Diseases of Tropical Fruit Crops

Diseases of Tropical Fruit Crops

Edited by

Randy C. Ploetz

University of Florida, IFAS,
Tropical Research and Education Center
Homestead, Florida,
USA

CABI Publishing

CABI Publishing is a division of CAB International

CABI Publishing
CAB International
Wallingford
Oxon OX10 8DE
UK

Tel: +44 (0)1491 832111
Fax: +44 (0)1491 833508
E-mail: info@cabi.org
Web site: www.cabi-publishing.org

CABI Publishing
745 Atlantic Avenue
8th Floor
Boston, MA 02111
USA

Tel: +1 (617)682-9015

E-mail: cabi-nao@cabi.org

A catalogue record for this book is available from the British Library, London, UK.

A catalogue record for this book is available from the Library of Congress, Washington, DC, USA.

ISBN 978 0 85199 390 4

Typeset in 9/11 pt Palatino by Columns Design Ltd, Reading, UK
Printed and bound by CPI Group (UK) Ltd, Croydon, CR0 4YY

Contents

The colour plate section can be found preceding p. 443

Contributors

Patricia Broadbent (formerly P. Barkley), Elizabeth Macarthur Agricultural Institute, Private Mail Bag 8, Camden, New South Wales 2570, Australia. e-mail: patricia_barkley_at_emai@smtpgwy.agric.nsw.gov.au

Carl W. Campbell, University of Florida, IFAS, Tropical Research and Education Center, 18905 SW 280th Street, Homestead, FL 33031-3314, USA

John B. Carpenter (Deceased), USDA Date and Citrus Experiment Station, Indio, California, USA

Lindy M. Coates, Queensland Horticulture Institute, Department of Primary Industries, 80 Meiers Road, Indooroopilly, Queensland 4068, Australia. e-mail: coateslm@dpi.qld.gov.au

Anthony W. Cooke, Queensland Horticulture Institute, Indooroopilly Research Centre, Queensland Department of Primary Industries, 80 Meiers Road, Indooroopilly, Queensland 4068, Australia. e-mail: cookea@dpi.qld.gov.au

Antonio de Goes, Campus Jaboticabal, Via de Acesso Prof. Paulo Donato Castellane s/n, 14.884–900 Jaboticabal. SP, Brazil

Steve M. Garnsey, University of Florida, Citrus Research and Education Center, 700 Experiment Station Road, Lake Alfred, FL 33850, USA. e-mail: garnsey@magicnet.net

Nigel A. Harrison, University of Florida, IFAS, Ft Lauderdale Research and Education Center, 3205 SW College Avenue, Ft Lauderdale, FL 33314-7799, USA. e-mail: naha@ufl.edu

Greg I. Johnson, Australian Centre for International Agricultural Research, GPO Box 1571, Canberra, ACT, Australia 2601. e-mail: johnson@aciar.gov.au

Phil Jones, Department of Crop and Disease Management, Rothamsted Experimental Station, Harpenden, Hertfordshire AL5 2JQ, UK. e-mail: phil.jones@bbsrc.ac.uk

Bernardo A. Latorre, Facultad de Agronomía e Ingeniería Forestal, Pontificia Universidad Católica de Chile, Av. Vicuna Mackenna 4860, Casilla 306-22, Santiago, Chile. e-mail: blatorre@puc.cl

T.-K. Lim, Biosecurity Australia, Department of Agriculture, Fisheries and Forestry Australia, GPO Box 858 Canberra, ACT 2601, Australia. e-mail: TK.Lim@affa.gov.au

Barry Q. Manicom, Institute of Tropical and Subtropical Crops, 1 River Street, Private Bag X11208, 1200, Nelspruit, Republic of South Africa. e-mail: barry@itsg2.agric.za

John A. Menge, Department of Plant Pathology, University of California, Riverside, CA 92521, USA. e-mail: menge@ucr.edu

Themis J. Michailides, University of California, 9240 S. Riverbend Avenue, Kearney Agriculture Center, Parlier, CA 93648, USA

Howard D. Ohr (Retired), Department of Plant Pathology, University of California, Riverside, CA 92521, USA. e-mail: hdohr@vcn.com

Henry Pak, The Horticulture and Food Research Institute of New Zealand Ltd, Mt Albert Research Centre, Private Bag 92 169, Auckland, New Zealand. e-mail: pakh@hort.cri.nz

Denis M. Persley, Plant Protection Unit, Queensland Department of Primary Industries, Plant Pathology Building, 80 Meiers Road, Indooroopilly, Queensland 4068, Australia

Yaacov Pinkas (Deceased), ARO, The Volcani Center, PO Box 6, Bet-Dagan 50250, Israel

Randy C. Ploetz, University of Florida, IFAS, Tropical Research and Education Center, 18905 SW 280th Street, Homestead, FL 33031-3314, USA. e-mail: rcp@mail.ifas.ufl.edu

Kenneth G. Rohrbach (retired), University of Hawaii, College of Tropical Agriculture, 3050 Maile Way, Gilmore 310B, Honolulu, HI 96822-2231, USA. e-mail: rohrbach@hawaii.edu. New address: 40900 South Applefield Circle, Elizabeth, CO 80107, USA

Carlos Ruggiero, Campus Jaboticabal, Via de Acesso Prof. Paulo Donato Castellane s/n 14.884–900 Jaboticabal. SP, Brazil. e-mail: ruggiero@fcav.unesp.br

Somsiri Sangchote, Department of Plant Pathology, Kasetsart University, Bangkok-10900, Thailand. e-mail: agrsrs@nontri.ku.ac.th

Donald Schmitt, University of Hawaii, Plant Pathology Department, 3190 Maile Way, St John 313, Honolulu, HI 96822-2279, USA. e-mail: schmitt@hawaii.edu

Sepiah, Faculty of Resource Science and Technology, Universiti Malaysia Sarawak (UNIMAS), 94300 Kota Samarahan, Sarawak, Malaysia. e-mail: msepiah@frst.unimas.my

C. Sittigul, Department of Plant Pathology, Chiang Mai University, Chiang Mai-50200, Thailand. e-mail: agppi004@chiangmai.ac.th

Walter R. Slabaugh, Agraquest Inc., 1050 Echo Avenue, Parma, ID 83660-6122, USA. e-mail: wslabaugh@agraquest.com

John E. Thomas, Queensland Department of Primary Industries, Agency for Food and Fibre Sciences, Queensland Horticulture Institute, Plant Pathology Building, 80 Meiers Road, Indooroopilly, Queensland 4068, Australia. e-mail: thomasje@dpi.qld.gov.au

'Pete' L.W. Timmer, University of Florida, Citrus Research and Education Center, 700 Experiment Station Road, Lake Alfred, FL 33850, USA. e-mail: lwt@lal.ufl.edu

Jacqui G. Wright, Secretariat of the Pacific Community, Private Mail Bag, Suva, Fiji Islands. e-mail: jacquiw@spc.org.fj

Dedication

Randy C. Ploetz[1] and John A. Menge[2]
[1] *University of Florida, Tropical Research and Education Center, Homestead, Florida USA; and* [2] *Department of Plant Pathology, University of California, Riverside, California, USA*

There have been many plant pathologists who have contributed to our understanding of the diseases that are described in the following chapters. Even the most obscure fruits in this book have had the causes and management of their associated diseases examined by several different researchers. Although far fewer individuals have made truly exceptional contributions, it would be a daunting task to mention all who deserve recognition. Rather than list all of these people, we highlight the careers of two outstanding figures in tropical fruit pathology, Robert Harry Stover (Fig. 1) and George Aubrey Zentmyer (Fig. 2). Their research on banana and avocado, respectively, was unparalleled in that it extended well beyond their roles as authorities on the diseases that affect these crops. To recognize their exceptional contributions, we dedicate this book to Drs Stover and Zentmyer.

Robert Harry Stover

Robert Harry Stover was born on 2 December 1926 in Chatham, Ontario, Canada, far from the tropics he came to love. In 1947, he obtained a BSc in Agriculture, specializing in bacteriology, from the University of Guelph. From 1949 to 1951, he was plant pathologist and head of investigations on tobacco diseases with the Canadian Department of Agriculture in Harrow, Ontario. During the winter months, he attended graduate school, and in 1950 he graduated from the University of Toronto with a PhD in plant pathology and mycology under Professor D.L. Bailey.

Dr Stover began his distinguished career as an authority on banana and tropical agriculture in La Lima, Honduras, soon after receiving his PhD. He was hired by the United Fruit Company in 1951 to investigate, and devise strategies for managing, Panama disease. At the time, this disease was decimating export plantations throughout the tropics and threatened the very existence of the trades. It is hard to imagine a more difficult challenge for a young scientist but, in a short time, Harry made major contributions to our understanding of the control and epidemiology of this disease, and the biology, ecology and pathology of its causal agent, *Fusarium oxysporum* f. sp. *cubense*. This research culminated in post-doctoral studies at Cambridge University in the laboratory of Dr S.D. Garrett, during which he wrote, *Fusarial Wilt (Panama Disease) of Bananas and Other* Musa *Species* (Stover, 1962).

After the trades converted to the resistant Cavendish cultivars, Dr Stover's focus shifted to Sigatoka leaf spot, caused by *Mycosphaerella musicola*. Although this disease did not have the lethal impact of Panama disease, it had a major influence on fruit yield and quality. His work on control of the disease witnessed a heightened sense of urgency when, in the early 1970s, black Sigatoka was first detected in the Americas. Black Sigatoka, caused by *Mycosphaerella fijiensis*, had a more serious impact on production than Sigatoka leaf spot, and quickly became the primary concern for export producers throughout the humid tropics. As he had done with Panama disease, Dr Stover established himself as the world authority on the Sigatoka leaf spots and, as an employee or consultant, numerous companies and organizations have relied on his advice on these diseases. In 1990, he co-edited the proceedings of an international conference in Costa Rica, *Sigatoka Leaf Spot Diseases of Bananas* (Fullerton and Stover, 1990), which remains a useful reference.

Dr Stover was an authority on all diseases that affect this crop, as was testified by his text, *Banana, Plantain and Abaca Diseases* (Stover, 1972). This book was the foremost reference on these diseases until the recent publication of *Diseases of Banana, Abacá and Enset* (Jones, 2000). He was also a world expert on banana, and was the senior author of a major treatise on this crop, the 3rd edition of *Bananas* (Stover and Simmonds, 1987). The following passage by his co-author, the late, eminent Professor N.W. Simmonds, alludes to the wealth of knowledge that Dr Stover possessed about this important crop. In the book's preface, Simmonds wrote:

> This book is almost entirely due to Dr. Stover, very little to me. The only justification for my name appearing as an author are that a few chapters of my *Bananas* remain more or less intact, that Longman urged joint authorship and that Dr. Stover was kind enough not to disagree. So this work is really 'Stover on Bananas' and I can't imagine a more suitable author. He has lived with those delightful plants for more than 30 years…and the fact is everywhere apparent. If I can claim any credit, it is that I acted as godfather to the present work in advising Longman that Dr. Stover was their man…and that he knew far more about the crop than I did.

Despite his record as a researcher, Dr Stover actually spent much of his career as an administrator. He served as the Chief of the Department of Plant Pathology of the United Fruit Company from 1961 to 1974, and Director of the Division of Tropical Research of the United Fruit Company between 1975 and 1977. In 1978, the subsidiary of the United Brands Company was changed to the Division of Tropical Research in the Services for Tropical Agricultural Research, SA (SIATSA), with the purpose of providing consultative and investigative services to private clients. Dr Stover was its director until 1984, when the Fundación Hondureña de Investigación Agrícola (FHIA) was created with facilities donated by United Brands. In 'retirement', Dr Stover started and directed a laboratory in which consultative work was conducted for banana companies, agrochemical firms, agencies and individual producers in Mexico, Central America, South America, the Caribbean, Africa, Australia and Asia.

In addition to the above books, Dr Stover wrote over 110 articles in refereed journals, and numerous book chapters, reviews and technical publications. He was on the governing board of the College of Agricultural Science Professionals of Honduras in 1969 and 1970, and was one of the original members of the Board of Trustees for the International Network for the Improvement of Banana and Plantain (INIBAP). In 1977, he was named a Fellow of the

American Phytopathological Society (APS), and in 1983 he received the Award for Contributions to Banana Research of the Asociacion para la Cooperacion en Investagaciones Bananeras en el Caribe y en America Tropical (ACORBAT). In 1985, he received the Gold Medal of the College of Professionals in Agricultural Sciences of Honduras (COLPROCAH), and the Medal of Honor for Merit from the University of San Pedro Sula. In 2001, the research library of FHIA, which houses one of the world's premier collections of literature on banana, plantain and tropical agriculture, was named in his honour to recognize contributions he made during his career and the decades-long support and direction he provided for the library. He was an active member of APS, the Canadian Society of Phytopathology, the American Association for the Advancement of Science, COLPROCAH and the Association of Researchers on Banana and Plantain (ACORBAT).

No account of Dr Stover's career would be complete without mentioning his love for his adopted country of Honduras. With Gene Ostmark and other United Fruit and FHIA employees, he frequently fished and hunted in southeastern Honduras, home of the Misquita people. He also frequented Lake Tioga and was responsible for stocking it with largemouth bass, resulting in an ongoing sport fishing industry on this body of water. Most important was the assistance and concern he demonstrated for the people of Honduras. He contributed in ways large and small to students and schools throughout the country, and conducted diverse activities to improve the health and living conditions of his countrymen, especially the poor. Of relevance to this book, he and other scientists from FHIA disseminated the Moko- and black Sigatoka-tolerant cooking banana 'Pelipita' among the impoverished Misquita people. Clearly, tropical fruit pathology and the world in general benefited from Dr Stover's efforts and compassion. On 25 February 2003, 17 days after the passing of Dr Zentmyer, he died in his adopted home city of La Lima.

George Aubrey Zentmyer

George Aubrey Zentmyer was born in North Platt, Nebraska, in 1913. He moved to Los Angeles, California, where he attended public schools and obtained his AB degree in Botany from the University of California, Los Angeles, in 1935. He enrolled in the University of California, Berkeley, and received his MSc degree in Plant Pathology in 1936 and his PhD in Plant Pathology in 1938.

Dr Zentmyer began his career as an Assistant Pathologist with the Division of Forest Pathology, USDA in San Francisco, working on diseases of forest trees. In 1940, he accepted a position with the Connecticut Agricultural Experiment Station where he worked on chemotherapy of vascular plants and studied the toxin production of the fungus causing Dutch elm disease. In 1941, George Zentmyer was married to Dorothy Anne Dudley, who worked side by side with him for his entire career. They have three daughters.

In 1944, Dr Zentmyer joined the Plant Pathology Department of the Citrus Experiment Station in Riverside California, which later became the University of California Riverside (UCR). He spent the rest of his career there, holding the rank of Professor and serving as Chairman of the Department of Plant Pathology. He retired in 1981 and remained an emeritus Professor in the department.

At UCR, Dr Zentmyer began his work on the diseases of avocado and other subtropical crops. During his career, he specialized in root-rotting pathogens, especially species of *Phytophthora*, and became one of the leading authorities on this destructive genus. He was instrumental in proving that the cause of avocado root rot is *P. cinnamomi*, and that this pathogen alone attacks more than 1000 host species (Zentmyer, 1980). He studied diverse aspects of avocado root rot, including: control with fumigants, irrigation effects, oxygen relations, effects of organic amendments, host plant nutrition, mechanisms of host resistance, control by soil fungicides and spread of the pathogen.

Dr Zentmyer was keenly interested in the taxonomy of the avocado and initiated studies in the 1960s with the late Eugenio Schieber of Guatemala. For three decades, Schieber and Zentmyer collected avocado germplasm throughout the tropics and named several new species. This material serves as breeding stock for rootstocks of the future. In 1975, Dr Zentmyer discovered the 'Duke 7' rootstock, which was the first commercial rootstock resistant to avocado root rot. 'Duke 7' has been credited with saving the avocado industry in many countries, and is still used to combat this important disease.

Dr Zentmyer established the world's foremost collection of *Phytophthora* at UCR, collecting more than 3000 isolates of 47 species during his career. His classic work on the genus included work on: spore formation; mating types; variability; effects of light, nutrition, temperature; and the induction of mating structures. Dr Zentmyer was the first to show that zoospores of *Phytophthora* are chemotactically attracted to root exudates prior to encystment and infection.

In addition to avocado and *Phytophthora,* Dr Zentmyer made major contributions to the science of other tropical crops, including cacao, coffee and macadamia, and other diseases, including Armillaria root rot, Verticillium wilt, and those caused by *Botryodipodia* (recently reclassified as *Diplodia*) and *Fusicoccum*. He was instrumental in determining that epidemic jarrah dieback in Australia, which affected huge areas of eucalyptus forest, was caused by *P. cinnamomi*.

During his career, he had a long and fruitful association with the California Avocado Society. He served the society in various capacities, and receiving its Award of Honor in 1954 and the Special Award of Merit in 1981. In 1964, the University of California elected Dr Zentmyer as an All-University Lecturer and Faculty Research Lecturer. The following year, he was awarded a Guggenheim Fellowship for a study leave in Australia, and in 1972 he received a NATO Senior Postdoctoral Fellowship for study in England. The same year, he was given a Special Award of Merit from the Caribbean Division of the American Phytopathological Society (APS). He is a Fellow and former President of the APS, and a Fellow of the American Association for the Advancement of Science (AAAS). In 1978, he became a member of the prestigious National Academy of Sciences of the USA. On the occasion of his retirement in 1981, an international symposium was convened in California to recognize his career. Over 300 participants from 24 countries attended the meeting, and from solicited papers the first comprehensive text on *Phytophthora* was published in his honour: Phytophthora: *Its Biology, Ecology and Pathology* (Erwin *et al.*, 1983). At the 75th anniversary of APS in 1983, he became only the fifth person to receive the society's highest award, the Award of Distinction.

Dr Zentmyer served the following academies, societies and foundations in numerous capacities: AAAS, American Institute of Biological Sciences, APS, the International Society of Plant Pathology, the National Research Council and the National Science Foundation. His association with the *Annual Review of Phytopathology* lasted for 30 years, becoming a member of the Editorial Committee in 1967 and serving for 25 years as an Associate Editor. Dr Zentmyer published more than 200 articles in refereed journals, and trained more than 15 PhD students, many of whom went on to research in the tropics. He served as a consultant for the Trust Territory of the Pacific Islands, the Commonwealth of Australia, AID-Ghana, Nigeria, the Government of South Africa, the Government of Israel, the Government of Western Australia, Ministry of Agriculture, Spain, and the Government of Costa Rica.

Despite Dr Zentmyer's tremendous achievements, he remained a humble friend of farmer, student and scientist alike. Until a recent illness, he was still active and would discuss his favourite subjects, plant pathology, avocados and the American Phytopathology Society, with all who were interested. Dr Zentmyer leaves a rich legacy. Sadly, he passed away on 8 February 2003 while proofs for this book were being edited.

Selected References

Erwin, D.C., Bartnicki-Garcia, S. and Tsao, P.H. (eds) (1983) Phytophthora: *Its Biology, Taxonomy, Ecology, and Pathology*. APS Press, St Paul, Minnesota.

Fullerton, R.A. and Stover, R.H. (1990) *Sigatoka Leaf Spot Diseases of Bananas*. INIBAP, Montpellier, France.

Jones, D.R. (ed.) (2000) *Diseases of Banana, Abacá and Enset*. CAB International, Wallingford, UK.

Stover, R.H. (1962) *Fusarial Wilt (Panama Disease) of Bananas and other* Musa *Species*. Phytopathological Paper No. 14, Commonwealth Mycological Institute, Kew, UK.

Stover, R.H. (1972) *Banana, Plantain and Abaca Diseases*. Commonwealth Mycological Institute, Kew, UK.

Stover, R.H. and Simmonds, N.W. (1987) *Bananas*, 3rd edn. Longman, London.

Zentmyer, G.A. (1952) Collecting avocados in Central America for disease resistance tests. *California Avocado Society Yearbook* 37, 107–111.

Zentmyer, G.A. (1959) Avocado diseases in Latin America. *Plant Disease Reporter* 43, 1229.

Zentmyer, G.A. (1961) Chemotaxis of zoospores for root exudates. *Science* 133, 1595–1596.

Zentmyer, G.A. (1965) Bacterial stimulation of sporangium production in *Phytophthora cinnamomi*. *Science* 150, 1178–1179.

Zentmyer, G.A. (1980) *Phytophthora cinnamomi* and the diseases it causes. *Monograph No. 10*, American Phytopathological Society.

Zentmyer, G.A. and Mitchell, D.J. (1985) Phytophthora diseases of fruit trees in the tropics. *Review of Tropical Plant Pathology* 2, 287–309.

Preface

Fruit are among the most important and interesting food crops that are produced in the tropical regions of the world (Martin *et al.*, 1987; Nakasone and Paull, 1998). In many ways, they exemplify the exotic nature of the tropics. Few who have eaten an excellent mango, seen a hairy rambutan or had the courage to sample a malodorous durian would deny that these fruit possess an incredible range of appearances, textures, smells and tastes.

Tropical fruit crops are valuable agricultural commodities (Table 1). Perhaps the best example of their importance is banana. It is the world's fourth most valuable food after rice, wheat and milk (FAO, 2001). In export trade, banana ranks fourth among all agricultural products and is the most significant fruit, with world trade totalling US$4.8 billion in 1999. (World trade in citrus fruit totalled US$4.6 billion in 1999. The higher figure in Table 1 also includes citrus juice and products.) Yet, only 17% of the global production of banana enters international commerce. Poor subsistence farmers in Africa, the Americas and Asia consume much of the remaining harvest. For many of these producers, banana is a staple food.

Although banana and citrus predominate, several other tropical fruit are also important (Table 1). The percentages of these fruit that reach export markets are quite variable, and range from 67% for kiwifruit to less than 5% for mango and papaya (FAO). For many of these crops, by-products or processed portions of the fruit are more important than the fruit itself. For example, the exported tonnage of coconut oil is six times higher than that for coconut, and more pineapple is exported in cans than as fresh fruit. In total, the value of tropical fruit and their associated products that are exported each year approaches US$20 billion.

In addition to the monetary importance of these fruit is their place in the everyday lives of people in the tropics (Martin *et al.*, 1987). They possess a wide array of nutritional qualities, and may contain significant amounts of vitamins, minerals, oils, starches and protein. They are dessert items that add flavour and variety to the diets of many.

In the future, demand for tropical fruit will increase in and outside the tropics. As production increases to meet these demands, information on the diseases that impact the health of these crops will be vital. Diseases often are the most important constraint to the production of tropical fruit. They indirectly reduce yields by debilitating the plant, and directly reduce the yield or quality of fruit before and after they are harvested. Failure to recognize and manage these diseases successfully can result in catastrophic losses. Thus, a primary objective of this book is to provide accurate information on the diagnosis and control of these important problems.

Table 1. 1999 production statistics for the world's major tropical fruit crops.[a]

Crop	Total production (Mt)[b]	Export production: Quantity (Mt)[b]	Export production: Value (millions US$)
Citrus, mainly *Citrus* spp.[c]	104.78	13.35	8655
Banana, *Musa* spp.[d]	89.44	15.1	4835
Coconut, *Cocos nucifera*[e]	48.15	2.59	1319
Mango, *Mangifera indica*[f]	25.0	0.61	397
Pineapple, *Ananas comosus*[g]	13.26	2.47	1818
Papaya, *Carica papaya*	5.23	0.14	90
Date, *Phoenix dactylifera*	5.21	0.47	249
Avocado, *Persea americana*	2.22	0.28	361
Fig, *Ficus* spp.[h]	1.14	0.09	139
Passion fruit, *Passiflora* spp.	1	n/a	n/a
Kiwifruit, *Actinidia deliciosa* var. *deliciosa*	0.95	0.64	539
Carambola, *Averrhoa carambola*	0.5	n/a	n/a
Tropical fruit[i]	n/a	0.13	93

[a]With two exceptions, figures are from the online database of the Food and Agricultural Organization of the United Nations (http://apps.fao.org/default.htm). Production estimates for carambola and passion fruit were made in this book based on figures from diverse sources.
[b]Production figures are in millions of metric tonnes.
[c]Includes fruit of all *Citrus* species, their hybrids and kumquat (*Fortunella* spp.), a citrus relative. The export figure is for fresh fruit, fruit products, juice and juice concentrate.
[d]Includes both banana and plantain, which is a type of banana. The export figure is for fresh fruit only.
[e]The export figure is the total for coconuts, coconut cake, desiccated coconut, coir, copra and coconut oil.
[f]The export figure is the total for fresh fruit, juice and pulp.
[g]The export figure is the total for canned fruit, fresh fruit, juice and juice concentrate.
[h]The export figure is the total for fresh and dried fruit.
[i]The export figure is the total for fresh fruit of several different crops, including *Annona* spp., breadfruit (*Artocarpus altilis*), carambola (*Averrhoa carambola*), durian (*Durio zibethinus*), guava (*Psidium guajava*), jackfruit (*Artocarpus heterophyllus*), longan (*Dimocarpus longan*), mangosteen (*Garcinia mangostana*), passion fruit (*Passiflora* spp.) and rambutan (*Nephelium lappaceum*). Global production figures for these and other less important tropical fruit crops are not available.

Several different criteria were used when choosing the fruit that are included in this book. Annual production figures were employed as primary indicators of importance, and in general those for which an excess of ~500,000 t year^{-1} are produced are found in the following pages (Table 1). Also included are crops that are widely known and appreciated in the tropics, but are produced in lower or unknown quantities. Among the latter fruit are species of *Annona*, guava and three sapindaceous fruit, longan, lychee and rambutan. Several other fruit with narrow distributions are included due to their exceptional flavour and potential for wider cultivation; these include durian, mangosteen and *Artocarpus* spp. (particularly breadfruit and jackfruit). Finally, although greater production of date, fig and kiwifruit occurs in the subtropics (i.e. north of the Tropic of Cancer and south of the Tropic of Capricorn), they are also produced in the tropics; they are included due to their major importance and wide distribution.

The need for this book became apparent while teaching a class on diseases of tropical crops at the University of Florida. When compiling the syllabus for tropical fruit diseases, it was evident that no comprehensive, up-to-date text was available on the diseases that impact these crops. Although significant coverage is available for two of the important crops in this book, banana (Jones, 2000) and citrus (Timmer *et al.*, 2000), and to a lesser extent for several other hosts (Lim and Khoo, 1985; Ridgeway, 1989; Broadley, 1991; Persley, 1993; Ploetz *et al.*, 1994; Dodd *et al.*, 1997; Johnson *et al.*, 1997; Ploetz and Prakash, 1997), the present book represents the most complete and diverse treatment on tropical fruit diseases that has been published since Cook's (1975) *Diseases of Tropical and Subtropical Fruits and Nuts*.

Every attempt has been made to ensure that the information in this book is current and accurate. I am responsible for any mistakes that occur in the following pages, and recognize that some of the book will be out of date soon after it is published. For example, new genetic and sexual data have delineated new species among taxa in the genus *Fusarium* and among other anamorphs that have *Botryosphaeria* teleomorphs (Leslie, 1995; Koenig *et al.*, 1997; O'Donnell *et al.*, 1998a,b; Crous and Palm, 1999; Denman *et al.*, 2000; Steenkamp *et al.*, 2000; Slippers *et al.*, 2001; Zhou and Stanosz, 2001; Britz *et al.*, 2002). Changes in the taxonomy and nomenclature of these and other pathogens undoubtedly will continue to occur in the future. Recent changes are recorded in Appendix I.

Note is made of recent changes in the taxonomy of oomycetes. Although species of *Peronophythora*, *Phytophthora* and *Pythium* have fungus-like lifestyles and a long prior history during which they were considered fungi, they are in the Kingdom *Chromista* and are more closely related to algae than true fungi (*Eumycota*) (Erwin and Ribiero, 1996). Thus, they are distinguished from fungi in this book. Also mentioned is the use of italics for virus and viroid agents, a practice that was advocated recently by the International Committee for the Taxonomy of Viruses and has been adopted by international journals that publish on these pathogens (Mayo and Horzinek, 1998; Van Regenmortel *et al.*, 1999).

To conserve space, the authorities of pathogens, hosts, insects and acarids are not included in the chapters, but are listed in three appendices. Likewise, descriptions of some of the most common and important pathogens are included in Chapter 1.

This book would not have been possible without the hard work and assistance of many colleagues. I am grateful to all of the authors for their effort and patience during the lengthy preparation of this book. I thank Carl Campbell, Monica Elliott, John Hu, Greg Johnson, David Jones, Bob Knight, John Thomas, Ken Pegg, Denis Persley and Dov Prusky for reviewing portions of the book, Cornelia Büchen-Osmond for assistance with the current taxonomy and nomenclature of the virus pathogens, and Paul Bridge, Monica Elliott and Julie Flood for information on the taxonomy of *Ganoderma* spp. In addition to the authors, I also thank Pedro Crous, Michel Dollet, J. De Filippis, Julie Flood, K. Gerber, Jim Graham, R. Inserra, David Jones, Kerry O'Donnell, Jeri Ooka, Jorge Parrado (deceased), Dov Prusky, Harry Stover, Janice Uchida, Tony Whiley and George Zentmyer for the use of colour slides, line drawings and black and white images. A special thanks goes to Tony Cooke, an author of the carambola chapter, who provided many excellent images of diseases in this book, and Ian Maguire, who scanned most of the line drawings and many of the black and white images. Finally, I am indebted to the Australian Centre for International Agricultural Research (ACIAR), Del Monte Fresh Fruit, John Menge, Pete Timmer and the Office of the Dean for Research of the University of Florida, IFAS. Their generous donations made possible the publication of the colour plate section in this book.

Randy C. Ploetz

References

Britz, H., Steenkamp, E.T., Coutinho, T.A., Wingfield, B.D., Marasas, W.F.O. and Wingfield, M.J. (2002) Two new species of *Fusarium* section *Liseola* associated with mango malformation. *Mycologia* 94, 722–730.

Broadley, R.H. (ed.) (1991) *Avocado Pests and Disorders*. Department of Primary Industries Queensland, Brisbane, Australia.

Cook, A.A. (1975) *Diseases of Tropical and Subtropical Fruits and Nuts*. Hafner Press, New York.

Crous, P.W. and Palm, M.E. (1999) Reassessment of the anamorph genera of *Botryoplodia*, *Dothiorella* and *Fusicoccum*. *Sydowia* 52, 167–175.

Denman, S., Crous, P.W., Taylor, J.E., Kang, J.-C., Pascoe, I. and Wingfield, M.J. (2000) An overview of the taxonomic history of *Botryosphaeria*, and a re-evaluation of its anamorphs based on morphology and ITS and rDNA phylogeny. *Studies in Mycology* 45, 129–140.

Dodd, J.C., Prusky, D. and Jeffries, P. (1997) Fruit diseases. In: Litz, R.E. (ed.) *The Mango: Botany, Production and Uses*. CAB International, Wallingford, UK, pp. 257–280.

Erwin, D.C. and Ribiero, O.K. (1996) *Phytophthora Diseases Worldwide*. APS Press, St Paul, Minnesota.

FAO (2001) Online database of the Food and Agricultural Organization of the United Nations (http://apps.fao.org/default.htm).

Johnson, G.I., Sharp, J.L., Milne, D.L. and Oosthuyse, S.A. (1997) Postharvest technology and quarantine treatments. In: Litz, R.E. (ed.) *The Mango: Botany, Production and Uses*. CAB International, Wallingford, UK, pp. 447–508.

Jones, D.R. (ed.) (2000) *Diseases of Banana, Abacá and Enset*. CAB International, Wallingford, UK.

Koenig, R., Ploetz, R.C. and Kistler, H.C. (1997) *Fusarium oxysporum* f. sp. *cubense* consists of a small number of divergent and globally distributed lineages. *Phytopathology* 87, 915–923.

Leslie, J.F. (1995) *Gibberella fujikuroi*: available populations and variable traits. *Canadian Journal of Botany* 73 (Suppl. 1), S282–S291.

Lim, T.K. and Khoo, K.C. (1985) *Diseases and Disorders of Mango in Malaysia*. Tropical Press, Kuala Lumpur.

Martin, F.W., Campbell, C.W. and Ruberté, R.M. (1987) *Perennial Edible Fruits of the Tropics*. USDA Agricultural Handbook No. 642.

Mayo, M.A. and Horzinek, M. (1998) A revised version of the international code of virus classification and nomenclature. *Archives of Virology* 143, 1645–1654.

Nakasone, H.Y. and Paull, R.E. (1998) *Tropical Fruits*. CAB International, Wallingford, UK.

O'Donnell, K., Cigelnik, E. and Nirenberg, H I. (1998a) Molecular systematics and phylogeography of the *Giberella fujikuroi* species complex. *Mycologia* 90, 465–493.

O'Donnell, K.O., Kistler, H.C., Cigelnik, E. and Ploetz, R.C. (1998b) Multiple evolutionary origins of the fungus causing Panama disease of banana: concordant evidence from nuclear and mitochondrial gene genealogies. *Proceedings of the National Academy of Sciences* USA 95, 2044–2049.

Persley, D. (ed.) (1993) *Diseases of Fruit Crops*. Department of Primary Industries Queensland, Brisbane, Australia.

Ploetz, R.C. and Prakash, O. (1997) Foliar, floral and soilborne diseases. In: Litz, R.E. (ed.) *The Mango: Botany, Production and Uses*. CAB International, Wallingford, UK, pp. 281–326.

Ploetz, R.C., Zentmyer, G.A., Nishijima, W., Rohrbach, K. and Ohr, H.D. (eds) (1994) *Compendium of Tropical Fruit Diseases*. APS Press, St Paul, Minnesota.

Ridgeway, R. (ed.) (1989) *Mango Pests and Disorders*. Department of Primary Industries Queensland, Brisbane, Australia.

Slippers, B., Johnson, G.I., Cooke, A.W., Crous, P.W., Coutinho, T.A., Wingfield, B.D. and Wingfield, M.J. (2001) Taxonomy of *Botryosphaeria* spp. causing stem end rot of mango. In: *Proceedings of 13th Biennial Australasian Plant Pathology Conference*. Cairns, Australia, September 24–27, 2001.

Steenkamp, E., Britz, H., Coutinho, T., Wingfield, B., Marasas, W. and Wingfield, M. (2000) Molecular characterization of *Fusarium subglutinans* associated with mango malformation. *Molecular Plant Pathology* 1, 187–193.

Timmer, L.W., Garnsey, S.M. and Graham, J.H. (eds) (2000) *Compendium of Citrus Diseases*, 2nd edn. APS Press, St Paul, Minnesota.

Van Regenmortel, M.H.V., Fauquet, C.M., Bishop, D.H.L., Carstens, E.R. and Maniloff, J. (1999) *Virus Taxonomy. Seventh Report of the International Committee for the Taxonomy of Viruses*. Academic Press, New York.

Zhou, S. and Stanosz, G.R. (2001) Relationships among *Botryosphaeria* species and associated anamorphic fungi inferred from the analyses of ITS and 5.8S rDNA sequences. *Mycologia* 93, 516–527.

Foreword

Tropical and subtropical fruit crops generally have received less attention from plant pathologists than have temperate fruit crops. Although exceptional crops such as banana and pineapple have been investigated extensively, much of the research has been done by private organizations for their own use, and the results have not been widely available to other producers and handlers.

This book, edited by Dr Randy Ploetz, includes comprehensive information on diseases of a variety of fruit crops. Causal organisms are identified, disease symptoms are given and effects of environmental factors are discussed. Strategies of management and control are presented.

The authors of the chapters have been chosen for their thorough knowledge of the crops and the diseases that affect them. Dr Ploetz is well qualified to do this because of the extensive research he has done in Florida and his wide experience with diseases of tropical fruit crops. The crops range from well-known ones such as citrus and fig, to those such as mango, which only recently has emerged as an important commodity in the world market, and *Artocarpus* species, which are relatively little known but have real promise for future expansion.

A large audience of researchers and fruit producers in the tropics should welcome the publication of this book.

Carl W. Campbell, PhD
Professor of Horticulture, Emeritus
University of Florida
Tropical Research and Education Center
Homestead, Florida, USA

Acknowledgements

Generous donations made by the Australian Centre for International Agricultural Research (ACIAR), Del Monte Fresh Fruit, John Menge (University of California at Riverside), Pete Timmer (University of Florida, Citrus Research and Education Center) and the Office of the Dean for Research of the University of Florida, IFAS, made possible the publication of the colour plate section in this book.

ACIAR
Australian Centre for International Agricultural Research

Del Monte
Quality

UNIVERSITY OF FLORIDA
CREC
Lake Alfred, FL
Institute of Food and Agricultural Sciences

UNIVERSITY OF CALIFORNIA
BERKELEY · DAVIS · IRVINE · LOS ANGELES · RIVERSIDE · SAN DIEGO · SAN FRANCISCO · SANTA BARBARA · SANTA CRUZ

UNIVERSITY OF
FLORIDA

Institute of Food and Agricultural Sciences

1 Common Pathogens of Tropical Fruit Crops

R.C. Ploetz[1], T.-K. Lim[2], John A. Menge[3], Kenneth G. Rohrbach[4]* and Themis J. Michailides[5]

[1] *University of Florida, Tropical Research and Education Center, Homestead, Florida, USA;* [2] *Biosecurity Australia, Department of Agriculture, Fisheries and Forestry Australia, Canberra, Australia;* [3] *Department of Plant Pathology, University of California, Riverside, California, USA;* [4] *University of Hawaii, College of Tropical Agriculture, Honolulu, Hawaii, USA;* [5] *University of California, Kearney Agricultural Center, Parlier, California, USA*

Introduction

Most of the pathogens that are described in this book cause specific diseases on one or a few hosts. However, some of the pathogens are generalists that affect several hosts, often in several different ways. For example, *Ceratocystis paradoxa* causes butt rot, decline, foot rot, fruit rot and leaf spot on five different host crops. Rather than repeat much of the information that would be included in the following chapters, attributes that these pathogens share are summarized below. Hosts on which these pathogens cause significant problems are listed in Table 1.1. With the exception of *Cephaleuros virescens* and the *Phytophthora* species, all of these pathogens are fungi. For detailed information, the reader is referred to the indicated chapters.

Common Pathogens

Alternaria alternata

Alternaria alternata causes fruit rots and leaf spots on carambola and mango, and surface mould on fig fruit (Chapters 6, 11 and 15). The fungus is widespread and has a large host range (Neergaard, 1945; Domsch *et al.*, 1980).

A. alternata is in Neergaard's (1945) *Longicatenatae* group that contains species that produce conidia in long chains of ten or more spores (*'A. alternata'* from citrus produces solitary conidia and is now viewed as a different species; see Chapter 7). Conidia and conidiophores of *A. alternata* are medium golden brown. Conidia have short beaks not exceeding one-third the length of the entire spore. They are ovoid, obclavate or obpyriform and 18–63 × 7–18 μm. They can be smooth walled or warty, and have three to eight transverse septa and one or two longitudinal septa towards the base (Fig. 1.1). Conidiophores are produced singly or in small groups; and are usually simple, straight or curved, two- to four-celled, up to 50 μm long and 3–6 μm wide. Cardinal temperatures for growth are 2.5–6.5, 25–28 and 31–32°C.

Armillaria spp.

Armillaria mellea causes root rot on avocado, cherimoya, fig, kiwifruit, lychee and soursop (Chapters 2, 3, 11, 13 and 14). It forms a

*New address: 40900 South Applefield Circle, Elizabeth, CO 80107, USA.

Table 1.1. Common pathogens of tropical fruit crops and some of the most important diseases they cause.

Pathogen	Hosts	Disease(s)
Alternaria alternata	Carambola, mango	Fruit rot, leaf spot
	Fig	Surface mould
Armillaria mellea	Avocado, cherimoya, fig, kiwifruit, lychee, soursop	Root rot
Armillaria socialis	Avocado, atemoya, lychee, soursop	Root rot
Botryosphaeria dothidea	Avocado	Fruit rot
(anamorph: *Fusicoccum aesculi*)	Mango	Decline, fruit rot
Botryosphaeria rhodina (anamorph: *Diplodia theobromae*)	Atemoya, carambola, durian, longan, lychee, mangosteen, rambutan, soursop, sugar apple	Fruit rot
	Avocado, breadfruit, jackfruit	Fruit rot, stem canker
	Mango	Decline, fruit rot
Botryotinia fuckeliana	Fig	Dieback, fruit rot
(anamorph: *Botrytis cinerea*)	Kiwifruit	Fruit rot
Cephaleuros virescens	Avocado, breadfruit, carambola, citrus, durian, longan, lychee, mango, mangosteen, rambutan	Leaf spot
Ceratocystis paradoxa (anamorph: *Chalara paradoxa*)	Pineapple	Butt rot, fruit rot, leaf spot
	Coconut	Butt rot
	Carambola	Fruit rot
	Fig	Foot rot
	Mango	Decline
Erythricium salmonicolor (anamorph: *Necator decretus*)	Breadfruit, carambola, citrus, durian, jackfruit, mango, mangosteen, rambutan	Pink disease
Glomerella cingulata (anamorph: *Colletotrichum gloeosporioides*)	*Annona marcgravii*, avocado, biriba, breadfruit, carambola, citrus, cherimoya, citrus, custard apple, durian, fig, guava, ilama, jackfruit, lychee, mango, mangosteen, papaya, passion fruit, soursop, sugar apple, rambutan	Anthracnose
Phytophthora cinnamomi	Avocado	Root rot, cankers
	Cherimoya, kiwifruit	Root rot, collar rot
	Pineapple	Root rot, heart rot
Phytophthora citricola	Avocado	Cankers
	Kiwifruit	Root rot, crown rot
	Fig	Fruit rot
	Guava	Fruit rot, leaf blight
Phytophthora citrophthora	Chempedek	Root rot, fruit rot
	Citrus	Root rot, gummosis, fruit rot
	Kiwifruit	Root rot, crown rot
Phytophthora nicotianae	Sugar apple	Root rot, fruit rot
	Pineapple	Root rot, heart rot
	Carambola	Root rot
	Fig, rambutan	Fruit rot
Phytophthora palmivora	Atemoya, breadfruit, papaya, pond apple, soursop	Root rot, fruit rot
	Sugar apple	Root rot
	Fig, longan	Fruit rot
	Pineapple	Heart rot
	Avocado	Cankers
	Mango	Root and crown rot
	Coconut	Bud rot
	Durian	Canker, fruit rot, leaf blight, root rot
Rigidoporus lignosus	Carambola, durian, mango	Root rot
Rosellinia necatrix (anamorph: *Dematophora necatix*)	Avocado, cherimoya, citrus, fig, kiwifruit	Root rot

Although some of these pathogens cause diseases on other hosts or on other organs of the listed hosts, only those that are significant are listed.

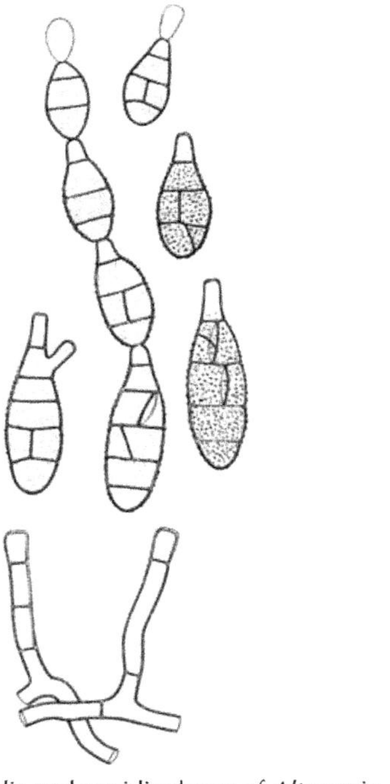

Fig. 1.1. Conidia and conidiophores of *Alternaria alternata* (CMI drawing).

number of structures including basidiomes (mushrooms) (Plate 1), basidiospores, mycelia, white mats or plaques between the bark and wood (Plate 2), pseudosclerotial tissue and rhizomorphs (brown–black, cylindrical fungal strands that look somewhat like roots) on the surface of infected roots. Infected tissues have a distinct mushroom odour when moist. Basidiomes are often produced around the base of affected trees after rainfall, but have no known role in the disease cycle (Shaw and Kile, 1991). They are usually clustered and have caps that are tan to bay in colour and are 1–15 cm in diameter. The caps often have a knob or bump in the centre and are covered with small yellow-green tufted scales. The stem is buff to brown, 3–30 cm long, 3–40 mm in diameter and has a shaggy, skirt-like ring surrounding it just below the cap. Below the ring, the stem often has concentric scales. The gills are pale yellow to dark tan. The basidiospores that are borne on the gills leave a white spore print and are 6–9 × 4–7 μm.

A. mellea is sometimes referred to as the honey mushroom, honey agaric, oak root fungus or shoestring fungus (Shaw and Kile, 1991). Its wide host range and its ability to survive as a saprophyte make it a difficult pathogen to control. Infection results when roots contact infected plants or rhizomorphs of the fungus in the soil. Infected tree stumps and large roots of dead hosts are common sources of inoculum, and the fungus may persist saprophytically for 10 years or more. After roots are penetrated, the pathogen grows up the root along the cambial layer, eventually girdling the root crown and causing tree death.

There currently are no fungicides for the control of this disease. When new orchards are planted into previously affected areas (including areas cleared of native vegetation), soil preparation is very important. After all tree stumps are cleared from the site, deep cultivation to remove as many diseased roots as possible is recommended. Soil fumigation following removal of infected plant material is another strategy, and carbon bisulphide or methyl bromide (most effective) can be used. However, fumigants may be unable to penetrate large infected roots. Replanting in established orchards should be done some distance away from affected trees. When orchards are thinned, tree stumps should be removed or their cut surfaces sealed to limit opportunities for infection.

A related species, *Armillaria socialis*, causes a root rot of atemoya, avocado, lychee and soursop in Florida (Chapters 2, 3 and 14). It is distinguished from *A. mellea* by the lack of a ring on the stem and the rare production of mushrooms and rhizomorphs.

Aspergillus niger

Several different species in this genus cause fruit rots on crops discussed in this book. They are cosmopolitan species that appear on a wide array of substrates in nature. Of these, *Aspergillus niger* is most important. Its colonies are effuse and blackish brown to black (Ellis, 1971). Mycelium is hyaline to pale yellow and 2–4 μm wide. Conidiophores are erect, straight or flexuous, often up to 3 mm long, 15–20 μm wide and hyaline with the upper portions becoming brown and swollen at the apex into a spherical, 40–70 μm diameter vesicle (Fig. 1.2). Phialides occur in groups at the apices of tightly packed, 20–30 μm long and 5–6 μm wide branches. They are flask-shaped, 7–10 μm long and 3–3.5 μm wide. Conidia are usually globose, brown, verrucose or echinulate, sometimes with warts or spines arranged in discontinuous bands, and 3–5 μm in diameter.

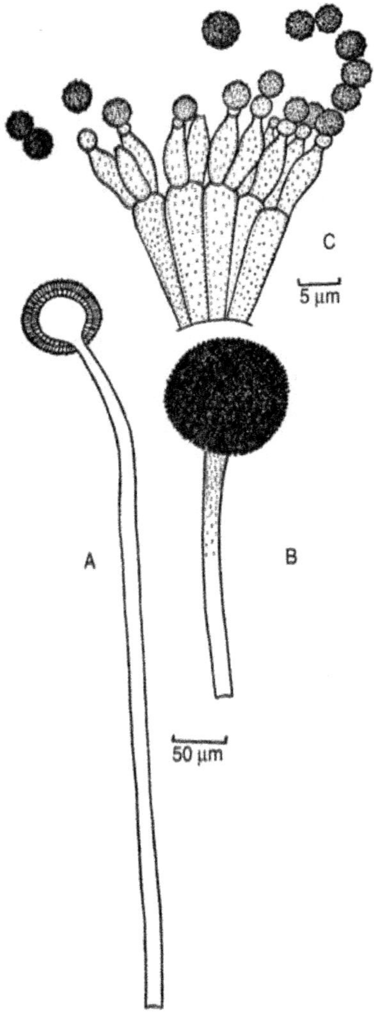

Fig. 1.2. (A) Conidiophore terminated in spherical vesicle, (B) vesicle surface with tightly packed branches and (C) detail of branches, flask-shaped phialides and echinulate conidia of *Aspergillus niger* (from Ellis, 1971).

Botryosphaeria spp.

Several of the pathogens in this book have *Botryosphaeria* teleomorphs, two of which, *B. rhodina* and *B. dothidea*, are described in this chapter. Based on internal transcribed spacer (ITS) and 5.8 rDNA sequence data, and conidial morphology, Zhou and Stanosz (2001) divided *Botryosphaeria* into two groups. Section *Brunnea* contained *B. rhodina* and species that have *Diplodia*, *Lasiodiplodia* and *Sphaeropsis* anamorphs. They produce conidia that are light to dark brown when mature and usually >10 μm in width. Members of the Section *Hyala* have *Fusicoccum* anamorphs (several fungi that were formerly in the genus *Dothiorella* recently were moved to *Fusicoccum* (Denman *et al.*, 2000; Slippers *et al.*, 2001)). Section *Hyala* includes *B. dothidea* and has anamorphs that produce conidia that are usually <10 μm in width and hyaline, but can darken to light brown with age.

B. dothidea (anamorph: *Fusicoccum aesculi*) causes a rot of avocado fruit, and fruit rot and decline of mango (Chapters 3 and 15). It produces fluffy grey mycelium with discrete pycnidia on potato dextrose agar (PDA) or stromatic multilocular fruiting bodies on oatmeal agar (OA). Discrete, immersed pycnidia are produced on mango. Conidia are fusiform to navicular, 12–25 × 4–6 μm, hyaline and single-celled (Fig. 1.3). The teleomorph is occasionally produced on OA and has been found in litter beneath avocado and mango trees (Johnson, 1994; Michailides *et al.*, 1999). On twigs, pseudothecia are subglobose to pyriform, 210 × 120 μm, and immersed beneath the epidermis. On OA, ascostromata are hemi-lenticular and up to 10 mm wide. Asci are eight-spored, bitunicate and irregularly biseriate. Ascospores are hyaline, single-celled, fusiform and 16–25 × 4.5–9.5 μm.

B. rhodina (anamorph: *D. theobromae*) is a common pathogen in the tropics that causes diverse diseases on atemoya, avocado, banana, breadfruit, carambola, durian, jackfruit, longan, lychee, mango, mangosteen, rambutan, soursop and sugar apple (Chapters 2–6, 10 and 14–16). *D. natalensis* and *Lasiodiplodia theobromae* recently were reduced to synonymy with *D. theobromae* (see Appendix I) (Denman *et al.*, 2000), and Crous

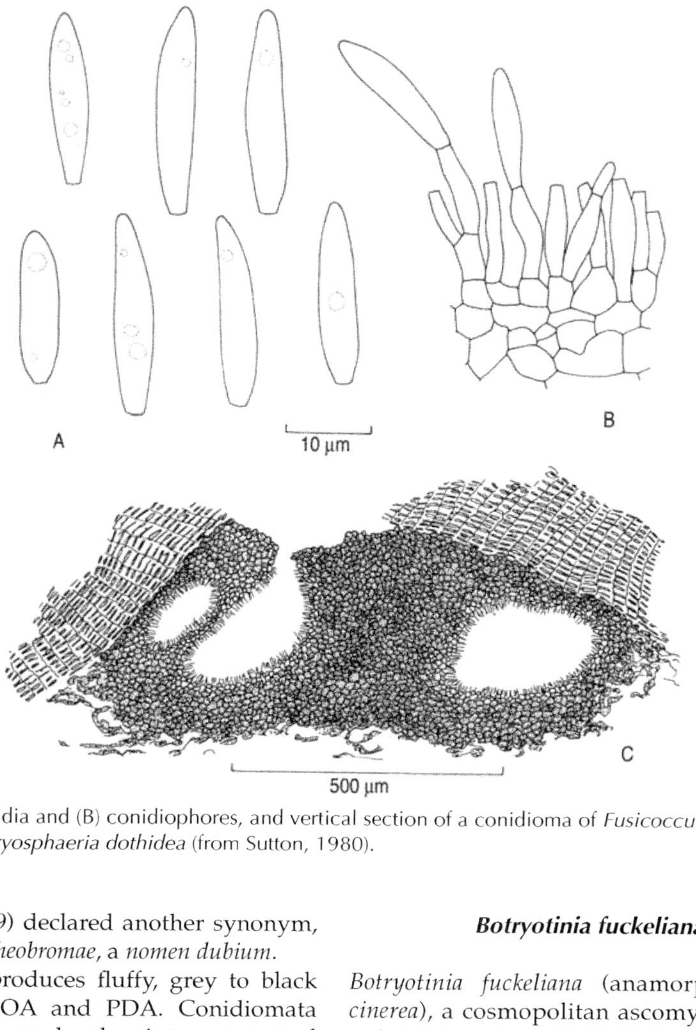

Fig. 1.3. (A) Conidia and (B) conidiophores, and vertical section of a conidioma of *Fusicoccum aesculi*, anamorph of *Botryosphaeria dothidea* (from Sutton, 1980).

and Palm (1999) declared another synonym, *Botryodiplodia theobromae*, a *nomen dubium*.

B. rhodina produces fluffy, grey to black mycelium on OA and PDA. Conidiomata may be simple or develop into aggregated stromatic bodies (Fig. 1.4). Cirri of conidia may ooze from ostioles. Conidia are initially hyaline, aseptate, granular, ovoid to ellipsoid and thick-walled. Mature conidia are two-celled, 20–30 × 10–15 µm, and brown-walled with numerous longitudinal striations. Paraphyses are usually present.

B. rhodina is usually not very aggressive, and attacks trees that are weakened by extreme temperatures, drought and other factors. It infects through wounds, and causes symptoms on fruit as they ripen. It is often an endophyte, but can also be found in soil, on dead twigs, on mummified fruit and on organic debris beneath trees.

Botryotinia fuckeliana

Botryotinia fuckeliana (anamorph: *Botrytis cinerea*), a cosmopolitan ascomycete, causes a dieback and fruit rot of fig, and fruit rot of kiwifruit (Chapters 11 and 13). The pathogen is a generalist and saprophyte, and is common in cooler regions (Holliday, 1980). Conidia are produced in dichotomously branched conidiophores on mycelia and sclerotia (Fig. 1.5). Conidia are 6–15 (11.7) × 1–12 (9.3) µm, hyaline or pigmented, ellipsoid–obovoid, globoid and aseptate. They are often abundant on affected plant tissue, resulting in the common grey mould sign. Sclerotia are small, hard, black survival structures and often are observed firmly attached to the outside of decayed fig fruits. The fungus is heterothallic, and carpogenic germination to

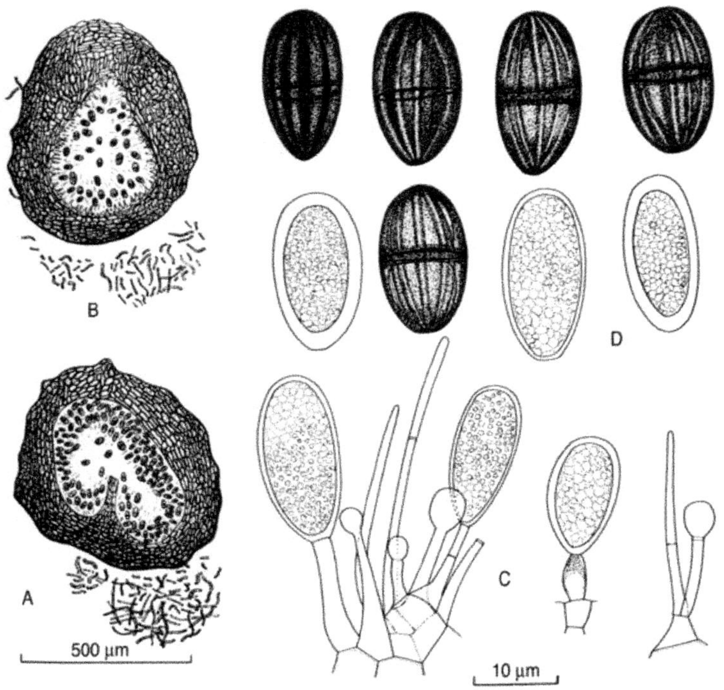

Fig. 1.4. Pycnidia (A and B), conidiogenous cells and immature conidia (C), and mature and immature conidia (D) of *Diplodia theobromae*, anamorph of *Botryosphaeria rhodina* (from CMI description no. 519).

Fig. 1.5. Conidia and conidiophores of *Botrytis cinerea*, anamorph of *Botryotinia fuckeliana* (from Ellis, 1971).

produce apothecia can occur after a dormant period of 2–6 months. The teleomorph is usually uncommon or not present in the field.

Cephaleuros virescens

Cephaleuros virescens is a green alga that causes algal leafspot (red rust) on avocado, breadfruit, carambola, citrus, durian, longan, lychee, mango, mangosteen and rambutan (Chapters 2, 3, 5–7, 10 and 14–16). The algal thallus is orange to rust coloured and develops below the host cuticle (Lim and Khoo, 1985). It produces 32 × 25 μm sporangia on the terminals of erect stalks (Fig. 1.6). Biflagellate zoospores are produced in the sporangia. Flask-shaped gametangia that are responsible for sexual reproduction are also formed in the thallus. Gametangia release biflagellate gametes in free water, which fuse in pairs to produce sporophytes.

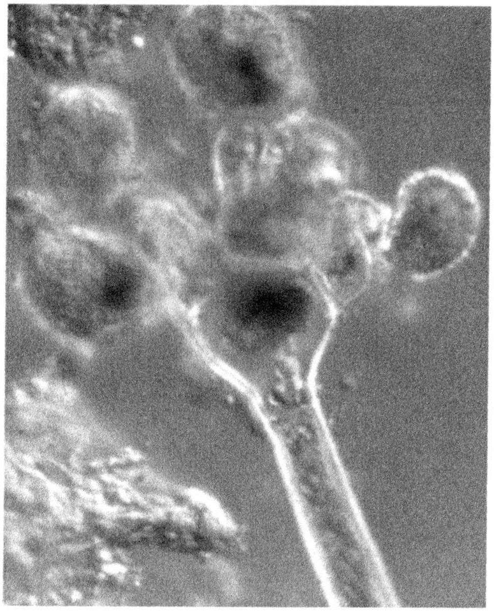

Fig. 1.6. Branch of the thallus of *Cephaleuros virescens* that has terminated in several oval sporangia (photo: T.-K. Lim).

Algal leafspot is a common problem in the tropics and subtropics (Joubert and Rijkenberg, 1971), but is usually serious only in poorly managed orchards (Lim and Khoo, 1985). In these situations, mites, insects and other foliar diseases can increase the severity of the disease. Algal leafspot requires a humid environment to establish and spread. The alga's zoospores are the primary infective propagules, and they are dispersed by rainsplash and wind.

Pruning, reduced planting densities and mowing beneath trees increases ventilation and helps reduce conditions that favour the disease. Also beneficial are measures that increase tree vigour, including adequate fertilization and irrigation, control of insect and mite pests, and management of other foliar diseases. Algicides or fungicides that contain copper are effective, but are needed only in extreme cases.

Ceratocystis paradoxa

Ceratocystis paradoxa (anamorph: *Chalara paradoxa*) causes butt rot, fruit rot and a leaf spot of pineapple, stem bleeding of coconut, foot rot of fig, and fruit rot of carambola (Chapters 6, 8, 11 and 19). Another species in the genus, *C. fimbriata* (anamorph: *Chalara* sp.), causes seca of mango in Brazil (Chapter 15).

C. paradoxa is an ascomycete in the order *Microascales*, and is widespread in the tropics. Hyphal growth occurs between 21 and 32°C, with optimum conditions of 25°C and pH 6 (Frossard, 1964). Conidia, which are also called endoconidia, and chlamydospores are hyaline to slightly brown, cylindrical to slightly oval, 6–24 × 2–5.5 μm, and are extruded through the end of the conidiophore (Fig. 1.7). Chains of thick-walled chlamydospores are produced in older cultures that are smooth and oval, brown to black, and average 10–25 × 7.5–20 μm (Dade, 1928). Strains with light and dark coloured mycelium have been reported from sugarcane and pineapple (Rashid, 1975; Simone, 1994). The perithecia and ascospores of the teleomorph are produced only occasionally.

Erythricium salmonicolor

Erythricium salmonicolor (anamorph: *Necator decretus*) causes pink disease on a wide range of agricultural crops including breadfruit, carambola, citrus, custard apple, durian, jackfruit, mango, mangosteen and rambutan (Chapters 5–7, 10 and 14–16). It produces basidiospores on sterigmata that arise from narrowly clavate to cylindrical basidia (Holliday, 1980; Lim and Khoo, 1985). They are hyaline, broadly ellipsoidal, thin-walled and 8–10 × 5–7 μm (Fig. 1.8). The anamorph produces ellipsoidal conidia that are unicellular, 10–18 × 6–12 μm and hyaline, but appear orange *en masse* in the erumpent or superficial sporodochia.

Pink disease is most important under high rainfall, tropical conditions, and basidiospores are known to infect rubber (Holliday, 1980). Successful management depends on early and accurate identification of the disease followed by prompt application of a fungicide. The following fungicides have been found to control pink disease on mango and other fruit crops: triadimefon,

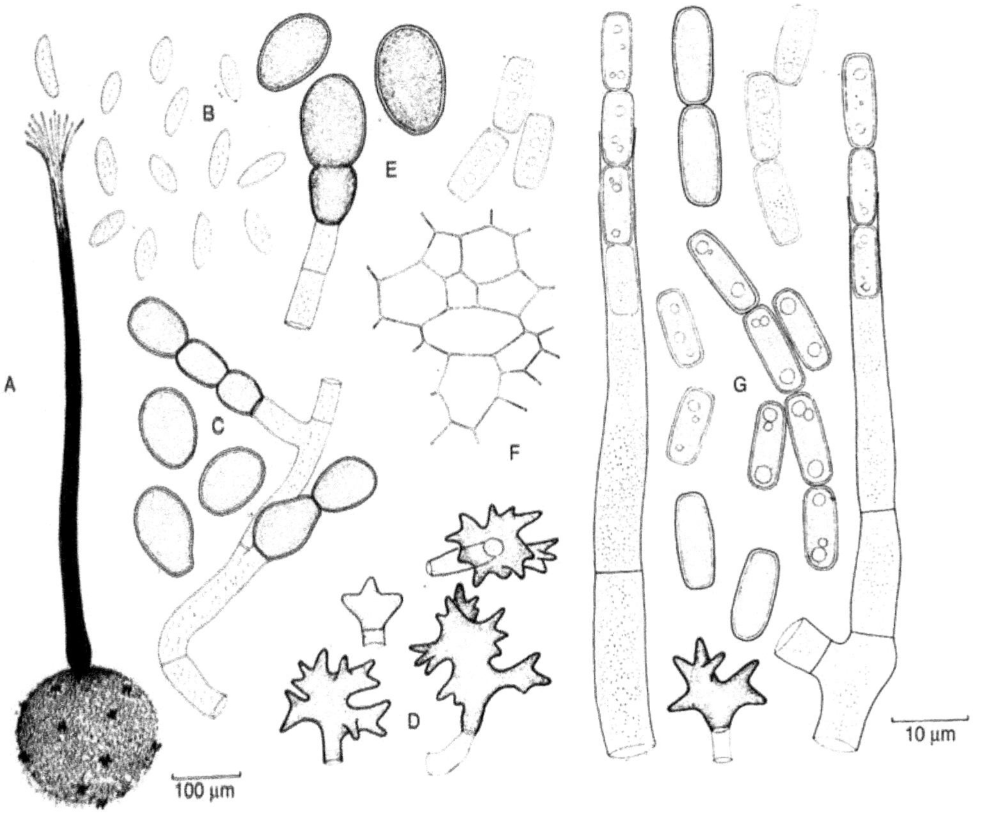

Fig. 1.7. (A) Perithecium, (B) ascospores, (D) perithecial appendages and (E) surface of the perithecial wall of *Ceratocystis paradoxa*, and (C) chlamydospores, (F) conidiophores and (G) conidia of its anamorph, *Chalara paradoxa* (from CMI description no. 143).

tridemorph, oxycarboxin, flusilazol and protectant fungicides such as copper hydroxide, copper oxide, copper oxychloride and Bordeaux mixture. They are applied by spraying or painting the infected bark.

Fusarium oxysporum

Special host-adapted forms of *Fusarium oxysporum*, the formae speciales, cause vascular wilts of banana (f. sp. *cubense*), citrus (f. sp. *citri*), date (f. sp. *albedinis*) and passion fruit (f. sp. *passiflorae*) (Chapters 4, 7, 9 and 18). In culture, colonies are fast growing (4–7 mm diameter on PDA at 24°C), with sparse to abundant aerial mycelium, and white, pink, salmon or purple pigmentation (Gerlach and Nirenberg, 1982; Nelson *et al.*, 1983). When formed, sporodochia are tan to orange, and sclerotia are blue and submerged (Fig. 1.9). Micro- and macroconidia form on branched and unbranched monophialides. Microconidia are one- or two-celled, oval- to kidney-shaped, and are borne in false heads. Macroconidia are four- to eight-celled, sickle-shaped, thin-walled and delicate, with foot-shaped basal and attenuated apical cells. Dimensions of the micro- and macroconidia are 5–16 × 2.4–3.5 µm and 27–55 × 3.3–5.5 µm, respectively (Gerlach and Nirenberg, 1982). Terminal and intercalary chlamydospores are usually globose, and are formed singly (7–11 µm) or in pairs in hyphae or conidia. The species has no teleomorph.

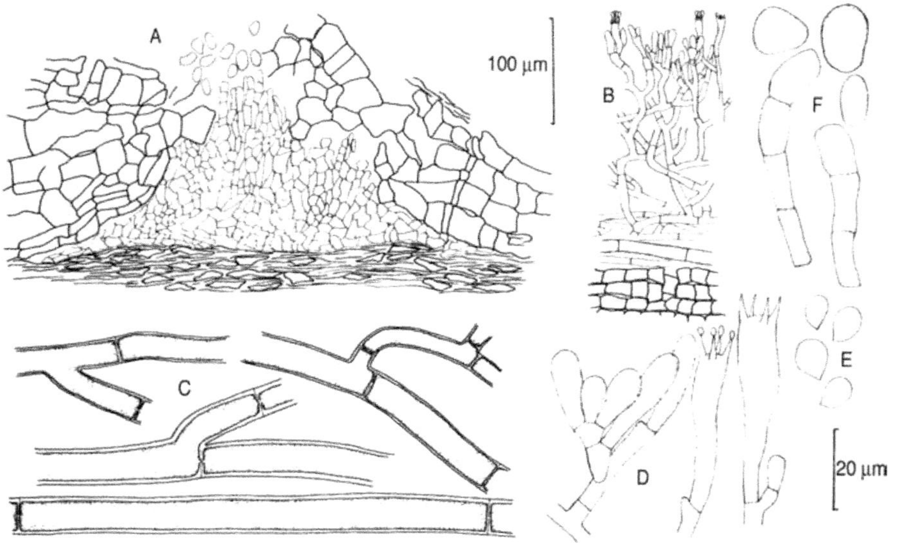

Fig. 1.8. (A) Conidium-bearing pustule and (F) conidiogenous cells and conidia of *Necator decretus*, and (B) hymenium, (C) basal hyphae, (D) immature and mature basidia, and (E) basidospores of its teleomorph, *Erythricium salmonicolor* (from CMI description no. 511).

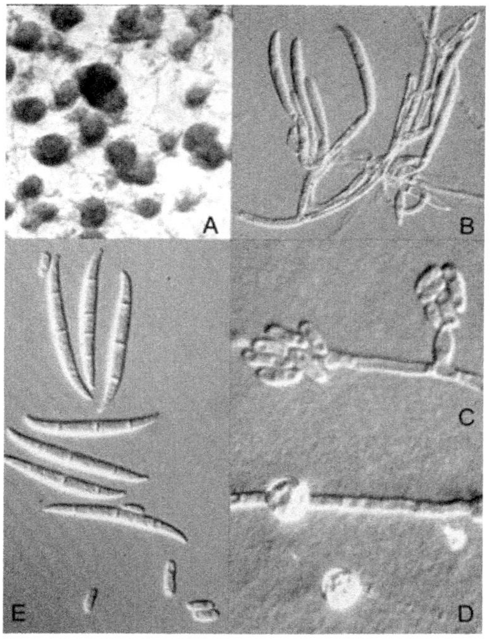

Fig. 1.9. (A) Sporodochia, (B) macroconidia, (C) microconidia in false head on monophialide, (D) terminal and intercalary chlamydospores, and (E) macroconidia and microconidia of *Fusarium oxysporum* (photo: K. O'Donnell, USDA).

In general, the plant pathogenic members of the species possess excellent saprophytic capabilities, and can survive for long periods as chlamydospores in host tissues, and as non-pathogenic parasites on alternative hosts. They damage their hosts by systemically colonizing and occluding the host xylem. The diseases they cause are difficult to control. No effective chemical means exist, and in most situations they can only be managed with host resistance.

Glomerella acutata

Glomerella acutata (anamorph: *Colletotrichum acutatum*) affects avocado, breadfruit, carambola, citrus, fig, guava, kiwifruit, lychee, mango and papaya (Chapters 3, 5–7, 11–15 and 17). It causes anthracnose primarily on fruit, but is usually less important than *G. cingulata*. It is also responsible for a serious fruit set disease on citrus, postbloom fruit drop.

Colonies are effuse, white becoming pale orange then greenish grey or black, often with a pink or reddish purple reverse (Dyko and Mordue, 1979). Conidia are hyaline, one-celled, straight, smooth, fusiform, 8–16 × 2.5–4 μm and salmon-coloured *en masse* (Fig. 1.10). Conidiophores are hyaline, septate, branched infrequently near the base and smooth. Appressoria are sparse, mostly light to medium brown, clavate to obvate, 6.5–11 × 4.5–7.4 μm, with smooth margins. Acervuli are superficial to subcuticular, up to 0.5 mm in diameter, and may or may not have setae that are brown, smooth, septate, straight to slightly curved, tapered to apices and 46.5–85 × 3–4 μm. *C. acutatum* differs from *C. gloeosporioides* in its orange to pink colony coloration during the first few weeks of growth and its fusiform conidia.

The fungus's teleomorph was described recently (Guerber and Correll, 2001). It is heterothallic and strains from avocado, kiwifruit and papaya formed perithecia in the original study. Whether strains from other hosts are fertile has not been reported. Perithecia have not been observed in the field. They are globose to obpyriform, generally ampuliform, 125–312 μm wide, ostiolate, periphysate and black–brown. Asci contain up to eight spores, are narrowly clavate, unitunicate, fasciculate, attached by a short pedicle and terminated by an annular pore. Ascospores generally are straight, oblong to ellipsoidal, hyaline, single-celled and 8.5–25.1 × 3.1–8.1 μm.

Glomerella cingulata

Glomerella cingulata (anamorph: *Colletotrichum gloeosporioides*) has the widest host range of any of the pathogens that are covered in this

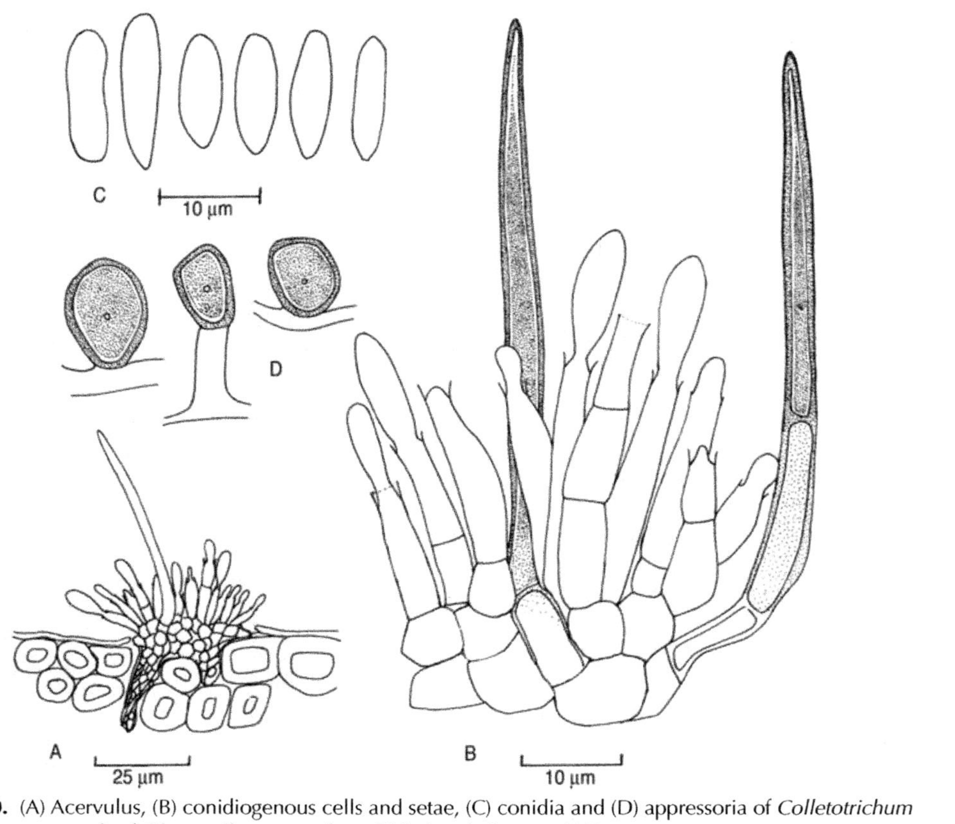

Fig. 1.10. (A) Acervulus, (B) conidiogenous cells and setae, (C) conidia and (D) appressoria of *Colletotrichum acutatum*, anamorph of *Glomerella acutata* (from CMI description no. 630).

book. It causes significant problems on avocado, biriba, breadfruit, carambola, cherimoya, citrus, custard apple, durian, fig, guava, ilama, jackfruit, lychee, mango, mangosteen, papaya, passion fruit, soursop, sugar apple and rambutan (Chapters 2, 3, 5–7, 10–12 and 14–18). Although it is most important as a fruit pathogen, it also causes branch and leaf diseases (Jeffries *et al.*, 1990). It is also a common endophyte and saprophyte.

On PDA, colonies are whitish to dark grey with thick to sparse lawns of aerial mycelium (Holliday, 1980; Jeffries *et al.*, 1990). Conidia are hyaline, one-celled, 7–20 × 2.5–5 μm and either cylindrical with obtuse ends or ellipsoidal with a rounded apex and a narrow, truncate base (Fig. 1.11). They form on light brown conidiophores in irregular acervuli and, upon maturity, appear orange and slimy *en masse*. Acervuli develop in lesions on leaves, branches and fruit, and conidia in acervuli remain viable for long periods, even under adverse climatic conditions. Setae that form in acervuli are brown, 4–8 × 200 μm, and two- to five-celled. The fungus is heterothallic and, although the teleomorph can be induced readily in culture (e.g. Correll *et al.*, 2000), it is observed rarely in the field. Perithecia are subspherical, dark brown to black, 90–220 μm in diameter and contain hyaline, unitunicate asci (Fig. 1.11) (Cedeño *et al.*, 1993; Wolcan and Larran, 2000). Ascospores are unicellular, curved, hyaline and 14–20 × 5–6 μm.

Conidia are the most important type of inoculum (Jeffries *et al.*, 1990). They are produced on virtually all host tissues and are usually dispersed by rainsplash. Moderate temperatures (25–30°C) and free moisture are needed for optimum production, germination and infection. New leaf flushes usually are most susceptible. Although fruits can be infected at any stage of development, infections that occur before ripening usually progress no further than the formation of appressoria. Once ripening commences, infection pegs form that forcefully penetrate the host. The biochemical events that control dormancy and infection are complex but are beginning to be understood for some of the crops in this book, including avocado and mango (Latunde-Dada, 2001).

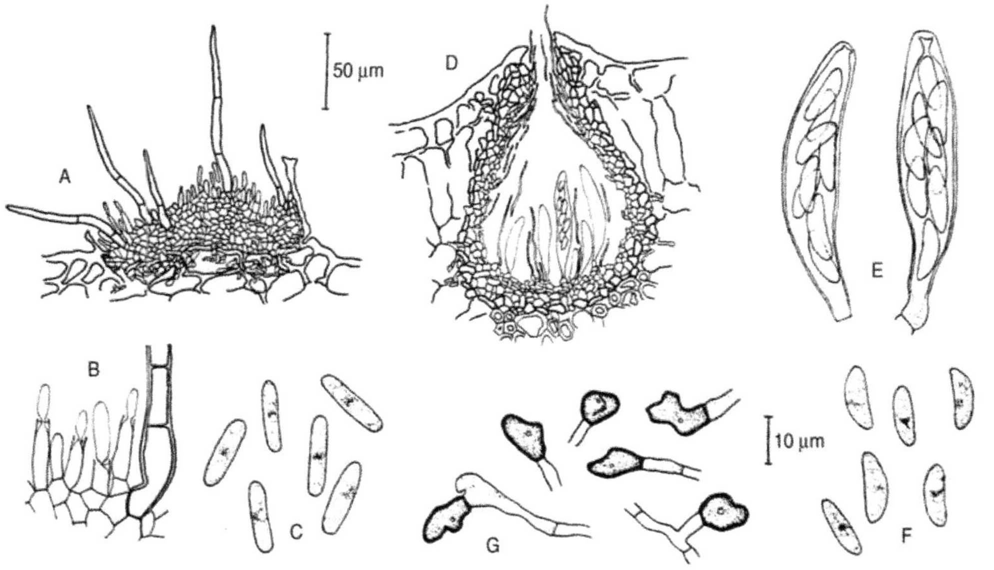

Fig. 1.11. (A) Acervulus and emergent setae, (B) conidiophores, (C) conidia, (D) perithecium, (E) asci, (F) ascospores and (G) appressoria at hyphal termini of *Glomerella cingulata* (anamorph: *Colletotrichum gloeosporioides*) (from CMI description no. 315).

Fungicide application focuses on reducing damage to fruit and, for some crops, inflorescences (Jeffries *et al.*, 1990). Trees may also require protection in crowded nurseries that receive overhead irrigation. Postharvest control usually utilizes hot water treatments and fungicide dips (Johnson *et al.*, 1997).

Phytophthora spp.

Members of this genus are among the most important of all tropical fruit pathogens. They have fungal-like lifestyles but are in the Kingdom *Chromista*, rather than the *Eumycota* (the true fungi). As oomycetes, they have diploid, coenocytic, vegetative hyphae, and cell walls that are made primarily of cellulose, rather than the chitin true fungi possess. Although selective media or specialized techniques may be required, these pathogens are fairly easy to recover from soil or infected host tissue (Tsao, 1990).

These pathogens produce a variety of propagules including chlamydospores, hyphal swellings, oospores, sporangia and zoospores (Erwin and Ribiero, 1996). Oospores are sexual spores that form after the fusion of antheridia (male structure) and oogonia (female). The position of the antheridium on the oogonium is an important morphological feature. If the oogonial incept grows through the antheridium, the antheridium is amphigynous, but if the antheridium is appressed laterally to the oogonial incept it is paragynous. If the oospore entirely fills the oogonium, it is plerotic, whereas those that are incompletely filled are aplerotic. Most of the species below are heterothallic and, thus, usually require isolates of both the A1 and A2 mating types for oospore formation. Oospores may germinate directly or indirectly by forming a sporangium. Chlamydospores and hyphal swellings are produced by some species and, unlike sporangia and zoospores, they and the oospores are capable of long-term survival in soil and plant tissues. Sporangia of some species are caducous (deciduous), and pedicel length in these species is a useful diagnostic character. Whether sporangia possess papilla is also a useful feature for species identification.

These pathogens prefer wet conditions and require free moisture for the production and motility of zoospores (Erwin and Ribiero, 1996). Thus, the diseases they cause are most severe in flooded or poorly drained sites. Motile zoospores have no cell wall, are kidney-shaped, 11×18 μm, and have both a posterior and anterior flagellum that propel the spore through water-filled pores in soil. Zoospores are attracted chemotactically to host products such as amino acids and carbohydrates that exude from roots, particularly at the zone of elongation behind the root tip. Upon attachment to the root surface, zoospores encyst, form a cell wall and infect their host, all within an hour. Many sporangia may be produced from a single root infection, and each can give rise to numerous zoospores that initiate new root infections. For these reasons, sporangia and zoospores are responsible for the extremely rapid spread of disease.

Although these pathogens are primarily soilborne, several of the important diseases discussed in this book occur on aerial portions of the host. In the latter cases, inoculum can be moved by rainsplash and irrigation water to aboveground portions of the host, and by wind-driven rain within and among plants.

Readers interested in detailed information on the species below, as well as the pathology, ecology, physiology and genetics of the genus, are referred to *Phytophthora Diseases Worldwide* (Erwin and Ribiero, 1996).

P. cinnamomi affects cherimoya, kiwifruit and pineapple, and is by far the most important pathogen of avocado (Chapters 2, 3, 13 and 19). It affects well over 1000 species of plants and, due to its unique suite of morphological characters, is one of the most easily identified species in the genus (Erwin and Ribiero, 1996). *P. cinnamomi* produces distinctive corraloid mycelium (Fig. 1.12). Its non-papillate, non-caducous sporangia are elliptical to ovoid, but are rarely formed in culture (Fig. 1.12). Their dimensions range dramatically (11–123 μm $\times$ 11–63 μm), depending on the host and reporting authors. Terminal and intercalary chlamydospores, 31–50 (40) μm in diameter, are abundant in culture and usually formed in botryose clusters. Their cell walls are much thinner than

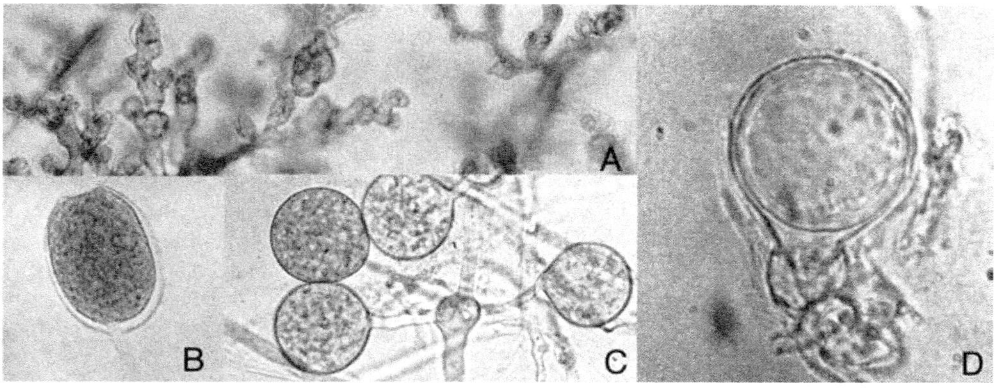

Fig. 1.12. (A) Corraloid mycelium, (B) non-papillate sporangium, (C) hyphal swellings and (D) oogonium and amphigyrous antheridium of *Phytophthora cinnamomi* (from CMI description no. 113).

those that are produced by other species. Hyphal swellings can also be abundant. *P. cinnamomi* is heterothallic. Oogonia are 21–58 (40) μm in dia, antheridia are amphigynous, and oospores are plerotic. The cardinal temperatures for growth are 5–15, 20–32.5 and 30–36°C. Detailed information on the epidemiology and management of Phytophthora root rot of avocado is given in Chapter 3.

P. citricola causes diverse diseases of avocado, fig, guava and kiwifruit (Chapters 3 and 11–13). It produces non-caducous sporangia that vary from obovoid, obclavate and obpyriform to slightly flattened on one side (Erwin and Ribiero, 1996). They are semipapillate and can have a single apex or be deeply bifurcated with two apices, or irregularly shaped with three or four apices (Fig. 1.13). Sporangia are 30–75 (47) μm × 21–44 (34) μm. Chlamydospores are rare. *P. citricola* is homothallic. Antheridia are paragynous and oogonia are 18–35 (26) μm in diameter. Oospores are 16–30 (22) μm in diameter and almost plerotic. Its cardinal temperatures for growth are 3, 25–28 and 31°C.

P. citrophthora affects chempedek, citrus and kiwifruit (Chapters 5, 7 and 13). It produces variable, non-caducous sporangia (Fig. 1.14) (Mchau and Coffey, 1994; Erwin and Ribiero, 1996). They range in shape from spherical, ovoid, obpyriform, obturnate, ellipsoidal to extremely distorted, and in size from 23 to 90 μm in length and from 18 to 60 μm in width. They are persistent, mostly papillate, and often have two or more papilla. Sporangiophores are irregularly

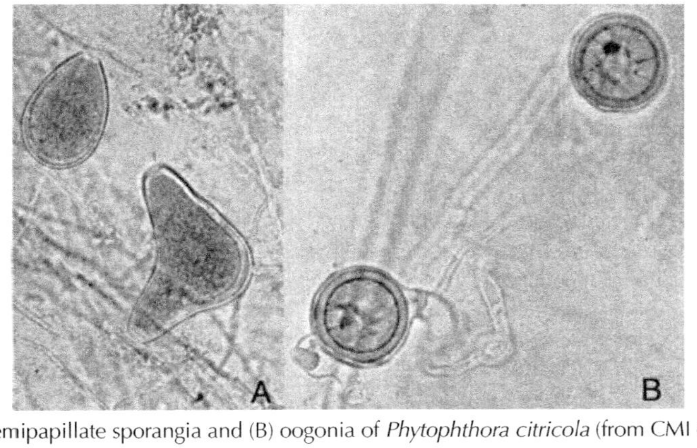

Fig. 1.13. (A) Semipapillate sporangia and (B) oogonia of *Phytophthora citricola* (from CMI description no. 114).

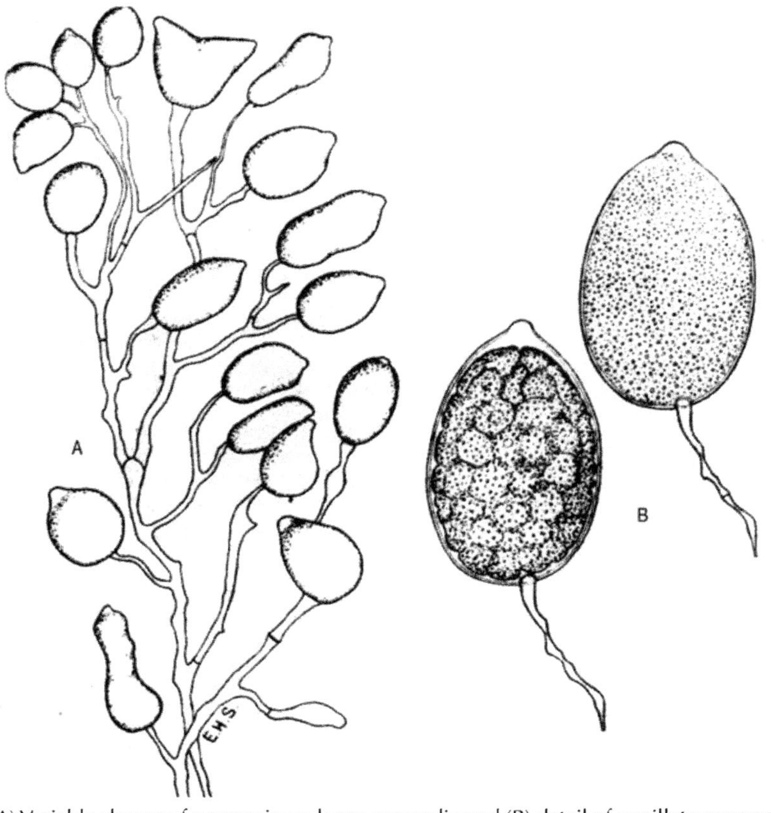

Fig. 1.14. (A) Variable shapes of sporangia on loose sympodia and (B) detail of papillate sporangia of *Phytophthora citrophthora* (from Waterhouse, 1956).

branched, some singly and some in loose sympodia with swellings at the branch points. Chlamydospores are uncommon for isolates from citrus, and sex organs do not occur in nature although oospores can be induced when some isolates are paired on carrot agar. The minimum temperature for growth is <5°C, the optimum is between 24 and 28°C, and the maximum is 32–33°C.

P. nicotianae causes fruit, heart and root rots on carambola, fig, pineapple, rambutan and sugar apple (Chapters 2, 6, 11, 14 and 19). It forms non-caducous ellipsoid, ovoid, pyriform to spherical sporangia with usually a single papillum (Fig. 1.15) (Erwin and Ribiero, 1996). They are produced either singly or in sympodia on stalks that range from 100 to 595 μm in length, and are 11–60 (40) μm × 20–45 (29) μm, with a length : breadth ratio of 1.1 : 1.7 (1.34). The pathogen forms intercalary and terminal chlamydospores that are 13–60 μm in diameter. Most isolates are heterothallic. Antheridia are amphigynous and spherical or oval, and oogonia are smooth, spherical and 15–64 μm in diameter. Oospores are aplerotic. Its cardinal temperatures for growth are 5–7, 27–32 and 37°C.

P. palmivora causes bud, crown, fruit, heart and root rots of atemoya, avocado, breadfruit, coconut, durian, fig, longan, mango, papaya, pineapple, pond apple and soursop (Chapters 2, 3, 5, 8, 10, 11, 14, 15, 17 and 19). It is a ubiquitous pathogen in the tropics with a very wide host range (Erwin and Ribiero, 1996). Its hyphae are often irregular and up to 7 μm in dia. The sporangia are prominently papillate, caducous, and ovoid, limoniform or ellipsoid. They are 40–60 × 25–35 μm with

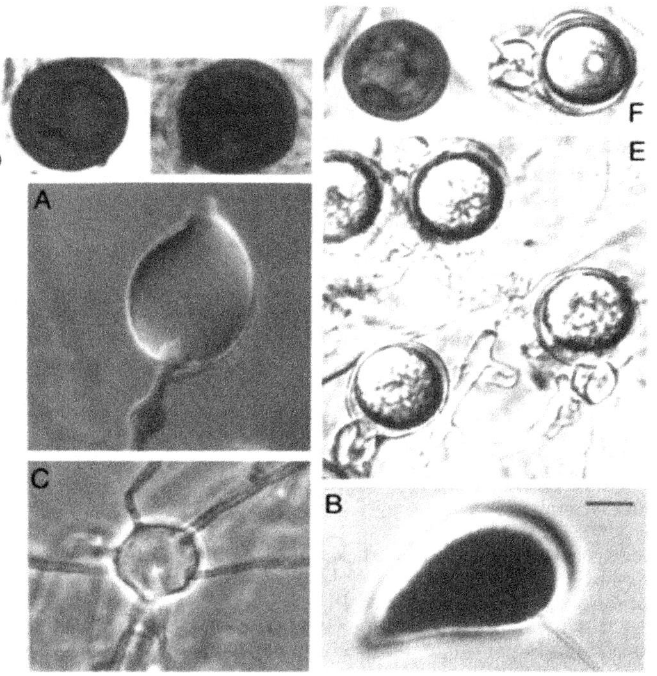

Fig. 1.15. (A) Discharged sporangium showing papilla, basal plug, swelling on the sporangiophore and external proliferation, (B) sporangium with lateral attachment, (C) hyphal swelling with radiating hyphae, (D) chlamydospores, (E) young oogonia with amphigynous antheridia and (F) oogonium with amphigynous antheridium and thick-walled oospore of *Phytophthora nicotianae*. Bar = 15 μm (from CMI description no. 1200).

length : breadth ratios of 1.3 : 1.8 (1.5) that may be lower on natural substrates (Fig. 1.16). Pedicels are <5 μm in length. Chlamydospores are formed by most isolates, and are spherical to subspherical, terminal or intercalary, and average 37 μm in diameter. *P. palmivora* is heterothallic, forming oospores either when crosses of the A1 and A2 mating types of the species are made or when they are crossed with the opposite mating type of several other species in Waterhouse's Group II (Waterhouse, 1956). Oospores are spherical and 22–24 μm in diameter, and antheridia are amphigynous. Its cardinal temperatures for growth are 11, 27.5–30 and 35°C.

Rigidoporus lignosus

Rigidoporus lignosus causes white root rot of carambola, durian and mango (Chapters 6, 10 and 15). It is a basidiomycete that is a common soil inhabitant in the humid tropics of Africa and Asia (Holliday, 1980). It has also been reported in the western hemisphere, but this identification may be in error. *R. lignosus* has a large host range on woody perennials that contains many important crops, including rubber, the host on which the pathogen was first reported (Nandris *et al.*, 1987).

R. lignosus produces white rhizomorphs on the surfaces of roots and root crowns which later darken to a yellowish and then a reddish colour (Lim and Khoo, 1985; Nandris *et al.*, 1987). The leading edge of the rhizomorph is well defined and seldom appears above ground. It must undergo a morphogenic change to produce infectious hyphae which are responsible for penetration of the host epidermis and the subsequent extracellular, enzymatic degradation of wood. *R. lignosus* produces a non-differentiated white rot and, as its specific name implies, degrades the lignin in host cell walls.

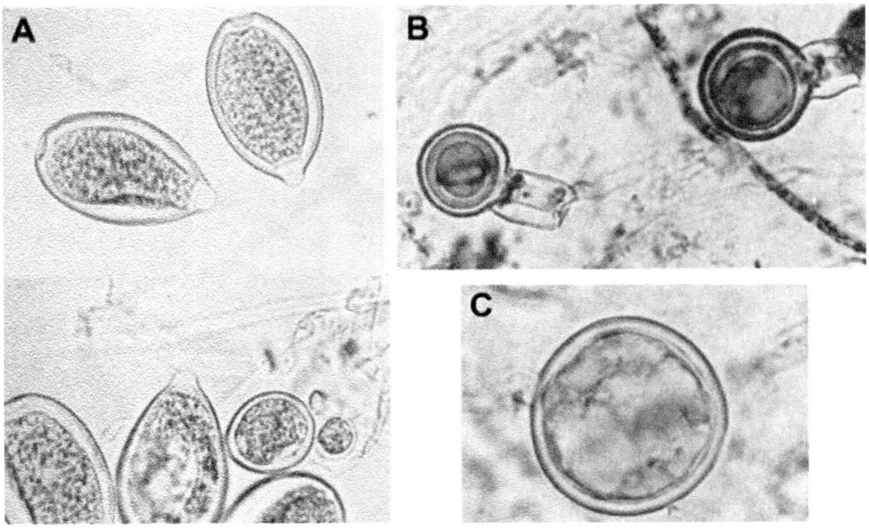

Fig. 1.16. (A) Sporangia, (B) oogonia with amphygynous antheridia and oospores and (C) chlamydospores of *Phytophthora palmivora* (from CMI description no. 831).

The fungus is most damaging if orchards are established in old rubber plantations or newly cleared jungle sites (Lim and Khoo, 1985). Previously colonized stumps and infected woody debris of rubber and other hosts are primary sources of inoculum. Orange–yellow, bracket-like sporophores are produced during the rainy season on the root collar, trunk or exposed roots (Fig. 1.17). Basidiospores produced on the sporophores are viable, and are thought to play a secondary role in disseminating the disease; at most, they probably colonize exposed stump surfaces. Rhizomorphs are more significant epidemiologically, since they grow rapidly and can advance great distances in soil in the absence of woody substrates. The most effective means for controlling white root rot rely on eliminating or avoiding colonized woody debris when new orchards are established (Lim and Khoo, 1985).

Rosellinia necatrix

Rosellinia necatrix (anamorph: *Dematophora necatrix*) causes root rot on avocado, cherimoya, citrus, fig and kiwifruit (Chapters 2, 3, 7, 11 and 13). The rarely seen teleomorph appears as swarms of spherical, black, 1–2 mm in diameter perithecia that have papillate ostioles (Hanlin, 1990). They are embedded in a mat of septate, brown hyphae. Asci are cylindrical, 250–380 × 8–12 μm and contain a single row of ascospores that are dark brown and one-celled (Fig. 1.18). They are 30–50 × 5–8 μm, elongate, often laterally compressed, with a longitudinal germ slit. Conidiophores of the more commonly found anamorph are produced on brown, ropey, rigid synnemata composed of intertwined, laterally cemented hyphae. The synnemata are up to 1.5 mm high and 40–300 μm thick. Conidia that are produced on the synnemata are one-celled, solitary, elliptical to ovoid, colourless to pale brown, smooth and 3–4.5 × 2–2.5 μm. Peculiar pear-shaped swellings often occur near the septa on the hyphae, especially on older hyphae. The fungus also produces scattered, black, rough, irregular masses of 98 × 130 μm microsclerotia that often unite to form irregular, flattened masses or sheets.

R. necatrix can be isolated from infected avocado roots on malt agar or it can be baited from soil using avocado leaf discs. It can survive for long periods in wood, roots

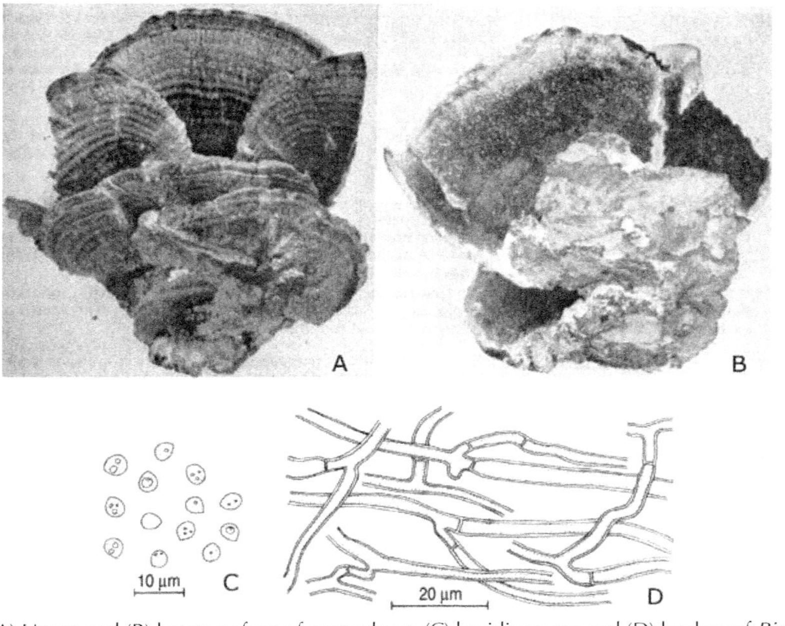

Fig. 1.17. (A) Upper and (B) lower surface of sporophore, (C) basidiospores and (D) hyphae of *Rigidoporus lignosus* (from CMI description no. 198).

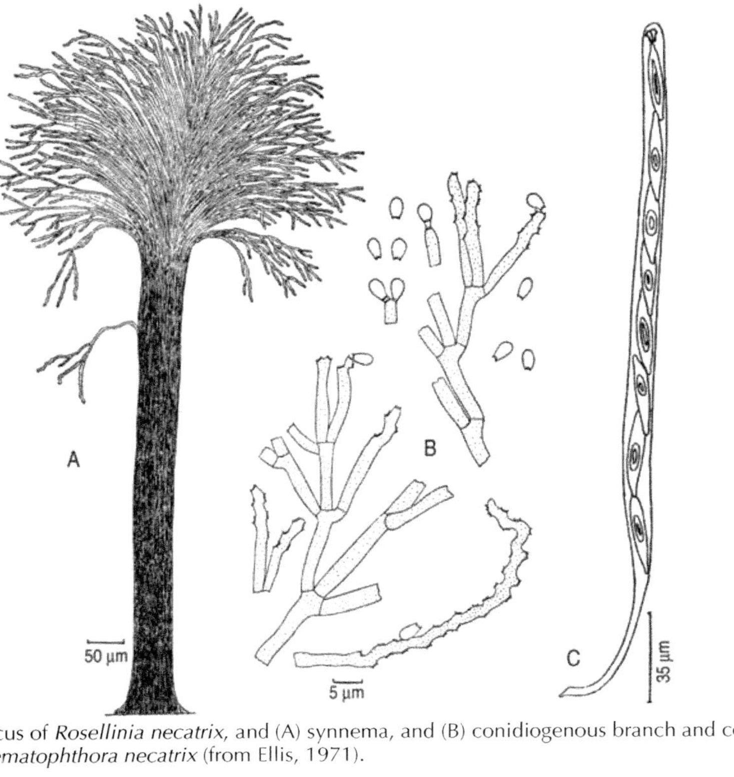

Fig. 1.18. (C) Ascus of *Rosellinia necatrix*, and (A) synnema, and (B) conidiogenous branch and conidia of its anamorph, *Dematophthora necatrix* (from Ellis, 1971).

and soil, primarily as microsclerotia. Feeder roots are directly infected when they contact hyphae or microsclerotia. The infection spreads into woody roots and may spread from tree to tree in this manner. Neither ascospores nor conidia appear to play a role in spreading the disease. The fungus is spread from orchard to orchard in soil contaminated with microsclerotia, infested roots and organic matter.

References

Cedeño, L., Mohali, S. and Palacios-Prü, E. (1993) Antracnose causada por dos cepas de *Glomerella cingulata* em frutos de parchita. *Fitopatologîa Venezolana* 6, 30–33.

Correll, J.C., Guerber, J.C., Wasilwa, L.A., Sherrill, J.F. and Morelock, T.E. (2000) Inter- and intra-species variation in *Colletotrichum* and mechanisms which affect population structure. In: Prusky, D., Freeman, S. and Dickman, M.B. (eds) Colletotrichum. *Host Specificity, Pathology, and Host–Pathogen Interaction*. APS Press, St Paul, Minnesota, pp. 145–179.

Crous, P.W. and Palm, M.E. (1999) Reassessment of the anamorph genera of *Botryoplodia, Dothiorella* and *Fusicoccum*. *Sydowia* 52, 167–175.

Dade, H.A. (1928) *Ceratostomella paradoxa*, the perfect stage of *Thielaviopsis paradoxa* (De Seynes) Von Hohnel. *Transactions of the British Mycological Society* 13, 184–194.

Denman, S., Crous, P.W., Taylor, J.E., Kang, J.-C., Pascoe, I. and Wingfield, M.J. (2000) An overview of the taxonomic history of *Botryosphaeria*, and a re-evaluation of its anamorphs based on morphology and ITS and rDNA phylogeny. *Studies in Mycology* 45, 129–140.

Domsch, K.H., Gams, W. and Anderson, T.H. (1980) *Compendium of Soil Fungi*. Academic Press, New York.

Dyko, B.J. and Mordue, J.E.M. (1979) Colletotrichum acutatum. *CMI Descriptions of Pathogenic Fungi and Bacteria No. 630*. Commonwealth Mycological Institute, Kew, UK.

Ellis, M.B. (1971) *Dematiaceous Hypomycetes*. Commonwealth Mycological Institute, Kew, UK.

Erwin, D.C. and Ribiero, O.K. (1996) *Phytophthora Diseases Worldwide*. APS Press, St Paul, Minnesota.

Frossard, P. (1964) Influences of temperature and acidity on the growth in culture of *Thielaviopsis paradoxa*, a pineapple parasite. *Fruits* 19, 461–463.

Gerlach, W. and Nirenberg, H. (1982) *The Genus* Fusarium *– a Pictorial Atlas*. Paul Parey, Berlin.

Guerber, J.C. and Correll, J.C. (2001) Characterization of *Glomerella acutata*, the teleomorph of *Colletotrichum acutatum*. *Mycologia* 93, 216–229.

Hanlin, R.T. (1990) *Illustrated Genera of Ascomycetes*. APS Press, St Paul, Minnesota.

Holliday, P. (1980) *Fungus Diseases of Tropical Crops*. Cambridge University Press, Cambridge, UK.

Jeffries, P., Dodd, J.C., Jeger, M.J. and Plumbley, R.A. (1990) The biology and control of *Colletotrichum* species on tropical fruit crops. *Plant Pathology* 39, 343–366.

Johnson, G.I. (1994) Stem-end rot. In: Ploetz, R.C., Zentmyer, G.A., Nishijima, W.T., Rohrbach, K.G. and Ohr, H.D. (eds) *Compendium of Tropical Fruit Diseases*. APS Press, St Paul, Minnesota, pp. 39–41.

Johnson, G.I., Sharp, J.L., Milne, D.L. and Oosthuyse, S.A. (1997) Postharvest technology and quarantine treatments. In: Litz, R.E. (ed.) *The Mango: Botany, Production and Uses*. CAB International, Wallingford, UK, pp. 447–508.

Joubert, J.J. and Rijkenberg, F.H.J. (1971) Parasitic green algae. *Annual Review of Phytopathology* 9, 45–64.

Latunde-Dada, A.O (2001) *Colletotrichum*: tales of forcible entry, stealth, transient confinement and breakout. *Molecular Plant Pathology* 2, 187–198.

Lim, T.K. and Khoo, K.C. (1985) *Diseases and Disorders of Mango in Malaysia*. Tropical Press, Kuala Lumpur.

Mchau, G.R.A. and Coffey, M.D. (1994) An integrated study of morphological and isozyme patterns found within a worldwide collection of *Phytophthora citrophthora* and a redescription of the species. *Mycological Research* 98, 1269–1299.

Michailides, T.J., Morgan, D.P. and Felts, D. (1999) Other hosts of *Botryosphaeria dothidea* and their relation to Botryosphaeria pistachio blight. In: *California Pistachio Production Reports*. California Pistachio Commission. Fresno, California, USA.

Nandris, D., Nicole, M. and Geiger, J.P. (1987) Root rot diseases of rubber trees. *Plant Disease* 71, 298–306.

Neergaard, P. (1945) *Danish species of* Alternaria *and* Stemphylium. *Taxonomy, Parasiticism, Economical Significance*. Einar Munksgaard, Copenhagen.

Nelson, P.E., Toussoun, T.A. and Marasas, W.O. (1983) Fusarium *Species. An Illustrated Guide for Identification.* Pennsylvania State University Press, University Park, Pennsylvania.

Rashid, A.R. (1975) Ecological studies of *Ceratocystis paradoxa* (De seynes) Moreau in pineapple and sugarcane soils in Hawaii. Thesis, Department of Plant Pathology, University of Hawaii, Honolulu, Hawaii.

Shaw, C.G. and Kile, G.A. (1991) Armillaria *Root Disease.* United States Department of Agriculture, Forest Service, Agricultural Handbook 691.

Simone, G.W. (1994) Stem bleeding disease. In: Ploetz, R.C., Zentmyer, G.A., Nishijima, W.T., Rohrbach, K.G. and Ohr, H.D. (eds) *Compendium of Tropical Fruit Diseases.* APS Press, St Paul, Minnesota, pp. 27–28.

Slippers, B., Johnson, G.I., Cooke, A.W., Crous, P.W., Coutinho, T.A., Wingfield, B.D. and Wingfield, M.J. (2001) Taxonomy of *Botryosphaeria* spp. causing stem end rot of mango. In: *Proceedings of the 13th Biennial Australasian Plant Pathology Conference.* 24–27 September, 2001, Cairns, Australia.

Sutton, B.C. (1980) *The Coelomycetes.* Commonwealth Mycological Institute, Kew, UK.

Tsao, P.H. (1990) Why many phytophthora root rots and crown rots of tree and horticultural crops remain undetected. *OEPP/EPPO Bulletin* 20, 11–17.

Waterhouse, G.M. (1956) *The Genus* Phytophthora. *Diagnoses (or Descriptions) and Figures from the Original Papers.* The Commonwealth Mycological Institute, Kew, UK.

Wolcan, S. and Larran, S. (2000) First report of anthracnose cause by *Glomerella cingulata* on passion fruit in Argentina. *Plant Disease* 84, 706.

Zhou, S. and Stanosz, G.R. (2001) Relationships among *Botryosphaeria* species and associated anamorphic fungi inferred from the analyses of ITS and 5.8S rDNA sequences. *Mycologia* 93, 516–527.

2 Diseases of Atemoya, Cherimoya, Soursop, Sugar Apple and Related Fruit Crops

Randy C. Ploetz
University of Florida, Tropical Research and Education Center, Homestead, Florida, USA

Introduction

The *Annonaceae* is a large neotropical family of trees and small shrubs that contains ~75 genera (Nakasone and Paull, 1998). Of these, *Annona* is the most important. It contains ~100 species and the most significant fruit crops. Martin *et al.* (1987) listed ten species or hybrids of major importance in the family, but only seven of these are grown commercially.

The Significant Species

Cherimoya, *A. cherimola*, originated in the highlands of Ecuador and Peru, but is now widely grown in the tropics and subtropics (Nakasone and Paull, 1998). Since the species requires cool night temperatures for appreciable fruit set, it is restricted to tropical highland and cool subtropical environments. Like many species in the family, it requires hand pollination for good production. The fruit weighs from 200 to 2000 g, is sweet and is eaten out of the hand or used in ice creams and sherbets. Propagation is by either seed or grafting, and superior clones have been selected in Australia, Chile, Ecuador, New Zealand, Spain and the USA (California) (Nakasone and Paull, 1998).

Ilama, *A. diversifolia*, originated in western Mexico and Central America (Nakasone and Paull, 1998). It requires hot, humid conditions and is quite sensitive to frost. It is often a poor producer of fruit and, since it is not widely adapted, it is not widely grown. Its fruit are comparable with those of cherimoya and weigh between 455 and 650 g.

Soursop, *A. muricata*, originated in the Caribbean and now has a pantropical distribution (Nakasone and Paull, 1998). It is the most tropical of these crops and demands hot, high rainfall conditions. It is very sensitive to temperatures below ~5°C, and freezing temperatures usually kill plants. Fruit set, which can be poor, presumably is related to frequency of pollination. Fruit weigh from 1 to 4 kg, are sweet to subacid in taste, and are eaten fresh and used in drinks and desserts. Propagation is primarily by seed, but superior genotypes have been selected from seedling populations for vegetative propagation in Malaysia and Mexico.

Custard apple or Bullock's heart, *A. reticulata*, originated in tropical America, but is now found throughout the tropics (Martin *et al.*, 1987). In some locations, such as Australia, the annonaceous fruits are referred to generically as custard apples. To avoid confusion in this chapter, custard apple will refer specifically to *A. reticulata*. Custard apple grows under the same types of conditions that are favoured by soursop, but it tolerates light frost. Its fruit weigh between 400 and 1000 g, and are used in the same ways as cherimoya, although their flavour is not as good. Cracking of the fruit surface at maturity, a common problem with annonaceous fruit, is particularly severe in this species and further limits the marketability of these fruit (Paull, 1996). Custard apple is propagated by seed.

Sugar apple or sweetsop, *A. squamosa*, is from tropical America and widely distributed in the eastern and western hemispheres. Its environmental requirements mirror those of the custard apple. It produces a sweet, widely appreciated fruit that is used in a variety of drinks and desserts. Propagation is by seed or grafting, and selected cultivars are produced in Cuba, India and Taiwan.

Atemoya, *A. cherimola* × *A. squamosa*, is the only important interspecific hybrid in this family, and may be the most widely adapted of the annonaceous fruit crops (Martin *et al.*, 1987). The common name is derived from the Brazilian names for the parents, ate for *A. squamosa* and moya for *A. cherimola*. The first hybrids were made in Florida in 1908, and subsequent selections were made in Australia, India, Israel and South Africa (Nakasone and Paull, 1998). Several named cultivars are now propagated by grafting in the tropics and subtropics. Like sugar apple, but unlike cherimoya, it bears fruit in warm tropical and subtropical situations. It tolerates a wide range of soils and climatic conditions, and survives light frosts. Some clones require hand pollination for adequate fruit set. Its fruit weigh between 300 and 900 g, have an exceptionally good, sweet taste, and are eaten fresh or used in a variety of desserts.

Biriba, *Rollinia deliciosa*, originated in Brazil and is grown primarily there where it is also known as fruta de condessa (Martin *et al.*, 1987). It has the same environmental requirements and sensitivities as soursop.

Pond apple, *A. glabra*, is native to southern Florida, the Caribbean and the lowlands of tropical America. It is not an important fruit crop, but is mentioned here due to its potential as a rootstock (Nakasone and Paull, 1998). Most commercial annonas are sensitive to waterlogged soils and must be planted in well-drained sites for optimum production. Since pond apple grows naturally in swampy areas, its use as a rootstock in poorly drained areas has been suggested.

Núñez-Elisea *et al.* (1999) reported that scions of the normally flood-sensitive custard apple, 'Gefner' atemoya and '49–11' ('Gefner' atemoya × custard apple) tolerated flooding (survived and grew) when grafted on rootstocks of pond apple, but that plants grafted on to rootstocks of sugar apple and custard apple often died. Seedlings of pond apple and soursop were also flood tolerant, but seedlings of sugar apple were not. Based on these results, the authors suggested that the use of pond apple and soursop rootstocks might enable production in marginal, flood-prone sites. However, this recommendation should be made with caution since pond apple and soursop are both susceptible to *Phytophthora palmivora*, a pathogen that is most damaging in poorly drained soils (see below).

Fruit Attributes

The annonaceous fruits are composites of many fused, single-seeded fruitlets. They are climacteric and must be picked when mature in order to ripen properly (Snowdon, 1990). Their storage life is very limited and, once fruit begin to soften, it is usually a matter of days before they must be consumed. They are sensitive to chilling injury, and are stored optimally at temperatures as high as 20°C. Atemoyas can be stored for up to 2 weeks at 12–15°C. Ethylene is used to facilitate uniform ripening in some commercial situations.

In general, these fruit tend to split between fruitlets once ripening has advanced, and this promotes the development of postharvest diseases (Paull, 1996). Wound sites that are associated with annona seed borers, *Brephratelloides* spp., are additional points of entry for these pathogens.

Major and Minor Diseases

A dozen diseases of these crops are discussed below. Some, such as anthracnose and bacterial wilt, can have major impacts on production and require adequate management to ensure commercially viable yields. Others, such as black canker, Diplodia fruit rot and purple blotch, are important problems only under certain conditions, and specific measures to control them usually are not indicated. The geographic distributions of these diseases are also quite variable, and range from global (anthracnose) to a single state or country (bacterial wilt and yellow blotch).

Minor diseases of these crops are listed at the end of the chapter (Table 2.1). Diseases on this list rarely, if ever, cause important problems, and scant descriptive information on them exists. The causal agents, affected taxa and associated literature references for these diseases are listed in Table 2.1.

Anthracnose

Anthracnose is among the most common and damaging diseases of annonaceous fruit crops. It can severely limit fruit production, and has been reported from Australia, the Azores, Bangladesh, Brazil, China, Dominican Republic, Egypt, Mozambique, the Philippines, Puerto Rico, Sierra Leone, Uganda and the USA (Florida and Hawaii) (Snowden, 1921; Li, 1936; Deighton, 1939; Aruda, 1940; DeCarvalho, 1948; Alvarez-García, 1949; Batista, 1953; Abo-El-Dahab and El-Goorani, 1971; Cook, 1975; Raabe *et al.*, 1981; Brown *et al.*, 1988; Snowdon, 1990; Alfieri *et al.*, 1994). In all likelihood, anthracnose occurs wherever these crops are grown in warm, humid environments. Atemoya, biriba, cherimoya, custard apple, ilama, soursop, sugar apple and *A. marcgravii* are affected.

Symptoms

Affected flower petals exhibit dark-brown lesions that enlarge, become black and cause flowers to shed, thereby reducing fruit set (Cook, 1975). Infections on young fruit cause either rotting and mummification, or the formation of a hardened plug and no external symptoms of rotting on mature fruit. Alternatively, maturing or harvested fruit can be severely affected by a black, dry rot.

Infections on leaves are first visible as light green dots that coalesce and darken to a chestnut or black colour, become circular or elongated, and eventually impart a scorched appearance to the canopy (Cook, 1975). In severe cases, trees defoliate prematurely. Lesions on seedlings can girdle the stem and cause damping off. Those on shoots in established trees cause desiccation above the lesion and the initiation of adventitious shoots below, which, themselves, may then be affected. This recurring damage can lead to a witches'-broom appearance in the canopy. Cankers that do not exude gum form in chronically damaged areas.

Causal agents

Glomerella cingulata (anamorph: *Colletotrichum gloeosporioides*) is the primary or perhaps only causal agent. *C. anonicola* (Ciferri and González-Fragoso, 1927; Deighton, 1939) and *Gloeosporium anonae* (Batista, 1953) are also reported causes, but were not compared with *C. gloeosporioides* to determine whether they were synonymous. *G. cingulata* and *C. gloeosporioides* are described in Chapter 1.

Epidemiology

Although little has been reported on this disease in the above species, it can be assumed that pathogen behaviour resembles that on other crops. Conidia are disseminated in rainsplash and wind-driven rain, and germinate and infect under warm, moist conditions. Infections are usually latent on fruit, and do not develop until ripening begins. Wet conditions promote disease development and the eventual production of conidia on all host tissues. Although the teleomorph has been found in badly rotted fruit, it is not known whether or to what extent ascospores play a role in the disease cycle.

The pathogen survives rather well in dead host debris. Thus, leaves, fruit and old flowers are probably important sources of inoculum.

Management

Frequent applications of fungicides are needed to control the disease, especially during wet weather. Applications may be needed from flower initiation until harvest, depending on prevailing weather conditions.

Armillaria (mushroom) root rot

This is a widespread disease that is caused by at least three different species of *Armillaria*. In the USA, *A. mellea* affects cherimoya in California and *A. socialis* affects cherimoya and soursop in Florida (Rhoads, 1942; Farr *et al.*, 1989). Both species are described in Chapter 1. Although *A. mellea* was also reported on soursop in Uganda (Small, 1926), recent studies have shown that *A. mellea senso stricto* (heterothallic) is not present in Africa, but a species that culturally resembles *A. mellea* is (Shaw and Kile, 1991). It is homothallic and partially sexually compatible with isolates of *A. mellea senso stricto* but, based on isozyme analyses, is distinct enough to be considered a different species (Mwenje and Ride, 1997). Although Kile *et al.* (1994) suggested the epithet *A. mellea* ssp. *africana* should be used to identify this taxon, Mwenje and Ride (1997) felt that it should be referred to as the African *A. mellea* group until more definitive studies were conducted. In Australia, *A. luteobubalina* is the causal agent (Vock, 1978). Based on the disease's presence on three different continents and the diversity of causal agents, it is probable that this disease occurs in other areas.

Shaw and Kile (1991) listed the following chronological order in which symptoms of Armillaria root rot develop: reduced canopy growth; chlorosis, stunting of foliage and defoliation; canopy dieback; stress-induced increase in fruit production; appearance of symptoms and signs at root collar; death. Unless trees are small, this is usually a slow progression. Confirmation of the disease relies on signs of the pathogen and its isolation from the host. Mycelial fans are often found in the cambium of dying or recently killed trees. Rhizomorphs also form beneath the bark and along the roots of affected trees, as do basidiomes (mushrooms or basidiocarps) on the soil surface or trunk of affected trees. Basidiome morphology is distinct for each of the described species (Shaw and Kile, 1991). This and interfertility studies are used to delineate most of the *Armillaria* species. Unfortunately, rhizomorphs and basidiomes may not be produced in tropical and warm subtropical environments. Identification is complicated further in homothallic species in which individuals produce basidiomes.

Armillaria root rot is difficult to control, especially now that soil fumigants are becoming increasingly scarce. The pathogens can survive long periods as saprophytes and, thus, contaminate soil long after a host has died. In addition, they can move considerable distances in soil via their rhizomorphs. Although removal of all host materials from fields prior to planting is a common recommendation, in practice it is often not possible to do this effectively (Shaw and Kile, 1991).

Bacterial wilt

Bacterial wilt causes widespread damage and tree mortality in Queensland, Australia (Mayers and Hutton, 1987). Seedlings in nurseries and grafted trees in the field are affected. The disease had been recognized as basal and root rot since 1918, but its cause remained elusive until the early 1980s. The above authors speculated that this disease might be found in other annona-producing areas since the causal agent is widely spread.

Symptoms

Two disease syndromes are recognized (Persley, 1993). Young plants or seedlings wilt rapidly and die during the summer and autumn. This occurs mainly on plants within 3 years of planting in the field. New growth stops, and foliage wilts and dries on the tree without becoming chlorotic.

Older trees decline more slowly and may linger for 2 years or more before dying. Affected limbs defoliate and new growth is stunted and pale green. As foliage becomes sparser, branches and fruit become increasingly susceptible to sunburn. Flowering and fruit set is increased, but fruit are undersized and non-marketable. Internally, the tree's vascular system becomes dark brown to black (Plate 3). This symptom may be limited to below the graft union, but usually extends into the canopy. Oozing of the causal bacterium from affected woody tissue is evident under the light microscope. In the disease's terminal phases, the root collar and surrounding roots rot and disintegrate.

Causal agent

Until 1983, the most probable cause of both wilt syndromes was assumed to be a species of *Phytophthora* (Mayers and Hutton, 1987). That year, *Ralstonia solanacearum* was recovered from root and trunk tissue of an affected tree and shown to be pathogenic on sugar apple and the 'African Pride' cultivar of atemoya, as well as *Casuarina equisetifolia*, potato, pepper and three different cultivars of tomato. The bacterium is a Gram-negative, aerobic rod. The agent is in biovar 3, the most nutritionally diverse biovar in this species. It is thought to have originated and evolved in Asia (Hayward, 1991).

Epidemiology

Up to 50% of the trees were killed by this disease when they were propagated on rootstocks of atemoya or sugar apple (Mayers and Hutton, 1987). The disease occurs most commonly in wet areas or those with poorly drained soils.

Management

No chemical control measures are available. Free-draining soils and pathogen-free planting materials should be used when new orchards are planted (Persley, 1993). Sugar apple is very susceptible, and its use as a rootstock should be avoided wherever there is a threat from this disease (George *et al.*, 1987). Cherimoya is much less susceptible, and some cultivars are more resistant than others. Alternative host crops of the pathogen should not be interplanted with trees, and weed hosts should be controlled.

Black canker

This disease, caused by the fungus *Phomopsis annonacearum*, is a relatively minor problem in Australia (Purss, 1953; Persley, 1993). It may also occur in California and Florida where *Phomopsis* sp. was reported to cause a fruit spot (Farr *et al.*, 1989; Alfieri *et al.*, 1994).

Spots on fruit originally are small and purple, and eventually enlarge, crack and darken to a brown colour (Plate 4) (Cook, 1975). Lesions are superficial (<0.5 cm deep); their surfaces become hard and embedded with black pycnidia of the fungus. In pycnidia, α-conidia, $2–3 \times 5–8\ \mu m$ in size, form. Infection and disease development are facilitated by wet conditions. Since the fungus survives on mummified fruit and leaves beneath trees, these tissues should be removed from the orchard and destroyed.

Botryodiplodia fruit rot

This disease has been reported in Australia, Egypt, India, Mauritius and the USA (Florida), and is usually a problem only in poorly managed orchards (Rao, 1964b; Lutchmeah, 1988; Snowdon, 1990; Persley, 1993; Alfieri *et al.*, 1994). Due to the wide geographic range of the causal agent, the disease may be found in other areas.

Small purple lesions enlarge and blacken and eventually are covered with black pycnidia (Plate 5) (Snowdon, 1990; Persley, 1993). Unlike black canker lesions, these penetrate the flesh that, depending on the presence of secondary microbes, eventually softens or hardens and cracks. Fruit usually remain attached to the tree. The pathogen also causes a blight and dieback of small branches (Cook, 1975; Farr *et al.*, 1989).

Botryodiplodia fruit rot is caused by *Diplodia theobromae* which is described in Chapter 1. The disease's common name comes from the genus in which the pathogen formerly was placed. It survives on dead host tissue, especially mummified fruit. The disease is managed by removing these reservoirs of inoculum from orchards and destroying them.

Cylindrocladium leaf and fruit spot

This disease was described in Queensland, Australia on atemoya and in Brazil on sugar apple (Figueiredo and Namekata, 1967; Hutton and Sanewski, 1988). It is usually an unimportant disease, but can cause serious fruit losses during periods of heavy rainfall.

Symptoms

Symptoms on leaves and fruit begin as black spots 1–2 mm in diameter (Hutton and Sanewski, 1988). Spots on fruit are skin deep, not sunken and eventually dry out and crack (Plate 6). As they enlarge, they coalesce to form large, irregular, brownish-black lesions that may elongate during wet periods down the sides of the fruit. The centres of lesions on leaves become light tan as they enlarge and coalesce (Plate 7). Stems and roots were also infected experimentally.

Causal agents

Cylindrocladium colhounii causes Cylindrocladium leaf and fruit spot in Queensland (Hutton and Sanewski, 1988). The fungus's teleomorph, *Calonectria colhounii*, was not reported. In Brazil, *Calonectria leguminum* (anamorph: *Cylindrocladium leguminum*) causes the disease (Figueiredo and Namekata, 1967).

C. colhounii produces cylindrical, straight, two- to four-celled, 45–80 × 4–6 μm macroconidia that are held in parallel clusters by a colourless slime (Fig. 2.1) (Crous, 2002). Macroconidiophores consist of a septate 80–190 × 6–7 μm stipe, penicilliate arrangement of fertile branches, a septate 160–280 × 3–4 μm stipe extension and a clavate 3–4 μm diameter vesicle. Its cardinal temperatures for growth are >5, <25 and 35°C.

C. leguminum produces orange to red-brown, subglobose to ovoid perithecia that are 360–580 μm high and 300–440 μm in width (Fig. 2.2) (Crous, 2002). Asci are clavate, 76–126 × 13–22 μm, taper to a long, thin stalk, and contain eight ascospores that are hyaline, fusoid with rounded ends, slightly curved, two- to four-celled and 30–100 × 4–8 μm. The anamorph produces hyaline, cylindrical, two- to seven-celled, 45–90 × 4–7 μm macroconidia that are held in parallel clusters by colourless slime. Macroconidiophores consist of a septate 80–200 × 6–7 μm stipe, penicilliate arrangement of fertile branches, a septate 120–240 × 3–4 μm stipe extension and a narrowly clavate 2–3 μm diameter vesicle.

Epidemiology

Since the causal fungi are soilborne, damage is most often prevalent in lower portions of the canopy (Hutton and Sanewski, 1988). In Australia, the disease has also been observed higher in the canopy as a result of soil that has been transported by the coastal brown ant, *Pheidole megacephala*, when building protective tunnels.

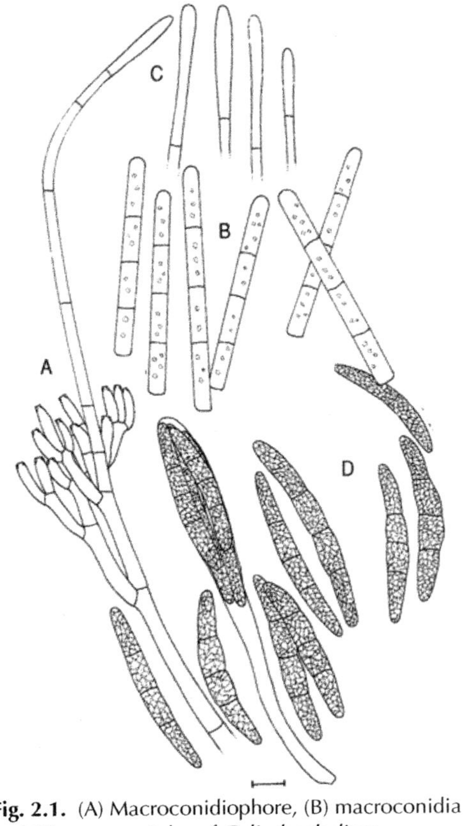

Fig. 2.1. (A) Macroconidiophore, (B) macroconidia and (C) macrovesicles of *Cylindrocladium colhounii*, and (D) ascus and ascospores of its teleomorph, *Calonectria colhounii*. Bar = 10 μm (from Crous, 2002).

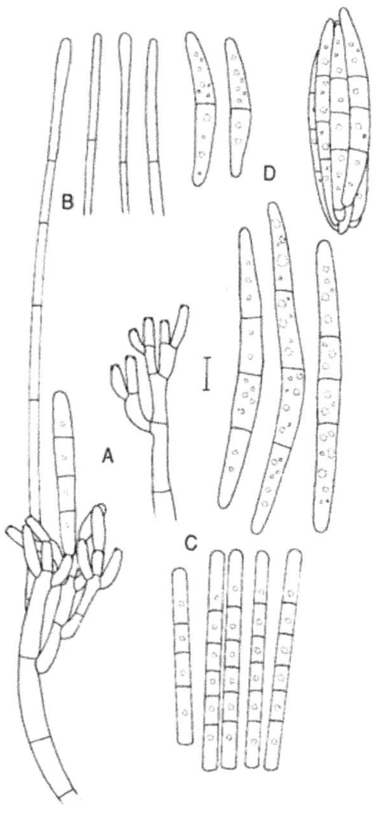

Fig. 2.2. (A) Macroconidiophores, (B) macrovesicles and (C) macroconidia of *Cylindrocladium leguminum*, and (D) ascospores of its teleomorph, *Calonectria leguminum*. Bar = 10 μm (from Crous, 2002).

Management

Low-lying branches should be removed to avoid fruit and foliage contact with soil and increase air circulation. Mulch under trees and permanent sod in the middles will reduce dispersal of inoculum, as will the control of ants and the mealybugs that they service. Of two popular cultivars of atemoya in Australia, 'Pinks Mammoth' was less susceptible than 'African Pride'.

Pestalotia fruit spot

Pestalotia fruit spot is a minor disease of sugar apple. It was reported in India where it was observed at low levels (3–5% incidence) in the markets of Poona (Rao, 1964a). The disease usually originated at the stem end of fruit as purplish brown, irregular spots that turned brown with age and eventually covered large areas of the fruit surface. The affected areas remained firm, and affected fruit failed to ripen. The causal fungus, *Pestalotia* sp., produced fusiform, five-celled conidia with the three central cells coloured a deep brown; they were 21–28 × 8–10 μm and had three or four divergent setulae at the apex. After artificial inoculation, the fungus affected sugar apple, but not soursop and fruit of several other species. Based on its morphology and host range, Rao (1964a) felt that the fungus was sufficiently different to be considered a new species.

Phytophthora root and fruit rot (purple blotch)

Phytophthora root and fruit rot have been reported in Australia, India, Malaysia, the Philippines, Spain and the USA (California and Hawaii) (Reinking, 1923; Purss, 1953; Rao *et al.*, 1962; Chee, 1969; Gomez, 1983; French, 1987; Persley, 1993; Tsao *et al.*, 1994; J.Y. Uchida, University of Hawaii, personal communication). Based on the wide host ranges of its causal agents, it probably occurs in other locations in which these pathogens are found.

Symptoms

Aboveground symptoms appear first on immature fruit as small, water-soaked lesions, especially on fruit that are near the soil surface. Affected fruit abscise, and excessive losses of all ages of fruit can occur. As symptoms progress, they become purplish to black, extend over large areas of the fruit surface, and extend as a brownish decay into the fruit pulp (Plate 8). Fruits eventually harden and mummify, the fruit interior rots completely and the seeds shrivel. The causal agents are soilborne and also cause dark-coloured root decay.

Causal agents

Phytophthora palmivora is the primary cause of Phytophthora root and fruit rot

(Reinking, 1923; Purss, 1953; Chee, 1969; Tsao *et al.*, 1994). It affects atemoya, pond apple, soursop and sugar apple. *P. nicotianae* has been reported on sugar apple in India (Rao *et al.*, 1962), and *P. cinnamomi* causes a root and collar rot on cherimoya in Spain (Gomez, 1983). *Phytophthora* sp. has been reported on sugar apple in California (French, 1987).

Recent reports have indicted two additional species. Weinert *et al.* (1998) reported that an isolate from sugar apple in Queensland that had been identified as *P. palmivora* (UQ3691) was actually *P. capsici*. Aragaki and Uchida (2001) examined 100 isolates from herbaceous and woody hosts that included *A. cherimola* from the islands of Hawaii and Oahu; based on the broad concept of the species (Erwin and Ribiero, 1996), all were considered *P. capsici*. Based on molecular data from others and morphological and pathological data of their own, Aragaki and Uchida (2001) proposed that the isolates from woody hosts comprised a distinct taxon that they named *P. tropicalis*. How widespread *P. tropicalis* is on annonas and whether diseases that previously were attributed to either *P. capsici* or *P. palmivora* were actually caused by it is not known.

Features of *P. cinnamomi*, *P. nicotianae* and *P. palmivora* are found in Chapter 1. In the broad sense, *P. capsici* possesses the following traits (Erwin and Ribiero, 1996). It produces ovoid to ellipsoid, papillate to semipapillate, caducous sporangia that have long pedicels (Fig. 2.3). Some sporangia have more than one papillum. Average dimensions of the UQ3691 strain from sugar apple were 40 × 24 μm with pedicels that exceeded 50 μm in length. *P. capsici* is heterothallic. Oogonia are 23–50 μm in diameter and spherical or subspherical, and antheridia are amphigynous. Oospores are usually plerotic with cell walls between 2 and 6 μm thick. Its cardinal temperatures for growth are 10, 28 and >35°C. *P. tropicalis* was distinguished from *P. capsici* by its production of chlamydospores by most isolates (27–33 μm in diameter; chlamydospores are generally not formed by strains of *P. capsici* from herbaceous hosts), no or poor growth at 35°C, weak or no virulence on *Capsicum*, and distinct isozyme and mitochondrial DNA profiles (Aragaki and Uchida, 2001).

The above species can be distinguished by the following sporangial characteristics. Those of *P. palmivora*, *P. tropicalis* and *P. capsici* are papillate and caducous, but the pedicels of *P. palmivora* are usually ~10 times shorter than those of *P. tropicalis* and *P. capsici* (5 versus 50 μm or more). *P. tropicalis* produces sporangia that have a predominantly tapered base and are narrower (≤26 μm in diameter) and have a greater length : breadth ratio (≥1.8) than *P. capsici* (Fig. 2.4). Sporangia of *P. cinnamomi* and *P. nicotianae* are non-caducous but, unlike those of *P. cinnamomi*, those of *P. nicotianae* are papillate.

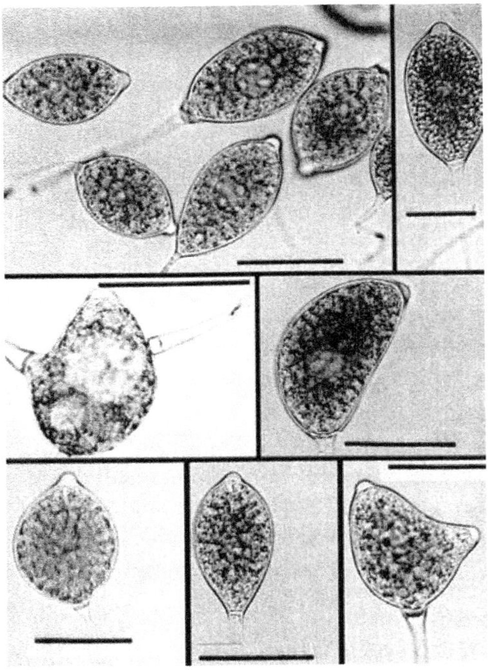

Fig. 2.3. Variable sporangia of *Phytophthora capsici*. Clockwise from the upper right: ellipsoid, bilaterally asymmetrical, bipapillate, ellipsoid with tapered base, nearly spherical, intercalary deciduous, and typical sporangia from pepper. Bars = 40 μm except for the first micrograph where it is 25 μm (from Aragaki and Uchida, 2001).

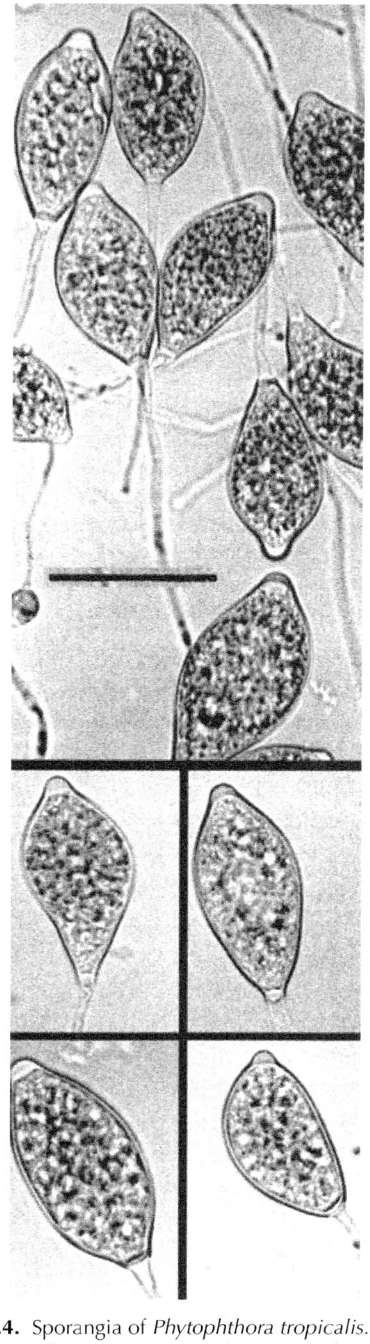

Fig. 2.4. Sporangia of *Phytophthora tropicalis*. Clockwise from the top: typical sporangia from anthurium, ellipsoid with tapered base, bilaterally asymmetrical, ellipsoid and pyriform. Note the tapered base of most of these sporangia. Bar = 40 μm (from Aragaki and Uchida, 2001).

Epidemiology

All of the above species have wide host ranges. Erwin and Ribiero (1996) reported ~200 hosts for *P. palmivora*, and Reinking (1923) demonstrated that isolates from coconut palm and cacao both caused severe disease on seedlings of soursop. Erwin and Ribiero (1996) listed more than 300 hosts for *P. nicotianae*, and Rao *et al.* (1962) showed that isolates from sugar apple caused either fruit or root rots on ~20 different species.

Rainfall favours disease development. Fruit usually abscise within 4 days of the appearance of purple lesions.

Management

Low-lying branches in the canopy should be removed. Mulches can reduce disease on fruit by reducing rainsplash dispersal of inoculum. Grass or other living mulches should be maintained between rows. Fungicidal control has not been reported.

Pink disease

The disease is not common, and appears during wet weather in mainly tropical areas (Persley, 1993). The causal fungus, *Erythricium salmonicolor*, has a wide host range and is described in Chapter 1. It forms light pink mats of mycelium on trunk and branch surfaces (Plate 9). Eventually, affected bark cracks and oozes gum. Affected areas should be pruned from trees and treated with a fungicide.

Pseudocercospora fruit and leaf spot

This disease is caused by a non-described species of *Pseudocercospora*, and may be widely distributed. It is a serious problem in Queensland where up to 50% of the fruit in a given orchard may be affected (Persley, 1993; although the affected taxa were not reported in this publication, symptoms of the disease were shown on custard apple). A leafspot caused by *Pseudocercospora* sp. has been reported on pond apple in Florida (Alfieri *et al.*, 1994), and there are numerous old reports of species of *Cercospora* causing damage on custard apple and sugar apple in Brazil (*C. anacardii* and *C. anonae*) and India (*C. anonae* and *C. caracasensis*) (Muller and Chupp, 1935; Mundkur and Ahmad, 1946; Holliday, 1980). Since the latter reports were published before *Pseudocercospora* was widely recognized as a distinct segregate of *Cercospora*, it is possible that all of the above diseases are caused by a single pathogen or closely related pathogens.

Symptoms begin in natural crevices on the fruit surface as purple-grey spots, 1–5 mm in diameter (Persley, 1993). Spots coalesce, increase to 10–15 mm in diameter, and darken with age (Plate 10). Badly affected fruit may crack, and be badly disfigured and unmarketable. On leaves, irregular dark red to brown lesions, 1–5 mm in diameter, form on the adaxial surface. The causal fungus eventually sporulates in necrotic areas in the centre of these lesions.

The pathogen's large conidia are disseminated by wind, and infection requires free moisture. Affected fruits from the previous season should be removed from the orchard before fruit set in the current season begins. Fungicides should be applied at the onset of fruit set.

Pythium root rot

Although this disease has only been reported in California and Florida, it can be quite damaging. Atemoya, cherimoya, custard apple and sugar apple are affected (Farr *et al.*, 1989; Ploetz, 1991; Alfieri *et al.*, 1994).

Aboveground symptoms include a thin, often chlorotic canopy and a general reduction in tree size and vigour (Fig. 2.5). Although damage is usually most severe in low-lying or poorly drained areas in a field, healthy trees may surround affected individuals. The cortical tissues of roots are brown and can be easily removed from the stele. When aboveground symptoms are evident, root systems are often so severely rotted that trees can be removed from the soil by hand.

Pythium splendens causes the disease in Florida (Ploetz, 1991), and is described in detail in Chapter 6. *Pythium* sp. has been reported in California (Farr *et al.*, 1989).

This disease most often starts in the nursery when contaminated soil or potting mix is used or where plants are placed on the ground. Clean pots and media should be used whenever possible. Metalaxyl was effective in pot studies, but its performance in the field has not been documented (Ploetz, unpublished results).

Rust

Cummins (1941) reported that this disease was common from Florida to Ecuador. It affects atemoya, cherimoya, custard apple, ilama and sugar apple, and can cause considerable defoliation in summer and autumn. However, specific control measures are usually not indicated.

The causal fungus, *Phakopsora cherimoliae*, produces telia in the base of old uredia or in separate subepidermal reddish brown crusts, 0.1–1 mm in diameter and 3–6 spores in thick-

Fig. 2.5. Aboveground symptoms of Pythium root rot, caused by *Pythium splendens*, on a 'Gefner' atemoya tree. Note the chlorotic, sparse canopy and the healthy tree in the background (photo: R.C. Ploetz).

ness. Teliospores, the outer layers of which are golden or chestnut-brown, are cubical, oblong or oblong–ellipsoid and 7–13 μm × 13–23 μm. Their cell walls are 1–2 μm thick.

Scab

This disease was reported in Brazil and Venezuela on cherimoya and sugar apple (Bitancourt and Jenkins, 1942). Dark spots with brown margins form on leaves and may coalesce into large irregular patches. They are less noticeable on the abaxial surface, and their centres eventually lighten and dry out. Ascocarps of the causal fungus, *Elsinoë annonae*, develop in lesion centres. They are 30–50 × 50–150 μm in diameter, and irregular, dark and erumpent. They contain spherical asci, 20 μm in diameter, which in turn contain eight hyaline, three-celled ascospores that are 5–8 × 12–15 μm. Acervuli of the anamorph, which presumably is a species of *Sphaceloma*, may form in old lesions.

Yellow blotch

Yellow blotch was reported in Brazil (Katijima *et al.*, 1993). It appears to be a minor problem on soursop in the state of Ceara. Affected leaves display vein clearing and diffuse, chlorotic areas that may be associated with some distortion of the laminar surface. Artificially inoculated seedlings were stunted, but rarely died. The authors felt that yield losses could occur if plants were infected at an early age.

A rhabdovirus was observed in field and experimentally infected plants (Katijima *et al.*, 1993). In leaf-dip preparations, it measured 60–70 nm × 250–300 nm and contained a 40–50 nm wide inner component with a transverse periodicity of 5 nm. In transmission electron micrographs of leaf thin sections, the virus was observed to accumulate in the endoplasmic reticulum, indicating that it belonged in the type I class of rhabdoviruses.

The virus was transmitted mechanically to seedlings of soursop, but not to biriba and sugar apple. It was transmitted most effectively to soursop via grafting, and could also be transmitted to biriba and sugar apple in this manner, although the virus did not appear to move beyond the graft regions in the latter plants. Spread of the disease in the field suggested that an unidentified insect vector transmitted the virus. No control measures were indicated.

Table 2.1. Miscellaneous minor diseases of fruit crops in the *Annonaceae*.

Disease	Affected taxa[a]	Cause(s)	Reference(s)
Cherimoya cambium disease	ch	*Ascochyta cherimolae*	Garmedia (1945)
Fruit rot	sa	*Botrytis* sp.	Alfieri *et al.* (1994)
	sa	*Phylctema* sp.	Alfieri *et al.* (1994)
Gliocladium rot	sa	*Gliocladium roseum*	Chaudry *et al.* (1985)
Leaf necrosis	il	*Corynespora* sp.	Alfieri *et al.* (1994)
Leaf spot	sa	*Alternaria alternata*	Kamal and Chandra (1980)
	ns, sa	*Alternaria* sp.	Farr *et al.* (1989); Alfieri *et al.* (1994)
	ns, pa	*Diplodia* sp.	Alfieri *et al.* (1994)
	ca, ch, ns, sa, pa	*Gloeosporium* sp.	Alfieri *et al.* (1994)
	sa	*Helminthosporium* sp.	Alfieri *et al.* (1994)
	sa	*Mycosphaerella* sp.	Farr *et al.* (1989)
	so	*Phyllostictina anonicola*	Batista (1952)
	ca, ch, ns, sa	*Phyllosticta* sp.	Alfieri *et al.* (1994)
	ns	*Stemphylium* sp.	Alfieri *et al.* (1994)

Continued

Table 2.1. *Continued.*

Disease	Affected taxa[a]	Cause(s)	Reference(s)
Not specified	ch	*Dothiorella* sp.	Farr *et al.* (1989)
	ch	*Penicillium* sp.	Farr *et al.* (1989)
	so	*Phyllachora anonicola*	Chardon *et al.* (1940)
	ch	*Phytomonas* sp.	Fernández-Becerra *et al.* (1996)
	ch	*Rhizoctonia solani*	Farr *et al.* (1989)
	ch	*Verticillium dahliae*	Farr *et al.* (1989)
Pink mould rot	sa	*Trichothecium roseum*	Chaudry *et al.* (1985)
Root rot	sa	*Phymatotrichopsis omnivora*	Farr *et al.* (1989)
	ch	*Rosellinia necatrix*	Montemartini (1934)

[a] at = atemoya, *A. cherimola* × *A. squamosa*; ca = custard apple or Bullock's heart, *A. reticulata* L.; ch = cherimoya, *A. cherimola* Mill.; il = ilama, *A. diversifolia* Saff.; ns = not specified; pa = pond apple, *A. glabra* L.; so = soursop, *A. muricata* L.; sa = sugar apple or sweetsop, *A. squamosa* L.

References

Abo-El-Dahab, M.K. and El-Goorani, M.A. (1971) Market and storage diseases of *Annona squamosa* L. fruits in U.A.R. (Egypt). *Phytopatologia Mediterranea* 10, 107–109.

Alfieri, S.A., Jr, Langdon, K.R., Kimbrough, J.W., El-Gholl, N.E. and Wehlburg, C. (1994) *Diseases and Disorders of Plants in Florida.* Bulletin No. 14, Florida Department of Agriculture and Consumer Services, Contribution No. 680.

Alvarez-García, L.A. (1949) Anthracnose of the Annonaceae in Puerto Rico. *Journal of Agriculture University of Puerto Rico* 33, 27–43.

Aragaki, M. and Uchida, J.Y. (2001) Morphological distinctions between *Phytophthora capsici* and *P. tropicalis* sp. nov. *Mycologia* 93, 137–145.

Aruda, S.C. (1940) Antracnose e cancro das Anonaceas. *Biologico* 6, 224–225.

Batista, A.C. (1952) Algumas novas espécies de *Phyllosticta* e *Phyllostictina. Boletin da Secretaria da Agricultura da Pernambuco* 19, 3–4.

Batista, A.C. (1953) Novos fungos agents de antracnose. *Annual Congress of the Society of Botany Brasil.* pp. 142–144.

Bitancourt, A.A. and Jenkins, A.E. (1942) New discoveries of Myriangiales in the Americas. *Proceedings of the Eighth American Science Conference* 1940, 3, 149–172.

Brown, B.I., Wong, L.S., George, A.P. and Nissen, R.J. (1988) Comparative studies on the postharvest physiology of fruit from different species of *Annona* (custard apple). *Journal of Horticultural Science* 63, 521–528.

Chardon, C.E., Miller, J.H. and Miller, A.S. (1940) Ascomycetes from the state of Minas Geraes (Brazil). *Mycologia* 32, 172–204.

Chaudry, A.S., Singh, G.N. and Singh, A.R. (1985) Effect of wrapping materials and ripening media on physico-chemical compositions of custard apple (*Annona squamosa* Linn.). *Indian Journal of Agricultural Research* 19, 90–92.

Chee, K.H. (1969) Hosts of *Phytophthora palmivora. Review of Applied Mycology* 48, 337–344.

Ciferri, R. and González-Fragoso, R. (1927) Hongos parásiticos y saprofitos de la República Dominicana. (9th Serie.). *Boletin Real Sociedad Española Histologia* 27, 68–81.

Cook, A.A. (1975) *Diseases of Tropical and Subtropical Fruits and Nuts.* Hafner Press. New York.

Crous, P.W. (2002) *Taxonomy and Pathology of* Cylindrocladium (Calonectria) *and Allied Genera.* APS Press, St Paul, Minnesota.

Cummins, G.B. (1941) Descriptions of tropical rusts. IV. *Bulletin of the Torrey Botanical Club* 68, 467–472.

DeCarvalho, T. (1948) Relacão preliminary de doenas encontradas em plantes e insectos com anatações fitopatológicas. Colónia de Moçambique, Repartição de Agricultura, Secção de Micologia.

Deighton, F.C. (1939) *Mycological Work*. Report of the Department of Agriculture Sierra Leone, 1938, pp. 64–66.

Erwin, D.C. and Ribiero, O.K. (1996) *Phytophthora Diseases Worldwide*. APS Press, St Paul, Minnesota.

Farr, D.F., Bills, G.F., Chamuris, G.P. and Rossman, A.Y. (1989) *Fungi on Plant and Plant Products in the United States*. APS Press, St Paul, Minnesota.

Fernández-Becerra, C., Osuna, A., Muller, E., Dollet, M. and Sánche-Moreno, M. (1996) Characterization of *Phytomonas* isolated from fruits by electrophoretic isoenzymes and kinetoplast DNA analysis. *FEMS Microbiology Letters* 145, 463–468.

Figueiredo, M.B. and Namekata, T. (1967) Constatação de *Calonectria quinqueseptata* n.sp., forma perfeita de *Cylindrocladium quinqueseptata* Boedijn and Reitsma, sôbre *Anonna squamosa* L. e *Eucalyptus* sp. *Arquivos do Instituto Biologico, São Paolo* 34, 91–96.

French, A.M. (1987) *California Plant Disease Host Index. Part 1: Fruit and Nuts*. California Department of Food and Agriculture.

Garmendia, J.O. (1945) La enfermedad del cambium en chirimoyos. *Simiente* 14, 38.

George, A.P., Nissen, R.J. and Brown, B.I. (1987) The custard apple. *Queensland Agricultural Journal* 287–296.

Gomez, M.C. (1983) The cherimoya. In: Chambers, C.M. (ed.) *California Rare Fruit Growers Yearbook* 15, 5–29.

Hayward, A.C. (1991) Biology and epidemiology of bacterial wilt caused by *Pseudomonas solanacearum*. *Annual Review of Phytopathology* 29, 65–87.

Holliday, P. (1980) *Fungus Diseases of Tropical Crops*. Cambridge University Press. Cambridge, UK.

Hutton, D.G. and Sanewski, G.M. (1988) Cylindrocladium leaf and fruit spot of custard apple in Queensland. *Australasian Plant Pathology* 18, 15–16.

Kamal, and Chandra, G. (1980) On sclerotial rot of *Nigelia sativa* and Alternaria leaf spot of *Annona squamosa* – new to India. *Indian Journal of Mycology and Plant Pathology* 9, 88.

Katijima, E.W., Martins, C.R.F. and Santos, A.A. (1993) Identification of a rhabdovirus in soursop (*Annona muricata*). *Plant Disease* 77, 276–278.

Kile, G.A, Guillaumin, J.J., Mohammed, C. and Watling, R. (1994) Biogeography and pathology of Armillaria. In: Johanson, M. and Stenlind, J. (eds) *Proceedings of the Eighth International Conference on Root and Butt Rots*. IUFRO, Uppsala, pp. 411–436.

Li, L.Y. (1936) Anthracnose of Hwangpee, *Clausena lansium* (Lour.) Skeels, in South China. *Lingnan Science Journal* 15, 113–117.

Lutchmeah, R.S. (1988) *Botryodiplodia theobromae* causing fruit rot of *Annona squamosa* in Mauritius. *Plant Pathology* 37, 152.

Martin, F.W., Campbell, C.W. and Ruberté, R.M. (1987) *Perennial Edible Fruits of the Tropics: an Inventory*. Agriculture Handbook No. 642, US Department of Agriculture, Washington, DC.

Mayers, P.E. and Hutton, D.G. (1987) Bacterial wilt, a new disease of custard apple: symptoms and etiology. *Annals of Applied Biology* 111, 135–141.

Montemartini, L. (1934) I parassiti e le malattie delle piante coltivate nella Sicilia occidentale durante il biennio 1932–33. *Rivista di Patologia Vegetale* 24, 11–36.

Muller, A.S. and Chupp, C. (1935) Cercosporae de Minas Geraes. *Arquives da Instituto da Biologia Vegetal* 1, 213–220.

Mundkur, B.B. and Ahmad, S. (1946) Revisions of and additions to Indian fungi. II. *Mycological Papers, Imperial Mycological Institute* 18.

Mwenje, E. and Ride, J.P. (1997) The use of pectic enzymes in the characterization of *Armillaria* isolates from Africa. *Plant Pathology* 46, 341–354.

Nakasone, H.Y. and Paull, R.E. (1998) *Tropical Fruits*. CAB International, Wallingford, UK.

Núñez-Elisea, R., Schaffer, B., Fisher, J.B., Colls, A.M. and Crane, J.H. (1999) Influence of flooding on net CO_2 assimilation, growth and stem anatomy of *Annona* species. *Annals of Botany* 84, 771–780.

Paull, R.E. (1996) Postharvest atemoya fruit splitting during ripening. *Postharvest Biology and Technology* 8, 329–334.

Persley, D. (ed.) (1993) *Diseases of Fruit Crops*. Department of Primary Industries, Brisbane.

Ploetz, R.C. (1991) Species of *Pythium* as pathogens of perennial, woody fruit crops in south Florida (Abstract). *Phytopathology* 81, 699.

Purss, G.S. (1953) The fruit rots of the custard apple. *Queensland Journal of Agricultural Science* 10, 247–265.

Raabe, R.D., Conners, I.L. and Martinez, A.P. (1981) *Checklist of Plant Diseases in Hawaii*. Information Text Series no. 22. Hawaii Institute of Tropical Agriculture and Human Resources, College of Tropical Agriculture and Human Resources, University of Hawaii.

Rao, V.G. (1964a) A new fruit spot of *Annona squamosa* from India. *Plant Disease Reporter* 48, 399–401.

Rao, V.G. (1964b) Some new market and storage diseases of fruits and vegetables in Bombay-Maharashtra. *Mycopathologia et Mycologia Applicata* 23, 297–310.

Rao, V.G., Desai, M.K. and Kulkarni, N.B. (1962) A new Phytophthora fruit rot of *Annona squamosa* from India. *Plant Disease Reporter* 46, 874–876.

Reinking, O.A. (1923) Comparative study of *Phytophthora faberi* on coconut and cacao in the Philippine Islands. *Journal of Agricultural Research* 25, 267–284.

Rhoads, A.S. (1942) Notes on Clitocybe root rot of bananas and other plants in Florida. *Phytopathology* 32, 487–496.

Shaw, C.G., III and Kile, G.A. (1991) *Armillaria Root Disease.* Agricultural Handbook No. 691. Forest Service, US Department of Agriculture, Washington, DC.

Small, W. (1926) Report of the mycologist for the period January 1st to September 30th. Annual Report of the Ugandan Department of Agriculture for the Year Ended 31st December, 1925.

Snowden, J.D. (1921) Report of the Government Botanist for the period 1st April to 31st December, 1920. Annual Report for the Department of Agriculture, 1920, pp. 43–46.

Snowdon, A.L. (1990) *A Colour Atlas of Post-harvest Diseases and Disorders of Fruits and Vegetables. Vol. 1: General Introduction and Fruits.* Wolfe Scientific, London.

Tsao, P.H., Gruber, L.C., Potales, L.A., Gochangco, A.M., Luzaran, P.B., de los Santos, A.B. and Pag, H. (1994) Some new records of Phytophthora crown and root rots in the Philippines and in world literature. (Abstract) *Phytopathology* 84, 871.

Vock, N.T. (1978) *A Handbook of Plant Diseases*, Vol. 1. Queensland Department of Primary Industries, Brisbane.

Weinert, M.P., Smith, B.N., Wagels, G., Hutton, D. and Drenth, A. (1998) First record of *Phytophthora capsici* from Queensland. *Australasian Plant Pathology* 28, 93.

3 Diseases of Avocado

John A. Menge[1] and Randy C. Ploetz[2]
[1]*Department of Plant Pathology, University of California, Riverside, California, USA;*
[2]*University of Florida, Tropical Research and Education Center, Homestead, Florida, USA*

Introduction

Avocado, *Persea americana* (family: *Lauraceae*), is one of the major fruit crops in the tropics and subtropics. Worldwide, over 2.4 million tonnes (Mt) of fruit were produced in 2000 (FAO). Although most of the fruit were grown and consumed in developing countries, lucrative national and international markets exist for the fruit. For example, in 2000, the 146 Mkg crop in California was worth US$380 million (Crane, 2001).

The commercial avocado is little changed from its wild ancestors that were cultivated for centuries in tropical America. An excellent overview of the avocado as a food crop may be found in Ploetz *et al.* (1994). It is an evergreen, subtropical tree with close relatives of laurel or sweet bay, camphor, cinnamon, sassafras and the California bay. There are many *Persea* species, but nearly all have small fruit, large seeds and are unsuitable for commercial use. Most are not graft compatible with *P. americana*. Several primitive Guatemalan species, *P. nubigena*, *P. steyermarkii* and a species known as Aguacate de Mico, are primitive ancestors of *P. americana* and can hybridize with it.

There are three races of *P. americana*. The Mexican race, *P. americana* var. *drymifolia*, is cold hardy and produces small, thin-skinned fruit that mature in ~6 months. The foliage is anise scented. The Guatemalan race, *P. americana* var. *guatemalensis*, has small to large, usually ovoid fruit, with thick, leathery skin. Guatemalan types grow at mid to high elevations, exhibit moderate cold tolerance and require 12–15 months for fruit to mature. The West Indian race, *P. americana* var. *americana*, is salt tolerant but not cold hardy and thrives mainly in low altitude, tropical areas. It produces fruit in 6–7 months that are small to as large as 1 kg in weight. Their skin is intermediate in thickness between that of the Guatemalan and Mexican varieties and the fruit often have a long-necked appearance.

Indigenous criollo types have been cultivated for centuries in mountain forests, lowland tropics and in subtropical areas of the Caribbean, Mexico, and Central and South America. When the Spanish conquistadors invaded the new world in the 1500s, they took note of this strange fruit and sent seeds back to southern Europe, where trees existed for many years as curiosities. It was not until the 1800s (Florida in 1833 and California in 1856) that avocados were first introduced into the USA. In 1911, nurseryman F.O. Popenoe imported a green-skinned, superior quality, Mexican–Guatemalan hybrid avocado called 'Fuerte' into California. This cultivar has spread around the world and marked the beginning of worldwide commerce in avocado.

Although avocado is a speciality fruit, the world industry is large and growing rapidly. It is now grown in 59 countries, the most important of which are, in descending order, Mexico, the USA, Indonesia, South Africa, Chile, Brazil, Dominican Republic, China, Colombia and Peru (Table 3.1). Most Latin America countries consume much of what they produce, while countries in other parts of the world are major exporters. Per capita consumption in Mexico, for instance, is 6 kg per person, while it is only 0.5 kg in the USA.

Avocado fruits have exceptional food value and are important in the diet of many Latin Americans. They are rich in minerals (iron, magnesium and potassium), vitamins (A, C, E and four essential B vitamins) and protein, and are also rich in lutein, glutathione and β-sitosterol, phytochemicals that are thought to aid the body in fighting disease (Crane, 2001). The fruit has no sodium and contains 3–30% (depending on the variety) monounsaturated oil.

Botanically, avocado fruit are considered berries. They consist of a single, large seed with two cotyledons surrounded by the thick, fleshy, edible mesocarp and the rind. The fruit of different cultivars are highly variable and range from black to green in colour, large to small in size, smooth to pebbly skinned and nearly spherical in shape to long-necked. They also have different harvest, quality, storage and taste characteristics.

The bewildering number of cultivars that exists has led to great difficulties in marketing. Due to its high quality and easily recognized dark, pebbly skinned appearance, the world avocado industry has adopted a Guatemalan cultivar, 'Hass'. 'Fuerte' is still grown in many areas, whereas local cultivars, often with a West Indian heritage, are grown in more tropical areas such as Brazil, Dominican Republic, Ecuador, El Salvador, Florida, Haiti and Indonesia. Guatemalan criollo types are often grown in Central America, and some other cultivars of note are grown in Australia ('Shepard'), Brazil ('Geada', 'Fortuna', 'Ouro Verde', 'Quinatal' and 'Sclano'), Israel ('Bnei Darom', 'Ettinger', 'Horshim', 'Iriet', 'Rosh Hanikra 4' and 'Tova'), Mexico ('Colin V-33') and Spain ('Torrox 23'). Several new cultivars that resemble 'Hass' but have improved or special characteristics include 'Sir Prize' and 'Lamb Hass'.

Avocado has such enormous genetic variability that seed-propagated trees usually bear no resemblance to the mother tree; hence, cultivars must be maintained clonally. As scions, they are grafted on rootstock cultivars that are usually selected for vigour, resistance to cold, certain soil characteristics or root disease. Seedling rootstocks are used but, in order to guarantee beneficial rootstock qualities, clonal rootstocks have become commonplace. 'Duke 7' is a commonly used rootstock since it tolerates Phytophthora root rot. Other widely used rootstocks include 'Borchard', 'Edranol', 'Ein Shener', 'Lahavot Haviva', 'Merensky II', 'Thomas', 'Toro Canyon' and 'Velvick'. Mexican rootstocks are popular in California, Chile, Mexico, New Zealand, South Africa and Spain; West Indian rootstocks are common in the Canary Islands, Florida and Israel; and Guatemalan rootstocks are used in Australia and Hawaii.

Table 3.1. Worldwide avocado yield by country in the year 2000.[a]

Country	Yield	Country	Yield	Country	Yield
Mexico	939,118	Peru	75,000	Congo, Republic	25,000
United States	164,500	Israel	53,400	Guatemala	25,000
Indonesia	121,822	Cameroon	50,000	Australia	24,311
South Africa	104,000	Spain	47,000	Ecuador	24,049
Chile	100,000	Venezuela	45,853	Costa Rica	23,000
Brazil	85,000	Haiti	45,000	Madagascar	23,000
Dominican Republic	81,720	El Salvador	42,500	Morocco	13,000
China	78,000	Philippines	29,085	Paraguay	12,000
Colombia	75,000	Congo, Dem. Rep.	27,000		

[a]Yield in metric tonnes. Data provided by Food and Agriculture Organization Statistics (2000).

Avocado Diseases

The most important disease of avocado worldwide is Phytophthora root rot. It is the limiting factor for production in many regions. Other important diseases include anthracnose, Phytophthora cankers, Pseudocercospora spot, ringneck, Rosellinia root rot, scab, stem-end rot and sunblotch. Of lesser importance are Armillaria root rot, avocado black streak, bacterial canker, Dothiorella stem canker, Dothiorella fruit rot and Verticillium wilt. Algal leaf spot, bacterial soft rot, Duke 6 stem pitting, powdery mildew, root and butt rot, silver spot, sooty blotch, sooty mould and tar spot are usually unimportant. Minor diseases can be significant in certain areas or may have a greater impact in the future if they attack new cultivars or are introduced into new areas.

Diseases of Fruit, Foliage or the Entire Plant

Anthracnose

Anthracnose is the most common rot of mature avocado fruit and also affects leaves, stems and young fruit. The disease has been reported from most avocado-producing areas of the world. In many areas, it is the most important disease and may cause losses of fruit of up to 37% (Fitzell, 1987).

Symptoms

On leaves, chlorotic, then necrotic, brown spots form and coalesce to form a dead tip on the leaf. Occasionally, leaf symptoms are severe enough to cause defoliation. Brown or purple lesions can form on new shoots, and shoot dieback accompanies severe infections. Dark lesions can form on the inflorescence, causing its death or abortion of fruit.

Most important are the symptoms on fruit. Prior to harvest, small dark lesions <5 mm in diameter develop around the fruit lenticels. These lesions can reduce fruit quality and cause fruit drop. Larger, dark, spreading lesions develop after injury from insects or wind. These lesions may become sunken and rotted, and pinkish spore masses may form on the surface. After harvest, infections frequently appear during storage. These lesions begin as small, depressed, brown spots that quickly enlarge and become black circular spots (Plate 11). They may coalesce until the entire fruit surface and pulp are affected. In a moist environment, pink, slimy spore masses erupt through the surface of the peel. The areas of decay are firm at first, but eventually become soft and putrid.

Causal agents

The disease is caused by *Glomerella cingulata* (anamorph: *Colletotrichum gloeosporioides*) and *Glomerella acutata* (anamorph: *Colletotrichum acutatum*). The former species is found worldwide and both are described in Chapter 1. The latter species has been indicted only in New Zealand, although it is widespread on other hosts including citrus, mango and papaya (Dyko and Mordue, 1979). It is described in Chapters 1 and 7.

Epidemiology

In the field, these fungi produce conidia in acervuli on dead twigs, leaves and other host tissues (Fitzell, 1987). They spread via rainsplash and can infect virtually all aboveground portions of the host, although flowers and fruit are most susceptible. As fruit, twigs and leaves senesce, they are colonized and new conidia-bearing acervuli are produced. *G. cingulata* is rarely found and probably plays little role in the disease (Fitzell, 1987). Infection of fruit can take place at any time from fruit set to harvest (Peterson, 1978). Infection and spread is encouraged by high moisture and warm temperatures (28°C is optimal) (Shabi *et al.*, 1994). Fruit infection seems to be highly dependent on the numbers of conidia: large numbers will cause heavy infection, but lower numbers cause no damage (Coates, 1991). Anthracnose is usually not important in areas with low summer rainfall.

As on other hosts of *C. gloeosporoides*, infections of unripe avocado fruit usually progress no further than the formation of appressoria, and remain quiescent until ripening begins. Quiescence in avocado is caused by fungistatic concentrations of antifungal compounds called dienes. When concentrations of these compounds drop below fungistatic levels during the ripening process, latent infections are activated and hyphae invade the fruit (Prusky *et al.*, 1983). A similar process allows the fungus to invade senescent leaves and twigs. In wounds initiated by insects or the fissures caused by Pseudocercospora spot and scab, the tissues ripen prematurely and the concomitant reduction in dienes predisposes the tissue to anthracnose development.

Management

A combination of resistant cultivars, cultural practices in the field, preharvest and postharvest fungicide treatment, correct storage conditions and rapid marketing is effective. Pruning the lower limbs to lower humidity in the canopy, and removing dead twigs, branches, leaves and fruit to reduce inoculum all help to combat the disease.

Since the fungus usually is quiescent in the cuticle of unripe fruit, combinations of protective preharvest fungicides and postharvest treatments can prevent development of the disease. Preharvest treatments with captafol, copper fungicides, dithiocarbamates, promyl and triadimefon have proven effective (Prusky, 1994). In many locations, copper sprays are applied at 14- to 28-day intervals from fruit set to harvest, although the fungus has become resistant to copper fungicides in some areas. In Australia, procloraz treatments after each wet period of 2 or more days plus copper fungicides have provided control (Muirhead *et al.*, 1982; Vock, 2001). In South Africa, combinations of captafol and copper are effective.

Postharvest treatments that have controlled anthracnose successfully include thiobendazole dips, and 5 min procloraz dips and low concentrations of procloraz in wax. Postharvest treatments are not needed if fruit are marketed rapidly, but the disease increases greatly as the length of time in storage and transit increases (Ploetz *et al.*, 1994).

Fruit should be placed in cold storage as quickly as possible after harvest. Storage temperatures that range from 5 to 18°C reduce anthracnose development, but they may not be appropriate for ripening in all cultivars. Once ripening has occurred, fruit can be held at 2–4°C for extended periods without damage. Anthracnose development is severe when avocados are stored above 24°C (Fitzell and Muirhead, 1983; Pegg, 1991; Prusky, 1994).

Biological control has been achieved with several bacteria and yeasts and may well provide a suitable alternative for postharvest fungicide treatments (Korsten *et al.*, 1995; Stirling *et al.*, 1995). In Israel, treatment with antioxidants that delay the breakdown of dienes, and treatment with CO_2 or nonpathogenic strains of *Colletotrichum* that stimulate diene production have also been effective (Prusky, 1988; Yakoby *et al.*, 2001).

Some cultivars, such as 'Fuerte', 'Rincon' and 'Wurtz' are more susceptible to anthracnose than 'Hass'. Anthracnose spots also are not clearly visible against the dark skin of 'Hass'. Cultivars that are less susceptible to anthracnose ripen and are ready to eat before dienes drop to subfungicidal levels (Prusky, 1994). The rootstock also impacts anthracnose development. In Australia, anthracnose incidence and severity were significantly lower and the percentage of acceptable fruit was significantly higher ($P < 0.05$) when 'Hass' was grafted on 'Velvick' rather than 'Duke 6' rootstocks (Willingham *et al.*, 2001). These differences were correlated significantly with higher leaf diene levels and lower N:Ca ratios in fruit skin.

Bacterial soft rot

Bacterial soft rot can occur in the field, but most often is a postharvest problem in wet subtropical or tropical climates.

Portions of the fruit skin become dark with a metallic sheen and a soft or mushy texture. Internally, the fruit is brown, often liquefied and has a putrid odour.

Bacterial soft rot is caused by *Erwinia herbicola* and *E. carotovora* (Allen, 1985). These bacteria often are common, saprophytic epiphytes on leaves, stems and fruit. Under stressful conditions or after wounding, they can become pathogenic. They infiltrate the peel during rainy weather, especially if it follows a dry period. Injured fruit, fruit with rind diseases such as anthracnose, scab and Pseudocercospora spot, or overmature fruit are most susceptible. 'Fuerte', 'Reed' and 'Sharwil' are highly susceptible (Allen, 1985).

Care should be taken to prevent injuries during harvest. Fruit should be clipped from the tree and the pedicels should be left attached to the fruit. Fruit should not be harvested when they are wet and they should not be wetted prior to arriving at the packing house (Pegg, 1991; Vock, 2001).

Dothiorella fruit rot

Dothiorella fruit rot can be an economically important postharvest decay. Losses it causes are often lumped together with those due to anthracnose and stem-end rot, since these diseases can be hard to distinguish.

Symptoms

Dothiorella fruit rot first appears as superficial, irregular, amber to reddish-brown lesions on the peel when the fruit softens during ripening. It is a disease of the packing house and only rarely found on overmature fruit in the orchard. The decay often spreads along vascular bundles in the fruit. As the fruit ages, the lesions enlarge rapidly, and become sunken and black. At this stage, a watery decay spreads throughout the fruit. The fruit eventually shrivels and an unpleasant odour develops. Finally, grey mycelium may envelop the fruit. Dothiorella fruit rot is distinct from stem-end rot in that it affects parts of the fruit other than the stem end. The lesions can be mistaken for anthracnose lesions in young fruit, but they are more superficial. The grey mycelium and lack of pink sporulation on the fruit surface distinguishes Dothiorella rot from anthracnose.

Causal agents

Dothiorella rot is caused by several species of *Botryosphaeria*, several of which also cause Dothiorella stem canker (see below). The anamorphic genus of some of these species was formerly known as *Dothiorella*, hence the name Dothiorella rot. The disease is most commonly caused by *B. dothidea* (anamorph: *Fusicoccum aesculi*). However, *B. ribis* (anamorph: *F. parvum*) and *F. luteum* also cause the disease. *B. dothidea* is described in Chapter 1, and *B. ribis* is described in the section on Dothiorella stem canker. *F. luteum* has no known teleomorph, and it cannot be distinguished reliably from *F. parvum*. In the absence of a sexual stage, *F. parvum* and *F. luteum* can be differentiated by molecular techniques (Denman *et al*., 2000; Zhou and Stanosz, 2001). All of these fungi can be isolated on potato dextrose agar (PDA).

Epidemiology

Ascospores or conidia that are produced on dead bark, twigs, cankers, senescent fruit and dead leaves are infectious (Horne and Palmer, 1935). Ascospores are wind disseminated and are more likely to be produced during the winter or spring. The ascostromata and ascospores are overwintering stages that survive under unfavourable conditions. Conidia are usually disseminated by rain and are produced more or less continuously under appropriate conditions. Ascospores and conidia infect fruit through wounds and lenticels, although direct penetration can occur when high levels of inoculum are present (Darvas, 1982). Fruit infections usually remain quiescent until after harvest, although small, superficial lesions may develop in the orchard (Johnson, 1994). Drought stress often weakens trees, resulting in more inoculum on infected twigs and branches, but humid, rainy conditions are necessary to spread the fungus to the fruit.

Management

See measures reported for stem-end rots.

Duke 6 stem pitting

'Duke 6' was a rootstock cultivar propagated in California for its resistance to Phytophthora root rot. Many countries imported it before the stem-pitting malady was discovered in South Africa.

Stunting, rapid defoliation, branch death, tree decline and collapse occur in young trees affected by Duke 6 stem pitting. Severe stem pitting occurs on the wood under the bark on the rootstocks, and mosaic patterns were observed on leaves when 'Duke 6' was grafted on 'Duke 7' and 'G6' clonal rootstocks (Moll *et al*., 1987).

The causal organism is unknown. The symptoms and transmissibility suggested a virus or a fastidious bacterium (Moll *et al*., 1987). However, viral double-stranded RNA is associated inconsistently with the syndrome.

The disease has been shown to spread to nearby 'Hass' trees grown on 'Edranol' and 'Duke 7'. The disease is graft transmissible, but insect transmission has not been demonstrated. Drought stress apparently triggers the disease (Moll *et al*., 1987). Fortunately, after all known carriers of the disease were eradicated, the disease appears to have disappeared.

Powdery mildew

Powdery mildew is a minor disease in North America. Leaves show a dark, watery discoloration on the surface, especially along the midribs, and may be stunted or curled and distorted. On the undersurface of the leaf, white, powdery mycelium and spores of the pathogen are visible. Eventually, purplish, net-like or vein-like blotches appear on the underside of the leaves. Young succulent flushes are often affected, and terminal shoots may be killed.

The disease is caused by an *Oidium* sp. (Stevens and Piper, 1941). Barrel-shaped to elliptical conidia are formed in chains on the lower surface of the leaves. A teleomorph of this fungus has not been observed. Spores are apparently disseminated by wind and rain to the young leaves in the spring. Spores from these primary infections attack subsequent flushes.

Powdery mildew is severe in damp, shaded areas, especially nurseries. Adequate pruning to reduce humidity and increase transmission of sunlight may reduce the disease. Copper or sulphur applications are effective (Stevens and Piper, 1941; Zentmyer, 1984).

Pseudocercospora spot (blotch)

Pseudocercospora spot is known by several names, including blotch, Cercospora spot and black spot. It is a disease of warm, humid and rainy climates, and is found in Florida, Latin America, northern Australia and South Africa. Losses can be heavy, and at one time reached 69% in some unsprayed orchards in South Africa (Darvas, 1982).

Symptoms

Symptoms occur on leaves, stems and fruit (Pohronezny *et al*., 1994). Lesions first appear as small (1–5 mm), angular, purple to brown flecks or spots near the leaf margins. Older leaf spots are surrounded by chlorotic haloes. Signs of the fungus may appear under humid conditions as grey, felty mycelium in the centre of the lesions. Individual lesions may coalesce to form large, brown, dead areas on the leaf. The leaves become curled and deformed and the tree may defoliate.

On fruit, lesions begin as small, dark flecks that later expand or coalesce into a circular or angular shape, become slightly sunken, brown to black in colour and eventually become cracked or fissured, which may allow other pathogens to enter. In some cases, if the disease is arrested temporarily, the disease is manifested as minute, raised, shiny, black specks associated with the corking of lenticels. The disease normally is superficial but can invade the flesh during advanced stages. If defoliation occurs, fruit can become chlorotic, shrivel and drop.

On green twigs and fruit pedicels, dark brown to black, 2–10 mm lesions can develop which may result in fruit drop.

Causal agent

Pseudocercospora spot is caused by *Pseudocercospora purpurea*. *P. purpurea* produces dense fascicles of conidiophores on dark brown to black, spherical to irregular stromata, 15–125 μm in diameter, and immersed in leaves or fruit. The conidiophores may be tightly packed or divergent, 20–200 μm long, colourless to olive-brown, rarely branched, straight or with a zigzag growth, with scars produced on the tip or sides where conidia have dehisced. Conidia are club-shaped to cylindrical, with a truncate base, pale olive, 9- to 11-septate, straight or curved and 20–200 × 2–5 μm. The fungus is thought to produce a *Mycosphaerella* teleomorph that is rarely observed and not thought to be important in the disease cycle.

P. purpurea is slow growing and very difficult to isolate. Care must be taken to isolate from fresh young lesions and to surface disinfest the tissue thoroughly to remove other contaminating fungi that will outgrow *P. purpurea*. It can be isolated on standard media, such as PDA. It produces a tufted, leathery growth, which is initially grey but becomes brown to black.

Epidemiology

Conidia can be present on leaves year-round if humidity and moisture are favourable (Pohronezny *et al.*, 1994). They are carried by wind, rain, irrigation water or insects to infection courts. After penetration, the pathogen remains latent for ~3 months. Fruit from a quarter to three-quarter size are susceptible, and very young fruit and fruit near maturity are immune (Pohronezny *et al.*, 1994). Severe disease develops when warm, humid, rainy weather occurs when fruit are about a quarter size.

P. purpurea affects many *Persea* species and all cultivars of *P. americana*. Since different cultivars are reported to be most susceptible in different geographical regions, there may be physiological races of this pathogen. Late-maturing cultivars appear to be most susceptible due to the longer period during which they can be infected. Anthracnose often flourishes in blotch lesions, resulting in fruit rot that is far worse than when either disease appears alone.

Management

Timely applications of cyproconazole, flusilazol or zinc, manganese or copper fungicides are effective (Darvas, 1982; Lonsdale, 1992; Teliz, 2000). Sprays should begin when the floral buds begin to swell and continue at monthly intervals until just before harvest. Sprays can be reduced during dry periods. Biological control with *Bacillus subtilis* has been effective in field trials in South Africa (Korsten *et al.*, 1997). In Mexico, inoculum can develop abundantly on prunings. Thus, in areas where this disease is a problem, pruning should be done during dry periods or the residue should be ground in a shredder or removed from the orchard (Teliz, 2000).

Ringneck

Ringneck can be very important, especially in areas where irrigation is not possible. It is a major problem in Mexico, where losses of 10–15% occur in an average year (Teliz, 2000).

Symptoms

The typical symptom is a partial or completely dry, corky, reddish brown ring that encircles the fruit pedicel (Plate 12) (Teliz, 2000; Vock, 2001). The ring may be from 2 mm to 2 cm wide. Dead tissue may peel and flare. Sometimes there is cracking of the peel near the stem end. The seed coat may die prematurely, leaving a dark, skin-like layer attached to the flesh when the seed is removed. When the seed coat dies, the fruit ceases normal growth, becomes stunted and oval in shape, and the pericarp may become purple. Fruit often drop or dehydrate and mummify. In some cultivars such as 'Fuerte', the basal end of the fruit may crack, darken, desiccate and eventually decay. In advanced stages, the seed and flesh can have pockets of decay.

Causal agent

This disease is thought to be a physiological problem resulting from insufficient moisture or rapid drying caused by hot, dry winds (Vock, 2001). In Mexico, the disorder is exacerbated by a variety of secondary microorganisms that gain entrance after the fruit is damaged (Teliz, 2000).

Epidemiology

In many cases, the disease is caused by damaged roots that result in desiccation of the foliage during periods of water stress (Vock, 2001). Trees with Phytophthora root rot appear to be particularly susceptible, and zinc deficiency can exacerbate the problem.

Management

Adequate soil moisture should be maintained at all times, especially during peak demand periods such as flowering and fruit set (Vock, 2001). Soil moisture monitoring devices should be used to optimize irrigation efficiency. Trees should be mulched to reduce loss of soil moisture, and root systems should be protected against root disease.

Scab

Scab is a serious problem in humid avocado-growing regions such as Florida, Latin America, Morocco, the Philippines and South Africa. Severe losses result from fruit drop and the lowered market value of affected fruit.

Symptoms

Fruit spots initially are oval to irregular in shape, brown to purple–brown, and slightly raised with a sandpaper-like surface (Pohronezny and Simone, 1994). As the disease progresses, spots enlarge and coalesce. Often there are intersecting raised ridges, and large, rough, corky areas may form over the surface of the fruit (Plate 13).

Lesions on leaves are less conspicuous because they often are high in the tree canopy. They are initially <3.5 mm in diameter and become necrotic and brown to black. They are often concentrated along leaf veins and cause the leaves to become stunted, crinkled and distorted. Lesions may coalesce into star-like patterns, and shot holes develop in the leaves. Raised, corky, roughened, oval to elongate lesions also occur on twigs and pedicels.

Causal agent

Scab is caused by *Sphaceloma perseae* (Jenkins, 1934). Acervuli erupt from leaves or fruit lesions as small, white, cream to olive masses of clustered conidiophores and spores. Conidiophores are 12–100 μm in length and bear conidia along the tip or sides. Conidia are colourless, non-septate, ovoid to strongly curved and 2–30 × 2–5 μm. The fungus can be isolated on PDA, on which it produces white to dark grey mycelium.

Epidemiology

Conidia may be formed on infected leaves, twigs and fruit throughout the year when conditions are favourable. They are carried to infection courts by wind, rain and insects (Teliz, 2000). *S. perseae* is favoured by cool, moist weather. In Mexico, most spores were produced in the winter prior to active growth, while most lesions were found 6 months later in summer after fruit set and foliage flush (Teliz, 2000).

The fungus is a pathogen of young tissue. Leaves become resistant 1 month after emergence and fruit become resistant once they are half-grown (Pohronezny and Simone, 1994). Disease is most severe when heavy rains or foggy weather keep the humidity above 80% when host tissue is at a susceptible growth stage (Teliz, 2000). Injuries caused by thrips create entry wounds for *S. perseae* and greatly exacerbate scab development.

Cultivars vary in their reaction to scab. 'Lula' is now seldom planted in Florida because of its extreme susceptibility. 'Booth 3, 5, 7 and 8', 'Choquette', 'Fuerte', 'Hass', 'Monroe' and 'Trapp' are moderately susceptible, whereas 'Booth 1', 'Collins', 'Pollack' and 'Waldin' are somewhat resistant (Pohronezny and Simone, 1994).

The rind injuries that are caused by scab are often used as entry points by other fruit pathogens. When other fruit diseases interact with scab, fruit quality drops dramatically.

Management

Scabby fruit, which are left on the tree during harvest, often become the main inoculum source the following year. Thus, scabby fruit should be removed from the grove. Canopies should be pruned regularly and skirts removed to improve air movement and the penetration of sunlight (Teliz, 2000).

Sprays of copper fungicides are effective. Applications should be made as flower buds appear, near the main bloom period, and 3–4 weeks later (Pohronezny and Simone, 1994). More sprays may be required during periods of heavy rain or fog. If humidity remains below 60%, some of these sprays may be omitted (Teliz, 2000).

Effective thrips control must accompany any effective scab control programme (Teliz, 2000).

Silver spot

Silver spot is a minor disease of avocado in Mexico and Central America. Leaf spots are slightly depressed and light brown on the upper surface and darker brown on the lower surface (Teliz, 2000). A dark brown or black border often surrounds spots. Pycnidia and ascostroma frequently are observed in dead tissue in lesions. When leaves are heavily infected, the necrotic lesions may coalesce to cover large sections of the leaf. Defoliation may occur, but is usually restricted to older leaves and apparently does not reduce yields (Teliz, 2000).

Silver spot is caused by *Mycosphaerella perseae* (Teliz, 2000). The *Septoria* anamorph produces tiny, black pycnidia. The spores are colourless, acicular to filiform, with pointed ends, and non- to many-septate with transverse septa. The teleomorph develops as single to multiple locules in a black ascostromata. Ascospores are two-celled, with the apical cell slightly larger than the basal cell. The fungus can be isolated on malt glucose agar, but dark pycnidia may take 2 months to develop (Teliz, 2000).

Ascospores produced in dead leaves infect young leaves. Secondary spread may occur via conidia from older leaves or dead twigs. Dissemination occurs via wind, rain and insects. Rain, fog and humid conditions in the spring favour silver spot development.

Judicious pruning to reduce humidity in the canopy is effective. When the disease is severe, fallen leaves that produce inoculum should be removed from the grove (Teliz, 2000). Three applications of maneb, captan or copper fungicides at 30-day intervals were successful (Teliz, 2000).

Sooty blotch

This minor disease does not damage avocado fruit, but discolours it and lowers its market value. It escalated into a major problem in South Africa in 1985, when 3% of all exported fruit was rejected due to the disease (Pieterse, 1986).

Sooty or smoky blotches develop on the branches, stems, leaf veins and fruit. They become dark and thick, and the affected area turns sooty black with age (Pegg, 1991).

Sooty blotch is caused by *Akarapeltopsis* sp. (Smith *et al.*, 1985). Ascostromata are produced abundantly on affected avocado tissues. They are flattened against the plant cuticle and are dark brown, circular shield or fish scale-like with a pore in the centre. They measure 220–300 μm in diameter, are 35–45 μm high and are composed of tightly interwoven, radial hyphae. The asci are obclavate, bitunicate and each contains 16 ascospores. Ascospores are colourless, smooth, club-shaped, two-celled and measure 14–16 × 4–5 μm. Paraphyses are thread-like, colourless, extend beyond the asci and are 1–2 μm in diameter.

Ascostroma on old leaves or petioles release ascospores in the spring after rains (Smith *et al.*, 1985). The population builds up during the summer from new infections caused by ascospores or hyphal fragments. Haustoria have not been observed.

Warm, rainy or foggy weather, which creates long periods of moisture on the leaves and fruit, favours this disease. Copper fungicide sprays applied in the spring and at intervals during the summer will control sooty blotch (Lonsdale, 1991; Pegg, 1991). Treating the fruit with a 1 min dip in Ca-hypochlorite in the packing house removes the dark growth of the fungus (Vock, 2001).

Sooty mould

Sooty mould is a general term for the dark mouldy signs of a closely related group of fungi that grow on plant and fruit surfaces. They do not parasitize avocado, but grow on the honeydew excreted by various insects. Sooty mould lowers the market value of fruit.

Symptoms

Black, felty fungal growth develops on fruit, leaves and twigs (Teliz, 2000). The mycelium is associated with the sticky honeydew (excreta) of scale insects and aphids. Leaf surfaces can be covered to such an extent that photosynthesis is inhibited and the leaves become chlorotic.

Causal agents

Sooty mould is caused by *Capnodium* spp. and other fungi. *Capnodium* spp. produce spherical- to pear-shaped ascostromata that are borne on short pedicels. Ascospores are dark brown with both vertical and transverse septa. Many of the sooty mould species that are from non-tropical areas are not *Capnodium* species. Many form perithecia rarely, if at all.

Many of these sooty moulds are perpetuated via hyphal fragments or conidia, which are dark, thick-walled and multiseptate. Ascospores, conidia and hyphal fragments can be disseminated by wind, rain and insects. When they contact honeydew, they germinate and produce dark, felty mats that contain ascocarps and conidia.

Epidemiology

Sooty moulds are associated with honeydew that are produced by a variety of insects and aphids. The heart-shaped scale is a common associate in Mediterranean countries and South Africa (Du Toit and De Villiers, 1988). It shows a preference for the 'Collinson', 'Hass' and 'Ryan' cultivars. Dust appears to affect the natural, biological controls of some these insects, and the disease becomes more severe along roads.

Management

The insects and aphids that produce honeydew should be controlled, and natural and assisted biological measures are effective (Du Toit and De Villiers, 1988). Reduction of dust along roadways may reduce the problem.

Copper fungicides, bupofexin and chlorpirifos are also effective, but repeated copper sprays may interfere with the natural biocontrol of the honeydew insects (Du Toit and De Villiers, 1988; Teliz, 2000). Washing the fruit in the packing house with water or a Ca-hypochlorite solution usually removes these fungi.

Stem-end rot

Stem-end rot is caused by a variety of pathogens. It can become severe anywhere avocados are grown, especially when anthracnose is well controlled and storage conditions are not optimal.

Symptoms

Rot begins at the stem end as a slight shrivelling of the tissue around the stem button (Plate 14). Mycelium is often observed on the abscission scar when the stem button is removed. Conspicuous, dark-brown to black lesions, which have well-defined margins, advance from the stem end and eventually encompass the entire fruit as ripening progresses. The fruit eventually become shrivelled, watery soft and are often covered by fungal mycelium (Johnson and Kotzé, 1994). Some fungi also discolour the vascular bundles, and this symptom precedes decay of the flesh (Plate 14). Some stem-end rots are dry and corky in texture and only become watery as other organisms invade. Fruiting bodies of the pathogens may be evident on well-decayed fruit.

Causal agents

Different fungi cause stem-end rot in different avocado-growing regions (Johnson and Kotzé, 1994). The main cause of stem-end rot in Israel is *Botryosphaeria rhodina* (see Chapter 1), and in South Africa is *Fusicoccum luteum* (see Dothiorella fruit rot) and *Nectria pseudotrichia* (anamorph: *Tubercularia laterita*). The primary pathogens in Australia and New Zealand are *B. ribis* and *F. luteum*, but in the USA it is *B. dothidea* (see Chapter 1).

Colletotrichum gloeosporioides can also cause stem-end rot alone and in combination with other fungi (see anthracnose). Other fungi reported to cause stem-end rot include *Albonectria rigidiuscula* (see galls and corky bark in Chapter 15), *Alternaria* sp., *Drechslera setariae*, *Gibberella pulicaris* (anamorph: *Fusarium sambucinum*), *Pestalotiopsis versicolor*, *Phomopsis perseae* (see Dothiorella stem canker) and *Rhizopus stolonifer* (Darvas and Kotzé, 1987). *Dothiorella mangiferae*, which has been synonymized with both *Fusicoccum aesculi* and *Fusicoccum parvum*, has been implicated but is a doubtful agent on avocado. Below are descriptions of important stem-end rot fungi that are not covered elsewhere in this book.

P. versicolor produces thick cottony, white to yellow mycelium and acervuli that are 0.5 mm in diameter and exude conidia in glistening, greenish-black drops. Conidia are fusiform and 22–29 × 7–10 μm. They have three thick-walled, dark median cells and thin-walled, colourless apical and basal cells. Three bristles radiate from the apex of the spore, while a single bristle-like pedicel protrudes from the basal cell giving the spore an insect-like appearance.

N. pseudotrichia produces conidia that stick together in balls at the tips of phialids on hyphae in culture (Johnson and Kotzé, 1994). The conidia are aseptate and 4–7 × 2–3 μm. In older cultures or on host tissue, the phialids are arranged on synnemata of tufts or pads of vertical, laterally fused hyphae. The synnemata are orange–red to brown at the base, becoming yellow to buff at the tips, and are 150–300 μm high. Perithecia are produced in clusters that erupt through the bark of dead twigs or wood. They measure 200–600 μm in diameter and are bright orange–red and become brown to nearly black with age. Under dry conditions, they collapse and appear disk-shaped. Ascospores are colourless to pale brown, elliptical, often tapering or curved towards the apex and measure 15–40 × 15–17 μm. They initially are three-septate but later have both transverse and longitudinal septa.

All of the above fungi can be cultured on PDA.

Epidemiology

All of these fungi are saprophytes or weak pathogens that are present in soil or senescent avocado flowers, bark, twigs, fruit and leaves (Johnson and Kotzé, 1994). Many of them are present in the pedicel or fruit peel as latent or endophytic infections, whereas others sporulate on dead tissues and are transferred to the fruit or pedicel by wind or rain. When fruit is harvested, the fungi are stimulated to grow in the injured pedicel or damaged button tissue. The likelihood of stem-end rot increases if the stem end of fruit comes into contact with litter or soil, and as the time between harvest and consumption increases. As the fruit senesces, the decay moves into the stem end and decays the flesh.

Stem-end rot is most severe on the east and north (northern hemisphere) or south (southern hemisphere) sides of trees. Stem-end rot fungi readily colonize senescent tissue that is associated with ring-neck and lesions of fruit-spotting diseases, such as scab and Pseudocercospora spot.

Environmental conditions can dictate which stem-end rotting fungus predominates. Hot weather favours *B. rhodina*, whereas wet conditions favour *C. gloeosporioides* and *N. pseudotrichia*. Water stress appears to stimulate latent infections by several of the *Botryosphaeria* species. Storage temperature also has an effect: cool storage promotes *C. gloeosporioides* and *P. perseae* over *N. pseudotrichia*, and at 30°C *B. rhodina* will predominate over other *Botryosphaeria* and *Fusicoccum* species (Johnson and Kotzé, 1994).

Management

To date, stem-end rot has been treated as a single entity, regardless of the causal agents that are involved. More effective treatments might be developed if the stem-end rots, caused by diverse agents, were recognized as different diseases. Many of the following treatments may not be acceptable in all production areas.

Dead wood and old fruit should be removed to reduce opportunities for preharvest latent infections. Optimum irrigation and nutrition reduces stress that interferes with natural resistance and creates senescent or dead avocado tissues. Mulching under trees often hastens litter decomposition and, thus, reduces inoculum.

Minimizing the length of time between harvest and consumption also reduces stem-end rot. Therefore, prompt marketing may be the best method for controlling stem-end rot (Vock, 2001). Fruit should never be harvested when they are wet or allowed to remain in a wet condition before reaching the packing house, since these conditions stimulate germination of spores and latent structures (Johnson and Kotzé, 1994). Removing the pedicel can reduce stem-end rot if it does not damage the button or the fruit rind. Sealing the stem end with wax can eliminate stem-end rot caused by *N. pseudotrichia*, but this procedure often results in increased stem-end rot due to *Botryosphaeria* sp. and *C. gloeosporioides* (Johnson and Kotzé, 1994).

Preharvest sprays with copper compounds or captafol usually will reduce the incidence of stem-end rot (Horne and Palmer, 1935; Darvas *et al.*, 1987). Where permitted, a postharvest spray of procloraz within 24 h of harvest is effective (Vock, 2001). The antioxidant butylated hydroxytoluene in combination with ascorbic or citric acid can delay stem-end rot caused by *B. rhodina* when it follows cool storage at 2°C, storage for 10 days and ripening at 20°C (Johnson and Kotzé, 1994). Controlled ripening at 16–18°C combined with fungicide treatments usually provides good stem-end rot control. Ripe fruit should be stored promptly at 7°C (Fitzell and Muirhead, 1983).

Bacillus subtilis has provided effective biocontrol of stem-end rots caused by species of *Botryosphaeria* and *Colletotrichum* (Korsten *et al.*, 1995).

Sunblotch

Sunblotch occurs wherever avocados are grown. Because the causal viroid can latently infect avocado, sunblotch can become a severe problem if persistent efforts are not made to exclude the pathogen from propagation material.

Symptoms

The most consistent symptoms are red, yellow, pink or white streaks that are slightly indented and run lengthwise on green twigs or young stems (Ohr *et al.*, 1994). Fruit symptoms include white, yellow or red blotches that may or may not form in depressed or scar-like areas on the fruit (Plate 15). Leaves may have white or yellow chlorotic, variegated areas and may be deformed. Sometimes clusters of leaves are completely chlorotic. Leaf symptoms are rare in the field. Regular cracking and checking of the bark, which resembles an alligator's skin, occurs on the trunk and larger branches. Trees are often stunted, sparsely foliated and exhibit a sprawling, prostrate architecture. Trees can become asymptomatic quickly. Although these trees show no obvious symptoms, all infected trees, symptomless or not, usually have greatly reduced yields (Desjardins, 1987). Further complicating the picture is the fact that sunblotch symptoms can vary widely depending on host cultivars, the environment and the viroid strain (Semancik and Szychowski, 1994).

Causal agent

Sunblotch is caused by the *Avocado sunblotch viroid* (ASBVd), which is a single-stranded, non-encapsidated, circular RNA molecule. ASBVd ranges from 246 to 251 (247) nucleotides in size, and sequence variants are commonplace. In a recent study, Schnell *et al.* (2001) detected 60 variants among a total of 122 clones that they sequenced. Moreover, variants in a given tree often were unique. In the same study, 60–81% of the variants that were detected in four different trees (each of a different cultivar) were found in only one tree.

Sunblotch can be diagnosed by observing symptom development on young Mexican seedlings that are grafted with buds from suspect trees. More modern and rapid methods of indexing rely on DNA probes or reverse transcriptase–polymerase chain reaction (RT–PCR). The latter technique is very sensitive and capable of detecting 1 ng of ASBVd (Schnell *et al.*, 1997).

Epidemiology

The viroid is systemic in avocado trees, but it is often distributed non-uniformly (Desjardins, 1987). Thus, diagnosis can be problematic. Ironically, asymptomatic carriers of sunblotch transmit the disease via seed at a nearly 100% rate, whereas trees with symptoms only occasionally transmit the disease (Desjardins, 1987). Sunblotch is readily graft-transmissible, and natural root grafts are an important mechanism of spread in the field. Sunblotch is also transmitted, but less often, via pruning and injecting wounds, and from pollen of infected trees. Infected pollen affects the fruit it produces, but no other part of the tree. There is no evidence of insect transmission (Desjardins, 1987).

Management

The primary control measure for this disease is the careful selection of pathogen-free sources of budwood and seed that are used during propagation. To ensure the pathogen-free status of these materials, the above indexing procedures should be used (Desjardins, 1987; Schnell *et al.*, 1997). Since seed from symptomless carriers is more likely to transmit the disease than seed from trees with symptoms, indexing is imperative. Sources of avocado seed that are used for propagation should be separated from commercial groves or the surrounding trees should be indexed periodically. Infected trees in a grove should be removed and all suspect trees should be indexed. Pruning tools, harvesting clippers and injection equipment should be disinfested with 1.5% sodium hypochlorite between trees (Desjardins, 1987). In many orchards, symptomless carriers are left in place because yield does not seem to be affected and spread does not occur. In other orchards, especially those planted on spacings of <6 m, the disease can spread down the rows.

Tar spot

Tar spot is a relatively minor disease in the Americas, but it is very common and often is present on 50% of the leaves in a grove. It appears to affect old leaves more severely than young leaves and thus usually does not affect yield (Teliz, 2000).

Tar spot lesions initially are pale to dark brown, irregularly circular and up to 10 mm in diameter on the surface of the leaves. Eventually, fungal stromata form that are shiny, black, circular, convex, raised and measure 1–7 mm in diameter. The stromata are often surrounded by a chlorotic halo. When the disease is severe, all leaves become chlorotic and the tree may defoliate (Teliz, 2000).

Tar spot is caused by *Phyllachora gratissima*. Conidia are formed in locules within stromata. Conidiophores are up to 36 μm long and elongate percurrently. Conidia are lanceolate to filiform, 44–58 × 1–2 μm, colourless, non-septate and wider towards the base and pointed at the apex. Locules within stromata also give rise to colourless, aseptate, bean-shaped ascospores that measure 21–25 × 11–13 μm and have a button-like, 2 μm long appendage at each end. Another species, *P. perseae*, is found on avocados in cooler climates and has larger ascospores.

Young avocado leaves are apparently infected by ascospores from old or fallen leaves in the spring. Secondary infection may take place during the summer by conidia. Dissemination of spores is via wind and rain, and windy, rainy conditions, especially in the spring, favour the disease.

Diseased, fallen leaves should be removed from the orchard. Copper sprays should begin in the spring as new foliage emerges, and in severe conditions applications should be repeated every 30 days (Teliz, 2000).

Diseases of the Trunk or Main Branches

Algal leaf spot

This relatively minor disease is also known as velvety spot. It is restricted to tropical areas.

Round, orange to red, raised, velvety spots are produced on leaves, branches or stems (Plate 16). The spots turn to white, grey, dark brown to nearly black as they age. A yellow halo may surround the spots (Pegg, 1991).

Algal leaf spot is caused by *Cephaleuros virescens* in the Americas. It is described in Chapter 1. In other parts of the world, other algae may be involved.

Spores from spots on old leaves are disseminated by wind and rain to new leaves. Spores are produced continuously during high rainfall and prolonged humid conditions that encourage development of the disease.

Reducing canopy humidity by judicious pruning, and applying copper fungicides in the spring controls the disease (Pegg, 1991; Vock, 2001).

Avocado black streak

Although black streak is a minor problem, it can cause severe cankers in California, the Canary Islands and Florida.

Symptoms

Cankers first appear on the lower trunk and the undersides of lower branches, but can appear higher in the tree. The cankers appear as superficial cracks in the outer layers of bark (Ohr and Zentmyer, 1994b). They ooze sap that dries and accumulates as a cinnamon to white, powdery sugar deposit on the canker surface (Plate 17). Since the exudate is water-soluble, it is not present during rainy periods. Scraping the lesions reveals shallow, reddish brown, irregular necrotic areas in the bark that form mottled areas of dead and live tissue or coalesce into large dead areas (Ohr and Murphy, 1987). Lesions are restricted to the outer bark or extend into the cambium. The dead areas can often be easily 'popped out' with a knife but usually do not extend deep enough into the bark to injure the tree severely.

On young trees and green shoots, lesions are seen as black blotches with definite margins. Affected trees have: shortened twig internodes; bunchy, terminal leaf growth; small, blotchy, chlorotic leaves; necrotic leaf spots; partial or full defoliation; shoot dieback; early blooms of chlorotic flowers; zinc deficiency; wilted leaves; and reduced yield. Symptoms can occur on the entire tree or on one side or a single branch. Usually trees continue to decline until they die, and collapse is sometimes sudden (Ohr and Zentmyer, 1994b).

Causal agent

The causal agent of black streak is not known.

Epidemiology

The disease can spread in the field and is transmissible (Zentmyer, 1984). Poor irrigation and inadequate fertilization appear to exacerbate the disease.

Management

Little can be done to control black streak. It is thought to occur primarily on Guatemalan cultivars such as 'Hass', but it has been found on both Mexican and West Indian cultivars. Recommendations include maintaining adequate irrigation and fertilization. Diseased trees should be removed and the sites fumigated before replanting (Ohr and Zentmyer, 1994b). Severely pruning trees or stumping them often will encourage vigorous new growth, but the disease usually reappears within several years (Ohr and Murphy, 1987).

Bacterial canker

Bacterial canker and diseases with similar symptoms occur widely in Australia,

California, Mexico, South Africa and other avocado-growing areas. The disease is of minor importance because it rarely impacts yields.

Symptoms

Cankers, which are 2–10 cm in diameter, appear on trunks as slightly sunken darker areas on the bark (Plate 18). Under the sunken area is a necrotic pocket containing a watery liquid. The bark above the canker may split and as the liquid leaks out and dries, it deposits a white powdery residue around the periphery of the canker (Cooksey *et al.*, 1994). Reddish brown necrotic areas appear under the bark, and streaks of necrosis radiate above and below the canker. The necrotic streaks may be restricted to the bark cortex or may extend into the centre of branches or even the trunk. When numerous cankers occur on affected trees, they frequently are connected by necrotic streaks. Poor growth, defoliation and low yields can occur on such trees, but this is rare (Cooksey *et al.*, 1993). Lesions can also occur on nursery trees.

Causal agents

The disease is caused by at least two bacteria. In South Africa, *Pseudomonas syringae* was shown to cause bacterial canker (Ohr and Korsten, 1990), whereas in California the causal agent was found to be *Xanthomonas campestris* (Cooksey *et al.*, 1993). These bacteria can be isolated on yeast dextrose carbonate agar or Pseudomonas agar F. However, positive identification requires the use of monoclonal antibodies or Biolog GN plates (Biolog, Inc., Hayward, California) that use a set of biochemical tests for identification (Ohr and Korsten, 1990).

Epidemiology

The bacteria are favoured by humid, rainy conditions. They are epiphytes on leaves and green twigs, and infect and systemically colonize the vascular system through wounds and branch stubs (Cooksey *et al.*, 1994). Infections that occur in or near the phloem apparently damage the phloem and cause cankers. Boron deficiency appears to exacerbate the disease. The pathogens can be spread through nursery practices, but apparently not by insects (Cooksey *et al.*, 1994).

Management

Screening nursery trees for symptoms of the disease before planting should help reduce the incidence of disease. Removal of infected trees and the use of copper bactericides help reduce spread of the disease. However, in most cases, control measures are not warranted (Cooksey *et al.*, 1994).

Dothiorella stem canker

Dothiorella stem canker causes minor losses of limbs on small trees, and large trees that are stressed by drought or frost are also affected.

Symptoms

Cankers occur on twigs, branches or trunks. They normally exude brownish red sap that dries to form white to brownish powder (Johnson, 1994). In young trees or branches, the slightly sunken, dark brown cankers contrast sharply with the normally greenish bark. The bark and wood under the canker is killed and turns red–brown to brown (Doidge, 1922). In older cankers, the bark can be removed easily from the damaged area. If most of the xylem is involved, trees or limbs may collapse and leaves quickly turn brown but remain attached. Graft unions of top-worked stumps or young trees can be infected, and a canker may form around the graft union. In other cases, the fungus remains in the graft union after it heals and, when a later stress occurs, the tree collapses. The telltale brownish discoloration in the wood at the graft union is a key symptom (Plate 19).

Causal agents

Several different fungi cause this disease. Their taxonomy is extremely confused and may still undergo more changes before it is correct (Denman *et al.*, 2000; Slippers *et al.*, 2001; Zhou and Stanosz, 2001).

In Mexico, New Zealand, Peru, South Africa and the USA, the most common cause is *B. dothidea* (anamorph: *F. aesculi*). The teleomorph is rarely found and occurs in small pustules on the dead bark where it produces irregular shaped ascostromata. The anamorph is common and produces black, knob-like stromata. Microscopic characteristics of the pathogen are found in Chapter 1.

B. ribis (anamorph: *F. parvum*) has been reported in Chile. It has elliptical conidia with rounded ends that are 7–11 × 15–24 μm (Fig. 3.1). Average widths of conidia of *B. dothidea* are <8 μm, while those of *B. ribis* are >8 μm and the length : width ratio of conidia of the former is 2.7 : 5.3 while that of the latter is 1.7 : 2.9. *F. luteum* causes serious cankers in California. It cannot be reliably differentiated from *F. parvum* (see Dothiorella fruit rot).

B. disrupta causes avocado cankers in Mexico and Central and South America. It produces brown ascospores that measure 24–40 × 11–20 μm.

P. persae produces dark, often multiloculate pycnidia (400–500 × 200–225 μm) on white to buff mycelia or, more commonly, that are erumpent from the bark of dead twigs. Two types of spores are borne in the same pycnidium: α-conidia that are colourless, fusiform, non-septate, 7–10 × 2–3 μm and have oil droplets at the poles; and β-conidia that are curved and needle-like.

Other fungi that have been reported to cause similar branch cankers and diebacks

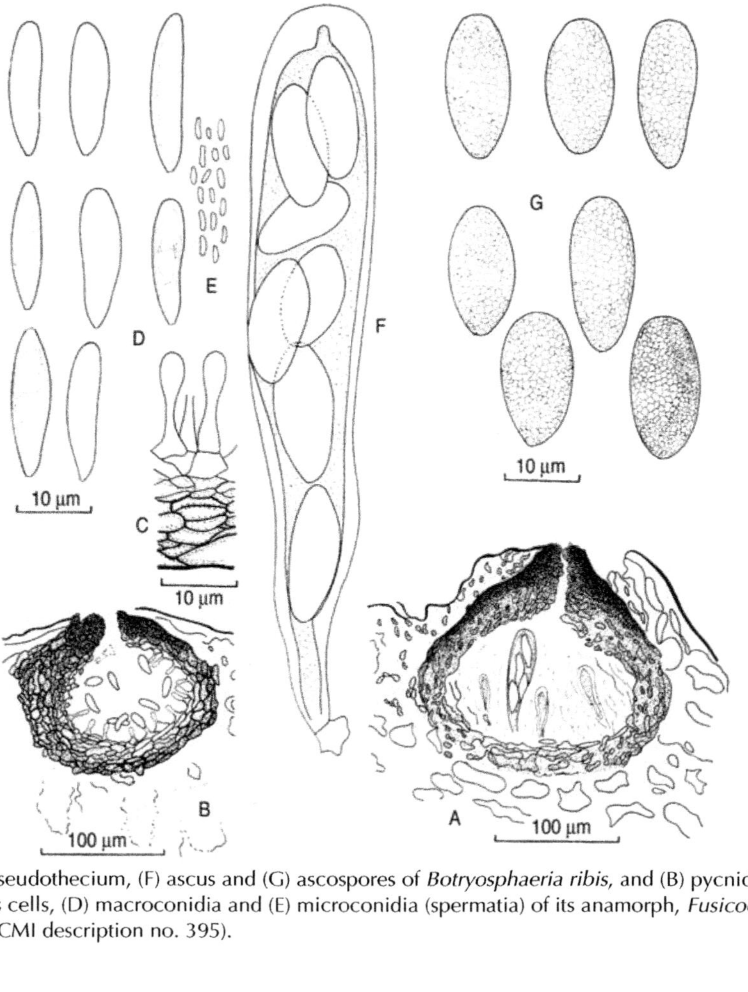

Fig. 3.1. (A) Pseudothecium, (F) ascus and (G) ascospores of *Botryosphaeria ribis*, and (B) pycnidium, (C) conidiogenous cells, (D) macroconidia and (E) microconidia (spermatia) of its anamorph, *Fusicoccum parvum* (from CMI description no. 395).

include *B. obtusa* (Mexico and the USA), *B. rhodina* (Chile, Mexico and the USA) and *B. quercuum* (Mexico and the USA). While their teleomorphs may be difficult to differentiate, their anamorphs are distinctive. *B. obtusa* has a *Sphaeropsis* anamorph with dark brown, rough-walled, one-celled conidia that measure 20–26 × 9–12 μm (Fig. 3.2). The anamorph of *B. rhodina*, *Diplodia theobromae*, is described in Chapter 1. *B. quercuum* has colourless, thick-walled, elliptical conidia that measure 18–25 × 12–16 μm.

All of these fungi produce rapidly growing colonies on PDA with tufted, greyish mycelium and dark coloured pycnidia.

Epidemiology

These pathogens are wound parasites (Doidge, 1922). In the field, ascospores are usually produced during the winter or spring, whereas conidia are produced more or less continuously under appropriate conditions. Ascospores are wind disseminated and conidia are usually disseminated by rain. They are produced on dead bark, twigs, cankers, senescent fruit and dead leaves, and both are infective (Horne and Palmer, 1935). They infect pruning wounds, branch splits from wind damage, frost damage, mechanical wounds and grafting wounds. Infection is far more likely when drought stress,

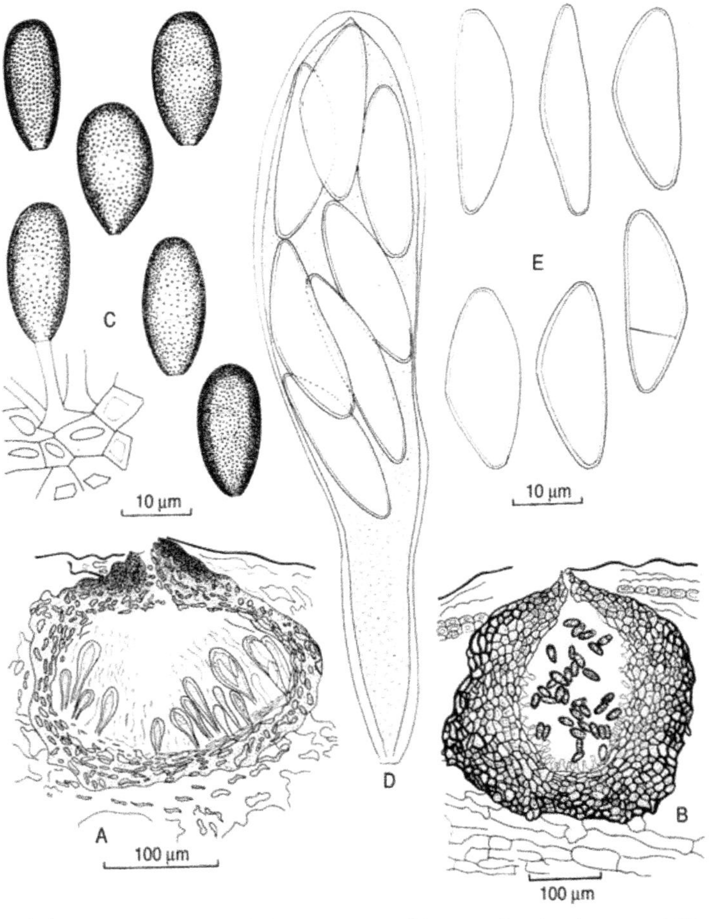

Fig. 3.2. (A) Pseudothecium, (D) ascus and (E) ascospores of *Botryosphaeria obtusa*, and (B) pycnidium and (C) conidia and conidiogenous cells of its anamorph, *Sphaeropsis malorum* (from CMI description no. 394).

flooding, insect attack, other diseases, nutrient deficiencies or chemical damage weakens trees. Infections can remain quiescent for long periods as sclerotial-like stromata until stress weakens the host (Johnson, 1994). Drought stress exacerbates the disease or triggers activity in latent infections. Heavy rainfalls under warm conditions spread the disease and increase infection (Horne and Palmer, 1935).

Management

Cankers, dead limbs, dead twigs and old fruit should be removed from trees (Doidge, 1922; Johnson, 1994). Pruning should be done during dry periods. Scraping cankers has been recommended as a way to encourage new bark development. If this method is employed, care should be taken not to injure healthy tissue. Correction of the stress factors that favour disease, such as drought or nutrient deficiency, will reduce disease severity. Field sprays with copper fungicides and treatment of graft unions with fungicides is beneficial (Horne and Palmer, 1935). Mexican rootstocks are more resistant than Guatemalan cultivars (Johnson, 1994).

Soilborne Diseases

Armillaria root rot

Armillaria root rot can kill avocado trees, but spreads slowly and is not as threatening as Phytophthora root rot. It occurs in many countries including Ecuador, Mexico, Spain and the USA (Florida and California).

Symptoms

Infected trees often decline slowly and exhibit chlorotic leaves and slow growth, or they may wilt and collapse suddenly and die with the dead leaves still attached (Ohr and Zentmyer, 1994a). Armillaria root rot attacks and rots the cambium and wood of the major roots and crowns of avocado. The pathogen may produce silky, white, stringy or fan-like plaques just under the bark in rotted areas that often are discoloured brownish or black. Dark structures known as pseudosclerotia often protrude through cracks in the bark, and slender, purplish-brown, shoestring-like rhizomorphs can often be found growing along the root surface or into the surrounding soil. The wood often becomes punky and soft as it decays. After rains, the fungus often produces clusters of characteristic mushrooms under infected trees.

Causal agents

The pathogen may be isolated from diseased tissue or surface-disinfested rhizomorphs on malt agar amended with *o*-phenylphenol (Shaw and Kile, 1991). The colonies are distinguished by their clamp connections on hyphae, red-brown crustose areas, rhizomorphs and lack of spores. Light inhibits growth of cultures.

Until recently, the causal agent was thought to be *Armillaria mellea*. This species was found to consist of several closely related species that are difficult to differentiate morphologically. Often genetic analysis or mating tests are the only way to differentiate species (Shaw and Kile, 1991). In many areas in the world, the true identity of the fungus causing Armillaria root rot is not known.

In California, the primary causal agent is the original *A. mellea*. In Florida, *A. socialis* causes the disease. It is distinguished from *A. mellea* by the lack of a ring on the stem and the rare production of mushrooms and rhizomorphs. Both species are described in Chapter 1.

Several other species may cause Armillaria root rot in other areas of the world, each of which have slightly different ecological niches, host ranges and virulence towards avocado (Shaw and Kile, 1991). Perhaps related to this proliferation of closely related species is the recent finding that nuclei in much of the vegetative hyphae of *Armillaria* species are diploid throughout most of the life cycle. This is rare among basidiomycetes. Confounding the fact that there may be many species of *Armillaria, A. mellea* has an extremely wide host range.

Epidemiology

A. mellea colonizes large roots, stumps or wood buried in soil. It cannot survive or infect healthy avocado without these food bases, and rarely colonizes wood or wood chips on the surface of the soil (Ohr and Zentmyer, 1994a). Healthy avocado roots are infected when they come in contact with rhizomorphs or when mycelial strands grow out from infected wood into the soil. Some *Armillaria* species, such as those in tropical areas, apparently produce no rhizomorphs and move only via mycelial strands.

Since *A. mellea* must colonize a tree thoroughly before it can move to another, the fungus moves very slowly. It often grows radially out from the infected tree, killing a new ring of trees each year. The fungus kills living roots by growing rapidly up the cambium. It kills the tree after it girdles the root or trunk, and survives for long periods as rhizomorphs or pseudosclerotia in these tissues.

A. mellea is present naturally in many forest ecosystems where it can be innocuous and inhabit large areas. If avocado orchards are planted in newly cleared forestland or in gullies or washes where trees or logs are buried during floods, they are often at risk from Armillaria root rot (Ohr and Zentmyer, 1994a).

Although *A. mellea* produces abundant basidiospores, they do not seem to infect avocado. It is thought that they may serve more as spermatia in carrying genetic material to already established colonies of the fungus (Ohr and Zentmyer, 1994a).

Management

Control of Armillaria root rot is difficult. Even the removal of all organic debris in new areas does not ensure freedom from this disease. In infested areas, dead trees and stumps should be pulled and replants should not be replanted in these locations. *A. mellea* is sensitive to drying, and removing soil from around the crowns of trees and major roots retards its spread (Ohr and Zentmyer, 1994a). *A. mellea* can be controlled with heavy applications of methyl bromide, although the fungus can still survive at great depths in large roots (Ohr *et al.*, 1973). Under these conditions, the fungus may be weakened by the fumigant and eventually killed by parasitic, soilborne fungi such as *Trichoderma* spp. Different approaches may be necessary to manage disease caused by different species of *Armillaria.*

Phytophthora cankers

Several species of *Phytophthora* cause cankers on the roots, crown, trunk and branches of avocado. These diseases exhibit different ecological and epidemiological characteristics and require different control strategies.

Symptoms

Phytophthora cankers usually originate below ground and may extend 3 m up the trunk and branches. The bark may be discoloured and is sometimes cracked or fissured (Zentmyer *et al.*, 1994). The cankers exude a brownish red, sugar-containing, viscous sap, which upon drying becomes a white to brownish powder and encrusts the lesion. The inner bark and the outer layer of wood are invaded and the phloem and cambium are killed (Plate 20). These tissues turn reddish brown when first invaded and then become brown. Cankers caused by *Phytophthora heveae* are often associated with a constriction of the trunk (Teliz, 2000), whereas those caused by *P. cinnamomi* are usually restricted to the root crown and do not extend far up the trunk.

Foliar symptoms in trees with advanced cases of trunk canker are often similar to those caused by Phytophthora root rot. Affected trees deteriorate, foliage becomes wilted and sparse, branches die back, and leaves are often small and yellowish (Zentmyer *et al.*, 1994). When lesions have completely girdled trees, the trees may collapse completely during hot, dry weather and the leaves turn brown but remain attached.

Causal agents

P. boehmeriae (Mexico), *P. cinnamomi* (Australia, Brazil, Cameroon, South Africa and the USA), *P. citricola* (Mexico and the USA), *P. heveae* (Guatemala and Mexico) and

P. palmivora (Honduras) have all been implicated as causes of Phytophthora canker (Teliz, 2000). These pathogens have fungal-like lifestyles, but are in the Kingdom *Chromista*, rather than the *Eumycota* (true fungi).

These pathogens can be isolated from new lesions in the inner bark or cambium on selective media, such as PARPH, or baits, such as avocado fruit and leaf discs (Erwin and Ribiero, 1996). *P. cinnamomi*, *P. citricola* and *P. palmivora* are described in Chapter 1. *P. heveae* produces numerous papillate, caducous sporangia in culture (Fig. 3.3). They are often irregular in shape, ovoid, ellipsoid or obpyriform, 29 (20–48) × 46 (27–66) μm, and with variable length : breadth ratios (1.1 : 2.9). Pedicels are up to 10 μm long. Although no chlamydospores are produced, abundant oospores form on agar, often in clusters (*P. heveae* is homothallic). Oogonia are spherical, 25–35 μm in diameter with thin, hyaline to yellowish walls. Antheridia are amphigynous and spherical (9 μm in diameter) or ellipsoidal (9 × 15 μm). *P. boehmeriae* is also homothallic and produces amphigynous oospores that are 19–41 μm in diameter (Fig. 3.4). Sporangia are prominently papillate, but are spherical, ovoid to pear-shaped, and 27–72 × 20–46 μm. Sporangia are deciduous, with pedicels <5 μm in length. Chlamydospores are common, spherical and 26–51 μm in diameter. The hyphae are also characteristically gnarled and irregular.

Epidemiology

Although these are primarily root pathogens, with the exception of *P. cinnamomi* they do surprisingly little damage to roots. Zoospores infect the feeder roots of avocado and other plants in avocado orchards. These infections produce sporangia and zoospores that allow additional infections to occur, whereas chlamydospores and oospores that are produced enable survival during unfavourable conditions. These propagules are often formed abundantly in the soil around infected trees and, when they are splashed or carried by insects to wounds in the bark, cankers are initiated.

The epidemiology of Phytophthora canker diseases that are caused by species other than *P. citricola* is not well known, and may be very different from what is described for *P. citricola*. *P. citricola* is present on feeder roots in nearly every avocado orchard that has been examined in California, but produces cankers on relatively few trees. It apparently is unable to infect non-wounded trunks (El Hamalawi and Menge, 1995a). However, there are many ways that trees can be wounded, including gophers, voles, pigs, sucker removal, staking, cultivation, pruning and violent winds. Wounds are susceptible to *P. citricola* for no more than 2 weeks after they occur (El Hamalawi and Menge, 1995a).

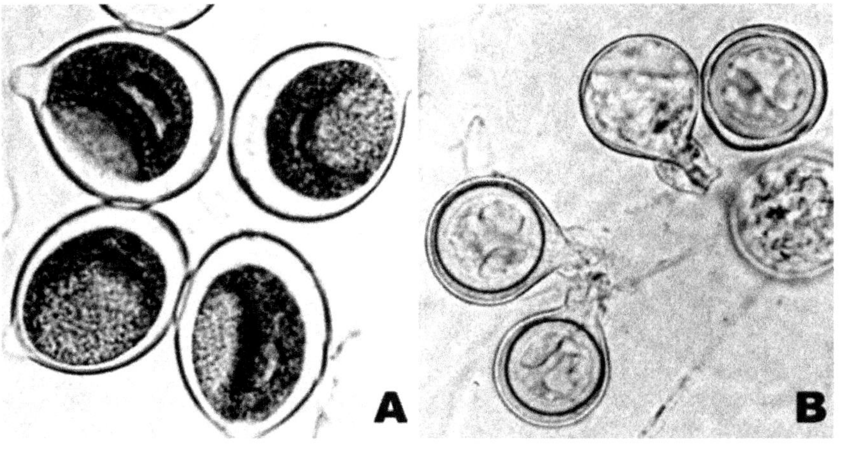

Fig. 3.3. (A) Papillate sporangia and (B) oospores of *Phytophthora heveae* (from CMI description no. 594).

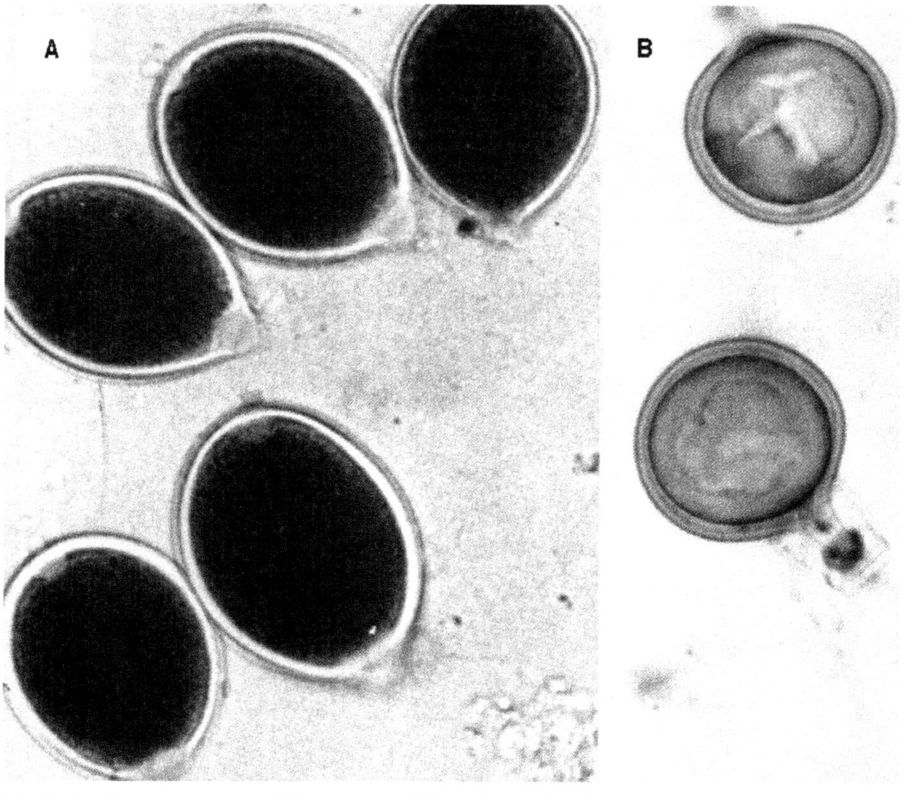

Fig. 3.4. (A) Papillate sporangia and (B) oospores of *Phytophthora boehmeriae* (from CMI description no. 591).

Disease development is stimulated where soil drainage is restricted and the soil remains saturated for long periods. Cankers also grow faster when they are constantly sprayed with irrigation water from sprinklers. Abundant oospores may be produced in sap that oozes from cankers, and they can spread by insects or birds to wounds on other trees. The accumulation of surface water also aids in the splash dissemination of zoospores.

Infection appears to be favoured by cool winter temperatures. Canker formation is also strongly affected by the host physiology (El Hamalawi and Menge, 1994). Cankers form readily during periods of nutrient storage in the bark during the winter or following leaf flushes, and pruning appears to enhance canker growth rates.

Avocado is often able to wall off infections and prevent girdling. In these cases, trees often make a full recovery. However, reinjuring the canker or pruning the tree at some later date can allow further development of the canker, spread to the cambium and ultimately death (El Hamalawi and Menge, 1994, 1995a).

Management

The best control measure is to prevent wounding of the trunk and roots. Unfortunately, many types of wounds cannot be prevented. Wounds from pruning and removal of suckers should be protected with fosetyl-Al or potassium phosphonate. Pruning tools should be disinfested periodically with bleach. When cankers are above ground, the associated bark should be cut or scraped and treated with fosetyl-Al or potassium phosphonate (El Hamalawi and Menge, 1995b). Injections or foliar sprays with these chemicals are also effective. However, since cankers are often below ground, combinations of aboveground applications and soil drenches may be required.

There are differences in susceptibility among rootstocks to Phytophthora canker. 'Duke 7' is among the most resistant, whereas 'Thomas', which is resistant to root rot caused by *P. cinnamomi*, is quite susceptible to Phytophthora canker caused by *P. citricola*.

Phytophthora root rot

Phytophthora root rot, which is also known as avocado root rot, is a serious disease and the most common limiting factor in avocado production worldwide (Coffey, 1992). It can be extremely destructive, spreading rapidly and killing most of the trees in a grove. The pathogen, *P. cinnamomi*, has spread throughout the world, and is found in more than 70 countries.

Symptoms

The pathogen invades the cortical tissue of avocado feeder roots, at first causing discrete brownish black lesions, which encircle individual roots. As the disease progresses, it may invade the entire feeder root system (Plate 21). Roots become black and brittle and, in advanced stages, feeder roots become scarce (Zentmyer, 1980). Phytophthora root rot reduces tree vitality since feeder roots require energy to produce and are needed for the uptake of water and nutrients. The reduced uptake of water in the root zone also causes the soil to remain abnormally wet. The disease normally does not progress into the larger woody roots, which remain functional until the tree dies.

Foliar symptoms include small, yellowish green leaves that wilt frequently, have brown necrotic tips and often drop prematurely. New growth is rare and entire limbs die back, leaving the tree sparsely foliated. Fruit yields decline, but large numbers of small fruit may be set. Eventually the tree dies. Trunk cankers may form occasionally, which are described under Phytophthora cankers.

Causal agent

P. cinnamomi is thought to have originated in Asia in the area encompassing Indonesia, Malaysia, New Guinea and Taiwan, although another theory indicates a South African origin (Erwin and Ribeiro, 1996). Since avocados originated in Central America, they have almost no natural resistance to this pathogen.

Microscopic features of *P. cinnamomi* are described in Chapter 1. It can be isolated from roots and soil as described for Phytophthora cankers.

Epidemiology

Wet soils favour zoospore release and movement (Zentmyer, 1980). They are only released at matric potentials of 0 to −0.4 bar (0 to −40 kPa) (Fig. 3.5). At −0.25 to −1 bar (−25 to −100 kPa), zoospores are unable to swim due to air that fills the soil pores, and the diffusion of root exudates from roots, which attracts the zoospores, is severely limited. Zoospores do not move well through clay soils because the pores are too small; however, because clay soils drain poorly, they can move for long distances in surface water runoff. It appears that any soil characteristic that maintains saturated conditions at the surface (where most avocado roots are located) will exacerbate this disease.

Sporangia are formed at matric potentials of −0.02 to −3 bars (−2 to −300 kPa) (Fig. 3.5). Mycelium is able to survive and grow in roots at 0 to −10 bars (0 to −1000 kPa), and chlamydospores and oospores form readily in the same range. Chlamydospores often are stimulated to form by dry conditions and they can survive for years under extremely dry conditions. Thus, drying the soil does not eliminate *P. cinnamomi*.

Soil water conditions also affect the host avocado. Flooding for as little as 3 days can predispose the roots to heavy infection due to the release of exudates. Drought stress can also injure roots and predispose them during the next wet period.

P. cinnamomi is an aerobic microorganism, and requires oxygen to grow and reproduce. While periodically wet soils enhance *Phytophthora* populations, they are reduced

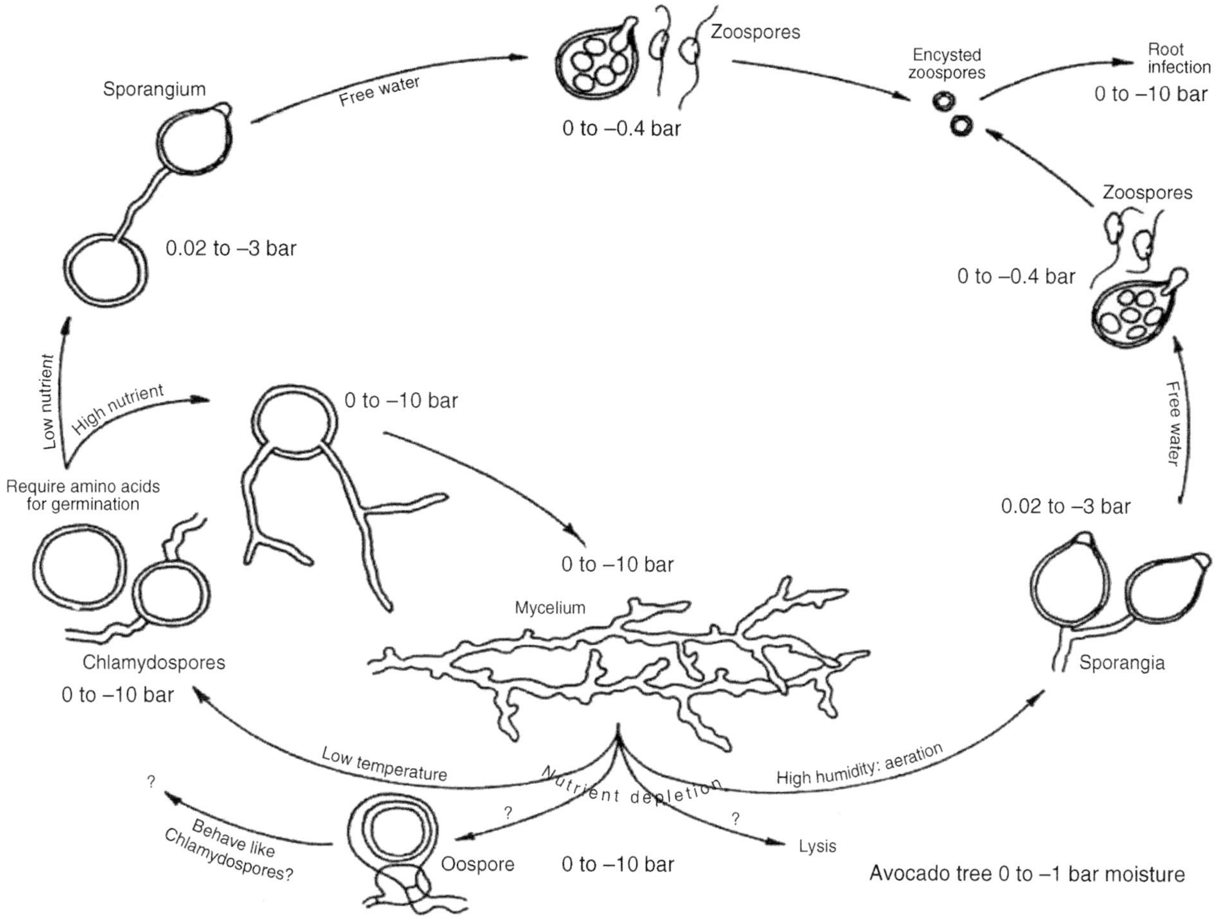

Fig. 3.5. The effect of soil matric potentials on the life cycle of *Phytophthora cinnamomi* (diagram: J. Menge, UCR).

when flooded soil becomes anaerobic. Unfortunately, avocado is even more sensitive to anaerobiosis than *P. cinnamomi*, and soil levels of 1–5% oxygen will damage or kill avocado roots. Air in well-drained, aerated avocado soil may contain 15% oxygen and 0.03% carbon dioxide, whereas poorly drained avocado soils may approach 1% oxygen and 16% carbon dioxide. Sporangial production, chlamydospore germination and zoospore activity require well-aerated conditions of 2.5–15% oxygen and 0–2% carbon dioxide (Fig. 3.6). However, zoospore germination, hyphal growth and chlamydospore formation are not restricted much under poorly aerated conditions and may occur at oxygen levels as low as 0.1% and carbon dioxide levels as high as 20%. Chlamydospore production is actually increased by high carbon dioxide levels. The impression that *P. cinnamomi* prefers waterlogged, anaerobic soils arises from its ability to survive, grow and cause root rot under decreased oxygen and increased carbon dioxide levels that are toxic to avocado roots.

Soil salinity is a major stress factor for avocado, and damage may occur when salinity levels reach 4 dS m^{-1}. Roots of avocado become more susceptible to root rot when they are stressed by salinity, presumably due to the increased exudation that occurs in damaged roots. When roots are rotted by *P. cinnamomi*, they lose the ability to exclude salt, and leaf margins often develop the brown, necrotic symptoms of salt damage.

Root rot is most severe between 9 and 21°C, and develops over a wide range of pHs (3.5–8.0), soil types and nutrient conditions. However, soils with high levels of organic matter, ammonia or calcium can inhibit root rot development.

Management

Although this disease has been studied for more than 60 years, no completely satisfactory control method has been found. However, many measures will reduce its impact. When they are packaged into a single, integrated strategy, they enable the continued economical production of avocados in the presence of *P. cinnamomi* (Coffey, 1987, 1992).

CLEAN NURSERY PRACTICES Diseased nursery stock has undoubtedly played a major role in the worldwide distribution of avocado root rot. Certification programmes that are run by growers or local governments help to ensure that clean nursery practices are followed. Clean seed must be used for propagation. Seed should not be collected from the ground, and should be treated with hot water (49–50°C for 30 min; ≥52°C damages seed) prior to use.

Soil and soil mixes must be disinfested (Zentmyer, 1980). Fumigation with methyl bromide (0.9 kg 11 m^{-3} for 24 h) is effective, as are heat treatment with steam (100°C for 30 min) and aerated steam (60°C for 30 min).

Irrigation water should come from deep wells since surface waters from rivers, canals and reservoirs are often contaminated with *P. cinnamomi.* Both copper sulphate (20 μg ml^{-1}) and chlorine (0.5 μg ml^{-1}) are effective disinfestants.

Sanitation is the single most important tool for preventing avocado root rot in the nursery (Zentmyer, 1980). Many nurseries require workers and visitors to dust their shoes with copper sulphate, and vehicles may be required to pass through a copper sulphate-treated water bath. Vehicles should be washed and disinfested before entering the nursery, especially if they have been in infested groves. In container nurseries, pots should never be placed on the ground, cement slabs or tarps, since *Phytophthora* spreads rapidly on these surfaces after irrigation and rain. All pots and liners should be sterilized before reuse.

Affected plants must be destroyed. Although fungicides are effective, they should not be used in nurseries because they mask symptoms and do not eliminate *P. cinnamomi* completely. Furthermore, continuous use of fungicides can select resistant isolates that could then be distributed to grower's fields. If good sanitation is practised, fungicides are not necessary in nurseries.

SITE SELECTION AND PREPARATION Poorly drained, saline and infested soils should be avoided when selecting a site for an avocado grove (Zentmyer, 1980). Soils with impervious subsoils can be improved by deep ripping and

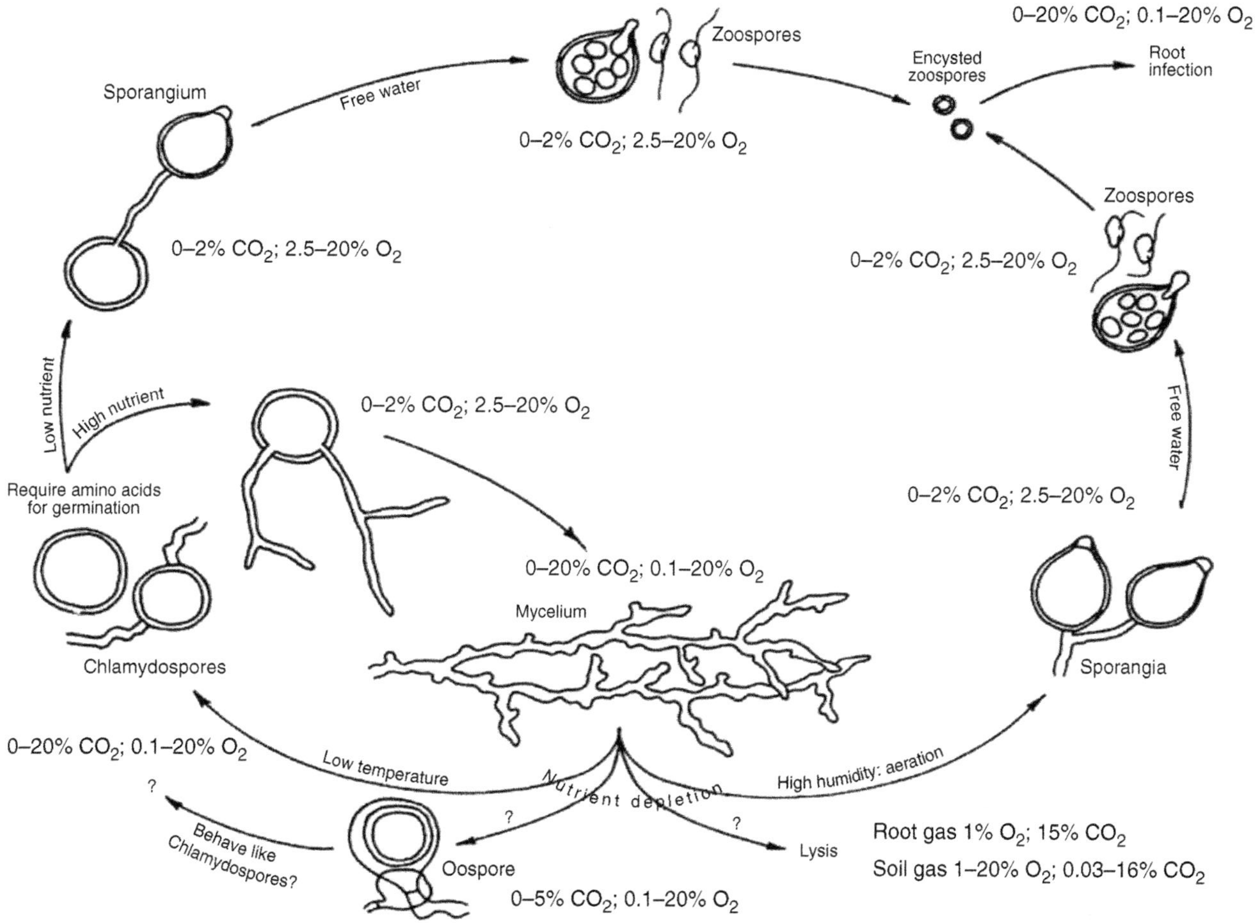

Fig. 3.6. The effect of oxygen and carbon dioxide levels on the life cycle of *Phytophthora cinnamomi* (diagram: J. Menge, UCR).

inserting subsurface tile drains. On sloped land, the construction of interception and diversion canals or watertight drainpipes may prevent the introduction of *P. cinnamomi*. In clay soils, planting trees on mounds (1–1.5 m in diameter and 0.5–1 m high) or ridges has been shown to increase survival and improve growth of avocado. Soil solarization has been found to reduce but not eliminate root rot in the Canary Islands (Gallo-Llobet *et al.*, 1995). Replanting in infested soil that has been fumigated is not recommended because complete eradication of *P. cinnamomi* is extremely difficult. Often, *P. cinnamomi* reinfests fumigated soil, and root rot becomes worse than during the original infestation.

GROVE SANITATION Excluding *P. cinnamomi* from clean groves is the most economical method of controlling the disease (Broadly, 1992). Groves should be fenced to protect them from human and animal traffic. Care should be taken to prevent movement of *P. cinnamomi* from one grove to another on cultivation equipment, vehicles, shovels, picking boxes, shoes, etc. (Zentmyer, 1980). Small pieces of equipment can be disinfested in 70% alcohol or chlorox. Vehicles should be washed and dried thoroughly before moving from a diseased grove to a healthy one.

P. cinnamomi affects >1000 species of plants, many of which are ornamentals (Zentmyer, 1980). Care should be taken when selecting and planting ornamentals and wind breaks to ensure that they do not carry *P. cinnamomi* into the grove. If diseased trees exist near healthy ones, a dry barrier of at least two rows of trees should separate them. Furrows should be dug to divert runoff from diseased trees away from healthy trees (Zentmyer, 1980). Once isolated, the diseased trees eventually should be removed and the soil fumigated. Irrigation water must be pathogen free.

RESISTANT ROOTSTOCKS Resistant rootstocks have the greatest potential for completely controlling Phytophthora root rot. Although no rootstock is yet immune, many selections are available that exhibit a high degree of tolerance, including 'Duke 7', 'D9', 'Merensky I', 'Merensky II' and 'Thomas' (Zentmyer, 1984).

IRRIGATION MANAGEMENT It is difficult to manage root rot with irrigation practices because avocado roots are very shallow and sensitive to drying. Tensiometers should be installed at depths of 15 and 30 cm near the dripline of one or two representative trees and used as a guide to determine when trees are receiving too much or too little water (Pegg, 1991; Broadly, 1992). Trees must not be over-watered (several days at 0 to -10 cb (-0 to -10 kPa)) or under-watered (several days at -50 to -70 cb (-50 to -70 kPa)), since these conditions predispose avocado roots. Irrigation should vary depending on local evapotranspiration. In saline soils, periodic leaching irrigations should be scheduled to force salt below the root zone.

SYSTEMIC FUNGICIDES Two fungicides have been very successful at reducing avocado root rot in many areas of the world. Metalaxyl (Ridomil) is highly soluble, moves readily in soil and is absorbed readily by avocado roots (Coffey, 1987). It may be applied as a granular formulation, or as a liquid drench or injection in irrigation water. A single application of metalaxyl provides 3 months of control. Isolates of some *Phytophthora* spp. have developed resistance to metalaxyl, and it is degraded rapidly in some soils although it is probably absorbed well before degradation begins. Metalaxyl will kill some, but not all, inoculum of *Phytophthora* in soil.

Fosetyl-Al (Aliette) is both phloem and xylem mobile. It and potassium phosphonate (which is the active metabolite of fosetyl-Al, phosphorous acid, that has been buffered with potassium hydroxide) can be applied as soil drenches, foliar sprays, trunk paints, trunk injections, and in irrigation water (Pegg *et al.*, 1985). All of these methods are effective if used properly, but some may work better than others under certain growing conditions.

Foliar sprays require more chemical and may not be practical on steep slopes. Since soil applications require sufficient numbers of functional roots for uptake, severely affected trees are difficult to rejuvenate in this manner. Heavy clay soils may also impede the uptake of this material from soil. Trunk paints are more effective for treating trunk lesions, but concentrations that are needed to rejuvenate roots often cannot be absorbed through the bark.

Injections are often the best way to rejuvenate severely affected trees (Whiley *et al.*, 1992). Fears have been expressed that trunk decay and damage may result from this practice, but there is little evidence to support either of these concerns.

The timing of trunk injections is critical (Whiley *et al.*, 1992). In order for effective concentrations of the phosphonate metabolite to be translocated to roots, injections should be made when foliar flushes are 75% complete. If injections are made during the start of a foliar flush or during flowering and fruit set, significant concentrations of phosphonate are translocated to the canopy and do not move to where they are needed, the root system.

Fosetyl-Al has little direct effect on soil populations of *Phytophthora*, but seems to increase the resistance of avocado roots to infection, thereby indirectly lowering populations in soil. A single application of fosetyl-Al or potassium phosphonate provides 3–4 months of control. For both, label directions should be heeded because rates, products and methods of application vary among countries.

CULTURAL PRACTICES Optimum fertility levels must be maintained via leaf analyses. Calcium is a particularly important nutrient, and applications of calcium carbonate, calcium nitrate and calcium sulphate have been shown to reduce avocado root rot (Zentmyer, 1980). Calcium may act as a weak fungicide by reducing the size and number of sporangia that are produced by *P. cinnamomi*, and it also may improve soil drainage. Applications of 1500–3000 kg of gypsum ha^{-1} are recommended, depending on tree size.

Animal manures reduce populations of *P. cinnamomi*, probably because they release ammonia, which is very toxic to *P. cinnamomi* (Zentmyer, 1984). However, avocado roots are also sensitive to ammonia and the roots it damages may be more susceptible to root rot. Therefore, animal manures should be broadcast sparingly and not on top of avocado roots. In Mexico, heavy applications of bovine manures are combined with severe pruning of infected trees (Teliz, 2000). This probably destroys infected roots as well as *P. cinnamomi* inoculum and allows trees to rejuvenate in soil that is relatively free of the pathogen.

MULCH In the presence of *P. cinnamomi*, mulches have been shown to stimulate avocado growth greatly (Broadly, 1992). Their use to control root rot originated in Australia. Certain Queensland rainforest soils were often free of *P. cinnamomi*-induced disease, even though the pathogen was present. This effect was attributed to high microbial populations, high levels of organic matter (>7%) and high exchangeable magnesium, calcium and nitrogen. These were labelled suppressive soils (Zentmyer, 1980).

A complex scheme called the Ashburner method was devised to simulate the naturally suppressive soils. It used bulky, organic mulches such as wheat straw, barley straw or sorghum stubble, together with fowl manure and dolomite to encourage breakdown of the mulch (Zentmyer, 1980). Today the practice has been modified to add only the key ingredients, which are organic mulches and gypsum. High populations of bacteria and actinomycetes or cellulose- and lignin-degrading microorganisms are stimulated. Layers of organic matter are recommended in the form of yard trimmings, avocado trimmings, corn stubble, sorghum stubble, wheat straw, lucerne straw and pine bark with a C:N ratio between 25:1 and 100:1. Layers should be 15–30 cm thick, placed under the canopy, and should be kept away from the trunk because animals that frequent the mulch may damage the trunk.

Tensiometers should be used to monitor soil moisture under the mulch. Since the mulch reduces water loss, it is easy to overwater mulched trees and eliminate much of the beneficial effects that the mulch produces (Pegg, 1991). Avocado roots that proliferate in the mulch and at the mulch–soil interface are relatively free of *P. cinnamomi*. Unfortunately, the beneficial effects of mulch do not extend very far into the soil, probably because enzymes that are produced in the mulch that are detrimental to *P. cinnamomi* are adsorbed and inactivated on soil particles.

BIOLOGICAL CONTROL Many soilborne microorganisms such as *Myrothecium roridum, Trichoderma harzianum, Epicoccum nigrum, Catenaria anguillae, Humicola fuscoatra, Anguillospora pseudolongissima, Hypochytrium catenoides, Myrothecium verrucaria, Streptomyces griseoalbus, Micromonospora carbonacea, Streptomyces violascens* and *Ceraceomyces tessulatus* have been shown to be inhibitory to *P. cinnamomi* via competition, antibiosis or parasitism (Erwin and Ribiero, 1996). Although several commercial products are available with *Trichoderma* or *Gliocladium* as the active agents, these products are mostly experimental at this time.

Root and butt rots

Root and butt rots are a loose group of diseases that are caused by several different fungi and are not well delineated. Many are considered to be of minor importance, while others can cause severe losses. Eventually the diseases must be separated and studied individually. Since each will have its own aetiology, different control measures may be needed.

Symptoms

Foliar symptoms range from a slow decline with branch death and a thinning canopy, to complete collapse of the tree, usually with the dead leaves remaining on the tree. Below ground, major roots can be rotted and punky and the disease may move up to the crown and rot it as well. In some cases, the trunk becomes hollow, and viable new sapwood is laid down each year as a ring around the hollow centre. In these cases, the tree may exhibit some chlorosis from toxins formed in the decay process, but may survive for many years with no noticeable reduction in yield. Often there are white or dark mycelial strands on the surface of the roots or penetrating the decayed wood.

Causal agents

Basidiomycetes cause these diseases (Zentmyer, 1984). They produce basidiospores on large basidiocarps that may appear as shelves jutting out from the affected avocado trunk, may encrust the trunk, especially at the soil surface, or may produce mushrooms.

Ganoderma applanatum, G. brownii, G. lucidum and *G. zonatum* have been implicated as root and butt rotters. Other members of the genus that attack avocado in Indonesia, Mexico and South America may represent different taxa since they are far more virulent than the rather benign species above. *Ganoderma* spp. are large, woody shelf fungi that produce conks with a smooth lacquered upper surface that is often marked with concentric ridges and grooves and a light coloured, actively growing margin (Figs 3.7 and 3.8). The undersurface is composed of light coloured pores. The basidiospores are very diagnostic and are brown, elliptical with a truncate base. The spores have two walls separated by spiny protuberances. When growing from buried roots, basidiocarps may be produced on an elongate, slender stem. *G. zonatum* is described in detail in Chapter 8.

Oxyporus latemarginatus, known formerly as *Poria latemarginatus*, often encrusts old stumps from which roots and crowns of healthy avocados are invaded. This fungus has pores, which are frilly or lacy and do not exhibit negative geotropism. The spores are 5–7 × 3–4 μm, and hymenial cystidia are present which are 20–28 × 4–6 μm, thin-walled and have encrusted apices.

Rigidoporus ulmarius and *R. vinctus* are similar pathogens of avocado. They are resupinate to shelf-like with very tiny pores on the pink, red, grey to black pore surface (Fig. 3.9). The spores are spherical. *R. vinctus* has club-like, encrusted, thick-walled cystidia, whereas *R. ulmarius* does not.

Coprinus micaceus is known as the inky cap mushroom. It deliquesces so that the spores and basidiocarp become a black, inky mass.

An unknown basidiomycete in Australia is very lethal. It infects trees and moves

Fig. 3.7. (A) Basidiocarp, and (B) basidiospores of *Ganoderma applanatum* (from CMI description no. 443).

down rows of avocados from root to root killing trees as it goes. The disease resembles Armillaria root rot in the way that it kills trees.

Root and butt rotting fungi usually can be isolated from freshly invaded wood on malt agar. However, identification is difficult since spores and fruiting bodies will not form. All

A

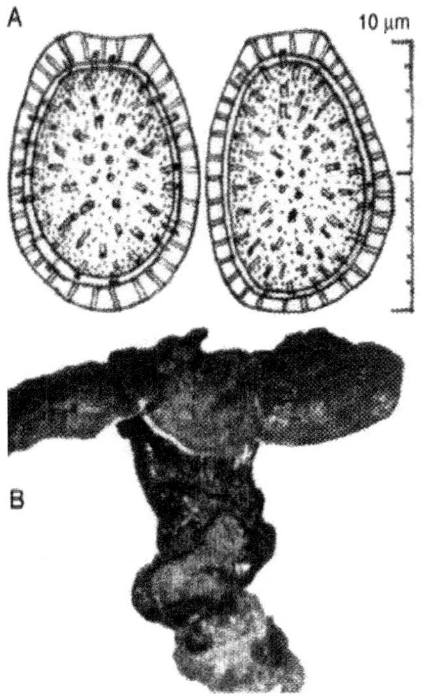

Fig. 3.8. (A) Basidiospores and (B) a basidiocarp of *Ganoderma lucidum* (from CMI description no. 445).

of these fungi are saprophytes and invade cut stumps or dead wood. From there, they often opportunistically invade nearby living avocado roots or trunks. These fungi probably invade more readily via mycelium growing through the soil than by spores.

Epidemiology

Since most of these fungi are saprophytes, they probably depend on stress or injuries to predispose avocado before they invade. Flooding or over-watering often favours attack by *C. micaceus*, and wounding trunks or roots may create a point of entry for many of these fungi. Most require significant inoculum on dead wood before they can invade healthy trees.

Management

Optimum irrigation and nutrition should be maintained, and wounding of the trunks and roots should be avoided. In severe cases, pruning or injection for *Phytophthora* control should be curtailed. Stumps, roots and dead wood should be removed before replanting.

Rosellinia (Dematophora) root rot

Rosellinia root rot, which is also known as white root rot, is a serious disease of avocado in Israel and Spain. It occurs in other countries, but does not appear to cause major losses.

Symptoms

Affected trees stop growing and develop yellow foliage and shrivelled fruit. As the disease progresses, trees defoliate and eventually die. The entire sequence takes from 1 to 3 years (Sztejnberg, 1994).

Below ground, white mycelium of the causal fungus can be seen colonizing and rotting small feeder roots. In later stages, the decay spreads through the cambium to larger roots, and finally to the crown or trunk. The surface of infected roots is covered with strands of white, cottony mycelium, and there is a continuous layer of white mycelium under the bark and in the surrounding soil (Sztejnberg, 1994). Later, the mycelium turns greenish grey or black, and patches of hyphae are found in the bark and extending into the soil. Infected roots become overgrown with characteristic white hyphae when they are placed in a moist chamber. The fungus eventually produces brown synnemata, which bear the asexual spores, and flattened sheets of black microsclerotia on the surface of infected tissues.

Causal agent

Rosellinia root rot is caused by *Rosellinia necatrix* (anamorph: *Dematophora necatrix*). It has been found in California, Israel, Mexico and Spain, can be isolated from infected avocado roots on malt agar or baited from soil using avocado leaf discs (Sztejnberg *et al.*, 1983a), and is described in detail in Chapter 1.

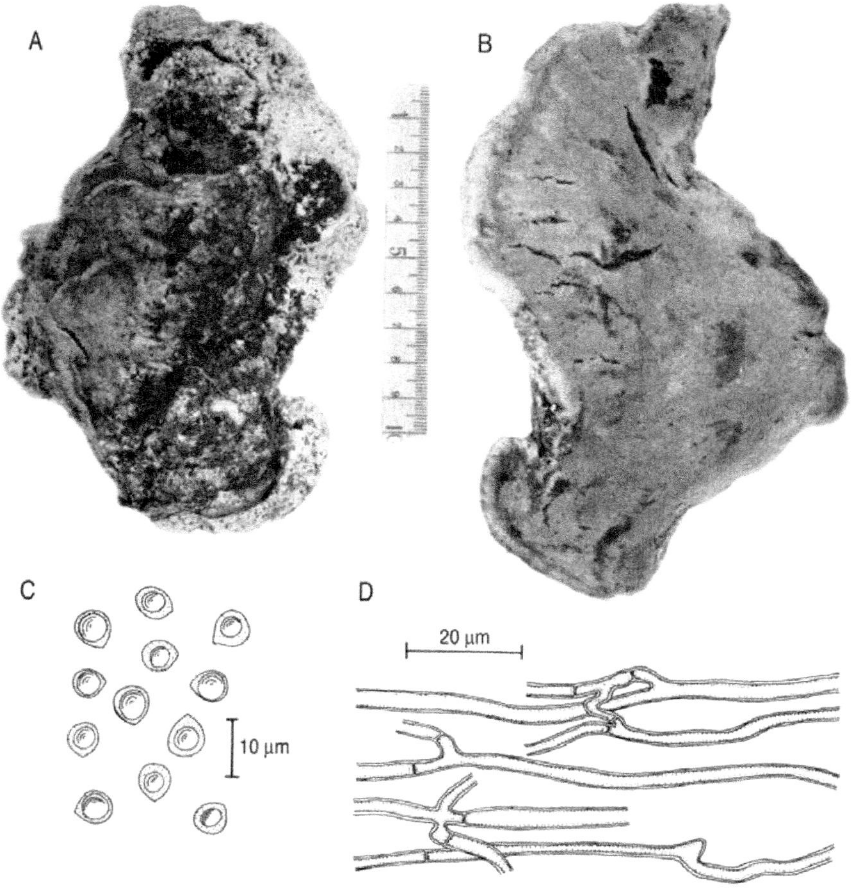

Fig. 3.9. (A) Upper and (B) lower surface of a basidiocarp, (C) basidiospores and (D) context hyphae of *Rigidoporus ulmarius* (from CMI description no. 199).

Other species of *Rosellinia* also affect avocado. In tropical areas *R. bunodes* (Fig. 3.10) and *R. pepo* (Fig. 3.11) cause Rosellinia black root rot, which is so named due to the appearance of black mycelial strands on infected roots (Sztejnberg, 1994).

Epidemiology

R. necatrix can survive for long periods in wood, roots and soil, primarily as microsclerotia. Avocado feeder roots are infected directly when they contact hyphae or microsclerotia (Sztejnberg, 1994). The infection spreads into woody roots and may spread from tree to tree in this manner. Neither ascospores nor conidia appear to play a role in spreading the disease. Instead, soil contaminated with microsclerotia, infested roots and organic matter appears to be the way in which the fungus is spread from orchard to orchard.

R. necatrix prefers wet, clay soils near saturation and soil temperatures of 20–25°C (Sztejnberg, 1994). The fungus spreads rapidly along drip irrigation lines, and can survive for years in roots or as microsclerotia.

Management

Since the fungus does not spread via spores, it is possible to stop its spread in orchards. Infected trees should be uprooted and destroyed, and trees surrounding the infec-

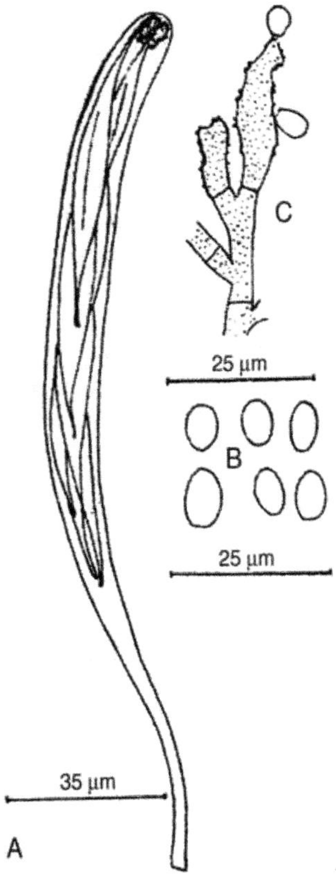

Fig. 3.10. (A) Ascus, (B) conidia and (C) conidiogenous branch of *Rosellinia bunodes* (from CMI description no. 351).

tion site should be removed to ensure that the pathogen is removed (Sztejnberg, 1994). Care should be taken to remove all roots. Trenching around the site to eliminate root bridges to other trees may isolate infection sites further.

Cultivation tools and other equipment should not be moved through infested areas. Irrigation of infected trees should be discontinued, and dry barriers should be established around infected sites to discourage growth of the fungus through the soil. Nurseries must make sure that their planting mixes and seedlings are not infested.

The fungus may be eliminated in soil by fumigation with deep placement of methyl bromide, heat treatment or soil solarization (Sztejnberg *et al.*, 1983b). In Israel, soil solarization has a long-term effect on inhibiting disease. In Spain, 6 weeks of solarization were necessary to eradicate *R. necatrix* to a depth of 60 cm.

Verticillium wilt

This widespread disease is not serious, and trees often recover with no lasting effects. The disease has also been called asphyxiation or apoplexy.

Symptoms

The entire tree or only one or several branches wilts suddenly. The leaves become brown but remain on the tree for several months (Zentmyer, 1994). Brown or grey–brown streaks are visible in the outer xylem of branches or roots if the bark is removed. Often, infected trees will send out new, vigorous shoots within a few months of the initial collapse. In many cases, the tree recovers completely and the disease does not recur.

Causal agent

Verticillium wilt is caused by *Verticillium dahliae* (Zentmyer, 1949). The pathogen affects avocado in Australia, California, Chile, Ecuador, Florida, Mexico, South Africa and Spain. The fungus is soilborne and affects a wide variety of hosts. It infects the feeder roots and invades the vascular tissue, moving upward in the xylem (Zentmyer, 1949). After killing tissues in the host, it produces abundant, verticillately branched (in a whorl) conidiophores. Three or four slender, elongate, phialides arise from each of the nodes and produce clusters of sticky phialospores (Fig. 3.12). The phialospores are colourless, elongate–elliptical to cylindrical, one-celled and 3–8 × 2–3 µm. The fungus also produces dark brown to black microsclerotia, 15–200 µm in dia, that are composed of torulose, swollen cells. Microsclerotia can survive for extended periods under unfavourable conditions. Dark, thickened, resting hyphae are often found in the host and in culture. The fungus is disseminated by water and wind-blown conidia as well as in soil and organic matter that are infested with microsclerotia and resting hyphae.

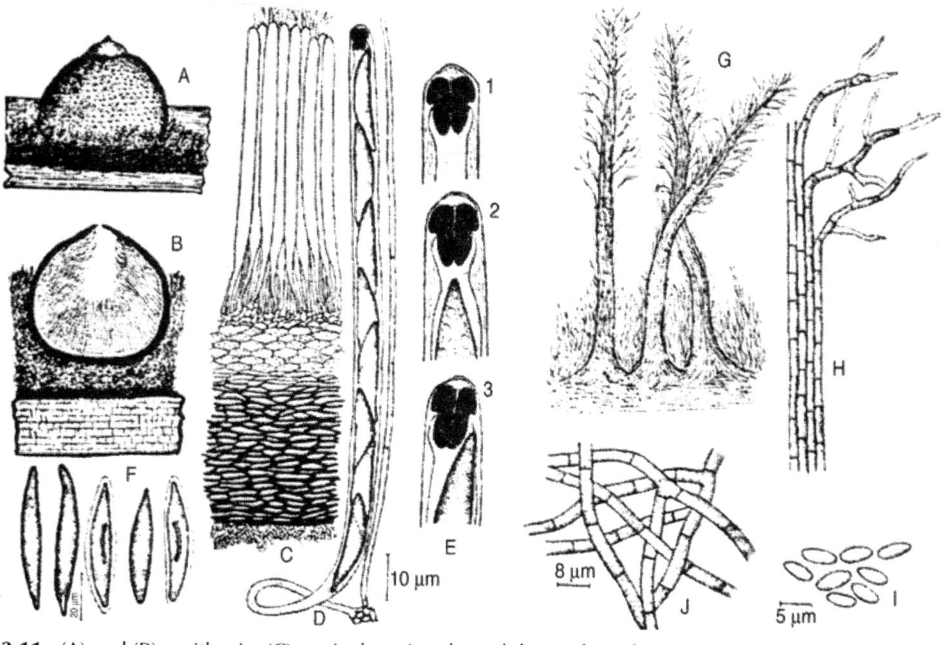

Fig. 3.11. (A) and (B) perithecia, (C) vertical section through base of perithecium, (D) ascus, (E) tips of mature asci, (F) mature ascospores, (G) synnemata, (H) sporogenous cells, (I) conidia and (J) conidia of *Rosellinia pepo* (from CMI description no. 354).

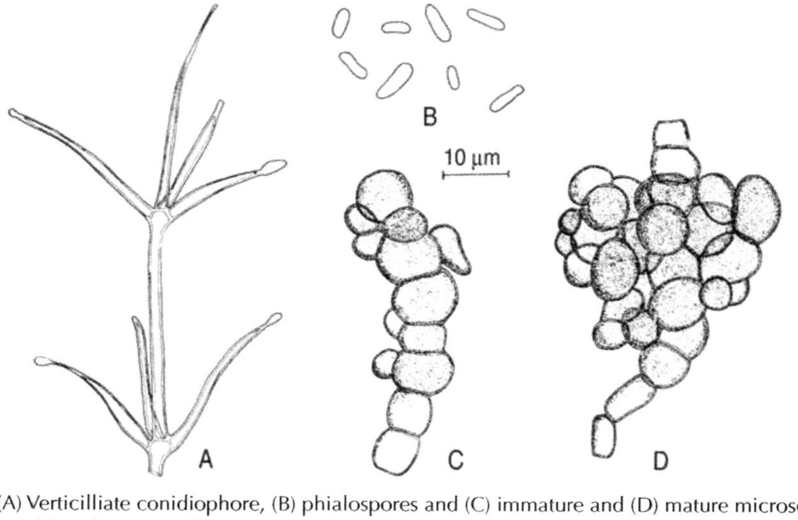

Fig. 3.12. (A) Verticilliate conidiophore, (B) phialospores and (C) immature and (D) mature microsclerotia of *Verticillium dahliae* (from CMI description no. 256).

V. dahliae can be isolated by placing infected, surface-disinfested twigs or roots in a moist chamber or on water agar or PDA that is amended with an antibiotic to discourage bacterial growth (Zentmyer, 1949). The characteristic conidiophores will develop on the plant parts. *V. dahliae* is often more difficult to isolate in the summer and winter than in the spring.

Epidemiology

This disease develops often in soil in which other hosts of *V. dahliae* have grown. These include tomato, pepper, aubergine, stone fruit trees, potato, strawberry, groundnut, olive, berries and many flower crops (Zentmyer, 1984). The fungus prefers cool (<25°C) and acidic conditions (Wilson and Ogawa, 1979). The disease often occurs in the spring after heavy rains. *V. dahliae* colonizes the water-conducting vessels of avocado in spring. Although the vessels of avocado are large, severe infections ultimately plug vessels and cause the canopy to wilt. High summer temperatures arrest development of infections and trees grow a new ring of water-conducting vessels and recover (Zentmyer, 1949).

Management

Trees should not be planted on land where crops susceptible to Verticillium wilt were grown previously (Zentmyer, 1949). These crops should never be interplanted with avocado. Budwood should never be taken from trees infected with Verticillium wilt. They can carry the disease, but more often die before they leave the nursery. Guatemalan rootstocks are far more susceptible than Mexican rootstocks, and the latter should be used in areas that are known to be infested with *V. dahliae* (Zentmyer, 1984). Usually no control measures are required. Dead branches should be pruned out after they die and new growth appears. Trees should be irrigated and fertilized optimally in order to promote rapid growth. If a tree dies, it should be removed and the hole fumigated with chloropicrin or methyl bromide before replanting (Zentmyer, 1994).

References

Allen, R.N. (1985) *Avocado Diseases*. Agfact H6. AB.5. Department of Agriculture, New South Wales.

Broadly, R.H. (1992) *Protect Your Avocados*. Information Series Q191031. Queensland Department of Primary Industry, Brisbane.

Coates, L.M. (1991) Latency of *Colletotrichum gloeosporioides* in avocado fruit. PhD thesis, University of Queensland, Brisbane, Australia.

Coffey, M.D. (1987) Phytophthora root rot of avocado: an integrated approach to control in California. *Plant Disease* 71, 1046–1052.

Coffey, M.D. (1992) Phytophthora root rot of avocado. In: Kumar, J., Chaube, H.S., Singh, U.S. and Mukhopadhyay, A.N. (eds) *Plant Diseases of International Importance*, Vol. III, *Diseases of Fruit Crops*. Prentice Hall, Englewood Cliffs, New Jersey, pp. 423–444.

Cooksey, D.A., Ohr, H.D., Azad, H.J., Menge, J.A. and Korsten, L. (1993) *Xanthomonas campestris* associated with canker of avocado in California. *Plant Disease* 77, 95–99.

Cooksey, D.A., Ohr, H.D. and Korsten, L. (1994) Bacterial canker. In: Ploetz, R.C., Zentmyer, G.A., Nishijima, W.T., Rohrbach, K.G. and Ohr, H.D. (eds) *Compendium of Tropical Fruit Diseases*. APS Press, St Paul, Minnesota, p. 75.

Crane, A. (2001) Future trends for the sales, marketing, packaging and consumption of avocados. Vision 2020, Australian and New Zealand Avocado Growers' Conference, Bundaburg, Australia.

Darvas, J.M. (1982) Etiology and control of some avocado fruit diseases. DSc (Agriculture) thesis, University of Pretoria, South Africa.

Darvas, J.M. and Kotzé, J.M. (1987) Fungi associated with pre- and postharvest diseases of avocado fruit at Westfalia Estate, South Africa. *Phytophylactica* 19, 83–85.

Darvas, J.M., Kotzé, J.M. and Wehner, F.C. (1987) Field occurrence and control of fungi causing postharvest decay of avocados. *Phytophylactica* 19, 453–455.

Denman, S., Crous, P.W., Taylor, J.E., Kang, J.-C., Pascoe, I. and Wingfield, M.J. (2000) An overview of the taxonomic history of *Botryosphaeria*, and a re-evaluation of its anamorphs based on morphology and ITS and rDNA phylogeny. *Studies in Mycology* 45, 129–140.

Desjardins, P.R. (1987) Avocado sunblotch. In: Diener , T.O. (ed.) *The Viroids*. Plenum Press, New York, pp. 299–313.

Doidge, E.M. (1922) A fungus of economic importance on the avocado. *Bothalia* 1, 179–186.
Du Toit, W.J. and De Villiers, E.A. (1988) Die hartvormige dopluis, *Protopulvinaria pyriformis* Cockerell, op avokado's. *South African Avocado Growers' Association Yearbook* 11, 79–80.
Dyko, B.J. and Mordue, J.E.M. (1979) Colletotrichum acutatum. *CMI Descriptions of Pathogenic Fungi and Bacteria No. 630*. Commonwealth Mycological Institute, Kew, UK.
El Hamalawi, Z. and Menge, J.A. (1994) Effect of leaf removal and plant pruning on the development of stem canker disease caused by *Phytophthora citricola* on *Persea americana* and *Persea indica*. *California Avocado Yearbook* 78, 131–142.
El Hamalawi, Z. and Menge, J.A. (1995a) Infection court and factors affecting the expansion of stem cankers caused by *Phytophthora citricola*. *Plant Disease* 79, 384–388.
El Hamalawi, Z. and Menge J.A. (1995b) Methods of fosetyl-Al application and phosphonate levels in avocado tissue needed to control stem canker caused by *Phytophthora citricola*. *Plant Disease* 79, 770–778.
Erwin, D.C. and Ribeiro, O.K. (1996) *Phytophthora Diseases Worldwide*. APS Press, St Paul, Minnesota.
Fitzell, R.D. (1987) Epidemiology of anthracnose diseases of avocados. *South African Avocado Growers' Association Yearbook* 10, 113–116.
Fitzell, R.D. and Muirhead. I.F. (1983) Reducing postharvest disease in Fuerte avocado by temperature management. *Australian Journal of Experimental Agriculture and Animal Husbandry* 23, 331–336.
Gallo-Llobet, L., Siviero, F. and Muñoz-Carpena, R. (1995) Influence of soil solarization on *Phytophthora cinnamomi* Rands in avocado (*Persea americana* Mill.). In: *Program and Book of Abstracts*. World Avocado Congress III, Tel Aviv, p. 57.
Horne, W.T. and Palmer, D.F. (1935) *The Control of Dothiorella Rot on Avocado Fruits*. Agricultural Experimental Station Bulletin 594. University of California, Berkeley.
Jenkins, A.E. (1934) *Sphaceloma perseae* the cause of avocado scab. *Journal of Agricultural Research* 49, 859–869.
Johnson, G.I. (1994) Dothiorella stem canker and fruit rot. In: Ploetz, R.C., Zentmyer, G.A., Nishijima, W.T., Rohrbach, K.G. and Ohr, H.D. (eds) *Compendium of Tropical Fruit Diseases*. APS Press, St Paul, Minnesota, p. 76.
Johnson, G.I. and Kotzé, J.M. (1994) Stem-end rot. In: Ploetz, R.C., Zentmyer, G.A., Nishijima, W.T., Rohrbach, K.G. and Ohr, H.D. (eds) *Compendium of Tropical Fruit Diseases*. APS Press, St Paul, Minnesota, pp. 81–83.
Korsten, L., De Jager, E.S., De Villiers, E.E., Lourens, A., Kotzé, J.M. and Wehner, F.C. (1995) Evaluation of bacterial epiphytes isolated from avocado leaf and fruit surfaces for biocontrol of avocado postharvest disease. *Plant Disease* 79, 1149–1156.
Korsten, L., De Villiers, E.E., Wehner, F.C. and Kotzé, J.M. (1997) Field sprays of *Bacillus subtilis* and fungicides for control of preharvest fruit diseases of avocado in South Africa. *Plant Disease* 81, 455–459.
Lonsdale, J.H. (1991) Control of preharvest fruit diseases of avocado. Part I: efficacy of various Triazole fungicides against Cercospora spot and Sooty blotch. *South African Avocado Growers' Association Yearbook* 14, 61.
Lonsdale, J.H. (1992) Evaluation of systemic fungicides as preharvest treatments of avocados. *South African Avocado Growers' Association Yearbook* 15, 35–38.
Muirhead, I.F., Fitzell, R.D., Davies, R.D. and Peterson, R.A. (1982) Post-harvest control of anthracnose and stem end rots of Fuerte avocados with prochloraz and other fungicides. *Australian Journal of Experimental Agriculture and Animal Husbandry* 22, 441–446.
Moll, J.N., Grech, N.M. and Van Vuuren, S.P. (1987) A lethal, transmissible stem-pitting of avocados associated with Duke 6 rootstocks. *South African Avocado Growers' Association Yearbook* 10, 122–123.
Ohr, H.D. and Korsten, L. (1990) Detecting bacterial canker. *California Grower* 14, 22–27.
Ohr, H.D. and Murphy, M.K. (1987) Blackstreak disease of avocado in California. *South African Avocado Growers' Association Yearbook* 10, 123–126.
Ohr, H.D. and Zentmyer, G.A. (1994a) Armillaria root rot. In: Ploetz, R.C., Zentmyer, G.A., Nishijima, W.T., Rohrbach, K.G. and Ohr, H.D. (eds) *Compendium of Tropical Fruit Diseases*. APS Press, St Paul, Minnesota, pp. 73–74.
Ohr, H.D. and Zentmyer, G.A. (1994b) Avocado black streak. In: Ploetz, R.C., Zentmyer, G.A., Nishijima, W.T., Rohrbach, K.G. and Ohr, H.D. (eds) *Compendium of Tropical Fruit Diseases*. APS Press, St Paul, Minnesota, pp. 74–75.
Ohr, H.D., Munnecke, D.E. and Bricker, J.L. (1973) The interaction of *Armillaria mellea* and *Trichoderma* spp. as modified by methyl bromide. *Phytopathology* 63, 965–973.

Ohr, H.D., Zentmyer, G.A. and Korsten, L. (1994) Sunblotch. In: Ploetz, R.C., Zentmyer, G.A., Nishijima, W.T., Rohrbach, K.G. and Ohr, H.D. (eds) *Compendium of Tropical Fruit Diseases*. APS Press, St Paul, Minnesota, p. 83.

Pegg, K.G. (1991) Causes of disease. In: Broadly, R.H. (ed.) *Avocado Pests and Disorders*. Queensland Department of Primary Industries. Brisbane, pp. 1–7.

Pegg, K.G., Whiley, A.W., Saranah, J.D. and Glass, R.J. (1985) Control of root rot of avocado with phosphorous acid. *Australasian Plant Pathology* 14, 25–29.

Peterson, R.A. (1978) Susceptibility of Fuerte avocado fruit at various stages of growth to infection by anthracnose and stem-end rot fungi. *Australian Journal of Experimental Agriculture and Animal Husbandry* 18, 158–160.

Pieterse, C.L. (1986) Afkeuringsfaktore by uitvoeravokado's. *South African Avocado Growers' Association Yearbook* 9, 14.

Ploetz, R.C., Zentmyer, G.A., Nishijima, W.T., Rohrbach, K.G. and Ohr, H.D. (eds) (1994) *Compendium of Tropical Fruit Diseases*. APS Press, St Paul, Minnesota.

Pohronezny, K.L. and Simone, G.W. (1994) Scab. In: Ploetz, R.C., Zentmyer, G.A., Nishijima, W.T., Rohrbach, K.G. and Ohr, H.D. (eds) *Compendium of Tropical Fruit Diseases*. APS Press, St Paul, Minnesota, p. 81.

Pohronezny, K.L., Simone, G.W. and Kotzé, J. (1994) Pseudocercospora spot (blotch). In: Ploetz, R.C., Zentmyer, G.A., Nishijima, W.T., Rohrbach, K.G. and Ohr, H.D. (eds) *Compendium of Tropical Fruit Diseases*. APS Press, St Paul, Minnesota, pp. 79–80.

Prusky, D. (1988) The use of antioxidants to delay the onset of anthracnose and stem end decay in avocado fruits after harvest. *Plant Disease* 72, 381–384.

Prusky, D. (1994) Anthracnose. In: Ploetz, R.C., Zentmyer, G.A., Nishijima, W.T., Rohrbach, K.G. and Ohr, H.D. (eds) *Compendium of Tropical Fruit Diseases*. APS Press, St Paul, Minnesota, pp. 72–73.

Prusky, D., Keen, N.T. and Eaks, I. (1983) Further evidence of the involvement of a preformed antifungal compound in the latency of *Colletotrichum gloeosporioides* on unripe avocado fruit. *Physiological Plant Pathology* 22, 189–198.

Schnell, R.J., Kuhn, D.N., Ronning, C.M. and Harkins, D. (1997) Application of RT–PCR for indexing avocado sunblotch viroid. *Plant Disease* 81, 1023–1026.

Schnell, R.J., Kuhn, D.N., Olano, C.T. and Quintanilla, W.E. (2001) Sequence diversity among avocado sunblotch viroids isolated from single avocado trees. *Phytoparasitica* 29 (5), 1–10.

Semancik, J.S. and Szychowski, J.A. (1994) Avocado sunblotch disease: a persistent viroid infection in which variants are associated with differential symptoms. *Journal of General Virology* 75, 1543–1549.

Shabi, E., Katan, T., Gera, H. and Elish, S. (1994) Taxonomic determination of pathogenic *Colletotrichum gloeosporioides* of almond, anemone and avocado according to fungicide sensitivity. *Phytoparasitica* 21, 130–131.

Shaw, C.G. and Kile, G.A. (1991) *Armillaria Root Disease*. Forest Service, US Department of Agriculture. Agricultural Handbook No. 691.

Slippers, B., Johnson, G.I., Cooke, A.W., Crous, P.W., Coutinho, T.A., Wingfield, B.D. and Wingfield, M.J. (2001) Taxonomy of *Botryosphaeria* spp. causing stem end rot of mango. In: *Proceedings of 13th Biennial Australasian Plant Pathology Conference*. Cairns, Australia, September 24–27, 2001.

Smith, E.M., Kotzé, J.M. and Wehner, F.M. (1985) Sooty blotch of avocado caused by *Akaropeltopsis* sp. *Phytophylactica* 17, 101–102.

Stevens, H.E. and Piper, R.B. (1941) *Avocado Diseases in Florida*. US Department of Agriculture Circular 582, 559–573.

Stirling, A.M., Coates, L.M., Pegg, K.G. and Hayward, A.C. (1995) Isolation and selection of bacteria and yeasts antagonistic to preharvest infection by avocado by *Colletotrichum gloeosporioides*. *Australian Journal of Agricultural Research* 46, 985–995.

Sztejnberg, A. (1994) Rosellinia (Dematophthora) root rot. In: Ploetz, R.C., Zentmyer, G.A., Nishijima, W.T., Rohrbach, K.G. and Ohr, H.D. (eds) *Compendium of Tropical Fruit Diseases*. APS Press, St Paul, Minnesota, pp. 80–81.

Sztejnberg, A., Omary, N. and Pinkas, Y. (1983a) Dematophthora root rot of avocado trees in Israel and development of a diagnostic method. *Phytoparasitica* 11, 238–239.

Sztejnberg, A., Omary, N. and Pinkas, Y. (1983b) Control of *Rosellinia necatrix* by deep placement and hot treatment with methyl bromide. *Bulletin Oepp* 13, 483–485.

Teliz, D.T. (2000) Enfermedades del aguacate. In: Teliz, D.T. (ed.) *El Aguacate y su Manejo Integrado*. Ediciones Mundi-Prensa, Mexico DF, pp. 139–181.

Vock, N. (2001) *Avocado Information Kit*. Agrilink series Qal 9906. DPI, Queensland.
Whiley, A.W., Saranah, J.B., Langdon, P.W., Hargreaves, P.A., Pegg, K.G. and Ruddle, L.J. (1992) Timing of phosphonate truck injections for *Phytophthora* root rot control in avocado trees. Vol. I. *Proceedings of the Second World Avocado Congress*, pp. 75–78.
Willingham, S.L., Pegg, K.G., Cooke, A.W., Coates, L.M., Langdon, P.W.B. and Dean, J.R. (2001) Rootstock influences postharvest anthracnose development in 'Hass' avocado. *Australian Journal of Agricultural Research* 52, 1017–1022.
Wilson, E.E. and Ogawa, J.M. (1979) *Fungal, Bacterial, and Certain Nonparasitic Diseases of Fruit and Nut Crops in California*. Division of Agricultural Sciences, University of California, Berkeley.
Yakoby, N., Zhou, R., Kobiler, I., Dinoor, A. and Prusky, D. (2001) Development of *Colletotrichum gloeosporioides* restriction enzyme-mediated integration mutants as biological control agents against anthracnose disease in avocado fruits. *Phytopathology* 91, 143–148.
Zentmyer, G.A. (1949) Verticillium wilt of avocado. *Phytopathology* 39, 677–682.
Zentmyer, G.A. (1980) Phytophthora cinnamomi *and the Diseases it Causes*. Monograph 10, American Phytopathological Society, St Paul, Minnesota.
Zentmyer, G.A. (1984) Avocado diseases. *Tropical Pest Management* 30, 388–400.
Zentmyer, G.A. (1994) Verticillium wilt. In: Ploetz, R.C., Zentmyer, G.A., Nishijima, W.T., Rohrbach, K.G. and Ohr, H.D. (eds) *Compendium of Tropical Fruit Diseases*. APS Press, St Paul, Minnesota, pp. 83–84.
Zentmyer, G.A., Ohr, H.D. and Menge, J.A. (1994) Phytophthora cankers. In: Ploetz, R.C., Zentmyer, G.A., Nishijima, W.T., Rohrbach, K.G. and Ohr, H.D. (eds) *Compendium of Tropical Fruit Diseases*. APS Press, St Paul, Minnesota, pp. 76–77.
Zhou, S. and Stanosz, G.R. (2001) Relationships among *Botryosphaeria* species and associated anamorphic fungi inferred from the analyses of ITS and 5.8S rDNA sequences. *Mycologia* 93, 516–527.

4 Diseases of Banana and Plantain

Randy C. Ploetz[1], John E. Thomas[2] and Walter R. Slabaugh[3]
[1] University of Florida, Tropical Research and Education Center, Homestead, Florida, USA; [2] Queensland Department of Primary Industries, Queensland Horticulture Institute, Indooroopilly, Queensland, Australia; [3] Agraquest Inc., 1050 Echo Avenue, Parma, Idaho, USA

Introduction

Banana is one of the most fascinating and important of all food crops. It arose in southern Asia, and may have been one of the first domesticated crops. In writing of the beginnings of agriculture in Southeast Asia, Simmonds (1966) concluded, 'It seems a reasonable assumption that the bananas evolved along with the earliest settled agriculture of that area and may therefore be some tens of thousands of years old'.

Due to its origins, diversity of banana is greatest in southern and Southeast Asia. Dissemination outside this region occurred at least 2500 years ago to Africa (Mbida Mindzie *et al.*, 2001) and ~2000 years ago to Polynesia and tropical South America (Langdon, 1993). Secondary centres of banana diversity developed later in the highland and coastal areas of East Africa, and of the plantains in West Africa and Latin America.

Taxonomy, Anatomy and General Attributes

The banana family, *Musaceae*, is one of eight in the order Zingiberales (Kress *et al.*, 2001). Most edible bananas arose from two species in the family, *Musa acuminata* and *M. balbisiana* (Daniells *et al.*, 2001b). They are diploid, triploid or tetraploid hybrids among subspecies of *M. acuminata*, and between *M. acuminata* and *M. balbisiana*. No subspecies of *M. balbisiana* exist, but there are nine of *M. acuminata*: *banksii*, *burmannica*, *burmannicoides*, *errans*, *malaccensis*, *microcarpa*, *truncata*, *siamea* and *zebrina*. Rare and unimportant hybrids also occur *M. acuminata* and *M. balbisiana*, and *M. schizocarpa* and *M. textilis* (Carreel, 1995). Also uncommon are the Fe'i bananas, which are complex hybrids among *M. lolodensis*, *M. maclayi* and *M. peekelii*. Information in this chapter focuses on bananas that are derived from *M. acuminata* and *M. balbisiana*.

The haploid contributions of *M. acuminata* and *M. balbisiana* to the cultivated bananas are indicated with an A and B, respectively (Simmonds and Shepherd, 1955). For example, the Cavendish and East African Highland cultivars are pure triploid *acuminata* and, thus, referred to as AAA, whereas the widespread cooking banana 'Bluggoe' is a triploid hybrid and ABB. The Linnaean species *M. paradisiaca* (the true plantains, which are AAB) and *M. sapientum* (the sweet dessert bananas, of which 'Silk' AAB is the type cultivar) are no longer considered valid.

Bananas are perennial herbs that develop from subterranean rhizomes (Purseglove, 1985; Stover and Simmonds, 1987). Suckers are sympodial shoots that develop laterally and in a radial fashion from the base of the rhizome. They form a pseudostem as soon as they clear the parent plant; it is composed of leaves and their fused petiole bases. Upon flowering, the true stem and associated apical meristem rise from within the rhizome and emerge from the top and through the centre of the pseudostem. Cold damage (chokethroat) and the banana streak disease can interrupt this sequence by causing the developing inflorescence to stall in, or break through the side of, the pseudostem (Fig. 4.1). As the stem emerges from the pseudostem, its growth usually becomes geotropic. Flowers are arranged on the stem (peduncle) in nodal clusters in a radial fashion. Each flower cluster is borne on a prominence called the crown or cushion, and is covered by a bract. Flowers in the proximal portion of the inflorescence are female, those in the distal portion are male, and there may be intervening clusters of intermediate flower structure. Since most cultivars are parthenocarpic, pollination is not required for fruit development. Individual fruit are called fingers, the contiguous, semicircular clusters of fruit are hands, and the entire infructescence is a bunch. Botanically, banana fruit are considered to be berries.

Fig. 4.1. Symptoms of chokethroat on 'Dwarf Cavendish' AAA at ~1000 m above sea level in Malawi. This disorder develops when temperatures of <15°C occur during bunch emergence (photo: R.C. Ploetz, UF).

Banana is propagated vegetatively, with either suckers or rhizome pieces. In addition, plantlets from meristem culture are now used widely (Vuylsteke, 1989). They enable the establishment of uniform plantings, and have the added advantage of being free of bacterial, fungal and nematode pathogens. Unfortunately, plants from meristem culture are more susceptible to Panama disease (Fusarium wilt), infectious chlorosis and possibly other diseases than those from traditional seed pieces (Ploetz *et al.*, 1994b; Smith *et al.*, 1998). Moreover, meristem culturing generally does not eliminate virus pathogens of banana. Thus, virus indexing has become increasingly important, especially during the international exchange of germplasm.

Banana is a climacteric fruit (Snowdon, 1990). After harvest, minute quantities of ethylene are produced that trigger a sharp increase in respiration. It is desirable to minimize postharvest injury of fruit since injuries are not only conspicuous upon ripening but will also cause premature ripening and enhance postharvest development of anthracnose and other diseases. Premature ripening also occurs in fruit from plants that are affected by the Sigatoka leaf spots. The preclimacteric life of fruit is increased by prompt refrigeration after harvest, preventing ethylene build-up in storage areas or ship holds with ethylene scrubbers, and adequate control of the Sigatoka diseases.

Significance as Agricultural Commodity and Food Source

Banana is now grown in almost all areas between 30°N and S latitude that have sufficient water. After citrus, it has the highest gross production of any fruit (97.7 Mt in 2001), and it is the most important fruit in world trade; in 2000, world commerce in banana was valued at US$4.8 billion (Anonymous, 2001). Although a handful of countries are responsible for most export production (Table 4.1), production in other countries is also significant. For example, bananas accounted for one-third of the total export revenues for Honduras, and about one-half of those for the Windward Islands, Guadeloupe and Martinique in 1982 (Stover and Simmonds, 1987).

Despite their importance, exported bananas represented only 17% of the total annual production of this crop in 2000 (Anonymous, 2001). Non-exported fruit are grown by diverse producers and are important commercial products, staple foods and dietary supplements. The latter bananas have diverse genomes, and are eaten raw; are baked, fried or boiled; or are brewed for beer. These include: AA (e.g. 'Pisang Mas') and AB (e.g. 'Ney Poovan') dessert bananas; AAA dessert bananas, e.g. members of the Cavendish subgroup and 'Gros Michel'; AAA highland cooking bananas that are endemic to East Africa; diverse AAB plantains that are most significant in Latin America and West Africa; AAB dessert bananas, of which 'Silk' and 'Pome' are prominent members; and ABB cooking bananas, including 'Bluggoe', 'Cardaba' and 'Pelipita'. In this chapter, all will be referred to as bananas.

Banana is a most important crop for millions of the world's poorest people. The fruit are nutritional and those of most cultivars are good sources of carbohydrates, potassium, calcium, phosphorus and vitamin C; in lower concentrations they also contain vitamins A, thiamine, riboflavin and niacin (Stover and Simmonds, 1987). Banana is prized in the developing world because it can be grown with minimal care. For example, a kilogram, 1000 calories or a hectare of these fruit can be produced more cheaply than any other important source of carbohydrates in West Africa.

Table 4.1. The major exporters of banana in 2000.

Country	Gross exports (1000 t)
Ecuador	4095
Costa Rica	2096
Colombia	1710
Philippines	1600
Guatemala	857
Panama	490

Data are from FAO, http://www.fao.org/default.htm

Impact of Banana Diseases

Diseases are major production constraints wherever banana is grown, and are the key reasons for the creation of the banana-breeding programmes (Buddenhagen, 1993). For example, Panama disease is considered one of the most destructive plant diseases in modern times (Stover, 1962; Simmonds, 1966). Currently, black Sigatoka is the most important disease on this crop. Export production of the Cavendish cultivars would not be possible without expensive and intensive applications of fungicides, and smallholder production of plantains and several other types of banana can be reduced by 50% or more.

In this chapter, we summarize current information on the most important diseases of this crop. They are listed alphabetically according to the causal agents and the organ of the plant that is affected. Minor diseases that could not be covered in the chapter are listed in Table 4.2. More thorough coverage on diseases of this crop is found in *Diseases of Banana, Abacá and Enset* (Jones, 2000).

DISEASES THAT ARE CAUSED BY BACTERIA

Bacteria cause serious diseases of the fruit, rhizome and pseudostem of banana. Although they are not usually as important as the major fungal diseases, they can cause considerable losses when insects vector the causal agents, and in export monocultures if adequate control measures are not used.

Diseases of Fruit

Bugtok

This disease, which is also known as tapurok, occurs in the Philippines. It was

Table 4.2. Non-infectious disorders and diseases of minor importance or unknown aetiology.

Disorder/disease	Cause
Alligator skin	Light abrasions on fruit peel caused by leaves or bracts
Armillaria corm rot	*Armillaria mellea* and *A. tabescens*
Banana dieback	Banana dieback virus
Black root rot	*Rosellinia bunodes*
Blue disease	Magnesium deficiency
Brown blotch	*Pestalotiopsis leprogena*
Ceratocystis fruit rot	*Ceratocystis paradoxa* (anamorph: *Chalara paradoxa*)
Choke	Low winter temperatures
Cladosporium speckle	*Cladosporium musae*
Corm dry rot	*Junghuhnia vincta*
Damping-off	*Deightoniella torulosa*
Dwarf Cavendish tip-rot	*Fusicoccum mangiferum*
Dwarfism	Genetic mutation
Elephantiasis	Unknown
Finger tip rot (gumming)	*Pseudomonas* spp.
Fruit chimera	Genetic mutation
Fruit rot	*Botryosphaeria ribis* (anamorph: *Fusicoccum parvum*)
Fungal root-rot	*Haematonectria haematococca* (anamorph: *Fusarium solani*) *Fusarium oxysporum* *Rhizoctonia* spp.
Fungal scald	*Colletotrichum musae*
Fused fingers	Genetic defect
Giantism	Genetic mutation
Heart leaf unfurling disorder	Unknown
High mat	Unknown
Javanese vascular wilt	*Pseudomonas* sp. (reports of this disease may actually have been of Fusarium wilt)
Leaf edge chlorosis	Unknown
Leaf spot	*Curvularia eragrostidis*
Leaf spot	*Drechslera musae-sapientum*
Leaf spot	*Leptosphaeria musarum*
Leaf spot	*Pestalotiopsis disseminata*
Leaf spot	*Pestalotiopsis palmarum*
Main stalk rot	*Ceratocystis paradoxa* (anamorph: *Chalara paradoxa*)
Maturity bronzing	High rainfall, humidity and temperature coupled with excessive fruit caliper
Peduncle rot	*Botryosphaeria rhodina* (anamorph: *Diplodia theobromae*) *Fusarium pallidoroseum* *Fusarium oxysporum* *Verticillium theobromae*
Pseudostem wet rot	*Erwinia chrysanthemi*
Rayadilla	Zinc deficiency
Rhizome and pseudostem soft rot	*Erwinia* sp.?
Root-knot	*Meloidogyne arenaria*, *M. incognita* and *M. javanica*
Rosetting	Nitrogen deficiency
Roxana	Unknown
Sclerotinia fruit rot	*Sclerotinia sclerotiorum*
Sheath rot	*Nectria foliicola*
Spike leaf	Low winter temperatures
Split peel	Rapid filling of pulp of fruit
Stem-end rot	*Colletotrichum musae*
Taiwan marginal scorch	Unknown
Trachysphaera finger rot	*Trachysphaera frutigena*
Underpeel discoloration	Chilling injury to fruit

Continued

Table 4.2. *Continued.*

Disorder/disease	Cause
Verticillium tip rot	*Verticillium theobromae*
Yellow mat	Adverse soil conditions
Yellow pulp	Delay in fruit filling, drought, excessive shading, magnesium deficiency, poor nutrition
Yellows	Lack of water, flooding, nutritional disorders

Major portions of this list are from the list of common names for banana diseases that was compiled by D.R. Jones (http://www.scisoc.org/resources/common/names/banana.htm). Additional entries are from Jones (2000).

first recorded in 1965, but may have been present long before this report (Roperos, 1965). Bugtok affects 'Cardaba', 'Saba' and other ABB cooking bananas and has no impact on export production of Cavendish cultivars.

Infection appears to occur initially via the male bud (Jones, 2000). The causal bacterium can ooze from and blacken bracts, and peduncles may shrivel. Bugtok causes grey to reddish yellow internal rot of fingers, and all or only a few fingers in a bunch may be affected. Unlike other bacterial diseases that affect fruit (e.g. blood disease and Moko), symptoms do not develop in the pseudostem's vascular system.

Bugtok is caused by *Ralstonia solanacearum* (general characteristics for the species are listed in the section on Moko disease). Based on different phenotypic, genetic and pathogenicity studies, strains of the bugtok bacterium are indistinguishable from those that cause Moko disease (Schaad *et al.*, 2001).

The pathogen is thought to be vectored by thrips and is not moved in planting material (Jones, 2000). Covering inflorescences with bags or nets controls the disease, but removing the male bud or managing the putative thrips vectors do not.

Diseases of the Rhizome and Pseudostem

Rhizome rot

This is the most important of the rhizome and pseudostem diseases that are caused by bacteria (Stover, 1972; Jones, 2000). New plantings of AA and AAA cultivars are affected most often. Newly planted rhizomes either do not germinate or, in older plants, decay can be so extensive that the pseudostem breaks off from the rhizome. There are few, if any, conspicuous external symptoms, but internally affected tissues are yellow to brownish water-soaked areas with black borders. These necrotic areas eventually blacken and can develop into extensive cavities that resemble those caused by the weevil borer, *Cosmopolites sordidus* (Fig. 4.2).

Both *Erwinia carotovora* ssp. *carotovora* and *E. chrysanthemi* have been indicted as causes of this disease (Jones, 2000). They are facultatively anaerobic, peritrichously flagellated Gram-negative rods. These pectolytic species are closely related, but can be separated using biochemical tests (Schaad *et al.*, 2001).

The pathogens are presumed to be opportunistic residents of banana soils that enter the rhizome through wounds (Jones, 2000). The disease is most prevalent in wet, humid areas. It is seldom a major problem but, where it is important, large, high quality rhizomes should be selected for planting purposes. In addition, knives that are used to prepare seed pieces should be disinfested frequently and the cut surfaces on rhizomes or rhizome pieces should be allowed to suberize thoroughly before planting.

Vascular Wilts

Blood disease

Blood disease is a lethal problem on dessert and cooking bananas, as well as wild *Musa* taxa (Jones, 2000). 'Penyakit Darah', as it is

Fig. 4.2. Extensive decay in the rhizome and pseudostem of a young 'Silk' AAB sucker caused by *Erwinia* sp. Note its resemblance to damage caused by the weevil borer, *Cosmopolites sordidus* (photo: R.C. Ploetz, UF).

known locally, was first reported from Sulawesi in 1906 where it destroyed new plantations of dessert banana. It was presumed to be an anomalous occurrence of Moko disease until work in the late 1980s demonstrated that an outbreak of blood disease on Java was caused by a bacterium that differed from that causing Moko (Eden-Green and Sastraatmadja, 1990).

Blood disease continues to spread and recently has been reported from Sumatra, Kalimantan, the Moluccan Islands and Irian Jaya. It has not been reported outside the Indonesian archipelago.

SYMPTOMS Leaves become chlorotic, necrotic and buckle, usually close to the pseudostem (Jones, 2000). Although the inflorescence may appear unaffected externally, it often develops the blackened shrivelled symptoms that occur when plants are affected by the SFR insect-vectored strain of the Moko pathogen (Ploetz *et al.*, 1994b). Internally, the fruit and vascular system become a reddish brown. Infection is systemic, and can extend into the rhizome and connected suckers in a mat. The causal bacterium exudes from severed vascular tissues as a cream to blackish ooze.

CAUSAL AGENT Gäumann named the blood disease bacterium *Pseudomonas celebensis* (Jones, 2000). None of his isolates remain and the name he used is no longer valid, but because his descriptions of the bacterium agree with more recent results he clearly worked on this disease.

Biochemical, genetic and physiological tests indicate that the pathogen is closely related to *Ralstonia solanacearum*. For example, restriction fragment length polymorphism (RFLP) and 16S rRNA sequence data place the pathogen firmly in the *R. solanacearum* species complex (Cook and Sequeira, 1994; Taghavi *et al.*, 1996).

Recently, it was suggested that the bacterium should be considered a separate species or a subspecies of *Ralstonia* (Thwaites *et al.*, 1999; Jones, 2000). Although random amplified polymorphic DNA (RAPD) and polymerase chain reaction (PCR) analyses clearly distinguish isolates of the pathogen from isolates of *R. solanacearum* that cause bugtok and Moko disease, the blood disease bacterium has not been described formally as a new taxon.

EPIDEMIOLOGY Despite the importance of blood disease, not much is known about the disease's epidemiology (Ploetz *et al.*, 1994b). The pathogen can reside in soil and in infested host tissue for more than a year, and in apparently healthy seed pieces and fruit. The latter traits enable long-distance spread of the pathogen; for example, its movement from Sulawesi to Java may have resulted from the movement of infected fruit. A rate of spread of 25 km year^{-1} was been reported recently, and this rapid movement is probably due, at least in part, to insect dissemination of the pathogen.

Gäumann (as reported in Jones, 2000) reported that plants were infected after he inoculated female flowers and, based on the symptoms of the disease and sequence in which they develop, it is clear that plant-to-plant movement occurs through the air. How the pathogen is disseminated and infects banana, and whether and what insects are involved should be researched.

MANAGEMENT No resistance has been reported among the edible clones, and dessert bananas appear to be most susceptible (Jones, 2000). Whether the removal of male buds would be as effective as it is against Moko disease has not been investigated. However, if female flowers are naturally infected, this measure might not be expected to be efficacious. Strict quarantine measures are needed to ensure that this destructive disease does not spread outside its current range. Under no circumstance should traditional seed pieces be moved from affected areas unless they are known to be pathogen-free.

Moko disease

Moko disease affects diverse dessert bananas, plantains and cooking bananas (Ploetz *et al.*, 1994b; Jones, 2000). 'Bluggoe' ABB is especially susceptible, and the disease is named after a synonym of this cultivar, 'Moko'. The disease has probably been present in South America for over a century, and was first recorded in the 1890s on Trinidad. In the western hemisphere, Moko disease is now recognized in an area extending from the Amazon Basin to Guatemala and southern Mexico, as well as on Grenada and Trinidad. Although the disease has been reported in India and several countries in Africa, the only verified occurrence in the eastern hemisphere is in the Philippines (Ploetz *et al.*, 1994b). Infested planting material from Honduras is thought to be responsible for the latter outbreak.

Moko is a classic 'new encounter' disease (Buddenhagen, 1960; French and Sequeira, 1970). Species of *Heliconia* are common understorey plants in rainforests in tropical America that are in the same taxonomic order as banana, the Zingiberales. Race 2 of *R. solanacearum*, the cause of Moko disease, co-evolved on these banana relatives in the New World. Thus, unlike the blood disease bacterium, the Moko pathogen did not evolve on banana.

SYMPTOMS Externally, the oldest leaves in the canopy become chlorotic, wilt, buckle and ultimately die (Ploetz *et al.*, 1994b). Younger leaves are then affected until the entire canopy is involved. Leaves remain attached to the pseudostem and eventually the pseudostem collapses. Suckers can also be affected, and in addition to the above symptoms their leaves may curve downwards as they die. If suckers are cut with infested machetes, they become blackened and stunted in 2–4 weeks. When insect-transmitted strains of the pathogen infect cushions on the peduncle, the male bud withers and darkens, and bacteria may ooze from the bud. The fruit may turn yellow and their peel split.

Internally, affected fruit pulp is firm but brown and later grey. On 'Bluggoe', the colour is more reddish brown and a red–brown liquid may occur at the fruit centre. The vascular system in the rhizome, pseudostem and peduncle is also discoloured light to dark brown (Plate 22). Severed vascular strands exude a milky discharge of the causal bacterium when placed in water.

Internal and external symptoms of Moko and Panama disease are quite similar.

However, only Moko affects fruit, suckers and plants younger than ~4 months.

CAUSAL AGENT *R. solanacearum* is an aerobic, Gram-negative, non-fluorescent, rod-shaped bacterium (Schaad *et al.*, 2001). It is a widespread and diverse pathogen that has been divided into five biovars based on carbohydrate utilization, and five races that are determined by host range.

Race 1 of *R. solanacearum* causes a vascular wilt on some *Musa* taxa, but not on the edible cultivars or heliconia (Jones, 2000). In contrast, strains that cause Moko disease are in biovar 1 and race 2; they are quite variable. They have restricted geographical distributions, occuring in a single country or region, and display varying levels of virulence on different banana cultivars and heliconia. They also display distinct colony phenotypes on Kelman's medium, and have disparate abilities to persist in soil and be vectored by insects. Virulent isolates of the bacterium produce extracellular polysaccharides and are not motile.

EPIDEMIOLOGY Root to root infection is possible, and moving water can disseminate the bacterium. However, spread usually involves insects or man (Jones, 2000). Trigona bees, wasps and other flying insects have been reported to disseminate certain strains of this pathogen (especially the SFR, and to a lesser extent B, strains) over 90 km. Insect-driven epidemics can move rapidly due to the strength and range of the insects that are involved and the rate at which plants become infectious. Within 15 days of flower infection, SFR strains begin to ooze from bracts and peduncles to initiate another cycle of infection. Contaminated farm machinery, machetes that are used for pruning, and infected fruit and rhizomes are all effective vehicles of dissemination.

MANAGEMENT Regular inspection and eradication programmes are essential wherever Moko disease is established. These include:

- early recognition of the disease;
- removal of the male bud;
- rigorous disinfestation of farm implements, especially machetes that are used for bud removal and mat maintenance; and
- destruction of affected and neighbouring plants with herbicides. These sites can be replanted 6–12 months after all host residues have decayed.

'Bluggoe' and other ABB cooking bananas with dehiscent bracts are most susceptible, and are also sources of inoculum for commercial bananas (Ploetz *et al.*, 1994b). In these situations, 'Pelipita' ABB, which has persistent bracts, or clones with aborting male buds could be used to replace the susceptible cultivars. Although alternative weed hosts have been reported, their importance in the Moko disease cycle is debated.

DISEASES THAT ARE CAUSED BY FUNGI

Fungi are the most important and prevalent pathogens of banana. They affect all organs of the host in the field and are the major causes of pre-harvest yield reductions and postharvest damage to and loss of fruit.

Diseases of the Foliage

Black cross

This leaf spot disease is found in Australia, Indonesia, the Philippines and southwestern Pacific, and is usually not a serious problem (Jones, 2000). Its symptoms are most conspicuous on the undersides of older leaves, and consist of black, cross-shaped lesions that are mature stroma of the causal fungus. Long axes of the crosses run up to 6 cm along leaf veins and short axes, up to 3 cm in length, and occur at right angles (Plate 23). On the upper leaf surface, lesions are usually restricted to black dots that silhouette the lesion on the underside.

The ascomycete *Phyllachora musicola* causes black cross leaf spot (Jones, 2000). It produces obpyriform to subglobose perithecia that are immersed in host leaf tissue. Their ostioles emerge from the upper surface of the leaf, and they form in stroma in groups of up to 40. Asci are cylindrical or somewhat clavate, unitunicate, 150–190 × 16.5–20 μm, and have a truncate apex with an apical ring (Fig. 4.3). Ascospores are one- or two-celled, hyaline, smooth and 42–52 × 8–10 μm. The fungus has no known anamorph.

Black Sigatoka

Black Sigatoka, which was known originally as black leaf streak, is currently the most important disease of banana. It was first reported on the Fijian island of Viti Levu in 1963 (Rhodes, 1964), but was probably widespread in the Southeast Asian/South Pacific region by that time (Jones, 2000). Its spread worldwide has occurred fairly recently: it first reached the western hemisphere (Honduras) in 1972 and Africa (Zambia) in 1973. By 1991, the disease had spread throughout the Americas and sub-Saharan Africa (Jones, 2000).

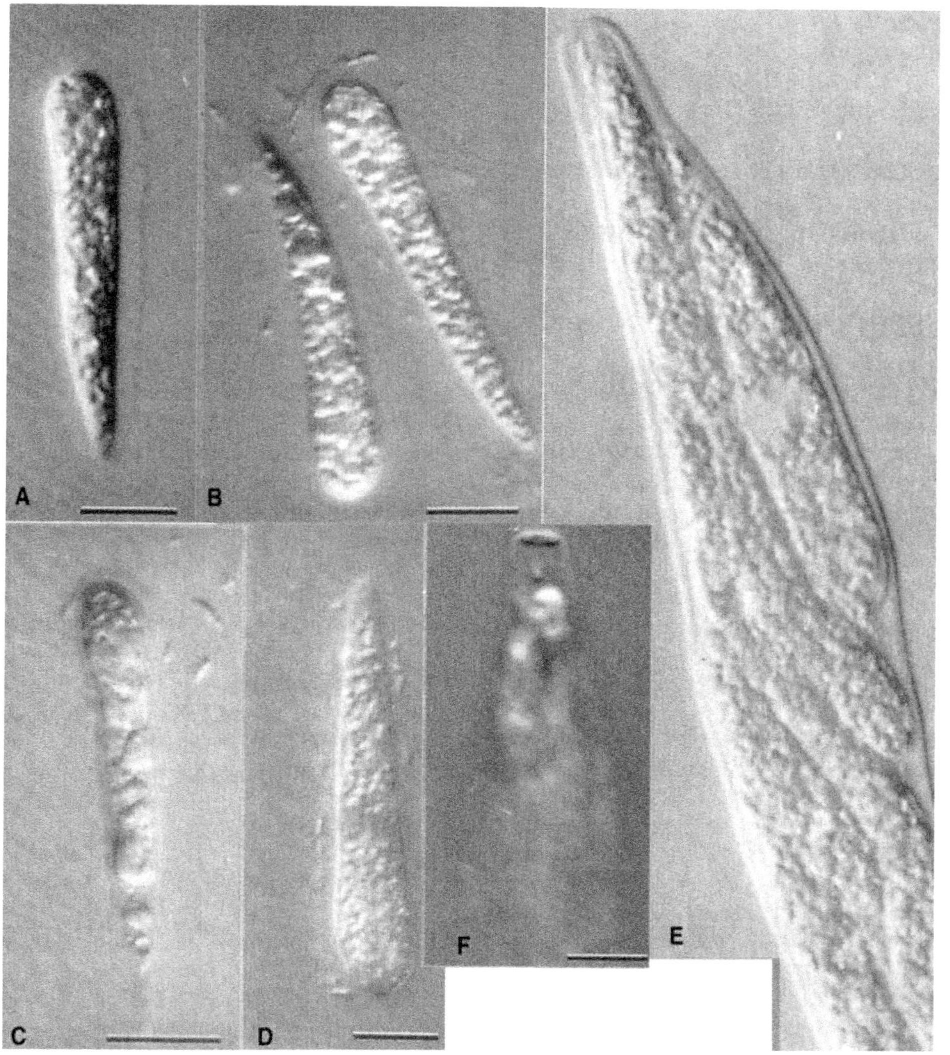

Fig. 4.3. (A–D) ascospores, (F) ascus tip with apical ring and (E) ascus of *Phyllachora musicola* (from CMI description no. 1127). Bars = 10 μm.

Another disease, yellow Sigatoka, is treated separately in this chapter, but is mentioned here due to its similar symptoms, the impact it and black Sigatoka have on banana, and the relatedness of the respective causal fungi. Yellow Sigatoka has a greater geographic distribution, but black Sigatoka is more aggressive and has a wider host range (Stover and Simmonds, 1987). On the Cavendish clones that are used by the export trades, fungicidal control measures for black Sigatoka are 3–6 times more expensive than those for yellow Sigatoka. In general, symptoms of black Sigatoka develop on the Cavendish cultivars 8–10 days earlier than those of yellow Sigatoka. Black Sigatoka affects plantain and banana cultivars that resist yellow Sigatoka, and causes greater defoliation and yield losses. In most of the lowland humid tropics, black Sigatoka has now displaced yellow Sigatoka as the predominant leaf spot on banana. Although yellow Sigatoka remains more important at elevations above 1000 m, there is evidence that the black Sigatoka pathogen adapts to higher elevations after it has been introduced into an area (Jones, 2000). In some areas in Latin America, the disease is now found as high as 1500 metres above sea level (masl).

SYMPTOMS Symptoms begin as minute reddish brown flecks on the lower leaf surface. As they progress, they become visible on the upper surface, elongate, darken and often develop a wet appearance (Plate 24). Dark borders and yellow haloes surround the spots as they exceed 1–2 cm in length, and as the spots mature their centres become grey and sunken. In susceptible cultivars, spots coalesce until the entire leaf surface is killed, and on plants that have flowered no healthy leaf tissue may remain by the time fruit mature (Plate 25). In severe cases, the bunch either does not develop fully or falls from the plant. The disease also causes fruit to ripen prematurely, which can be a serious problem when fruit are shipped long distances.

Two other diseases cause symptoms that resemble those of black Sigatoka. Phaeoseptoria leaf spot is found in portions of Asia, Australia, Africa and the Americas, and eumusae leaf spot (formerly called Septoria leaf spot) is found in Malaysia, Mauritius, Nigeria, southern India, Sri Lanka, Thailand and Vietnam (Jones, 2000). They are covered individually in this chapter.

CAUSAL AGENT Black Sigatoka is caused by the ascomycete, *Mycosphaerella fijiensis* (anamorph: *Pseudocercospora fijiensis*). A variant of the pathogen, *M. fijiensis* var. *difformis*, is no longer recognized. *M. fijiensis* is closely related to *M. eumusae*, the cause of eumusae leaf spot, and *M. musicola*, the cause of yellow Sigatoka. DNA sequence analyses suggest that these major leaf pathogens of banana may have evolved from a common ancestor (Carlier *et al.*, 2000).

Conidia of *M. fijiensis* are produced in streaks early in their development, mainly on the lower leaf surface (Jones, 2000). Conidiophores are pale brown, single- to six-celled, straight to geniculate, occasionally branched, subcylindric, 16.5–62.5 × 4–7 μm and in predominantly hypophyllous fascicles (Crous and Mourichon, 2002) (Fig. 4.4). Conidiogenous cells are up to 25 μm long, 2–4 μm wide at the apex, and have 1–3 minutely thickened scars. Conidia are subhyaline, obclavate to cylindric-obclavate, have an obclavate basal cell, usually six- to eight-celled, 10–120 × 2.5–5 μm, with hila that are slightly thickened and darkened along the rim. Although the presence of this basal scar was shown recently to be phylogenetically unimportant in cercosporoid fungi (the former anamorph genus for this pathogen, *Paracercospora*, was based on this feature) (Crous and Mourichon, 2002), it is a valuable diagnostic character that enables this pathogen to be distinguished from *M. eumusae* and *M. musicola*.

The teleomorph is virtually indistinguishable from that of *M. eumusae* and *M. musicola*. The pseudothecia in which ascospores are produced are mainly globose, 47–85 μm in diameter and immersed in the leaf tissue (Jones, 2000). They occur on both leaf surfaces although they are most common on the upper side. Their ostioles protrude above the leaf surface and are dark brown and conspicuous. Asci are bitunicate,

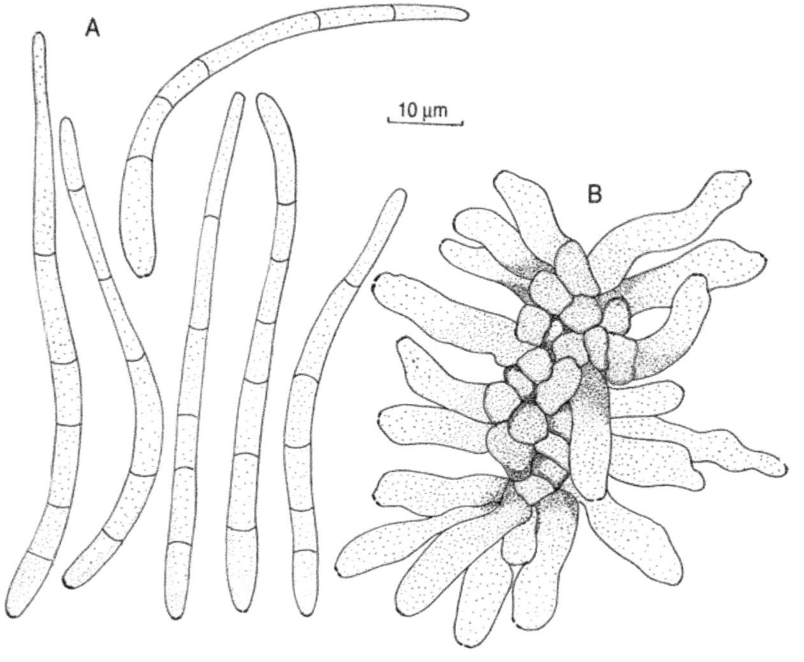

Fig. 4.4. Conidia and conidiophores of *Pseudocospora fijiensis*, anamorph of *Mycosphaerella fijiensis* (from CMI description no. 413).

obclavate and lack paraphyses, and the ascospores are colourless, 12.5–16.5 × 2.5–3.8 μm, and two-celled with a constricted septum.

M. fijiensis grows very slowly on artificial media. At optimal temperatures of between 24 and 28°C, single-conidium cultures reach 1 cm in diameter after 38 days on potato dextrose agar (PDA).

The fungus is heterothallic, and considerable pathogenic and genetic diversity has been reported among geographically dispersed isolates (Jones, 2000). Genetic analyses indicate that the centre of diversity for the pathogen is the Australasian–Southeast Asian region (Carlier *et al.*, 1996). There is evidence that isolates from some areas attack banana cultivars that are resistant in other regions, and that the pathogen is capable of adapting to resistant cultivars in a given location over time (Fullerton and Olsen, 1995). Meiotic recombination in the pathogen probably plays a major role in generating diversity in this pathogen.

EPIDEMIOLOGY Conidia and ascospores of the fungus are both infective. They are formed under high moisture conditions and are disseminated by wind (Rutter *et al.*, 1998). Ascospores are most important in spreading the disease within plants and plantations. Since they are killed after 6 h of UV radiation, Parnell *et al.* (1998) concluded that ascospores would probably not disperse more than a few hundred kilometres by wind. In contrast, infected planting material and leaves that are used as packing materials can move the pathogen great distances.

Leaves are infected indirectly via stomata and, due to the greater abundance of stomata there, the lower surface is the primary infection site (Washington *et al.*, 1998). The cardinal temperatures for ascospore germination are 12, 27 and 36°C, and substomatal penetration for conidia and ascospores occurs within 48–72 h on moist surfaces. Although free moisture is required for ascospore germination, conidia can also germinate under high humidity.

Further colonization of the leaf occurs via epiphyllic hyphae of the pathogen that emerge from stomata and grow along the lower leaf surface to other stomata. Under prolonged wet conditions, an extensive network of hyphae can cover and infect the leaf surface.

Transition periods (the length of time between the appearance of the first fleck symptom to the development of mature spots with grey centres) are affected by environment, host genotype and pathogen isolate, and ranged between 11 and 139 days. Important environmental factors include temperature (optimum: 25–28°C), moisture (optimum: free water on host surfaces) and shade (which reduces symptom development by as much as 50%). Development rate decreases dramatically as conditions become drier, or are hotter or cooler than optimum. Likewise, development is much slower on resistant cultivars and where less aggressive populations of the pathogen exist.

MANAGEMENT Management strategies vary according to the cultivars that are grown, the environment and the intended market for the fruit (Fullerton and Stover, 1990). In export plantations that produce dessert bananas or plantains, frequent applications of fungicides are usually needed. Depending on the sensitivity of local populations of the pathogen to the utilized fungicides, application frequencies can range as high as 36 year^{-1} for dessert bananas and 19 year^{-1} for plantains (Jones, 2000). Fungicide applications usually require airplanes or helicopters and permanent landing strips and facilities for mixing and loading the fungicides. When the high recurring expense of the spray materials themselves is considered, costs run as high as 26% of the total cost of production for exported dessert bananas (Stover and Simmonds, 1987).

Cultural practices are also used in commercial situations. These include the removal of leaves with mature spots, and reducing humidity within plantations by increasing spacing between plants and providing efficient drainage with permanent canals and pumps.

Since most smallholders cannot afford these control measures, they are affected most by black Sigatoka. For example, plantain yield losses of 33 and 76% were recorded during, respectively, the first and second cropping cycle in West Africa (Mobambo *et al.*, 1996). Under marginal conditions, production is often abandoned due to low yields (Jones, 2000).

Chemical control of first yellow, and then black, Sigatoka has evolved considerably over the last 60 years (Ploetz, 2000; Stover, 1990a). Bordeaux mixture, first used in the mid-1930s, was replaced by several succeeding generations of protectant and, later, systemic fungicides. Presently, a sterol biosynthesis inhibitor, tridemorph, several different sterol demethylation inhibitors, most importantly propiconazole, and different strobilurins are the most common systemics.

In general, insensitivity develops in *M. fijiensis* towards the systemic fungicides after prolonged use. Within 2–3 years of the introduction of benomyl in Central America, resistance was observed, and by the late 1970s it was no longer effective in many areas (Stover, 1990a). Recently, a 500-fold increase in resistance to a strobilurin fungicide, trifloxystrobin, was recorded after 4 years of 6–11 applications year^{-1} in Costa Rica (Chin *et al.*, 2001). In contrast, tolerance to propiconazole has developed quite differently. Reduced control with this fungicide developed in Central America in the 1990s. Romero and Sutton (1997) observed a rapid shift towards insensitivity in three different plantations in Costa Rica. However, unlike the situation with benomyl, they observed wide ranges in sensitivity in each plantation, and in no case was there evidence for the predominance of a highly tolerant population. Although they speculated that individuals with high levels of tolerance to propiconazole were less fit than sensitive strains, the cost of tolerance in *M. fijiensis* to this important fungicide has not been researched.

Due to resistance problems, the systemic fungicides usually are applied in combination or alternation with broad-spectrum, protectant fungicides, such as the dithiocarbamates and chlorothalonil. With the exception of chlorothalonil, these fungicides are usually mixed with petroleum-based spray oils. The oils themselves are fungistatic and retard the development of the pathogen in the leaf. When they are mixed in water emulsions with fungicides, the resulting 'cocktails' provide superior disease control.

Application schedules are determined with disease forecast systems that incorporate data on disease severity within the plantation and environmental factors that affect infection and disease development (Stover, 1990). The models are based on the use of oils and systemic fungicides only, since protectant fungicides have no curative effect on symptom development.

Although some cultivars resist black Sigatoka, resistance is poor among many important types of banana, including export dessert AAA, AAB plantain, highland AAA and AAB dessert (Table 4.3). Furthermore, clones that resist black Sigatoka may be susceptible to other problems such as Panama disease, nematodes and weevil borer (*Cosmopolites sordidus*).

As fungicides lose their effectiveness against black Sigatoka and subsistence production continues to deteriorate, the demand for resistant genotypes will increase. International breeding programmes, most notably that of the Fundación Hondureña de Investigación Agrícola (FHIA) in La Lima, Honduras and the International Institute of Tropical Agriculture (IITA) in Onne, Nigeria, have made significant progress in incorporating disease resistance in several breeding targets (Jones, 2000).

Cladosporium speckle

Cladosporium speckle affects older leaves of banana plants in humid environments (Stover, 1972). It is usually a minor problem, but yields may be impacted in some areas. The disease has been recorded in Africa, Asia, Australasia, Oceania and tropical America.

SYMPTOMS Symptoms reported from different regions are similar (Jones, 2000). Pale brown linear streaks, 0.3 × 1.5 mm, appear 3–4 weeks after leaves unfurl. These enlarge, coalesce and eventually turn brown (Plate 26). Blotchy necrotic areas develop where high concentrations of lesions are found, and they are usually associated with chlorotic areas that run parallel to the leaf veins. Alternatively, diffuse, brown to blackish blotches develop on the upper surface of older leaves without linear streaks. In both cases, large areas of the leaf surface can eventually die.

CAUSAL AGENT *Cladosporium musae* produces erect, dark brown conidiophores that have persistent branches at the apex, average 610 μm in length and have thick-walled bases that are 6–8 μm wide (Fig. 4.5) (David, 1988). Conidiogenous cells are terminal or intercalary. Conidia are ovate–cylindrical, ellipsoidal or fusiform, and may have a constricted middle and one or more scars at each end. They form in branched chains, are 6–22 × 3–5 μm, mainly single-celled, thin-walled and almost hyaline.

EPIDEMIOLOGY Conidia are carried by air currents and require free moisture to germinate. Disease development is favoured by high humidity.

MANAGEMENT Diverse AAA cultivars are susceptible, including those in the Cavendish subgroup, 'Gros Michel', the East African highland cultivars, 'Pisang Berangan' and 'Pisang Nangka'. Based on the susceptibility of other clones, in some regions there may be pathogenic specialization in the fungus. Although 'Sucrier' AA was not susceptible in Côte d'Ivoire, 'Kluai Khai' (a synonym of 'Sucrier') is the most susceptible cultivar in Thailand. Clones with AAB genomes were also not susceptible in Côte d'Ivoire, but 'Dwarf Horn' plantain AAB is susceptible in Panama.

Although control measures usually are not needed, fungicides that are used against the Sigatoka leaf spots are effective.

Cordana leaf spot (leaf blotch)

This is a common, but usually innocuous disease of banana. It causes its greatest damage during rainy weather on the lower

Table 4.3. Reaction of important banana cultivars to Panama disease and yellow and black Sigatoka.

			Reaction to[a]		
Genome	Subgroup	Cultivar	Panama disease	Yellow Sigatoka	Black Sigatoka
AA		'Sucrier'	R[b]	S	MS
		'Pisang Lilin'	R	R	R
AB		'Ney Poovan'	S	R	
AAA		'Ibota Bota'	R[b]	R	R
	Cavendish		R[b]	S	S
	Gros Michel		S	S	S
	Lujugira-Mutika		R[c]	R	S
AAB		'Silk'	HS	SS	MS
		'Mysore'	R	R	SS
	Maia-Maoli/Popoulu		S	S[d]	MS
	Plantain		R[b]	R[e]	MS
	Pome	'Pome'	MS	SS	S
ABB		'Bluggoe'	S	R	MS
		'Pisang Awak'	MS	R	SS
		'Saba'	R	R	SS
		'Cardaba'	R	R	SS
		'Pelipita'	R	R	MS
AAAA		'FHIA23'	R[b]		
		'I.C. 2'	S		MS
AAAB		'FHIA01'	R	R	R
AABB		'FHIA03'	R[b]	R	R

[a] Reactions are: HS (highly susceptible); S (susceptible); MS (moderately susceptible); SS (slightly susceptible); and R (resistant). From: Stover and Simmonds (1987); Ploetz *et al.* (1994b); Pegg (2000).
[b] 'Sucrier', 'Yangambi km 5', plantains, 'FHIA03' and 'FHIA 23' succumb to race 4 but are otherwise resistant. 'Dwarf Parfitt', a Cavendish cultivar, and somaclones of 'Giant Cavendish' from the Taiwan Banana Research Institute (i.e. the GCTCV series) also resist race 4.
[c] Reports that clones in the Lujugira-Mutika subgroup are susceptible to Fusarium wilt (Ploetz *et al.*, 1994a) may be in error (Kangire and Rutherford, 2001).
[d] Resistance to yellow Sigatoka may exist in some accessions in the Maia-Maoli/Popoulu subgroup (Jones, 2000).
[e] Resistance in plantains to yellow Sigatoka breaks down at high elevations.

leaves of plantains, and to a lesser extent on ABB cultivars.

SYMPTOMS Pale brown, oval patches, ranging from one to several centimetres in diameter, form towards the leaf margins and in association with wounds caused by other diseases and leaf tears (Plate 27). Lesions are surrounded by bright yellow haloes and have light grey, necrotic centres with concentric zonations that are most noticeable on the upper leaf surface. Lesions eventually may encompass entire leaf margins, and large portions of the leaf lamina can be affected, especially when the disease occurs in concert with the Sigatoka leaf spots.

CAUSAL AGENTS Two species cause Cordana leafspot. *Cordana musae* was first described in 1902 on Java (Jones, 2000). More recently, Ellis (1971) described a new species, *C. johnstonii*, from Irian Jaya, peninsular Malaysia and Tonga. It has also been recognized since in New South Wales and on Lord Howe and Norfolk Islands (Priest, 1990). Its occurrence in cooler climates and absence in warmer

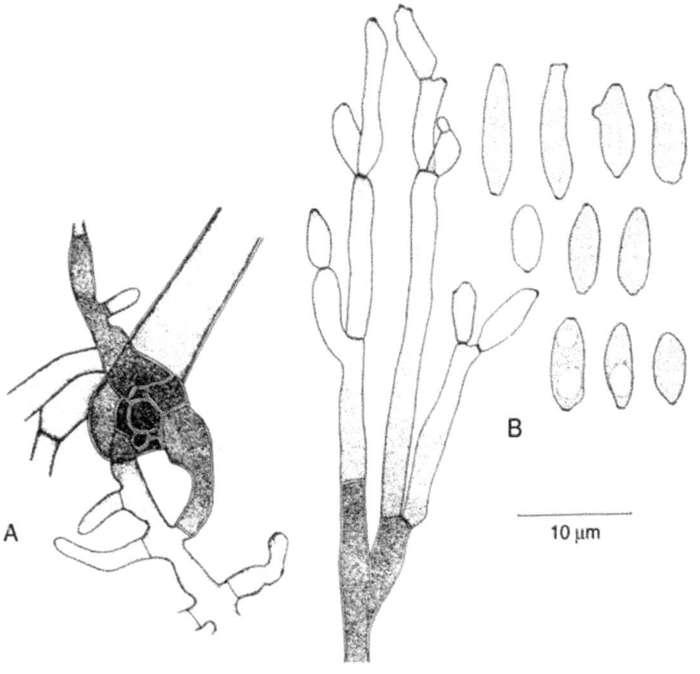

Fig. 4.5. (A) Base of conidiophore and (B) conidiophore and conidia of *Cladosporium musae* (from CMI description no. 958).

production areas led Jones (2000) to speculate that it tolerated cooler conditions than *C. musae*.

Conidiophores of each species are straight or flexuous and septate (Ellis, 1971; Ellis and Holliday, 1972). Those of *C. musae* are pale to mid-brown, smooth, up to 220 μm long and 4–6 μm in diameter with terminal and intercalary swellings of 6–8 μm, and basal swellings of 8–11 μm (Fig. 4.6). Conidia are solitary, arising from small pegs on the terminal end of the conidiophores, obovoid to pyriform, two-celled, subhyaline to pale brown, smooth, 11–18 μm long and 7–10 μm at their widest point, and with a thickened hilum. In contrast, conidiophores of *C. johnstonii* often are twisted at the base and up to 300 μm long, and its conidia, although quite similar to those of *C. musae*, are much larger, ranging from 19 to 26 μm in length and 14 to 16 μm in width (Fig. 4.7).

EPIDEMIOLOGY The pathogens sporulate abundantly on the undersides of lesions (Jones, 2000). Production occurs at night during periods of dew or rainfall, and spore release takes place as vapour pressure decreases; peak release of conidia occurs around 7 a.m. Under moist conditions, conidia germinate and form appressoria within 8 h of their deposition on the leaf surface. After another several hours, infection pegs form on appressoria and penetrate the epidermis directly. Healthy and necrotic banana tissue are infected. However, since barriers to infection are often created in healthy tissue shortly after invasion, this is primarily a disease of wounded and weakened tissue.

MANAGEMENT In export plantations, where the Sigatoka leaf spots are controlled with protectant or systemic fungicides, Cordana leaf spot is uncommon. Oil, used in combination with fungicides for control of the Sigatoka leaf spots, was reported to increase Cordana leaf spot when used alone. Also, oil is phytotoxic to plantain cultivars on which Cordana leaf spot is most severe. Specific

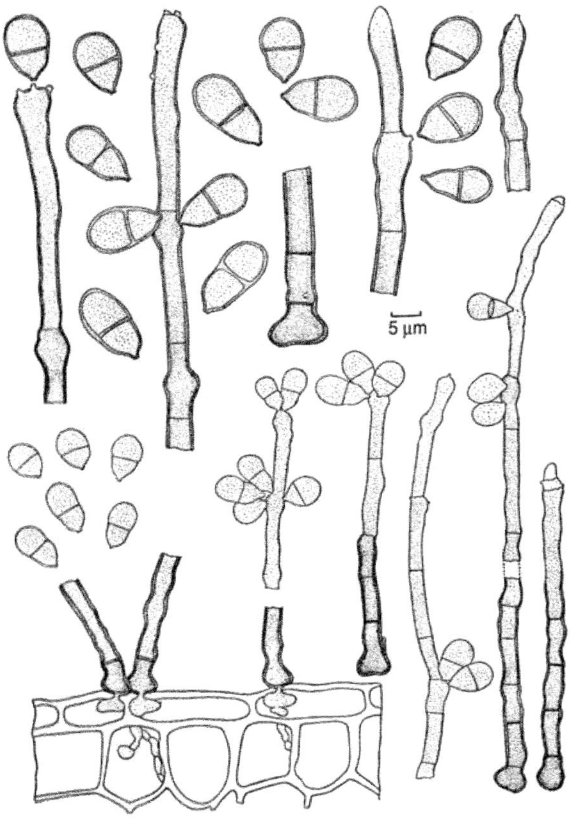

Fig. 4.6. Conidia and conidiophores of *Cordana musae* (from Ellis, 1971).

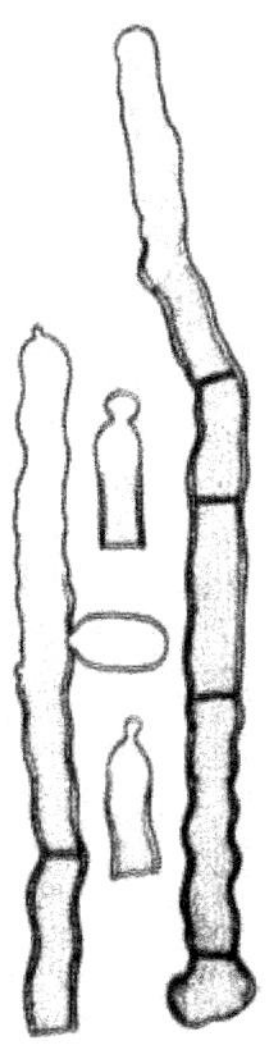

Fig. 4.7. Immature conidium and conidiophores of *Cordana johnstonii* (from Ellis, 1976).

control measures for this disease usually are not indicated.

Eumusae leaf spot

This disease is found in Malaysia, Mauritius, Nigeria, India, Sri Lanka, Thailand and Vietnam (Carlier *et al.*, 2000). It was first recognized during surveys in the 1990s during which the distribution of the Sigatoka leaf spots in South and Southeast Asia was investigated. Surprisingly, none of the samples yielded *M. musicola*, and *M. fijiensis* was isolated from only a few. From most of the samples, an undescribed species of *Mycosphaerella* was recovered that appeared to have a *Septoria* anamorph. The original common name of the disease, Septoria leaf spot, referred to this identification. Further examination of samples from other regions expanded the disease's range outside Asia, and recently has shown that the anamorph is a species of *Pseudocercospora* (Carlier *et al.*, 2000; Crous and Mourichon, 2002). Its disparate geographic distribution and the close resemblence of its symptoms to those of the Phaeoseptoria and Sigatoka leaf spots suggest that it may be present, but not recognized, in countries not on the above list. At this time, it appears to be the predominant leaf spot in Thailand, and is common in peninsular Malaysia, southern India and Sri Lanka (Jones, 2000).

SYMPTOMS The symptoms are very similar to those of the Sigatoka leaf spots (Carlier *et al.*, 2000; Jones, 2000). Faint brown streaks expand and become oval to elliptical (Plate 28). They darken and develop a grey centre and dark border as they mature. At this stage, the lesions are broader and larger than those of the Sigatoka leaf spots. However, they do resemble those caused by Phaeoseptoria leaf spot, a disease for which Eumusae leaf spot initially was mistaken. On severely affected leaves, lesions coalesce, large portions of the leaf die, and surrounding tissues yellow.

CAUSAL AGENT The causal fungus is *M. eumusae* (anamorph: *Pseudocercospora eumusae*) (Carlier *et al.*, 2000; Crous and Mourichon, 2002). The species name refers to the section of *Musa*, *Eumusae*, in which the host taxa reside.

The teleomorph of *M. eumusae* cannot be distinguished from that of *M. fijiensis* or *M. musicola* (Crous and Mourichon, 2002). Pseudothecia are globose, 42–51 μm in diameter, dark brown and have a short ostiole. Eight two-celled ascospores, 12–16.5 × 3–4.5 μm, are produced in oblong asci. The fungus produces predominantly epiphyllous sporodochia on dark brown substomatal stromata. Conidiophores are subhyaline to pale olivaceous and pale brown at the base, subcylindrical, single- to four-celled and 10–25 × 3–5 μm with truncate ends. Its sporodochia develop similarly to those of *M. musicola*, but its conidiophores are much longer. Conidia of *P. eumusae* are subhyaline to pale olivaceous, subcylindrical, 18–65 × 2–3 μm, four- to nine-celled, and have subtruncate ends without scars. Conidia can be distinguished from those of *M. musicola* by their more cylindrical shape, subtruncate ends and shorter dimensions.

EPIDEMIOLOGY AND MANAGEMENT These topics have not been researched, although there is evidence that chemicals that are used against the Sigatoka leaf spots are effective (Jones, 2000).

Eyespot

This disease, which is also known as Drechslera leaf spot, occurs where Bermuda grass, the primary host of the pathogen, is found beneath plants (Stover, 1972). It has been observed in Central America, Jamaica and Uganda, and is not economically important.

Symptoms develop on suckers <2 m in height. Lesions are oval or lenticular, as large as 16 × 8 mm, and develop a bleached greyish centre as they mature (Fig. 4.8). The causal agent, *Drechslera gigantea*, sporulates on Bermuda grass, but not banana.

Freckle

Freckle is common in Asia, Australia and the Pacific (Jones, 2000). Reports in Africa and the Caribbean may be in error. Freckle is one of the most serious problems affecting the Cavendish-based banana industry in Taiwan, is becoming more important in export production in the Philippines, and damages 'Pisang Berangan' AAA in plantations in Malaysia. The disease usually is not important in other situations.

SYMPTOMS The upper surface of older leaves is affected (Jones, 2000). Dark brown to black spots, <1 mm across, give the leaf a greyish caste. They can coalesce in bands parallel to the leaf veins or run lengthwise down the leaf, due apparently to infective spores dripping down the leaf while it was in the candela stage. Larger spots, up to 4 mm across, may also form and aggregate in large tan areas. Pycnidia of the causal fungus that form in necrotic areas feel rough and are

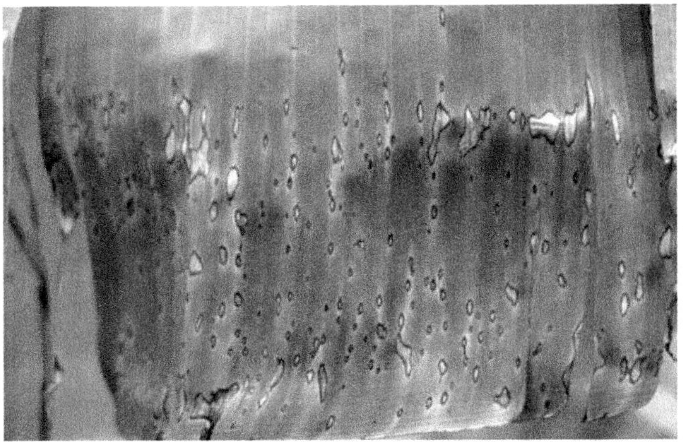

Fig. 4.8. Symptoms of eyespot, caused by *Drechslera gigantea* (photo: R.C. Ploetz, UF).

a diagnostic feature for the disease (Plate 29). Petioles, midribs, transition leaves, bracts and fruit can also be affected. Symptoms begin developing on 2- to 4-week old fruit, but are most conspicuous at harvest (Punithalingham and Holliday, 1975).

CAUSAL AGENT *Guignardia musae* (anamorph: *Phyllosticta musarum*) causes freckle. Pycnidia are most common on the host, and are 60–170 μm in diameter, globose, brown to black, and occur individually or in groups (van der Aa, 1973). Conidiogenous cells are cylindrical or conical and 4–11 × 2.5–5 μm. Conidia, 15–18 × 9–10 μm, are aseptate, obovoid, ellipsoidal or short cylindrical with a rounded apex and truncate base. They are indented, may have a single 8–16 μm long apical appendage, and are encased in a gelatinous 1–3 mm thick envelope. Perithecia are 70–220 μm in diameter, globose and papillate. Asci are 35–85 × 20–25 μm, eight-spored, clavate to cylindrical, and usually have a short stalk. Ascospores, 17–22 × 8–10 μm, are aseptate, and ovoid to oblong ovoidal. Spermatia, 6–10 × 0.5–2 μm, are aseptate, cylindrical, or dumb-bell-shaped.

Punithalingham and Holliday (1975) described as *G. musae* an ascomycete that differed somewhat from van der Aa's (1973) description. Chuang (1981) felt that it was a saprophyte on diseased tissue.

Based on host range and symptomatology (see below), Jones (2000) speculated that two different races or species might cause freckle in southern and Southeast Asia, one of which spread to Taiwan and the northern Pacific, and the other that spread to Australasia and the South Pacific. Whether the similar but distinct fungi that van der Aa (1973) and Punithalingham and Holliday (1975) described represent the taxa that Jones (2000) deduced is not known.

EPIDEMIOLOGY Although the role that ascospores play in spreading the disease is not known, conidia are important (Jones, 2000). They are dispersed in water and usually travel only a short distance. On moist host surfaces at 24°C, they germinate in 2–3 h, and after 12 h form irregular, hyaline appressoria in grooves between host epidermal cells. After 24–72 h, scattered cells discolour. After 96 h, >60% of the appressoria are associated with necrotic cells, and necrosis is more rapid and extensive where several appressoria develop in an area.

Fine penetration hyphae that develop under appressoria swell to a diameter of 3–5 μm upon entering the cell. Inter- and intracellular hyphae subsequently invade surrounding epidermal cells, but rarely penetrate beneath the fifth layer. In Taiwan, the incubation period varies from 20 days in warm, wet weather to 60 days when it is dry and cool.

Pycnidia develop in lesions of all sizes. Secondary infections intensify symptoms that then often extend laterally on leaves in streaks. Conidia are dispersed from leaves to fruit. Continuous infection in the presence of rain or dew results large areas eventually dying.

Freckle affects *M. acuminata* ssp. *banksii* and *M. schizocarpa* in Papua New Guinea, but *M. balbisiana* is not known to be affected anywhere (Jones, 2000). In southern Asia, a wide range of cultivars is affected, including: 'Pisang Berangan' and those of the Cavendish subgroup (both AAA); 'Silk', 'Horn' and 'French' plantain and Mysore (all AAB); and 'Pisang Awak', 'Bluggoe', 'Pelipita', 'Saba' and 'Kluai Teparot' (all ABB). In contrast, in Hawaii and Taiwan, Cavendish cultivars are susceptible but 'Bluggoe' is resistant, whereas in Australasia and the South Pacific, the Cavendish clones are resistant and 'Bluggoe' is susceptible (Jones, 2000; D.R. Jones, personal communication, 2002). The differential response of the latter cultivars in these regions suggests that different taxa may cause freckle in these areas. Studies to clarify this situation are warranted since the existence of different freckle pathogens would have important quarantine implications for countries in which Cavendish-based trades exist but are not affected seriously by this disease (e.g. Australia).

MANAGEMENT Where freckle is important on exported fruit (especially Taiwan), removal of diseased leaves and protection of bunches with bags is carried out (Jones, 2000). Several fungicides that are effective against black Sigatoka are also used against freckle, including mancozeb and two sterol demethylation inhibitors, propiconazole and flusilazole.

Leaf speckle

This disease is found in Southeast Asia, and is usually a minor problem on diverse cultivars (Jones, 2000). However, it has been serious in Cavendish AAA export plantations in Taiwan since 1981 after disease forecast systems were begun for black Sigatoka control. Severe leaf speckle developed when fungicides were not used after bunch initiation. Dithiocarbamates in oil are effective.

Lesions develop as brown specks on the underside of the leaf. As they enlarge, tan and brown blotches develop on the upper surface. Surrounding areas become chlorotic and then necrotic. *Acrodontium simplex* causes leaf speckle. Its brown, erect conidiophores are septate, single or sparsely branched and 41–90 μm long. Conidia are oval, 2.9–3.8 × 1.9 μm and cover the conidiophore terminus. Symptoms identical to those described above were associated with the tropical speckle agent, *Periconiella musae*, in Queensland and Vietnam.

Malayan leaf spot

This is usually a minor disease in Fiji, Tonga, Western Samoa and the highlands of peninsular Malaysia and Papua New Guinea (Stover; 1972; Jones, 2000). Severe development occurs in areas of Fiji with >1 m of annual rainfall when temperatures are below 24°C, and on some local cultivars in Papau New Guinea.

Symptoms vary in the different locations. In Fiji, lesions on the upper leaf surface are diamond-shaped, light grey, 2–4 × 3–5.5 mm, and have 0.5 mm wide black borders. Profuse growth of the causal fungus, *Haplobasidion musae*, occurs on the leaf undersurface. In Malaysia, lesions have dark purple borders, are pale grey on the upper and pale brown on the lower surface, and are either ellipsoid (2–4 × 3–12 mm) or round (2–5 mm in diameter). Lesions are similar in Papua New Guinea, and large patches of confluent necrosis can develop. Symptoms develop as early as the second or third leaf on local 'Mala' AA.

Single to six conidiophores of *H. musae* are produced at the ends of hyphae that arise through the epidermis on the leaf's lower surface (Fig. 4.9). They are straight or flexuous, pale brown, one- to four-celled and 50–110 × 4–6 μm. They terminate in a subglobose apex that bears spherical, 4–8 μm diameter, sporogenous cells. Conidia are spherical, brown, verrucose, 4–6 μm in diameter, and borne singly or in chains of two to five.

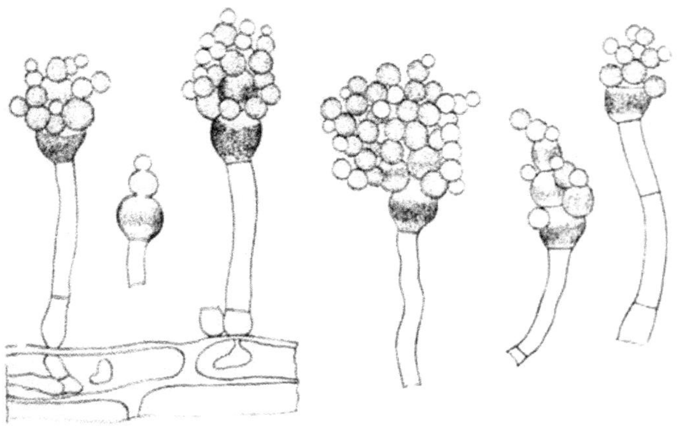

Fig. 4.9. Conidia and conidiophores of *Haplobasidium musae*, cause of Malayan leaf spot (from CMI description no. 496).

Although control measures usually are not needed, maneb was effective. Spray oil enhanced disease development.

Mycosphaerella speckle

This disease has a worldwide distribution, but is a significant problem only in subtropical Australia (Stover, 1972) and South Africa (Viljoen *et al.*, 2002). Symptoms occur below the fourth leaf, and consist of smoky patches on the upper leaf surface and tan irregular blotches on the lower surface (Plate 30). Under wet conditions, these appear water soaked and exude droplets of moisture. The blotches darken, becoming purple to black, speckled and visible on both leaf surfaces. Associated tissues yellow, and eventually the speckled areas coalesce and bleach to grey on the lower, and tan on the upper surface.

Pseudothecia of the causal fungus, *Mycosphaerella musae*, appear at this stage. They are 45–99 × 34–81 μm, scattered, globose, black and immersed in host tissue. Asci are eight-spored, obclavate and 24–44 × 8–12 μm, and ascospores are hyaline, obtuse to cylindrical, two-celled and 9–16 × 2–3 μm. Conidia are not produced on the host, but form on agar after 4–5 days. They are *Cercospora*-like, average 127 × 2.9 μm, are usually verrucose and have a basal scar.

Disease development is greatest under humid, warm conditions, and on old, senescing leaves. Diverse cultivars are affected, and control measures that are used for the Sigatoka leaf spots are effective.

Phaeoseptoria leaf spot

This disease damages 'Mysore' and 'Nendran' (both AAB) in Kerala State, India (Jones, 2000). It is also found in Cameroon, Colombia, Ghana, Guyana, Honduras, Kenya, Queensland, Sabah, Trinidad, Uganda and Zanzibar. This disparate distribution suggests that the disease may be found elsewhere.

The symptoms resemble those of the Sigatoka leaf spots. Lesions are elliptical to oval, 1–2 cm wide, and pale yellow with dark borders and chlorotic haloes (Punithalingham, 1983). As spots merge, they form large necrotic areas. Young leaves are not affected.

Phaeoseptoria musae produces light to dark brown, subglobose, 85–145 μm in diameter, ostiolate pycnidia that are immersed and later erumpent in host tissue (Punithalingham, 1983) (Fig. 4.10). Conidia are 22–30 × 2.5–3 μm, three- to five-celled, hyaline to pale brown, straight or slightly curved, cylindrical to slightly clavate, with a rounded or truncate base and gradually narrowed apex.

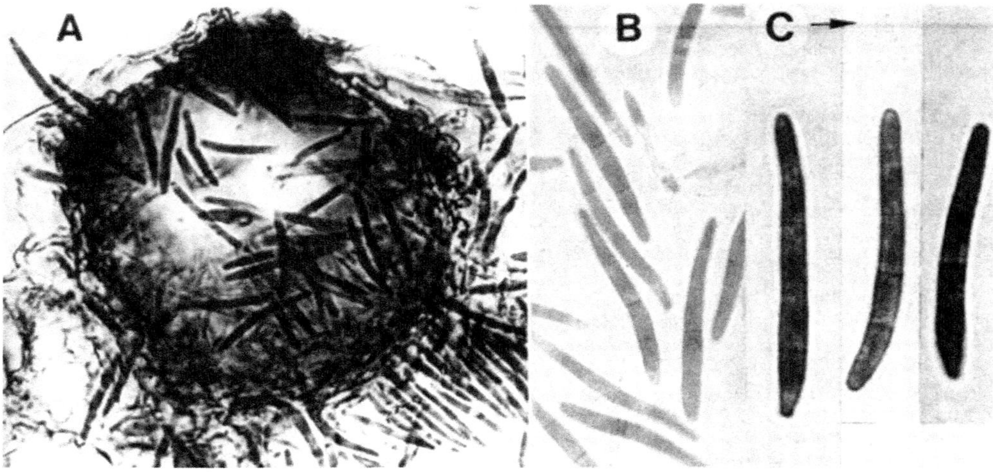

Fig. 4.10. (A) Vertical section of an immersed pycnidial conidioma, and (B) and (C) conidia of *Phaeoseptoria musae* (from CMI description no. 772).

Nothing is known about the disease's epidemiology or management.

Rust

This is generally a minor disease in the eastern hemisphere (Mulder and Holliday, 1971). It affects a wide range of AA, AAA, AAAA, AAB and ABB cultivars (Jones, 2000).

SYMPTOMS Small brown to black streaks develop primarily on the underside of older leaves. As they enlarge and coalesce, the associated tissues become chlorotic and then necrotic. Rusty brown masses of uredosori of the pathogens cover the lesion surfaces.

CAUSAL AGENTS Authenticated specimens of *Uromyces musae* are only known from Africa, and all collections in the Pacific islands have been referred to *Uredo musae*. (Firman, 1976).

Uredia of *Uromyces musae* are round and pulvinate (Mulder and Holliday, 1971). Urediospores are globose to subglobose or ellipsoid, finely echinulate, light brown, 20–28 × 17–24 μm, and have a 2.5 μm thick wall (Fig. 4.11). Telia are scattered or aggregated, oblong or ellipsoid, 0.5–1.0 mm in length to up to 3 mm when confluent. Teliospores are subglobose, ovoid or oblong, brown, 23–35 × 17–25 μm, and have a rounded or slightly acute apex. Pedicels are persistent, hyaline and ~60 μm in length.

The form genus *Uredo* includes all species in which only the uredinal stage is known (Alexopoulos *et al.*, 1996). *Uredo musae* differs from *Uromyces musae* in having more crowded uredia, urediospores with thinner walls (~1.5 μm) and more pronounced echinulations (Cummings, 1941).

EPIDEMIOLOGY AND MANAGEMENT Urediospores of *Uromyces musae* are wind-disseminated and germinate on wet leaf surfaces (Jones, 2000). No hosts other than *Musa* spp. are known. Control is usually not required, although severe outbreaks have occurred in Cavendish AAA plantations in Western Samoa.

Tropical speckle

This disease is found throughout the humid tropics. Although it affects younger leaves than the Sigatoka leaf spots (as young as the third leaf), it causes no economic damage or growth reduction.

SYMPTOMS Two types of symptoms have been described (Jones, 2000). One appears as

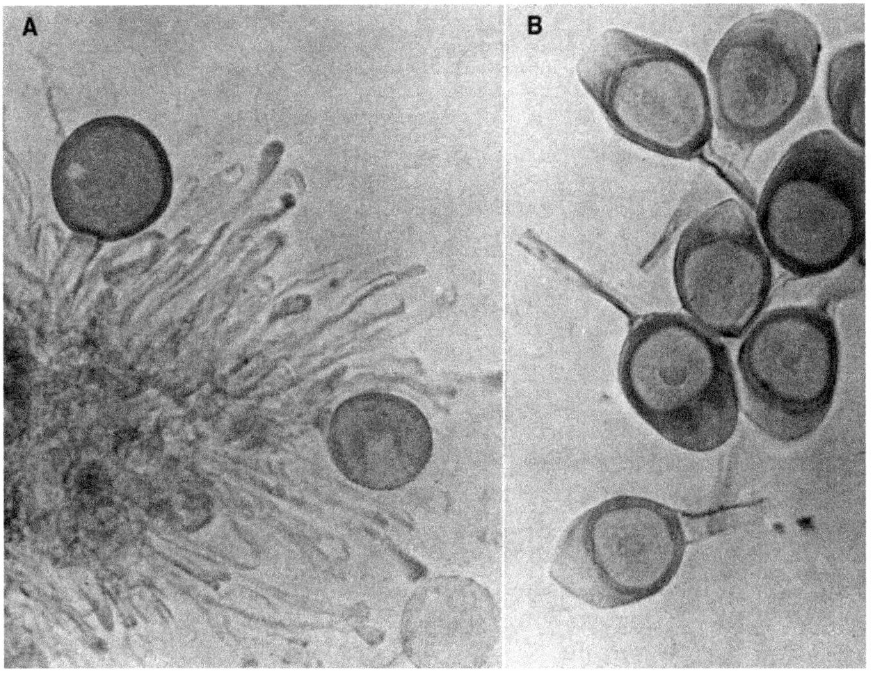

Fig. 4.11. (A) Urediospores and (B) teliospores of *Uromyces musae* (from CMI description no. 295).

circular blotches, ≤4 cm in diameter, which are chlorotic on the upper and tan on the lower leaf surface. Dense aggregations of dark brown or black pinpoint specks are evident on the upper surface, and erect groups of conidiophores of the causal fungus are visible when the lower surface is examined obliquely. Blotches can merge to form large tan areas, but the affected tissues generally are not killed.

The second symptom consists of dark grey to black patches on the lower leaf that, when viewed with a hand lens, are densely packed tiny black specks. Individual blotches are less distinct on the upper surface, and are smaller than those for the first symptom but can merge to involve large sections of the leaf. As for the first symptom, thick lawns of conidiophores are evident on the lower surface. Symptoms can also develop on leaf midribs, peduncles and tips of fingers, the latter of which can make 'Umalag' (AAA, Cavendish subgroup) fruit unmarketable in the Philippines.

CAUSAL AGENT There is some controversy over the taxonomy of the pathogen (Jones, 2000). Ellis (1971, 1976) indicated that two fungi were involved. *Periconiella musae* was associated with the second symptom described above, and *Veronaea musae* with the first. Although the fungi were quite similar, the conidiophores of *P. musae* were reported to be branched and those of *V. musae* were unbranched. However, de Hoog (1977) indicated that specimens of *V. musae* often produced branched conidiophores, and that the two fungi should be considered a single species, *Ramichloridium musae* (Fig. 4.12). In the absence of recent studies that clearly distinguish *P. musae* and *V. musae*, it will be assumed that *R. musae* is the correct epithet for this pathogen.

Conidia of *R. musae* are 5.5–8.5 × 2–2.6 μm, hyaline to subhyaline, thin-walled, ellipsoidal and have inconspicuous basal scars (de Hoog, 1977). Conidiogenous cells are vertical, pale brown, cylindrical and terminal on conidiophores that are up to 500 μm high, 1.8–2.5 μm wide, four- to ten-celled, golden brown and

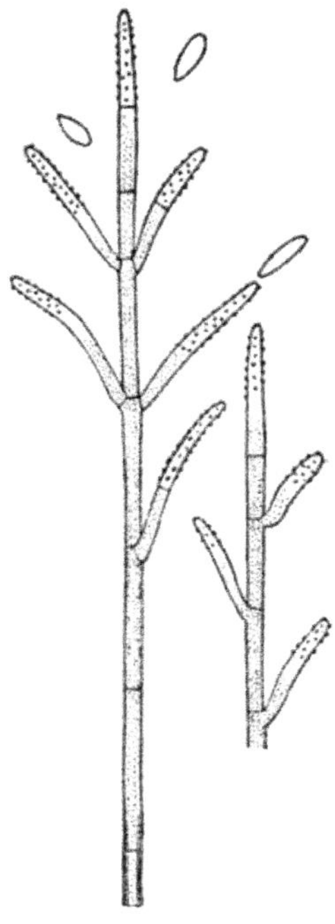

Fig. 4.12. Conidia and condiophores of *Ramichloridium musae* (from Ellis, 1971).

thick-walled. They arise from epiphyllic mycelium on the lower surface of leaves.

EPIDEMIOLOGY AND MANAGEMENT The pathogen is not very aggressive, and penetrates and kills only cells that are adjacent to stomata on the lower leaf surface. The disease affects diverse AA, AAA, AAB and ABB cultivars, as well as *M. schizocarpa* and *M. acuminata* ssp. *banksii*. Control measures are rarely needed.

Yellow Sigatoka

Yellow Sigatoka, which is also known as Sigatoka, was the most important leaf spot disease of banana before the spread of black Sigatoka (Jones, 2000). It also originated in the Southeast Asian/South Pacific region, but global epidemics of yellow Sigatoka began decades before those of black Sigatoka. By the 1950s, the disease had spread to most producing regions. Very few production areas (e.g. Canary Islands, Egypt and Israel) are free of the disease.

SYMPTOMS Pale green flecks, <1 mm long, become chlorotic streaks, 3–4 × 1 mm (Mulder and Holliday, 1974). They broaden, lengthen, turn brown to rusty, and become surrounded by chlorotic haloes (Plate 30). As they mature, they enlarge to 12–15 (up to 50) × 2–5 mm, darken and then become grey with a dark brown or black border. Coalescence of spots can kill large areas. In general, these symptoms resemble those of black Sigatoka and Eumusae leaf spot.

CAUSAL AGENT Yellow Sigatoka is caused by *Mycosphaerella musicola* (anamorph: *Pseudocercospora musae*). Conidia are pale olivaceous brown, cylindric to obclavate-cylindric, straight, curved or undulate, 10–80 (up to 100) × 2–6 μm, and four- or more celled (Fig. 4.13) (Mulder and Holliday, 1974). They have a rounded or obtuse apex and, unlike those of *M. fijiensis*, do not have a basal scar. The two species can also be distinguished with PCR (Johanson and Jeger, 1993). Conidia form singly and terminally on conidiophores that are densely packed in dark brown to black stroma, 15–35 μm in diameter, that erupt through stomata. Conidiophores are pale olivaceous brown, single-celled, borne terminally on stromatal hyphae, straight or slightly curved, rarely branched, narrow towards the apex, rounded, mostly ampulliform and 5–25 × 2–6 μm.

Pseudothecia are formed primarily on the upper leaf surface in mature spots. They are 36.8–72 μm in diameter, dark brown or black, and erumpent with a short ostiole. Asci are 14.4–18 × 3–4 μm, and bear eight two-celled ascospores.

EPIDEMIOLOGY Both conidia and ascospores are infective. Ascospores and conidia are formed under high moisture conditions, and are disseminated by wind and, in the case of conidia, also by rain and irrigation water. Due to their small size, ascospores are more important than conidia in spreading the disease long distances.

Major infection occurs above 21°C, and the candela and the first fully opened leaf are primary infection sites (Stover, 1972). Several distinct patterns of symptoms can develop depending on the infective spore and when

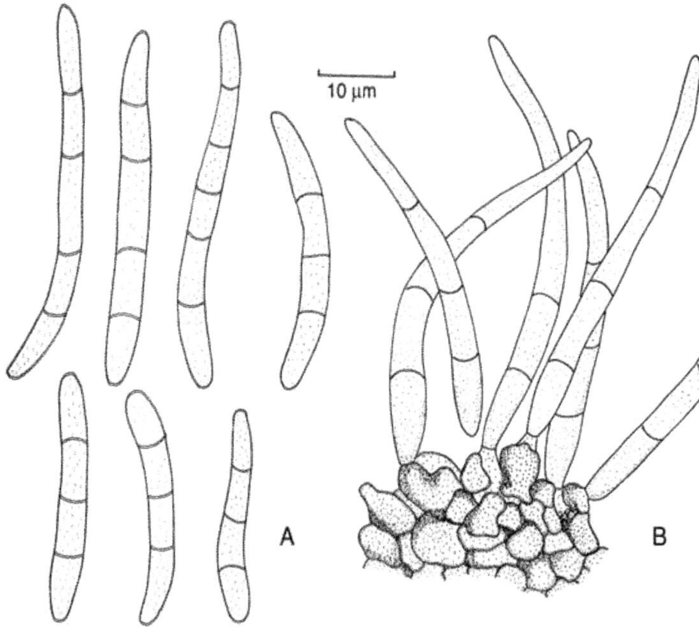

Fig. 4.13. Conidia and conidiophores of *Pseudocercospora musae*, anamorph of *Mycosphaerella musicola* (from CMI description no. 414).

infection occurs. Ascospore infection is evident as tip spotting where spots are confined to the apical third of the leaf, usually near the margins. When the opening, newly developed leaf is infected by conidia, line spotting occurs where lines of spots are found in more basal locations on the leaf. When streaks are evident on the left margin of the third and fourth leaf, infection by ascospores and conidia probably occurred during the candela stage. Finally, infection of the first leaf results in a more random spotting over the entire leaf.

Ascospore and conidium germination is sensitive to temperature. Optimum ranges are 25–26°C for ascospores and 25–29°C for conidia, and germination rates decline dramatically once temperatures approach 20°C. Optimum rates for the growth of ascospore germ tubes, 25°C, is ~2°C lower than for *M. fijiensis*, and this is thought to explain at least partially the prevalence of yellow Sigatoka at higher elevations.

Disease development is enhanced when infection densities and light intensities are high. On susceptible cultivars, the time between spore germination and the appearance of symptoms can range from 11 to >100 days, and between streak development and their maturation into brown spots from 2 to >100 days.

MANAGEMENT Wide ranges of susceptibility occur among different cultivars, and resistance increases as the relative proportion of the B genome increases (Table 4.3). However, there are exceptional situations where AA and AAA clones are resistant. Although the AAB plantains are resistant at low elevations, they are susceptible at high elevation in Puerto Rico, Colombia, West Africa and elsewhere.

The history of chemical control of this disease is found in the section on black Sigatoka. The same chemistries and strategies that are used against black Sigatoka are also highly effective against yellow Sigatoka.

Diseases of Fruit

Anthracnose

Anthracnose is one of the most common and important diseases of banana fruit (Jones,

2000). It is primarily a postharvest disease that affects the peel and crown surface area (crown rot), but can also develop when green fruit are injured.

SYMPTOMS Small brown spots begin to appear on all parts of the peel as fruit ripen. These expand and coalesce to form large depressed, black to brown areas of decay that often are covered with orange to salmon-coloured masses of conidia of the causal fungus (Plate 31). The pulp is affected only after fruit become overripe.

Lesions that develop on wounded green fruit are black and in the shape of the wound. These lesions may be surrounded by chlorotic haloes, and can invade the pulp under warm conditions.

CAUSAL AGENT The causal fungus, *Colletotrichum musae*, resembles in many ways the more common cause of anthracnose on other fruit, *C. gloeosporioides*. The identity and correct name of its teleomorph is confused. Sutton (1992) indicated that there is some doubt over whether *Glomerella musarum*, as described by Petch in 1917, is the actual teleomorph of *C. musae*. A more recent report in which the teleomorph was produced in the laboratory used the epithet *Glomerella musae*, although no formal description of it was made (Rodriguez and Owen, 1992). Until such time, *Glomerella* sp. should be used when referencing the teleomorph (personal communication, Amy Rossman).

C. musae can be distinguished from *C. gloeosporioides* by its longer and wider conidia and faster growth rate on artificial media at 24°C. *C. gloeosporioides*/*G. cingulata* can be recovered from decaying bananas, but *C. musae* has adapted to, and is the most predominant cause of anthracnose on, banana.

Conidia of *C. musae* are hyaline, single-celled, elliptical to oval and 11–17 × 3–6 μm (Fig. 4.14).

EPIDEMIOLOGY *C. musae* is a common component of the microflora within the banana

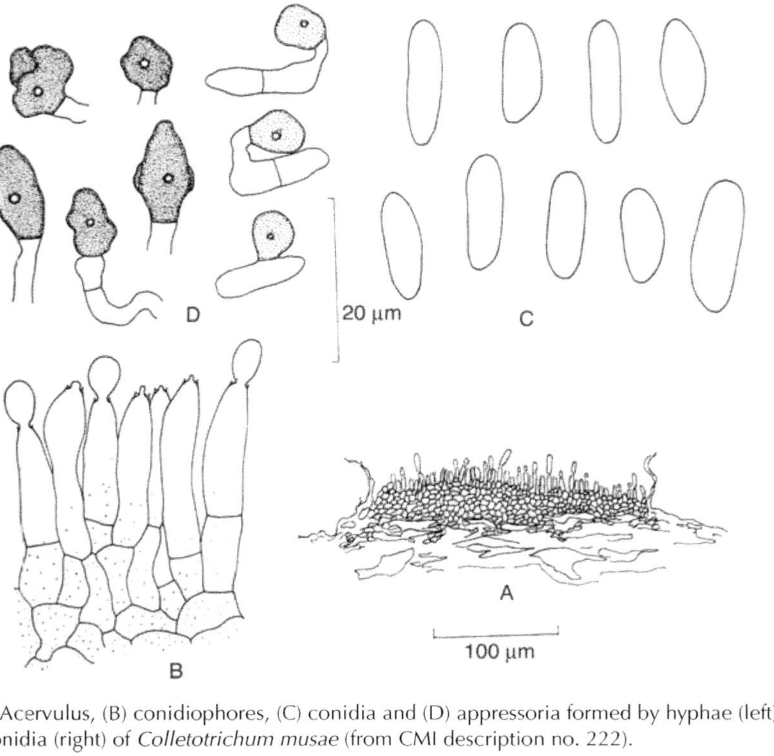

Fig. 4.14. (A) Acervulus, (B) conidiophores, (C) conidia and (D) appressoria formed by hyphae (left) and germinated conidia (right) of *Colletotrichum musae* (from CMI description no. 222).

canopy. Work in Guadeloupe indicated that floral parts and bracts are most important sources of inoculum, and that a threefold reduction in the severity of this disease was realized when these sources were removed at flowering (de Lapeyre de Bellaire *et al.*, 2000). Conidia are formed under wet conditions and are dispersed in rain, primarily within the same plant or bunch. Rain is also required for infection. In Guadeloupe, disease severity was significantly correlated with cumulative rainfall during the first 35 days after flowering, and the disease was controlled completely if fruit were protected from rain with covers.

C. musae forms melanized appressoria on the fruit peel shortly after conidia germinate. These infections usually remain latent until ripening begins.

MANAGEMENT Anthracnose can be a serious problem when shipped fruit ripen prematurely in the ship's hold. Minute quantities of ethylene, that are produced by both the pathogen and the host, initiate the climacteric ripening process. The importance of ethylene control and shipping green fruit from plantations in which the Sigatoka leaf spots have been well controlled was mentioned in the introduction to this chapter.

Removal of flower debris and other tissues that harbour the pathogen is beneficial, as is the protection of fruit from injury. In export production, fruit are routinely washed and dipped in fungicide before they are packed and cooled for shipment. Where such measures are not available (e.g. most domestic and smallholder production), anthracnose can cause significant losses.

Botryodiplodia finger rot

This postharvest decay is common when shipped bananas are in transit for more than 14 days (Stover, 1972). The disease has been reported from Central and South America, the Caribbean, India, Taiwan and the Philippines (Ploetz *et al.*, 1994b).

SYMPTOMS Symptoms usually begin at the flower-end of the finger or at a wound site (Fig. 4.15) (Ploetz *et al.*, 1994b). The decay spreads uniformly, causing a brownish black discoloration of the peel and a softening of the pulp. The affected area of the peel becomes wrinkled and encrusted with pycnidia of the pathogen. The pulp is reduced to a soft, rotten mass and a dark grey mould grows on the peel surface under high humidity. The rate of disease development increases during ripening and can spread to adjacent fingers. Infected clusters tend to ripen prematurely, and fully mature fruit are most susceptible. Microscopic examination of spores may be necessary to distinguish this disease from tip rot caused by *Botryosphaeria dothidea* (anamorph: *Fusicoccum aesculi*).

CAUSAL AGENT The disease's common name comes from the former name of the pathogen's anamorph, *Botryodiplodia theobromae*. *Botryosphaeria rhodina* (anamorph: *Diplodia theobromae*) is described in Chapter 1.

EPIDEMIOLOGY *B. rhodina* is a common inhabitant of decaying banana trash (Ploetz *et al.*, 1994b). Conidia are disseminated by

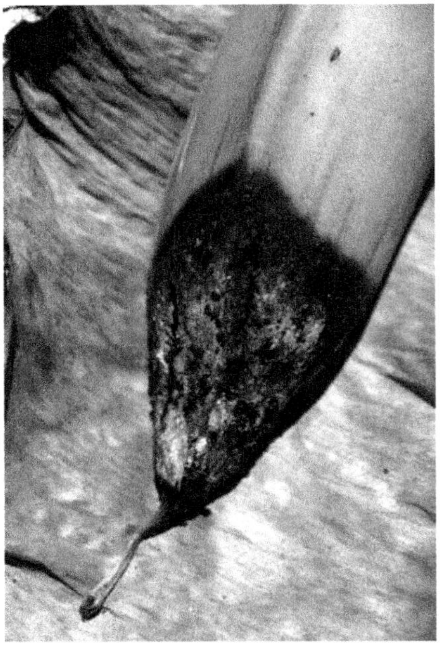

Fig. 4.15. Symptoms of Botryodiplodia finger rot at the flower end of a 'Bluggoe' fruit (photo: R.C. Ploetz, UF).

wind and water, and infection occurs through tissues at the flower-end of fingers and wounds. The fungus grows very slowly at <20°C, and optimum growth and rotting occurs between 25 and 30°C.

MANAGEMENT This disease can be controlled by minimizing fruit injury, treating with systemic fungicides, rapidly reducing fruit temperatures following harvest and excluding over-mature fruit from shipment. Reducing the caliper (grade) and age of fruit that are shipped will probably suppress disease development when transit times exceed 14 days. Measures utilized against crown rot are also helpful.

Brown spot

Brown spot was first described in 1965, but was probably present in commercial plantations much earlier (Kaiser and Lukezec, 1965). Brown spot was common during warm, rainy weather in the western hemisphere, where its presence varied greatly but was most severe in Mexico, Honduras and Guatemala. Although losses of entire bunches and up to 20% of the hands occurred at packing stations in these countries, brown spot is no longer important. Strategies used against pitting disease are effective.

Brown spot occurs on peduncles, fruit crowns and fingers. Spotting is generally more prevalent on inner whorl fingers, and only occurs >50 days after fruit emergence. The spots are irregular, light to dark brown with irregular margins, average 5–6 mm in diameter, and are surrounded by a water-soaked halo. They are not as sunken as those caused by pitting disease and do not increase in size or number during ripening. Brown spots are centred on stomata and no aerial mycelium or fructifications are produced within the spot.

Brown spot is caused by *Cercospora hayi*. Conidia are hyaline, 3–4 × 90–150 μm, with truncate bases and acute tips and more than five-celled. Dead banana and weed foliage are the principal sources of inoculum, and conidia form on dead leaf trash within 16 h at 23–26°C and saturated humidities (Kaiser and Lukezic, 1966a). Conidia are disseminated by wind, and can remain viable for at least 5 weeks when exposed to fluctuating humidity and temperature. *C. hayi* can also survive in dried foliar tissues for 15 weeks as mycelium. Spore release peaks during afternoon hours that usually coincide with the highest daily wind velocities. Spore densities and spotting incidence are the highest during periods with high rainfall.

Cigar-end rot

Cigar-end rot is an economically important disease in Central and West Africa (Ploetz *et al.*, 1994b). It also occurs in the Canary Islands, Egypt, India, Iran, South Africa, South America and the West Indies.

SYMPTOMS One or all fingers on a hand may be affected (Fig. 4.16). The first symptoms are a localized darkening and wrinkling of the peel at the tip (Ploetz *et al.*, 1994b). A black band and a narrow chlorotic region between infected and healthy tissues border the darkened area. In Trachysphaera tip-rot, the surface of the lesion becomes covered with white spores that turn pink or brown as they mature, giving the finger tip

Fig. 4.16. Cigar-end rot on fingers of 'Dwarf Cavendish' AAA (photo: R.C. Ploetz, UF).

the greyish, ashen appearance usually associated with cigar-end rot. The pulp undergoes a dry rot and becomes mummified. A wet rot can occur when secondary organisms are also present. In Verticillium tip-rot, the pulp is characteristically dry and fibrous with grey, powdery spore masses occurring on the lesion.

CAUSAL AGENTS *Trachysphaera fructigena* and *Verticillium theobromae* cause cigar-end rot in Central and West Africa (Ploetz *et al.*, 1994b). *T. fructigena* has not been reported in the western hemisphere, but *V. theobromae* has been reported in both hemispheres.

Conidiophores of *T. fructigena* are erect, usually have a terminal vesicle, and conidia arise singly or in whorls from its apex (Holliday, 1970). Conidia are 13–48 μm in diameter, echinulate, spherical, hyaline and borne on 10–30 μm long pedicels. Oogonia are somewhat pyriform and 24 × 40 μm. Oogonial walls are thick and have irregular, sac-like outgrowths (Fig. 4.17). The antheridia are amphigynous and completely surround the oogonial stalk.

Solitary or small groups of conidiophores of *V. theobromae* are produced on infected tissues. Conidia are hyaline, ellipsoidal to subcylindrical, 3–8 × 1.5–3 μm, and are borne terminally on tapering phialides and aggregate into rounded, mucilaginous, transluscent heads (Fig. 4.18) (Hawksworth and Holliday, 1970). Conidiophores are erect, hyaline to brownish, verticillately branched and 4–6 × 150–400 μm; phialides are 14–37 × 1.5–5 μm.

EPIDEMIOLOGY The frequency of cigar-end rot increases during periods of high humidity and rainfall. Conidia of *V. theobromae* are wind disseminated and infect dying flower parts (Meredith, 1965). *V. theobromae* is a common colonizer of banana flowers and leaf trash. The source of *T. fructigena* inoculum is unknown. In West Africa, cigar-end rot incidence is highest along plantation borders and on plants grown at higher elevations (Tezenas du Montcel and Laville, 1977). Optimum growth of *T. fructigena* occurs at 24°C, whereas cigar-end rot caused by the fungus is favoured by moderate (20°C) followed by higher (<27°C) temperatures. Optimal growth of *V. theobromae* occurs at 25°C.

In fruit infected only with *V. theobromae*, premature ripening and postharvest rotting do not occur. In contrast, fruit infected with *T. fructigena* continues to rot after harvest. New *T. fructigena* infections can occur either in dehanding and delatexing tanks or in contaminated ripening rooms. These new infections typically occur at freshly cut crown surfaces and peel injury sites caused by improper handling.

MANAGEMENT Cigar-end rot control begins in the field with the frequent removal of dead

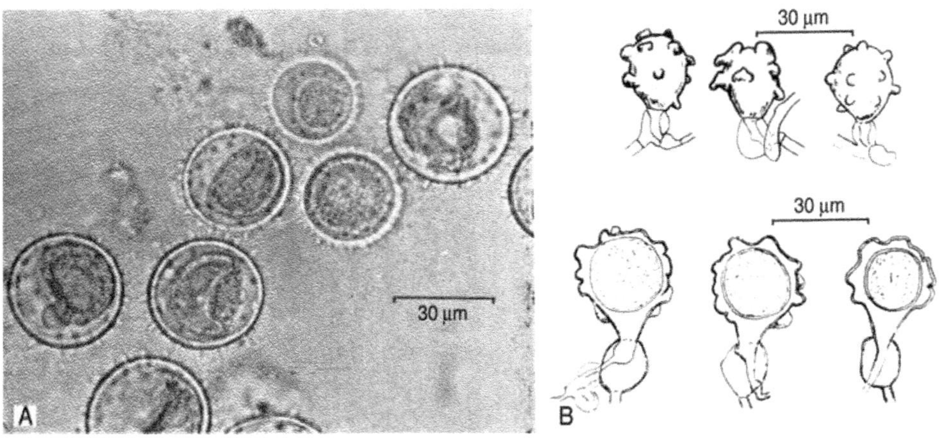

Fig. 4.17. (A) Oospores and (B) oogonia and antheridia of *Trachysphaera fructigena* (from CMI description no. 229).

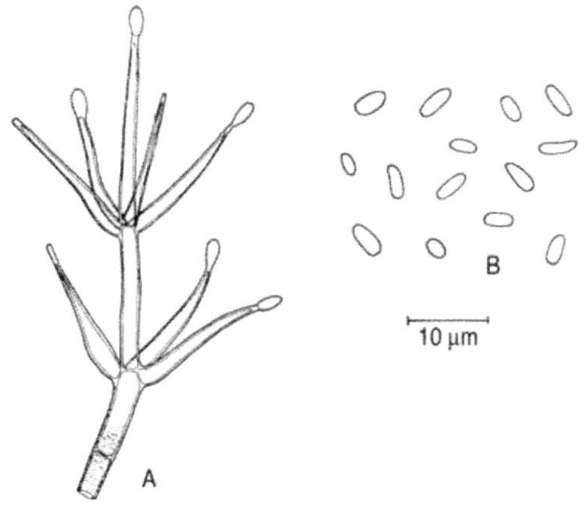

Fig. 4.18. (A) Verticillate conidiophore and (B) conidia of *Verticillium theobromae* (from CMI description no. 259).

flowers followed by bagging bunches with perforated polyethylene sleeves (Ploetz *et al.*, 1994b). Bracts and dead flower parts accumulate in the fruit bags and should be removed a few weeks after bagging. Field sanitation, as practised for fruit spot control, is helpful in reducing *V. theobromae* inoculum pressure and subsequent cigar-end rot. Packing station sanitation is essential to reduce postharvest infection and rot caused by *T. fructigena*. Infected fruit should be culled prior to dehanding to avoid contamination of dehanding and delatexing tank water with *T. fructigena* spores. Fungicide sprays may be necessary during some peak cigar-end rot seasons (Tezenas du Montcel, 1981).

Crown and pedicel rot

Crown and pedicel rot can be major problems on exported fruit, and are present in all banana-growing regions (Jones, 2000). These diseases were not important when the export trades relied on fruit of 'Gros Michel' AAA since these fruit were shipped as intact bunches. With the trades' conversion to the Cavendish AAA cultivars in the 1960s came the need to remove and pack individually hands of its more fragile fruit, thereby opening the crown area to invasion by various fungi.

Crown and pedicel rot cause significant aesthetic damage and drop of fingers. In addition, some of the causal fungi have been reported to produce mycotoxins (Jiménez *et al.*, 1997) or cause changes in the quantity and quality of soluble sugars in fruit (Odebode and Sanusi, 1996). Damage is most severe when fruit are in transit for more than 14 days. Losses can be substantial, especially in Europe where fruit is displayed on hooks.

SYMPTOMS Crown rot is first observed on fruit that are in transit for more than 7 days. It begins as a softening and blackening of tissues at the cut crown surface (Plate 31). Greyish white, pink or white mould that may be present on the cut surface at this early stage is known as crown mould. Crown rot spreads rapidly during ripening, and can spread into the pedicel, and ultimately the pulp when severe or if *Ceratocystis paradoxa* is present.

A water-soaked, dark band of creased and bruised tissue becomes evident on pedicels that are flexed excessively during handling and transit. The crease turns black, enlarges and withers as pedicel rot progresses. When *Colletotrichum musae* is present, pink spore masses may appear on the blackened tissue under moist conditions. Pedicel rot can also result from infection at mechanical injury sites, peduncle rot and severe pitting disease.

CAUSAL AGENTS A complex of fungi causes crown rot (Greene and Goos, 1963; Kaiser and Lukezic, 1966b; Griffee, 1976; Knight *et al.*, 1977; Johanson and Blazquez, 1992). The pathogens vary based on location, time of year and other factors. *C. musae* (Fig. 4.14) is most important, but the following species have also been indicted: *Acremonium* sp., *C. paradoxa*, *Diplodia theobromae*, *Fusarium moniliforme*, *F. pallidoroseum*, *F. subglutinans* (Fig. 4.19), *Nigrospora sphaerica* and *V. theobromae* (Fig. 4.18). Snowden (1990) indicated that *N. sphaerica* and *N. oryzae* (Fig. 4.20) were considered by some to be synonymous. Recently, O'Donnell *et al.* (1998) redescribed isolates of *F. subglutinans* from banana as a new species, *F. concentricum*, based on DNA sequence analyses.

The mycotoxigenic species include *F. concentricum*, *F. moniliforme* and *F. subglutinans* (reports of the latter species may be of *F.*

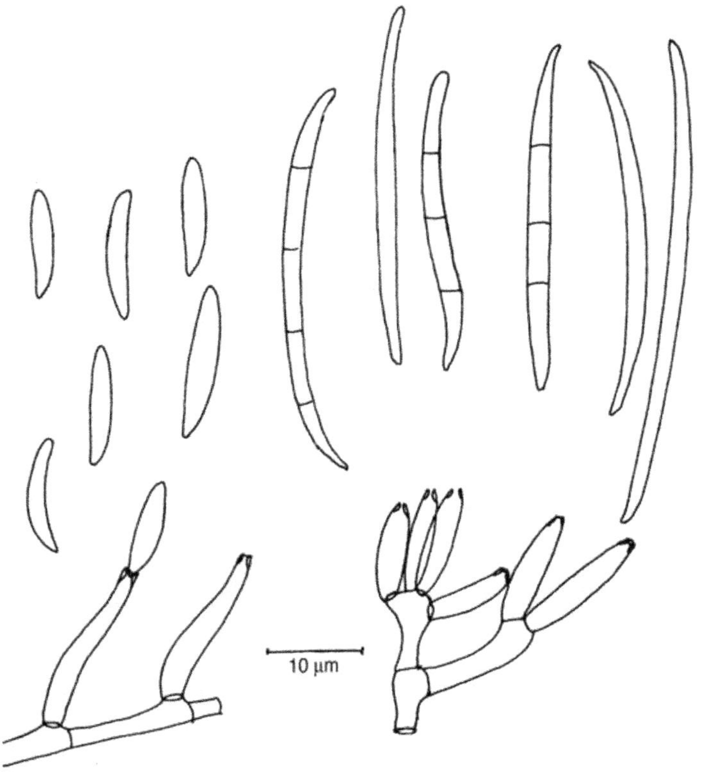

Fig. 4.19. Counterclockwise from upper left: microconidia, macroconidia and conidiophores of *Fusarium subglutinans* (from CMI description no. 23).

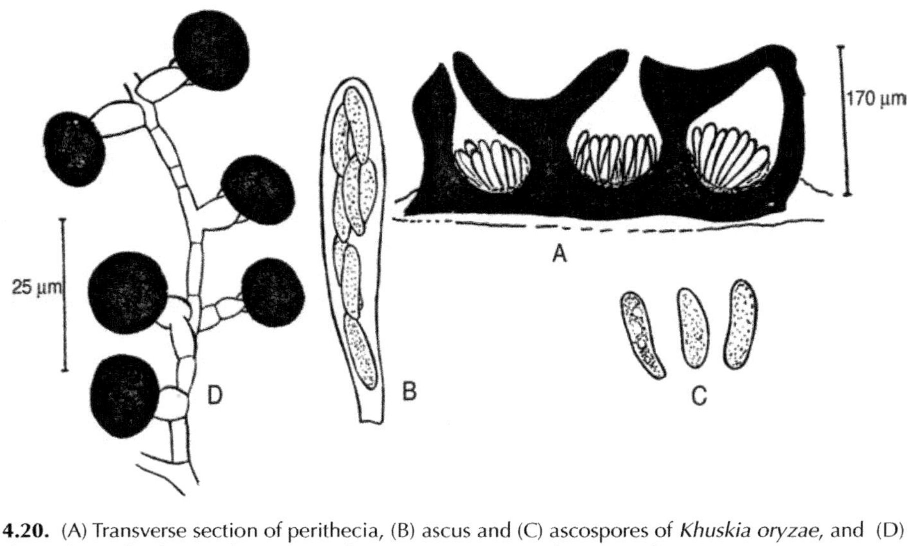

Fig. 4.20. (A) Transverse section of perithecia, (B) ascus and (C) ascospores of *Khuskia oryzae*, and (D) conidia and conidiophores of its anamorph, *Nigrospora oryzae* (from CMI description no. 311).

concentricum) (Chakrabarti and Ghosal, 1986; Jiménez *et al.*, 1997; Logrieco *et al.*, 1998). Unfortunately, toxigenicity of these species has been tested in maize cultures, rather than banana. Whether these species produce mycotoxins in banana and colonize fruit to the extent that they represent *bone fide* human health risks has not been investigated.

EPIDEMIOLOGY The causal fungi are common colonizers of banana leaves, flowers, bracts and transitional leaves (Stover, 1972). They sporulate on decaying debris and are disseminated by wind and water-splash to all parts of the fruit. Infection occurs primarily at the cut surfaces of crowns during either dehanding with contaminated knives or contact with infested wash water. Spores from fruit surfaces and decaying flower parts that remain attached to the fingers tend to accumulate in wash water. Spores present in the wash water can be drawn several millimetres into the vascular system at the wound site where they germinate and cause a rot (Greene and Goos, 1963). Other spores may germinate at the cut surface and invade the adjacent cells. Susceptibility to crown rot is increased by desiccation of crowns, poor crown-trimming techniques and poor dehanding practices that result in crushed subsurface crown tissues. The use of dull knives during trimming causes ragged edges and favours entry of crown rot organisms into adjacent, sound tissue.

Crown rot incidence and severity vary depending on the organisms that gain entry and climatic conditions, and certain groups of fungi and bacteria appear to act synergistically in boxed bananas. Hot, dry conditions prior to harvest tend to favour crown rot development.

MANAGEMENT Crown rot control begins in the field with good sanitation, as practised for pitting disease control (Stover, 1972). Good hygiene in packing stations and clean water in dehanding and delatexing tanks are essential. Trimmed crowns should have bevelled edges since sharp edges are sensitive to handling injury and can result in increased disease in otherwise good quality fruit. Prompt (<48 h after harvest) and rapid fruit temperature reduction also aids crown rot control.

Diverse measures for managing this disease were reviewed recently (Krauss and Johanson, 2000). Postharvest treatment with fungicides usually is essential for exported bananas (Shillingford and Sinclair, 1978). Fungicides can be applied with dips, sprays or recirculating drenches. Those approved for use depend on the country of destination. Fruit destined for the USA and Europe can be treated with thiabendazole or imazalil. Other measures that reduce disease development include hot water (50°C) dips for 20 min, retrimming the crown surface once it is in the packing house, and using controlled atmospheres, various fruit coatings or natural products (Reyes *et al.*, 1998; Krauss and Johanson, 2000). In general, control with the later products is either inconsistent or does not achieve the levels realized with fungicides.

Deightoniella fruit speckle and black tip

Deightoniella torulosa causes Deightoniella speckle, which is also known as swamp spot, and black tip, which is also known as tip-end rot (Stover, 1972). The same fungus also causes Deightoniella leaf spot. All three diseases are relatively minor problems wherever banana is grown.

Speckles occur on fruit at all stages of maturity and consist of reddish brown to black spots, up to 2 mm in diameter, with a dark green halo. The disease is important only during the wet season, and control measures for pitting disease are also effective against it. Cultural practices that reduce moisture in the plantation are also helpful.

Black tip symptoms consist of a slowly advancing black lesion at the flower-end of one or more fingers. A single side of the finger is usually affected, and the diseased area is bordered by a pale yellow or narrow, grey margin. Old lesions tend to rupture and a pale brown mould can develop under moist conditions. *V. theobromae* can infect fruit through black tip lesions and transform black tip into cigar-end rot. Black tip can be controlled by removing leaf trash and improving drainage to reduce humidity in the field.

D. torulosa is a common component of the airborne microflora within plantations. Conidiophores arise singly or in small groups from hyphae or swollen cells (Fig. 4.21) (Subrumanian, 1968). They are septate, brown, 6–10 μm wide, are swollen to 13–16 μm in diameter at the apex, and can have up to six successive proliferations of the same dimension. Conidia are blastospores and are produced singly. They are straight or slightly curved, obpyriform to obclavate, subhyaline to smoky olive, usually four- to six-celled, and 35–70 × 13–25 μm.

Diamond spot

Diamond spot was first described in 1968 and was common in parts of the Americas and the Philippines (Berg, 1968). It is no longer an important disease.

Diamond spot first appears on the peel of green fruit as slightly raised, yellow spots, 3–5 mm in diameter (Fig. 4.22). Since infected cells do not grow at the same rate as adjacent healthy cells, a longitudinal crack develops through the spot that is surrounded by a yellow halo. The tissue exposed by the crack and the yellow halo becomes necrotic, collapses, turns black and appears as a sunken, diamond-shaped lesion, 1.0–3.5 × 0.5–1.5 cm. Large spots occasionally expose the pulp. Diamond spots begin to appear as the fruit approaches harvest grade, and latent infection is common. Lesion size and disease incidence can increase during transit and ripening.

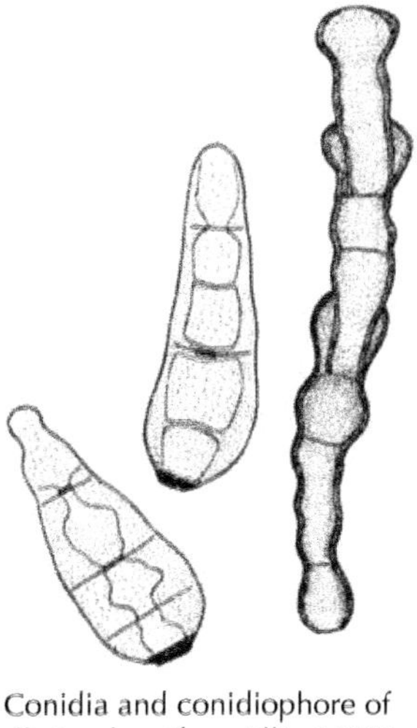

Fig. 4.21. Conidia and conidiophore of *Deightoniella torulosa* (from Ellis, 1971).

A complex of fungi causes diamond spot. A strain of *C. hayi*, which differs from that involved with brown spot, is accompanied by *F. pallidoroseum*, *F. solani* and, less commonly, other *Fusarium* species. The species in the complex are prevalent on decaying and dead leaf trash where they sporulate under moist conditions. They are part of the airborne microflora commonly found in banana plantations.

Latent infections are common, and lesion development frequently occurs during transit and ripening. The same procedures used against pitting disease control diamond spot.

Pitting

Pitting disease, which is also known as Johnston spot, is found in Asia, Africa, Australia, the Canary Islands, the Caribbean, and Central and South America (Stover, 1972; Snowdon, 1990). Losses of up to 50% of the fruit in packing stations occurred in some Central American areas in the 1960s. Pitting was once considered the most important fruit spot disease in Central America, but is now less important. Cavendish cultivars are more susceptible than 'Gros Michel' AAA and plantains.

SYMPTOMS Small, reddish sunken spots develop on maturing fruit (Stover, 1972; Snowdon, 1990). As fruit begin to ripen, they enlarge to 4–6 mm in diameter, become brown and are surrounded by a reddish brown border. The fruit pulp is not damaged even when lesion centres crack. Smaller spots that develop on the pedicle and crown can cause finger drop. Lesions can also develop on leaves of young water suckers and transition bracts. They are shallow, sunken and larger than those on fruit and, unlike those on fruit, support sporulation of the pathogen.

CAUSAL AGENT Pitting disease is caused by *Pyricularia grisea* (its teleomorph,

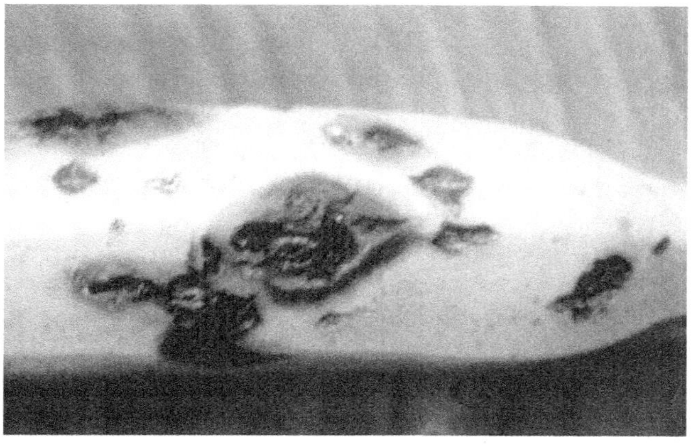

Fig. 4.22. Severe symptoms of diamond spot in the Philippines (photo: R.H. Stover).

Magnaporthe grisea, is uncommon) (Stover, 1972). It is isolated readily from pitting disease lesions or as conidia from recently collapsed, dry banana foliage when placed in a moist chamber.

Conidia are attached singly at the broader end in scarpioid cymes, ovate to pyriform with a small basal apiculus, three-celled, essentially hyaline and 17–19 × 6.5–8.5 μm (Fig. 4.23). Although *P. grisea* closely resembles *P. oryzae* (a rice pathogen), these fungi are host specific (Meredith, 1963). *P. grisea* sporulates poorly on most agar media, but sporulates abundantly on autoclaved *Commelina erecta* leaves.

EPIDEMIOLOGY Pitting disease is most prevalent during periods of high rainfall. The principal source of inoculum is dead leaves and bracts. Conidia are disseminated by wind currents and are present throughout the year in Central American banana plantations. They germinate and form appressoria on green fruit within 4–8 h under high humidity. The optimum temperature for appressorial formation and infection is 24–26°C.

In the laboratory, lesion development can occur on green fruit within 2–3 weeks of inoculation. In contrast, fruit in the field seldom develop lesions until 3 weeks before harvest. The fungus may remain dormant in infected tissues until the fruit nears maturity or until after harvest. Latent infections are common and can result in unacceptable levels of pitting disease during transit and ripening. However, since severe pitting occurs only after prolonged periods of high

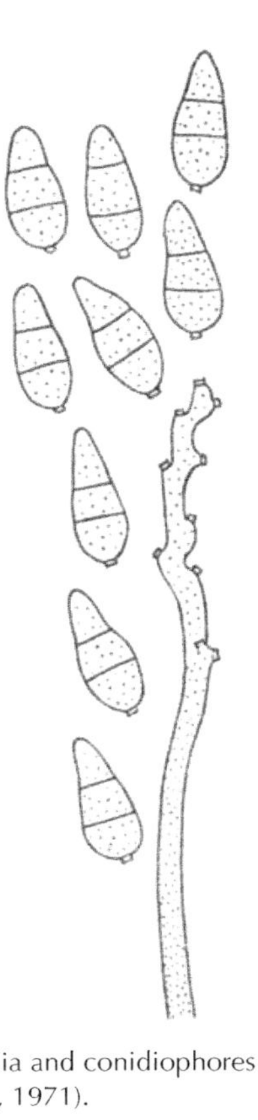

Fig. 4.23. Conidia and conidiophores of *Pyricularia grisea* (from Ellis, 1971).

rainfall and where control measures are inadequate, it has become a minor problem in export plantations where intensive control of foliar diseases is practised.

MANAGEMENT All symptomatic fruit are culled at packing stations. Frequent removal of inoculum reservoirs in the field is essential (Halmos, 1970). All collapsed and dying banana foliage, transition leaves and bracts should be removed at regular intervals during the rainy weather. Fruit can be protected with fungicide sprays prior to bagging with perforated polyethylene film or with perforated polyethylene film dusted with fungicide. Fungicides used for fruit spot control depend on tolerances established in import countries; maneb, mancozeb, benomyl and thiophanate methyl are common (Guyon, 1970).

Sooty blotch and sooty mould

These diseases produce similar diffuse, smoky grey to black areas on the finger and crown surface. They are superficial, do not affect the pulp and often are not distinguished (e.g. Stover, 1972). Sooty mould requires excreta (honeydew) of mealybugs and aphids to develop, whereas sooty blotch does not.

Sooty blotch occurs in Australia, Costa Rica and probably other locations (Jones *et al.*, 1993; T. Sutton, NCSU, personal communication, 2001; R. Ploetz, personal observations). It is caused by *Chaetothyrina musarum* and does not warrant specific management.

Stover (1972, 1975) reported that the most common cause of sooty mould in the Americas and the Philippines was *Cladosporium cladosporioides*. It occurs most frequently during cool, rainy weather, is controlled with insecticides or by covering bunches with insecticide-impregnated polyethylene bags, and is described in Chapter 6.

Squirter disease

This disease occurs when single fingers are packed (Stover, 1972; Snowdon, 1990). It is caused by *Khushia oryzae* (anamorph: *Nigrospora oryzae*) (Fig. 4.20). The pathogen infects the cut pedicel surface and proceeds to colonize and liquefy the pulp. Externally, the peel becomes bluish tan during ripening. Packing fruit only in hands and treatment with standard postharvest fungicides controls the disease.

Diseases of the Rhizome and Pseudostem

Cylindrocladium root rot

This disease can cause significant root damage in its own right, but is most serious when it interacts with *Radophlous similis*, the burrowing nematode (Jones, 2000). It has been reported on several different cultivars in Cameroon, Costa Rica, Côte d'Ivoire, Guadeloupe and Martinique.

Cortical tissues of banana roots are killed and blackened by at least two species, *Cylindrocladium gracile* and *Calonectria spathiphylli* (anamorph: *Cylindrocladium spathiphylli*). *Cy. gracile* has no known teleomorph. Macroconidiophores consist of a stipe, penicilliate fertile branches, a stipe extension and a terminal vesicle (Fig. 4.24) (Crous, 2002). Stipes are septate, hyaline, smooth and 50–150 × 5–6 μm. The stipe extensions are septate, straight to flexuous, 140–350 μm long, 2.5–3 μm at the apical septum above which is a clavate 2–6 μm wide vesicle. The conidiogenous apparatus is 35–75 long, 15–60 wide with 10–25 × 3–5 μm single- or two-celled primary branches and 10–18 × 3–5 μm single-celled secondary branches. Conidia are cylindrical, rounded at the ends, 38–65 × 3.5–6 μm and straight.

Ca. spathiphylli produces orange to red perithecia that are subglobose to ovoid, 380–655 μm high and 340–650 μm in diameter (Fig. 4.25) (Crous, 2002). Asci are two- to eight-spored, clavate and 120–230 × 7–25 μm, and ascospores are aggregated in the upper third of the ascus, are hyaline, guttulate, fusoid with rounded ends, one- to three-septate and 22–65 × 3–7 μm. Macroconidiophores are composed of a stipe, penicilliate fertile branches, a stipe extension and a terminal vesicle. Stipes are septate, hyaline, smooth and 120–150 × 6–8 μm. The stipe extensions are septate, straight to flexuous,

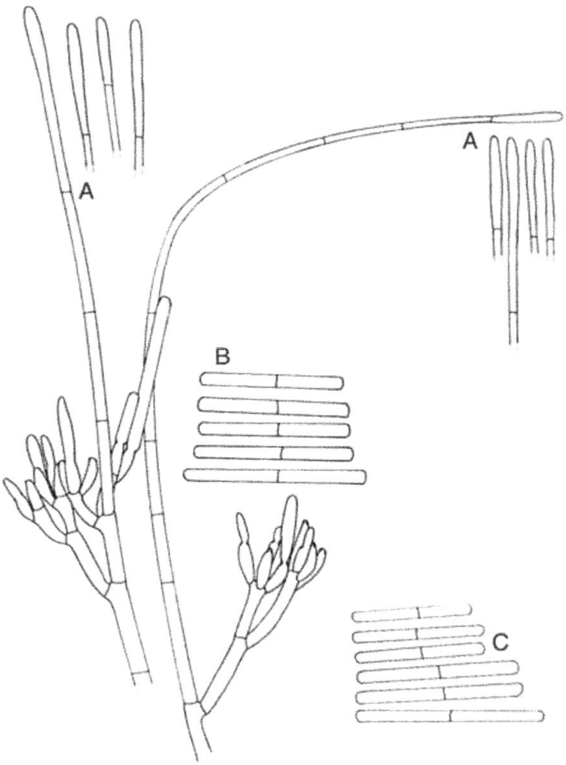

Fig. 4.24. (A) Macroconidiophores and vesicles and (B) conidia of *Cylindrocladium gracile* (from CMI description no. 1152).

170–326 μm long, 3–4 μm at the apical septum and terminate in a globose or ellipsoid to obpyriform 8–15 μm wide vesicle. The conidiogenous apparatus is 60–150 μm long, 40–90 wide with 18–40 × 4–6 μm single- or two-celled primary branches and 18–30 × 4–6 μm single-celled secondary branches. Conidia are cylindrical, rounded at the ends, 45–120 × 5–7 μm and straight.

Risède and Simoneau (2001) examined the morphology and genetic variation of isolates from banana in Cameroon, the Caribbean and Costa Rica. RFLP analyses of the intergenic spacer region (IGS) confirmed the presence of *Cy. gracile* and *Ca. spathiphylli*, and the *Calonectria* teleomorph was formed in crosses of *Ca. spathiphylli* isolates, but not *Cy. gracile*.

Two additional species have been reported on *Musa*. Risède (1994) tentatively identified isolates from Cameroon, the Caribbean and Costa Rica as *Cy. pteridis* (Fig. 4.26). Although Risède and Simoneau (2001) later concluded that these isolates were more appropriately assigned to *Cy. gracile*, Crous (2002) indicated that this conclusion should be re-examined since the isolates were larger, 69–86 × 5.6–6.2 μm, than was typical for *Cy. gracile* (conidia of *Cy. pteridis* are 50–130 μm × 4–6 μm). *Cy. musae* was reported from Costa Rica as a new species (Semer *et al.*, 1987), but recent results indicate that it is synonymous with *Cy. spathiphylli* (Risède and Simoneau, 2001).

The epidemiology and management of this disease have not been reported.

Marasmiellus rot

This widespread disease occurs in neglected plantings, and in sandy, gravelly or poorly

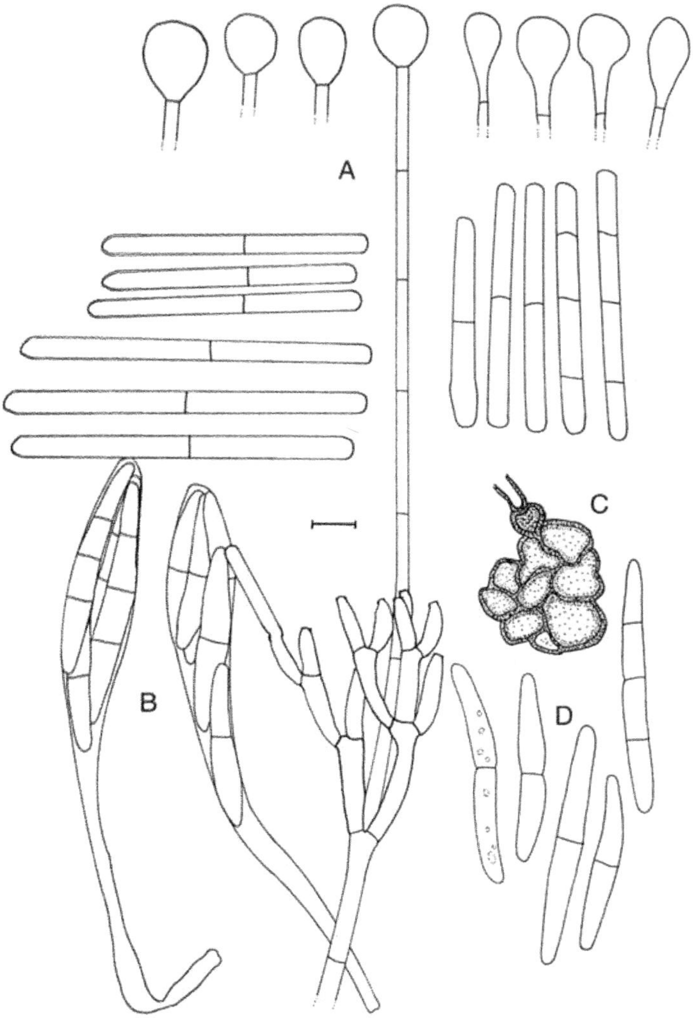

Fig. 4.25. (A) Conidiophore, vesicles and conidia, and (C) chlamydospores of *Cylindrocladium spathiphylli*, and (B) asci and (D) ascospores of its teleomorph, *Calonectria spathiphylli*. Bar = 10 μm (from CMI description no. 1175).

drained soils (Stover, 1972). The pathogen *Marasmiellus inoderma* colonizes dead and dying banana trash and is a pathogen on weakened plants. Outer leaf sheaths dry and plants become stunted. Water-soaked, brown oval patches of necrosis develop on the inner leaf sheaths and may extend to the pseudostem's centre, but not the rhizome. Roots are covered in white mycelium, develop a blackened tip and ultimately shrivel and have a brown cortex. Whitish mycelium and rhizomorphs that may develop between leaf sheaths have a mushroom odour, and basidiocarps of the pathogen form on the pseudostem and soil surface under moist conditions. The fruiting structures are brownish to salmon coloured on the top, turning pale later, are 5–15 mm in diameter and have widely spaced gills on the bottomside.

Basidiospores of the fungus are wind disseminated. Where the disease is a problem, improved plantation sanitation and manage-

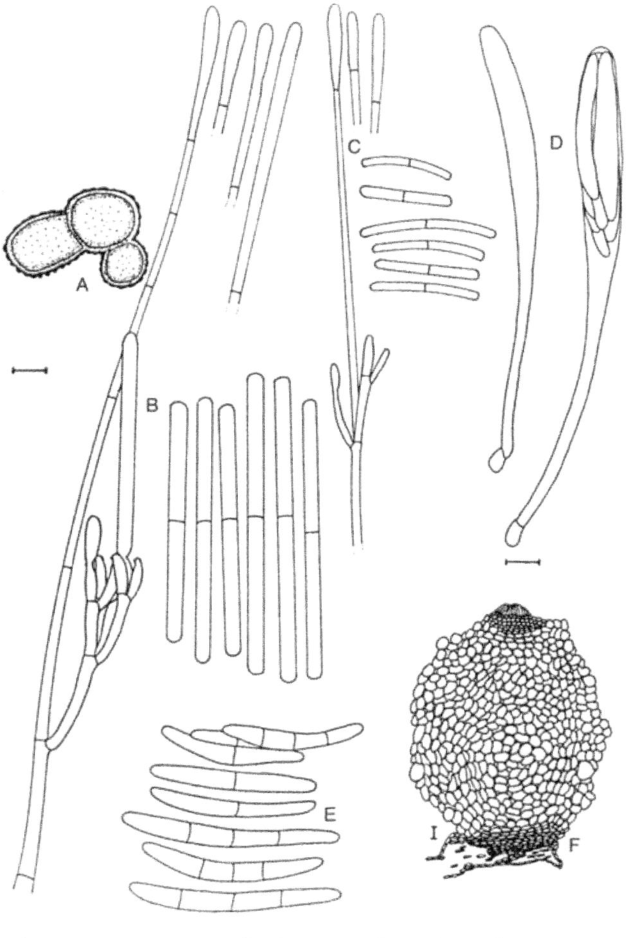

Fig. 4.26. (A) Chlamydospores, (B) macroconidiophore, vesicles and conidia, (C) microconidiophore, vesicles and conidia of *Cylindrocladium pteridis*, and (D) asci, (E) ascospores and (F) perithecium of its teleomorph, *Calonectria pteridis*. All bars except that for the perithecium (20 μm) equal 10 μm (from CMI description no. 1153).

ment are indicated. In particular, dead banana tissue and other trash should be removed.

Panama disease (Fusarium wilt)

Panama disease, which is also known as Fusarium wilt, originated in Southeast Asia, but was first reported in Australia in 1876 (Ploetz and Pegg, 1997). It was responsible for destructive epidemics in export plantations of 'Gros Michel' AAA before the 1960s that caused the trades to convert to the Cavendish cultivars. It is widely spread, and is now found in all banana-producing regions except islands in the South Pacific, the Mediterranean, Melanesia and Somalia.

SYMPTOMS The first internal symptom, a reddish brown discoloration of the xylem, develops in feeder roots, the initial sites of infection. It progresses to the rhizome and is most prominent where the stele joins the cortex. As the pseudostem is colonized, faint brown streaks or flecks become evident on and within older leaf sheaths. Eventually, large portions of the xylem turn brick red to brown (Plate 32).

The first external symptoms of Panama disease are a yellowing of the oldest leaves or a longitudinal splitting of the lower portion of the outer leaf sheaths on the pseudostem of plants that are usually more than 4 months old. This is followed by wilting and buckling of leaves at the petiole base. In some cases, these leaves remain green. As the disease progresses, younger and younger leaves collapse, until the entire canopy consists of dead or dying leaves.

Symptoms of Moko disease, caused by race 2 of the bacterium *R. solanacearum*, can be confused with those caused by Panama disease. However, Moko causes wilt and chlorosis on plants that are younger than 4 months old, and also discolours internal portions of fruit.

Two disorders of unknown aetiology can be confused with Panama disease. The term 'false Panama disorder' was used by Deacon *et al.* (1985) to identify a condition of Cavendish cultivars in South Africa that resembled Panama disease. The disorder also occurs on Cavendish clones in the Canary Islands, and elsewhere on other dessert and plantain cultivars (AAA, AAB and ABB genomes), including Colombia, Grenada, Panama and Suriname (summarized by de Beer *et al.*, 2001). External symptoms are similar to those of Panama disease, but internally affected plants exhibit only discontinuous wine-coloured vascular strands without the browner, more extensive and continuous vascular discoloration caused by Panama disease. The disorder is thought to be caused by a combination of stress factors such as drought, cold temperatures, soil compaction and nutrient imbalance (de Beer *et al.*, 2001).

In the highlands of western Uganda, another disorder, matooke wilt, occurs at elevations above 1300 m. Originally reported on the highland AAA cultivars (Lujugira-Mutika subgroup), it is now also recognized on introduced cultivars (not specified) (Kangire and Rutherford, 2001). Leaves on affected plants generally are healthy, but may be smaller and exhibit marginal necrosis. Affected pseudostems are thin, often dry in appearance, and may buckle, especially after the fruit bunch has emerged. Fruit are small and do not develop fully. Internal symptoms in the pseudostem are similar to those of Panama disease and include conspicuous brown to purple vascular strands (Fig. 4.27). In a previous study, *Fusarium oxysporum* was recovered from 59% of the symptomatic highland AAA samples that were examined, but only 14% of these (three of 22) were vegetatively compatible with the causal agent of Panama disease, *F. oxysporum* f. sp. *cubense* (Ploetz *et al.*, 1994a). Kangire and Rutherford (2001) reported that they were unable to recover *F. oxysporum* from affected plants. Thus, the true cause of this problem is not clear. Since affected plants are usually found in areas where household or animal wastes have been discarded, current hypotheses focus on an edaphic aetiology.

CAUSAL AGENT *F. oxysporum* f. sp. *cubense* is a soilborne hyphomycete (Domsch *et al.*, 1980). It is one of more than 100 formae speciales (special forms) of *F. oxysporum* that

Fig. 4.27. Internal symptoms of matooke wilt in Uganda. Note their resemblance to symptoms of Panama disease (photo: R.C. Ploetz, UF).

cause vascular wilts of flowering plants. It contains pathogenic and saprophytic strains that cannot be distinguished morphologically. Colonies are white to purple on PDA, grow 4–7 mm day^{-1} at 24°C, and have slight to copious aerial mycelium. Sporodochia are tan to orange, and sclerotia are blue and submerged.

Micro- and macroconidia are produced on branched and unbranched monophialides. Microconidia are 5–16 × 2.4–3.5 μm, one- or two-celled, oval- to kidney-shaped, and are borne in false heads. Macroconidia are 27–55 × 3.3–5.5 μm, four- to eight-celled and sickle-shaped with foot-shaped basal cells. Chlamydospores are 7–11 μm in diameter, usually globose and are formed singly or in pairs in hyphae (terminal and intercalary) or conidia. Atypically for the species, chlamydospores are not produced by isolates in vegetative compatibility group (VCG) 01214.

Four races of *F. oxysporum* f. sp. *cubense* have been described, only three of which affect banana (race 3 is a pathogen of *Heliconia* spp.). Race 1 caused the epidemics on 'Gros Michel' and also affects 'Maqueño' AAB, 'Silk' AAB, 'Pome' AAB, 'Pisang Awak' ABB and the hybrid 'I.C.2' AAAA. Race 2 affects cooking bananas, such as 'Bluggoe' ABB, and some bred tetraploids.

Race 4 is most destructive since it affects race 1 and race 2 suscepts as well as the Cavendish cultivars, plantains and other bananas that are resistant to races 1 and 2. Until recently, it had been reported only in subtropical regions where cold winter temperatures are thought to be predisposing factors (Moore *et al.*, 1993). However, considerable damage has occurred in Cavendish monocultures in tropical Southeast Asia within the last decade. A distinct population of the pathogen, VCG 01213–01216, is responsible for the latter outbreaks. It affects the same clones as race 4 in the subtropics, but does not require predisposing factors. VCG 01213–01216/tropical race 4 is found currently in Australia (Northern Territory), Indonesia (Halmahera, Irian Jaya, Java, Sulawesi and Sumatra), peninsular Malaysia and Papua New Guinea. Were it to spread to the Cavendish-dependent export trades in the western hemisphere or plantain production areas in Western Africa and Latin America, it could cause significant damage.

Vegetative or somatic compatibility has been used extensively to characterize this pathogen, and >20 VCGs have been reported to date (Jones, 2000). Phylogenetic work indicates that the pathogen originated in Southeast Asia, and probably co-evolved with its banana host in this region (Ploetz and Pegg, 1997).

EPIDEMIOLOGY Rhizomes ('suckers') are used traditionally as vegetative seedpieces for banana. Because they are usually free of symptoms when they are infected by *F. oxysporum* f. sp. *cubense*, they often are responsible for its dissemination. *F. oxysporum* f. sp. *cubense* can also spread in soil and running water, and on farm implements and machinery. Work in the early export plantations indicated that susceptible clones could not be replanted successfully in infested sites for up to 30 years due to the long-term survival of *F. oxysporum* f. sp. *cubense* in soil and as a parasite of non-host weed species (Waite and Dunlap, 1953; Stover, 1962).

Root tips are the initial sites of infection, and wounded rhizome surfaces are apparently minor infection courts. In most cases, root tip infections are stopped shortly after the pathogen reaches the xylem due to the formation of gels and tyloses, and vascular collapse. However, some of these infections are not recognized early enough in susceptible cultivars, and the colonization of the xylem and associated parenchymal tissues continues unabated. Macroconidia and chlamydospores that form on dead or dying plants are significant survival structures of the pathogen.

MANAGEMENT Few effective options exist for managing this lethal disease. In work in South Africa, methyl bromide significantly reduced disease incidence, but was effective for less than 5 years since the fumigated areas were recolonized by the pathogen (Herbert and Marx, 1990). Pseudostem/rhizome injections of carbendazim and potassium phosphonate have provided erratic or unrepeatable control (Lakshmanan *et al.*,

1987; Herbert and Marx, 1990; Davis *et al.*, 1994). Heat treatment of soil was used recently to control the spread of the pathogen in the Philippines, but this method will probably suffer the fate described for methyl bromide-treated soil.

Soils that suppress the development of Panama disease are found in several different locations (Stover, 1962, 1990b; Chuang, 1988; Peng *et al.*, 1999; Domínguez *et al.*, 2001). Diverse attributes have been associated with this trait. Stolzy and co-workers (reviewed by Toussoun, 1975) associated disease suppression with chemical and physical factors and found its closest association with soils in which a clay fraction of the montmorillonoid type was found. In the Canary Islands, suppression was associated with electrical conductivity, Na content and the structural stability of soil aggregates (Domínguez *et al.*, 2001). In studies in Australia, the addition of $CaCO_3$, $Ca(OH)_2$, $CaSO_4$ or Fe-EDDHA to soil reduced germination of chlamydospores of the pathogen as well as disease severity under controlled conditions (Peng *et al.*, 1999). Whether these treatments would be effective in the field was not examined. Unfortunately, the conversion of large-scale tracts of conducive field soil to a suppressive condition has not been reported.

Studies on the biological and cultural control of this disease have begun only recently. Arbuscular mycorrhizal fungi have been shown to reduce disease severity in short-term greenhouse studies, but results from long-term field studies have not been reported. Soil amendments, endophytic fungi and rhizosphere bacteria currently are being examined in Australia and South Africa. Achieving success with these or other approaches is a daunting task due to the high susceptibility of the cultivars for which protection is sought (usually Cavendish clones) and the perennial nature of the crop in most areas. The success of these strategies should be greater in areas where banana is treated as an annual or single-cycle crop (e.g. Taiwan).

Susceptible clones can be grown if pathogen-free propagation material is used in non-infested soil. Tissue-cultured plantlets are free of bacterial, fungal and nematode pathogens, and should be used to establish new plantings whenever possible. It should be noted, however, that plants grown from plantlets are more susceptible to Panama disease than those that are grown from rhizomes (Smith *et al.*, 1998). In addition, since meristem culturing usually does not eliminate the important viral pathogens, such planting material should be virus indexed or otherwise known to be free of these pathogens. Although the expense of plantlets may make their use in subsistence agriculture impractical, plantlets could be used to initiate nurseries for producing pathogen-free, conventional seedpieces.

Genetic resistance offers the greatest promise for managing this disease in infested soils. To date, pre-existing cultivars have been identified that perform well in different regions and against different populations of the pathogen (Table 4.3). Resistant hybrids have also been produced in the breeding programmes. Although these generally lack the flavour or postharvest attributes that are found in the susceptible cultivars, progress is being made.

Pseudostem heart rot

This disease was important on injured 'Gros Michel' AAA, but has become unimportant after the trades coverted to the Cavendish clones (Stover, 1972). It is uncommon on other clones, and is usually associated with wounding.

The centre whorl of leaves darkens and becomes necrotic. Initially the decay is firm, but later softens and develops a foul odour due to secondary colonization by bacteria. The associated leaves become yellow or brown upon emergence and may or may not open. In extreme instances, the entire canopy and pseudostem collapse. Damage is not systemic, and only one or a few pseudostems in a mat are affected. Plants can recover and throw a new set of leaves.

The cause of this disease may be confused, since two different fungi have been reported. Stover (1972) indicted *Fusarium moniliforme* (Fig. 4.28) and listed a *Gibberella fujikuroi* teleomorph. In contrast, Reinking

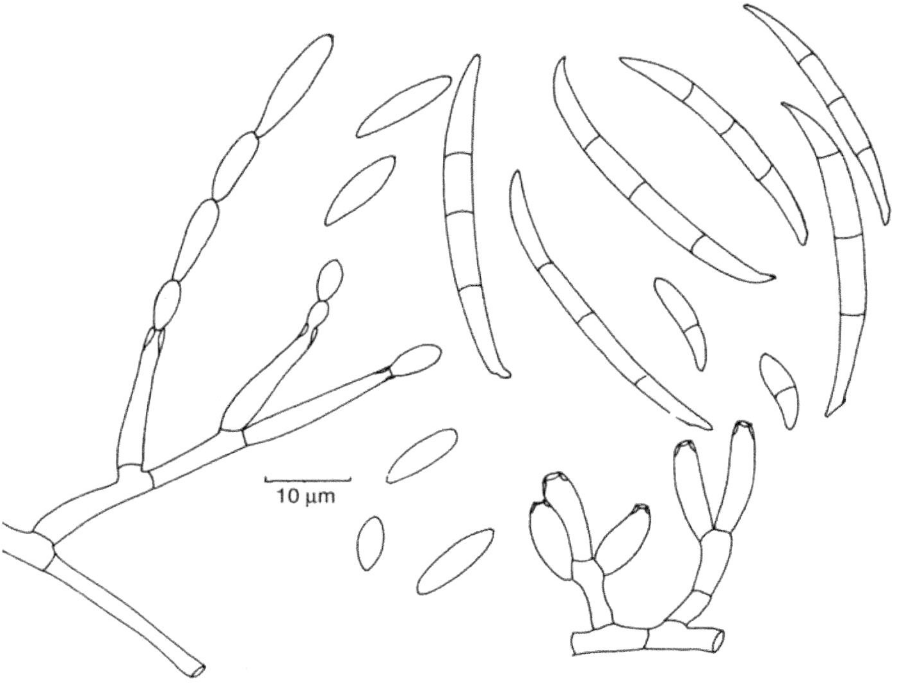

Fig. 4.28. Counterclockwise from upper left: microconidia, macroconidia and conidiophores of *Fusarium moniliforme* (from CMI description no. 22).

(1937) mentioned both *F. moniliforme* and *F. subglutinans* (Fig. 4.19), and suggested that further studies were warranted to clarify whether both or only one of the species caused heart rot. Over six decades after this paper was published, the situation is still not clear. Whether the heart rot *F. subglutinans* is synonymous with the newly described *F. concentricum* (see section on crown rot in this chapter) is also not known.

Heart rot is not a major concern unless plantations either are poorly maintained or have been damaged by flooding, wind or other agents.

DISEASES THAT ARE CAUSED BY NEMATODES

Worldwide, 151 species of nematodes in 43 genera have been reported on *Musa* spp. (Gowen and Quénéhervé, 1990). The most important are *Radopholus similis*, *Pratylenchus coffeae*, *P. goodeyi* and *Helicotylenchus multicinctus*, although *Rotylenchus reniformis* and several species of *Meloidogyne* are are also common. In general, these pests damage roots and the structural integrity of the banana mat, and provide entry points for root-rotting fungi. They have also been associated with an increased susceptibility to the banana weevil, *Cosmopolites sordidus* (Speijer *et al.*, 1993).

Burrowing nematode root rot (blackhead toppling disease)

In general, the burrowing nematode, *R. similis*, is the most important nematode pest of banana (Gowen and Quénéhervé, 1990; Sarah *et al.*, 1996). It has a pantropical distribution and is found in all banana-producing areas except the Canary Islands, Cape Verde Islands, Cyprus, Crete, Israel, Mauritius, Taiwan and the highlands of East Africa. It has a serious impact on export production of the Cavendish clones, and is also common on cooking bananas and plantains in Puerto Rico and in lowland central and eastern Africa. The nematode has a relatively wide host range that includes weed and crop species. Its wide distribution undoubtedly is due to the movement of infected hosts. For example, Marin *et al.* (1998) indicated that at least three separate introductions of the species occurred as different banana clones were disseminated in the New World.

Symptoms

R. similis causes dark brown to black necrotic patches on root and rhizome surfaces. The entire cortex of roots can be killed, but the stele is not affected; in cross-section, a sharp demarcation between the cortex and stele is evident. Damage is enhanced in the presence of several fungal pathogens, including *Cylindrocarpon musae* (Fig. 4.29), *Cylindrocladium* spp. (see section on Cylindrocladium root rot in this chapter), *Acremonium stromaticum* and *Rhizoctonia solani* (Jones, 2000).

The most conspicuous symptom is the blackhead toppling syndrome in which pseudostems lodge and expose the blackened remains of roots and the blackened, but intact, rhizome. It results from the reduced mechanical strength of the root system and can occur on plants of any age. It is most common, however, on plants with a bunch and during excessive rain and wind. Uprooted plants often fall with attached suckers. Blackhead toppling differs from rhizome breakage caused by *C. sordidus*, in which the lower portion of the rhizome remains in the soil.

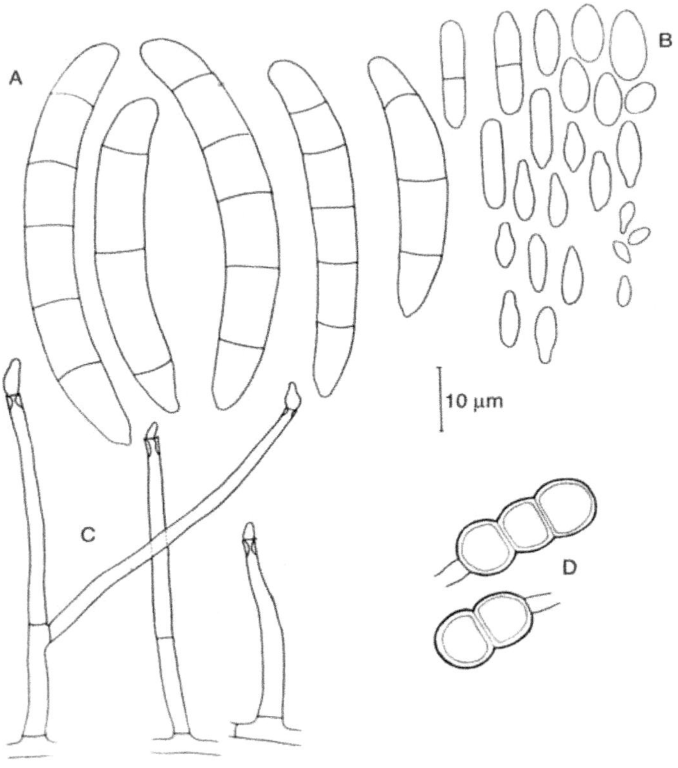

Fig. 4.29. (A) Macroconidia, (B) microconidia, (C) conidiogenous cells and (D) thick-walled resting structures of *Cylindrocarpon musae* (from CMI description no. 927).

Causal agent

R. similis is a migratory, endoparasitic nematode that completes its life cycle in 20–25 days, and whose cardinal temperatures are 17, 30 and 33°C (Sarah *et al.*, 1996). Its relatively high optimum temperature for growth generally restricts it to the lowland tropics.

Pathogenic and genetic variation occur among populations from different regions and from other hosts. On banana, some populations from Africa are more pathogenic than others in the Caribbean, Queensland and Sri Lanka, and another pathotype from Puerto Rico is more pathogenic on plantain than other bananas (Sarah *et al.*, 1996). Although differences in various genetic characters exist, the species generally is uniform.

Epidemiology

All life stages are vermiform and mobile. Females are infective, but males have an atrophied stylet and are not. Eggs are laid in the cortex, and the entire life cycle can be completed within the root.

Management

Meristem cultured plantlets should be used in new production areas where banana has not been grown previously (Gowen and Quénéhervé, 1990; Sarah *et al.*, 1996). Traditional suckers can be treated with a nematicide or in hot water (52–55°C for 15–20 min) after all symptomatic tissues have been removed. However, since the latter measures are not totally effective, there is some risk when they are used in non-infested soil.

Where sites are infested with the nematode, fallowing for 12 months or rotation with a non-host is preferred, but often not desirable. In addition to taking land out of production this practice can be a challenge since it is difficult to eliminate all suckers of the previous crop. Flooding is also effective, but requires level areas with access to large quantities of water. Effective fumigants that had been used to disinfect fields in the past generally are no longer used since they are either no longer registered or are too expensive.

In established banana plantations, successful control is limited to the use of organophosphate and carbamate nematicides. They are applied as granules at the base of plants or as emulsifiable concentrates via drip irrigation. As for the broad-spectrum soil fumigants, the number of nematicides that retain registration is being reduced continually.

Many important banana cultivars are susceptible to *R. similis*. Stoffelen *et al.* (2000) tested the resistance of 14 clones to this nematode that varied in their resistance to to Panama disease. Only three clones that resist Panama disease, two accessions of 'Pisang Jari Buaya' AA and one of 'Yangambi Km5' AAA, were resistant to *R. similis*. None of the cultivars in this study were resistant to the lesion nematode, *Pratylenchus coffeae*.

Lesion nematode damage

P. coffeae and *P. goodeyi* are serious pests of bananas (Bridge *et al.*, 1997). They cause lodging and lesions on roots and rhizomes of banana that are identical to those caused by *R. similis*. They usually appear in roots in concert with several different fungi, including *Fusarium oxysporum*, *F. redolens*, *F. sambucinum*, *Nigrospora musae* and *Rhizoctonia solani*.

P. coffeae has a pantropical geographic distribution and a wide host range that includes important crop species (Bridge *et al.*, 1997). Its worldwide occurrence is traced to the dissemination of infected planting materials. It is the most important nematode species on banana in many areas in Central and South America, as well as Southeast Asia. In Africa, its more localized distribution suggests that it was introduced there only recently.

In contrast, *P. goodeyi* is only known on banana and occurs in the Canary Islands and the cooler banana-growing regions of Africa. In Africa, it is usually only found in smallholder production, which would explain why it has not been distributed as widely as *P. coffeae* and *R. similis*.

These species can be misidentified as *R. similis* (Bridge *et al.*, 1997). They are migratory endoparasites, and both juvenile and adult stages are capable of invading roots. Like *R. similis*, they can complete their entire life cycle in roots. Control measures are similar to those listed for *R. similis*, although rotation and fallowing are less effective for *P. coffeae* due to its wide host range.

Spiral nematode

The spiral nematode, *Helicotylenchus multicinctus*, occurs in most regions where banana is grown, but is most important in the subtropics and at high elevations in the tropics (McSorley and Parrado, 1986). In subtropical areas where *R. similis* is uncommon, *H. multicinctus* often is the most important nematode on this crop. It causes significant damage in Argentina, Cuba, Cyprus, Florida, Israel, Lebanon and South Africa.

H. multicinctus causes lesions on root surfaces and destruction of small feeder roots. The damage is superficial, and penetrates the root cortex only slightly and not the rhizome. However, root damage can be severe, resulting in yield loss and lodging.

H. multicinctus is an endoparasite that can complete its life cycle in the root cortex. Over 40 plant species are reported to be hosts. Nematode-free sites and planting material are the best ways to combat this problem. Otherwise, the measures listed for *R. similis* can be used. Due to its less invasive colonization, seedpieces can be disinfested by removing all root tissue from rhizomes (McSorley and Parrado, 1983). Treatment with hot water is also effective, and it need not be as severe as that used for *R. similis* (50–55°C for 7.5 minutes).

DISEASES OF BANANA THAT ARE CAUSED BY VIRUSES

Bunchy top is by far the most serious virus-induced disease of banana worldwide (Dale, 1987). However, other virus-induced diseases are also damaging and/or widespread. The most significant of these diseases are covered below alphabetically.

Banana mild mosaic

Filamentous, virus-like particles have been noted frequently in germplasm collections and field specimens (Rivera *et al.*, 1992; Anonymous, 1993; Lockhart, 1994; Belalcázar *et al.*, 1998; Caruana and Galzi, 1998). Recent research indicates that a single virus, *Banana mild mosaic virus* (BanMMV), is present in most cases (Gambley and Thomas, 2001). All *Musa* genotypes appear to be susceptible. The economic impact of mild mosaic is not known, but observations in Latin America indicate that banana and plantain yields may be affected (B.E.L. Lockhart, Minnesota, 1998, personal communication).

Symptoms

Symptoms of infection are inconsistent and in many cases may not develop. 'Ducasse' (ABB, syn. 'Pisang Awak') has been observed with chlorotic mosaic and streak symptoms in the field, but these symptoms usually disappear as plants mature. Transitory chlorotic leaf streaks were noted on some leaves of 'Daluyao' (AAB, Plantain subgroup) under glasshouse conditions, and 'Pisang Seribu' (AAB) displays silvery streaks on the leaf lamina. Plants of 'Gros Michel' (AAA) from Latin America have displayed large chlorotic blotches on the leaves, stunting and delayed bunch development (Jones, 2000).

Mixed infections of BanMMV and *Banana streak virus* (BSV) are common. In such cases, symptoms are reminiscent of those caused by BSV alone (Jones, 2000). In mixed infections with *Cucumber mosaic virus* (CMV), BanMMV seems to cause additional leaf necrosis (Caruana and Galzi, 1998).

Causal agent

BanMMV virions are flexuous filaments ~580 nm long and 14 nm wide (Jones, 2000). The genome is single-stranded (ss) RNA of ~7.4 kb, contains five open reading frames, 3′- and 5′-untranslated regions and a poly (A) tail. BanMMV is related to, but distinct from, the *Carlavirus*, *Foveavirus* and *Potexvirus* genera (Gambley and Thomas, 2001).

Polyclonal antisera that have been prepared against isolates from 'M'bouroukou' (AAB, Plantain subgroup), 'Pisang Seribu' AAB (B.E.L. Lockhart, Minnesota, 2001, personal communication) and 'Ducasse' (C.F Gambley and J.E. Thomas, unpublished) work well in immunosorbent electron microscopy (ISEM), but the 'Pisang Seribu' antiserum is best for enzyme-linked immunosorbent assay (ELISA). All isolates that have been tested with polyclonal antibodies are serologically related. Monoclonal antibodies (mAbs) have been prepared to an isolate from 'Cardaba' ABB, and these have been used in tissue blot and western blot assays (Caruana *et al.*, 1995). A sensitive triple antibody sandwich ELISA has been developed which uses polyclonal coating antibodies and the mAbs (M.L. Iskra-Caruana, Montpellier, 1999, personal communication). BanMMV can also be detected by reverse transcription (RT)–PCR in total nucleic acid extracts from infected plants and by immunocapture RT–PCR using crude sap extracts (Gambley *et al.*, 1999).

Epidemiology

Vegetative propagation through conventional planting material and tissue culture is the only known means of transmission of BanMMV, although field transmission in mixed infections with the aphid-transmitted CMV was suspected in Guadeloupe (Caruana and Galzi, 1998). BanMMV has not been transmitted mechanically to banana or herbaceous indicators, or with the banana aphid, *Pentalonia nigronervosa* (Caruana *et al*, 1995).

Management

The use of indexed, virus-free vegetative planting material is recommended.

Banana mosaic

Banana mosaic was first described in New South Wales, Australia, in 1930 (Magee, 1930, 1940b) and now occurs in most banana-growing areas worldwide. It is also known as infectious chlorosis, heart rot, virus sheath rot, cucumber mosaic and banana mosaic (Magee, 1930; Wardlaw, 1961; Stover, 1972).

Symptoms

Symptoms are affected by environment and the strain of CMV. Although infected leaves may be symptomless, they typically have chlorotic stripes, stippling or line patterns, or a more general mosaic (Plate 33) (Yot-Dauthy and Bové, 1966; Ploetz *et al.*, 1994b). Fruit may be distorted and display chlorotic streaks or a mosaic. At temperatures below 24°C, symptoms are often more severe and may include heart rot (Ploetz *et al.*, 1994b). Severe strains of the virus can cause leaf distortion and necrosis. In the past, symptoms were often confused with those caused by BSV (e.g. Wardlaw, 1961; Stover, 1972).

Causal agent

Banana mosaic disease is caused by CMV (Yot-Dauthy and Bové, 1966), a member of the genus *Cucumovirus* (Family *Bromoviridae*). The virus has a tripartite ssRNA genome encapsidated in isometric virions 28–30 nm in diameter (Francki *et al.*, 1979; Palukaitis *et al.*, 1992). A satellite RNA is encapsidated with some isolates from Israel, and is responsible for symptom attenuation (Gafny *et al.*, 1996).

CMV isolates have been divided into subgroups I and II based on serology and nucleic acid hybridizations (Devergne and Cardin, 1973; Piazolla *et al.*, 1979; Owen *et al.*, 1990; Palukaitis *et al.*, 1992). Subgroup I predominates in the tropics and subtropics (Hasse *et al.*, 1989; Ploetz *et al.*, 1994b) and includes most, but not all, isolates from banana (Hu *et al.*, 1995; Singh *et al.*, 1995; Gafny *et al.*, 1996).

A number of methods are available for the diagnosis of CMV in banana, including serology, nucleic acid hybridization and PCR (Diekmann and Putter, 1996). Inconsistent symptom expression and the similarity of symptoms induced by other viruses, such as BSV and *Banana bract mosaic virus* (BBrMV), limit the value of symptoms for diagnosis. Serological detection is reliable, and a number of ELISA kits are available commercially. A range of polyclonal and monoclonal antibodies detects a wide range of virus strains, or differentiates subgroups I and II (Hasse *et al.*, 1989; Diekmann and Putter, 1996). Sensitive nucleic acid-based methods, especially PCR, are being used increasingly in banana (Hu *et al.*, 1995; Singh *et al.*, 1995; Sharman *et al.*, 2000a).

Epidemiology

CMV infects at least 800 plant species. Resistance is not known to occur in *Musa*, although Stover (1972) noted that *M. balbisiana* was free of symptoms in the field. CMV is transmitted in a non-persistent manner by at least 75 aphid species (Palukaitis *et al.*, 1992), and often is seed transmitted in other hosts. CMV may also be seed transmitted in banana (Gold, 1972), so seedlings need to be tested if they come from a plant whose virus status is unknown. Although CMV can be sap transmitted experimentally, this probably is not important in the field.

Weeds or nearby crop plants are the primary sources of inoculum, and spread from banana to banana appears to be less common (Stover, 1972). *Commelina* is a common weed host in banana plantations, as are vegetable crops, such as cucurbits, tomatoes and peppers, which often are intercropped with bananas.

Management

Alternative hosts should be removed from within and around banana plantings. Growing bananas next to non-hosts (e.g. rice) instead of susceptible vegetable crops reduced mosaic in Taiwan (Tsai *et al.*, 1986). Roguing infected banana plants in plantations is advisable, as it eliminates these plants as potential sources of infection. These plants are also likely to have a poor yield and to bear symptomatic fruit.

Most aphid vectors do not colonize bananas, but high transient populations of inefficient species can still result in the transmission of CMV to banana. Insecticides are unlikely to be effective in controlling the disease in banana because these non-colonizing species can transmit the virus after feeding only for a few seconds. A combination of roguing and insecticides has been reported to give good control in commercial plantations (Adam, 1962), but is probably not cost-effective (Jeger *et al.*, 1995). Roguing alone has given adequate control (Stover, 1972).

Pentalonia nigronervosa is not usually considered a significant vector of CMV. However, large populations have been observed on shoot tips of *Commelina diffusa* in West Africa, and may play a role in infection of banana and subsequent within-crop spread (Jones, 2000).

As CMV is transmitted in vegetative propagules, including tissue-cultured plantlets, it is important that planting material comes from virus-free sources. It has been suggested that a severe strain of CMV in Morocco was imported in planting material (Bouhida and Lockhart, 1990). Mother plants should be free of symptoms and indexed to ensure freedom from CMV. The virus has been eliminated from banana by heat treatment of rhizomes followed by apical meristem culture (Berg and Bustamante, 1974), and through cryopreservation (Helliot *et al.*, 2002). However, it is generally more cost- and time-effective to obtain virus-free mother plants from another source. Tissue-cultured plantlets should also be protected from infection in nurseries during the acclimatization phase.

Banana streak

Banana streak (mosïque à tirets) was first described as distinct from 'typical' mosaic caused by CMV in Côte d'Ivoire (Lassoudière, 1974, 1979). Illustrations of what appears to be banana streak had been presented as CMV-induced mosaic in earlier publications (Wardlaw, 1961; Yot-Dauthy and Bové, 1966; Stover, 1972).

Banana streak has a worldwide distribution (Jones, 2000), although disease incidence varies greatly between districts and in different cultivars. 'Mysore' AAB, a popular dessert banana from India, appears to be universally infected with BSV (Ploetz *et al.*,

1994b). The economic impact of BSV depends not only on its direct effect on fruit yield and quality, but also on the quarantine restrictions imposed on many improved, but infected hybrids.

Symptoms

Symptom expression is very variable, and influenced by virus strain, host cultivar and physiology, and the environment (Gauhl and Pasberg-Gauhl, 1994; Ploetz *et al.*, 1994b; Dahal *et al.*, 1998b; Daniells *et al.*, 2001a). The most common symptoms are continuous or discontinuous chlorotic streaks running parallel to the leaf veins, and range from prominent to very sparsely distributed (Plate 34). The streaks darken over time, and may become brown or black (Plate 35). In some cases, spindle-shaped lesions or chlorotic blotches occur (Jones, 2000). A range of other symptoms are sometimes associated with the disease, including splitting of the pseudostem, heart rot, leaf and pseudostem necrosis, petiole and pseudostem streaks, aberrant bunch emergence and altered leaf phyllotaxy (Lassoudière, 1979; Gauhl and Pasberg-Gauhl, 1994; Jones, 2000; Daniells *et al.*, 2001a). Young, infected, suckers usually display few, if any, symptoms (Lassoudière, 1979; Daniells *et al.*, 2001a).

Symptom expression is sporadic over time, and not all leaves on an infected plant may display symptoms. Similarly, not all infected plants in a stand will have symptoms at the same time. The factors responsible for inconsistent symptom expression are unclear. Lower temperatures, or possibly temperature fluctuations, have been correlated with symptom expression (Lockhart, 1995; Dahal *et al.*, 1998b, 2000). However, the developmental stage of the plant may also be important, as the proportion of plants with symptoms was shown to increase progressively during each cropping cycle (Lassoudière, 1979; Daniells *et al.*, 2001a). Daniells *et al.* (2001a) also noted greater expression during rapid plant growth at bunch initiation during warm periods.

Streak can deform fruit and bunches, lengthen the cropping cycle and reduce bunch size, and fruit number, size and quality (Lassoudière, 1974, 1979; Dahal *et al.*, 2000; Daniells *et al.*, 2001a). Yield depression increases through successive cropping cycles, and is more apparent under suboptimal growth conditions. Reported losses have ranged from 6 to 45% (Lassoudière, 1979; Daniells *et al.*, 2001a).

Causal agent

The causal virus, BSV, was first identified in Morocco (Lockhart, 1986). BSV (genus *Badnavirus*) has bacilliform virions ~30 nm × 130–150 nm. It has a double-stranded (ds) DNA genome of ~7.4 kbp and replicates via reverse transcription (Lockhart and Olszewski, 1993). Isolates of BSV are highly heterogeneous, and induce a range of symptoms. They also differ serologically and genomically (Lockhart and Olszewski, 1993; Geering *et al.*, 2000). Banana can also be infected experimentally with a related badnavirus, *Sugarcane bacilliform virus* (SCBV) (Bouhida *et al.*, 1993). Since some isolates of SCBV and BSV have closely related genomes, the distinction between these two species is blurred (Braithwaite *et al.*, 1997; Geering and Thomas, 2002).

Sequences of BSV are integrated into the genomes of *Musa* and *Ensete* (LaFleur *et al.*, 1996). These sequences are very variable, and probably include many 'dead' and partial inserts of a number of virus strains (LaFleur *et al.*, 1996; A.D.W. Geering, Brisbane, 2001, personal communication). At least one integrated BSV sequence appears to give rise to a circularized, transcriptionally active episomal form of the virus after excision from the *Musa* genome and two homologous recombinations (Harper *et al.*, 1999b; Ndowora *et al.*, 1999; Hull *et al.*, 2000). This 'active' integrant of strain BSV-OL is linked to the B genome of *Musa* (Geering *et al.*, 2001). Hybridization and propagation by tissue culture are likely triggers for episomal expression of integrants (Ndowora *et al.*, 1999; Dallot *et al.*, 2001), and BSV-OL is a common contaminant in new hybrids that are produced by banana-breeding programmes. Other strains of BSV might also be integrated and give rise to episomal infections (A.D.W. Geering, Brisbane, 2001, personal communication).

Epidemiology

BSV is transmitted in a semi-persistent manner by the mealybugs *Planococcus citri*, *Saccharicoccus sacchari* and possibly other species (Dahal *et al.*, 1998a; Jones, 2000; Kubiriba *et al.*, 2001b). Field spread of BSV is slow and appears to be of limited significance in most locations (Lockhart, 1995; Daniells *et al.*, 2001a; Kubiriba *et al.*, 2001a). However, this may not be the case in some commercial plantations of Cavendish cultivars in Ecuador, where BSV is a serious problem (Jones, 2000).

The major means of spread is in vegetative planting material. The virus is neither mechanically transmissible nor soilborne. Although there is some evidence that BSV is seed transmitted (Daniells *et al.*, 1995; Jones, 2000), this was before integration of BSV in the host genome was recognized. The natural and experimental host range of BSV is restricted to *Musa* and *Ensete* species, and a wide range of *Musa* cultivars and genotypes are susceptible (Jones, 2000).

Management

The use of virus-free planting material is crucial since the virus is spread primarily by vegetative propagation. However, the selection of suitable planting material is hampered by the difficulties in visually diagnosing BSV infection and identifying such a heterogeneous virus with diagnostic assays. Visual inspections should be conducted on all leaves and on a number of occasions during the cropping cycle.

Fluctuations in virus titre and the serological heterogeneity of BSV combine to make ELISA assays unreliable (Dahal *et al.*, 1998a,b; Daniells *et al.*, 2001a). With the currently available antisera, some BSV isolates are also poorly detected by ELISA (Ndowora, 1998). Immunosorbent electron microscopy using partially purified virus preparations (Bouhida *et al.*, 1993) and broad-spectrum antiserum (Ndowora, 1998) seems to detect all isolates and is generally more sensitive than ELISA (Lockhart, 1995; Thottappilly *et al.*, 1998).

Nucleic acid-based assays are compromised by the genomic heterogeneity of BSV isolates and the presence of integrated BSV sequences (Lockhart and Olszewski, 1993; Geering *et al.*, 2000). False positives due to integrated viral sequences can be avoided with immunocapture PCR (Harper *et al.*, 1999a). Both BSV strain-specific (Geering *et al.*, 2000) and degenerate (Lockhart and Olszewski, 1993; Ahlawat *et al.*, 1996; N.E. Olszewski and B.E.L. Lockhart, Minnesota, 2000, personal communication) PCR primers are available.

Bract mosaic

Symptoms of bract mosaic were first noted on Mindanao in 1979 (Magnaye and Espino, 1990). The disease is widespread in the Philippines, and has also been recorded from India and Sri Lanka (Rodoni *et al.*, 1997; Thomas *et al.*, 1997). Although *Banana bract mosaic virus* (BBrMV) has also been recorded in Thailand and Vietnam, symptoms there were more characteristic of CMV-induced mosaic. In Western Samoa, a plant doubly infected by BBrMV and BSV showed symptoms of streak only (Rodoni *et al.*, 1999).

Limited data are available on the economic impact of bract mosaic. In one study conducted in Mindanao, yield losses of up to 40% were noted in 'Cardaba' ABB and 'Lakatan' AAA (Kenyon *et al.*, 1996; Thomas and Magnaye, 1996).

Symptoms

Mosaic patterns on bracts are diagnostic and distinct from symptoms caused by all other known viruses of banana. Mosaic patterns, stripes and spindle-shaped streaks may also be visible on the pseudostem when outer leaf sheaths are removed and on the petioles and midribs. The colour of symptoms depends on host pigmentation, and can vary from chlorotic or yellow, through red, brown and even black. Leaf symptoms, consisting of spindle-shaped lesions and streaks running parallel to the veins, are not always evident, but can occur on young plants that have been infected recently. Suckering is also suppressed, and suckers that do emerge are distorted and deeply pigmented (Anonymous, 1995).

In the Philippines, chlorotic streaks may be present on peduncles, and a high disease incidence is associated with increased levels of malformed fruit in commercial plantations. In India, petioles and peduncles of 'Nendran' AAB become brittle and fruit is only rarely carried to maturity. If fruit does mature, it is undersized. Mosaics can be seen on the fruit of other cultivars (Jones, 2000).

Initial symptoms in aphid-inoculated plants include broad, chlorotic patches along the major leaf veins, surrounded by a rusty red border and green or reddish streaks or spindle-shaped lesions on the petioles. Leaf symptoms, consisting of spindle-shaped lesions and streaks running parallel to the veins, are not always evident, but can occur on young plants that have been infected recently.

Causal agent

BBrMV has flexuous virions, 750 × 11 nm (Muñez, 1992; Bateson and Dale, 1995; Thomas *et al.*, 1997). Virions contain a major coat protein of ~39 kDa, a buoyant density in caesium chloride of 1.29–1.31 g cm^{-3} and an $A_{260/280}$ of 1.17 (Thomas *et al.*, 1997). Nucleotide sequence analysis indicates that BBrMV is a distinct potyvirus (Bateson and Dale, 1995; Rodoni *et al.*, 1997, 1999; Thomas *et al.*, 1997). All isolates of BBrMV tested are closely related serologically (Thomas *et al.*, 1997; Rodoni *et al.*, 1999). Identity at the nucleotide level within the coat protein gene was >87% for isolates from the Philippines, India, Western Samoa and Vietnam (Thomas *et al.*, 1997; Rodoni *et al.*, 1999).

Although bract mosaic-affected banana plants from India and the Philippines frequently also contain BanMMV (M.L. Iskra-Caruana and J.E. Thomas, 1998, Montpellier and Brisbane, personal communication), BBrMV alone appears to be the causal of bract mosaic. BBrMV virions have been aphid transmitted to healthy banana test plants that subsequently developed bract mosaic (Caruana and Galzi, 1998).

Both serological and nucleic acid-based assays are now available for BBrMV. The virus can be detected by ELISA using polyclonal (Thomas *et al.*, 1997; Rodoni *et al.*, 1999) and/or monoclonal (J.E. Thomas, Brisbane, 1996, unpublished) antibodies. It can also be detected by PCR in total nucleic acid extracts from infected plants, using either virus-specific or degenerate potyvirus group primers (Bateson and Dale, 1995; Thomas *et al.*, 1997), and by immunocapture PCR using crude sap extracts (Sharman *et al.*, 2000b). Virion concentration in infected plants is relatively low, and the virions usually are not detected readily by direct electron microscopy of sap.

Epidemiology

BBrMV is transmitted by at least three species of aphids: *Aphis gossypii*, *Rhopalosiphum maidis* and *Pentalonia nigronervosa* (Magnaye and Espino, 1990; Muñez, 1992; Diekmann and Putter, 1996). *P. nigronervosa* transmitted BBrMV after an acquisition access period of 1 min, indicating that transmission is of the non-persistent type (Muñez, 1992). Efficiency of transmission with the latter species was <10% (Caruana and Galzi, 1998).

Attempts to transmit BBrMV by sap inoculation to herbaceous indicator plants so far have been unsuccessful (Magnaye and Espino, 1990; Muñez, 1992; Diekmann and Putter, 1996; S. Cohen, Montpellier, 1996, personal communication). However, occasional sap transmission from banana to banana has been reported (L.V. Magnaye and L. Herradura, Davao, 1998, personal communication). The virus can be transmitted through vegetative planting material including suckers, bits, rhizomes and micropropagated plantlets.

The natural and experimental host range of BBrMV appears to be restricted to *Musa*. The virus has been detected in a wide range of naturally infected banana cultivars and genotypes, and no resistance to the virus is known.

Management

The use of indexed, virus-free planting material is the best means of control. Roguing and sanitation programmes have been introduced into commercial production areas in the Philippines (Magnaye, 1994).

Bunchy top

It is probable that bunchy top originated in the centre of origin of the genus *Musa* in the south and southeast Asian–Australasian region. Banana bunchy top was first reported in Fiji in 1889, but was almost certainly present as early as 1879 (Darnell-Smith, 1924; Magee, 1953). Interest in the disease arose due to its effects on the Cavendish-based export industry that began there in 1877. Its peak production of 788,000 bunches in 1892 declined to 147,000 bunches by 1895, mainly because of bunchy top. Other early records include Egypt in 1901 (Fahmy, 1924, in Magee, 1927) and Australia and Sri Lanka in 1913 (Magee, 1953).

Due to rapid expansion of the banana industry in Australia and use of infected planting material, bunchy top extended along a 300 km range from central New South Wales to southern Queensland between 1913 and 1927 (Magee, 1927). Production in the industry peaked in 1922, but 3 years later had collapsed, with production in the most affected areas reduced by 90–95% (Magee, 1927). Recently, severe outbreaks have occurred in Pakistan (Khalid and Soomro, 1993) and Hawaii (Ferreira *et al.*, 1989).

The countries in which bunchy top has been authenticated are listed in Table 4.4. Reports from East Malaysia (Sabah), West Malaysia and Papua New Guinea (Magee, 1953; Wardlaw, 1961) need verification. Significantly, Hawaii is the only location in the western hemisphere where the disease is present.

Symptoms

Symptoms of bunchy top are distinctive (Magee, 1927). Plants can become infected at any stage of growth. The first leaf to emerge from infected suckers can develop severe symptoms. The leaves are rosetted and small with very chlorotic margins that tend to turn necrotic (Plate 36). Dark green streaks are usually evident in the leaves. In contrast, symptoms usually appear in the second leaf to emerge after aphid inoculation. They consist of a few dark green streaks or dots on the minor veins on the lower portion of the lamina. The streaks form 'hooks' as they enter the midrib and are best seen from the underside of the leaf in transmitted light. The 'dot-dash' symptoms can sometimes also be seen on the petiole (Plate 37). Successive leaves become shorter and narrower and often have chlorotic, upturned margins. The leaves become brittle and erect, giving the plant a bunched appearance.

Once infected, plants rarely produce a bunch and do not fruit in subsequent years. When infected late in the growing cycle, they may fruit, but the bunch stalk and fruit are small and distorted. On plants infected very late, only a few dark green streaks may appear on the tips of the flower bracts (Thomas *et al.*, 1994).

Mild disease symptoms are expressed in some banana cultivars and *Musa* species. The dark green leaf and petiole streaks, so diagnostic in most cultivars in the Cavendish subgroup, can be rare or absent in other cultivars (Magee, 1953). Severely affected plants of 'Veimama' (AAA, Cavendish subgroup) have been observed to recover to later produce few if any symptoms (Magee, 1948). In addition, symptomless strains and mild strains that produce only limited vein clearing and dark green flecks have been reported from Taiwan (Su *et al.*, 1993).

Causal agent

In only 3 years, Magee (1927) determined that the causal agent was a virus transmitted by *Pentalonia nigronervosa* and in infected planting material. He also proposed management strategies that still form the basis of Australia's control programmes today.

Banana bunchy top virus (BBTV) is presumed to be the causal agent of bunchy top, even though the disease has not been produced via inoculation with purified virions or cloned genomic components. BBTV virions are associated intimately with the disease (Harding *et al.*, 1991; Thomas and Dietzgen, 1991) and are always detected in symptomatic plants (Dietzgen and Thomas, 1991; Thomas, 1991; Thomas and Dietzgen, 1991; Karan *et al.*, 1994). Virions are icosahedra, ~18–20 nm in diameter, have a coat

Table 4.4. Countries in which banana bunchy top disease has been recorded.[a]

Region/country	Year first recorded	BBTV detected
Pacific		
Australia	1927	+
Fiji	1927	+
Guam	1982	
Hawaii (USA)	1991	+
Kiribati (formerly Gilbert Islands)	1980	
New Caledonia	2001	+
Samoa (American)	1967	
Samoa (Western)	1967	+
Tonga	1967	+
Tuvalu (formerly Ellice Islands)	1926	
Wallis Island	1933	
Asia		
China	1996	+
India	1953	+
Indonesia	1978	+
Japan (Bonin Islands)	1926	
Japan (Okinawa)	1993	+
Malaysia (Sarawak)	1995	+
Pakistan	1993	+
Philippines	1961	+
Sri Lanka	1921	+
Taiwan	1961	+
Vietnam	1969	+
Africa		
Burundi	1988	+
Central African Republic	1996	
Congo	1961	+
Democratic Republic of Congo (formerly Zaire)	1982	
Egypt	1927	+
Gabon	1982	+
Malawi	1997	+
Rwanda	1988	

[a] References for specific records are found in Thomas and Iskra-Caruana (2000).

protein of Mr ~20,000 Da, a sedimentation coefficient of ~46S and a buoyant density of 1.29–1.30 g cm^{-3} in caesium sulphate (Wu and Su, 1990c; Dietzgen and Thomas, 1991; Harding *et al.*, 1991; Thomas and Dietzgen, 1991). Purified preparations have an $A_{260/280}$ of 1.33 (Thomas and Dietzgen, 1991). The virus possesses a multicomponent genome, consisting of at least six circular ssDNA components each ~1000–1100 nucleotides long (Burns *et al.*, 1995; Xie and Hu, 1995; Karan *et al.*, 1997).

Although biologically similar, isolates have been divided into two groups based on nucleotide sequences of components 1, 3 and 6 (Karan *et al.*, 1994, 1997; Wanitchakorn *et al.*, 2000). The 'South Pacific' group comprises isolates from Australia, Fiji, Western Samoa, Tonga, Burundi, Egypt and India, whilst the 'Asian' group comprises isolates from the Philippines, Taiwan and Vietnam.

Polyclonal and monoclonal antibodies are now used routinely in ELISAs to detect BBTV in field and tissue culture plants, as well as in single viruliferous aphids (Wu and Su, 1990b; Dietzgen and Thomas, 1991; Thomas and Dietzgen, 1991; Thomas *et al.*, 1995; Geering and Thomas, 1996). All isolates tested from Africa, Australia, Asia and the Pacific region are related serologically (Thomas, 1991). BBTV is also detected in plant tissue and viruliferous aphids with DNA and RNA probes, labelled either non-radioactively or with ^{32}P (Hafner *et al.*, 1995; Xie and Hu, 1995), and by PCR (Xie and Hu, 1995). Substances in banana sap that inhibit PCR can be circumvented by immunocapture PCR (Sharman *et al.*, 2000).

Epidemiology

BBTV is transmitted by *P. nigronervosa* and in vegetative planting material (conventional and micropropagated), but not by mechanical inoculation (Magee, 1927; Drew *et al.*, 1989; Ramos and Zamora, 1990; Wu and Su, 1991). Magee (1927) showed that banana bunchy top was systemic and concluded that the virus was restricted to the phloem tissue (Magee, 1939). Following aphid inoculation, symptoms generally do not appear until two or more leaves are produced (Magee, 1927). This period can vary between 19 days in summer and 125 days in winter (Allen, 1978a). The virus can only be recovered by aphids from the first or subsequent symptomatic leaves (Magee, 1940a). Suckers produced in infected mats generally develop symptoms before reaching maturity (Magee, 1927).

Using RNA probes and PCR, Hafner *et al.* (1995) demonstrated that BBTV replicates for a short period at the site of aphid inoculation, then moves down the pseudostem to the basal meristem, and ultimately to the rhizome, roots and newly formed leaves. Trace levels of virus were detected eventually by PCR in leaves formed prior to inoculation, but replication was not demonstrated, consistent with the inability to transmit the virus by aphids from such leaves (Magee, 1940a).

BBTV is transmitted by *P. nigronervosa* in a circulative, non-propagative manner (Magee, 1927). The transmission parameters reported from Hawaii (Hu *et al.*, 1996) and Australia (Magee, 1927) are, respectively: minimum acquisition access period of 4 and 17 h; minimum inoculation access period of 15 min and of 30 min to 2 h; and retention of infectivity after removal from virus source of 13 and 20 days. No evidence was found for multiplication in or transmission of BBTV to the parthogenetic offspring of the aphid vector (Magee, 1940a; Hafner *et al.*, 1995; Hu *et al.*, 1996). Reported transmission efficiencies for individual aphids range from 46 to 67% (Magee, 1927; Wu and Su, 1990a; Hu *et al.*, 1996), and the virus is acquired more efficiently by nymphs than adults (Magee, 1940a).

P. nigronervosa is found worldwide on banana, *M. textilis*, and other species in the *Musaceae* and related families (Wardlaw, 1961; R.N. Allen, Brisbane, 1996, personal communication). Some host preference is displayed and it is transferred with difficulty between some host species.

Spread of the virus over long distances is by infected planting material, and it is by this means that new plantings in isolated areas usually become infected. Dissemination over short distances from these foci is by the banana aphid. Allen (1978b, 1987) showed that in Australia the average distance of secondary spread of the disease by aphids was only 15.5–17.2 m. Nearly two-thirds of the new infections were within 20 m of the nearest infected plant, and 99% were within 86 m. Allen and Barnier (1977) showed that if a new plantation was adjacent to a diseased plantation, there was an 88% probability that the disease would move into the new plantation within 12 months. This was reduced to 27% if the plantations were separated by 50–1000 m, and to <5% if they were 1000 m apart. The average interval between infection of a plant and spread of the disease from it by aphids to other plants (the disease latent period) equalled the time needed for 3.7 leaves to emerge. The maximum rate of leaf emergence occurred during the summer (Allen, 1987).

BBTV infects a wide range of cultivars in the *Eumusa* and *Australimusa* series of edible banana, *Ensete ventricosum*, and the following taxa: *M. balbisiana* (Magee, 1948; Espino *et al.*, 1993), *M. acuminata* ssp. *banksii* and *M. textilis* (Magee, 1927), *M. velutina* (Thomas and Dietzgen, 1991), *M. coccinea*, *M. jackeyi*, *M. ornata* and *M. acuminata* ssp. *zebrina* (A.D.W. Geering and J.E. Thomas, Brisbane, 1998, personal communication). Although hosts outside the *Musaceae* have been reported, these results have not been corroborated (for a summary, see Jones, 2000).

Management

Although there are no confirmed reports of immunity to bunchy top, differences in susceptibility between cultivars have been noted (Magee, 1948; Jose, 1981; Muharam, 1984; Espino *et al.*, 1993). All AA and AAA cultivars evaluated by Espino *et al.* (1993) were highly susceptible. However, mild or no symptoms developed on some cultivars containing the B genome including, 'Bungaoisan' AAB (plantain subgroup) and 'Abuhon' (ABB). Although cultivars in the Cavendish subgroup generally are highly susceptible, not all cultivars with an AAA genome are so susceptible. For example, the concentration of BBTV virions and the proportion of plants infected by aphid inoculation is higher in the Cavendish clone 'Williams' than in 'Gros Michel'. Symptoms are also slower to develop and less severe on 'Gros Michel' (A.D.W. Geering and J.E. Thomas, Brisbane, 1997, personal communication).

Magee (1927) proposed a range of measures for the control of bunchy top. They involved two major components, exclusion of the disease from non- or slightly affected areas, and eradication of infected plants. In Australia, the following control measures were legislated (Magee, 1936) and remain in force to this day:

- registration of all banana plantations;
- establishment of quarantine zones (Fig. 4.30);
- restrictions on the movement and use of planting material;
- regular inspections of all banana plantations for bunchy top;
- ongoing education and extension programmes for growers; and
- prompt destruction of all infected plants.

The last of these measures is critical and involves first killing aphids on the plant with kerosene or insecticide, and then killing all contiguous plants in a mat and preventing regrowth (Beaver, 1982). In total, these measures enabled the complete rehabilitation of the Australian banana industry. Attempts to control bunchy top by controlling the aphid vector have met with limited success.

Fig. 4.30. Signpost at the entry to a banana plantation in Queensland, Australia. Strict quarantine measures that were enacted to combat banana bunchy top in Australia in the late 1920s were responsible for rejuvenating the banana industry in this country (photo: R.C. Ploetz, UF).

Acknowledgements

The authors thank David Jones for reviewing the chapter, and Anne Desjardins for information on mycotoxological species of *Fusarium*. R. Harry Stover is thanked for use of his contributions to our understanding of banana diseases during the last 50 years.

References

Adam, A.V. (1962) An effective program for the control of banana mosaic. *Plant Disease Reporter* 46, 366–370.

Ahlawat, Y.S., Pant, R.P., Lockhart, B.E.L., Srivastava, M., Chakraborty, N.K. and Varma, A. (1996) Association of a badnavirus and citrus mosaic disease in India. *Plant Disease* 80, 590–592.

Alexopoulos, C.J., Mims, C.W. and Blackwell, M. (1996) *Introductory Mycology*. 4th edn. John Wiley & Sons, New York.

Allen, R.N. (1978a) Epidemiological factors influencing the success of roguing for the control of bunchy top disease of bananas in New South Wales. *Australian Journal of Agricultural Research* 29, 535–544.

Allen, R.N. (1978b) Spread of bunchy top disease in established banana plantations. *Australian Journal of Agricultural Research* 29, 1223–1233.

Allen, R.N. (1987) Further studies on epidemiological factors influencing control of banana bunchy top disease, and evaluation of control measures by computer simulation. *Australian Journal of Agricultural Research* 38, 373–382.

Allen, R.N. and Barnier, N.C. (1977) The spread of bunchy top disease between banana plantations in the Tweed River District during 1975–76. *NSW Department of Agriculture, Biology Branch Plant Disease Survey (1975–76)*, pp 27–28.

Anonymous (1993) Risks involved in the transfer of banana and plantain germplasm. In: *Bananas, Plantains and INIBAP, Annual Report 1993*. INIBAP, Montpellier, France, pp. 39–47.

Anonymous (1995) *Musa*News. *Infomusa* 4 (2), 26–30.

Anonymous (2001) FAOSTAT online database at: http://www.fao.org/default.htm

Bateson, M.F. and Dale, J.L. (1995) Banana bract mosaic virus: characterisation using potyvirus specific degenerate PCR primes. *Archives of Virology* 140, 515–527.

Beaver, R.G. (1982) Use of picloram for eradication of banana diseased with bunchy top. *Plant Disease* 66, 906–907.

Belalcázar, S., Reichel, H., Pérez, R., Gutierréz, T., Múnera, G. and Arévalo, E. (1998) Banana streak badnavirus infection in *Musa* plantations in Colombia. In: Frison, E. and Sharrock, S.L. (eds) *Banana Streak Virus: a UniqueVirus–Musa Interaction? Proceedings of a Workshop of the PROMUSA Virology Working Group*, Montpellier, France. January 19–21, 1998. INIBAP, Montpellier, France.

Berg, L.A. (1968) Diamond spot of bananas caused by *Fusarium roseum* 'Gibbosum'. *Phytopathology* 58, 388–389.

Berg, L.A. and Bustamante, M. (1974) Heat treatment and meristem culture for the production of virus-free bananas. *Phytopathology* 64, 320–322.

Bouhida, M. and Lockhart, B.E. (1990) Increase in importance of cucumber mosaic virus infection in greenhouse-grown bananas in Morocco. *Phytopathology* 80, 981.

Bouhida, M., Lockhart, B.E.L. and Olszewski, N.E. (1993) An analysis of the complete, nucleotide sequence of a sugarcane bacilliform virus genome infection of banana and rice. *Journal of General Virology* 74, 15–22.

Braithwaite, K., Geijskes, J., Geering, A., McMichael, L., Thomas, J. and Smith, G. (1997) Genetic variation in sugarcane bacilliform virus and banana streak virus in Australia. In: *Abstracts of the Pathology and Molecular Biology Workshop*. International Society of Sugar Cane Technologists, Umhlanga Rocks, Keva Zulu-Natel, South Africa.

Bridge, J., Fogain, R. and Speijer, P. (1997) The root lesion nematodes of banana. Musa *Pest Fact Sheet No. 2*. INIBAP, Montpellier, France.

Buddenhagen, I.W. (1960) Strains of *Pseudomonas solanacearum* in indigenous hosts in banana plantations in Costa Rica, and their relationship to bacterial wilt of banana. *Phytopathology* 50, 660–664.

Buddenhagen, I.W. (1993) Whence and whither banana research and development. In: *Biotechnology Applications for Banana and Plantain Improvement*. INIBAP, Montpellier, France, pp. 12–26.

Burns, T.M., Harding, R.M. and Dale, J.L. (1995) The genome organization of banana bunchy top virus: analysis of six ssDNA components. *Journal of General Virology* 76, 1471–1482.

Caruana, M.L. and Galzi, S. (1998) Identification of uncharacterised filamentous viral particles on banana plants. *Acta Horticulturae* 490, 323–335.

Caruana, M.L., Galzi, S., Séchet, H., Bousalem, M. and Bringaud, C. (1995) Etiologie de la maladie de la mosaqque des bractées des bananiers pour un diagnostic cible. *Laboratoire de Phytovirologie des Régions Chaudes, CIRAD, ORSTOM, Rapport D'Activities 1994–1995*. pp. 6–9.

Carlier, J., Lebrun, M.H., Zapater, M.-F., Dubois, C. and Mourichon, X. (1996) Genetic structure of the global population of banana black leaf streak fungus *Mycosphaerella fijiensis*. *Molecular Ecology* 5, 499–510.

Carlier, J., Zapater, M.-F., Lapeyre, F., Jones, D.R. and Mourichon, X. (2000) Septoria leaf spot of banana: a newly discovered disease caused by *Mycosphaerella eumusae* (anamorph *Septoria eumusae*). *Phytopathology* 90, 884–890.

Carreel, F. (1995) Etude de la diversité génétique des bananiers (genre *Musa*) á l'aide des marqueurs RFLP. PhD thesis, Institut National Agronomique, Paris-Grignan, France.

Chakrabarti, D.K. and Ghosal, S. (1986) Occurrence of free and conjugated 12,13-epoxytrichothecenes and zearalenone in banana fruits infected with *Fusarium moniliforme*. *Applied and Environmental Microbiology* 51, 217–219.

Chin, K.M., Wirz, M. and Laird, D. (2001) Sensitivity of *Mycosphaerella fijiensis* from banana to trifloxystrobin. *Plant Disease* 85, 1264–1270.

Chuang, T.Y. (1981) Isolation of *Phyllosticta musarum*, causal organism of banana freckle. *Transactions of the British Mycological Society* 77, 670–671.

Chuang, T.-Y. (1988) Studies on the soils suppressive to banana fusarium wilt. II. Nature of suppression to race 4 of *Fusarium oxysporum* f. sp. *cubense* in Taiwan soils. *Plant Protection Bulletin (Taiwan)* 30, 125–134.

Cook, D. and Sequeira, L. (1994) Strain differentiation of *Pseudomonas solanacearum* by molecular genetic methods. In: Hayward, A.C. and Hartman, G.L. (eds) *Bacterial Wilt: the Disease and its Causative Agent,* Pseudomonas solanacearum. CAB International, Wallingford, UK, pp. 77–93.

Crous, P.W. (2002) *Taxonomy and Pathology of* Cylindrocladium (Calonectria) *and Allied Genera*. APS Press, St Paul, Minnesota.

Crous, P.W. and Mourichon, X. (2002) *Mycosphaerella eumusae* and its anamorph *Pseudocercospora eumusae* spp. nov.: causal agent of eumusae leaf spot disease of banana. *Sydowia* 54, 35–43.

Cummings, G.B. (1941) Uredinales of New Guinea – III. *Mycologia* 33, 143–154.

Dahal, G., Hughes, J. d'A. and Lockhart, B.E.L. (1998a) Status of banana streak disease in Africa: problems and future research needs. *Integrated Pest Management Reviews* 3, 85–97.

Dahal, G., Hughes, J. d'A., Thottappilly, G. and Lockhart, B.E.L. (1998b) Effect of temperature on symptom expression and reliability of banana streak badnavirus detection in naturally infected plantain and banana (*Musa* spp.). *Plant Disease* 82, 16–21.

Dahal, G., Ortiz, R., Tenkouano, A., Hughes, J.A., Thottapilly, G., Vuylsteke, D. and Lockhart, B.E.L. (2000) Relationships between natural occurrence of banana streak badnavirus and symptom expression, relative concentration of virus antigen, and yield characteristics of some micropropagated *Musa* spp. *Plant Pathology* 49, 68–79.

Dale, J.L. (1987) Banana bunchy top: an economically important tropical plant virus disease. *Advances in Virus Research* 33, 301–325.

Dallot, S., Acuña, P., Rivera, C., Ramírez, P., Côte, F., Lockhart, B.E.L. and Caruana, M.L. (2001) Evidence that the proliferation stage of micropropagation procedure is determinant in the expression of *Banana streak virus* integrated into the genome of the FHIA 21 hybrid (*Musa* AAAB). *Archives of Virology* 146, 2179–2190.

Daniells, J., Thomas, J.E. and Smith, B.J. (1995) Seed transmission of banana streak confirmed. *Info*Musa 4 (1), 7.

Daniells, J.W., Geering, A.D.W., Bryde, N.W. and Thomas, J.E. (2001a) The effect of *Banana streak virus* on the growth and yield of dessert bananas in tropical Australia. *Annals of Applied Virology* 139, 51–60.

Daniells, J., Jenny, C., Karamura, D. and Tomekpe, K. (2001b) Musa*logue. A Catalog of Musa Germplasm. Diversity in the Genus* Musa. INIBAP, Montpellier, France.

Darnell-Smith, G.P. (1924) 'Bunchy top' disease in banana. *Queensland Agricultural Journal* 21, 169–179.

David, J.C. (1988) Cladosporium musae. *CMI Descriptions of Pathogenic Fungi and Bacteria No. 958*. International Mycological Institute, Kew, UK.

Davis, A.J., Say, M., Snow, A.J. and Grant, B.R. (1994) Sensitivity of *Fusarium oxysporum* f. sp *cubense* to phosphonate. *Plant Pathology* 43, 200–205.

Deacon, J.W., Herbert, J.A. and Dames, J. (1985) False Panama disorder of bananas. *ITSC Information Bulletin* 149, 15–18.

de Beer, Z. Hernández, J.M. and Sabadel, S. (2001) False Panama disorder on banana. Musa *Factsheet No. 9*. INIBAP, Montpellier, France.

de Hoog, G.S. (1977) *Rhinocladiella* and allied genera. *Studies in Mycology* 15, 1–140.

de Lapeyre de Bellaire, L., Chillet, M., Dubois, C. and Mourichon, X. (2000) Importance of different sources of inoculum and dispersal methods of conidia of *Colletotrichum musae*, the causal agent of anthracnose, for fruit contamination. *Plant Pathology* 49, 782–790.

Devergne, J.C. and Cardin, L. (1973) Contribution à l'étude du virus de la mosaïque du concombre (CMV). IV. Essai de classification de plusieurs isolats sur la base de leur structure antigénique. *Annales de Phytopathologie* 73, 409–430.

Diekmann, M. and Putter, C.A.J. (eds) (1996) *FAO/IPGRI Technical Guidelines for the Safe Movement of Germplasm. No. 15*, Musa, 2nd edn. Food and Agriculture Organization of the United Nations/International Plant Genetic Resources Institute, Rome.

Dietzgen, R.G. and Thomas, J.E. (1991) Properties of virus-like particles associated with banana bunchy top disease in Hawaii, Indonesia and Tonga. *Australasian Plant Pathology* 20, 161–165.

Domínguez, J., Negrín, M.A. and Rodríguez, C.M. (2001) Aggregate water-stability, particle size and soil solution properties in conducive and suppressive soils to Fusarium wilt of banana from Canary Islands. *Soil Biology and Biochemistry* 33, 449–455.

Domsch, K.H., Gams, W. and Anderson, T.-H. (1980) *Compendium of Soil Fungi*, Vol. 1. Academic Press, New York.

Drew, R.A., Moisander, J.A. and Smith, M.K. (1989) The transmission of banana bunchy-top virus in micropropagated bananas. *Plant Cell, Tissue and Organ Culture* 16, 187–193.

Eden-Green, S.J. and Sastraatmadja, H. (1990) Blood disease of bananas in Sulawesi and Java. *FAO Plant Protection Bulletin 38*.

Ellis, M.B. (1971) *Dematiaceous Hyphomycetes*. Commonwealth Mycological Institute, Kew, UK.

Ellis, M.B. (1976) *More Dematiaceous Hyphomycetes*. Commonwealth Mycological Institute, Kew, UK.

Ellis, M.B. and Holliday, P. (1972) Cordana musae. *CMI Descriptions of Pathogenic Fungi and Bacteria No. 350*. Commonwealth Mycological Insititute, Kew, UK.

Espino, R.C., Johns, A.P., Juanillo, C. and Magnaye, L.V. (1993) Evaluation of Philippine banana cultivars for resistance to bunchy-top and fusarium wilt. In: Valmayor, R.V., Hwang, S.C., Ploetz, R., Lee, S.C. and Roa, N.V. (eds) *Proceedings: International Symposium on Recent Developments in Banana Cultivation Technology*. Taiwan Banana Research Institute, Chiuju, Pingtung, Taiwan, December 14–18, 1992. INIBAP/ASPNET, Los Baños, Philippines, pp. 89–102.

Ferreira, S.A., Trujillo, E.E. and Ogata, D.Y. (1989) *Bunchy Top Disease of Bananas*. Information leaflet prepared by the Hawaii Cooperative Extension Service, Hawaii Institute of Tropical Agriculture and Human Resources, University of Hawaii at Manoa. Commodity Fact Sheet BAN-4(A), FRUIT.

Firman, I.D. (1976) Banana rust in Fiji and other Pacific islands. *Fiji Agricultural Journal* 38, 85–86.

Francki, R.I.B., Mossop, D.W. and Hatta, T. (1979) Cucumber mosaic virus CMI/AAB. *Descriptions of Plant Viruses No. 213*. Commonwealth Mycological Institute and Association of Applied Biologists, Kew, UK.

French, E.R. and Sequeira, L. (1970) Strains of *Pseudomonas solanacearum* from Central and South America: a comparative study. *Phytopathology* 60, 506–512.

Fullerton, R.A. and Olsen, T.L. (1995) Pathogenic variability in *Mycosphaerella fijiensis* Morelet, cause of black Sigatoka in banana and plantain. *New Zealand Journal of Horticultural Science* 23, 39–48.

Fullerton, R.A. and Stover, R.H. (eds) (1990) *Sigatoka Leaf Spot Diseases of Banana*. Proceedings of an International Workshop held at San José, Costa Rica, March 28–April 1, 1989. INIBAP, Montpellier, France.

Gafny, R., Wexler, A., Mawassi, M., Israeli, Y. and Bar-Joseph, M. (1996) Natural infection of banana by a satellite-containing strain of cucumber mosaic virus: nucleotide sequence of the coat protein gene and the satellite RNA. *Phytoparasitica* 24, 49–56.

Gambley, C.F., Sharman, M., Thomas, J.E., Ndowora, T.R.C. and Lockhart, B.E.L. (1999) Detection and relationships of a new filamentous virus in banana. 12th Biennial Conference, Australasian Plant Pathology Society. Canberra, September 27–30, 1999.

Gambley, C.F. and Thomas, J.E. (2001) Molecular characterisation of banana mild mosaic virus, a new filamentous virus in *Musa* spp. *Archives of Virology* 146, 1369–1379.

Gauhl, F. and Pasberg-Gauhl, C. (1994) *Symptoms Associated with Banana Streak Virus (BSV)*. Plant Health Management Division, International Institute of Tropical Agriculture, Ibadan, Nigeria.

Geering, A.D.W. and Thomas, J.E. (1996) A comparison of four serological tests for the detection of banana bunchy top virus in banana. *Australian Journal of Agricultural Research* 47, 403–412.

Geering, A.D.W. and Thomas, J.E. (2002) Banana streak virus. *Association of Applied Biologists Description of Plant Viruses*. Description No. 390, AAB, Wellesbourne, UK.

Geering, A.D.W., McMichael, L.A., Dietzgen, R.G. and Thomas, J.E. (2000) Genetic diversity among *Banana streak virus* isolates from Australia. *Phytopathology* 90, 921–927.

Geering, A.D.W., Olszewski, N.E., Dahal, G., Thomas, J.E. and Lockhart, B.E.L. (2001) Analysis of the distribution and structure of integrated *Banana streak virus* DNA in a range of *Musa* cultivars. *Molecular Plant Pathology* 2, 207–213.

Gold, A.H. (1972) Seed transmission of banana viruses. *Phytopathology* 62, 760.
Gowen, S.R. and Quénéhervé, P. (1990) Nematode parasites of bananas, plantains and abaca. In: Luc, M. Sikora, R.A. and Bridge, J. (eds) *Plant Parasitic Nematodes in Subtropical and Tropical Agriculture.* CAB International, Wallingford, UK, pp. 431–460.
Greene, G.L. and Goos, R.D. (1963) Fungi associated with crown rot of boxed bananas. *Phytopathology* 53, 271–275.
Griffee, P.J. (1976) Pathogenicity of some fungi isolated from diseased crowns of banana hands. *Phytopathologische Zeitschrift* 85, 206–216.
Guyon, M. (1970) Essais de lutte chimique contre la Johnson fruit spot au Nicarauga. *Fruits* 25, 685–691.
Hafner, G.J., Harding, R.M. and Dale, J.L. (1995) Movement and transmission of banana bunchy top virus DNA component one in bananas. *Journal of General Virology* 76, 2279–2285.
Halmos, S. (1970) Inoculum sources of *Pyricularia grisea*, the cause of pitting disease of bananas. *Phytopathology* 60, 183–184.
Harding, R.M., Burns, T.M. and Dale, J.L. (1991) Virus-like particles associated with banana bunchy top disease contain small single-stranded DNA. *Journal of General Virology* 72, 225–230.
Harper, G., Dahal, G., Thottappilly, G. and Hull, R. (1999a) Detection of episomal banana streak badnavirus by IC-PCR. *Journal of Virological Methods* 79, 1–8.
Harper, G., Osuji, J., Heslop-Harrison, J.S. and Hull, R. (1999b) Integration of banana streak badnavirus into the *Musa* genome: molecular and cytogenetic evidence. *Virology* 255, 207–213.
Hasse, A., Richter, J. and Rabenstein, F. (1989) Monoclonal antibodies for detection and serotyping of cucumber mosaic virus. *Journal of Phytopathology* 127, 129–136.
Hawksworth, D.L. and Holliday, P. (1970) Verticillium theobromae. *CMI Descriptions of Pathogenic Fungi and Bacteria No. 259.* Commonwealth Mycological Insititute, Kew, UK.
Helliot, B., Panis, B., Poumay, Y., Swennen, R., Lepoivre, P. and Frison, E. (2002) Cryopresevation for the elimination of cucumber mosaic and banana streak viruses from banana (*Musa* spp.). *Plant Cell Reports* 20, 1117–1122.
Herbert, J.A. and Marx, D. (1990) Short-term control of Panama disease of bananas in South Africa. *Phytophylactica* 22, 339–340.
Holliday, P. (1970) Trachysphaera fructigena *CMI Descriptions of Pathogenic Fungi and Bacteria No. 229.* Commonwealth Mycological Institute, Kew, UK.
Hu, J.S., Li, H.P., Barry, K. and Wang, M. (1995) Comparison of dot blot, ELISA and RT–PCR assays for detection of two cucumber mosaic virus isolates infecting banana in Hawaii. *Plant Disease* 79, 902–906.
Hu, J.S., Wang, M., Sether, D., Xie, W. and Leonhardt, K.W. (1996) Use of polymerase chain reaction (PCR) to study transmission of banana bunchy top virus by the banana aphid (*Pentalonia nigronervosa*). *Annals of Applied Biology* 128, 55–64.
Hull, R., Harper, G. and Lockhart, B. (2000) Viral sequences integrated into plant genomes. *Trends in Plant Sciences* 5, 362–365.
Jeger, M.J., Eden-Green, S., Thresh, J.M., Johanson, A., Waller, J.M. and Brown, A.E. (1995) Banana diseases. In: Gowen, S. (ed.) *Bananas and Plantains.* Chapman and Hall, London, pp. 317–381.
Jiménez, M., Huerta, T. and Mateo, R. (1997) Mycotoxin production by *Fusarium* species isolated from banana. *Applied and Environmental Microbiology* 63, 364–369.
Johanson, A. and Blazquez, B. (1992) Fungi associated with banana crown rot on field-packed fruit from the Windward Islands and assessment of their sensitivity to fungicides thiabendazole, prochloraz and imazilil. *Crop Protection* 11, 79–83.
Johanson, A. and Jeger, M.J. (1993) Use of PCR for detection of *Mycosphaerella fijiensis* and *M. musicola*, the causal agents of Sigatoka leaf spots in banana and plantain. *Mycological Research* 97, 670–674.
Jones, D.R. (ed.) (2000) *Diseases of Banana, Abacá and Enset.* CAB International, Wallingford, UK.
Jones, D.R., Pegg, K.G. and Thomas, J.E. (1993) Banana. In: Persley, D. (ed.) *Diseases of Fruit Crops.* Department of Primary Industries Queensland, Brisbane, Australia, pp. 25–35.
Jose, P.C. (1981) Reaction of different varieties of banana against bunchy top disease. *Agricultural Research Journal of Kerala* 19, 108–110.
Kaiser, W.J. and Lukezec, F.L. (1965) Brown spot disease of banana caused by *Cercospora hayi. Phytopathology* 55, 977–980.
Kaiser, W.J. and Lukezic, F.L. (1966a) Influences of certain environmental conditions on spore dispersal and survival of *Cercospora hayi* from banana. *Phytopathology* 56, 1290–1293.
Kaiser, W.J. and Lukezic, F.L. (1966b) Occurrence, sporulation and pathogenicity studies with *Glomerella cingulata* associated with crown rot of boxed bananas. *Mycologia* 58, 397–405.

Kangire, A. and Rutherford, M.A. (2001) Wilt-like disorder of bananas in Uganda. *Musa Factsheet No. 10.* INIBAP, Montpellier, France.

Karan, M., Harding, R.M. and Dale, J.L. (1994) Evidence for two groups of banana bunchy top virus isolates. *Journal of General Virology* 75, 3541–3546.

Karan, M., Harding, R.M. and Dale, J.L. (1997) Association of banana bunchy top virus DNA components 2 to 6 with bunchy top disease. *Molecular Plant Pathology* online. http://www.bspp.org.uk/mppol/1997/0624karan

Kenyon, L., Magnaye, L., Warburton, H. and Herradura, L. (1996) Epidemiology and control of banana virus diseases in the Philippines. NRI-Department for International Development Crop Protection Programme Project A0217/X0258 Final Technical Report.

Khalid, S. and Soomro, M.H. (1993) Banana bunchy top disease in Pakistan. *Plant Pathology* 42, 923–926.

Knight, C., Cutts, D.F. and Colhoun, J. (1977) The role of *Fusarium semitectum* in causing crown rot of bananas. *Phytopathologische Zeitschrift* 89, 170–176.

Krauss, U. and Johanson, A. (2000) Recent advances in the control of crown rot of banana in the Windward Islands. *Crop Protection* 9, 151–160.

Kress, W.J., Prince, L.M., Hann, W.J. and Zimmer, E.A. (2001) Unraveling the evolutionary radiation of the families of the Zingiberales using morphological and molecular evidence. *Systematic Biology* 50, 926–944.

Kubiriba, J., Legg, J.P., Tushemereirwe, W. and Adipala, E. (2001a) Disease spread patterns of *Banana streak virus* in farmers' fields in Uganda. *Annals of Applied Biology* 139, 31–36.

Kubiriba, J., Legg, J.P., Tushemereirwe, W. and Adipala, E. (2001b) Vector transmission of *Banana streak virus* in the screenhouse in Uganda. *Annals of Applied Biology* 139, 37–43.

LaFleur, D.A., Lockhart, B.E.L. and Olszewski, N.E. (1996) Portions of the banana streak badnavirus genome are integrated in the genome of its host *Musa* spp. *Phytopathology* 86, 11, S100.

Lakshmanan, P., Selvaraj, P. and Mohan, S. (1987) Efficacy of different methods for the control of Panama disease. *Tropical Pest Management* 33, 373–374.

Langdon, R. (1993) The banana as a key to early American and Polynesian history. *Journal of Pacific History* 28, 15–35.

Lassoudière, A. (1974) La mosaique dite 'a tirets' du bananier 'Poyo' en Côte d'Ivoire. *Fruits* 29, 349–357.

Lassoudière, A. (1979) Mise en évidence des répercussions économiques de la mosaqque en tirets du bananier en Côte d'Ivoire. Possibités de lutte par éradication. *Fruits* 34, 3–34.

Lockhart, B.E.L. (1986) Purification and serology of a bacilliform virus associated with banana streak disease. *Phytopathology* 76, 995–999.

Lockhart, B.E.L. (1994) Development of detection methods for banana streak virus. In: *The Global Banana and Plantain Network, INIBAP Annual Report 1994.* INIBAP, Montpellier, France, pp. 20–21.

Lockhart, B.E.L. (1995) Banana streak badnavirus infection in *Musa*: epidemiology, diagnosis and control. *Food and Fertilizer Technology Center Technical Bulletin 143.* Food and Fertilizer Technology Center, Taipei, Taiwan.

Lockhart, B.E.L. and Olszewski, N.E. (1993) Serological and genomic heterogeneity of banana streak badnavirus: implications for virus detection in *Musa* germplasm. In: Ganry, J. (ed.) *Breeding Banana and Plantain for Resistance to Diseases and Pests, Proceedings of the International Symposium on Genetic Improvement of Bananas for Resistance to Diseases and Pests,* organized by CIRAD-FLHOR, Montpellier, France, September 7–9, 1992. CIRAD, Montpellier, France, pp. 105–113.

Logrieco, A., Moretti, A., Castella, G., Kostecki, M., Golinski, P., Ritieni, A. and Chelkowski, J. (1998) Beauvericin production by *Fusarium* species. *Applied and Environmental Microbiology* 63, 364–369.

Magee, C.J.P. (1927) *Bulletin No. 30. Investigation on the Bunchy Top Disease of the Banana.* Council for Scientific and Industrial Research, Melbourne.

Magee, C.J.P. (1930) A new virus disease of bananas. *Agricultural Gazette of New South Wales* XLI, 929.

Magee, C.J. (1936) Bunchy top of bananas – rehabilitation of the banana industry of NSW. *Journal of the Australian Institute of Agricultural Science* 2, 13–16.

Magee, C.J.P. (1939) Pathological changes in the phloem and neighbouring tissues of the banana (*Musa cavendishii* Lamb.) caused by the bunchy-top virus. In: *Science Bulletin.* Department of Agriculture, New South Wales, Sydney, pp. 4–32.

Magee, C.J.P. (1940a) Transmission studies on the banana bunchy-top virus. *Journal of the Australian Institute of Agricultural Science* 6, 109–110.

Magee, C.J.P. (1940b) Transmission of infectious chlorosis or heart-rot of the banana and its relationship to cucumber mosaic. *Journal of the Australian Institute of Agricultural Science* 6, 44–47.

Magee, C.J.P. (1948) Transmission of banana bunchy top to banana varieties. *Journal of the Australian Institute of Agricultural Science* 14, 18–24.
Magee, C.J. (1953) Some aspects of the bunchy top disease of banana and other *Musa* spp. *Journal and Proceedings of the Royal Society of New South Wales* 87, 3–18.
Magnaye, L.V. (1994) Virus diseases of banana and current studies to eliminate the virus by tissue culture. In: Tangonan, N.G. (ed.) *Towards Making Pest and Disease Management Relevant to Big and Small Banana Growers. Proceedings of the 1st PPS-SMD National Symposium on Pests and Diseases in the Philippines,* Davao City, Philippines, April 23–24, 1993. Phytopathological Society Inc., Southern Mindanao Division, pp. 38–43.
Magnaye, L.V. and Espino, R.R.C. (1990) Note: banana bract mosaic, a new disease of banana. I. Symptomatology. *The Philippine Agriculturist* 73, 55–59.
Marin, D.H., Sutton, T.B. and Barker, K.R. (1998) Dissemination of bananas in Latin America and the Caribbean and its relationship to the occurrence of *Radopholus similis. Plant Disease* 82, 964–974.
Mbida Mindzie, C., Doutrelepont, H., Vrydaghs, L., Swennen, R., Swennen, R.J., Beeckman, H., DeLanghe, E. and DeMaret, P. (2001) First archaeological evidence of banana cultivation in central Africa during the third millennium before present. *Vegetation History and Archaeobotany* 10, 1–6.
McSorley, R. and Parrado, J.L. (1983) The spiral nematode, *Helicotylenchus multicinctus*, on bananas in Florida and its control. *Proceedings of the Florida State Horticultural Society* 96, 201–207.
McSorley, R. and Parrado, J.L. (1986) *Helicotylenchus multicinctus* on bananas: an international problem. *Nematropica* 16, 73–91.
Meredith, D.S. (1963) *Pyricularia grisea* (Cooke) Sacc. causing pitting disease of bananas in Central America. I. Preliminary studies on pathogenicity. *Annals of Applied Biology* 52, 453–463.
Meredith, D.S. (1965) Tip rot of banana fruits in Jamaica. 2. *Verticillium theobromae* and *Fusarium* spp. *Transactions of the British Mycological Society* 48, 327–336.
Mobambo, K.N., Gauhl, F., Swennen, R. and Pasberg-Gauhl, C. (1996) Assessment of the cropping cycle effects on black leaf streak severity and yield decline of plantain and plantain hybrids. *International Journal of Pest Management* 42, 1–8.
Moore, N., Pegg, K.G., Langdon, P.W., Smith, M.K. and Whiley, A.W. (1993) Current research on Fusarium wilt of banana in Australia. In: Valmayor, R.V., Hwang, S.C., Ploetz, R.C., Lee, S.W. and Roa, V.N. (eds) *Proceedings: International Symposium on Recent Developments in Banana Cultivation Technology*, Taiwan Banana Research Institute, Chiuju, Pingtung, Taiwan, December 14–18, 1992. INIBAP/ASPNET, Los Baños, Laguna, Philippines, pp. 270–284.
Muharam, A. (1984) Test for resistance of some banana cultivars to banana bunchy-top disease. *Bulletin. of Penel. Horticulture* 11, 16–19.
Mulder, J.L. and Holliday, P. (1971) Uromyces musae. *CMI Descriptions of Pathogenic Fungi and Bacteria No. 295*. Commonwealth Mycological Institute, Kew, UK.
Mulder, J.L. and Holliday, P. (1974) Mycosphaerella musicola. *CMI Descriptions of Pathogenic Fungi and Bacteria No. 414*. Commonwealth Mycological Insititute, Kew, UK.
Muñez, A.R. (1992) Symptomatology, transmission and purification of banana bract mosaic virus (BBMV) in 'Giant Cavendish' banana. MSc thesis. University of the Philippines, Los Baños, Philippines.
Ndowora, T., Dahal, G., LaFleur, D., Harper, G., Hull, R., Olszewski, N.E. and Lockhart, B. (1999) Evidence that badnavirus infection in *Musa* can originate from integrated pararetroviral sequences. *Virology* 255, 214–220.
Ndowora, T.C.R. (1998) Banana streak virus: development of an immunoenzymatic assay for detection and characterization of sequences that are integrated in the genome of the host. PhD thesis, University of Minnesota.
Odebode, A.C. and Sanusi, J. (1996) Influence of fungi associated with bananas on nutritional content during storage. *Zeitschrift für Lebesmittel-Untersuchung und -Forschung* 202, 471–473.
O'Donnell, K., Cigelnik, E. and Nirenberg, H.I. (1998) Molecular systematics and phylogeography of the *Giberella fujikuroi* species complex. *Mycologia* 90, 465–493.
Owen, J., Shintaku, Aeschleman, P., Ben Tahar, S. and Palukitis, P. (1990) Nucleotide sequence and evolutionary relationships of cucumber mosaic virus (CMV) strains. CMV RNA 3. *Journal of General Virology* 71, 2243–2249.
Palukaitis, P., Roossink, M.J., Dietzgen, R.G. and Francki, R.I.B. (1992) Cucumber mosaic virus. *Advances in Virus Research* 41, 281–348.
Parnell, M., Burt, P.J.A. and Wilson, K. (1998) The influence of exposure to ultraviolet radiation in simulated sunlight on ascospores causing black Sigatoka disease of banana and plantain. *International Journal of Biometerology* 42, 22–27.

Pegg, K.G. (2000) Results and discussion, IMTP Fusarium, Wamuran, Australia. In: Orjeda, G. (ed.) *Evaluating Bananas: a Global Partnership. Results of IMTP III.* INIBAP, Montpellier, France, pp. 231–259.

Peng, H.X., Sivasithamparam, K. and Turner, D.W. (1999) Chlamydospore germination and Fusarium wilt of banana plantlets in suppressive and conducive soils are affected by physical and chemical factors. *Soil Biology and Biochemistry* 31, 1363–1374.

Piazzolla, P., Diaz-Ruiz, J.R. and Kaper, J.M. (1979) Nucleic acid homologies of eighteen cucumber mosaic virus isolates determined by competition hybridization. *Journal of General Virology* 45, 361–369.

Ploetz, R.C. (2000) Management of the most important disease of banana and plantain, black Sigatoka. *Pesticide Outlook* 11, 19–23.

Ploetz, R.C. and Pegg, K.G. (1997) Fusarium wilt of banana and Wallace's line: was the disease originally restricted to his Indo-Malayan region? *Australasian Plant Pathology* 26, 239–249.

Ploetz, R.C., Jones, D.R., Sebasigari, K. and Tushemerirewe, W. (1994a) Fusarium wilt (Panama disease) on East African highland cultivars of banana. *Fruits* 49, 253–260.

Ploetz, R.C., Zentmyer, G.A., Nishijima, W.T., Rohrbach, K.G. and Ohr, H.D. (eds) (1994b) *Compendium of Tropical Fruit Diseases.* APS Press, St Paul, Minnesota.

Priest, M.J. (1990) Distribution of *Cordana* spp. on *Musa* in Australia. *Mycological Research* 94, 861–863.

Punithalingham, E. (1983) Phaeoseptoria musae. *CMI Descriptions of Pathogenic Fungi and Bacteria No. 772.* Commonwealth Mycological Institute, Kew, UK.

Punithalingham, E. and Holliday, P. (1975) Guignardia musae. *CMI Descriptions of Pathogenic Fungi and Bacteria No. 467.* Commonwealth Mycological Institute, Kew, UK.

Purseglove, J.W. (1985) *Tropical Crops. Monocotyledons.* Longman, London.

Ramos, C.S. and Zamora, A.B. (1990) Elimination of banana bunchy top infection from banana (*Musa* sp cv Lakatan) by heat pretreatment and meristem culture. *Philippine Journal of Crop Science* 15, 119–123.

Reinking, O.A. (1937) Isolations made from heart rot of banana in Honduras. *Phytopathology* 27, 853–854.

Reyes, M.E.Q., Nishijima, W. and Paull, R.E. (1998) Control of crown rot in 'Santa Catarina Prata' and 'Williams' banana with hot water treatments. *Postharvest Biology and Technology* 14, 71–75.

Rhodes, P.L. (1964) A new banana disease in Fiji. *Commonwealth Phytopathological News* 10, 38–41.

Risède, J.-M. (1994) Partial characterization of *Cylindrocladium* sp., a root pathogen of banana in Martinique. *Fruits* 49, 167–178.

Risède, J.-M. and Simoneau, P. (2001) Typing *Cylindrocladium* species by analysis of ribosomal DNA spacers polymorphism: application to field isolates from the banana rhizosphere. *Mycologia* 93, 494–504.

Rivera, C., Ramírez, P. and Pereira, R. (1992) Preliminary characterization of viruses infecting banana in Costa Rica. In: *Biotechnology Applications for Banana and Plantain Improvement.* INIBAP, Montpellier, France, pp. 63–68.

Rodoni, B.C., Ahlawat, Y.S., Varma, A., Dale, J.L. and Harding, R.M. (1997) Identification and characterization of banana bract mosaic virus in India. *Plant Disease* 81, 669–672.

Rodoni, B.C., Dale, J.L. and Harding, R.M. (1999) Characterisation and expression of the coat protein-coding region of banana bract mosaic potyvirus, development of diagnostic assays and detection of the virus in plants from five countries in southeast Asia. *Archives of Virology* 144, 1725–1737.

Rodriguez, R.J. and Owen, J.L. (1992) Isolation of *Glomerella musae* [teleomorph of *Colletotrichum musae* (Berk. & Curt.) Arx.] and segregation analysis of ascospore progeny. *Experimental Mycology* 16, 291–301.

Romero, R.A. and Sutton, T.B. (1997) Sensitivity of *Mycosphaerella fijiensis*, causal agent of black Sigatoka of banana, to propiconazole. *Phytopathology* 87, 96–100.

Roperos, N.I. (1965) Note on the occurrence of new disease of cooking banana in the Philippines. *Coffee and Cacao Journal* 8, 135–136.

Rutter, J., Burt, P.J.A. and Ramirez, F. (1998) Movement of *Mycosphaerella fijiensis* spores and sigatoka disease development on plantain close to an inoculum source. *Aerobiology* 14, 201–208.

Sarah, J.L., Pinochet, J. and Stanton, J. (1996) The burrowing nematode of bananas, *Radopholu similis* Cobb, 1913. Musa *Pest Fact Sheet No. 1.* INIBAP, Montpellier, France.

Schaad, N.W., Jones, J.B. and Chun, W. (eds) (2001) *Laboratory Guide for the Identification of Plant Pathogenic Bacteria.* APS Press, St Paul, Minesota.

Semer, C.R., Mitchell, D.J., Mitchell, M.E., Martin, F.N. and Alfenas, A.C. (1987) Isolation, identification and chemical control of *Cylindrocladium musae*, a new species associated with toppling disease of banana. *Phytopathology* 77, 1729.

Sharman, M., Thomas, J.E. and Dietzgen, R.G. (2000a) Development of a multiplex immunocapture PCR with colorimetric detection for viruses of banana. *Journal of Virological Methods* 89, 75–88.

Sharman, M., Gambley, C.F., Oloteo, E.O., Abgona, R.V.J. and Thomas, J.E. (2000b) First record of natural infection of abaca (*Musa textilis*) with banana bract mosaic potyvirus in the Philippines. *Australasian Plant Pathology* 29, 69.

Shillingford, C.A. and Sinclair, J.B. (1978) Uptake and translocation of systemic fungicides by banana fruits as determined by assay. *Plant Disease Reporter* 62, 1107–1111.

Simmonds, N.W. (1966) *Bananas*, 2nd edn. Longmans, London.

Simmonds, N.W. and Shepherd, K. (1955) Taxonomy and origins of cultivated bananas. *Journal of the Linnaean Society of Botany (London)* 55, 302–312.

Singh, Z., Jones, R.H.C. and Jones, M.G.K. (1995) Identification of cucumber mosaic subgroup I isolates from banana plants affected by infectious chlorosis disease using RT-PCR. *Plant Disease* 79, 713–716.

Smith, M.K., Whiley, A.W., Searle, C., Langdon, P.W., Schaffer, B. and Pegg, K.G. (1998) Micropropagated bananas are more susceptible to fusarium wilt than plants grown from traditional material. *Australian Journal of Agricultural Research* 49, 1133–1139.

Snowdon, A.L. (1990) *A Color Atlas of Postharvest Diseases and Disorders of Fruits and Vegetables*. Vol. 1. *General Introduction and Fruits*. Wolfe Scientific Publishers, London.

Speijer, P.R., Budenberg, W.J. and Sikora, R.A. (1993) Relationship between nematodes, weevils, banana and plantain cultivars and damage. *Annals of Applied Biology* 123, 517–525.

Stoffelen, R., Verlinden, R., Pinochet, J., Swennen, R. and DeWaele, D. (2000) Screening of Fusarium wilt resistant bananas to root-lesion nematodes. *Info*Musa 9, 6–8.

Stover, R.H. (1962) *Fusarial wilt (Panama Disease) of Bananas and other* Musa *species*. Commonwealth Mycological Institute, Kew, UK.

Stover, R.H. (1972) *Banana, Plantain and Abaca Diseases*. Commonwealth Mycological Institute, Kew, UK.

Stover, R.H. (1975) Sooty moulds of bananas. *Transactions of the British Mycological Society* 65, 328–330.

Stover, R.H. (1990a) Sigatoka leaf spots: thirty years of changing control strategies: 1959–1989. In: Fullerton, R.A. and Stover, R.H. (eds) *Sigatoka Leaf Spot Diseases of Banana*. INIBAP, Montpellier, France, pp. 66–74.

Stover, R.H. (1990b) Fusarium wilt of banana: some history and current status of the disease. In: Ploetz, R.C. (ed.) *Fusarium Wilt of Banana*. APS Press, St Paul, Minnesota, pp. 1–7.

Stover, R.H. and Simmonds, N.W. (1987) *Bananas*, 3rd edn. Longmans, London.

Su, H.J., Wu, R.Y. and Tsao, L.Y. (1993) Ecology of banana bunchy-top virus disease. In: Valmayor, R.V., Hwang, S.C., Ploetz, R., Lee, S.C. and Roa, N.V. (eds) *Proceedings: International Symposium on Recent Developments in Banana Cultivation Technology*, Taiwan Banana Research Institute, Chiuju, Pingtung, Taiwan, December 14–18, 1992. INIBAP/ASPNET, Los Baños, Philippines, pp. 308–312.

Subrumanian, C.V. (1968) Deightoniella tourlosa. *CMI Descriptions of Pathogenic Fungi and Bacteria No. 165*. Commonwealth Mycological Institute, Kew, UK.

Sutton, B.C. (1980) The *Coelomycetes*. Commonwealth Mycological Institute, Kew, UK.

Sutton, B.C. (1992) The genus *Glomerella* and its anamorph *Colletotrichum*. In: Bailey, J.A. and Jeger, M.J. (eds) Colletotrichum: *Biology, Pathology and Control*. CAB International, Wallingford, UK, pp. 1–26.

Taghavi, M., Hayward, C., Sly, L.I. and Fegan, M. (1996) Analysis of the phylogenetic relationships of strains of *Burkholderia solanacearum*, *Pseudomonas syzygii*, and the blood disease bacterium of banana based on 16S rRNA gene sequences. *International Journal of Systematic Bacteriology* 46, 10–15.

Tezenas du Montcel, H. (1981) Perspectives nouvelles dans la lutte chimique contre *Trachysphaera fructigena* du bananier au Cameroun. *Fruits* 36, 3–8.

Tezenas du Montcel, H. and Laville, E. (1977) Influence des conditions climatiques sur developpement du *Trachysphaera fructigena* sur bananier dans le sudouest du Cameroun. *Fruits* 32, 77–85.

Thomas, J.E. (1991) Virus indexing procedures for banana on Australia. In: Valmayor, V.V., Umali, B.E. and Besjosano, C.P. (eds) *Banana Diseases in Asia and the Pacific: Proceedings of a Technical Meeting on Diseases affecting Banana and Plantain in Asia and the Pacific*, Brisbane, Australia, April 15–18, 1991, INIBAP, Montpellier, France, pp. 144–157.

Thomas, J.E. and Dietzgen, R.G. (1991) Purification, characterization and serological detection of virus-like particles associated with banana bunchy top disease in Australia. *Journal of General Virology* 72, 217–224.

Thomas, J.E. and Magnaye, L.V. (1996) *Banana Bract Mosaic Disease*. Musa *Disease Fact Sheet No. 7*. INIBAP, Montpellier, France.

Thomas, J.E., Iskra-Caruana, M.L. and Jones, D.R. (1994) *Banana bunchy top disease*. Musa *Disease Fact Sheet No. 4*. INIBAP, Montpellier, France.

Thomas, J.E., Smith, M.K., Kessling, A.F. and Hamill, S.D. (1995) Inconsistent transmission of banana bunchy top virus in micropropagated bananas and its implication for germplasm screening. *Australian Journal of Agricultural Research* 46, 663–671.

Thomas, J.E., Geering, A.D.W., Gambley, C.F., Kessling, A.F. and White, M. (1997) Purification, properties and diagnosis of banana bract mosaic potyvirus and its distinction from abaca mosaic potyvirus. *Phytopathology* 87, 698–705.

Thottappilly, G., Dahal, G. and Lockhart, B.E.L. (1998) Studies on a Nigerian isolate of banana streak badnavirus. I. Purification and enzyme-linked immunosorbent assay. *Annals of Applied Biology* 132, 253–261.

Thwaites, R., Mansfield, J., Eden-Green, S.J. and Seal, S. (1999) RAPD and rep PCR-based fingerprinting of vascular bacterial pathogens of *Musa* spp. *Plant Pathology* 48, 121–128.

Toussoun, T.A. (1975) Fusarium-suppressive soils. In: Bruehl, G.W. (ed.) *Biology and Control of Soil-borne Plant Pathogens.* APS Press, St Paul, Minnesota, pp. 145–151.

Tsai, Y.P., Hwang, M.T., Chen, S.P. and Liu, S.S. (1986) An ecological study of banana mosaic. *Plant Protection Bulletin (Taiwan ROC)* 28, 383–387.

van der Aa, H. (1973) Studies in Phyllosticta I. *Studies in Mycology* 5, 1–110.

Viljoen, A., Surridge, A.K.J. and Crous, P.W. (2002) The impact of minor *Mycosphaerella* pathogens on bananas (*Musa*) in South Africa. In: *2nd International Workshop on Mycosphaerella Leaf Spot Diseases of Banana*, San Jose, Costa Rica May 20–23.

Vuylsteke, D. (1989) Shoot-tip culture for the propagation, conservation, and exchange of *Musa* germplasm. *Practical Manuals for Handling Crop Germplasm* in vitro 2. International Board for Plant Genetic Resources. Rome, Italy.

Waite, B.H. and Dunlap, V.C. (1953) Preliminary host range studies with *Fusarium oxysporum* f. sp. *cubense*. *Plant Disease Reporter* 37, 79–80.

Wanitchakorn, R., Harding, R.M. and Dale, J.L. (2000) Sequence variability in the coat protein gene of two groups of banana bunchy top isolates. *Archives of Virology* 145, 593–602.

Wardlaw, C.W. (1961) *Banana Diseases Including Plantains and Abaca.* Longmans, Green and Co. Ltd, London.

Washington, J.R., Cruz, J., Lopez, F. and Fajardo, M. (1998) Infection studies of *Mycosphaerella fijiensis* on banana and the control of black Sigatoka with chlorothalonil. *Plant Disease* 82, 1185–1190.

Wu, R.Y. and Su, H.J. (1990a) Transmission of banana bunchy top virus by aphids to banana plantlets from tissue culture. *Botanical Bulletin of Academia Sinica* 31, 7–10.

Wu, R.Y. and Su, H.J. (1990b) Production of monoclonal antibodies against banana bunchy top virus and their use in enzyme-linked immunosorbent assay. *Journal of Phytopathology* 128, 203–208.

Wu, R.Y. and Su, H.J. (1990c) Purification and characterization of banana bunchy top virus. *Journal of Phytopathology* 128, 153–160.

Wu, R.Y. and Su, H.J. (1991) Regeneration of healthy banana plantlets from banana bunchy top virus-infected tissues cultured at high temperature. *Plant Pathology* 40, 4–7.

Xie, W.S. and Hu, J.S. (1995) Molecular cloning, sequence analysis, and detection of banana bunchy top virus in Hawaii. *Phytopathology* 85, 339–347.

Yot-Dauthy, D. and Bové, J.M. (1966) Mosaique du bananier. Identification et purification de diverses souches du virus. *Fruits* 21, 449–466.

5 Diseases of Breadfruit, Jackfruit and Related Fruit Crops

Somsiri Sangchote[1], Jacqui G. Wright[2] and G.I. Johnson[3]
[1]Department of Plant Pathology, Kasetsart University, Bangkok, Thailand; [2]Secretariat of the Pacific Community, Suva, Fiji Islands; [3]ACIAR, Canberra, Australia

Introduction

Artocarpus spp. (family: *Moraceae*) are tropical trees that are native to South and Southeast Asia and the Pacific (Table 5.1) (Pareek *et al.*, 1998). They are in the same family as fig, *Ficus* spp., and the genus contains ~50 species, several of which produce edible fruit. Best known are jackfruit, *Artocarpus heterophyllus*, and breadfruit, *A. altilis* (Purseglove, 1968). Lesser known species with regional importance include

Table 5.1. Common names, uses and putative centres of origin for species of *Artocarpus*.[a]

Species	Common name(s)	Putative origin(s)	Primary use(s)	Other uses
A. altilis	Breadfruit	Pacific islands	Fruit cooked or raw	Flour made from pulp, seeds roasted and eaten as nuts. Fermented flesh traditional off-season food in Polynesia.
A. camansi	Kamansi	Philippines	Young fruit boiled	Seeds boiled or roasted
A. gomezianus	Tapang, Tampang	Malaysia	Fruit eaten cooked	As jam or in brine
A. heterophyllus	Jackfruit	India (Western Ghats, Assam)	Flowers, fruit and seeds cooked or raw	Timber
A. mariannensis	Dugdug Cheibiei	Palau and northern Mariana Islands	Fruit Seeds	
A. hirsuta	Aini	India	Fruit pulp	
A. lakoocha	Monkey jack	India	Fruit	Also pickled
A. integer	Chempedek	Southeast Asia, Malaysia	Fruit	Young fruit boiled in soup
A. lingnanensis	Kwai muk	China	Fruit	
A. odoratissimus	Marang, Tarap, Morang	Borneo, Sulu in South Philippines	Fruit	Rind and seeds roasted
A. rigidus	Monkey jack	Tropical Asia	Fruit aril	Seed roasted
A. rotundus	Monkey jack	Malaysia	Fruit	Fruit pulp tastes like honey

[a]From Arora (1998) and Pareek *et al.* (1998).

chempedek, *A. integer*, monkey jack, *A. lakoocha* and morang, *A. odoratissima*, in Asia, and the dugdug (or chebiei) in the Pacific (Table 5.1) (Ragone, 1997; Arora, 1998; Pareek *et al.*, 1998).

Each compound fruit, the sorosus or syncarpa, is formed from numerous enlarged fleshy connate calyces and carpels, arising from a peduncle, and bearing hardened tips. The achenes fuse during development and are deeply embedded in the axis of the inflorescence. The fruit has a thick verrucose rind, and a layer of pith containing seed cavities (Spjut, 1994). Immature fruit contain sticky latex. Mesocarp textures may be firm and starchy, firm and rubbery, or soft, depending on the degree of ripeness and species or cultivar (Piper, 1989).

The jackfruit is a popular garden tree throughout tropical Asia, growing up to 20 m in height and bearing fruit 3 years after planting. The compound fruit can reach up to 70 cm in length, are barrel or pear shaped, and can weigh up to 30 kg (Nakasone and Paull, 1998). The rind is pale green to dark yellow with short hexagonal fleshy spines. Flesh of each fruitlet is golden yellow and contains one or no seed. Seeds have a thick gelatinous brown covering, and fruitlets are separated by soft, fibrous mesocarp (Fong and Hoi-Sen, 1987; Tankard, 1987). Jackfruit is cross-pollinated and mostly grown from seed. Although there has been little selection, clonal selections with superior size and flesh characteristics are grown locally for fresh consumption or cooking (Ghosh, 1998).

Polynesians have cultivated the breadfruit as an important staple for centuries. There are many documented uses for its seeds, fruits, leaves, timber and other products. Throughout the Pacific, breadfruit exhibits great morphological variability, and fruits may be seedless, or contain numerous, minute aborted seeds, one to a few viable seeds, or numerous seeds. The breadfruit is propagated from root cuttings. Breadfruit trees may grow up to 20 m in height, and clonally propagated trees start fruiting in 3–6 years (Ragone, 1997). The compound fruit are round to oval in shape, 10–20 cm in diameter, and have a moist creamy white or pale yellow mesocarp that is mainly carbohydrate.

Several diseases of *Artocarpus* spp. have been reported. However, in comparison with other tree fruit crops, losses are not significant. Little has been reported on disease resistance mechanisms, but the hard skin on the fruit of many species (except breadfruit) would reduce susceptibility to some wound pathogens. A potentially antifungal stilbene has been reported from *A. incisus* (sic. = *A. altilis*) (Shimizu *et al.*, 1997).

Aboveground Diseases

Anthracnose

Breadfruit and jackfruit are affected. Symptoms can develop on leaves, twigs and fruit. Lesions on leaves begin as small dark brown spots that expand gradually to become grey at the centre with dark brown margins. Concentric rings of acervuli may be produced on lesions and, under favourable conditions, spots can coalesce to cause blighting of young leaves. Severe twig disease and defoliation can also develop on breadfruit (Abraham *et al.*, 1988).

On immature breadfruit, anthracnose may develop in association with injuries that are caused by insects (Plate 38) (Trujillo, 1971b). Lesions also develop on mature breadfruit and jackfruit as they start to ripen. Small, round, dark brown spots develop on the rind (Plate 39) and expand gradually, coalescing to form larger lesions. Anthracnose is caused by *Glomerella cingulata* (anamorph: *Colletotrichum gloeosporioides*) (Fig. 1.11), whereas *G. acutata* (anamorph: *C. acutatum*) (Fig. 1.10) causes premature fruit fall of breadfruit (McKenzie and Jackson, 1990). See Chapter 1 for a detailed description of both species. Conidia of *C. gloeosporioides* are produced on affected areas in the tree canopy and on ripe fruit. They are spread by water, rain and wind, and high relative humidity and temperature encourage disease development.

Pruning diseased parts to remove inoculum and open the canopy can reduce the incidence and severity of anthracnose. Strategic sprays with mancozeb can also reduce losses (Sirayoi, 1993).

Bacterial dieback

Bacterial dieback, caused by *Erwinia carotovora*, has been recorded on chempedek (Agrolink, 1999). The disease causes leaf yellowing and gummy exudate from stems and branches.

Corynespora leaf spot

Corynespora cassiicola causes irregular spots on breadfruit leaves and, less often, stems, roots and flowers (Dingley *et al.*, 1981; Macfarlane, 1997). The lesions are up to 2 cm in diameter, often have an undulate border and display a zonate pattern that darkens with age. Shot holing and defoliation may occur.

Conidia of the pathogen are airborne, and it persists saprophytically in plant debris (Macfarlane, 1997). Benomyl, copper fungicides and mancozeb have been reported to be effective.

Diplodia fruit rot and collar rot

Lesions develop on jackfruit and breadfruit fruit, twigs and branches. A collar rot has also been recorded on the trunk of breadfruit trees (Kohler *et al.*, 1997).

Lesions develop as small brown spots on the sides or stem ends of fruit. They expand to become large, dark brown, soft lesions that can extend through the entire fruit. Under humid conditions, dark grey to black mycelium can develop on the lesion surface (Plate 40). Twig and branch dieback symptoms can also develop, and a dry collar rot has been reported on breadfruit trunks where the wood beneath the bark shows white patches with dark brown margins (Kohler *et al.*, 1997).

Diplodia theobromae is a common pathogen of tropical tree fruits. Its teleomorph, *Botryosphaeria rhodina*, apparently has not been reported on these hosts. Both are described in Chapter 1. *D. theobromae* occurs widely in soil, on dead twigs, on mummified fruit and on organic debris. It infects through wounds, and causes symptoms on fruit as they ripen. Kuthubutheen and Muid (1985) noted that while *Colletotrichum*, *Pestalotiopsis* and *Fusarium* spp. were isolated from symptomless tissue of newly opened panicles and remained common throughout the life of leaves, *D. theobromae* did not establish until 25–30 days after leaf expansion.

Careful handling of fruit to avoid soil contact and wounds can reduce losses. Cultural control methods can reduce collar rot in breadfruit. Efforts should be made to prevent or protect wounds. Pruning wounds may be sealed with tar mixed with copper fungicides.

Miscellaneous fruit rots

Several fruit rots are of minor importance. *Aspergillus niger* (Fig. 1.2) causes a soft black rot, and *Drechslera rostrata* causes irregular white and brown spots on monkey jack (Roy *et al.*, 1982). *Botrytis* sp. and *Botryosphaeria* sp. caused fruit rot of jackfruit, and *Botrytis cinerea* was associated with blossom blight and premature fall of young jackfruits (Pandey *et al.*, 1981). *Dothiorella* sp. has been reported on breadfruit (Wall, 1989).

Miscellaneous leaf diseases

The following leaf spots generally are not significant. *Septoria artocarpi* affects jackfruit (Butani, 1978) and *Septoria eburnea* affects breadfruit (Firman, 1975). *Pseudocercospora artocarpi* (Fig. 5.1) was reported on jackfruit and breadfruit (Shaw, 1984), and McKenzie (1996) associated *Pseudocercospora* sp. with leaf blotch and circular to irregular black mould on the lower surface of older breadfruit leaves. Rust, caused by *Uredo artocarpi*, has been recorded on jackfruit, chempadek and *Artocarpus* sp. (Vevai, 1971; Singh, 1980; Shaw, 1984). On jackfruit, brown leafspot is caused by *Nigrospora sphaerica* and *Pestalotiopsis versicolor* (Basak, 1992), and grey leaf blight is caused by *Pestalotiopsis elasticola* (Vevai, 1971; Zhang *et al.*, 2003). On breadfruit, a leafspot is caused by *Pestalotia* sp. (Hammes and Chant, 1989), and algal leaf spot is caused by *Cephaleuros virescens* (Zaiger and Zentmyer, 1966; Gerlach, 1988; McKenzie and Jackson, 1990).

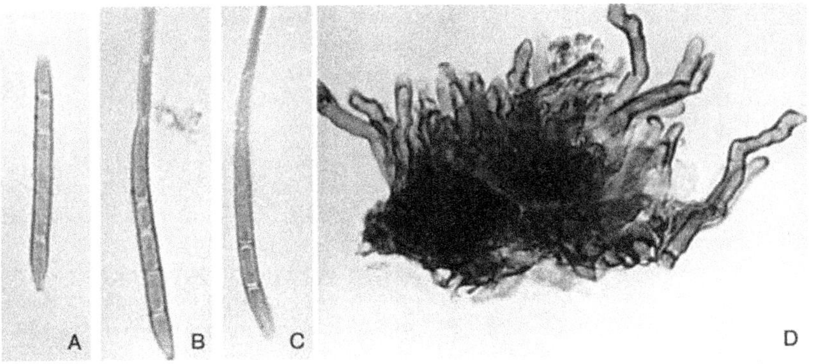

Fig. 5.1. (A–C) Conidia and (D) conidiophores and stroma of *Pseudocercospora artocarpi* (from CMI description no. 1128).

Phomopsis leaf spots and fruit rot

Phomopsis leaf spots and fruit rot are minor problems on *Artocarpus* spp. *Phomopsis artocarpina* has been associated with leaf spots on jackfruit (Butani, 1978) and *P. artocarpi* has been recorded as causing leaf spot and branch dieback of breadfruit (Zaiger and Zentmyer, 1966; McKenzie and Jackson, 1990). *Phomopsis* sp. causes small, brown spots on jackfruit that expand, become soft and dark brown, and penetrate the mesocarp (Fig. 5.2).

The epidemiology of these diseases has not been studied. On other hosts, stem-end infections of fruit arise from infected floral remnant tissue or by endophytic colonization of the peduncle (Johnson, 1997). Surface lesions on fruit probably result from conidial infections by direct penetration or infection through style remnants, and sporulation occurs on mature lesions on infected fruit. Careful handling to avoid damage and spraying with carbendazim 2 weeks before harvest can reduce fruit losses.

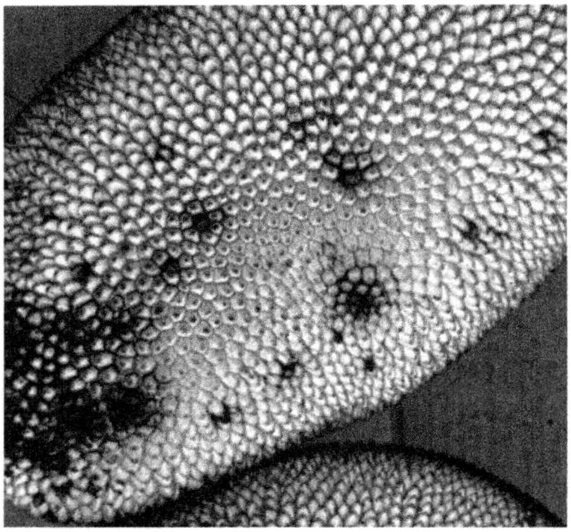

Fig. 5.2. Soft, dark brown spots on jackfruit caused by *Phomopsis* sp. (photo: Somsiri Sangchote).

Phytophthora fruit, stem and root rot

Several organs of breadfruit and chempedek are affected. Small, water-soaked lesions are the first visible symptoms on the fruit surface. As the lesions enlarge, the centres become light brown and, at the margin, white mycelial growth and sporulation occur. Rotted fruit mummify on the tree, and infections that spread from these fruit can lead to defoliation and branch dieback. Root and collar rot and stem lesions have been recorded on chempedek (Agrolink, 1999).

Fruit rot is caused by *Phytophthora palmivora* (Fig. 1.16; Gerlach and Salevao, 1984), and *P. citrophthora* (Fig. 1.14) causes root and stem rot on chempedek (Agrolink, 1999). Both are described in Chapter 1. Zoospores of *P. palmivora* are spread in rain and water splash, and insects and snails have been implicated on other hosts (Lim, 1990). The pathogen infects green, mature fruits directly. Lesion enlargement occurs during the day when temperatures are high, and sporulation occurs at night under high relative humidity and lower temperatures

(Trujillo, 1971b). Diurnal growth often results in the formation of concentric bands of decay in older lesions.

Regular removal of diseased plant parts, removal of weeds and the use of a mulch to cover the soil are important techniques that can be used to help control this disease. New plantings should be established in well-drained soils. Contact with soil or plant debris at harvest should be avoided. Copper fungicides, copper oxide plus metalaxyl, mancozeb plus metalaxyl, or potassium phosphonate may provide some control (Kohler *et al.*, 1997).

Pink disease

This disease has a wide host range on tropical woody plants. It is mostly a trunk, branch and twig problem in the rainy season especially on plants with a dense canopy. Initial symptoms on jackfruit, breadfruit and chempedek appear as a white patch on branches, often in the crook between branches, or between the branch and trunk (see Plate 9). Symptoms expand rapidly to encircle the branch, appearing as an extensive superficial salmon pink sheath extending ≥20 cm from the crook down the trunk. Infected branches may wither or defoliate (Visarathanonth and Jermsiri, 1998).

Pink disease is caused by *Erythricium salmonicolor* (anamorph: *Necator decretus*). The diagnostic features of this fungus are listed in Chapter 1. Conidia of the pathogen are spread by wind and rain, and symptoms are more likely to develop beneath shady canopies. Affected tissue should be pruned and remaining branches sprayed or painted with copper oxychloride or mancozeb (Sirayoi, 1993). Pruning to improve air circulation and light penetration in the canopy reduces the likelihood of reinfection.

Rhizopus rot or transit rot

This disease develops on flowers and young fruits and is favoured by hot and humid conditions (Butani, 1978; Pandey *et al.*, 1979). It can also be severe on ripening fruit that have not been harvested, or develop on wounded fruit during storage or shipment after harvest.

Symptoms

Lesions develop as soft, watery brown spots and expand rapidly. Affected areas become covered with a black mass of mycelia and sporangia (Johnson, 1993) (Plate 41). Moisture that condenses on fruit after removal from cold storage favours mycelium growth and sporulation on fruit tissue, soil particles and organic debris, and leads to a rapid increase in infection (Trujillo, 1971a,b; Johnson, 1993). Fruit-to-fruit spread can occur when they are in contact during shipment.

Causal agents

Three *Rhizopus* species have been recorded on these hosts, *R. oryzae* (Singh and Singh, 1989), *R. artocarpi* (McMillan, 1975) and *R. stolonifer* (Fig. 5.3) (Almeida and de Landim, 1980). They rapidly produce sporangia and sporangiophores on young fruit and on organic debris. Spores are spread by insects and wind; however, infection most commonly arises after wounds are contaminated by mycelium or debris.

Management

Pruning to remove organic debris and encourage good ventilation, and the application of copper fungicides or dithiocarbamates to flowers and young fruit have reduced losses of jackfruit and breadfruit (Almeida and de Landim, 1980; Gupta and Pandey, 1985; Singh and Singh, 1989; Katuruak *et al.*, 1990; Johnson, 1993). Removal and destruction of diseased fruit from trees and the ground may reduce the risk of postharvest losses.

At harvest, fruit should not come in contact with soil. Adhering organic debris should be cleaned from fruit prior to packing, and skin damage and the use of dirty wash water should be avoided. Fruit should not be packed with organic materials such as leaves, coconut fibre or wood shavings. These materials can wound the skin and allow infection to spread directly from the organic materials into fruit.

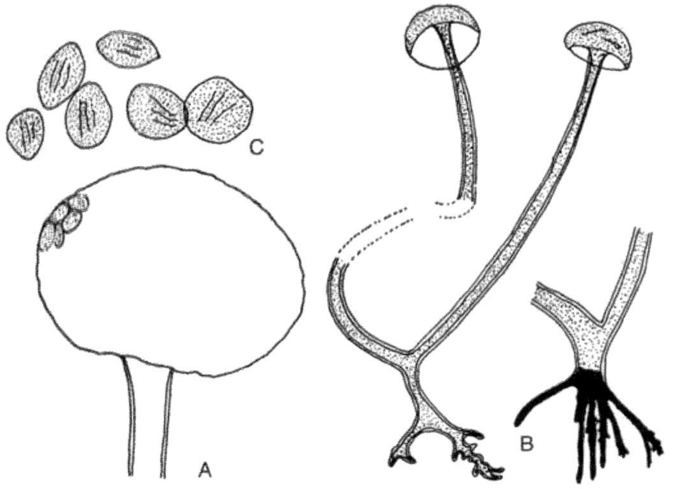

Fig. 5.3. (A) Sporangium, (B) branched sporangiophore with rhizoids and (C) sporangiospores of *Rhizopus stolonifer* (from CMI description no. 110).

Root and Dieback Diseases

Brown root and crown rot

This disease affects many tree species, including breadfruit.

Symptoms

Roots become encrusted with soil that is held together by a brown mycelial crust with a white margin; it may run up to 1.5 m above the soil line. Cracking of bark and gummosis may occur. The wood is discoloured, and later becomes dry, friable and honeycombed (Putter, 1998). Basidiocarps of the pathogen are produced at the base of the tree on the bark surface, and as it develops the tree is girdled, resulting in sudden death. Wilting, yellowing and necrosis of leaves occurs and leads to defoliation and premature death. Diseased trees occur in pockets and appear to die slowly (Trujillo, 1971b).

Causal agents

This disease is caused by *Phellinus noxius* (Fig. 5.4) (Dingley *et al.*, 1981). It is found in the eastern tropics, and causes important diseases of rubber, tea and cacao (Holliday, 1980). *P. lamaoensis* has also been reported (Huguenin, 1964).

Epidemiology

P. noxius has a very wide host range and spreads from diseased to healthy trees by root contact. Infections can also arise from root contact with decaying stumps buried in the ground after land has been cleared. More important, however, are its airborne basidiospores that initiate infections on pruning wounds (Holliday, 1980).

Management

Brown root and crown rot is difficult to control. Occasionally, trees can be cured if diseased parts are removed immediately after symptoms appear. Tree-to-tree spread has been prevented by removing diseased trees and roots more than 2.5 cm in diameter, and by exposing the base of the trunk and major roots of adjacent trees to determine if they are infected (Kohler *et al.*, 1997). Painting pruning wounds with proprietary wound sealants, Trichoderma preparations or house paint may also reduce risk of infection.

Miscellaneous root and trunk diseases

Rhizoctonia solani causes feeder root dieback of breadfruit (Trujillo, 1971a), and seedling blight of *A. elastica* (Singh, 1980). *Ganoderma*

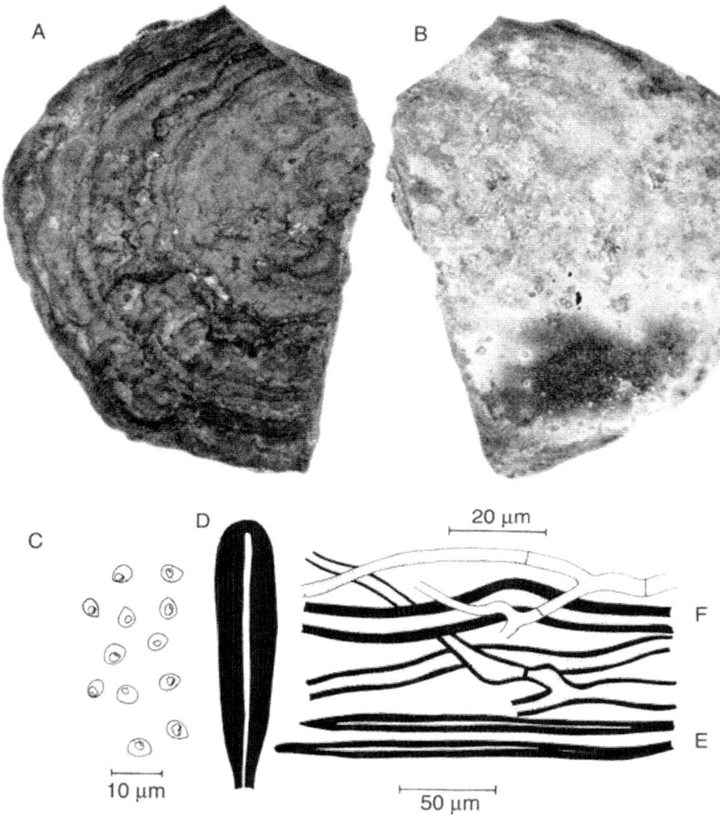

Fig. 5.4. (A) Dorsal and (B) ventral surface of basidiocarp, (C) basidiospores, (D) tramal setal-hypha, (E) context setal-hyphae and (F) context hyphae of *Phellinus noxius* (from CMI description no. 195).

lucidum (Fig. 3.8) causes trunk rot on the latter host.

Nematodes

The following nematodes have been reported as pathogenic on jackfruit and breadfruit: *Aphelenchoides* sp., *Helicotylenchus dihystera*, *Helicotylenchus multicinctus*, *Hemicriconemoides cocophilus*, *Meloidogyne* sp., *Paratylenchus* sp. and *Xiphinema brevicolle* (Orton Williams, 1980; Grandison, 1990).

Crossonema malabaricum and *Neolobocriconema palamiensis* have been reported from soil around the roots of jackfruit (Reddy, 1998). Other nematodes associated with breadfruit include: *Achlysiella williamsi*, *Aphelenchoides bicaudatus*, *Criconemella denoudeni*, *C. onoensis*, *Ditylenchus* sp., *Gracilacus* sp., *Helicotylenchus erythrinae*, *H. indicus*, *H. microcephalus*, *H. pseudorobustus*, *Helicotylenchus* spp., *Hemicriconemoides mangiferae*, *Heterodera* sp., *Meloidogyne incognita*, *Pratylenchus coffeae*, *P. loosi*, *Pratylenchus* sp., *Sphaeronema* sp., *Xiphinema basiri*, *X. ensiculiferum* and *Xiphinema* sp. (Kirby *et al.*, 1980; Orton Williams, 1980; Grandison, 1996).

Pingelap disease

Pingelap has caused near total destruction of breadfruit in several Pacific Island nations (Zaiger and Zentmyer, 1966). The disease affected numerous cultivars of seeded and seedless breadfruits of *A. altilis* and *A. mariannensis*. It attacked trees of bearing age, but in severe epidemics it affected trees of all ages.

The diseased trees exhibited two disease syndromes with a variety of intergradations: a dieback syndrome and a wilt syndrome.

The cause of Pingelap is not known. Several fungi were isolated from dead and dying trees including *Fusarium solani* (now recognized to be *Rhizoctonia solani*), *Phomopsis* sp., *Phyllosticta* sp. and *Pythium* sp., but none were shown to be pathogenic (Zaiger and Zentmyer, 1966). It was concluded that a number of factors contributed to these devastating epidemics, including drought stress and salt damage due to seawater inundation of the atolls by typhoons (Trujillo, 1971a). These could have predisposed the trees to colonization by weak pathogens such as *Phomopsis* sp. Although Trujillo (1970) noted that symptoms described for Pingelap also resembled those caused by *Phytophthora palmivora*, it is not known whether it is involved in the development of this disease.

An epidemic in the late 1990s and early 2000s with similar symptoms was observed in Northern Pacific countries where droughts had been severe (2+ years). Fungi isolated from affected fruits and tree cankers include *C. gloeosporioides*, *D. theobromae*, *F. solani* and *Rhizopus* sp. *Phytophthora* was not isolated from fruits or trees. These microbes were recognized as secondary pathogens that attacked drought-stressed and salinity-affected trees to cause a variety of symptoms including cankers, dieback to the point of death and fruit rot. Replanting with drought- and salinity-tolerant cultivars, top working trees that suffer from dieback, mulching to conserve water, and removal and destruction of infected plant parts and fruits may all be beneficial.

References

Agrolink (1999) Department of Agriculture Malaysia website: http://agrolink.moa.my/doa/english/croptech/che_pes.html

Abraham, M., Padmakumary, G. and Nair, M.C. (1988) Twig blight (die-back) of *Artocarpus incisa*. *Indian Phytopathology* 41, 629–630.

Almeida, R.T. and de Landim, C.M.U. (1980) *Rhizopus stolonifer* (Ehrenb. ex Fr.) Vuill., causal agent of soft rot of jackfruit (*Artocarpus heterophyllus* Lam.) and breadfruit (*Artocarpus altilis* Fosberg) in Ceara State. *Fitossanidade* 4, 23–24.

Arora, R.K. (1998) Genetic resources of native tropical fruits in Asia. In: Arora R.K. and Ramanatha Rao, V. (eds) *Tropical Fruits in Asia, Diversity, Maintenance, Conservation and Use. Proceedings of the IPGRI-ICAR-UTFANET Regional Training Course on the Conservation and Use of Germplasm of Tropical Fruits in Asia*. Bangalore, India, May 18–31, 1997, pp. 42–53.

Basak, A.B. (1992) Brown leaf spot of jack fruit tree caused by *Nigrospora spaherica* (Sacc.) Mason and *Pestalotiopsis versicolor* (Sperg) Steyaert. *Bangladesh Journal of Forest Science* 21, 68–70.

Butani, D.K. (1978) Pests and diseases of jackfruit in India and their control. *Fruits* 33, 351–357.

Dingley, J.M., Fullerton, R.A. and McKenzie, E.H.C. (1981) *Records of Fungi, Bacteria, Algae, and Angiosperms Pathogenic on Plants in Cook Islands, Fiji, Kiribati, Niue, Tonga, Tuvalu, and Western Samoa*. South Pacific Commission, Noumea, New Caledonia.

Firman, I. (1975) Plant diseases in the area of the South Pacific Commission 2. American Samoa. *South Pacific Commission Information Document* 38. Noumea, New Caledonia.

Fong, C.H. and Hoi-Sen, Y. (1987) *Malaysian Fruits in Colour*. Tropical Press SDN.BHD, Kuala Lumpur.

Gerlach, W.W.P. (1988) *Plant Diseases of Western Samoa*. Samoan German Crop Protection Project, Apia, Western Samoa.

Gerlach, W.W.P. and Salevao, F. (1984) Fruit rot of breadfruit, *Artocarpus altilis*, caused by *Phytophthora palmivora* in Western Samoa. *Alafua Agricultural Bulletin* 9, 21–26.

Ghosh, S.P. (1998) Fruit wealth of India. In: Arora, R.K. and Ramanatha Rao, V. (eds) *Tropical Fruits in Asia, Diversity, Maintenance, Conservation and Use. Proceedings of the IPGRI-ICAR-UTFANET Regional Training Course on the Conservation and Use of Germplasm of Tropical Fruits in Asia*. Bangalore, India, May 18–31, 1997, pp. 3–15.

Grandison, G.S. (1990) *Report on a Survey of Plant Parasitic Nematodes in the Cook Islands (Southern Group)*. South Pacific Commission, Noumea, New Caledonia.

Grandison, G.S. (1996) *Plant Parasitic Nematodes of American Samoa*. South Pacific Commission, Noumea, New Caledonia.

Gupta, J.H. and Pandey, I.C. (1985) Chemical control of fruit rot disease of jack-fruit. *Progressive Horticulture* 17, 361–362.

Hammes, C. and Chant, H. (1989) *Manuel de Défense des Cultures en Polynésie Française*. Institut Français de Recherche Scientifique pour le Développement en Coopération, Service de L'économie Rurale de Polynésie Français, Entomologie Agricole, Notes et documents No. 3.

Holliday, P. (1980) *Fungus Diseases of Tropical Crops*. Cambridge University Press. Cambridge.

Huguenin, B. (1964) *FAO Plant Protection Committee for the South East Asia and Pacific Region*. First Quarterly Report, 1964, January–March 8–9.

Johnson, G.I. (1993) Jackfruit. In: Persley, D. (ed.) *Diseases of Fruit Crops*. Information Series Q192023 Department of Primary Industries, Queensland p. 59. Plate 3,10.

Johnson, G.I. (1997) Mango disease losses: balancing economy and ecology. *Acta Horticulturae* 455, 575–586.

Katuruak, C.,Vichitranonth, S. and Leelasartrakul, K. (1990) *Studies on Jackfruit Diseases*. Progress Report, Division of Plant Pathology and Microbiology, Department of Agriculture, Bangkok, pp. 36–37 (in Thai).

Kirby, M.F., Kirby, M.E., Siddiqi, M.R. and Loof, P.A.A. (1980) *Fiji Nematode Survey Report: Plant Parasitic Nematode Distributions and Host Associations*. Ministry of Agriculture and Fisheries, Fiji, Report No. 68.

Kohler, F., Pellegrin, F., Jackson, G. and McKenzie, E. (1997) *Diseases of Cultivated Crops in Pacific Island Countries*. South Pacific Commission, pp. 16–18, 163–164, 169–170.

Kuthubutheen, A.J. and Muid, S. (1985) Fungus succession of fruit trees and its importance to the control of fruit-tree diseases in Malaysia. *Proceedings of the 2nd Asian Conference on Technology for Rural Development*. Kuala Lumpur, Malaysia, December 4–7, 1985. World Scientific Publishing Company Pte Ltd, Singapore pp. 683–699.

Lim, T.K. (1990) *Durian Diseases and Disorders*. Tropical Press Sdn Bhd, Malaysia.

Macfarlane, R. (1997) *Corynespora cassiicola*. Global Plant Protection Information System, FAO, Rome, Italy.

McKenzie, E.H.C. (1996) *Fungi, Bacteria and Pathogenic Algae on Plants in American Samoa*. South Pacific Commission, Noumea, New Caledonia.

McKenzie, E.H.C. and Jackson, G.V.H. (1990) The fungi, bacteria and pathogenic algae of the Federated States of Micronesia. *SPC Technical Paper* 199.

McMillan, R.T., Jr (1975) *Rhizopus artocarpi* rot of jackfruit (*Artocarpus heterophyllus*). *Proceedings of the Florida State Horticultural Society* 87, 392–393.

Nakasone, H.Y. and Paull, R.E. (1998) *Tropical Fruits*. CAB International, Wallingford, UK.

Orton Williams, K.J. (1980) *Plant Parasitic Nematodes of the Pacific*. UNDP/FAO-SPEC Survey of Agricultural Pests and Diseases in the South Pacific, Technical Report 8.

Pandey, R.S., Bhargava, S.N., Shukla, D.N. and Khati, D.V.S. (1979) Control of *Rhizopus* rot of jack-fruit. *Indian Phytopathology* 32, 479–480.

Pandey, P.C., Puri, Y.N. and Rehill, P.S. (1981) Blossom and fruit blight of jackfruit and its control. *Indian Journal of Forestry* 4, 5–7.

Pareek, O.P., Sharma, S. and Arora, R.K. (1998) *Under-utilised Edible Fruits and Nuts: an Inventory of Genetic Resources in Their Regions of Diversity. Part II Inventory of Genetic Resources*. IPGRI Office for South Asia, New Dehli, India, pp. 33, 51–52; Appendix II Under-utilised edible fruits and nuts p. 208.

Piper, J.M. (1989) *Fruits of South-East Asia Facts and Folklore*. Oxford University Press, Singapore, pp. 22–29.

Purseglove, J.W. (1968) *Artocarpus altilis*. In: *Tropical Crops. Dicotyledons*. Longman, London, pp. 379–384.

Putter, C.A.J. (1998) *Phellinus noxius*. Global Plant Protection Information System, FAO, Rome, Italy.

Ragone, D. (1997) *Breadfruit.* Artocarpus altilis *(Parkinson) Fosberg. Promoting the Conservation and Use of Underutilised and Neglected Crops. 10.* Institute of Plant Genetics and Crop Plant Research, Gatersleben/International Plant Genetic Resources Institute, Rome, Italy.

Reddy, P.P. (1998) Nematodes and their control in tropical fruit crops. In: Arora, R.K. and Ramanatha Rao, V. (eds) *Tropical Fruits in Asia, Diversity, Maintenance, Conservation and Use. Proceedings of the IPGRI-ICAR-UTFANET Regional Training Course on the Conservation and Use of Germplasm of Tropical Fruits in Asia*. Bangalore, India, May 18–31, 1997, pp. 246–250.

Roy, A.N., Sharma, R.B. and Sharma, K.G. (1982) New market diseases of barhal fruit. *Current Science* 51, 143.

Shimizu, K., Kondo, R. and Sakai, K. (1997) A stilbene derivative form *Artocarpus incisus*. *Phytochemistry* 45, 1297.

Shaw, D.E. (1984) *Microorganisms in Papua New Guinea*. Research Bulletin No. 33. Department of Primary Industry, Port Moresby.

Singh, K.G. (1980) A check list of hosts and diseases in Malaysia. *Ministry of Agriculture Bulletin No. 154*, 14–15.

Singh, N.I. and Singh, K.U. (1989) Efficacy of certain fungicides against *Rhizopus* rot of jackfruit. *Indian Phytopathology* 42, 465–466.

Sirayoi, A. (1993) *Diseases of Fruit Crops, Spices and Control*. Takansak Foundation, Kasetsart University.

Spjut, R.W. (1994) A systematic treatment of fruit types. *Memoirs of the New York Botanical Garden* 70, 112–115.

Tankard, G. (1987) *Exotic Tree Fruit for the Australian Home Garden*. Thomas Nelson, Melbourne, Australia.

Trujillo, E.E. (1970) *Preliminary Report on the Causes of Breadfruit Diseases in the Trust Territory*. Office of the High Commissioner, Trust Territory of the Pacific Islands, Saipan.

Trujillo, E.E. (1971a) *A List of Diseases of Economic Plants in the Trust Territory of the Pacific Islands*. Trust Territory of the Pacific Islands Department of Resources and Development, Division of Agriculture.

Trujillo, E.E. (1971b) *The Breadfruit Diseases of the Pacific Basin*. Information Document No. 27. South Pacific Commission, Noumea, New Caledonia.

Vevai, E.J. (1971). Know your crop its pest problems and control – minor tropical fruits. *Pesticides* 5, 33–54.

Visarathanonth, N. and Jermsiri, J. (1998) *Diseases of Fruit Crops*. Department of Agriculture, Bangkok.

Wall, G.C. (1989) *Plant Diseases Reported on Guam*. University of Guam, Mangilao, Guam.

Zaiger, D. and Zentmyer, G.A. (1966) A new lethal disease of breadfruit in the Pacific Islands. *Plant Disease Reporter* 50 892–896.

Zhang, J., Tong, X. and Qixin, G. (2003) Notes on *Pestalotiopsis* from southern China. *Mycotaxon* 85, 91–99.

6 Diseases of Carambola

Sepiah[1], Randy C. Ploetz[2] and Anthony W. Cooke[3]
[1]*Faculty of Resource Science and Technology, University Malaysia Sarawak, Sarawak, Malaysia;* [2]*University of Florida, Tropical Research and Education Center, Homestead, Florida, USA;* [3]*Queensland Horticulture Institute, QDPI, Indooroopilly, Queensland, Australia*

Introduction

Carambola, *Averrhoa carambola* (family: *Oxalidaceae*), is a small tree, growing to 12 m in height (Sedgley, 1984; Vijaysegaran, 1988). It has compound leaves and pink or light red flowers with purple hearts that are borne on leafy twigs in clusters. Its fruit are ellipsoid, whitish yellow to light orange or golden red, and ribbed (Martin *et al.*, 1987; Vijaysegaran, 1988).

Carambola requires tropical conditions for optimal growth and production. It will grow on almost any type of soil provided that it is not waterlogged (Sedgley, 1984; Abidin, 1987; Vijaysegaran, 1988). It will establish in sandy tin tailings, and peat and mineral soils, but grows best in acidic to slightly acidic soils. It thrives in areas with high rainfall and temperatures, partial shade and protection from wind (Martin *et al.*, 1987; Marler and Zozor, 1992; Marler *et al.*, 1994). Minor element deficiencies often plague the crop when it is grown in alkaline soil, and in the subtropics trees often defoliate during the winter.

Carambola is thought to have originated in Southeast Asia. The crop has been cultivated in the region since ancient times, but only recently elsewhere (Martin *et al.*, 1987; Ngah *et al.*, 1989). It was recorded in the western hemisphere in 1854 (Rio de Janeiro Botanical Garden), and has now naturalized and developed a secondary centre of diversity in northern South America (Knight, 1989). Carambola is now grown to 30°S in Australia and 32°N in Israel.

The name 'carambola' came from Malabar (Popenoe, 1924). The Portuguese adopted this name shortly after their first encounter with the fruit, and it is the common name of the fruit in English and Spanish. Other names include 'belimbing manis' in Malaysia, 'five-finger' in Guyana and, due to its silhouette in cross-section, 'starfruit' in the USA (Martin *et al.*, 1987; Ramsammy, 1989). Unfortunately, the last name confuses the fruit with *Dasmasonium stellatum*, a water plant whose common name is starfruit. A close relative, the bilimbi, *Averrhoa bilimbi*, is a minor fruit crop and will not be considered in this chapter.

Carambola is usually self-incompatible (Knight, 1966). Thus, seedling progeny are often out-crossed hybrids that do not retain the full suite of characters that are possessed by the female parent. In most commercial situations, superior selections are clonally propagated, usually by graftage on to seedling rootstocks (Campbell, 1989; Crane, 1992). However, in some areas, commercial production is mainly from seedlings (e.g. Guyana).

Current Status

The popularity of carambola has increased since the early 1980s for two reasons. First, the sour types of carambola, which were the first types to be widely grown outside Asia, have been replaced by sweet, low-acid cultivars that have far broader appeal. Secondly, as disposable incomes and the awareness and acceptance of exotic foods have increased in many parts of the world, so too have the demands for fruit such as carambola (Green, 1989).

Improved carambola germplasm has been developed in several areas (Knight, 1989). The Malaysian Agricultural Research and Development Institute (MARDI) has a series of selected clones, the B-series, which are used widely in that country and are beginning to be used in other areas. The most important are 'B10' and, to a lesser extent, 'B17' (Green, 1989; Knight, 1989). Several different cultivars are grown in Taiwan, of which one known variously as 'Cheng-Tsey', 'Chun-Choi' or 'Cheng-Chui' is probably the most important. 'Fwang Tung' is the best known of the selections from Thailand, and 'Arkin' (Florida) and 'Kary' (Hawaii) are two of the most popular selections outside Asia.

In Australia, Florida, Malaysia and Taiwan, pruning is practised to remove branches and control the height and growth of the canopy (Crane, 1992; Izham *et al.*, 1992; Samson, 1992; Vijaysegaran, 1988; Watson *et al.*, 1988). Recent work in Florida has also shown that pruning and fruit removal promote flowering (Nuñéz-Elisea and Crane, 2000). Pruning can also be used to reduce the inoculum of several carambola pathogens and is used in many disease control strategies.

Management on commercial farms usually includes pruning and spraying. In some locations, young fruit are bagged to exclude insect pests and increase fruit quality. In Malaysia, heavy pruning is usually conducted to keep the tree at a height that facilitates bagging and harvesting. Carambola is non-seasonal in the tropics and usually produces 3–5 peak crops of fruit per year. However, in subtropical areas, there are usually only two seasons due to cool, winter temperatures. Depending on the climate, fruit mature from 40 to 70 days after they set. The fruit is picked green or mature, depending on cultivars and markets, and can be stored effectively for >6 weeks at 5°C without chilling injury (Campbell *et al.*, 1989).

Production figures for carambola are scarce. In descending order, the following locations recently had the largest areas devoted to production: Taiwan (3140 ha), Guyana (2450 ha), Malaysia (900 ha), Brazil (300 ha) and Florida (300 ha) (Donadio, 1989; Green, 1989; Ngah *et al.*, 1989; Ramsammy, 1989; Crane, 1997). Carambola is also grown commercially in Australia, Hawaii, Indonesia, Israel and Thailand. Annual production ranges from over 38,000 million tonnes (Mt) in Taiwan and 24,000 Mt in Malaysia to a few thousand tonnes or less in other production areas. In total, global output probably does not exceed 0.5 Mt year^{-1}.

PESTS

Insects are serious fruit pests. In Florida, fruit are affected by: stinkbugs, *Nezaera* sp.; squash bugs, *Acanthocephala* sp.; fruit blotch miner; red-banded thrips, *Selenothrips rubrocinctus*; and soft brown scale, *Coccus hesperidum* (Crane, 1992, 1993). The weevil, *Diaprepes abbreviatus*, damages roots and can cause canopy dieback. Important pests in Malaysia are: fruit fly, *Bactrocera* sp.; fruit borer, *Diacrotricha fasciola*, which attacks flowers; *Porthesia (Euproctis) scientillans*, which attacks flowers and fruit; *Archips tabescens*, which attacks fruit and mature green leaves; *Adoxophyes privatana*, which attacks shoots and young leaves; and stem borer, *Indarbela disciplaga* (Ooi, 1984; Vijaysegaran, 1988; Ithnin *et al.*, 1992). A fruit fly, *Bactrocera dorsalis* (*Dacus dorsalis* complex), is the most serious pest in Southeast Asia (Samson, 1992). Since it can cause losses of 100% (Vijaysegaran, 1988), some form of fruit protection is necessary. Birds also attack fruit. Pest control is achieved by bagging the fruit, spraying with insecticides, use of protein bait sprays and removal of infested fruit or stems.

DISEASES

In general, diseases are not limiting factors in carambola production in most countries. This is due, in part, to the short history of cultivation in most areas, and the fact that large, homogeneous plantings that would be conducive to the development of diseases are uncommon. In many areas, fruit are produced in small plots, <1 ha, or on backyard trees that are interplanted with other crops.

As the volume and intensity of carambola production increases around the world and new cultivars are introduced, it is logical to assume that diseases will increase in importance. Minor diseases that rarely impact small or interplanted plots may become damaging as large monocultures of this crop are established. Furthermore, as carambola's popularity increases, so too will the marketplace's demand for high quality, blemish-free fruit. Finally, the increased movement of people and agricultural commodities across international borders will undoubtedly result in the movement of carambola pathogens into new production areas. Although it is not possible to predict which will be important, it is clear that some will cause problems when they find a conducive niche.

Foliar and Canopy Diseases

Leaf spots, and stem and branch diseases are usually not severe problems. Pink disease potentially is the most debilitating of these problems. Many of the diseases that are listed below can be serious during rainy weather and when canopy management is not practised.

Algal disease

Algal disease is usually not a major problem, but can be damaging in unmanaged plantings where it reduces host photosynthesis and causes dieback of twigs and branches (Joubert and Rijkenberg, 1971).

SYMPTOMS Algal disease is characterized by the presence of orange, rust-coloured spots on the surfaces of leaves, stems, twigs, limbs and fruits. Leaf spots can be up to 5 mm in diameter, and they often merge to form large, irregular spots.

The alga can penetrate cortical tissues of the host, causing these tissues to swell and crack as the pathogen filaments grow and expand inside the host. The leaves on affected twigs wilt, turn yellow and fall, and shoots gradually die back.

CAUSAL AGENT Algal disease of carambola is caused by the alga, *Cephaleuros virescens*. It is the most important of a few algae that cause diseases on fruit and plantation crops in the tropics and subtropics (Joubert and Rijkenberg, 1971). It is described in Chapter 1.

EPIDEMIOLOGY Algal disease is found in humid tropical and subtropical regions, and is most prevalent during the rainy season and on plants with dense canopies. The pathogen's primary infective propagules are biflagelle zoospores that are dispersed by water splash and wind. A wet, humid environment, and poor ventilation within the plant canopy are favourable for its establishment and spread (Manners, 1993).

MANAGEMENT Good plantation management is the most effective means for controlling the disease. Proper fertilization schemes, good irrigation and pruning systems are needed to make conditions less favourable for the pathogen, and copper-based fungicides can be used to control the disease in severely affected plantations.

Cercospora leaf spots

Cercospora leaf spots are common diseases in warm, humid production areas. They have been reported in Australia (Watson *et al.*, 1988), Canary Islands (Galan-Sauco and Menini, 1991), Florida (McMillan, 1986; Campbell, 1989), India (Mukerji and Bhasin, 1986) and Malaysia (Ting and Tai, 1971; Ithnin *et al.*, 1992). With the exception of Malaysia, they usually are not important (Ting and Tai, 1971).

SYMPTOMS Symptoms occur on both young and mature leaves of seedlings and mature trees, and may also affect twigs, branches and leaf petioles. Initially, they appear as tiny necrotic or chlorotic spots on leaflets. As they enlarge, they develop greyish white centres, definite dark reddish brown margins, and chlorotic haloes. Spots may enlarge to 5 mm in diameter, and adjacent spots often coalesce to form large, irregular lesions. Sporulation of the causal fungus is evident in lesion centres as dark masses, and heavily affected leaves become yellow and fall prematurely.

CAUSAL AGENTS Cercospora leaf spot is caused by *Cercospora averrhoae* in Australia, the Canary Islands, Florida and Malaysia (Ting and Tai, 1971; McMillan, 1986; Watson *et al.*, 1988; Campbell, 1989; Galan-Sauco and Menini, 1991; Ithnin *et al.*, 1992), and *C. wellesiana* in India (Mukerji and Bhasin, 1986).

The pathogens are hyphomycetes that produce conidia in the small greyish centres of old lesions on the lower leaf surface. Conidiophores are unbranched, light brown and two-celled, and conidia are 71–284 × 2.5–4.3 μm, hyaline, multiseptate (8–23), and straight to slightly curved with a truncate base and tapered apex (Ting and Tai, 1971).

EPIDEMIOLOGY Conidia are produced during periods of rainfall and high humidity, and are dispersed by rainsplash, wind, insects and irrigation. Infection requires high moisture conditions, and under favourable conditions can occur on successive new leaf flushes.

MANAGEMENT Certain cultivars, such as 'B17', are more susceptible to the disease, and may suffer considerable defoliation. Less susceptible cultivars should be considered in conducive disease environments. Cercospora leaf spot can be controlled with regular applications of several different fungicides.

Pink disease

Pink disease is a problem in Southeast Asia (Samson, 1992). It usually does not cause major damage, but severely affected branches may wilt and die, especially in high-density plantations or in shaded areas. The disease occurs in areas with high rainfall and on plants that are more than 2 years old and have dense canopies.

SYMPTOMS The most common and conspicuous symptoms of pink disease are patches of silky pale pink fungal growth that develop along twigs, branches and trunks (see Plate 9). Under high humidity, mycelial threads of the pathogen rapidly form a thin, rough, pink encrustation on the bark surface. The pathogen penetrates the bark and wood, eventually causing tissue to dry and crack. Twigs and branches above the infected area may be killed.

CAUSAL AGENT *Erythricium salmonicolor* (anamorph: *Necator decretus*) causes pink disease on many woody plants in the tropics (Hawksworth *et al.*, 1983). It is described in Chapter 1.

EPIDEMIOLOGY During wet conditions, the pathogen forms basidiospores on white–pinkish hymenial layers. They are released and dispersed by rainsplash and wind. The fungus can also spread through contact with infected limbs, and wounded plants are most susceptible.

MANAGEMENT Early detection of the disease is important. Affected branches should be removed immediately and destroyed. Trees should be pruned regularly to provide good air circulation and penetration of sunlight in the canopy.

Miscellaneous foliar and canopy diseases

Leaf spots caused by *Corynespora cassiicola*, *Diplodia* sp., *Gloeosporium* sp., *Phomopsis* sp. and *Phyllosticta* sp. have been reported in Florida and Guyana, but they are not important (Campbell, 1989; Ramsammy, 1989). Leaf spots can also be caused by a bacterium, *Xanthomonas* sp. In the Canary Islands, carambola trees can also be affected by *Rhizopus* sp. (Galan-Sauco and Menini, 1991).

Fruit Diseases

Diseases can occur on fruit during every stage of development, but are most common after harvest. Since they impact the quantity and quality of the final product, they exert a most important influence on the production of these fruit.

Several precautions are needed in order to avoid disease problems and maintain the quality of fruit during storage and marketing (Watson *et al.*, 1988). Symptomatic fruit must be discarded since some of these diseases can spread to healthy fruit during storage, and fruit with physical damage should also be rejected at the packing plant since wounding enhances their development. Many of these diseases are latent, and thus are not evident upon packing. Whenever possible, fruit should be pre-cooled and stored at 5–10°C to reduce the rate at which these diseases develop.

Alternaria black spot (brown spot)

Alternaria black spot, which is also known as brown spot, has been reported in India and observed on stored fruit in Malaysia (Tandon and Verma, 1964; Jain and Saksena, 1984; Singh, 1992; Sepiah, Malaysia, personal observation). The disease can be an important problem during transit and storage of fruit.

SYMPTOMS Small, light brown or black circular spots develop on the fruit skin. Lesion centres are slightly sunken and, if the fruit are kept moist, olive–brown spores of the pathogen will develop. The spots are more limited, darker and firmer than those of anthracnose. The associated decay is firm and does not penetrate deeper than 1 or 2 mm until later stages, when the flesh becomes discoloured and partially softened.

CAUSAL AGENT *Alternaria alternata* is described in Chapter 1.

EPIDEMIOLOGY The pathogen colonizes dead and dying plant material, and leaves and twigs are significant sources of inoculum. Conidia are dispersed by wind and rainwater, and germinate between 4 and 35°C. Although penetration of the fruit surface via lenticels can occur, it is greatest via wounds. Infections occur during all stages of fruit development, but usually remain latent until fruit ripening begins.

MANAGEMENT Dead or dying plant material should be removed from trees. Recommended fungicides should be applied regularly to trees and young fruit prior to bagging. Injury of fruit should be avoided during postharvest handling, and fruit should be stored in a cool environment.

Anthracnose, fruit speckle, black spot and scab

Several different diseases on carambola fruit are associated with species of *Colletotrichum*. The most important, anthracnose, probably occurs in all countries that produce fruit. It is present in Australia, the Canary Islands, Florida, India, Southeast Asia and Taiwan (Srivastava and Tandon, 1968; Rana and Upadhyaya, 1971; Watson *et al.*, 1988; Campbell, 1989; Duan *et al.*, 1991; Galan-Sauco and Menini, 1991; Singh, 1992; Crane, 1993; Persley, 1993). It is usually a major postharvest disease, but substantial losses can also occur prior to harvest.

Speckle, black spot and scab have been reported in Malaysia, and are the most important fruit diseases in this country. They can appear at any stage of fruit development, particularly during the rainy season, and can result in serious losses if infection occurs early in development since fruit are then distorted. All cultivars are affected, but 'B17' is particularly susceptible. Speckle also affects leaves.

SYMPTOMS Anthracnose first appears on fruit as tiny, slightly sunken, light brown or dark spots. These eventually expand and soften the flesh. Spots often coalesce to form large, irregular necrotic lesions (Plate 42). Lesions may develop anywhere on the fruit surface, but often are associated with physical damage on the ribs (Fig. 6.1).

When conditions are favourable, orange masses of spores of the pathogen develop on affected tissue. In storage, greyish white or

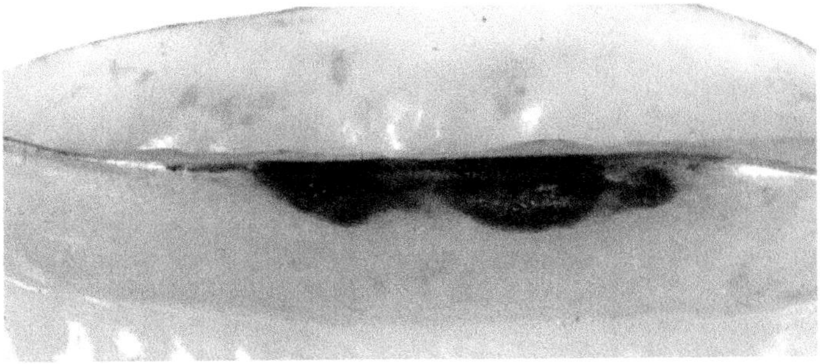

Fig. 6.1. Anthracnose symptoms on a 'Leng Bak' carambola fruit. Anthracnose often develops on the outer portions of the ribs of carambola fruit (photo: A.W. Cooke).

dark mycelium may develop, especially at temperatures above 15°C. During extended storage, the entire fruit surface may be covered with mycelium and become rough due to the formation of dark fruiting bodies that produce pinkish or whitish conidia.

Oval to irregular brown anthracnose lesions of variable sizes can also develop on leaves. They may be at the tip, margin or the midrib of the leaf, and they enlarge and coalesce under moist conditions. Young leaves are more susceptible than older leaves.

Speckle appears on fruit as tiny brown to dark brown spots that can develop soon after fruit set and eventually reach 1 mm in diameter (Fig. 6.2). The spots are composed of dark brown masses of fungal mycelium that can be removed from the fruit skin without leaving a visible indication of infection until a few hours after harvest. Thereafter, spots cannot be removed and the surrounding tissue turns light brown. These symptoms do not develop further.

On leaves, speckle appears as numerous tiny pinkish red or red spots on the lower surface. Where spots are concentrated, the symptoms also appear on the upper surface of the leaf as irregular reddish spots surrounded by yellowing tissue. In severe cases, leaves yellow and abscise.

Black spot appears as dark, irregular, dry spots on mature or ripening fruit. Only a few layers of surface cells are affected. Spots may be up to 5 mm in diameter and do not increase in size after harvest.

Scab causes raised, light brown, irregular patches that contain numerous tiny dark brown spots. They develop on any part of the fruit, and in severe cases enlarge to cover the entire fruit. Young fruit crack and their development is distorted. Scab develops when speckle is severe.

CAUSAL AGENTS *Glomerella cingulata* (anamorph: *Colletotrichum gloeosporioides*) is the major cause of anthracnose. *C. acutatum* also

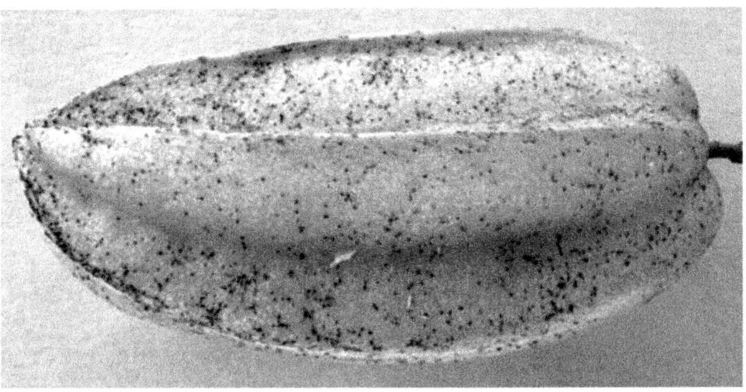

Fig. 6.2. Speckle symptoms on a carambola fruit (photo: Sepiah).

causes anthracnose in Australia (Persley, 1993). Both are described in Chapter 1. In Malaysia, postharvest anthracnose on fruits of 'B2' and 'B10' is due to *C. gloeosporioides*, whereas that on 'B17' is caused mainly by *C. crassipes* and *C. gloeosporioides*, and less frequently by *C. acutatum* and *C. capsici* (Sepiah, Malaysia, personal observations).

The pathogens that are associated with speckle, black spot and scab on carambola fruit are *C. crassipes* and *C. gloeosporioides*.

EPIDEMIOLOGY Conidia of the pathogens are produced on dead stems, twigs, branches, leaves, flowers and fruits, and are disseminated by wind, rainsplash and insects. Optimum conidium germination and appressorium formation for *C. gloeosporioides* occurs, respectively, between 16 and 36°C and 16 and 24°C. Infection occurs on most aboveground tissues of the host, and requires free moisture. In water, conidia form germ tubes within 6–8 h and appressoria within 10 h. Infection pegs are formed beneath appressoria, and wounding is not required.

In Taiwan, Duan *et al.* (1991) reported that the primary inoculum of *C. gloeosporioides* for fruit infection was conidia that were produced on fruit lesions. Several scenarios are possible. Conidia may germinate to produce small, dark brown colonies of mycelia that are superficial and removed easily by hand. Conidia can also infect, grow inside and kill epidermal cells soon after germination to form tiny brown spots, or may damage bigger areas and deeper layers of fruit tissue to form black spots. If lesions are dry and no fungal structures appear on the surface, they may not expand until after harvest. Most commonly, fruit are infected in a latent fashion; these infections remain quiescent on fruit for months and cause perceptible damage only after ripening begins. Postharvest anthracnose can be severe and develops mainly during long storage periods or on fruit that is overripe. Fruit-to-fruit spread after harvest probably is uncommon.

MANAGEMENT To protect fruit, effective control measures must be initiated during flowering and fruit development. Good plantation hygiene should be practised during the wet season. Pruning dead twigs and branches, and removing dead leaves and infected fruit helps reduce inoculum levels in the canopy. Several systemic and non-systemic fungicides control anthracnose, but the timing and frequency of application are critical. They should begin shortly before fruit are either set or bagged, and subsequent applications should be made to reduce pre- and postharvest disease on fruit. Careful handling of fruit to minimize injury, storing fruit at 5°C, and avoiding long storage periods are also helpful.

Aspergillus fruit rot

Aspergillus fruit rot is a minor disease that occurs during storage. The disease has been observed in Florida and Malaysia (Alfieri *et al.*, 1994; Sepiah, Malaysia, personal observation). Symptoms appear initially as a discoloration and softening of fruit tissue, and lesions are almost circular, enlarge slowly, and may be covered with white mycelium and dark coloured conidia of the pathogen, *Aspergillus niger*. *A. niger* is a cosmopolitan fungus with significant saprophytic capabilities. It is described in Chapter 1. Its conidia are dispersed by wind, rainwater and insects. During storage and transit, the fungus enters the flesh via wounds. Applying fungicides in the plantation can reduce inoculum, and careful handling of fruit to minimize injury, and storage at 5–15°C reduces the occurrence of the disease.

Black rot

Black rot occurs on fruit during storage. It can cause significant damage in India (Subramanian and Rao, 1981), but is unimportant in Malaysia (Sepiah, Malaysia, personal observation). Initially, lesions are small, dark brown and dry. As they enlarge, they become black, slightly sunken and irregular in shape. Small, black pycnidia of the pathogen, *Phoma averrhoae*, form at the centre of lesions (Subramanian and Roa, 1981). Lesions may develop slowly in green fruit or remain latent until fruit ripening. The fungus survives on branches, twigs and leaves of the plant, and releases conidia from pycnidia during rain. Good ventilation within the plant canopy should be maintained, and fungicides should be applied when needed.

Diplodia rot

Diplodia rot causes serious damage to fruit during storage and transit. The disease has been reported in Australia, Florida and India (Mukerji and Bhasin, 1986; Watson *et al.*, 1988; Singh, 1992; Alfieri *et al.*, 1994) and on fruit imported into Britain (Snowdon, 1990). It is common on fruit of 'B17' in Malaysia.

Symptoms begin as light brown lesions that enlarge relatively quickly, especially above 20°C. Infected tissue becomes dark, soft and watery, and covered with dark greyish mycelium of the pathogen, *Diplodia theobromae* (Fig. 6.3). Pycnidia that exude tendrils of white conidia are produced eventually on the fruit surface. The teleomorph of the fungus, *Botryosphaeria rhodina*, has not been reported on this host. The pathogen is described in Chapter 1 (Fig. 1.1).

Rainwater and insects disperse conidia and mycelium to new infection sites. The pathogen grows on fruit surfaces and infects via wounds that occur during handling, or at the stem or stylar end. The entire fruit can be rotted within 2–3 days at 25–30°C. Diplodia rot can be reduced by regular applications of fungicides, particularly before young fruit are bagged. Plants should be pruned regularly and all dead materials in the canopy removed. Fruit should be stored between 5 and 15°C and handled carefully to avoid damage.

Ceratocystis fruit rot (black rot)

Ceratocystis fruit rot is also known as black rot. To avoid confusion with the disease that is caused by *Phoma averrhoae*, the former name will be used in this chapter. Ceratocystis fruit rot is usually a postharvest problem, and can occur in the field if fruit are wounded or overripe. The disease occurs in Malaysia and Queensland, and is usually a minor problem (Watson *et al.*, 1988; Sepiah, Malaysia, personal observation).

Initial, water-soaked lesions enlarge rapidly and turn greyish black. The entire fruit can become soft and watery, and dark coloured mycelium and chlamydospores of the causal fungus, *Ceratocystis paradoxa* (anamorph: *Chalara paradoxa*), form on the fruit surface. The pathogen is described in Chapter 1 (Fig. 1.1). It survives as chlamydospores on plant debris in the plantation. Conidia are splashed on to fruit by rain, and infection of fruit may occur before, during and after harvest via wounds.

Fruit damage should be avoided at all stages of development. Regular applications of fungicides, and pruning and removing plant residues from the plantation are beneficial. Fruit should be stored between 5 and 10°C.

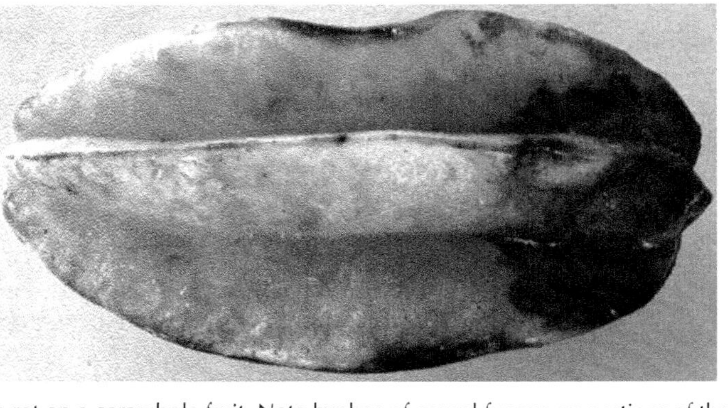

Fig. 6.3. Diplodia rot on a carambola fruit. Note hyphae of causal fungus on portions of the lesion surface (photo: Sepiah).

Cladosporium spot

Cladosporium spot occurs on fruit before and after harvest. It is present in India and Malaysia (Sharma and Khan, 1978; Jain and Saksena, 1984; Campbell, 1989; Singh, 1992; Sepiah, Malaysia, personal observation). The disease is most important during postharvest storage, especially on fruit stored between 5 and 15°C. In Malaysia, the fruit of 'B 17' are most susceptible.

SYMPTOMS Tiny watery spots form on the fruit surface, and masses of greyish brown mycelium may develop at their centre. Symptoms are most common at the stem end of fruit.

CAUSAL AGENTS *Cladosporium cladosporioides* (Jain and Saksena, 1984; Snowdon, 1990) and *C. herbarum* (Sharma and Khan, 1978; Singh, 1992) cause this disease. On fruit, growth of the fungi is limited to just a few millimetres, and small masses of conidiophores and conidia are often formed on the spots. Conidiophores are olivaceous brown and bear conidia from the upper to middle portion (Domsch *et al.*, 1980). Those of *C. cladosporioides* are 2–6 μm wide and up to 350 μm long (although usually much shorter), and bare ellipsoid, single-celled, olivaceous brown conidia that are 3–7 × 2–4 μm (Fig. 6.4). Conidiophores of *C. herbarum* are 3–6 μm wide and up to 250 μm long, and bear golden brown conidia, 5.5–13 × 3.8–6.0 μm, that are usually single-celled (Fig. 6.5).

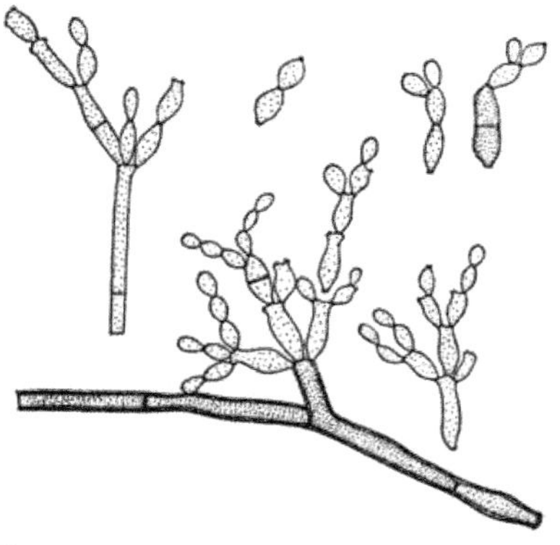

Fig. 6.4. Conidia and conidiophores of *Cladosporium cladosporioides* (from Ellis, 1971).

EPIDEMIOLOGY The pathogens are cosmopolitan colonists of plant surfaces and litter. They normally grow as saprophytes. Their conidia are common on live and dead plants and other organic materials, and are dispersed to other fruits by wind, rainwater or insects.

Cladosporium spot is more common during the rainy season, and it can reduce the quality of fruit in storage, especially between 5 and 15°C. At higher storage temperatures, diseases that are caused by *D. theobromae*, *Colletotrichum* spp. and *Phomopsis* sp. are more prevalent due to their faster growth rates.

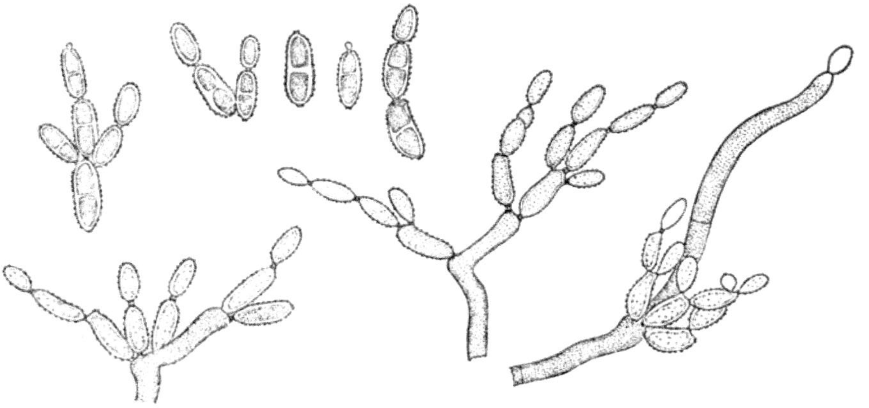

Fig. 6.5. Conidia and conidiophores of *Cladosporium herbarum* (from Ellis, 1971).

MANAGEMENT Cladosporium fruit disease can be controlled with good plantation management and by avoiding prolonged cold storage. Strategic pruning after flowering, to force new growth, may also reduce postharvest losses.

Dothiorella rot

Symptoms of this disease appear as fruit ripen. A brown soft decay starts usually at the stem end (Plate 43), but may also develop on other portions of the fruit surface that have been injured. As the disease spreads, fruit collapse and may split open. A straw-coloured fluid drains from the stem end or from splits in the side of the fruit, and steel-grey mycelium of the causal fungus, *Dothiorella* sp., may cover the surface of the fruit. The pathogen may spread from diseased to healthy fruit by physical contact, and flesh of affected fruit has an 'off' flavour.

Pre-harvest sprays for anthracnose control may reduce the incidence of this disease. The disease is less prevalent in young plantations where leaf litter and prunings have not accumulated. Water stress during fruit development and maturation should be avoided, and immature fruit should not be harvested. Harvested fruit should not be 'bled' on the soil surface, since this can be a source of infection. Fruit should be cooled and stored in well-ventilated containers immediately after harvest.

Flyspeck

This is a relatively unimportant fruit disease in Florida; apparently, it has not been reported previously on carambola (R.C. Ploetz, UF, personal observation). The most conspicuous symptoms of flyspeck are small black dots that are formed in roughly circular patterns on the fruit surface; they resemble marks made by flies (Plate 44). They are superficial, and can be rubbed off. The disease is often found in association with sooty blotch, but is far less common.

Flyspeck is caused by the ascomycete, *Schizothyrium pomi* (anamorph: *Zygophiala jamaicensis*). It has a wide host range, including apple, carnation and many other temperate and tropical hosts. Baker *et al.* (1977) indicated that although the epithet *Leptothyrium pomi* was often used when reporting the flyspeck agent, this fungus has not been demonstrated to be a part of the flyspeck disease cycle.

On the fruit surface, *S. pomi* forms minute pseudothecia (flyspecks) and sparse networks of interconnecting dark hyphae that produce undulating conidiophores ~20 μm in length (Fig. 6.6). Each conidiophore produces a pair of two-celled conidia.

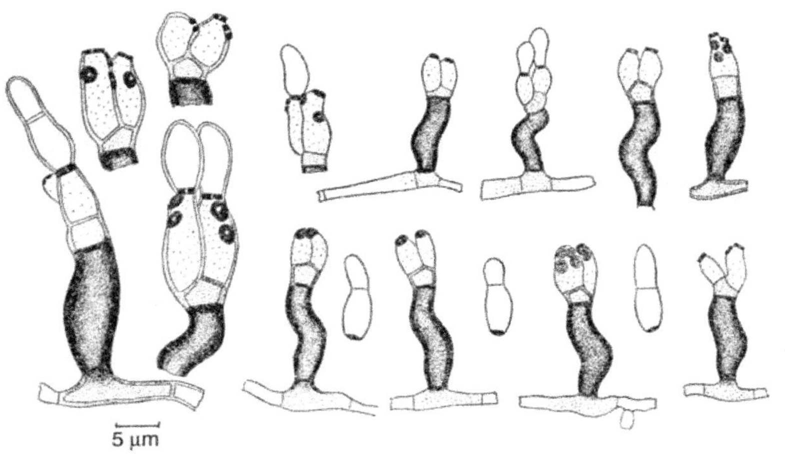

Fig. 6.6. Conidia and conidiophores of *Zygophiala jamaicensis*, anamorph of *Schizothyrium pomi* (from Ellis, 1971).

The disease is observed during warm, wet periods. Specific control measures for it are not indicated.

Fusarium fruit rot

Fusarium fruit rot has been observed infrequently on stored fruit of 'B17' in Malaysia (Sepiah, Malaysia, personal observation). The disease is caused by *Fusarium pallidoroseum* and *F. moniliformae* (Fig. 4.28). It causes a softening and darkening of fruit tissue. Lesions up to 15 mm in diameter develop in fruit grooves or on wounded areas. Pinkish white or yellowish mycelium of the causal fungi is always present on lesions. The pathogens usually survive as saprophytes in the field on leaves, branches and fruits. They produce micro- and macroconidia on dead or live plant materials, which are dispersed by rainwater. The fungi grow on the fruit surface, and infection is increased on wounded fruit. Fusarium fruit rot can be controlled with regular applications of fungicides, especially before fruit are bagged. Plants that have a dense canopy should be pruned.

Penicillium spot

Penicillium spot occurs on fruit of 'B17' in Malaysia (Sepiah, Malaysia, personal observation). It develops before and after harvest, but is not important until fruit are stored. The disease can cause substantial losses, particularly when fruit are harvested in the rainy season and kept for long periods at 5–15°C.

SYMPTOMS In the field, tiny dark, dry spots, up to 1 mm in diameter, appear on the fruit surface. They usually do not enlarge until storage. In storage, lesions may expand or be initiated at new sites. Lesions can develop at any place on the fruit surface, and begin as tiny, pale brown spots that become darker and increase slowly up to 2 mm in diameter as time in storage increases. The pathogen often produces whitish mycelium on spots that turns bluish as conidia form.

CAUSAL AGENT *Penicillium expansum* causes Penicillium spot. It produces spherical to oval, slightly greenish or bluish conidia, that are 3.0–3.5 μm in diameter and are formed in chains on 400 μm long conidiophores (Fig. 6.7) (Domsch *et al.*, 1980). In culture, the fungus produces a fruity odour that is reminiscent of apples.

EPIDEMIOLOGY The fungus occurs commonly on plants, and in air and soil, and saprophytically colonizes dead and dying plant materials. Conidia are released and dispersed by wind and rainwater. In wet, warm weather, the fungus may penetrate fruit tissue and form spots, or it remains latent until conditions are favourable. Infection may also occur via wounds that occur during harvesting and handling.

MANAGEMENT Inoculum of the pathogen should be reduced by pruning trees, and by applying recommended fungicides on a regular basis. Treating young fruit with fungicides before they are bagged is also effective. Wounding and prolonged storage of fruit should be avoided.

Pestalotiopsis rot

Pestalotiopsis rot occurs during storage. It affects fruit of 'B17' in Malaysia, and only rarely causes serious losses (Sepiah, Malaysia, personal observation).

Initially, light brown, circular lesions enlarge up to 2 cm in diameter and soften. White mycelium of the pathogen, *Pestalotiopsis guipini*, covers lesions and eventually produces dark masses of conidia. It is a weak pathogen that usually requires wounding in order to infect carambola.

The disease is controlled with regular applications of fungicides, by pruning trees and removing the debris from the plantation, and by storing fruit at 5–10°C.

Phomopsis rot

Phomopsis sp. causes leaf and fruit diseases of carambola. Phomopsis rot of fruit is most important, particularly after harvest. It can cause substantial damage during storage and transit, and is present in Australia,

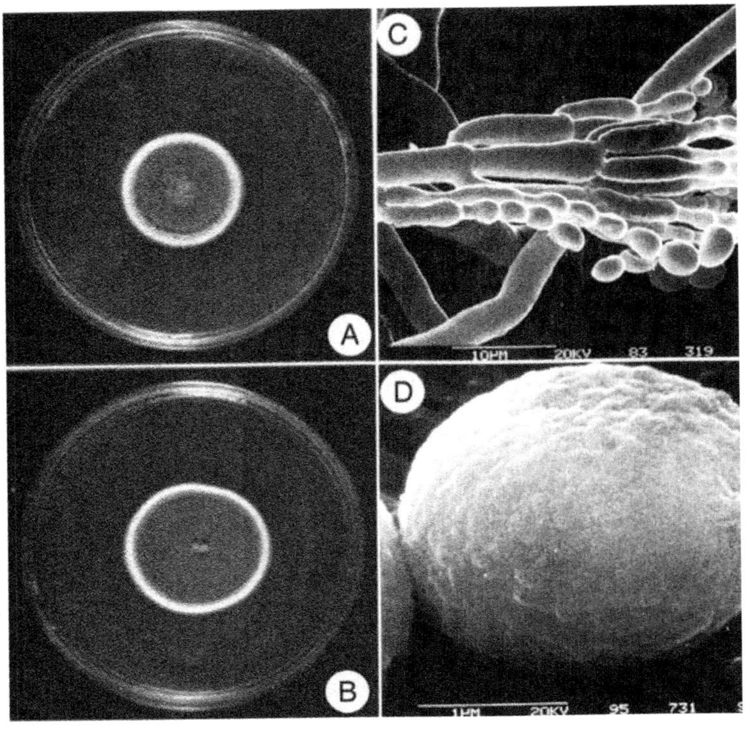

Fig. 6.7. (A) Czapek agar culture after 2 weeks at 25°C, (B) malt agar culture after 2 weeks at 25°C, (C) scanning electron micrograph of conidial head and (D) conidium of *Penicillium expansum* (from CMI descriptions nos 97 and 1258).

Florida, India and Malaysia (Mukerji and Bhasin, 1986; Watson *et al.*, 1988; Campbell, 1989; Singh, 1992; Sepiah, Malaysia, personal observation).

SYMPTOMS Phomopsis rot may occur while fruit are still in the field, or during transit or storage. Symptoms usually develop near the stem and stylar ends of fruit (Plate 45), but also occur on wounded areas, particularly on the ribs. Affected areas become discoloured, soft and watery. Depending on the strain of the pathogen, the whole fruit can be affected within a few days at temperatures higher than 20°C. Greyish or yellowish white mycelium of the pathogen often grows on affected tissue, and produces black pycnidia that release yellowish masses of conidia.

CAUSAL AGENT A coelomycete, *Phomopsis* sp., causes Phomopsis rot. It produces two types of hyaline conidia in pycnidia, α-conidia that are ovate or ellipsoidal, and β-conidia that are long, filamentous and sickle shaped (Sutton, 1982). Conidiophores are simple, and pycnidia are dark, ostiolate, immersed, erumpent and nearly globose.

EPIDEMIOLOGY Pycnidia are formed on dead parts of the plant. They ooze conidia in tendrils that are dispersed by splashing water, and can also be spread by insects. Wounding enhances infection, and growth of the pathogen during storage is promoted above 20°C.

MANAGEMENT Precautions are needed when wet weather follows a dry season, since stressed plants are more prone to growth cracks. It is also important to treat young fruit with recommended fungicides before bagging. Harvested fruit should be handled properly to avoid unnecessary wounding, and should be stored at a low temperature as soon as possible.

Sooty blotch

Sooty blotch can be an important problem. Severely affected fruit are unsightly and are either downgraded at the packing plant or require cleaning before they can be sold.

This disorder has been confused with sooty mould, a distinctly different problem, in at least two producing areas. A recent illustration and description of 'sooty blotch' in Malaysia was clearly of sooty mould, and in Florida the problem was reported initially as sooty mould (Campbell, 1989; Crane, 1992). The problem in Florida is now recognized as sooty blotch due to its distinct symptoms and the absence of insect associations that are needed for sooty mould development (Ploetz *et al.*, 1995). In Queensland, Watson *et al.* (1988) reported 'a surface mould' on fruit that could be rubbed off and was not associated with insect feeding. Thus, sooty blotch may also occur in Australia.

SYMPTOMS In general, symptoms appear on the fruit and leaf surface as smoky grey to black, irregular splotches of varying intensity (Plate 44). They are finely webbed networks of hyphae of the causal fungi, are usually superficial, and in severe cases cover >50% of the fruit surface. They possess diverse microscopic, hyphal morphologies that presumably relate to the complex aetiology of the disease (Ploetz *et al.*, 2000). Most often, they are aborescent and fernlike with few plectenchymal bodies; the latter mature to become pycnothyria.

CAUSAL AGENT The cause of this disorder is somewhat confused. Although Watson *et al.* (1988) clearly described sooty blotch in Australia, they mentioned no causal agent. In Florida, *Leptothyrium* sp. was cited originally (Campbell, 1989; Crane, 1992) (this genus had been associated erroneously with flyspeck in the past, but not sooty blotch (Baker *et al.*, (1977)). *Leptothyrium* sp. was later recovered from a leafspot of an *Averrhoa* species other than carambola (not specified) (Alfieri *et al.*, 1994), and *Microthyrium* sp. was indicted as the cause of the fruit blemish (G. Simone, personal communication). It was determined subsequently that this problem was related to sooty blotch of apple (Ploetz *et al.*, 1995).

Johnson *et al.* (1996, 1997) recently determined that sooty blotch of apple is a disease complex. They associated a ramose symptom with either *Geastrumia polystigmatis* or *Peltaster fructicola*, a fuliginous symptom with *Leptodontium elatius*, and a punctate symptom with a fungus that resembled, but was distinct from, *P. fructicola*. Preliminary work indicates that the most prevalent symptom on carambola in Florida is of the ramose type (Ploetz *et al.*, 2000). These symptoms are very similar to the ramose symptoms on apple, as are the cultural appearances of *P. fructicola* and '*Peltaster* sp.' (carambola). However, they differ in the dimensions of their brown pycnothyria (81–113 μm versus 35–64 μm, respectively) and single-celled conidia (4–6 × 2 μm versus 1.5–2 × 3.2–4.8 μm, respectively), and DNA restriction fragment length polymorphisms (RFLPs) (Ploetz *et al.*, 2000; S. Williamson, personal communication).

EPIDEMIOLOGY In Florida, symptoms begin during the summer rainy season, and moist microclimates appear to favour sooty blotch development (Ploetz *et al.*, 2000). The disease's latent period was estimated to range between 41 and 53 days.

MANAGEMENT Variation in the susceptibility of fruit of different cultivars has been noted in different locations. In Australia, 'B8', 'B10', 'B16' and 'Fwang Tung' were reported to be susceptible during some seasons, but not others, whereas 'Arkin' was not affected (Watson *et al.*, 1988). In contrast, 'Arkin' is severely affected in Florida (Ploetz *et al.*, 2000).

Canopy size and density should be managed to increase airflow and the rate of evaporation. Likewise, when new plantations are planted, an E–W, rather than a N–S, row orientation should be considered. In Florida, ferbam, and less so copper fungicides, are somewhat effective. Superior control in the field will probably depend upon an integrated approach that utilizes fungicides and the cultural information above. After harvest, severely affected fruit must be washed with dilute bleach.

Sooty mould

Sooty mould is usually not a major problem (Sepiah, Malaysia, personal observation). It appears as light brown to dark blotches or encrustations on leaves and twigs. On fruit, the stem end is affected most often. The causal fungi do not penetrate the host and can be removed by hand.

The disease is caused by dark coloured, saprophytic fungi. They grow on plant surfaces in the presence of honeydew of ants, aphids, mealybugs and scale insects. The fungi produce spores on these substrates that are spread by rain and insects. The disease does not spread during storage. Systematic pruning to provide good ventilation and reduce humidity within the canopy is helpful, but the most effective means by which this problem is overcome is to apply recommended insecticides, especially before fruit are bagged.

Miscellaneous fruit diseases

Botryosphaeria ribis causes an uncommon postharvest fruit rot in Malaysia (Fig. 3.4) (Sepiah, Malaysia, personal observation). *Khuskia oryzae* (anamorph: *Nigrospora oryzae*) (Fig. 4.20) has been reported to cause postharvest disease of carambola in India (Jain and Saksena, 1984). In Malaysia, *Rhizopus* sp. damages fruit during storage, but is not common. In Florida, a fruit rot caused by the bacterium *Pseudomonas* sp. has been reported (Alfieri *et al.*, 1994).

Root Diseases

Root disease can kill trees, but is most often responsible for reducing vigour of the carambola host. In many cases, these diseases are inconspicuous and may not be recognized unless roots are excavated. To avoid these problems, it is imperative that clean nursery stock and non-infested planting sites be used.

Pythium root rot

This can be a serious problem in Florida in low-lying areas and when infected trees are planted in the field (Ploetz *et al.*, 1991a).

SYMPTOMS Aboveground symptoms of this disease include a general sparseness of the canopy, which is smaller than that of healthy trees of a similar age. Foliage often displays symptoms of nutritional deficiencies, especially iron and manganese, and, during periods of water stress, wilts.

Root systems can be dramatically reduced in size and mechanical strength to the extent that badly affected trees can be pulled from the ground by hand. The pathogen kills apical portions of roots, especially in the region of the root cap.

CAUSAL AGENTS The oomycetes, *Pythium splendens*, and, to a lesser extent, *P. ultimum* var. *ultimum* cause Pythium root rot. They produce coenocytic hyphae and have very wide host ranges (van der Plaats-Niterink, 1981).

Unlike most members of this genus, *P. splendens* does not produce sporangia or zoospores (van der Plaats-Niterink, 1981). It produces globose, hyphal swellings, 25–49 μm in diameter, which are usually terminal and germinate with one to six germ tubes (Fig. 6.8). Although homothallic strains are known, the fungus is primarily heterothallic and requires complementary strains to form oogonia. Antheridia are diclinous and usually terminal, and one to eight form per terminal or intercalary oogonia. Oospores are aplerotic, 20–33 (25) μm in diameter, with 1–3 μm thick walls. Its cardinal temperatures for growth are 5, 25 and 34°C. Sporangia are not produced.

P. ultimum var. *ultimum* produces sporangia and zoospores infrequently. Hyphal swellings are globose, 20–29 μm in diameter, and usually intercalary. Oogonia are smooth, globose, 14–25 (21.5) μm in diameter, usually terminal, each with one to three mostly monoclinous antheridia. Oospores are aplerotic, globose, 12–21 (18) μm in diameter, and have cell walls of 2 μm or more in thickness. Its cardinal temperatures for growth are 5, 25–30 and 35°C.

EPIDEMIOLOGY *P. splendens* is an opportunistic pathogen that causes its greatest

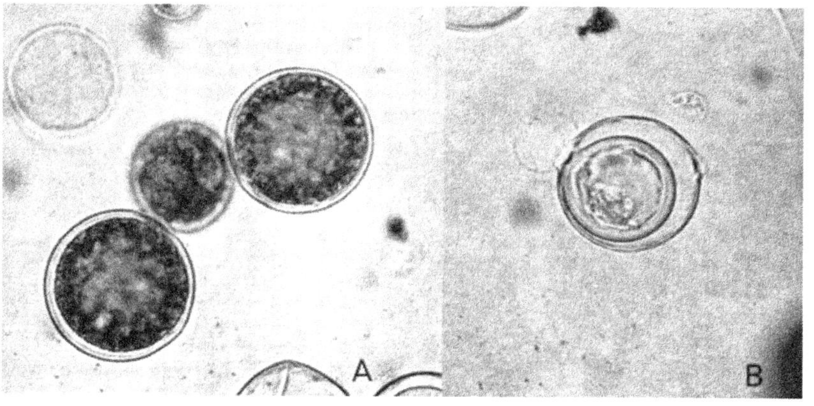

Fig. 6.8. (A) Globose hyphal swellings and (B) an oospore of *Pythium splendens* (from CMI description no. 120).

damage during cool weather that does not favour the carambola host. In controlled temperature studies, Ploetz (2003) demonstrated that it caused its greatest root damage and reduction in canopy growth on carambola at 15 and 20°C, far below the pathogen's optimum temperature for growth (Fig. 6.9). Nutritional deficiencies that were induced by the pathogen were eliminated by either treating soil with metalaxyl or by pasteurizing soil (Ploetz, 1991b).

MANAGEMENT Disease-free nursery stock and non-infested planting sites should be utilized whenever possible. Low-lying areas should also be avoided. Although metalaxyl controlled the disease in pot studies, its efficacy on mature, bearing trees in the field has not been demonstrated (Ploetz, 1991b; R.C. Ploetz, USA, personal observation).

White root disease

White root disease of carambola has been reported in Malaysia, where it is a minor problem (Ithnin *et al.*, 1992). It occurs in limited areas of a plantation, and appears to be related to the plantation's history. The disease occurs where rubber (Ithnin *et al.*, 1992) and cassava (B.S. Lee, Kuala Lumpur, 1998, personal communication) were grown previously.

SYMPTOMS The primary symptom of white root disease is rot of the major roots that is covered with white rhizomorphs of the pathogen (Nandris *et al.*, 1987). Affected trees wilt, yellow, defoliate prematurely, and eventually die.

CAUSAL AGENT *Rigidiporus lignosus*, the basidiomycete that causes white root disease, is a common soil inhabitant in the humid tropics of Africa and Asia (Holliday, 1980). It is described in Chapter 1.

EPIDEMIOLOGY Previously colonized stumps and infected woody debris of rubber and other hosts are primary sources of inoculum. Orange–yellow, bracket-like sporophores are produced during the rainy season on the root collar, trunk or exposed roots. Basidiospores produced by these sporophores are viable, but are thought to play a secondary role in disseminating the disease. Rhizomorphs are more significant, since they grow rapidly and can advance great distances in soil in the absence of woody substrates.

MANAGEMENT Colonized woody debris should be eliminated when new plantations are established. Affected trees in pre-existing plantations should be removed and destroyed, and the soil in the surrounding area treated with suitable fungicides (Tan and Hashim, 1992; Lam and Chew, 1993; Jayasinghe *et al.*, 1995).

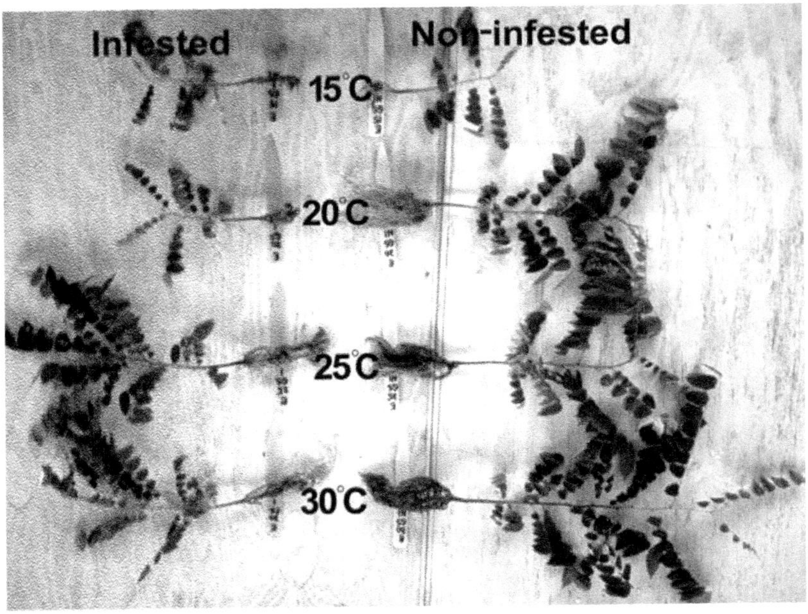

Fig. 6.9. Influence of Pythium root rot and temperature on 'Golden Star' carambola seedlings. Seedlings on the left were grown in soil infested with *Pythium splendens* and those on the right were grown in pathogen-free soil. Note the smaller root systems and canopies of seedlings from infested soil and the interaction of low temperatures (15 and 20°C) and root rot on plant size (photo: R.C. Ploetz).

Miscellaneous root diseases

Phytophthora citrophthora, *P. nicotianae* and *Rhizoctonia solani* cause root rots of carambola in the field (Ho, 1983; Alfieri *et al.*, 1994). Collar rot, caused by *D. theobromae*, has been seen on carambola in Malaysia, and causes wilting and plant death (B.S. Lee, Kuala Lumpur, 1998, personal communication). All but *R. solani* are described in Chapter 1.

References

Abidin, M.I.B.Z. (1987) *Cultivation of Tropical Fruits*. Hi-Tech Enterprise, Kuala Lumpur.

Alfieri, S.A., Jr, Langdon, K.R., Kimbrough, J.W., El-Gholl, N.E. and Wehlburg, C. (1994) *Diseases and Disorders of Plants in Florida*. Bulletin No. 14. Florida Department of Agriculture and Consumer Services, Contribution No. 680.

Baker, K.F., Davis, L.H., Durbin, R.D. and Snyder, W.C. (1977) Greasy blotch of carnation and flyspeck of apple: diseases caused by *Zygophiala jamaicensis*. *Phytopathology* 67, 580–588.

Campbell, C.A., Huber, D.J. and Koch, K.E. (1989) Postharvest changes in sugars, acids, and color of carambola fruit at various temperatures. *HortScience* 24, 472–475.

Campbell, C.W. (1989) Propagation and production system for carambola. *Proceedings of the Interamerican Society for Tropical Horticulture* 33, 67–71.

Crane, J.H. (1992) *The Carambola (Star Fruit)*. Factsheet FC 12. Florida Cooperative Extension Service, University of Florida, IFAS, Gainesville, Florida.

Crane, J.H. (1993) Commercialization of carambola, atemoya, and other tropical fruits in South Florida. In: Janick, J. and Simon, J.E. (eds) *New Crops*. John Wiley & Sons, New York, pp. 448–460.

Crane, J.H. (1997) Tropical fruit crops acreage in Florida. *Extension Handout*. University of Florida, Homestead, Florida.

Domsch, K.H., Gams, W. and Anderson, T.-H. (1980) *Compendium of Soil Fungi*, Vol. 1. Academic Press, London.

Donadio, L.C. (1989) Carambola growing in Brazil. *Proceedings of the Interamerican Society for Tropical Horticulture* 33, 26–29.
Duan, C.H., Tasi, W.H. and Tu, C.C. (1991) Pathogenicity and inoculum sources of *Colletotrichum gloeosporioides* in carambola. *Journal of Agricultural Research of China* 40, 425–432.
Ellis, M.B. (1971) *Dematiaceous Hyphomycetes*. Commonwealth Mycological Institute, Kew, UK, 608pp.
Galan-Sauco, V. and Menini, U.G. (1991) Carambola and its cultivation. *FAO Plant Production and Protection Paper* 108. FAO, Rome.
Green, J.G. (1989) Carambola production in Malaysia and Thailand. *Proceedings of the Florida State Horticultural Society* 100, 275–278.
Hawksworth, D.L., Sutton, B.C. and Ainsworth, G.C. (1983) *Dictionary of the Fungi*, 7th edn. Commonwealth Mycological Institute, Kew, UK.
Ho, H.H. (1983) The nature of paragyny in *Phytophthora*. *Mycopathologia* 83, 119–123.
Holliday, P. (1980) *Fungus Diseases of Tropical Crops*. Cambridge University Press, Cambridge.
Ithnin, B., Lim, W.H., Vijaysegaran, S., Nik Masdik, N.M. and Lee, S.A. (1992) Perosak, penyakit dan rumpai. In: Abd. Rahman, M., Izham, A. and Raziah, M.L. (eds) *Panduan Pengeluaran Belimbing*. MARDI, Kementerian Pertanian Malaysia, Kuala Lumpur, pp. 26–35.
Izham, A., Zainudin, J., Chang, S.T. Masri, M., Lim, S.P., Syed Mohd, S.I. and Ithnin, B. (1992) Amalan kultur. In: Abd. Rahman, M., Izham, A. and Raziah, M.L. (eds) *Panduan Pengeluaran Belimbing*. MARDI, Kementerian Pertanian Malaysia, Kuala Lumpur, pp. 11–25.
Jain, S. and Saksena, S.B. (1984) Three new soft rot diseases of *Averrhoa carambola* from India. *National Academy of Sciences, India* 7, 327–328.
Jayasinghe, C.K., Jayasuria, K.E. and Fernado, T.H.P.S. (1995) Pentachlorophenol– effective and economical fungicide for management of white root disease caused by *Rigidiporus lignosus* in Sri Lanka. *Journal of the Rubber Research of Sri Lanka* 75, 64–70.
Johnson, E.M., Sutton, T.B. and Hodges, C.S. (1996) *Peltaster fructicola*: a new species in the complex of fungi causing apple sooty blotch disease. *Mycologia* 88, 114–120.
Johnson, E.M., Sutton, T.B. and Hodges, C.S. (1997) Etiology of apple sooty blotch in North Carolina. *Phytopathology* 87, 88–95.
Joubert, J.J. and Rijkenberg, F.H.J. (1971) Parasitic green algae. *Annual Review of Phytopathology* 9, 45–64.
Knight, R.J. (1966) Heterostyly and pollination in carambola. *Proceedings of the Florida State Horticultural Society* 78, 72–78.
Knight, R.J. (1989) Carambola cultivars and improvement programs. *Proceedings of the Interamerican Society for Tropical Horticulture* 33, 72–78.
Lam, C.H. and Chew, S.B. (1993) Hexaconazole (Anvil 5 Sc), a cost effective fungicide for controlling white root disease in immature rubber. *Planta* 69, 465–474.
Manners, J.G. (1993) *Principles of Plant Pathology*, 2nd edn. Cambridge University Press, Cambridge.
Marler, T. E. and Zozor, Y. (1992) Carambola growth and leaf gas-exchange responses to seismic or wind stress. *HortScience* 27, 913–915.
Marler, T.E., Schaffer, B. and Crane, J.H. (1994) Developmental light level affects growth, morphology, and leaf physiology of young carambola trees. *Journal of the American Society for Horticultural Science* 119, 711–718.
Martin, F.W., Campbell, C.W. and Ruberte, R.M. (1987) *Perennial Edible Fruits of the Tropics: an Inventory USDA Handbook* No. 642.
McMillan, R.T. Jr (1986) Serious diseases of tropical fruits in Florida. *Proceedings of the Florida State Horticultural Society* 99, 224–227.
Mukerji, K.G. and Bhasin, J. (1986) *Plant Diseases of India. A Source Book*. Tata McGraw Hill Publisher Comp., New Delhi.
Nandris, D., Nicole, M. and Geiger, J.P. (1987) Root rot diseases of rubber trees. *Plant Disease* 71, 298–306.
Ngah, A.W.B., Ahmad, I. and Hassan, A. (1989) Carambola production, processing and marketing in Malaysia. *Proceedings of the Interamerican Society for Tropical Horticulture* 33, 30–43.
Núñez-Elisea, R. and Crane, J.H. (2000). Selective pruning and crop removal increase early-season fruit production of carambola (*Averrhoa carambola* L.). *Scientia Horticulturae* 86, 115–126.
Ooi, P.A.C. (1984) A fruit fly survey in a star fruit orchards in Serdang Selangor. *Journal of Plant Protection in the Tropics* 1, 63–65.
Persley, D. (ed.) (1993) *Diseases of Fruit Crops*. Department of Primary Industries, Brisbane.
Ploetz, R.C. (1991a) Species of *Pythium* as pathogens of perennial, woody fruit crops in south Florida. *Phytopathology* 81, 699.

Ploetz, R.C. (1991b) Effects of fungicides and supplemental applications of iron on the control of *Pythium splendens*-induced root rot of carambola (*Averrhoa carambola*). *Abstracts, XIIth International Plant Protection Congress*. Poster 30, August 15.

Ploetz, R.C. (2003) *Pythium splendens* is an opportunistic pathogen of carambola, *Averrhoa carambola. Mycopathologie* (in press).

Ploetz, R.C., Dorey, A.J. and Benscher, D. (1995) Observations on the epidemiology of sooty blotch on carambola (*Averrhoa carambola*) in south Florida. *Phytopathology* 85, 1195.

Ploetz, R.C., Dorey, A.J. and Benscher, D. (2000) The cause, epidemiology and control of sooty blotch of carambola, *Averrhoa carambola* L., in South Florida. *Fruits* 55, 241–252.

Popenoe, W. (1924) *Manual of Tropical and Subtropical Fruits*. MacMillan, New York.

Ramsammy, P. (1989) The carambola in Guyana. *Proceedings of the Interamerican Society for Tropical Horticulture* 33, 12–25.

Rana, O.S. and Upadhyaya, J. (1971) A new record of anthracnose on carambola fruits (*Averrhoa carambola*). *Science and Culture* 37, 529.

Samson, J.A. (1992) *Averrhoa* L. In: Verheij, E.W.M. and Corond, R.E. (eds) *Prosea: Plant Resources of South East Asian Edible Fruits and Nuts*. Prosea Foundation, Bogor, pp. 96–98.

Sedgley, M. (1984) *Oxalidaceae*. In: Page, P.E. (ed.) *Tropical Tree Fruits for Australia*. Queensland Department of Primary Industries, Brisbane, pp. 125–128.

Sharma, N. and Khan, A.M. (1978) Fruit rot in *Averrhoa carambola* L. *Geobios* 5, 48.

Singh, S.P. (1992) *Fruit Crops for Wasteland*. Scientific Publisher, Jodhpur, India.

Snowdon, A.L. (1990) *A Colour Atlas of Post-harvest Diseases and Disorders of Fruits and Vegetables. Vol. 1: General Introduction and Fruits*. Wolfe Scientific, London.

Srivastava, M.P. and Tandon, R.M. (1968) Some storage diseases of fruits. *Current Science* 37, 292.

Subramanian, V. and Rao, V.G. (1981) Three new storage diseases of fruits and vegetables from India. *Journal of the University of Poona, Science and Technology* 54, 145–149.

Sutton, B.C. (1982) *Coelomycetes*. Commonwealth Mycological Institute, Kew, UK, 253 pp.

Tan, A.M. and Hashim, I. (1992) Fungicide drenching for white-root disease control. *Planters Bulletin Rubber Research Institute of Malaysia* 87–93, 212–213.

Tandon, R.N. and Verma, A. (1964) Some new storage diseases of fruits and vegetables. *Current Science* 33, 625–627.

Ting, W.P. and Tai, L.H. (1971) Occurrence of Cercospora leaf spot of starfruit (*A. carambola*) in Selangor. *Malaysian Agriculture Journal* 48, 25–27.

van der Plaats-Niterink, A.J. (1981) *Monograph of the Genus Pythium. Studies in Mycology No. 21*. Centraalbureau voor Schimmelcultures, Baarn.

Vijaysegaran, S. (1988) Notes on carambola production in Peninsular Malaysia. *Workshop Paper on Postharvest Loss Assessment Methodology*. MARDI, Kuala Lumpur.

Watson, B.J., George, A.P., Nissen, R.J. and Brown, B.I. (1988) Carambola: a star of the horizon. *Queensland Agriculture Journal* 114, 45–51.

7 Diseases of Citrus

L.W. Timmer[1], S.M. Garnsey[1] and P. Broadbent[2]

[1] *University of Florida, Citrus Research and Education Center, 700 Experiment Station Road, Lake Alfred, Florida, USA;* [2] *formerly P. Barkley, New South Wales Agriculture, Elizabeth Macarthur Agricultural Institute, Private Mail Bag 8, Camden, 2570 New South Wales, Australia*

Introduction

World production, important production areas and cultural practices

Citrus originated in the region between northeast India, South China and Indonesia. Citrus fruit are now produced widely in tropical and subtropical countries in a belt from 35° north latitude to 35° south latitude. In subtropical climates, e.g. California and the Mediterranean, where there is definite seasonality, flowering and growth are controlled by temperature. In distinctly tropical areas near the equator, flowering occurs following the replenishment of soil moisture after drought. Variable rind colour and maturity, low fruit yields, and fruit blemishes caused by diseases and pests limit the full exploitation of citrus in tropical areas (Reuther *et al.*, 1967) where fruit are grown largely for local consumption. Some species, such as sweet orange and lemon, grow best in the subtropics, whereas limes and pummelo are produced primarily in the lowland tropics.

The world produces more citrus than any other kind of fruit, an average of 87 million tonnes (Mt) year^{-1} in recent years (Anonymous, 2001). Global annual earnings from exports of citrus, a major cash crop for many developing countries, amount to an average of more than US$8 billion. Citrus also contributes to the nutritional requirements of many lower income countries.

The major citrus production areas are shown in Fig. 7.1. While production is widespread, 15 countries account for 84% of world production. Total world citrus production in 2000 was estimated at 106 Mt. In 2000–2001, total production of fresh citrus fruit in major citrus-producing countries in the northern hemisphere was 45.9 Mt and in the southern hemisphere 27.6 Mt. Total citrus used for processing was 29.9 Mt (Anonymous, 2001).

Orange production increased in Mexico and most Mediterranean countries. Tangerine production increased in China, Morocco, Spain and Japan. Lemon and lime production grew globally, with increases in US production more than making up for declines in Spain. Brazil is the world's largest citrus producer (20% of world production) and the world's largest exporter of frozen concentrated orange juice (50% of world exports). Spain is the world's largest exporter of fresh citrus (tangerines and oranges) with an all time record of 1.4 Mt in 2000–2001. Total citrus production in China increased more than fourfold over the last decade to 9.1 Mt, two-thirds of which are tangerine types consumed by the domestic market.

Cultural practices vary considerably in citrus areas around the world, depending on

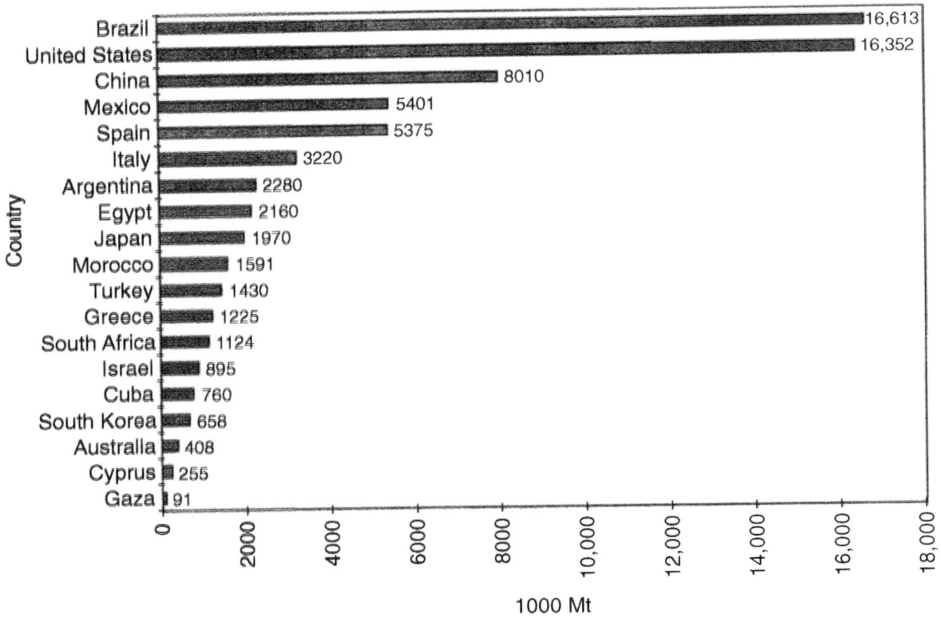

Fig. 7.1. Citrus production in the most important producing countries in 1997/1998. World Horticultural, Trade and Export Opportunities, US Department of Trade, FHORT 2-99, February 1999.

soil factors, climate, destination of the fruit, and local economic and labour situations. In the cooler winter-rainfall areas, such as the Mediterranean, parts of California, Australia, and South Africa, the majority of fruit is produced for fresh market. Diseases and pests that blemish the fruit are less of a problem in these low rainfall regions, but citrus trees in these areas require irrigation. In more recent years, most orchards have been converted from flood or furrow irrigation to drip or microsprinkler irrigation. Fertilizer and some pesticides are delivered via the irrigation system. In arid, subtropical areas, spacing is usually closer and there is less need for pruning since trees grow more slowly. Weeds are controlled by herbicides applied across the entire orchard or more commonly in strips down the tree row, especially in more hilly terrain.

In high rainfall areas, such as Florida, Brazil and Mexico, fruit is often destined for processing, and cultural practices differ considerably. Trees come into bearing more quickly and grow quite large. The exterior quality of fruit grown for processing is of little importance, and only pests affecting yield are controlled. Tree spacing tends to be wider and, once trees reach full size, they are often hedged and/or topped to maintain tree size. In some areas, such as Brazil, irrigation is not common and yields may suffer from periodic droughts. In other areas, such as Florida, irrigation is mostly by microsprinklers that have replaced overhead irrigation. Weed control is mostly by application of herbicides in the tree rows or in some cases by tillage with some manual control on young trees.

In Asia, trees may be grown on mounds or berms in and around rice paddies. In many low-lying areas, ditches are needed for proper drainage and it may be necessary to grow trees on raised beds. Care is often minimal in small plantings of fruit for local consumption. Weeds are removed manually, fertilization may consist of organic waste from farm animals or other crops, and spraying of pests is done only when there is an urgent need. Production is not usually very high, but costs are low and these small orchards often are economically viable and provide variety in the local diet. These plantings may consist of budded trees but, in some areas, rooted cuttings, marcotts or seedlings continue to be used.

Botany, taxonomy and cultivars of commercial importance

Nearly all cultivars of citrus grown for commercial purposes are members of three closely related genera, *Citrus*, *Fortunella* and *Poncirus* (Reuther *et al.*, 1967). These belong to the subtribe *Citrinae* of the tribe *Citreae*, which is in the subfamily *Aurantioideae* of the family *Rutaceae*. The most commonly propagated citrus species are sweet orange, *Citrus sinensis*, mandarin, *C. reticulata*, lemon, *C. limon*, grapefruit, *C. paradisi*, Mexican lime, *C. aurantifolia*, Persian lime, *C. latifolia*, and pummelo, *C. grandis*. Minor taxa include citron, *C. medica*, combava, *C. hystrix*, and calamondin, *C. mitis*. Some popular hybrids include tangelo, *C. reticulata* × *C. paradisi*, and tangor, *C. reticulata* × *C. sinensis*. Numerous named cultivars or selections of oranges, mandarins and grapefruit are cultivated. Some of the more widely grown varieties and cultivars include 'Valencia', 'Pera', 'Hamlin' and navel oranges; 'Satsuma' and 'Clementine' mandarins; 'Marsh', 'Ruby Red', 'Rio Red', 'Flame' and 'Star Ruby' grapefruits; and 'Eureka' and 'Lisbon' lemons. Trifoliate orange, *Poncirus trifoliata*, and hybrids between it and citrus are widely used as commercial rootstocks, but the fruit is not edible. Kumquats, which include several species in the genus *Fortunella*, are grown to a limited extent in many areas.

Most commercial citrus is propagated by budding or grafting the desired scion on a nucellar seedling of a cultivar that has been selected for its rootstock attributes (Castle *et al.*, 1993). A few species are propagated as rooted cuttings, or marcotts and dooryard citrus in tropical areas is frequently grown from seed. Some of the more common rootstocks include sour orange, *C. aurantium*, rough lemon, *C. jambhiri*, Rangpur lime, *C. limonia*, 'Cleopatra' and 'Sunki' mandarins, trifoliate orange, citranges, *C. sinensis* × *P. trifoliata*, and citrumelos, *C. paradisi* × *P. trifoliata*.

Nucellar embryony is common among many citrus cultivars, and nucellar seedlings theoretically are identical to the female parent genetically. However, many citrus species are highly heterozygous and zygotic seedlings are generally quite variable (Reuther *et al.*, 1968).

Commercially cultivated citrus species apparently originated in Asia where citrus culture has an ancient history. The first movement of citrus to other countries was primarily by seed. With the advent of more rapid forms of transportation, movement of budwood and whole plants became common, but this unfortunately was also responsible for the inadvertent dissemination of many pests and diseases.

New cultivars have arisen in different citrus-growing areas via selection of naturally occurring mutants or variants and through breeding programmes. Fruit quality and yield are affected by climate and soil conditions, and cultivars have been selected for adaptation to different local conditions. Market factors, local preferences, and disease and pest tolerance also affect what is cultivated in different citrus-growing areas. Rootstocks are selected for tolerance to soilborne diseases, adaptability to local soil conditions, and for high yield or influence on fruit quality.

Overview of citrus diseases

This chapter is divided into four sections: Diseases of Fruit and Foliage, Root and Trunk Diseases, Postharvest Decays and Systemic Diseases. Most diseases of fruit and foliage do not cause direct loss of trees and are correctable without replacing trees. The degree to which they must be controlled depends on whether the crop is intended for the fresh market or for processing. Many of these diseases, such as melanose, scab and greasy spot rind blotch, cause only superficial blemishes and need not be controlled if the fruit is to be processed. Others, such as postbloom fruit drop, canker, Phytophthora brown rot and black spot, may result in loss of fruit without necessarily affecting the health of the tree. Yet others, such as greasy spot, Phaeoramularia spot and Alternaria brown spot, cause fruit blemishes and leaf drop and, if severe, can weaken trees and jeopardize future production.

Root diseases cause substantial losses in citrus, and *Phytophthora* spp. are the most widespread causal agents. These pathogens attack virtually every part of the citrus tree, from germinating seedlings to fruit after harvest. Foot rot and gummosis may cause tree death or debilitation by girdling the trunk, and fruit drop and decay in storage and transport can be important. Nematodes also attack root systems, debilitate trees, and reduce yield and fruit size.

Postharvest decays are caused mostly by pathogens that are less aggressive and cause little or no damage to the tree. Nevertheless, the losses from postharvest decays are economically more significant than those caused by many other diseases. By the time fruit have been produced, harvested, transported, and washed and waxed, investments are high and losses more significant.

Diseases caused by systemic pathogens arguably are the most severe of all citrus diseases. Many result in tree losses and most are not curable. To solve these problems, trees must be removed and replaced with trees of tolerant varieties or rootstocks. *Citrus tristeza virus* (CTV) has caused losses of trees on sour orange rootstock in many areas of the world including Brazil, Argentina, California, Spain, Israel, Florida and Venezuela. Huanglongbing (greening) has resulted in extensive loss and debilitation of trees in Asia and Africa. Citrus blight causes severe tree losses, especially in Florida and Brazil. Citrus variegated chlorosis has emerged recently as a major problem in Brazil. Many other diseases, such as psorosis, those caused by viroids, and tristeza stem-pitting, may reduce production and orchard life.

The major diseases of citrus are covered below. Those that have a minor or local impact are listed in Table 7.1. More detailed information on citrus diseases has been published elsewhere (Klotz, 1973; Knorr, 1973; Reuther *et al.*, 1978; Roistacher, 1991; Browning *et al.*, 1995; Barkley, 1998; Timmer and Duncan, 1999; Timmer *et al.*, 2000a).

Diseases of Fruit and Foliage

Alternaria brown spot

Alternaria brown spot was first described in Australia in 1903. It occurs in the Mediterranean area, the Caribbean basin, in South Africa and probably in many other areas where susceptible cultivars are grown. It causes serious losses of susceptible tangerine and tangerine hybrids. A similar leaf and fruit spot affects rough lemon and Rangpur lime.

Symptoms

The disease affects young leaves, twigs and fruit, and produces brown to black lesions which vary in size from small dots to large, expanding lesions. Diseased fruit may abscise, and lesions on remaining fruit may vary from small spots to larger lesions (Whiteside, 1976).

Causal agent

Alternaria brown spot is caused by a host-specific, small-spored strain of *Alternaria*, but the specific classification of these fungi is uncertain at present. Isolates from tangerines and their hybrids do not affect rough lemon, and most isolates from rough lemon do not affect tangerines (Kohmoto *et al.*, 1979).

Epidemiology

Alternaria produces thick-walled, pigmented conidia that are dispersed by the wind. Conidia are produced under moist conditions on lesions on attached or recently fallen mature leaves (Timmer *et al.*, 1998). Moderate to high temperatures and rainfall favours the disease but, since heavy dews are sufficient for infection, fruit blemishes occur even in semi-arid areas where no rainfall occurs after flowering (Timmer *et al.*, 2000b).

Management

Minimizing the period of leaf wetness of the tree canopy can reduce disease incidence. Nursery trees free of the disease should be used for new plantings, and overhead irrigation should be avoided. Excessive nitrogen fertilization and irrigation that promote abundant growth flushes should be avoided. Foliar fungicide applications are needed in

Table 7.1. Citrus diseases of local or minor importance.

Disease	Causal agent	Occurrence
Diseases of fruit and foliage		
Algal spot	*Cephaleuros virescens*	Humid tropics – all citrus
Areolate leaf spot	*Pellicularia filamentosa*	Humid tropics – all citrus
Bacterial blast and black rot	*Pseudomonas syringae*	Winter rainfall areas – lemons
Bacterial spot	*Xanthomonas axonopodis* pv. *citrumelo*	Florida – trifoliate orange and hybrids, grapefruit
Botrytis blight	*Botrytis cinerea*	Winter rainfall areas – lemons
Pink disease and thread blight	*Corticium* spp.	Humid tropics – all citrus
Sclerotinia twig blight	*Sclerotinia sclerotiorum*	Winter rainfall areas – lemons
Septoria spot	*Septoria citri*	Winter rainfall areas – grapefruit, lemons
Sphaeropsis knot	*Sphaeropsis tumefaciens*	Caribbean – mostly Mexican lime
Root and trunk diseases		
Black root rot	*Thielaviopsis basicola*	Most citrus, nurseries with artificial potting mixes
Damping-off	*Phytophthora* spp., *Rhizoctonia solani*, *Pythium* spp.	All citrus, seedbeds
Fusarium wilt	*Fusarium oxysporum* f. sp. *citri*	Mostly Mexican lime, greenhouses
Mushroom root rot	*Armillaria* and *Clitocybe* spp.	All citrus – recently cleared land
Rosellinia root rot	*Rosellinia* spp.	All citrus – recently cleared land
Ustulina root and collar rot	*Ustulina deusta*	All citrus – recently cleared land
Nematodes		
Lesion nematode	*Pratylenchus coffeae*, *P. vulnus* and *P. brachyurus*	Brazil, Taiwan, Florida
Sheath nematode	*Hemicycliophora* spp.	California, Australia
Spreading decline	*Radopholus similis*	Florida, sandy soils
Sting nematode	*Belonolaimus longicaudatus*	Florida, sandy soils
Postharvest decays		
Anthracnose	*Colletotrichum gloeosporioides*	Humid areas – ethylene-degreened fruit
Aspergillus rot	*Aspergillus niger*	Worldwide, but uncommon
Cottony rot	*Sclerotinia sclerotiorum*	Winter rainfall areas – lemons
Grey mould	*Botrytis cinerea*	Winter rainfall areas – lemons
Trichoderma rot	*Trichoderma viride*	Winter rainfall areas – lemons
Systemic diseases		
Australian citrus dieback	Fastidious procaryote	Australia
Chlorotic dwarf	Whitefly-transmitted virus	Turkey
Citrus variegation, crinkly leaf and leaf rugose	Ilarviruses	Worldwide, seldom damaging
Concave gum	Virus?	Primarily Mediterranean
Cristacortis	Virus?	Primarily Mediterranean
Gummy pitting	Unknown	Australia
Gum pocket	Viroid?	South Africa – trifoliate orange
Impietratura	Virus?	Primarily Mediterranean
Shell bark	Viroid?	Worldwide – lemons
Vein enation-woody gall	Luteovirus	Widespread, minor damage
Yellow mosaic	Badnavirus	India
Yellow vein	Virus?	Pakistan – lemons

most affected orchards, with frequency based on disease severity. Iprodione, maneb and copper fungicides are the most commonly used products (Timmer and Zitko, 1997).

Anthracnose diseases

Three anthracnose diseases affect citrus. Postbloom fruit drop (PFD) and lime anthracnose are the most important, whereas postharvest anthracnose is only significant locally and not covered below (Timmer and Brown, 1999). PFD affects all citrus species and causes severe crop losses, especially in humid, tropical areas. Lime anthracnose affects only Mexican lime and precludes its commercial production in humid areas. Both diseases occur primarily in the Americas, but lime anthracnose is also found in Zanzibar.

Symptoms

PFD produces orange brown lesions on flower petals (Plate 46). Fruitlets on affected flowers abscise, leaving persistent buttons, consisting of the peduncle, calyx and floral disc, that are diagnostic for the disease. Persian lime and navel, Natal and Valencia oranges are the most susceptible cultivars.

Lime anthracnose affects flowers, young leaves and fruit, and lesions range from small spots to large expanding lesions. Leaves and fruit often abscise and twigs are killed, resulting in the 'withertip' symptoms.

Causal agent

PFD and lime anthracnose are caused by strains of *Colletotrichum acutatum* that differ in pathogenicity, but are similar in morphology (Brown *et al.*, 1996). It is described in Chapter 1.

Epidemiology

Conidia are produced in acervuli on petals that are affected by PFD and on all tissues that are affected by lime anthracnose. Spores are splash dispersed, and susceptible tissues are infected rapidly under moist conditions. Conidia splashed to mature tissues germinate to form appressoria that serve as survival structures (Timmer *et al.*, 1994). PFD is most severe in areas where flowering occurs more than once a year or is prolonged.

Management

Fungicide application may be needed during bloom to control PFD. A predictive model has been developed to aid in timing fungicide applications (Timmer and Zitko, 1996b). Benomyl and captafol are effective, and other products such as captan, maneb and ferbam provide some control. Lime anthracnose is difficult to control in humid areas, but applications of mancozeb and captafol may be effective.

Black spot

This disease causes fruit loss and a serious external blemish of citrus fruit. Black spot is widespread in the humid to semi-arid citrus-growing areas in the southern hemisphere that have summer rainfall. The disease is more serious on lemons and late oranges than on early oranges, grapefruit and tangerines.

Symptoms

Black spot produces lesions on fruit varying from small brown to black spots to large sunken lesions (Plate 47). Symptoms may appear in the orchard on fruit, and cause premature fruit drop, or infections may remain quiescent until harvest. Lesions that occur on leaves do not affect the tree greatly, but are important for reproduction of the pathogen, *Guignardia citricarpa* (anamorph: *Phyllosticta citricarpa*).

Causal agent

The pathogen produces pycnidia in lesions on fruit and leaves, and ascocarps are formed in decomposing leaves on the orchard floor. Pycnidia are solitary, sometimes aggregated, globose, immersed, mid- to dark brown, 115–190 μm in diameter, with a papillate,

12–14.5 μm diameter ostiole (Sutton and Waterston, 1966). Conidia are obovate to elliptical, hyaline, single-celled, 8–10.5 × 5.5–7 μm, and form as blastospores on hyaline, cylindrical conidiophores (Fig. 7.2). Ascocarps are solitary or aggregated, globose, immersed, dark brown to black and 95–125 μm in diameter. Asci are clavate–cylindrical, eight-spored and 40–65 × 12–15 μm. Ascospores are 12.5–16 × 4.5–6.5 μm, single-celled, hyaline, multiguttulate, cylindrical but swollen in the middle, with obtuse ends and a colourless appendage.

Epidemiology

Most infections result from ascospores that are dispersed by wind from the orchard floor (Kotzé, 1981, 1997). Infections usually occur from early to mid-summer and remain latent for some time. Rainfall is required for release of ascospores, and moisture is essential for infection.

Management

Fungicide applications are the primary means for managing black spot. Often a single, late-summer spray of benomyl will provide sufficient disease control except where resistant strains occur. In the latter cases, two to three preventive sprays with copper or dithiocarbamate fungicides in early to mid-summer are used.

Canker

Citrus canker is a serious bacterial disease in humid tropical and subtropical areas. The disease causes external blemishes on fruit, making them unsuitable for the fresh market, and may cause fruit drop. The disease is widespread in Asia and is spreading in southern South America and in South Florida (Gottwald *et al.*, 1997), but generally is absent from areas with Mediterranean climates.

Symptoms

Canker affects young leaves, stems and fruit of most citrus species, producing water-soaked lesions of variable size (Plate 48). As lesions age, they form a raised pustule surrounded by a chlorotic halo.

Causal agent

Citrus canker is caused by the bacterium, *Xanthomonas axonopodis* pv. *citri*. The Asiatic or A strain seriously affects grapefruit and early oranges and is less severe on late oranges and mandarins (Gottwald *et al.*, 1993). The B strain in southern South

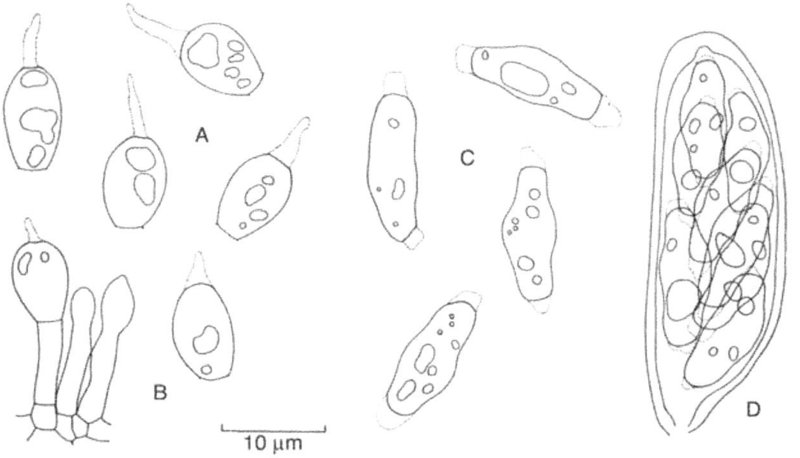

Fig. 7.2. (A) Conidia and (B) blastic conidiophores of *Phyllosticta citricarpa*, and (C) ascospores and (D) ascus of its teleomorph, *Guignardia citricarpa* (from CMI description no. 85).

America affects primarily lemons, whereas the C strain in Brazil affects Mexican lime (Stall and Civerolo, 1991) (also see bacterial spot, Table 7.1).

Epidemiology

The pathogen reproduces extensively only in young lesions on fruit and leaves. Bacteria ooze out of such lesions and are dispersed primarily by wind-blown rain. The disease is dependent on storms and wind-blown rain not only for dispersal, but also to force the bacteria into wounds and stomata. Canker is most serious in areas with severe thunderstorms, hurricanes and typhoons. The bacterium survives for relatively short periods outside host tissue (Timmer *et al.*, 1996b). The presence of leaf miners exacerbates canker because tunnels provide entry points for the bacterium and expose additional tissue in which it multiplies.

Management

Citrus canker is controlled by quarantine and eradication in countries in which it is absent or has a limited distribution. Movement of citrus fruit and budwood from infested areas is restricted (Graham and Gottwald, 1996; Schubert *et al.*, 2001). Diseased trees are burned in place, and the area is kept free of citrus root sprouts for 6–12 months.

Windbreaks are quite effective in reducing spread of the disease and in limiting the amount of infection. Copper fungicides are effective in preventing fruit infection if applied frequently (Stall *et al.*, 1981). Antibiotics have been less effective because of their short residual activity and the development of resistant strains.

Greasy spot

Greasy spot occurs throughout the Caribbean basin, and diseases with similar symptoms have been reported in South America, Japan, Australia and other areas. In areas with moist, warm summers, greasy spot causes substantial leaf drop and consequent reductions in yield and fruit size. Greasy spot is more severe on grapefruit, lemons and early oranges, than on late oranges and mandarins. Greasy spot rind blotch occurs on grapefruit and reduces the external quality of fresh market fruit.

Symptoms

Symptoms on leaves first appear on the underside of the leaf as yellow to tan slightly raised areas. Lesions eventually become brown to black but rarely become necrotic (Plate 49). Infections on fruit occur through stomata, forming minute black lesions. Chlorophyll is retained in areas surrounding these lesions as fruit matures, and an unsightly blemish results.

Causal agents

The disease in the Caribbean basin is caused by the fungus *Mycosphaerella citri* (anamorph: *Stenella citri-grisea*) (Sivanesan and Holliday, 1969; Whiteside, 1977b). Another species, *M. horii*, has been reported in Japan, but agents for the other greasy spot-like diseases have not been described.

M. citri produces pseudothecia that are immersed in decomposing leaves on the orchard floor. They are densely grouped, up to 90 μm in diameter, and ostiolate (Fig. 7.3) (Sivanesan and Holliday, 1969). Asci are obclavate, bitunicate, eight-spored and 25–35 × 8–10 μm. Ascospores are hyaline, fusiform, two-celled, straight or slightly curved and 6–12 × 2–3 μm. Conidiophores of *S. citri-grisea* are sparse, simple, erect, deep, septate, lightly rough walled, 12–40 × 2–3.5 μm, and arise from extramatricular hyphae. Conidia are 10–40 × 2–3.5 μm, pale olivaceous, verrucose, cylindric, straight or slightly bent, rounded at the apex, tapered to a truncate base with a thick hilum.

Epidemiology

Most infections are caused by ascospores. Spores germinate and hyphae penetrate through stomata and spread slowly in the leaf. Symptoms appear 4–6 months after infection. Conidia are produced from epiphytic mycelia on the underside of the leaf,

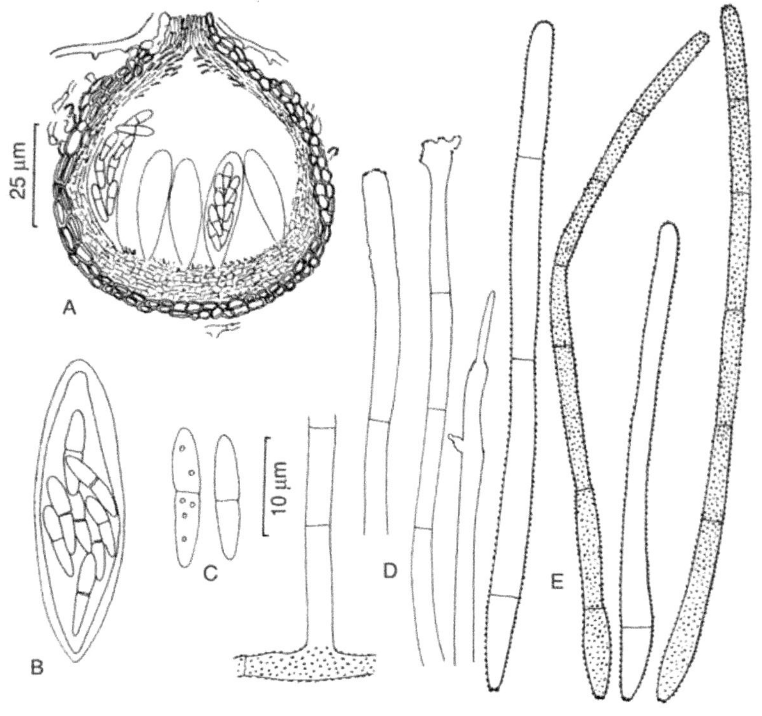

Fig. 7.3. (A) Pseudothecium, (B) ascus and (C) ascospores of *Mycosphaerella citri*, and (D) conidiophores and (E) conidia of its anamorph, *Stenella citri-grisea* (from CMI description no. 510).

but are considered of minor importance for infection. Rainfall is important for spore release, and warm humid nights are necessary for epiphytic growth of the fungus and infection (Whiteside, 1974).

Management

The disease is controlled by fungicide applications in mid-summer (Whiteside, 1983). Copper products, petroleum oil and sterol biosynthesis-inhibiting fungicides are the most effective. Sprays are timed to coincide with the epiphytic growth of the pathogen prior to penetration of the leaf.

Mal secco

This disease is primarily a problem of lemons, but can also affect tangerines and their hybrids. Oranges and grapefruit are seldom affected. Mal secco can result in losses of tree limbs or, in severe cases, of the entire tree. It occurs primarily in the Mediterranean Basin and in the Mid-East. Infected leaves develop a veinal chlorosis. As the infection proceeds downward in the vascular system, leaves wilt and the shoots die back. Eventually, limbs or the tree may die. When the bark of affected branches is removed, the wood shows a characteristic orange or orange–red discoloration.

Mal secco is caused by the fungus *Phoma tracheiphlia*. Pycnidia of the fungus are produced on dead twigs and branches. Conidia produced in pycnidia are splash dispersed or carried by wind-blown rain (Solel, 1976). Infection occurs primarily through wounds, but infection through stomata may occur. Infection requires long periods of moisture (36–48 h) and relatively cool temperatures (20°C). Infection of root sprouts can result in rapid disease development and tree loss.

Use of disease-free propagation material helps to reduce the dissemination of mal secco. Diseased trees and branches should be removed and burned to reduce inoculum. Foliar sprays of benomyl or copper fungicides in the spring and autumn reduce new infections. Resistant cultivars of lemon are available, but these are not planted widely since they are not acceptable commercially (Solel, 1977; Recupero *et al.*, 1997).

Melanose

External blemishes of melanose on fruit reduce its value for the fresh market, but usually do not affect yield. Melanose is most severe on lemons and grapefruit. It is an important disease of fruit produced for the fresh market in humid subtropical areas, but is not of major concern in Mediterranean climates or in high-rainfall tropical areas.

Symptoms

Melanose appears as raised, brick red to brown pustules on the leaves, twigs and fruit. Spores carried down the side of the fruit by water may cause lesions to form in a tearstain or droplet pattern. In severe cases, lesions may coalesce to form a mud cake symptom on fruit.

Causal agent

Melanose is caused by *Diaporthe citri* (anamorph: *Phomopsis citri*) (Ruehle and Kuntz, 1940). Pycnidia are ostiolate, scattered or clustered, immersed but later erumpent, black, conical to lenticular, and up to 600 μm in diameter (Fig. 7.4) (Punithalingam and

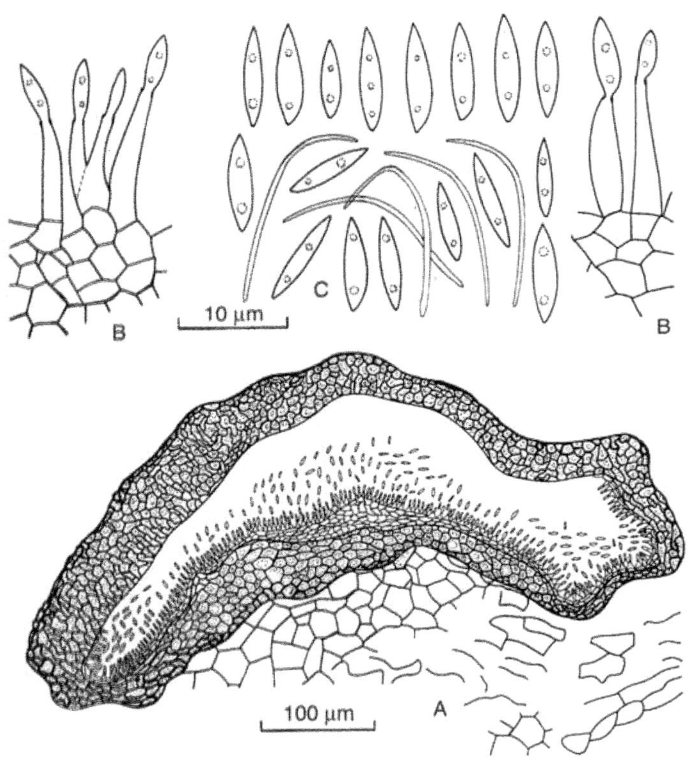

Fig. 7.4. (A) Pycnidium, (B) pycnidial wall, conidiogenous cells and α-conidia, and (C) α- and β-conidia of *Phomopsis citri*, anamorph of *Diaporthe citri* (from CMI description no. 396).

Holliday, 1969). α-Conidia are hyaline, single-celled, fusiform to elliptisoidal, biguttulate and 6–10 × 2–3 μm. β-Conidia are 20–30 × 0.5–1 μm, hyaline, single-celled, filiform, curved and often strongly hooked.

Epidemiology

The teleomorph contributes little to disease severity, but airborne ascospores are responsible for long distance spread of the disease. Pycnidia of the fungus are produced on dead twigs. Conidia produced in pycnidia are dispersed by rainsplash and washed over the fruit. Relatively long periods of wetting (12–18 h) are required for infection even at high temperatures. Dead twigs decay rapidly in tropical, high-rainfall areas, and serve as a source of inoculum for only brief periods.

Management

Copper fungicides are the most widely used means to control melanose (Timmer and Zitko, 1996a) because they are highly effective and have a long residual (Whiteside, 1977a). However, they must be applied frequently when fruit growth is rapid. Copper fungicides can produce necrotic lesions and darken existing blemishes when applied in hot weather (Timmer *et al.*, 2000a). Other fungicides are effective, but often have shorter residuals or are more costly.

Phaeoramularia leaf and fruit spot

This leaf and fruit spot was first reported in 1952 in Angola and Mozambique and now occurs through most of southern Africa and in Yemen. Virtually all citrus species are affected, but damage is less severe on lemons and limes (Brun, 1972). It is a devastating disease that causes considerable leaf and fruit drop and blemishes fruit that remain on the tree. Lesions up to 1 cm in diameter, which usually have light grey centres and a large chlorotic halo, develop on leaves. The centre of lesions may drop out, causing a shot-hole effect. Similar lesions occur on fruit.

This disease is caused by *Phaeoramularia angolensis*. Hyaline, cylindrical, multiseptate conidia are produced on tufts of conidiophores on leaf lesions. They are wind dispersed and the disease is most prevalent under warm, humid conditions. Preventive applications of fungicides are the only means of control, and copper fungicides, chlorothalonil, flusilazole and propineb are the most effective products (Seif and Hillocks, 1997).

Powdery mildew

Powdery mildew occurs throughout the humid areas of Asia, from India to the Philippines, and reduces yield by debilitating trees and causing fruit drop. Whitish, powdery patches of mildew occur on the upper surface of leaves, especially at the edges and on young fruits. Immature leaves and entire shoots may shrivel and drop, and infected young fruit falls prematurely.

In some areas, powdery mildew is caused by *Oidium tingitaninum* (Roy and Ghosh, 1991). Conidia are colourless, barrel shaped with rounded ends and are produced in chains. In other areas, *O. citri* has been identified as the causal agent. Conidia of this species are longer and are formed singly.

Spores are produced on the surface of infected leaves and dispersed by wind. Moderate temperatures (18–20°C) and high humidity, dews or light rains favour infection. When conditions are favourable, fungicide applications are needed every 10–14 days for good control. Wettable sulphur, benomyl and many of the sterol biosynthesis-inhibiting fungicides are effective (Narasimhan *et al.*, 1984).

Scab

Scab diseases affect only the external quality of the fruit of susceptible citrus and are important primarily on fruit that are grown for fresh market. Sweet orange scab affects the fruit only of sweet oranges and mandarins, and occurs primarily in southern South America (Timmer, 1997). Citrus scab affects many mandarins and their hybrids, lemons and grapefruit, and occurs in all areas where conditions are favourable.

Symptoms

The first symptoms are clear to slightly pink, water-soaked areas on leaves or fruit. These grow rapidly to raised pustules that become warty and grey with age (Plate 50). Lesions on fruit tend to flatten with age, especially on grapefruit, and lesions of sweet orange scab tend to be flatter than those of citrus scab.

Causal agents

Sweet orange scab is caused by *Elsinoë australis* (anamorph: *Sphaceloma australis*), whereas citrus scab is caused by *E. fawcettii* (anamorph: *S. fawcettii*). At least four pathotypes of *E. fawcettii* been described based on host range, two in Florida and two in Australia (Tan *et al.*, 1996; Timmer *et al.*, 1996a).

Epidemiology

All citrus scab infections originate from conidia of *S. fawcettii* that are produced directly on scab lesions. Brief periods of wetting trigger production of hyaline conidia, which are spread by rainsplash. Only a brief period of wetting (4–6 h) is required for infection (Whiteside, 1975). Spindle-shaped, coloured conidia are also produced and can be airborne for short distances.

The disease cycle for sweet orange scab is similar, but *S. australis* does not produce spindle-shaped conidia. Ascospores may play a role since *E. australis* does not affect leaves and there would appear to be no source of infection once fruit are harvested.

Management

Fruit are susceptible to scab until they reach ~3 cm in diameter. Fungicide applications during this period are effective in controlling the disease. The most effective products include the sterol biosynthesis-inhibiting fungicides, benomyl, ferbam and copper materials (Timmer and Zitko, 1997).

Disorders that resemble diseases

Fungal colonization of insects or their products are some of the most obvious disease disorders. Sooty mould, which appears as a grey to black film on the leaf, results from several different fungi colonizing honeydew of homopteran insects. Sooty mould can be eliminated by controlling the responsible insects and by spraying fruit with petroleum oil to control the responsible fungi. Species of *Aschersonia*, fungi that attack whitefly nymphs, form bright yellow or red colonies on the leaf surface. While the effect is dramatic, these fungi cause no damage to the plant.

Injuries on fruit and leaves that result from the application of insecticides, fungicides and herbicides can be mistaken for diseases. Complex tank mixes, especially those that include acidic products, can cause fruit burn. This damage is usually more severe on the outside of exposed fruit and is minimal on interior fruit. When copper fungicides are applied in hot weather, they may produce black specks on fruit that closely resemble melanose damage. Contact herbicides applied to soil can drift and produce necrotic spots on leaves that resemble fungal leaf spots.

Damage caused by environmental factors can also resemble diseases. Hail can produce scars on fruit and twigs that are similar to fungal and bacterial diseases. In desert areas, sunscald may affect exposed fruit, and this damage may be exacerbated by application of petroleum oil sprays. Excessive salt can cause burn of leaf edges as well as defoliation. Although wind scar damage may be difficult to distinguish from scab symptoms on fruit, the latter are more raised and occur as individual lesions.

Certain pest problems also cause damage that is similar to disease symptoms. Rust mite damage on fruit resembles greasy spot rind blotch. Rind blotch tends to produce many pinpoint flecks that are black, whereas rust mite damage is rusty and more confluent. Bud mites and broad mites cause distortion of leaves that can be confused with symptoms of viral diseases. Thrips produce lesions on fruit that could also be confused with diseases such as scab or rind blotch.

Diseases of the Roots and Trunk

Phytophthora root rot, foot rot and brown rot

Phytophthora spp. cause some of the most economically important diseases of citrus worldwide (Erwin and Ribeiro, 1996). Losses occur in seedbeds from damping-off; in nurseries due to root or foot rot (gummosis and collar rot); in orchards due to foot rot, fibrous root rot and brown rot; and in the packing house due to brown rot (Feld *et al.*, 1979; Graham and Timmer, 1992).

Symptoms

Aboveground symptoms of root rot are a thinning of the canopy, failure to form vigorous new growth, and a reduced yield. These symptoms result from deterioration of the fibrous root system and failure of the tree to replace roots rapidly enough to maintain tree health. The root cortex is lost and only the central stele remains on affected fibrous roots. Under favourable conditions, lateral roots may also develop frogeye lesions.

On fruit, lesions are light tan and firm. Under humid conditions, white mycelium of the pathogen grows on the fruit surface.

Gummosis or collar rot is characterized by gum exudation from the trunk, usually above the bud union. The affected area has damp, sometimes soft bark, with olive to brown or black wood underneath. Later the bark dries and cracks and the wood underneath hardens and dries (Plate 51). The affected area is often surrounded by callus tissue.

Causal agents

Phytophthora citrophthora, *P. palmivora* and *P. nicotianae* cause root rot and gummosis. Each is described in Chapter 1. In Mediterranean climates with winter rains and dry summers, *P. citrophthora* is more important, whereas, in warmer areas (e.g. Queensland, Australia and Florida), *P. nicotianae* is most important. In California, both pathogens occur; *P. citrophthora* is active in winter and *P. nicotianae* in summer. In Florida, *P. palmivora* causes a serious decline, especially of trees that are also attacked by root weevils.

Epidemiology

Phytophthora spp. survive unfavourable periods as chlamydospores or oospores in soil or in decaying roots (Lutz and Menge, 1986). Chlamydospores germinate readily when moisture is present and quickly form sporangia. Infection is usually by zoospores that are released from sporangia when free moisture is abundant. Zoospores are attracted to wounds or to the region of elongation of root tips, where they encyst, germinate and penetrate. A wound is necessary for infection of suberized bark.

All species of *Phytophthora* that affect citrus can cause brown rot. This disease is caused most commonly by *P. citrophthora* where rainfall occurs during the winter. In humid, subtropical and tropical areas, it is often caused by *P. palmivora*. Both species produce abundant sporangia on the fruit surface that can be spread through the tree by wind-blown rain. In contrast, *P. nicotianae* is mostly soilborne and propagules must be splashed from the soil; consequently, most infections are within 1 m of the ground (Graham *et al.*, 1998).

Management

Management strategies for foot and root rots and gummosis include improving soil drainage via enhanced surface runoff, installation of underground tile drains above impervious soil layers, improved irrigation practices; applying fungicides; or replanting with trees on a more tolerant rootstock. The grove should not be over-irrigated, and mini-sprinkler irrigation systems, which accurately control water delivery, are recommended. If registered for use on citrus in a production area, the following chemicals can be used: copper-containing compounds applied as sprays to the skirts of trees; metalaxyl delivered through the irrigation system or as a soil surface spray or drench; fosetyl-Al as a foliar spray, drench or trunk injection; or phosphorous acid as a trunk injection or foliar spray. However, the most effective measure for all but *P. palmivora* is the use of resistant rootstocks such as trifoliate orange and its hybrids.

Control of gummosis or collar rot requires additional measures, as the scion is usually more susceptible than the rootstock. The bud union must be kept well above soil through high budding, careful planting and care to keep soil away from the trunk. Mechanical damage to the trunk must be avoided since such wounds serve as infection points.

These diseases can be controlled in nurseries by only using seed from fruit high in the canopy, or those that have been treated with hot water (52°C for 10 min) or steam-aerated (60°C for 30 min). Pathogen-free potting mixes and good nursery practices to avoid contamination should be used. In field nurseries, soils may be fallowed or fumigated chemically to reduce populations of the pathogens.

Foliar applications of copper fungicide or fosetyl-Al prior to the onset of the rainy season usually provide good control in winter rainfall areas. In humid areas, applications are made just prior to the maturation of the variety affected.

Slow decline

The citrus nematode is found worldwide in citrus orchards. It causes a disease referred to as slow decline that debilitates trees and reduces yields and fruit size. It usually is most serious in areas with fine-textured soils where trees are grown on susceptible rootstocks.

Symptoms

The symptoms of slow decline are not diagnostic. Affected trees are usually weak and have thin foliage, small leaves and many dead twigs. Deficiencies of micronutrients are common, and plants often appear to be slightly wilted or water-stressed. Rootlets appear coarse, and soil clings to fibrous roots because of the gelatinous egg masses that are produced by the female nematode.

Causal agent

Slow decline is caused by the citrus nematode, *Tylenchulus semipenetrans*. It is a semi-endoparasitic nematode that feeds in the cortex of the fibrous roots. Each female nematode (Fig. 7.5) produces from 75 to 100 eggs that hatch in 2–3 weeks (Van Gundy, 1958). There are three biotypes of this nematode, all of which can attack citrus (Inserra *et al.*, 1980).

Epidemiology

Multiplication of citrus nematodes is favoured by fine-textured or organic soils, and this nematode is not usually a problem in sandy soils. Population growth is also favoured by salinity, and high nematode populations increase salt damage to trees.

Management

Resistant rootstocks can be used effectively for control wherever trifoliate orange and its hybrids can be grown. Most other commonly used citrus rootstocks are moderately to highly susceptible. The key to reducing problems with citrus nematodes is to produce nursery stock that is free of the pest and plant orchards in areas that are free of the nematode. Native soils are usually free of

Fig. 7.5. A swollen female of *Tylenchulus semipenetrans* dissected from a fibrous root of citrus. Bar = 120 μm (photo: R. Inserra).

nematodes that attack citrus. Old orchard soils may be freed of nematodes by chemical fumigation or fallowing for 1–2 years. Post-plant nematicides reduce nematode populations and increase yields and fruit size. However, some of these products have caused environmental problems and microbes rapidly degrade others.

Diseases of local or minor importance

Other nematodes reported on citrus in addition to those in Table 7.1 are root knot nematodes, *Meloidogyne* spp., stubby root nematodes, *Paratrichodorus* spp., spiral nematodes, *Helicotylenchus* spp., lance nematodes, *Hoplolaimus* spp., and dagger nematodes, *Xiphinema* spp. (Timmer *et al.*, 2000a).

Root and trunk disorders that resemble diseases

Dry root rot or sudden death is a disorder in which apparently healthy trees with a normal crop suddenly collapse and die (Broadbent *et al.*, 1972). Affected trees have blackened, rotted roots, but no gumming or pitting. The disorder is usually associated with fine-textured soils, excess moisture and poor aeration, and at times with root injury. Several fungi, such as *Fusarium* spp. and *Coprinus* sp., are associated with the disease, but are not able to induce the disorder unless trees are under severe stress. The incidence of the problem can be reduced by improved drainage, good irrigation practices and careful selection of planting sites.

Marcott or tangerine collapse and various twig diebacks affect some varieties of mandarins. Collapse occurs when trees set excessive amounts of fruit, leaving little stored carbohydrate in the leaves and twigs. Trees may recover, but produce little fruit the following year, or may die. Varieties such as 'Robinson' and 'Fallglo' tangerines have a genetic propensity for twig dieback. Fungi such as *Colletotrichum* or *Diplodia* spp. are readily isolated from affected twigs, but do not cause the disease if twigs or branches are inoculated. Removing dead limbs and twigs and application of benomyl may help to reduce the damage.

Root rots may have abiotic causes. Anaerobic conditions caused by flooding can produce root necrosis resembling damage by *Phytophthora* spp. Excessive nitrogen fertilizer, or other chemicals, can also kill roots. Mechanical damage to roots or trunks can produce symptoms that resemble diseases.

Postharvest Decays

Penicillium decays

Penicillium spp. cause the most significant postharvest problems in areas with Mediterranean climates and important problems in virtually all other citrus areas. The causal species are cosmopolitan and cause millions of dollars in losses every year.

Symptoms

Initial symptoms are a soft, watery area on the fruit surface. The spots grow and gradually become covered with white mycelium of the pathogen. If the decay is caused by green mould, the affected area is somewhat irregular and covered with green spores with a narrow, white mycelial border. With blue mould, the affected area expands more slowly, is covered by blue spores and surrounded by a broader, white margin (Plate 52). Areas affected by whisker mould are similar to those affected by blue mould, but develop more slowly and may have concentric circles of coremia. Penicillium decays cause fruit to become mushy and to disintegrate under humid conditions.

Causal agents

Green mould is caused by *Penicillium digitatum*, blue mould by *P. italicum*, and whisker mould by *P. ulaiense* (Holmes *et al.*, 1994).

Colonies of *P. digitatum* are restricted and thin on Czapek agar and other synthetic media (Fig. 7.6) (Onions, 1966a). On malt agar, they grow rapidly, are velvety, dull yellow–green to greyish olive with age, have

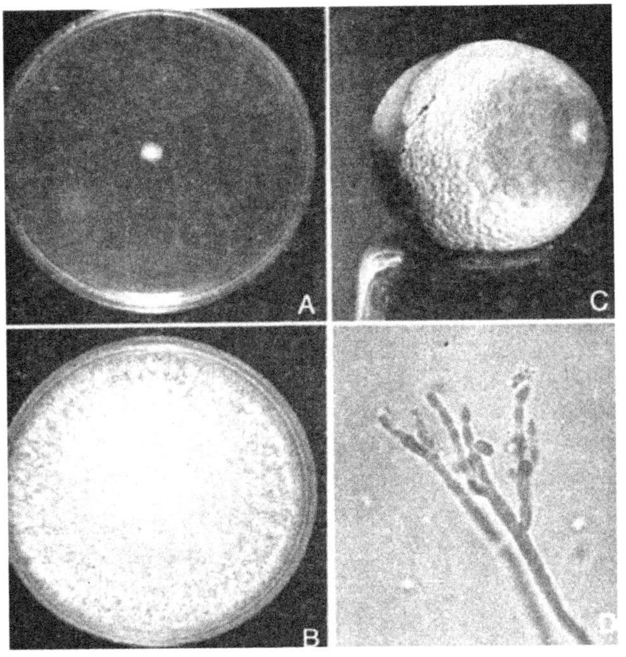

Fig. 7.6. (A) Czapek agar culture after 2 weeks at 25°C, (B) malt agar culture after 2 weeks at 25°C, (C) infected orange after 7 days at room temperature and (D) conidial head of *Penicillium digitatum* (from CMI description no. 96).

a colourless to pale brown reverse, and a strong odour of decaying citrus. The conidial apparatus is very fragile and tends to break apart. Conidiophores are 30–100 × 4–5 μm and smooth walled, metulae are 15–30 × 4–6 μm, phialides are sparse and 15–29 × 3.5 μm, and conidia are smooth, subglobose to cylindrical but usually elliptical, and 3.4–12 × 3–8 μm.

Colonies of *P. italicum* are also restricted on synthetic media but grow more than *P. digitatum* (Fig. 7.7) (Onions, 1966b). They are aromatic, pale grey–green to graphalium green, with a zonate, grey to yellowish brown reverse. The conidial apparatus consists of asymmetric penicilli bearing tangled chains of conidia. Conidiophores are up to 250 × 4–5 μm and smooth, and penicilli are 50–70 μm and usually consist of a main axis and one to three branches, 1–25 × 2.8–4.4 and are 50–70 μm, but are occasionally much longer and rebranched. Metulae and phialides are sparse, and conidia are cylindrical becoming elliptical or subglobose, smooth and 4–5 × 2.5–3.5 μm.

Epidemiology and management

Penicillium spp. penetrate only through wounds, and decays progress most rapidly at 22–27°C. Thus, damage to fruit during harvesting and handling should be avoided. Since these fungi sporulate abundantly, sanitary measures are essential. Rotted fruit must be removed from the packing house area and equipment disinfested with chlorine or quaternary ammonium products (Timmer and Duncan, 1999). Postharvest fungicides that are effective against these decays include benzimidazole compounds, imazalil, prochloraz and guazatine (Gutter, 1975). Benomyl, applied as a preharvest spray, is also effective in controlling decay. Refrigeration of packed fruit is important in slowing decay and preventing spread in cartons.

Stem-end rots

There are three important stem-end rots: Diplodia stem-end rot and Phomopsis stem-end rot, which are most important in humid

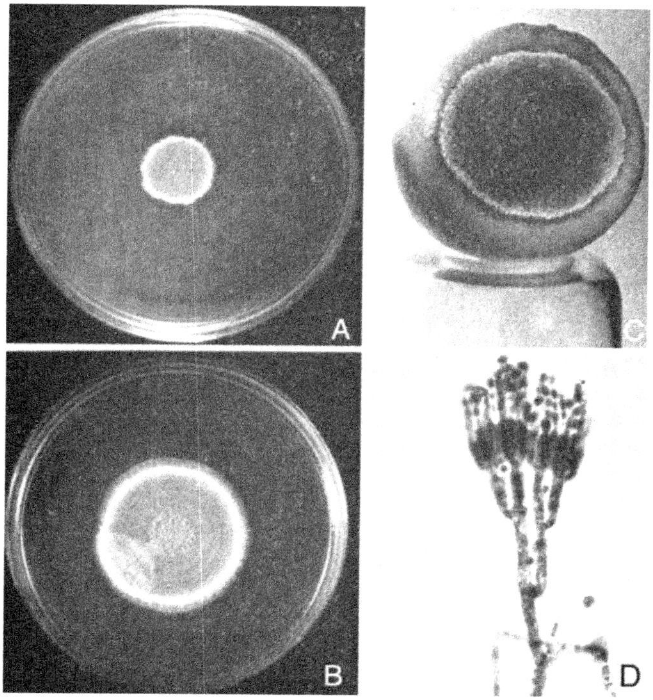

Fig. 7.7. (A) Czapek agar culture after 2 weeks at 25°C, (B) malt agar culture after 2 weeks at 25°C, (C) infected orange after 7 days at room temperature and (D) conidial head of *Penicillium italicum* (from CMI description no. 99).

areas, and Alternaria black rot, which is more prevalent in drier regions. The fungal species that are involved are common in many citrus areas, and disease development is restricted primarily by environmental conditions.

Symptoms

All stem-end rots produce rather firm brown decays that usually proceed from the button end of the fruit down the centre columella and the peel (Plate 53). Diplodia stem-end rot progresses rapidly at the junctures between fruit segments, whereas Phomopsis stem-end rot progresses more uniformly. The decayed portion of the fruit surface is depressed with Phomopsis rot, but not with Diplodia rot. Alternaria black rot progresses down the core from the stem or stylar end, often with no external symptoms. Affected fruit colour prematurely and, when cut, reveal a dark black, rotted core.

Causal agents

Botrysphaeria rhodina (anamorph: *Diplodia theobromae*) causes Diplodia stem-end rot and is described in Chapter 1. The cause of Phomopsis stem-end rot, *Diaporthe citri*, also causes melanose on citrus and is described under the section on this disease in this chapter.

Alternaria sp. causes Alternaria black rot (although the agent was formerly named *A. citri*, new research will result in its renaming). In culture, its colonies are grey, olivaceous brown or black, and sometimes zonate (Ellis, 1971). Conidiophores are simple or branched, straight or flexuous, septate, pale to mid- or olivaceous brown, up to 300 μm long, 3–5 μm wide, with a terminal and sometimes one or two lateral scars (Fig. 7.8). Conidia are colourless or pale brown, solitary or in branched chains of 2–7, straight or slightly curved, usually obclavate or oval,

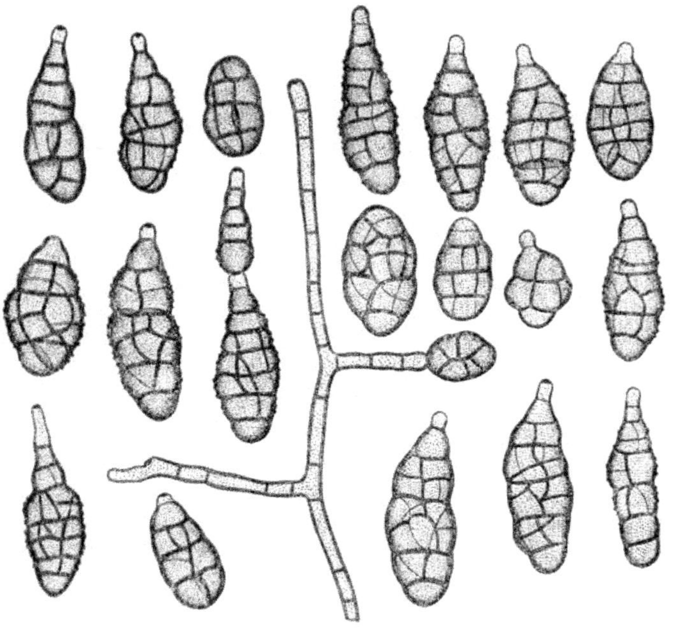

Fig. 7.8. Conidia and conidiophores of *Alternaria* sp. (formerly *A. citri*) (from Ellis, 1971).

pale to mid or sometimes dark or olivaceous brown, smooth to verruculose with up to eight transverse and numerous longitudinal or oblique septa, constricted at the septa, 8–60 μm long with the beak present and 6–24 μm wide. Beaks usually are >8 μm long and 2.5–4 μm wide.

Epidemiology

All of the causal agents are common saprophytic fungi in citrus orchards. Spores of these organisms germinate and form quiescent infections on the buttons of the fruit. After fruit are harvested and buttons senesce, the fungi grow into and rot the fruit.

Management

Application of benzimidazole fungicide prior to harvest or in the packing house is effective in controlling Diplodia and Phomopsis stem-end rots, but not Alternaria black rot. Application of 2,4-D to harvested fruit helps delay button senescence and slows development of black rot.

Sour rot

This decay is a common problem in most citrus areas of the world. Fully mature or overripe fruit are attacked more frequently than less mature fruit. The problem is common on stored fruit.

Lesions begin as moist, light coloured areas on the surface of the fruit. Fruit rapidly decomposes into a watery, slimy mess of peel and sections. Under moist conditions, the rotted fruit are covered with a white layer of mycelium of the causal fungus, *Galactomyces citri-aurantii* (anamorph: *Geotrichum citri-aurantii*). Arthrospores of the fungus are formed by segmentation of the mycelium of the fungus. The pathogen is a common soil inhabitant in citrus orchards that is carried with soil on fruit, harvesting bags, boxes and equipment (Brown, 1979). Since infection occurs through wounds, it is important to handle fruit carefully (Baudoin and Eckert, 1982).

Equipment and packing areas should be maintained free of soil and plant debris. None of the commonly used postharvest fungicides is highly effective for control of sour rot, but washing harvested fruit with sodium *o*-phenylphenate will reduce disease incidence.

Phytophthora brown rot

Brown rot was addressed as a field problem earlier in the chapter (see Diseases of Fruit and Foliage). As a postharvest problem, it occurs sporadically depending on its incidence in the field. None of the commonly used packing house fungicides controls the disease. Affected fruit must be eliminated prior to packing by delaying harvest to allow diseased fruit to fall and by culling in the packing house. Refrigeration of fruit slows brown rot development and prevents its spread to other fruit.

Fruit Disorders that Resemble Diseases

Many disorders of harvested fruit do not involve pathogens, but result from mishandling of fruit or from field conditions that lead to physiological breakdown after harvest (Reuther *et al.*, 1989; Timmer *et al.*, 2000a).

Oleocellosis is caused by peel oil that damages the fruit surface. The surface of the fruit becomes brown and slightly sunken and the disorder may occur in small spots or cover large areas of the fruit. Oleocellosis results when fruit is turgid at harvest and pressure within the fruit causes oil to extrude on to the surface. It can be avoided by delaying harvest until the fruit is dry and turgor pressure has declined.

Stylar-end breakdown of limes begins as a water-soaked, light brown area at the stylar end of the fruit. It is caused by juice that is released into the rind from juice vesicles that rupture due to high turgor pressures inside the fruit. Pressures are highest in large turgid fruit that are exposed to high temperatures after harvest. Incidence can be reduced by harvesting fruit late in the day, harvesting only small fruit, and by avoiding exposure of fruit to direct sun after harvest.

Chilling injury produces small sunken spots on the surface of the peel and occurs when fruit is stored below 10°C. Damage is more common on grapefruit, lemons and limes than on oranges.

Postharvest pitting of grapefruit appears as small, sunken spots on the surface of fruit and may superficially resemble chilling injury. Clusters of oil glands collapse on the peel surface. This type of pitting results from storage of waxed fruit at high temperatures. Rapid cooling after waxing will reduce the incidence of pitting.

Zebra skin, a pattern of darkened rind over the fruit segments, occurs mostly on tangerines that are harvested immediately after heavy rainfall or irrigation proceeded by drought. Delays in harvest to allow fruit turgidity to decrease will minimize zebra skin.

Stem-end rind breakdown is the darkening and collapse of the rind around the stem end that is caused by desiccation and ageing of the peel. Minimizing desiccation by prompt washing and waxing of fruit after harvest and storage at high humidity will delay the onset of rind breakdown.

Creasing (albedo breakdown) is a common disorder of thin-skinned tangerines and navel oranges and consists of grooves or furrows in the peel due to incomplete development of the albedo. It is thought to be related to potassium deficiency.

Granulation occurs when the juice sacs shrivel after gels form in the juice. It is more prevalent at the stem end of more mature fruit. The cause is unknown but it is more common in arid areas.

DISEASES THAT ARE CAUSED BY SYSTEMIC PATHOGENS

Diseases that are Caused by Virus and Virus-like Agents

A large number of virus and virus-like agents affect citrus (Roistacher, 1991). Some of these cause serious and widespread diseases, whereas others are less harmful or are restricted to certain cultivars or geographic regions. In some cases, the effects of virus infection, such as tristeza quick decline, are dramatic and easily recognized. More often, the effects of virus infection are subtle and easily overlooked. Often only certain cultivars or combinations of scions and rootstocks are visibly affected, whereas infection of others is symptomless. A number of citrus viruses have been isolated and their basic properties are well characterized. Some of these are members of recognized groups of plant viruses such as badna-, capillo-, clostero- and ilarviruses. Others, such as satsuma dwarf and psorosis, are apparently members of previously unrecognized virus groups. The agents of several long recognized virus-like diseases remain uncharacterized, and the agents for several recently reported diseases have not been investigated.

Some of the more notable virus and virus-like diseases are described briefly below. Comments on procedures for detection of virus and virus-like pathogens and control of the diseases that they cause are included in a separate section.

Leprosis

Leprosis has been described in Florida and various countries in South America. It remains a major disease problem in parts of Brazil, but has not been seen in Florida in recent years. Zonate chlorosis, also reported from Brazil, apparently is a similar disease.

Leprosis causes chlorotic to necrotic areas on the fruit, leaves and twigs of susceptible cultivars (Plate 54). Initial symptoms are chlorotic lesions that often become necrotic and gum-impregnated and show concentric patterns. A chlorotic zone around the lesion may remain. Leaf and fruit drop occurs when infections are abundant.

Leprosis is caused by a bacilliform virus that is vectored by mites in the genus *Brevipalpus* (Rodrigues *et al.*, 2000). The virus has a rhabdovirus-like morphology (Lovisolo *et al.*, 1996). A similar mite-transmitted virus is found in coffee, but the relationship between the two remains undetermined.

Infections are localized and apparently associated with feeding activity of mites that carry the causal virus. The virus has been transmitted experimentally by mites, and with difficulty by mechanical inoculation and by grafting. The virus does not infect citrus systemically and trees do not develop symptoms on new growth after infective mites are removed.

Psorosis and citrus ring spot

Psorosis was the first recognized virus-induced disease of citrus. The term psorosis is now retained only for diseases previously described as psorosis A, psorosis B, citrus necrotic ringspot and naturally spread psorosis.

SYMPTOMS Affected trees exhibit various types of transient chlorotic leaf patterns plus large bark scaling lesions on the trunks of susceptible cultivars, especially sweet orange and grapefruit. Trees with large lesions on the trunk or scaffold limbs become debilitated, and fruit production is severely affected. An especially severe form of psorosis is designated as psorosis B. Other diseases that cause symptoms on immature leaves, but do not induce bark scaling (see concave gum, Table 7.1) were included in what became known as the psorosis complex (Reuther *et al.*, 1978).

Leaf symptoms include various types of chlorotic patterns in young leaves that may, in some cases, persist in mature leaves. A necrotic shock reaction may occur in young shoots following infection, especially in seedling indicator plants. Ring spot patterns are seen sometimes on fruit, but are not common. Most, but not all, isolates of the causal virus produce a scaling and flaking of the bark in infected sweet orange and grapefruit trees that may not appear for a number of years. Lesions vary in size and may coalesce to encircle the trunk or large branches and reduce canopy vigour. Callus tissue forms under the sloughing bark and some gum impregnation may occur in the wood under lesion areas. While bark scaling is commonly associated with psorosis, it is sometimes observed when no transmissible agent is present. Rio Grande gummosis, a disease of unknown aetiology, also produces lesions on the trunk and limbs that can be confused with psorosis, but gumming is more profuse and the bark usually dies around the site of the initial lesion (Childs, 1978).

CAUSAL AGENT Psorosis is associated with a bi-component virus that has an unusual flexible rod-like structure for which the names spirovirus and ophiovirus have both been proposed (Garcia *et al.*, 1997; Barthe *et al.*, 1998).

EPIDEMIOLOGY AND MANAGEMENT Study of the causal agents is complicated because trees frequently are co-infected by several different pathogens that cannot be separated readily. Some isolates of the virus, especially those that produce chlorotic ring spot symptoms in leaves or fruit, are mechanically transmissible to non-citrus hosts, a step that facilitates their characterization (da Graça *et al.*, 1991).

The virus can be transmitted mechanically to herbaceous plants, including *Chenopodium quinoa*. Natural spread by unknown means has been observed in several countries, most notably in Argentina, where the disease remains a major problem. The incidence of psorosis in most citrus-growing areas has been reduced greatly by propagation of psorosis-free budwood.

Satsuma dwarf

Satsuma dwarf was first described in Japan and has become a significant problem. Affected trees are stunted and yields are reduced, but the disease is not lethal. Affected 'Satsuma' mandarin trees often have cupped, boat-shaped leaves. Citrus mosaic, natsudaidai dwarf and navel orange infectious mottling are related diseases in Japan that produce various chlorotic leaf symptoms in sensitive cultivars, and citrus mosaic may also cause fruit symptoms (Miyakawa and Yamaguchi, 1981). Satsuma dwarf has been reported from several other countries where it was introduced via infected budwood, but no information on crop losses in these areas is available.

The *Satsuma dwarf virus* (SDV) particle is isometric and has two RNA components. Sequencing studies indicate that it is phylogenetically related to both cucumo- and nepoviruses (Iwanami *et al.*, 1998).

Natural infections of SDV have been found in China laurestine, *Viburnum odoratissimum*, and the virus can be transmitted experimentally by mechanical inoculation to a number of non-citrus hosts. Natural spread occurs in the field, and the pattern of spread suggests a soilborne vector, but none has been identified. Once an area becomes infested, SDV apparently remains in the soil since new trees replanted in these sites become infected (Miyakawa and Yamaguchi, 1981). Long-distance spread has occurred via movement of infected budwood within Japan and to other countries.

Tatterleaf

The *Tatterleaf* (*Citrange stunt*) *virus* (TLV) was discovered originally in symptomless 'Meyer' lemon trees in California. It subsequently was discovered that TLV is widespread in citrus in China where 'Meyer' lemon originated. It has been spread to a number of countries via movement of infected budwood.

SYMPTOMS Although most citrus cultivars do not develop symptoms when infected with TLV, a severe bud union incompatibility occurs in trees that are propagated on trifoliate orange or its hybrids (Miyakawa, 1980). The extensive use of these rootstocks in modern plantings and the continued widespread use of trifoliate orange as a breeding parent for new rootstock varieties results in widespread genetic susceptibility that is of concern.

The incompatibility symptoms include formation of a deep crease in the wood of the trunk at the bud union that may also be accompanied by a brown stain or gum. The leaves become chlorotic, and trees are stunted and may break off in a severe windstorm. Several genetic incompatibilities also produce similar symptoms, and indexing is required for accurate diagnosis. Citrange seedlings graft-inoculated with TLV develop chlorotic spotting and distortion on some leaves, and stems often show a zig-zag growth habit. Some diffuse chlorotic symptoms and leaf distortion also occur in recently inoculated Mexican lime and *Citrus excelsa*.

CAUSAL AGENT The TLV is rod shaped and a member of the capillovirus group. A very close sequence homology with viruses isolated from apples, pears and lilies suggests that they have a common origin (Magome *et al.*, 1997).

EPIDEMIOLOGY No vector has been described for TLV or its related viruses, but the presence of such similar viruses in distinct hosts suggests that some form of natural spread occurs. The unresolved epidemiology of TLV remains a concern in assessing its potential impact on citrus production.

Tristeza decline and tristeza stem pitting

Tristeza decline has resulted in the loss of well over 50 million trees on sour orange rootstock since the 1930s, and is a continuing threat to the >200 million citrus trees on this rootstock that remain worldwide (Rocha-Peña *et al.*, 1995) (Plate 55). In addition to the decline that CTV causes in trees grafted on sour orange rootstock, some CTV isolates also cause serious stem pitting diseases of limes, grapefruit and sweet orange (Fig. 7.9). Because CTV causes several serious diseases and is easily spread by aphid vectors, it is the most serious virus problem on citrus production worldwide (Bar-Joseph *et al.*, 1989).

CTV apparently originated in Southeast Asia. It has been disseminated via infected budwood and spread naturally by aphid vectors so that it is now widespread. Areas where it is not widespread, such as Mexico and parts of the Mediterranean Basin, are under an increasing threat. Differences in virulence of strains, efficiency of aphid vectors and susceptibility of different cultivars have resulted in a complex disease situation. A broad range of effects has been observed, ranging from minimal injury to devastating crop losses that prevent economic cultivation of sensitive cultivars.

SYMPTOMS CTV causes a variety of symptoms in different hosts. Foliar symptoms include vein clearing, leaf cupping and chlorosis on sensitive species such as Mexican lime. Canopy symptoms of tristeza decline include wilting, leaf chlorosis and premature flowering. Trees may decline gradually or collapse and die in a matter of weeks (quick decline). CTV-induced bud union phloem necrosis that results in tree decline may appear as pinhole pitting in the bark of the sour orange rootstock just below the union or as a yellow to brown stain at the bud union. Nursery trees propagated on sour orange seedlings with budwood of trees infected with decline isolates are stunted and chlorotic, but rarely show a quick decline reaction. Trees affected by decline typically set abnormally large numbers of fruit which are small and may ripen prematurely.

Stem pitting typically consists of well-defined pits in the wood that are observed when the bark is removed (Fig. 7.9). Mexican lime is especially sensitive, but a wide variety of cultivars can be affected by certain CTV isolates. Commercial production of grapefruit, pummelo and some sweet oranges is affected in many areas. The degree of pitting can vary from a few small pits that apparently have no effect on tree vigour to numerous pits that may coalesce and deform stems. In severe cases, the bark of infected plants is thickened, and the wood is brittle and has a porous texture. Trees that are severely

Fig. 7.9. Symptoms of stem-pitting on grapefruit, caused by *Citrus tristeza virus*, in the Dominican Republic (photo: S.M. Garnsey).

affected by stem pitting produce small and often misshapen fruit of poor quality.

CAUSAL AGENT CTV is a flexuous, rod-shaped virus and a member of the closterovirus group. The genome is encoded in a single RNA species that is the largest plant viral RNA described to date. The sequence of nearly all isolates of CTV is similar in the 3′ portion of the genome that encodes the coat protein. Considerable variation that is present in the 5′ half of the genome indicates variation in the origin of isolates (Hilf *et al.*, 1999).

EPIDEMIOLOGY CTV is transmitted by several species of aphids in a semi-persistent manner. The brown citrus aphid, *Toxoptera citricida*, is considered the most efficient vector, and tristeza decline and stem pitting problems have been most severe in areas where it has become endemic (Rocha-Peña *et al.*, 1995). The cotton or melon aphid, *Aphis gossypii*, and the spirea aphid, *A. spiraecola*, are other important vectors.

Diseases that are Caused by Viroids

Viroids are infectious, single-stranded circular RNA molecules that lack a capsid protein and mRNA activity. Viroids are the smallest plant pathogens and cause diseases in a range of crops. A complex of viroids affects citrus, two of which cause the economically important diseases, exocortis and cachexia–xyloporosis (Duran-Vila *et al.*, 1998).

Exocortis, caused by *Citrus exocortis viroid* (CEVd) is a bark-scaling disease of trifoliate orange rootstock (Fig. 7.10). The viroid is present in almost all citrus-growing regions, and numerous sequence variants are known. Most scion cultivars are symptomless carriers and disease symptoms are only expressed when infected scions are propagated on sensitive rootstocks, e.g. trifoliate orange, citranges and 'Rangpur' lime.

A disease causing gum impregnation of the bark and pitting of the wood with corresponding bark pegs was described on 'Palestine' lime in Israel as xyloporosis, and in USA on 'Orlando' tangelo as cachexia. Cachexia–xyloporosis also affects mandarins, mandarin hybrids, *C. macrophylla* and 'Rangpur' lime, and is caused by specific variants, CVd-IIb and CVd-IIc, of the *Hop stunt viroid*, HSVd. Another HSVd variant, CVd-IIa, is sometimes associated with dwarfing of trees on trifoliate orange rootstock.

Other viroids are: *Citrus viroid* I (CVd-I), which is also known as *Citrus bent leaf viroid* (CBLVd); *Citrus viroid III* (CVd-III), which is associated with dwarfed trees on trifoliate orange and is related to the *Apple scar skin viroid* (ASSVd) family; and *Citrus viroid IV* (CVd-IV), a putative hybrid, which shares sequence homology with CEVd. Other possible viroid-induced diseases are listed in Table 7.1.

Fig. 7.10. Bark scaling symptoms on trifoliate orange rootstock, caused by *Citrus exocortis viroid*, in Australia (photo: P. Broadbent).

Viroid-induced dwarfing of trees is now used commercially in Australia, Israel and California, and for 'Tahiti' limes in Brazil. Australian field trials have shown that inoculation of trees with CVd-IIIb, with and without CVd-IIa, reduced tree size of selected sweet orange scions on Australian trifoliate orange and citrange rootstocks without affecting tree health (Gillings *et al.*, 1991; Hutton *et al.*, 2000). Based on the planted area, closely spaced, dwarfed trees produced higher yields than conventional plantings without affecting fruit quality. No effects on canopy development become apparent until 4 years after inoculation. Rapid, early canopy development of viroid-inoculated trees ensures a high bearing volume per hectare and promotes early orchard productivity.

Citrus viroids are distributed principally by the propagation of infected scion wood and subsequently by natural root grafting and mechanical transmission on budding knives and hedging equipment, especially from lemon to lemon. Citrus viroids are not known to be seed or insect transmitted.

The detection of viroid pathogens and control of the diseases they cause are included in a separate section.

Diseases that are Caused by Fastidious Prokaryotes

Citrus variegated chlorosis

Citrus variegated chlorosis (CVC) has become a serious and widespread disease in Brazil and parts of Argentina (Donadio and Moreira, 1998). Affected trees have mottled leaves on one or more branches, and in chronic stages may be stunted and show twig dieback. Leaves with interveinal chlorosis may have brown areas on the lower surface that enlarge (Plate 56). Fruits are small and hard and change colour prematurely. They frequently are sunburned, and also may have sunken brown areas on the surface of the rind. Movement of the pathogen in xylem, especially basipetally, is slow following infection.

The disease is caused by a strain of the bacterium *Xylella fastidiosa* that inhabits xylem and impairs its normal function. It can be cultured from CVC-affected trees, and Koch's postulates have been fulfilled (Chang *et al.*, 1993; Hartung *et al.*, 1994). The strain of *X. fastidiosa* that causes CVC is apparently different from other strains that cause other diseases, including Pierce's disease of grape, plum leaf scald, phony peach and leaf scorch diseases in coffee, oaks and sycamore. It is most closely related to strains that cause coffee leaf scorch.

Extensive spread has occurred unknowingly via propagation of infected budwood sources and by leafhopper vectors. Control measures include avoiding propagation of CVC-infected budwood for new plantings, removing infected limbs from recently affected trees and removal of affected trees in young plantings. Mandarins, grapefruit and lemons appear to be less sensitive to CVC than sweet orange and are more suitable for areas that are severely affected by CVC.

Huanglongbing (greening)

Huanglongbing (HLB) was reported in mainland China in 1919, and in South Africa in 1937 as citrus greening disease (da Graça, 1991). Its name translates as 'yellow shoot disease', and it has been reported under different names in many Asian countries, in southern and eastern Africa and the southwestern part of the Arabian Peninsula. 'Greening' is the most common name in English-speaking countries. HLB has destroyed an estimated 60 million trees in Africa and Asia.

SYMPTOMS The most characteristic symptom of HLB is green patches on a pale green background that often begins in one part of the canopy. Zinc-like deficiency symptoms commonly are associated with HLB, resulting in its confusion with nutritional problems. Leaf yellowing and leaf drop result in twig dieback (Plate 57). Fruit on severely affected trees are small, lopsided and poorly coloured, hence the name greening. Juice is bitter, low in soluble solids and high in acid. Nursery trees are stunted, terminal leaves are yellowed, new leaves are small, leathery and upright, and old leaves are mottled. As these symptoms take 4–6 months to appear; symptomless trees may be distributed from affected nurseries.

Rutaceous hosts include *Citrus* spp. and possibly other citrus relatives. *Murraya paniculata* (L.) Jack and other citrus relatives are good hosts for the psyllid vectors. Among commercial citrus, limes and lemons are more tolerant than oranges, mandarins, tangors, tangelos and grapefruit.

CAUSAL AGENTS Although Koch's postulates have not been completed, bacteria are consistently associated with this disease. They have a filamentous morphology and are polymorphic, varying in both length and diameter. In 1992, the bacteria were characterized taxonomically by studying the nucleotide sequence of its 16S ri RNA gene (Villechanoux *et al.*, 1993). They are Gram-negative members of the α-subdivision of the proteobacteria. Two *Candidatus* species have been recognized: Liberibacter asiaticus in Asia and L. africanus in Africa (Jagoueix *et al.*, 1996). Polymerase chain reaction (PCR) and DNA–DNA hybridization techniques can now be used to detect the two species.

EPIDEMIOLOGY Transmission of HLB occurs by grafting and by the African citrus psyllid, *Trioza erytreae*, and the Asian psyllid, *Diaphorina citri*. Each psyllid is able to transmit L. africanus and L. asiaticus.

MANAGEMENT HLB in Asia can only be controlled when coordinated efforts are made to eliminate infected trees, control the vector and ensure that nurseries distribute only pathogen-free trees. South Africa has shown that it is possible to live with HLB, but disease pressure is probably lower there than it is in Asia (Buitendag and von Broembsen, 1993). Most of the nursery trees planted in South Africa originate from budwood that is obtained from the Outspan Foundation Block which is located in an area with little HLB. By pursuing effective biological and chemical psyllid control programmes after planting, growers can raise greening-free trees even in areas where psyllids and HLB are endemic. Biological control of the psyllids and eradication of HLB-affected trees has eliminated the disease from Reunion. However, in many countries, the effectiveness of psylla parasites is reduced by hyperparasites. Eradication of all citrus and replanting with pathogen-free trees has been attempted in Indonesia (Bové *et al.*, 1998) and in coastal China (Aubert, 1993) as a means of controlling HLB, but these efforts appear to have failed.

Stubborn

Stubborn is an important disease of citrus in arid regions such as parts of California, North Africa and the Middle East (Gumpf and Calavan, 1981; Bové, 1995). The disease is especially severe in young plantings. The pathogen and vectors both have extensive non-citrus host ranges that create problems for disease management.

SYMPTOMS Stubborn can affect most citrus cultivars and relatives. Leaves typically are small, and internodes are shortened, giving the tree canopy a bushy appearance. Some leaf chlorosis may be present and fruit is sparse. Fruit often show seed abortion, are acorn shaped, and colouring of the stylar end may be delayed. Symptoms in trees that are infected when mature often are less conspicuous. In contrast to many citrus diseases, symptoms are most severe under high temperature conditions. The pathogen has a wide host range and, indeed, citrus is probably a secondary host. Natural or experimental infections have been observed or induced in plants in at least 19 families.

CAUSAL AGENT Stubborn is caused by *Spiroplasma citri* (Salgio *et al.*, 1973), a mollicute that lacks a cell wall and is readily culturable (Bové *et al.*, 1983). Cells of *S. citri* frequently have a filamentous form with a distinct helical morphology. The pathogen is found only in the phloem of infected plants, and symptoms reflect its pathogenic effects on that tissue.

EPIDEMIOLOGY Several species of leafhopper can vector stubborn, and epidemics of stubborn seem to be associated with large-scale periodic migrations of leafhoppers into citrus from other crops or native vegetation (Calavan and Bové, 1989).

MANAGEMENT A primary focus for control of stubborn is to avoid infection in nurseries and in young plantings. Use of stubborn-free budwood is essential. Tolerant or resistant cultivars have not been selected.

Witches' broom disease of limes

Witches' broom disease of limes (WBDL) is a lethal disease that occurs in Oman and the United Arab Emirates, and has been reported recently in India and Iran. Symptoms in the early stages of infection are one or more witches' brooms that form from repeated abnormal proliferation of axillary buds. Leaves on stems in the broom are small and paler than normal. As broom formation increases, twig and limb dieback occurs and trees decline rapidly and become unproductive.

The disease is caused by a phloem-restricted phytoplasma that has not been cultured (Garnier *et al.*, 1991). It can be visualized in infected trees by electron microscopy and has no cell wall.

A leafhopper vector is suspected, although none has been confirmed experimentally. The primary reservoir of the pathogen may be another host. Secondary spread from infected limes is suspected since the disease becomes epidemic once it enters a grove. Some spread to new areas via movement of infected plant materials apparently has occurred. The disease is confined primarily to Mexican limes, but some other cultivars have been infected experimentally. Sweet orange apparently is resistant (Bové *et al.*, 1996).

Detection and Control of Systemic Pathogens

The ability to rapidly and accurately detect viroid, virus and virus-like pathogens is critical to most types of control efforts (Roistacher, 1991). Initially, identification was based on symptoms in field trees or symptoms in graft-inoculated indicator plants that were chosen for their ability to express diagnostic symptoms. Prior to biological indexing for CEVd on 'Etrog' citron, it was recommended that budwood only be taken from trees >10 years old on trifoliate orange that had no symptoms of bark scaling on the rootstock or dwarfing of the tree. Biological indexing for cachexia–xyloporosis was based on the observation of 'Orlando' tangelo, as a seedling indicator or as a rootstock, 'Parson's Special' mandarin on a vigorous rootstock, or 'Ellendale' tangor on trifoliate or citrange rootstocks. Inoculated indicators must be kept at 27–32°C for 3–6 months and 9–18 months for good symptom expression of exocortis and cachexia–xyloporosis, respectively. 'Etrog' citron is widely used for rapid detection of CEV and group I, III and IV viroids, but does not react strongly to group II viroids.

The use of citrus indicators is still the only method to detect uncharacterized virus-like pathogens such as the concave gum agent, and is also the only reliable method to identify specific pathotypes of characterized pathogens such as CTV.

Purification and characterization of several citrus viruses has allowed virus-specific antisera to be produced and serological assays to be developed (Gonsalves *et al.*, 1978; Kawai *et al.*, 1996). Enzyme-linked immunosorbent assay (ELISA) is used routinely for several citrus viruses, including CTV and SDV, and it provides a rapid and sensitive assay for viruses for which good antisera are available (Roistacher, 1991). Tissue imprint assays allow rapid and sensitive serological detection of viruses with simple equipment (Garnsey *et al.*, 1993).

All citrus viroids replicate in 'Etrog' citron, even when plants are incubated at suboptimal temperatures for symptom expression. Consequently, citron is used as an amplification host from which RNA extracts are subjected to sequential polyacrylamide gel electrophoresis (sPAGE) for viroid detection (Roistacher, 1991). Nucleic acid analyses by electrophoresis and molecular hybridization have been proposed for viroid detection directly from test plants without a citron amplification step, but low titre and uneven distribution of viroids have hampered applications.

Electron microscopy has been used to detect CTV and, when grids are sensitized with virus-specific antibodies, can provide a rapid and sensitive method for identification. Inclusions formed by CTV are also easily visualized by light microscopy in sections of bark or petioles stained with azure A (Garnsey *et al.*, 1980).

As sequence information have been obtained for different viroids and viruses, nucleic acid and PCR-based assays have been developed for their detection (Yang *et al.*, 1992; Cevik *et al.*, 1996). PCR is a more expensive and difficult procedure than ELISA, but can be more sensitive and has the potential to differentiate strains of a single pathogen that would be difficult to separate by ELISA. PCR currently is feasible for CTV, SDV, TLV and CVV, as well as several prokaryotic pathogens and citrus viroids. Hybridization assays that use labelled cDNA probes have been developed, but have not been widely adopted for practical applications.

The basic control strategies for most graft-transmissible diseases of citrus are to prevent infection where possible. All budwood distributed from quarantine and multiplication programmes should be tested for these pathogens. Citrus viruses and viroids can be eliminated by shoot-tip grafting *in vitro* or by use of nucellar budlines.

Quarantine measures are important for preventing movement of viruses via infected propagation materials (Frison and Taher, 1991). Most citrus-growing countries now restrict importation of citrus germplasm from other countries to avoid the introduction of new pathogens or more virulent strains. Programmes to obtain and promote propagation of pathogen-free budwood for new plantings are especially effective against pathogens that have no vectors, and may also be an important component of programmes to manage those that have vectors. Disinfestation of pruning tools is also used to prevent movement of mechanically transmitted pathogens.

Cultivation of tolerant or resistant cultivars often is the only practical control strategy when vector-borne viruses are endemic and the vectors are abundant (Timmer and Duncan, 1999). Unfortunately, sources of resistance or tolerance are not available for some host–virus combinations, such as grapefruit and stem pitting isolates of CTV (Garnsey *et al.*, 1998). Scion–rootstock combinations that tolerate a particular pathogen may be susceptible to others with equally severe consequences (Castle *et al.*, 1993). In these cases, the choices are either to avoid growing the sensitive cultivar or attempt control via mild strain cross-protection. Cross-protection has ameliorated crop loss from CTV-induced stem pitting in several locations, but is not always effective and risk-free. Cross-protecting isolates themselves may cause some damage to the host. Genetic engineering offers some possibilities for the development of resistant cultivars (see Chapter 20).

A Disease of Unknown Aetiology

Blight

Citrus blight is one of the most economically important diseases in Brazil and Florida, and also occurs in other areas in the Americas, as well as in Australia and South Africa. Losses in Florida exceed 500,000 trees annually and are probably greater in Brazil. Plants on rough lemon, 'Rangpur' lime, trifoliate orange and most of the citrange rootstocks are most susceptible.

SYMPTOMS Blight is a wilt and decline disease of citrus trees (Smith, 1977) (Plate 58). The disease does not occur on non-bearing trees, and trees may be 8–10 years old when decline is first noted. The first symptom of the disease is usually zinc deficiency in the leaves and a mild wilt and greyish cast to the canopy. These symptoms are followed by more severe wilt, leaf drop and twig dieback. Trees with blight seldom die, but usually become unproductive within a year. As the canopy declines, fibrous roots are lost and eventually major roots may be rotted as well. Symptoms are the results of blockage of xylem vessels with amorphous plugging material.

CAUSE AND EPIDEMIOLOGY The cause of blight is unknown. Many causes have been suggested, such as soil and nutritional factors, soil pathogens, viruses and *X. fastidiosa*, but a causal role has not been confirmed for any of these. Blight has been transmitted by root-piece and tree-to-tree root grafts, but not by budwood or branch-to-branch grafts. Thus, it is an infectious disease with a root-associated agent (Timmer *et al.*, 1992). However, neither is it spread by soil taken from around blighted trees nor is blight more common in replanted areas.

Disease occurrence initially is random, but clusters of affected trees may develop or severity may become greater in certain parts of the block. Disease increase is often linear which is not typical of infectious diseases.

MANAGEMENT Trees affected by citrus blight must be replaced since there is no known cure. Before trees are removed, the disease should be diagnosed accurately to differentiate it from other decline diseases. Blight-affected trees have the following unique characteristics that can be used to identify the disease: (i) they fail to take up water when it is injected into the trunk with a syringe (Lee *et al.*, 1984); (ii) they have a higher content of zinc in the wood or bark than healthy trees or those that are affected by other diseases (Wutscher *et al.*, 1977); and (iii) they produce a characteristic 12 kDa and larger proteins in leaves and other tissues that can be assayed serologically (Derrick *et al.*, 1990).

Blight-affected trees should be replaced with trees on tolerant rootstocks. Sweet orange, sour orange, 'Swingle' citrumelo and, to a lesser extent, 'Cleopatra' mandarin are more tolerant to the disease. Tolerance to the disease is expressed as the age at which symptoms usually appear and as reduced incidence. Once trees on any rootstock develop the disease, they usually decline within a short time.

Disorders that resemble diseases

Symptoms that are caused by some viruses and phloem-limited prokaryotes are often not specific and reflect a plant response to disruption of vascular function. Deficiency symptoms due to inadequate nutrient levels can be confused with symptoms of huanglongbing, stubborn and severe stem pitting forms of tristeza. Mechanical girdling of the trunk or limbs can produce chlorotic symptoms similar to those produced by huanglongbing or tristeza decline. Chimeras and other genetic abnormalities can also be confused with symptoms produced by viruses. Herbicide applications can also induce growth abnormalities and leaf chlorosis suggestive of infections by viruses and prokaryotic agents. Bark scaling symptoms closely resembling those induced by psorosis have been observed in grapefruit, but these apparently are not associated with an infectious agent. Bud union incompatibilities of genetic origin can look very similar to those that are induced by tatterleaf virus and an unidentified infectious agent that induces bud union symptoms in trees grafted on rough lemon. Pesticides applied to young leaves, especially in glasshouses, can induce virus-like ringspot patterns.

Acknowledgements

Florida Agricultural Experiment Station Journal Series No. R-06953.

References

Anonymous (2001) Food and Agricultural Organization Committee. Economic and Social Dept. Commodities and Trade Division. http://www.fao.org/es/ESC/esce/cmr/cmmots/CMRcie.htm

Aubert, B. (1993) Citrus greening disease, a serious limiting factor for citriculture in Asia and Africa. *Proceedings of the IV World Congress of the International Society of Citrus Nurserymen.* pp. 134–142.

Bar-Joseph, M., Marcus, R. and Lee, R.F. (1989) The continuous challenge of citrus tristeza virus control. *Annual Review of Phytopathology* 27, 291–316.

Barkley, P. (1998) *Citrus Diseases and Disorders*. New South Wales Agriculture, Camden, NSW, Australia, 24 pp.

Barthe, G.A., Ceccardi, T.L., Manjunath, K.L. and Derrick, K.S. (1998) Citrus psorosis virus: nucleotide sequencing of the coat protein gene and detection by hybridization and RT–PCR. *Journal of General Virology* 79, 1531–1537.

Baudoin, A.B.A.M. and Eckert, J.W. (1982) Factors influencing the susceptibility of lemons to infection by *Geotrichum candidum*. *Phytopathology* 72, 1592–1597.

Bové, J.M. (1995) *Virus and Virus-like Diseases of Citrus in the Near East Region*. Food and Agriculture Organization, Rome.

Bové, J.M., Whitcomb, R.F. and McCoy, R.E. (1983) Culture techniques for spiroplasmas from plants. In: Tully, J.G. and Razin, S. (eds) *Methods in Mycoplasmology*, Vol. 2. Academic Press, New York, pp. 225–234

Bové, J.M., Navarro, L., Bonnet, P., Zreik, L. and Garnier, M. (1996) Reaction of citrus cultivars to graft-inoculation of Phytoplasma aurantifolia-infected lime shoots. In: Moreno, P., da Graça, J.V. and Timmer, L.W. (eds) *Proceedings of the 13th Conference of the International Organization of Citrus Virologists*. Riverside, California, pp. 249–251.

Bové, J.M., Erti Dwiastuti, M., Triwiratno, A., Supriyanto, A., Nasli, E., Becu, P. and Garnier, M. (1998) Incidence of huanglongbing and citrus rehabilitation in North Bali, Indonesia. In: da Graça, J.V., Lee, R.F. and Yokomi, R. (eds) *Proceedings of the 14th Conference of the International Organization of Citrus Virologists*. Riverside, California, pp. 200–206.

Broadbent, P., Fraser, L.R. and Waterworth, Y. (1972) Sudden death of citrus on *Poncirus trifoliata* rootstock. *Plant Disease Reporter* 56, 81–84.

Broadbent, P., Bevington, K.B. and Coote, B.G. (1991) Control of stem-pitting of grapefruit in Australia by mild strain cross protection. In: Brlansky, R.H., Lee, R.F. and Timmer, L.W. (eds) *Proceedings of the 11th Conference of the International Organization of Citrus Virologists*, Riverside, California, pp. 64–70.

Brown, A.E., Sreenivasaprad, S. and Timmer, L.W. (1996) Molecular characterization of slow-growing orange and Key lime anthracnose strains of *Colletotrichum* from citrus as *C. acutatum*. *Phytopathology* 86, 523–527.

Brown, G.E. (1979) Biology and control of *Geotrichum candidum*, the cause of citrus sour rot. *Proceedings of the Florida State Horticultural Society* 92, 186–189.

Browning, H.W., McGovern, R.J., Jackson, L.K., Calvert, D.V. and Wardowski, W.F. (1995) *Florida Citrus Diagnostic Guide*. Florida Science Source, Lake Alfred, Florida, 244 pp.

Brun, J. (1972) Citrus leaf spot caused by *Cercospora angolensis*. *Fruits* 27, 539–541.

Buitendag, C.H. and von Broembsen, L.A. (1993) Living with citrus greening in South Africa. In: Moreno, P., da Graça, J.V. and Timmer, L.W. (eds) *Proceedings of the 12th Conference of the International Organization of Citrus Virologists*. Riverside, California, pp. 269–273.

Calavan, E.C. and Bové, J.M. (1989) Ecology of *Spiroplasma citri*. In: Whitcomb, R.F. and Tully, J.G. (eds) *The Mycoplasmas*, Vol. V. Academic Press, New York, pp. 425–485.

Castle, W.S., Tucker, D.P.H., Krezdorn, A.H. and Youtsey, C.O. (1993) *Rootstocks for Florida Citrus*. University of Florida Publication SP-42.

Cevik, B., Pappu, S.S., Pappu, H.R., Benscher, D., Irey, M., Lee, R.F. and Niblett, C.L. (1996) Application of bio-directional PCR to citrus tristeza virus: detection and strain differentiation. In: da Graça, J.V., Moreno, P. and Yokomi, R.K. (eds) *Proceedings of the 13th Conference of the International Organization of Citrus Virologists*, Riverside, California, pp. 17–24

Chang, C.J., Garnier, M., Zreik, L., Rossetti, V. and Bové, J.M. (1993) Culture and serological detection of the xylem-limited bacterium causing citrus variegated chlorosis and its identification as a strain of *Xylella fastidiosa*. *Current Microbiology* 27(3), 137–142.

Childs, J.F.L. (1978) Rio Grande gummosis of citrus trees. Part 1. A brief review of the history and occurrence of Rio Grande gummosis. *Plant Disease Reporter* 62, 390–394.

da Graça, J.V. (1991) Citrus greening disease. *Annual Review of Phytopathology* 29, 109–136.

da Graça, J.V., Lee, R.F., Moreno, P., Civerolo, E.L. and Derrick, K.S. (1991) Comparison of isolates of citrus ringspot, psorosis, and other virus-like agents of citrus. *Plant Disease* 75, 613–616.

Derrick, K.S., Lee, R.F., Brlansky, R.H., Timmer, L.W., Hewitt, B.G. and Barthe, G.A. (1990) Proteins associated with citrus blight. *Plant Disease* 74, 168–170.

Donadio, L.C. and Moreira, C.S. (eds) (1998) *Citrus Variegated Chlorosis*. Bebedouro, SP, Brazil.

Duran-Vila, N., Roistacher, C.N., Rivera-Bustamante, R. and Semancik, J.S. (1998) A definition of citrus viroid groups and their relationship to the exocortis disease. *Journal of General Virology* 69, 3069–3080.

Ellis, M.B. (1971) *Dematiaceous Hyphomycetes*. Commonwealth Mycological Insititute, Kew, UK.

Erwin, D.C. and Ribeiro, O.K. (1996) *Phytophthora Diseases Worldwide*. APS Press. St Paul, Minnesota.

Feld, S.J., Menge, J.A. and Pehrson, J.E. (1979) Brown rot of citrus: a review of the disease. *California Citrograph* 64, 101–106.

Frison, E.A. and Taher, M.M. (1991) *Technical Guidelines for the Safe Movement of Citrus Germplasm*. Food and Agricultural Organization/International Board for Plant Genetic Resources.

Garcia, M.L., Sánchez de la Torre, M.E., Dal Bó, E., Djelouah, K., Rouag, N., Luisoni, E., Milne, R.G. and Grau, O. (1997) Detection of citrus psorosis-ringspot virus using RT–PCR and DAS–ELISA. *Plant Pathology* 46, 830–836.

Garnier, M., Zreik, L. and Bové, J.M. (1991) Witches' Broom, a lethal mycoplasmal disease of lime trees in the Sultanate of Oman and the United Arab Emirates. *Plant Disease* 75, 546–551.

Garnsey, S.M., Christie, R.G., Derrick, K.S. and Bar-Joseph, M. (1980) Detection of citrus tristeza virus. II. Light and electron microscopy of inclusions and viral particles. In: Calavan, E.C., Garnsey, S.M. and Timmer, L.W. (eds) *Proceedings of the 8th Conference of the International Organization of Citrus Virologists*, Riverside, California, pp. 9–16

Garnsey, S.M., Permar, T.A., Cambra, M. and Henderson, C.T. (1993) Direct tissue blot immunoassay (DTBIA) for detection of citrus tristeza virus (CTV). In: Moreno, P., da Graça J.V. and Timmer, L.W. (eds) *Proceedings of the 12th Conference of the International Organization of Citrus Virologists*. Riverside, California, pp. 39–50.

Garnsey, S.M., Gottwald, T.R. and Yokomi, R.K. (1998) Control strategies for citrus tristeza virus. In: Hadidi, A., Khetarpal, R. and Koganezawa, H. (eds) *Plant Virus Disease Control*. APS Press, St Paul, Minnesota, pp. 639–658.

Gillings, M.R., Broadbent, P. and Gollnow, B.I. (1991) Viroids in Australian citrus: relationship to exocortis, cachexia and citrus dwarfing. *Australian Journal of Plant Physiology* 18, 559–570.

Gonsalves, D., Purcifull, D.E. and Garnsey, S.M. (1978) Purification and serology of citrus tristeza virus. *Phytopathology* 68, 553–559.

Gottwald, T.R., Graham, J.H., Civerolo, E.L., Barrett, H.C. and Hearn, C.J. (1993) Differential host range reaction of citrus and citrus relatives to citrus canker and citrus bacterial spot determined by leaf susceptibility. *Plant Disease* 77, 1004–1009.

Gottwald, T.R., Graham, J.H. and Schubert, T.S. (1997) An epidemiological analysis of the spread of citrus canker in urban Miami, Florida, and synergistic interaction with the Asian citrus leafminer. *Fruits* 52, 383–390.

Graham, J.H. and Gottwald, T.R. (1996) Research perspectives on eradication of citrus bacterial diseases in Florida. *Plant Disease* 75, 1193–1200.

Graham, J.H. and Timmer, L.W. (1992) Phytophthora diseases of citrus. In: Kumar, J., Chaube, H.S., Singh, U.S. and Mukhopodhyay, A.N. (eds) *Plant Diseases of International Importance, Vol. III, Diseases of Fruit Crops*. Prentice-Hall, Englewood Cliffs, New Jersey, pp. 250–269.

Graham, J.H., Timmer, L.W., Drouillard, D. and Peever, T.L. (1998) Characterization of *Phytophthora* spp. causing outbreaks of citrus brown rot in Florida. *Phytopathology* 88, 724–729.

Gumpf, D.J. and Calavan, E.C. (1981) Stubborn disease of citrus. In: Maramorosch, K. and Raychaudhuri, S.P. (eds) *Mycoplasma Diseases of Trees and Shrubs*. Academic Press, New York, pp. 97–134

Gutter, Y. (1975) Interrelationship of *Penicillium digitatum* and *P. italicum* in thiabendazole-treated oranges. *Phytopathology* 65, 498–499.

Hartung, J.S., Beretta, J., Brlansky, R.H., Spisso, J. and Lee, R.F. (1994) Citrus variegated chlorosis bacterium: axenic culture, pathogenicity, and serological relationship with other strains of *Xylella fastidiosa*. *Phytopathology* 84, 591–597.

Hilf, M.E., Karasev, A.V., Albiach-Marti, M.R., Dawson, W.O. and Garnsey, S.M. (1999) Two paths of divergence in the citrus tristeza virus complex. *Phytopathology* 89, 336–342.

Holmes, G.J., Eckert, J.W. and Pitt, J.I. (1994) Revised description of *Penicillium ulaiense* and its role as a pathogen of citrus fruits. *Phytopathology* 84, 719–727.

Hutton, R., Broadbent, P. and Bevington, K.B. (2000) Viroid dwarfing for high density citrus plantings. *Horticultural Reviews* 24, 277–317.

Inserra, R.N., Vovlas, V. and O'Bannon, J.H. (1980) A classification of *Tylenchulus semipenetrans* biotypes. *Journal of Nematology* 12, 283–287.

Iwanami, T., Kondo, Y., Makita, Y., Azeyanagi, C. and Ieki, H. (1998) The nucleotide sequence of the coat protein genes of Satsuma dwarf virus and navel orange infectious mottling virus. *Archives of Virology* 143, 405–412.

Jagoueix, S., Bové, J.M. and Garnier, M. (1996) PCR detection of the two liberobacter species associated with greening disease of citrus. *Molecular Cellular Probes* 10, 43–50.

Kawai, A., Tsukamoto, T., Namba, S. and Nishio, T. (1996) Citrus tatter leaf virus: a review of its properties and the development of a serological detection system. In: da Graça, J.V., Moreno, P. and Yokomi, R. (eds) *Proceedings of the 13th Conference of the International Organization of Citrus Virologists*, Riverside, California, pp. 339–342.

Klotz, L.J. (1973) *Color Handbook of Citrus Diseases*. University of California, Division of Agricultural Science, Berkeley, 362 pp.

Knorr, L.C. (1973) *Citrus Diseases and Disorders*. University of Florida Press, Gainesville, 122 pp.

Kohmoto, K., Scheffer, R.P. and Whiteside, J.O. (1979) Host selective toxins from *Alternaria citri*. *Phytopathology* 69, 667–671.

Kotzé, J.M. (1981) Epidemiology and control of citrus black spot in South Africa. *Plant Disease Reporter* 65, 945–950.

Kotzé, J.M. (1997) History and epidemiology of citrus black spot in South Africa. *Proceedings of the International Society of Citriculture 1996* 2, 1296–1299.

Lee, R.F., Marais, L.J., Timmer, L.W. and Graham, J.H. (1984) Syringe injection of water into the trunk: a rapid diagnostic test for citrus blight. *Plant Disease* 68, 511–513.

Lovisolo, O., Colariccio, A., Chagus, C.M., Rossetti, V., Kitajima, E.W. and Harakawa, R. (1996) Partial characterization of citrus leprosis virus. In: da Graça, J.V., Moreno, P. and Yokomi, R.K. (eds) *Proceedings of the 13th Conference of the International Organization of Citrus Virologists*, Riverside, California, pp. 179–188

Lutz, A. and Menge, J. (1986) Citrus root health. II: Phytophthora root rot. *Citrograph* 72, 33–39.

Magome, H., Yoshikawa, N., Takahashi, T., Ito, T. and Miyakawa, T. (1997) Molecular variability of the genomes of capilloviruses from apple, Japanese pear, European pear, and citrus trees. *Phytopathology* 87, 389–396.

Miyakawa, T. (1980) Occurrence and varietal distribution of tatter leaf-citrange stunt virus and its effects on Japanese citrus. In: Calavan, E.C., Garnsey, S.M. and Timmer, L.W. (eds) *Proceedings of the 8th Conference of the International Organization of Citrus Virologists*, Riverside, California, pp. 220–224.

Miyakawa, T. and Yamaguchi, A. (eds) (1981) *Citrus Diseases in Japan*. Japan Plant Protection Association, Tokyo, Japan.

Narasimhan, V., Subramanian, K.S., Shanmugan, N. and Jeyarajan, H. (1984) Efficacy of certain fungicides in the control of powdery mildew of mandarin. *Pesticides* 18(1), 61–62.

Onions, A.H.S. (1966a) Penicillium digitatum. *CMI Descriptions of Pathogenic Fungi and Bacteria No. 96*. Commonwealth Mycological Institute, Kew, UK.

Onions, A.H.S. (1966b) Penicillium italicum. *CMI Descriptions of Pathogenic Fungi and Bacteria No. 99*. Commonwealth Mycological Institute, Kew, UK.

Punithalingam, E. and Holliday, P. (1969) Diaporthe citri. *CMI Descriptions of Pathogenic Fungi and Bacteria No. 396*. Commonwealth Mycological Institute, Kew, UK.

Recupero, G.R., Gentile, A., Russo, M.P. and Domina, F. (1997) Genetic analysis of resistance to *Phoma tracheiphila* in three *Citrus* and *Poncirus* progenies. *Plant Breeding* 116, 198–200.

Reuther, W., Webber, H.J. and Batchelor, L.D. (eds) (1967) *The Citrus Industry*, Vol. I. University of California, Division of Agricultural Sciences, Berkeley.

Reuther, W., Batchelor, L.D. and Webber, H.J. (eds) (1968) *The Citrus Industry*, Vol. II. University of California Division of Agricultural Sciences, Berkeley.

Reuther, W., Calavan, E.C. and Carmen, G.E. (eds) (1978) *The Citrus Industry*, Vol. IV. University of California, Division of Agricultural Sciences, Berkeley.

Reuther, W., Calavan, E.C. and Carmen, G.E. (eds) (1989) *The Citrus Industry*, Vol. V. University of California, Division of Agriculture and Natural Resources, Berkeley.

Rocha-Peña, M.A., Lee, R.F., Lastra, R., Niblett, C.L., Ochoa Corona, F.M., Garnsey, S.M. and Yokomi, R.K. (1995) Citrus tristeza virus and its aphid vector *Toxoptera citricida*: threats to citrus production in the Caribbean and Central and North America. *Plant Disease* 79, 437–445.

Rodrigues, J.C.V., Machado, M.A., Kitajima, E.W. and Müller, G.W. (2000) Transmission of Citrus leprosis virus by *Brevipalpus phoenicis* (Acari: Tennipalpidae). In: da Graça, J.V., Lee, R.F. and Yokomi, R.K. (eds) *Proceedings of the 14th Conference International Organization of Citrus Virologists*, Riverside, California, pp. 174–178.

Roistacher, C.N. (1991) *Graft-transmissible Diseases of Citrus. Handbook for Detection and Diagnosis*. International Organization of Citrus Virologists and Food and Agricultural Organization, Rome.

Roy, S.D. and Ghosh, S.K. (1991) Global status of powdery mildew (*Oidium tingitaninum*) disease on *Citrus* spp. *Journal of Mycopathology Research* 2, 127–132.
Ruehle, G.D. and Kuntz, W.A. (1940) *Melanose of Citrus and its Commercial Control.* Florida Agricultural Experiment Station, University of Florida, Gainesville, Bulletin 349.
Salgio, P., L'hospital, M., Laflèche, D., Dupont, G., Bové, J.M., Tully, J.G. and Freundt, E.A. (1973) *Spiroplasma citri* gen. and n.: a mycoplasma-like organism associated with 'stubborn' disease of citrus. *International Journal of Systematic Bacteriology* 23(3), 191–204.
Schubert, T.S., Rizui, S.A., Sun, X., Gottwald, T.R., Graham, J.H. and Dixon, W.N. (2001) Meeting the challenge of eradicating citrus canker in Florida – again. *Plant Disease* 85, 340–356.
Seif, A.A. and Hillocks, R.J. (1997) Chemical control of *Phaeoramularia* fruit and leaf spot of citrus in Kenya. *Crop Protection* 16(2), 141–145.
Sivanesan, A. and Holliday, P. (1969) Mycospaerella citri. *CMI Descriptions of Pathogenic Fungi and Bacteria No. 510.* Commonwealth Mycological Institute, Kew, UK.
Smith, P.F. (1977) A review of the nature and history of citrus blight in Florida. *Proceedings of the International Society of Citriculture* 3, 881–884.
Solel, Z. (1976) Epidemiology of mal secco disease of lemons. *Phytopathology Zeitschrift* 85, 90–92.
Solel, Z. (1977) Control of mal secco disease of lemon trees. *Proceedings of the International Society of Citriculture* 3, 928–930.
Stall, R.E. and Civerolo, E.L. (1991) Research relating to the recent outbreak of citrus canker in Florida. *Annual Review of Phytopathology* 29, 399–420.
Stall, R.E., Miller, J.W., Marco, G.M. and Canteros de Echenique, B.I. (1981) Timing of sprays to control cancrosis of grapefruit in Argentina. *Proceedings of the International Society of Citriculture* 1, 414–417.
Sutton B.C. and Waterston, J.M. (1966) Guignardia citricarpa. *CMI Descriptions of Pathogenic Fungi and Bacteria No. 85.* Commonwealth Mycological Institute, Kew, UK.
Tan, M.K., Timmer, L.W., Broadbent, P., Priest, M. and Cain, P. (1996) Differentiation by molecular analysis of *Elsinoe* spp. causing scab diseases of citrus and its epidemiological implications. *Phytopathology* 86, 1039–1044.
Timmer, L.W. (1997) *Elsinoe fawcettii* and *E. australis. Crop Protection Compendium.* CAB International, Wallingford, UK.
Timmer, L.W. and Brown, G.E. (1999) Biology and control of anthracnose diseases of citrus. In: Prusky, D., Freeman, S. and Dickman, M. (eds) *Host Specificity, Pathology, and Host-Parasite Interaction of* Colletotrichum. APS Press, St Paul, Minnesota, pp. 300–316.
Timmer, L.W. and Duncan, L.W. (eds) (1999) *Citrus Health Management.* APS Press, St Paul, Minnesota.
Timmer, L.W. and Zitko, S.E. (1996a) Evaluation of copper fungicides and rates of metallic copper for control of melanose on grapefruit in Florida. *Plant Disease* 80, 166–169.
Timmer, L.W. and Zitko, S.E. (1996b) Evaluation of a model for prediction of postbloom fruit drop of citrus. *Plant Disease* 80, 380–383.
Timmer, L.W. and Zitko, S.E. (1997) Evaluation of fungicides for control of Alternaria brown spot and citrus scab. *Proceedings of the Florida State Horticultural Society* 110, 71–76.
Timmer, L.W., Lee, R.F., Brlansky, R.H., Graham, J.H., Albrigo, L.G., Derrick, K.S. and Tucker, D.P.H. (1992) The infectious nature of citrus blight. *Proceedings of the Florida State Horticultural Society* 105, 21–26.
Timmer, L.W., Agostini, J.P., Zitko, S.E. and Zulfiqar, M. (1994) Postbloom fruit drop of citrus, an increasingly prevalent disease of citrus in the Americas. *Plant Disease* 78, 329–334.
Timmer, L.W., Priest, M., Broadbent, P. and Tan, M.K. (1996a) Morphological and pathological characterization of *Elsinoe* spp. causing citrus scab diseases. *Phytopathology* 86, 1032–1038.
Timmer, L.W., Zitko, S.E. and Gottwald, T.R. (1996b) Population dynamics of *Xanthomonas campestris* pv. *citri* on symptomatic and asymptomatic citrus leaves under various environmental conditions. *Proceedings of the International Society of Citriculture* 2, 448–451.
Timmer, L.W., Solel, Z., Gottwald, T.R., Ibañez, A.M. and Zitko, S.E. (1998) Environmental factors affecting production, release, and field populations of conidia of *Alternaria alternata*, the cause of brown spot of citrus. *Phytopathology* 88, 1218–1223.
Timmer, L.W., Garnsey, S.M. and Graham, J.H. (eds) (2000a) *Compendium of Citrus Diseases*, 2nd edn. APS Press, St Paul, Minnesota.
Timmer, L.W., Darhower, H.M., Zitko, S.E., Peever, T.L., Ibañez, A.M. and Bushong, P.M. (2000b) Environmental factors affecting the severity of Alternaria brown spot of citrus and their potential use in timing fungicide applications. *Plant Disease* 84, 638–643.

Van Gundy, S.D. (1958) The life history of the citrus nematode, *Tylenchulus semipenetrans*. *Nematologica* 3, 283–294.

Villechanoux, S., Garnier, M., Laigret, F., Renaudin, J. and Bové, J.M. (1993) The genome of the non-cultured, bacterial-like organism associated with citrus greening disease contains the nusG–rplKAJL–rpoBC gene cluster and the gene for the bacteriophage type DNA polymerase. *Current Microbiology* 26(3), 161–166.

Whiteside, J.O. (1974) Environmental factors affecting infection of citrus leaves by *Mycosphaerella citri*. *Phytopathology* 64, 115–120.

Whiteside, J.O. (1975) Biological characteristics of *Elsinoe fawcettii* pertaining to the epidemiology of sour orange scab. *Phytopathology* 65, 1170–1175.

Whiteside, J.O. (1976) A newly recorded Alternaria-induced brown spot disease on Dancy tangerines in Florida. *Plant Disease Reporter* 60, 326–329.

Whiteside, J.O. (1977a) Sites of action of fungicides in the control of citrus melanose. *Phytopathology* 67, 1067–1072.

Whiteside, J.O. (1977b) Behavior and control of greasy spot in Florida citrus groves. *Proceedings of the International Society of Citriculture* 3, 981–986.

Whiteside, J.O. (1983) Timing of spray treatments for citrus greasy spot control. *Proceedings of the Florida State Horticultural Society* 96, 17–21.

Wutscher, H.K., Cohen, M. and Young, R.H. (1977) Zinc and water soluble phenolic levels in the wood for diagnosis of citrus blight. *Plant Disease Reporter* 61, 572–576.

Yang, X., Hadidi, A. and Garnsey, S.M. (1992) Enzymatic cDNA amplification of citrus exocortis and cachexia viroids from infected citrus hosts. *Phytopathology* 82, 279–284.

8 Diseases of Coconut

Nigel A. Harrison[1] and Phil Jones[2]
[1] *University of Florida, Fort Lauderdale Research and Education Center, 3205 College Avenue, Fort Lauderdale, FL 33314, USA;* [2] *Department of Crop and Disease Management, IACR Rothamsted, Harpenden, Herts AL5 2JQ, UK*

Introduction

The coconut, *Cocos nucifera,* is the most widely cultivated of all palms. It originated in Melanesia and has been spread by sea currents and man throughout the coastal tropics (Harries, 1992). The genus *Cocos* is monospecific, but many varieties and ecotypes are recognized. Coconut palms are classified as having either a tall or dwarf growth habit. Those of the tall type grow to 30–40 m in height, can live and produce fruit for >80 years, are usually cross-pollinated, and will thrive in diverse soils and environments. Tall palms mature slowly and usually do not bear fruit until they are at least 8 years old. In contrast, dwarfs generally grow no more than 10 m in height and are usually self-pollinated. They begin to bear after 3 years but do not achieve full production until they are >8 years old. Dwarfs are shorter lived than talls and typically have a productive lifespan of 35 years.

Coconut is the archetypal smallholder crop of the coastal tropics (Persley, 1992). It is a low input crop that provides food, drink, fuel, shelter and cash income for producers. Almost all parts of the plant are used. Nut yield varies greatly among varieties, and local selection occurs depending on whether production is for oil or sale on the fresh market as a source of food and drink. Copra, the dried endosperm and the source of edible oil, is the main economic product worldwide. Husks are retted to provide coir for fibre industries, and shells are burnt in kilns to produce charcoal. In some countries, shells and husks are chipped to make a peat mulch substitute that is exported to temperate countries for horticultural use.

Coconuts are rarely grown in pure stands. On smallholder farms, they are under- and interplanted with other crops, including cassava, citrus, coffee, cocoa, maize, mango, millet, sweet potato and spices. In plantation systems, coconuts are often underplanted with coffee and cocoa for which they provide necessary shade (Purseglove, 1972). Breeders are constantly striving to produce higher yielding cultivars that are suitable for a wide variety of tropical environments. Hybrids between talls and dwarfs are favoured by many plantation growers because they tend to mature quickly and are dwarf in stature, which makes nut harvest simpler (Been, 1995).

Coconut is grown on an estimated 11.6 Mha hectares in 86 countries, and ~96% of total world production comes from smallholdings. About 85% of the crop is produced in Asia and the Pacific region. In 1998, world production was ~47.7 million tonnes (Mt), and Indonesia and the Philippines were the leading producers with, 14.7 and 10.5 Mt, respectively (Table 8.1).

Table 8.1. Coconut production in 1998 in the most important countries in various regions and major diseases that constrain production.[a]

Country	Production (Mt)[b]	Major diseases
Indonesia	14.7	Bud rot, wilt diseases
Philippines	10.5	Cadang-cadang
India	10.0	Kerala (root) wilt
Sri Lanka	2.0	Leaf scorch decline
Mexico	1.1	Lethal yellowing
Brazil	0.65	
Mozambique	0.45	
Tanzania	0.34	Lethal disease

[a]Adapted from Persley (1992). Production data are from the FAOSTAT database http://apps.fao.org/default.htm
[b]Production is in millions of metric tonnes.

Coconut suffers from numerous diseases. The most widespread is bud rot, which is caused primarily by the chromistan, *Phytophthora palmivora*. Bud rot is distributed worldwide, and severe infestations can cause total loss of a crop. Fungi cause most of these diseases, and at least 38 genera have been reported on this host. More than one species may act in concert to reduce the vigour of the palm. For example, in India, a complex of 14 species of fungi, including *Colletotrichum gloeosporioides* and *Exserohilum rostratum*, have been isolated from palms with leaf rot, a syndrome that affects the spear leaf.

Coconut diseases caused by non-fungal agents are often restricted in their geographic distribution. For example, red ring disease, caused by the nematode *Bursaphelenchus cocophilus*, is limited to Central and South America; foliar decay, caused by the coconut foliar decay virus, occurs only on Vanuatu in the New Hebrides archipelago; and cadang-cadang and tinangaja, both caused by viroids, are found only in the Pacific region (Griffiths, 1987; Hodgson and Randles, 1999; Randles *et al.*, 1999). Lethal yellowing, caused by a phytoplasma, was first reported from the Caribbean, but diseases attributed to similar phytoplasmas are now recognized in Africa, Central America and Indonesia (Warokka, 1999; Cordova *et al.*, 2000).

In this chapter, we describe the most important diseases of coconut. Minor diseases that have limited distributions or small impact on the health of this host are listed in Table 8.2.

The Major Diseases of Coconut

Bud rot and nutfall

Bud rot is a serious disease of coconut that occurs throughout the warm moist tropics.

Symptoms

Bud rot causes a chlorosis and rapid necrosis of the youngest opened leaves (Plate 59). Bending and breaking of affected leaves can occur. Necrotic spotting occurs on leaf bases and pinnae when the youngest leaves are infected in the spindle (Menon and Pandalai, 1960) and hyphal mats of the causal agents may develop on lesions on leaf bases. As infection advances to encompass the newest, unopened spear leaf, it can be pulled out with minimal force, revealing the underlying rotted apical bud. A foul-smelling odour invariably accompanies exposure of these softened and pink to reddish brown tissues (Joseph and Radha, 1975). Characteristically, leaf necrosis progresses quickly throughout the central leaves of the upper crown, leaving a surrounding fringe of unaffected green leaves. In later stages, infection spreads downward into mature, woodier tissues and the upper trunk. Affected areas are water soaked, pinkish and delineated from healthy tissues by a dark border (Fig. 8.1). Although nut fall is not typical of bud rot, inflorescences may become infected, at which time attached nuts abort. Fallen nuts may exhibit superficial water-soaked lesions on their husks from which the causal agents can be readily cultured (Fig. 8.2). Affected palms rarely recover, but when they do they usually have a bitten leaf appearance.

Causal agents

Although *Fusarium moniliforme*, *F. solani* and *Graphium* sp. have been reported as causal agents, species of *Phytophthora* are more com-

Table 8.2. Miscellaneous diseases of coconut, *Cocos nucifera*.

Disease	Causal agent(s)
Algal leaf spot	*Cephaleuros virescens*
Anthracnose	*Glomerella cingulata* (anamorph: *Colletotrichum gloeosporioides*)
Bacterial bud rot	*Erwinia* spp.
Bitten leaf	*Ceratocystis paradoxa* (anamorph: *Chalara paradoxa*)
Bipolaris leafspot	*Bipolaris incurvata*
Black scorch	*Ceratocystis paradoxa* (anamorph: *Chalara paradoxa*)
Blast	Phytoplasma suspected
Bristle top	Not known
Catacauma leaf spot	*Catacauma mucosum*
Damping-off	*Fusarium* spp.
	Phytophthora spp.
	Pythium spp.
	Rhizoctonia solani
Dry basal rot	*Ceratocystis paradoxa* (anamorph: *Chalara paradoxa*)
Dry bud rot	Not known (possibly vectored by *Sogatella kolophon* and *S. yubana*)
Finschafen disease	Not known
Frond rot	Physiological disorder
Graphiola leaf spot	*Graphiola phoenicis*
Koleroga	*Phytophthora arecae*
Leaf blight	*Cytospora palmarum*
Leaf spots	*Alternaria* sp.
	Botryosphaeria disrupta
	Capitorostrum cocoes
	Cercospora sp.
	Curvularia lunata
	Cylindrocladium pteridis
	Drechslera gigantea
	Epicoccum nigrum
	Exserohilum rostrata
	Helminthosporium sp.
	Macrophoma sp.
	Macrosporium cocos
	Melanconium sp.
	Mycosphaerella palmicola
	Periconiella cocoes
	Pseudoepicoccum cocos
	Phomopsis sp.
	Phyllosticta palmetto
	Ramularia necator
Lixa grande	*Sphaerodothis acrocomiola*
Lixa pequena	*Phyllachora torrendiella*
Powdery mildew	*Oidium* sp.
Pudricion del cogollo	Phytoplasma
Queima das folhas	*Botryosphaeria cocogena* (anamorph: *Diplodia theobromae)*
Root rot	*Fusarium* spp.
	Phytophthora spp.
	Pythium spp.
	Rhizoctonia solani
Soccoro wilt	Not known
Stem necrosis	Phytoplasma suspected
Stigmina leaf spot	*Stigmina palmivora*
Thread blight	*Pellicularia filamentosa*, *P. koleroga*
	Corticium penicillatum

From Ploetz *et al.* (1999).

Fig. 8.1. Cross-section of coconut apex affected by bud rot. The affected areas are water-soaked, pinkish and delineated from healthy tissues by a dark border (photo: J.J. Ooka and J.Y. Uchida).

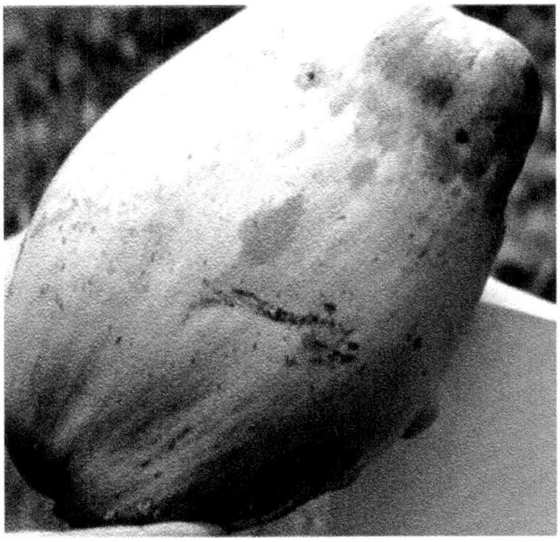

Fig. 8.2. Coconut from a tree affected by bud rot. Note the water-soaked, darkened areas at the stem end of the nut (photo: J. Flood).

mon. *P. palmivora* is most significant, and was first isolated from affected palms in India by Butler (1906). It and *P. nicotianae*, a less frequent cause of the disease, are described in Chapter 1. *P. heveae*, cause of bud rot and nutfall in Côte d'Ivoire, is described in Chapter 3 under Phytophthora cankers.

An unnamed species of *Phytophthora* (formerly referred to as *P. katsurae*) was described in Hawaii (Uchida *et al.*, 1992). It causes bud and fruit rot on Hawaii, Kauai, Maui and Oahu. It produces sporangia on artificial media that are 31 (23–42) × 40 (31–49) μm, non-caducous, papillate, ovoid and bilaterally symmetric. It is homothallic and produces oogonia with varying numbers of irregular surface protuberances and tapered, funnel-shaped bases. They are 27 (22–31) μm in diameter, antheridia are amphigynous, and oospores are aplerotic. The species has longer sporangia than *P. katsurae* (40 versus 28 μm) and fewer oogonial protuberances.

A fifth species, *P. arecae*, causes primarily nutfall and mahli (nut and flower) disease in India and Sri Lanka (Erwin and Ribiero, 1996). Chlamydospores are uncommon and 35–40 μm in diameter. Sporangia are 40–50 × 35–40 μm, broadly ellipsoid to nearly spherical, papillate and caducous, with 1–6 μm long pedicels (Fig. 8.3) (Stamps, 1985). The pathogen is heterothallic. Oogonia average 30 μm in diameter, and antheridia are amphigynous, 14 × 15 μm, and often wider than long.

Epidemiology

Bud rot affects palms of all ages, but those between 15 and 40 years old are most susceptible. It occurs in all coconut-growing regions, but is most severe in areas with high rainfall (Cook, 1971). Damage caused by high winds and hurricanes frequently is associated with the disease, and whether mature trees can be affected in the absence of such injury is debated. Spread occurs primarily via rainsplashed zoospores and caducous sporangia (*P. palmivora* and *P. heveae*). Oospores and chlamydospores can survive in soil and coconut debris such as cull piles, and infected roots are also reservoirs of inoculum.

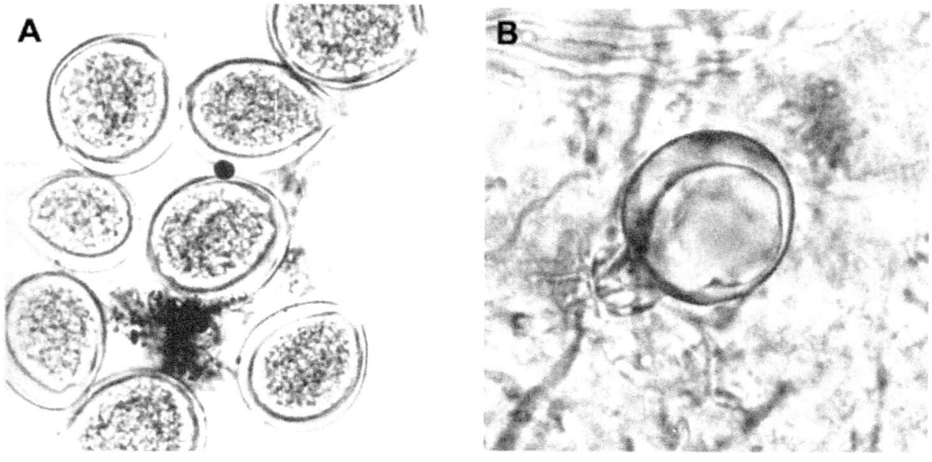

Fig. 8.3. (A) Sporangia and (B) an oospore of *Phytophthora arecae* (from CMI description no. 833).

Management

Few control measures are successful. Good cultural conditions are an essential first step. Infested debris and dead trees should be removed from plantations and nurseries and burned. To avoid prolonged periods of free moisture on host surfaces, nurseries should not be irrigated at night. Thevenin *et al.* (1995) reported that bud rot could be controlled by a stem injection of 20–25 ml of Foli-r-fos 400AS per palm at the onset of the rainy season.

Cadang-cadang and tinangaja

Cadang-cadang is the major cause of death of coconut palms in the Philippines. Although cadang-cadang-like symptoms were described as early as 1914 (Randles *et al.*, 1997), it was not until the 1950s and 1960s that the disease began to reach epidemic proportions. Randles *et al.* (1997) calculated losses due to cadang-cadang of US$39 million for 1978 and US$21 million for 1980.

The disease is confined to, and continues to spread on, Catanduanes, Masbate and Northern Samar Islands and the Bicol Peninsula of Luzon Island (Hanold and Randles, 1997). A similar disease, tinangaja, was reported from Guam in 1917, and by 1946 had destroyed the coconut industry on that island (Boccardo, 1985).

Causal agents

Cadang-cadang and tinangaja are caused by viroids, small, naked, single-stranded circular RNAs. The *Coconut cadang-cadang viroid*, CCCVd, was first isolated by Randles (1975). It has a basic 246 nucleotide structure consisting of a central conserved region of 44 nucleotides that is common to many viroids (Haseloff *et al.*, 1982). It predominates in the early stages of infection, but is replaced by larger forms (287–301 nucleotides) as the disease progresses to the mid- and late stages. This variation is due to the reiteration of 41, 50 or 55 nucleotides. The *Coconut tinangaja viroid*, CtiVd, has a sequence of 254 nucleotides with a 64% homology with the basic CCCVd molecule (Keese *et al.*, 1988). Differences in the sequences of CCCVd and CTiVd could account for the variation in symptoms of the respective diseases.

Symptoms

There are three main stages in the development of cadang-cadang, early (E), mid (M) and late (L) (Fig. 8.4) (Randles, 1997). At the E0 stage, CCCVd is detected in the youngest leaves, but plants are symptomless. After 1–2 years, the E1 stage develops as newly formed nuts become more rounded and exhibit equatorial scarifications, but no leaf symptoms (Fig. 8.5). At E2,

Fig. 8.4. Coconut palms in various stages of decline caused by the *Coconut cadang-cadang viroid* (CCCVd). Note the understorey of *Alpinia* sp., a symptomless host of CCCVd variants (photo: J.W. Randles).

more nuts are rounded and scarified, chlorotic spots appear on leaves, and inflorescences are stunted with tip necrosis and the loss of some male florets. By E3, leaf spots enlarge, fewer nuts are produced, and new inflorescences are stunted and sterile. Spathe, inflorescence and nut production declines and then ceases by the M stage. Leaf spots become more numerous, but appear only on leaves at or lower than positions 3 or 4 in the crown. By the L stage, leaves decline in size and number, and leaflets become brittle. Leaf spots coalesce, resulting in a general chlorosis, crown size is reduced, and ultimately the palm dies. The length of time that it takes for a palm to die varies with age. In 22-year-old palms, this averages 7.5 years, whereas in 44-year-old palms the average is 15.9 years (Zelazny and Niven, 1980).

Symptoms of tinangaja differ slightly from those of cadang-cadang (Hodgson and Randles, 1999). The earliest recognizable symptom is a reduction in canopy size, but this may be missed until yellow spots appear on the leaves and small elongate nuts that lack a kernel are produced (Fig. 8.6). As the disease progresses, palms decline, the crown shrinks and the upper trunk may show tapering. Ultimately, nut production ceases and the palm dies.

Fig. 8.5. Nuts from healthy (right) and cadang-cadang-affected coconut palms (photo: J.W. Randles).

Fig. 8.6. Spindle-shaped nuts from a tinangaja-affected coconut palm compared with one from a healthy palm (right) (photo: J.W. Randles).

Epidemiology

CCCVd has induced symptoms typical of cadang-cadang after artificial inoculation of healthy palms (Randles *et al.*, 1977; Mohamed *et al.*, 1985). The experimental host range of CCCVd is narrow and limited to the following palm species: *Adonidia merrilli, Areca catechu, Coryphya elata, Dypsis lutescens, Elaeis guineensis, Phoenix dactylifera, Ptychosperma macarthuri* and *Roystonea regia* (Imperial *et al.*, 1985). Experimentally infected palms all develop yellow leaf spots and often are stunted.

Production from affected palms falls below economic levels ~5 years before they die (Price, 1971). Studies on the epidemiology of cadang-cadang and tinangaja have been constrained by several problems. These include a long latent period between infection and symptom appearance; difficulties in recognizing early stages of the diseases; no knowledge of how natural transmission of the pathogens occurs; and low rates of disease spread. All of these factors have conspired against the development of sustainable control measures.

Although there have been many studies of the spread of cadang-cadang, few have been conducted on tinangaja since suitable diagnostic tools for it were developed only recently (Hodgson and Randles, 1998, 1999). The rate and pattern by which cadang-cadang spreads may vary from site to site, and disease incidence within the boundary of spread is variable (Zelazny and Niven, 1980; Zelazny and Pacumbaba, 1982; Zelazny *et al.*, 1982). Zelazny and Pacumbaba (1982) identified three beetle species that were more abundant in areas of high disease incidence, but no vector has been identified unequivocally.

Hanold and Randles (1997) analysed 1024 coconut samples from an extensive survey for the presence of viroid-like sequences that shared sequence similarity with CCCVd. Only India had no samples which tested positive for such sequences. In some countries in Southeast Asia, a high incidence of CCCVd-like sequences has been detected. Further characterization of the newly detected molecules should determine whether any of them warrant classification as new viroids, and the health risks, if any, that these molecules pose to coconut and other palms.

Management

Replanting reduces disease losses since the rate of spread in new plantings is not related to the closeness of affected palms (Bigornia, 1977). Since removal of symptomatic palms does not eradicate cadang-cadang, Randles *et al.* (1997) suggested that removing all palms that were positively indexed for CCCVd infection might be more successful.

Trials have been established with nuts from palms that have remained healthy in areas with a high incidence of cadang-cadang in the hope of finding tolerance. Palms from these trials are being tested by both natural and experimental inoculation. Also, in directed breeding, dwarf × tall crosses are being made to produce hybrid seed that are challenged by experimental inoculation with the viroid.

Mild strain cross-protection is used for a number of economically important viruses, and the lack of options to control cadang-cadang has raised hope that this strategy could work for this disease. Rodriguez and Randles (1993) found variations in the pathogenicity and conserved domain of CCCVd associated with the 'brooming' form of the disease, suggesting that other variants may still be found. Similarly, Rodriguez (1993), using molecular detection techniques, found variants of CCCVd in ginger (*Alpinia* sp.) and arrowroot from the Philippines (Fig. 8.4).

Coconut foliar decay

Coconut foliar decay, which is also known as New Hebrides coconut disease, affects introduced coconut palm varieties on, and is limited to, the island of Vanuatu.

Symptoms

The local 'Vanuatu Tall' variety remains symptomless after infection, but introduced varieties and progeny derived from hybridization between these varieties and 'Vanuatu Tall' develop a range of symptoms (Calvez *et al.*, 1980). These begin with partial yellowing of one or more of the 7th to 11th oldest leaves that spreads the length of the leaf and to adjacent leaves in the whorl. A lateral necrosis of the petiole then develops, which causes the leaf to die and hang from the canopy. This appearance of a normal apex and several yellowish leaves followed by death of young leaves that hang through green older leaves is characteristic of this disease (Randles *et al.*, 1999). Narrowing of the trunk can occur but, with symptom remission, as occurs in tolerant varieties, it may thicken once again. Susceptible varieties die between 1 and 2 years after symptoms first appear.

Causal agent

Foliar decay is caused by a small, 20 nm in diameter, isometric virus particle, *Coconut Foliar decay virus* (CFDV) (Randles and Hanold, 1989). It is classified as a nanovirus and contains a 1.76 kb genome of circular single-stranded DNA (Randles *et al.*, 1987). Purified preparations of CFDV or sap from infected coconut palms are not infective.

Epidemiology

CFDV is transmitted in nature by the planthopper, *Myndus taffini* (Julia, 1982). *M. taffini* breeds on the roots of a native tree species, *Hibiscus tiliaceous*, which grows prolifically on Vanuatu. It is a common reservoir for the vector in and around coconut plantations, but it has not been shown to host the virus. No other plant hosts of CFDV have been reported.

The development of molecular techniques for detection, such as polymerase chain reaction (PCR) (Rohde *et al.*, 1990; Randles *et al.*, 1992) and a cDNA probe (Randles *et al.*, 1999), have allowed the diagnosis of CFDV in palms in the absence of symptoms. This led to the discovery that some 'Vanuatu Tall' palms have symptomless infections. Typically, CFDV can be detected in young leaves and secondary roots where it is localized in vascular tissues, primarily phloem, in a non-systemic pattern (Randles *et al.*, 1992). CFDV can be detected in coconut embryos and husks, but not pollen (Randles, 1997). Although seed transmission of foliar decay has not been recorded, movement of nuts from infected palms could result in the transfer of CFDV to new sites.

Management

The local 'Vanuatu Tall' and 'Vanuatu Red Dwarf' coconuts are tolerant and offer the best means of controlling the disease (Persley, 1992).

Ganoderma butt rot

Ganoderma butt rot, which is also known as basal stem rot, is a lethal disease of coconut and other species of palm. In Florida (USA), Elliott and Broschat (2001) listed 57 palm taxa that were affected. Although the disease has its greatest economic impact on oil palm, *Elaeis guineensis*, it can also damage coconut palm in the landscape and in plantations. For example, in India, losses of up to 31% have been reported (Bhaskaran and Ramanathan, 1984), and complete plantations were destroyed within 7–8 years when control measures were not used (Bhaskaran, 2000).

Symptoms

Overall symptoms vary from region to region, but the one found in all areas is internal discoloration of lower stem tissue. Initially, older fronds begin to yellow, droop and appear drought stressed (Peries, 1974). They eventually die, collapse parallel to the trunk, and either remain attached to the tree or break off. In some regions, younger and younger leaves become pale green and chlorotic, and the size of new leaves decreases. Several unopened leaves may develop in the crown, and flowering and nut set decrease and then stop. Eventually, only a few stunted leaves remain in the canopy before the palm dies. From the onset of symptoms to death usually takes 1–4 years.

In India, stem bleeding (exudation of a reddish brown liquid), chocolate brown bands that extend up the trunk and extensive root rot may also occur (Bhaskaran, 2000). These symptoms are not found in all regions and may reflect differences among the different causal agents. Stem bleeding may extend to 10–15 m up the trunk, but the pathogen usually extends no higher than 1.5 m (Elliott and Broschat, 2001). Internally, colonized tissues are discoloured brown.

Basidiocarps of the pathogens may form on lower portions of the trunk at any stage in disease development, and may appear before external symptoms develop. Basidiocarps also develop on stumps of trees that were killed by the disease or by other causes.

Causal agents

The taxonomy of the causal agents is confused (Seo and Kirk, 2000). Traits that have been used to delineate species of *Ganoderma* often are ambiguous and can vary when a given species is observed in different environments. New species have been named based on one or a few specimens, and hundreds of these have been reduced to synonymy with well-established species upon further examination. Furthermore, assignment of isolates to a given species has often been indiscriminate, resulting in erroneous host ranges being published for a given species. Although different taxa appear to affect coconut palm, with one exception their identities are unclear.

G. zonatum is reported from Florida, South America, Java and tropical Africa (Steyaert, 1975b), and the name has also been used for collections from Southeast Asia and Oceania (P. Bridge, Royal Botanic Gardens, Kew, personal communication, 2002). It affects mainly palms, but has been reported on *Eucalyptus* spp. and as a rotter of dead wood. The fungus produces sessile, reniform to irregular, dimidiate basidiocarps up to 30–40 cm at their widest point and up to 9 cm thick at the base (Fig. 8.7) (Steyaert, 1975b; Ryvarden, 2000). The upper surface usually is tuberculate and sometimes sulcate, glossy (laccate) and reddish brown, but darkens with age. The lower pore surface is white. In cross-section, the cutis is 20 μm thick and underlined in yellow. The context extends through one-third to half of the depth of the bracket and is tan, whereas the lower tube layer is walnut brown and half to two-thirds the bracket thickness. Basidiospores are pale yellow, ellipsoid, with long and thick echinules, and 9–16 × 5.5–9 μm. Basidiocarps from previous years that remain on palms are infertile.

G. boninense occurs in Australia, Japan, Indonesia, Malaysia, the Philippines, Samoa, Sri Lanka and Tasmania (Steyaert, 1975a). Its host range includes mainly palms, but some dicot tree species as well. It causes basal stem rot of oil palm in Southeast Asia and Oceania (Abdullah, 2000). Although it is isolated commonly from coconut palms, it does not cause disease. Coconut and *G. boninense* appear to have co-evolved in the region to

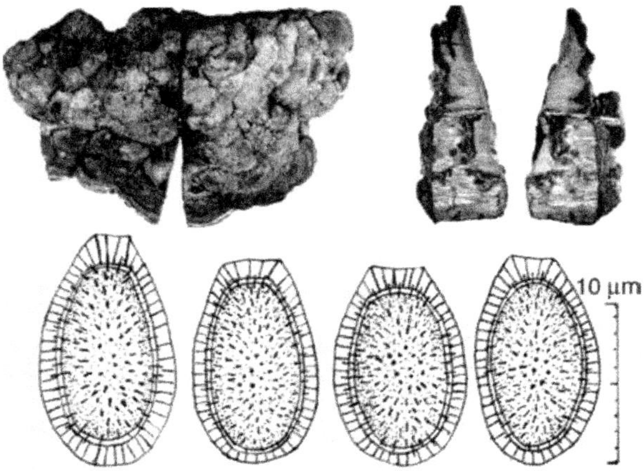

Fig. 8.7. Clockwise from top left-hand corner basidiocarp, sections of a basidiocarp and basidiospores of *Ganoderma zonatum* (from CMI description no. 448).

reach a state of benign equilibrium (J. Flood, CABI, personal communication, 2002).

A *Ganoderma* sp. kills palms in Sri Lanka and India (Peries, 1974; Bhaskaran, 2000). It has been called variously *G. applanatum*, *G. boninense* and *G. lucidum*, but resembles none of these species. Although it has a narrow genetic base, its true identity and affinity for these or other species are not known (P. Bridge, Royal Botanic Gardens, Kew, personal communication, 2002). Work to clarify its place in the genus is needed.

Other species of *Ganoderma* that have been reported on palms in general or specifically on coconut palm include *G. cupreum*, *G. miniatotinctum*, *G. tornatum* and *G. xylonoides* (Moncalvo, 2000). Their validity as species and importance and identification on coconut palm are uncertain.

Epidemiology

G. zonatum is thought to move primarily via basidiospores, but there are no experimental data to support this assumption (Elliott and Broschat, 2001). Recent work on another laccate species, *G. boninense*, indicates that basidiospores probably play a major role in disseminating it, either by directly infecting the host or by indirect infections from organic debris or dead host materials that they have colonized (Miller *et al.*, 1999). Somatic incompatibility tests demonstrated the presence of highly diverse populations in plantations to the extent that it was uncommon to find two isolates of the same somatic incompatibility group. Based on these results, root-to-root contact between palms, an avenue that previously was thought to be important, was concluded to be of no consequence. It is assumed that basidiospores infect wounds on trunks or cut ends of palm fronds, but experimental evidence is lacking.

Management

In general, this is a disease of palms with a woody trunk. Mortality begins at ~10 years of age (Venkatarayan, 1936). Where the disease is established, sanitation is most important. Basidiocarps and infected trees should be removed and destroyed whenever they are found. All root systems, stumps and trunks of palms that die from other causes should also be removed as quickly as possible to prevent their colonization by these pathogens. Wounding should be avoided.

Fumigation of infested sites may be beneficial (Simone, 1994a). Products containing dazomet, metam sodium or methyl bromide+chloropicrin are effective. Although treating stumps with creosote, copper-based fungicides and tridemorph did not inhibit colonization of stumps, urea appeared to be effective and enhanced microbial decay of stumps in Sumatra.

Grey leaf blight

Grey leaf blight was first recognized in Guyana in 1931 and is now known throughout the tropics (Cook, 1971; Holliday, 1980). It usually is a minor problem, but can be severe under crowded or wet conditions, or after leaves are damaged by insects. Young palms in nurseries are especially vulnerable.

Symptoms

Symptoms begin as small yellow to brown spots on leaflets and raches that develop grey centres with dark brown borders as they enlarge. Dark, globose acervuli of the causal fungus form in lesion centres on upper leaf surfaces (Plate 60). Lesions elongate parallel to veins, and may eventually coalesce to form large, irregular necrotic areas on leaves. In severe cases, leaflet tips and margins become grey and frizzled, imparting a blighted appearance to the canopy.

Causal agent

Pestalotiopsis palmarum causes grey leaf blight (Mordue and Holliday, 1971). Although a teleomorph in the genus *Rhynchosphaeria* has been reported for the pathogen in India (Agnihothrudu and Barua, 1965), it apparently is uncommon.

Conidiophores are hyaline, cylindrical to obovoid, 1–4 μm in diameter and 5–18 μm long (Fig. 8.8). Conidia emerge from acervuli in black cirri, are fusiform, usually straight, four-septate with slight constrictions at septa, and 20 (17–25) × 6 (4.5–7.5) μm. Three median cells in conidia are olivaceous and larger than the hyaline apical and basal cells. Apical cells have three (rarely two or four) cylindrical appendages that are 16 (5–25) μm in length. Basal appendages are straight and 2–6 μm long.

On potato dextrose agar (PDA), colonies initially are white and fluffy. Acervuli develop in yellowish clumps and produce greenish black masses of conidia. Colonies are usually diurnally zonate and, on the obverse side, exhibit little coloration.

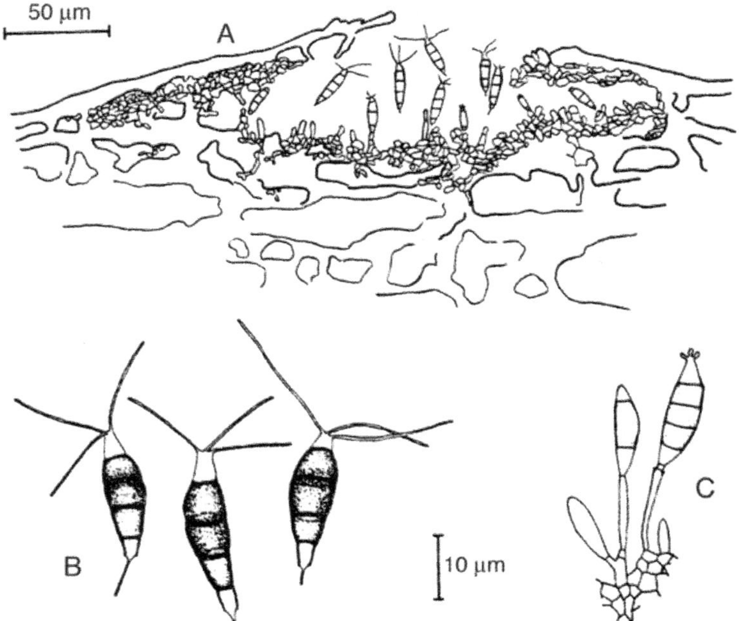

Fig. 8.8. (A) Acervulus, (B) conidia and (C) conidiophores of *Pestalotiopsis palmarum* (from CMI description no. 319).

Epidemiology

The pathogen does not exhibit much host specificity, and several species of palm, as well as some non-palm hosts, are affected. *P. palmarum* is a relatively non-aggressive pathogen, and young, wounded or otherwise weakened host tissues are infected preferentially, presumably by conidia. The disease develops slowly on palms, and symptoms are usually only seen on older leaves. Thus, older leaves and infested host debris are the primary sources of inoculum.

Infection events apparently have not been studied in coconut palm. However, in banana, the fungus was shown to form appressoria and infection pegs that penetrated host surfaces directly (Vakili, 1963). Rainfall and high humidity play important roles in the development of this disease.

Management

Control measures are usually needed only in nurseries. To prevent damage from occurring in these situations, optimal growth of young plants is required. In particular, insect damage should be suppressed, and overcrowded or unsanitary conditions should be avoided. If possible, dry foliage should be maintained.

In nurseries, or the rare instances where control in mature trees is required, broad-spectrum, protectant-type fungicides are efficacious.

Hartrot

Hartrot is a fatal wilt disease. Historically, it has also been known as bronze leaf wilt, fatal wilt, Coronie wilt, Cedros wilt or Marchitez. Hartrot was first recognized during the early 1900s in Surinam where it now prevents successful coconut production. During the mid-1970s, the disease killed an estimated 15,000 coconut palms in the Cedros region of Trinidad within 3 years. Currently, it appears to be restricted to Central America (Costa Rica, Honduras and Nicaragua), South America (Brazil, Colombia, Ecuador, Guyana, Peru, Surinam and Venezuela) and the West Indies (Grenada, Trinidad and Tobago).

Symptoms

The earliest symptom on coconut is a simultaneous yellowing or browning of the lowest two or three leaves, starting from the tips and spreading to the leaf base (Plate 61). Immature nuts are shed but mature nuts generally remain attached for longer. As discoloration advances to involve younger leaves, older leaves become necrotic, desiccate and may break near the petiole. The rachillae of newly opened and unopened inflorescences are partially rotten and blackened, while leaves of the upper crown are still green; they die and collapse when the disease reaches an advanced stage. Roots also show apical or complete necrosis and desiccate at this time. Finally, when most of the leaves are brown and nut drop is complete, a putrid wet rot of the basal spear leaf and apical meristem develops and the palm dies.

Coconut palms of all ages are equally susceptible, as are both tall and dwarf types. In general, symptoms are indistinguishable from those of red ring and lethal yellowing. Mortality occurs within 1–3 months of the onset of external symptoms, depending on the age, size and condition of the palm, and vigorously growing palms succumb most rapidly.

Causal agent

Hartrot disease is associated with the systemic colonization of coconut palm by a culturable, phloem-restricted, uniflagellate protozoan (phylum *Euglenozoa*, order *Kinetoplastida*, family *Trypanosomatidae*) (Fig. 8.9). Plant-associated trypanosomatids belong to the genus *Phytomonas*, which includes flagellates that parasitize latex cells of laticiferous plants, and intraphloemic flagellates that infect non-laticiferous host plants. Trypanosomatids from laticiferous plants are not considered to be pathogenic, although some damage fruit (Dollet, 1984). Speciation of those from non-laticiferous host plants has not been fully resolved, nor has the pathogenicity been demonstrated for those that are associated with coconut. Their role as phytopathogens is supported

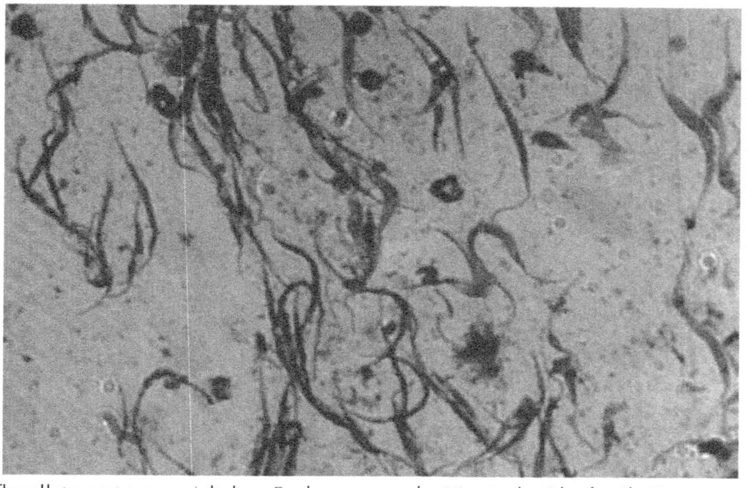

Fig. 8.9. Uniflagellate protozoans (phylum Euglenozoa, order Kinetoplastida, family Trypanosomatidae, genus *Phytomonas*) in the sap of a coconut palm affected by hartrot (photo: Michel Dollet).

by their consistent association with hartrot symptoms, and the correlation between the level of phloem colonization by *Phytomonas* and disease severity (Waters, 1978).

Although culturable, *Phytomonas* isolated from palms initially require several passages on complex media containing cultured insect cells before they can be grown in cell-free media (Menara *et al.*, 1988). The promastigote form predominates in coconut. These flagellates are between 1 and 1.5 μm wide and 12 and 27 μm long, including a single, 7 μm long flagellum. Within the cytoplasm, the base of the flagellum is associated with a kinetoplast composed of a reticulate network of electron-dense DNA. The ribosome-rich cytoplasm also contains a single nucleus and both electron-translucent and -dense, membrane-bound structures (Waters, 1978).

Isolates from palms are morphologically uniform. However, they can be distinguished by: their isoenzyme profiles (Guerrini *et al.*, 1992); the sizes, homology and restriction length polymorphisms of minicircle components of their kinetoplastid DNA (Muller *et al.*, 1995); the presence of double-stranded RNA or virus-like particles (Marche *et al.*, 1993); and the sequences of their small subunit 18S rDNA (Marche *et al.*, 1995).

Epidemiology

The epidemiology of hartrot is poorly understood. Although it is capable of rapid spread, as evidenced by the epiphytotic in the Cedros region of Trindad (Waters, 1978), spread is usually slower and more localized. Hartrot affects mostly bearing palms, although very young palms, 18–20 months old, are sometimes involved. To date, all coconut ecotypes that have been tested have been equally susceptible (Louise *et al.*, 1986). When symptoms first appear, 10–100% of mature sieve tubes may contain flagellates and many are completely occluded (Parthasarathy *et al.*, 1976), presumably obstructing transport of photosynthates. Sap expressed from coconut inflorescences or roots at this stage contain flagellates that can be resolved by light microscopy (Fig. 8.9) (Waters, 1978).

Several species of the pentatomid insect genera *Lincus* and *Ochlerus* are known to transmit the trypanosomatids responsible for hartrot (Desmier de Chenon, 1984; Dollet, 1984). Crepuscular feeders that do not fly, adults apparently acquire *Phytomonas* during feeding activity on diseased palms and vector it to adjacent palms. Other palms, such as *Bentinckia nicobarica*, *Maximiliana maripa* and *Roystonea regia*, may serve as alternative sources of the protozoan (Dollet, 1984; Kastelein and Parsadi, 1986).

Management

Control of hartrot relies primarily upon early detection, prompt removal and destruction of affected palms. Treatment of palms with insecticides such as lindane or deltamethrin arrests disease spread (Louise *et al.*, 1986), but whether this strategy is useful or practical over the long term is unproved. Technical guidelines have been established to prevent the inadvertent spread of hartrot during the transfer of coconut germplasm (Frison *et al.*, 1993).

Leaf scorch decline

Leaf scorch decline was first reported in southern Sri Lanka in 1955 (Kirthisinghe, 1966). However, there is evidence that it existed in the area considerably before this time (Bryce, 1921; Park, 1927). The disease continues to spread and is now found in almost all coconut-growing areas of the country (Fernando, 1999).

Symptoms

Scorching and drying of leaflets accompanied by a slight curling of the lowermost leaves are characteristic of the disease. Scorching gradually extends from the tips of leaflets to the leaf midrib and then progresses to successively younger leaves in the upper crown. On each leaf, distal leaflets are more severely affected than proximal ones. The oldest affected leaves usually remain attached to palms for an extended time as compared with leaves of similar age on healthy palms (Ekanayake, 1963). Tapering of the trunk just below the crown is apparent on palms that have reached mid-stage development of the syndrome. As symptoms advance, fewer and smaller inflorescences are produced and nuts are elongated in shape. In its advanced stage, new leaves are markedly reduced in length and width, and palms become completely unproductive as the crown diminishes in size and eventually topples from the trunk (Mahindapala and Chandrasena, 1975; Rajapakse and Fernando, 1997).

Foliar symptom development is accompanied by considerable root decay. Root systems of affected palms are more fibrous than those of healthy palms due to rootlet necrosis. Cross-sections of root tissues reveal blackening of cortical tissue and blockage of xylem vessels by tyloses (Peries, 1968; Jayasekara *et al.*, 1990). No consistent changes in trunk tissues have been reported. The full course of the syndrome averages between 5 and 10 years. Similar declines have been observed on several other palm species (Davis, 1962).

Epidemiology

An initial 1963 survey of the Southern Province of Sri Lanka revealed a disease incidence of 5.1%. Subsequent surveys of the major coconut-growing areas during 1985 and again in 1996 showed an increase of up to 12%, and incidence was highest in small holdings (<1 ha). The distribution pattern of affected palms was scattered, and there was no indication of focal origin or directional spread (Fernando, 1999). At this time, no cause has been identified for this disease.

Lethal bole rot

Lethal bole rot, caused by the basidiomycete *Marasmiellus cocophilus*, is an important disease of coconut palm in East Africa but is rarely found outside this region. The disease is often fatal to palms up to 8 years old. It spreads readily through the soil by root contact between palms and probably by airborne basidiospores. The first visible symptoms are a wilting of the oldest leaves that turn yellow or bronze. At an advanced stage, reddish dry rot and cavities lined with fungal mycelium appear in the tissue of the basal trunk (bole). Sometimes the entire bole region of palms is decayed due to invasion by secondary bacteria. Extensive destruction of the root system, death of the spear leaf and a progressive soft rot of the apical bud accompany basal trunk decay.

Seed infestation by *M. cocophilus* leading to a basal stem rot of seedling coconuts has also been described (Jackson and Firman,

1982; Jackson and McKenzie, 1988). Affected seedlings in field nurseries show premature death of the oldest two or three leaves, and often snap at the junction of the stem and nut. Younger leaves develop a brown rot at the leaf bases. Sporocarps of *M. cocophilus* are produced in abundance from seed husks and bases of dead leaves. After transplanting, leaves of surviving seedlings unfurl before becoming fully emerged. Decay of some areas of the bole and roots is usually evident. In some instances, affected seedlings eventually recover to produce normal roots and leaves.

M. cocophilus has been reported from several grasses including goosegrass, marsh grass and Bermuda grass (Frison *et al.*, 1993). They may serve as alternative hosts and sources of inoculum.

Lethal yellowing

The term lethal yellowing (LY) was first used in the mid-1950s to denote a fatal disease of uncertain aetiology that had affected coconut palms in western Jamaica since the 1800s (Nutman and Roberts, 1955). LY is found in the Bahamas, the Cayman Islands, Cuba, Dominican Republic, Haiti, Jamaica and the southern USA (Florida and Texas), and recently has spread to southern Mexico, Belize and Honduras (Ashburner *et al.*, 1996; Eden-Green, 1997a). It threatens global coconut production because it spreads rapidly, kills palms quickly and is incurable. About two-thirds of the world's coconut palms are susceptible (Harries, 1978).

Symptoms

Symptoms on highly susceptible mature, tall-type coconut palms begin with a premature drop of most or all the fruit. Aborted fruit often develop a brown–black, water-soaked appearance at the calyx end. Inflorescence necrosis follows, with flower stalks partially or totally blackened as they emerge from the surrounding spathe, in contrast to the creamy white to yellow flower stalks on healthy palms (Plate 62). The foliar yellowing phase initially involves the lowest, older leaves and eventually encompasses the entire crown (Plate 63). Occasionally, this symptom is first seen as a single yellow leaf (flag leaf) in the mid-crown. Yellowed leaves remain turgid before turning brown, then desiccate and hang down around the trunk for several days before falling. Death of the apical bud occurs midway or late in the foliar yellowing phase, at which time root necrosis is extensive. Bud death is accompanied by a putrid decay of these tissues, due to the invasion of secondary microbes, and collapse of the newly emerged spear (youngest) leaf that then hangs from the crown. Finally, the remaining crown withers and topples from the palm, leaving a bare trunk standing. Palms succumb to LY within 3–6 months of symptom onset.

On dwarf-type coconuts, symptoms also include premature fall of nuts and inflorescences followed by a pronounced browning of older leaves, rather than the yellowing that is typical on most tall varieties. Often, foliar symptoms on the 'Malayan Green Dwarf' more closely resemble those of a wilt during which inflorescences shrivel inside unopened spathes and fail to emerge. Pinnae of mid- and upper-crown leaves are noticeably flaccid and fold inward, and pronounced constrictions in the trunk may develop during the later stages of the disease.

Causal agent

Phytoplasmas (previously known as mycoplasma-like organisms or MLOs) were implicated as the probable cause of LY when these minute, wall-less prokaryotes were first observed by electron microscopy in phloem sieve tubes of symptomatic, but not healthy, coconut palms in the early 1970s (Plavsic-Banjac *et al.*, 1972). Unlike other phytopathogenic mollicutes, phytoplasmas are obligate parasites. Thus, evidence for their being the cause of LY is derived primarily from their consistent association with diseased palms, from symptom remission in palms after injection with tetracycline but not penicillin antibiotics, and from the absence of other pathogens.

Besides coconut, the LY phytoplasma affects at least 35 other palm species and the arborescent monocot *Pandanus utilis* (Harrison and Oropeza, 1997). In palms, phytoplasma profiles vary from ovoid to filamentous when viewed by electron microscopy. In coconut, non-filamentous forms average 295 nm in diameter, and filamentous forms average 142 nm in diameter and at least 16 μm in length (Waters and Hunt, 1980). Phytoplasma cells are enclosed by a trilaminar unit membrane and contain cytoplasm with DNA fibrils and ribosomes (Thomas, 1979; Thomas and Norris, 1980).

Molecular diagnostic assays that employ DNA probe hybridization and the PCR have been developed for the LY phytoplasma (Harrison *et al.*, 1992, 1994). Besides enhancing pathogen detection and identification in host plants or vector planthoppers, these assays have also shown that the phytoplasma exists as a group of closely related, possibly identical strains in the western Caribbean (Harrison and Oropeza, 1997). They are phylogenetically distinct from other coconut-infecting phytoplasmas in Africa (Tymon *et al.*, 1998), and are most closely related to a phytoplasma associated with decline-type diseases of coconut and the arborescent monocot, *Carludovica palmata*, in southern Mexico, and Canary Island date palm, *Phoenix canariensis*, in Texas (Harrison *et al.*, 2002).

Epidemiology

The onset of external symptoms follows a protracted incubation (pre-symptomatic) phase estimated as between 114 and 405 days (McCoy *et al.*, 1983). Generally, phytoplasma concentrations are highest in immature rather than mature tissues and may be found most reliably in phloem-rich unemerged leaf bases surrounding the apical meristem of affected hosts. They are less abundant in flower stalk bases or roots and are only occasionally present and sporadically distributed in mature leaves.

From primary disease foci that typically involve a few palms, two types of disease spread have been described. One involves spread in a random pattern around the focus, eventually claiming most of the susceptible palms within, and a second type occurs as a series of jumps of a few to 100 km or more to establish new disease foci from which the local pattern of spread is then repeated. McCoy *et al.* (1983) studied disease spread and apparent infection rates during the epidemic of LY in southeastern Florida during the early 1970s. After an initial lag phase, the number of diseased palms increased in a logarithmic fashion. It was estimated that each affected coconut palm in a primary focus was responsible for the infection of 4.6 new palms during the first 8 months of the epidemic, and 9.3 new palms within 2 years.

In Florida, the rates of spread generally have been lower in coastal than in inland sites. Although a higher level of cultural maintenance occurs at the inland sites, the actual reason for this relationship is unclear. Differences in the rate of spread at different geographical locations have also been noted. In 3 years, the disease spread over 128 km in southeastern Florida, and similar rates of dispersal have been reported in southern Mexico (Oropeza and Zizumbo, 1997). In contrast, LY required ~60 years to move across Jamaica, a distance of 238 km.

There is persuasive evidence implicating a planthopper, *Myndus crudus*, as a primary vector of LY (Fig. 8.10). In Florida, transmission of LY to coconut and other susceptible palm species occurred following their exposure to *M. crudus* collected from palms in areas with a high incidence of LY, but did not occur when palms were protected from the planthopper (Howard *et al.*, 1983, 1984). Furthermore, *M. crudus* and LY have coincident distributions in the Americas (Howard, 1983).

Management

Given our present understanding of the disease, use of resistant ecotypes and hybrids offers the only practical long-term solution to LY (Harries, 2001). Although a recent report (Broschat *et al.*, 2002) questions whether the 'Malayan Dwarf' and hybrid 'Maypan' ('Malayan Dwarf' × 'Panama Tall') possess high levels of resistance, they

Fig. 8.10. *Myndus crudus*, the probable vector of lethal yellowing in the Americas (photo: J. De Filippis).

have been used extensively for replanting in Jamaica and other countries that are affected by the disease. Promising levels of resistance have been identified in other coconut ecotypes that include 'Chowghat Green Dwarf', 'Fiji Dwarf', 'Red Spicata Dwarf', 'Sri Lanka Yellow Dwarf' and 'King' (Harries, 1995), but these ecotypes have not been exploited commercially.

To discourage the inadvertent spread of LY, commercial movement of living palms and palm seeds from LY-affected to disease-free areas is generally not permitted. However, quarantine requirements vary according to the localities that are involved. Technical guidelines for the safe movement of coconut germplasm from LY-affected areas for research but not for commercial purposes have been developed under the auspices of the International Board for Plant Genetic Resources (now known as the International Plant Genetic Resources Institute) (Frison *et al.*, 1993). Other measures, such as the eradication of affected palms, have not reduced the spread of LY, but slight reductions were achieved by insecticide suppression of the reputed vector (Howard and McCoy, 1980). Prophylactic antibiotic treatments, although effective (McCoy *et al.*, 1976), are not economical in commercial coconut production.

Lethal yellowing-like diseases

Recurrent LY-like diseases also affect coconut cultivation in the eastern hemisphere. These include Awka disease (Nigeria); Cape St Paul wilt (Ghana) (Plate 64); Kaïncopé disease (Togo); Kribi disease (Cameroon); lethal decline (Tanzania, Kenya and Mozambique) (Plate 65); Kalimantan wilt, Natuna wilt (Fig. 8.11) and Sulawesi yellows in Indonesia; leaf scorch decline in Sri Lanka; Malaysian wilt in peninsular Malaysia; and Tatipaka disease and root wilt in India (Eden-Green, 1997b). The symptoms and epidemiology of, and devastation that is caused by, these diseases closely resemble those of LY (Plates 64 and 65; Fig. 8.11) (Eden-Green, 1997b). However, it was not until molecular diagnostics became available that the associated phytoplasmas could be easily diagnosed (Tymon *et al.*, 1997) and relationships between the African and Caribbean LY diseases established (Tymon *et al.*, 1998; Mpunami *et al.*, 1999).

Awka wilt, Cape St Paul wilt, Kaïncopé and Kribi

These diseases spread rapidly and begin with premature nutfall. Varietal susceptibility has been studied most extensively in Ghana, where natural exposure of germplasm in replicated field trials has demonstrated that the 'Vanuatu Tall' and 'Sri Lankan Green Dwarf' varieties are most tolerant. Insect vectors of these diseases have not been identified. However, in Nigeria, two insects, *Meenoplus proximus* and *Malenia cocos*, were widespread on palms year round and considered the most promising candidate vectors of Awka wilt (Aisagbonhi and Kolade, 1994). *Myndus adiopodoumeensis* has been investigated intensively as a potential vector of St Paul wilt in Ghana, but no posi-

Fig. 8.11. Decline of a coconut plantation in Indonesia caused by Natuna wilt. Note the presence of healthy, non-bearing palms in the foreground and the absence of nuts on palms in the background (photo: P. Jones).

tive transmissions using this planthopper have been demonstrated to date (Eden-Green, 1997b).

Molecular studies have shown that phytoplasmas implicated as aetiological agents of coconut diseases in Ghana and Nigeria are distinct from, although related to, the LY phytoplasma (Tymon *et al.*, 1998).

Coconut lethal disease

Lethal disease (LD) was first identified in Tanzania, but subsequently has also been found affecting coconuts in the adjacent countries of Kenya and Mozambique. In Tanzania, disease incidence differs significantly from region to region. In the southern region, LD has killed >50% of the coconut palms since 1965, while in the northern region it has killed only 8.5% (Schuiling *et al.*, 1992). Using molecular diagnostics, Mpunami *et al.* (1999) showed the presence of two distinct phytoplasma strains amongst diseased East African coconuts. The most prevalent of these has been classified as the LD phytoplasma (Tymon *et al.*, 1998), while the second strain is most closely related to the phytoplasma associated with Cape St Paul wilt in West Africa.

Studies in Tanzania over many years have shown that coconut varieties resistant to LY in the Caribbean are highly susceptible to LD. Both Schuiling *et al.* (1992) and Kullaya *et al.* (1997) concluded that no exotic varieties have exploitable resistance to LD. However, the 'Pemba Red Dwarf' as well as a local tall ecotype show promising levels of resistance. Most recently, *Diastrombus mkurangai* and *Meenoplus* spp. (Homoptera: Meenoplidae) have been implicated as vectors of LD in Tanzania (Mpunami *et al.*, 2000).

Kalimantan wilt

This disease presently is restricted to the district of Kotawaringin Timur of Central Kalimantan where it threatens at least 17,000 ha of coconut grown by smallholders in settlements forming a 5–7 km wide strip along the Sampit river. Symptoms in the area of Samuda resemble those of Natuna wilt. Immature and middle-aged nuts already present on palms may be malformed and are often elongated in shape with poorly developed endosperm (Plate 66). Death of the palm occurs 6–18 months after symptoms first appear. The disease first occurs as a focus within a planting and then spreads outward to neighbouring palms. Some palms may escape infection as new foci develop some distance away. No systematic epidemiological study of Kalimantan wilt has been undertaken, but these preliminary observations suggest the involvement of an airborne vector (Warokka, 1999).

Jones *et al.* (1999) implicated the involvement of phytoplasmas with palms affected by Kalimantan wilt and Natuna wilt based on evidence obtained from diagnostic PCR assays. No comparative studies have been carried out as yet to determine the relationship between these putative aetiological agents and other phytoplasmas from coconut palm.

Natuna wilt

Lying 4° north of the equator between Peninsular Malaysia and Borneo, Natuna Besar is the largest of the Natuna Islands with an area of 1095 km^2. Natuna wilt was first recorded in 1965 in Kelarek in the Western District (Sitepu, 1983). This lethal disease has now spread throughout Natuna Besar and the offshore islands to the southwest (Warokka, 1999). Although Natuna wilt usually affects palms that are >10 years old, symptoms have been observed on palms as young as 7 years. Non-bearing palms apparently are unaffected (Fig. 8.11).

Symptoms first appear as a pronounced wilting of the lower leaves that progresses to involve successively younger leaves. Affected leaves droop and lose their sheen before changing from green to brown, without any pronounced yellowing stage as seen with LY. While necrosis of immature inflorescences does not occur, a cessation of nut production coincides with the onset of foliar symptoms (Fig. 8.11). Immature and middle-aged nuts already present on palms may be malformed and often are elongated, with poorly developed endosperm. Other nuts may fail to develop any endosperm and consist of solid fibrous husk only. Nutfall is inconsistent and therefore not a reliable disease symptom. An internal examination of bud tissues reveals no apparent abnormality of these immature tissues during the early stages but, as symptoms progress, brown lesions appear on both the basal region of the unemerged petioles and inflorescence sheaths.

According to a 1980 disease survey, as many as 500,000 palms were affected on Natuna Besar (Sitepu, 1983). In 1997, it was estimated that the regionwide incidence of Natuna wilt was ~6–7%, although the incidence within certain localized sites approached 40% (Warokka, 1999).

Premature decline

This disorder was first recognized during 1995 in the Western and North Western Provinces of Sri Lanka, but it may have been present there much longer. Palms >20 years of age are affected. In contrast to leaf scorch decline, palms die more rapidly, usually within 1–3 years of symptom onset. Symptoms begin as a slight drooping and pale green or yellowish green colour of the lower leaves. Due to the drooping, a noticeable gap forms between the lower and upper leaves of the crown. By this stage, an acute tapering of the trunk below the crown is evident. On palms with advanced symptoms, the size of the crown is greatly diminished as a result of a marked reduction in the length of newly formed leaves. Nut production ceases and is followed by death of the palm. The cause of this disease is not known.

Red ring

Red ring was probably first described in Trinidad in the early 1900s. Known locally as root disease or fever, it was eventually named red-ring disease by Nowell (1919) who noted the internal red-ring symptom in diseased palms in Grenada and Trinidad. Red ring is one of the most important diseases of coconut palm in the Neotropics. It is found in Central America, Dominican Republic, the Lesser Antilles, southern Mexico (Yucatan Peninsula) and parts of South America (Brazil, Colombia, Ecuador, French Guyana, Guyana, Peru, Surinam and Venezuela). Annual losses of 10–15% occur in some areas (Griffiths, 1987; Giblin-Davis, 1993).

Symptoms

Symptoms on tall-type coconut palms include premature nutfall of all but mature nuts, necrosis and withering of inflorescences followed by yellowing, and senescence (browning) of progressively younger leaves (Plate 67). Leaf discoloration begins at the tips of the pinnae, moving inward to the rachis and basal petiole. Senescent leaves often break at or near the petiole and hang downward. A transverse section of the lower stem invariably reveals a diagnostic orange to red–brown ring 2–6 cm in width within 2–6 cm of the stem periphery (Plate 67). In longitudinal section, internal stem discoloration is usually continuous throughout its length as two discrete, peripheral bands that fuse in the basal stem while forming discontinuous streaks or spots in the upper stem. The apical meristem is not affected, while the normally white cortical tissues of roots turn orange to light red and desiccate. An internal red–brown discoloration that extends for varying distances into leaf bases may also be evident. Full development of internal symptoms occurs before the appearance of any recognizable external symptoms (Griffiths, 1987; Griffiths and Koshy, 1990).

Symptoms vary widely according to the environmental conditions, age and palm type. For example, in dwarf coconut ecotypes and dwarf × tall hybrids, leaves become brown and desiccate rather than turn yellow. Also, internal discoloration is brown rather than the orange or red–brown colour evident in most tall ecotypes. Severe structural damage to the crown of diseased palms results from larval feeding by the weevil vector, often causing the crown to topple over under its own weight. Coconut palms 3–10 years old are affected most frequently and die within a matter of months. Infection of older palms leads to a more protracted decline marked by the appearance of progressively smaller leaves and a cessation of normal fruit production (Hoof and Seinhorst, 1962).

Causal organism

The disease is caused by the red-ring or coconut palm nematode, *Bursaphelenchus cocophilus*, a stylet-bearing member of the order *Aphelenchida*. This obligate parasite is vectored to coconut and other palm species mostly by adults of the American palm weevil, *Rhynchophorus palmarum*, and also by the sugarcane weevils *Dynamis borassi* and *Metamasius hemipterus* (Giblin-Davis, 1993).

Both sexes of mature *B. cocophilus* are <15.5 μm in diameter and 775–1369 μm in length. Stylet length is ~11–15 μm in adults (Gerber *et al.*, 1989). Survival of the nematode depends upon the dispersal stage juveniles (third stage) that may persist intercellularly in coconut tissues for as long as 3 months (Griffiths, 1987). Third-stage juveniles are 700–920 μm long and have a pointed tail with or without a mucron. Up to 11,000 nematodes can be recovered per gram of discoloured stem parenchyma tissues, leaf petioles and roots of coconut, or from *R. palmarum* (Blair, 1964; Gerber and Giblin-Davis, 1990). Survivorship of *B. cocophilus* is poor in soil and water (<7 days), but has been prolonged experimentally on the mesocarp of near mature coconut fruits or in excised leaf stalks (Giblin-Davis *et al.*, 1989).

Epidemiology

Palm and sugarcane weevils colonize affected palms. Their larvae become associated with the dispersal stage of *B. cocophilus*, and carry it through their metamorphosis. Adult weevils then vector the nematodes to other palms (Giblin-Davis, 1993). The incidence of *R. palmarum* females that are infested with nematodes upon emergence from diseased coconut palms can be high. In a study in Trinidad, all were found to contain nematodes and >47% harboured >1000 nematodes (Gerber and Giblin-Davis, 1990). Seasonally abundant, *R. palmarum* is a strong flier capable of rapid long-distance travel. Weevils are attracted to moisture in the leaf bases of palms and by semiochemicals emitted from palm wounds (Giblin-Davis *et al.*, 1996). Visiting males produce 7- to 9-carbon methyl alcohol aggregation pheromones that attract weevils of both sexes to the palm for mating and oviposition (Giblin-Davis *et al.*, 1996). Transmission of *B. cocophilus* seems to occur primarily during oviposition. The nematodes are highly motile and only 10–50 are needed in wounds for disease establishment (Goberdhan, 1964). Although nematodes are excluded from vascular tissues, occluding tyloses that develop in xylem vessels close to invaded tissues impair water uptake and translocation in palms (Giblin-Davis, 1993). In 4- to 5-year-old coconut palms, experimental inoculations with dispersal stage nematodes induced external symptoms in some palms within 28 days, with peak populations of nematodes occurring in tissues at 42 days. Test palms succumbed within 10 weeks (Goberdhan, 1964; Griffiths, 1987).

Management

Promising resistance has been identified in some coconut ecotypes (Dean, 1979) but has not been exploited. Disease control relies primarily on prompt removal and destruction of diseased or *R. palmarum*-infested palms. A rigorous phytosanitation programme coupled with mass trapping of *R. palmarum* using baits and a synthetic aggregation pheromone, Rhyncholure (racemic 6-methyl-2-hepten-4-ol), reduces weevil numbers and dispersion patterns while lowering incidence in oil palm plantations (Oehlschlager *et al.*, 1995). Whether a similar strategy would be effective in coconut palm plantations is untested.

Root (wilt) disease

Root (wilt) disease first appeared in the late 1800s (Koshy, 1999). According to a survey conducted during 1984–85, the disease was present in the eight southern districts of Kerala, and in small isolated pockets in some northern districts and Tamil Nadu. Disease incidence varies considerably, and is highest in Kottayam (75.6%) and lowest in Thiruvananthapuram (1.5%). Both pre-bearing and mature palms are affected and, although the disease is not fatal, it causes estimated losses of 968 million nuts annually (Solomon, 1994). Although the disease is thought to be caused by a phytoplasma, studies on it have yielded conflicting results.

Symptoms

The primary symptom is an abnormal bending of the leaflets that is termed ribbing or flaccidity. Yellowing and marginal necrosis of leaflets of the outer whorls also occur. In seedlings and juvenile palms, foliar yellowing is uncommon and symptoms begin with a rotting of the spear leaf followed by leaf flaccidity. In some cases, flaccidity may precede spindle rot. Root necrosis is extensive in affected palms, and the capacity of the palm to produce new roots is drastically reduced. Shedding of buttons and immature nuts, a reduction in the number and size of leaves, and drying of the spathes and a necrosis of spikelets from the tip down on unopened inflorescences may also occur. Leaves break at the tips, then yellow, desiccate and collapse before falling off. Abnormal structural changes occur in vascular tissues. Tyloses are induced in xylem vessels of roots on diseased palms. In its advanced stages, the palm crown is markedly smaller, fewer or no inflorescences are produced, and flowering on juvenile palms is delayed.

Causal agent

There have been numerous studies to identify a causal agent, and fungi, bacteria, viruses and nematodes have all been implicated (Solomon, 1997). Phytoplasmas have been observed in transmission electron micrographs of sieve elements of some affected palms (Solomon *et al.*, 1987). Further support for a phytoplasma aetiology was obtained following transmission of the disease by *Stephanitis typical*, a lacewing in the *Tingidae*. (Interestingly, all known phytoplasma vectors belong to genera of leafhoppers, planthoppers or psyllids (*Auchenorrhyncha*).) Furthermore, remission of symptoms occurred in 53% of the affected palms that were treated with oxytetracycline. The disease was also graft transmitted from coconut to periwinkle with dodder laurel (Sasikala *et al.*, 1988). A serodiagnostic test (Solomon *et al.*, 1987) and a physiological test based on stomatal resistance (Rajagopal *et al.*, 1988) have been standardized for detecting the disease, but attempts to verify the presence of phytoplasma in affected palms by molecular diagnostics such as PCR have not been successful.

Epidemiology

Spread is faster, although erratic, among palms grown in sandy loams or alluvial soils. Higher disease incidence has been observed in low-lying waterlogged areas bordering rivers and canals.

Management

Based on nutritional studies conducted at the Central Plantation Crops Research Institute (CPCRI) in Kerala, India, it is recommended that each affected palm be treated with 500 g N, 300 g P_2O_5, 1000 g K_2O and MgO annually to maintain its productivity. Viable economic yields can be obtained from diseased palms by spraying fungicides against leaf rot, adding of organic matter, raising green manure crops beneath trees, irrigating during summer months and adopting inter- and mixed-cropping and mixed farming (Nair *et al.*, 1991; Solomon, 1994). Management practices advocated by CPCRI include eradicating severely affected palms and those that are affected prior to flowering.

Stem bleeding disease

Stem bleeding disease is found worldwide, and old plantings are affected most. Symptoms on the trunk begin as a soft, yellow decay (Simone, 1994b). Affected areas are dark and blacken with age. A reddish brown or rust-coloured liquid may ooze from infections, and may extend several feet down the trunk discolouring it black as it dries. Pith tissues of affected roots decay. Infections may coalesce and cause heart rot and, as the disease progresses, lower leaf pinnae may die and defoliation and death may occur.

The causal fungus, *Chalara paradoxa*, is described in Chapter 1. Its teleomorph, *Ceratocystis paradoxa*, is uncommon on palms. The pathogen infects primarily the trunk and roots through wounds. Chlamydospores of the fungus are important for its survival, and it is moved via rainsplash and air currents, as well as in infested soil by birds or rodents.

Wounding of the host should be avoided, and affected palms should be removed and destroyed (Simone, 1994b). When infections are restricted in size, they can be removed and treated with benomyl and a wood preservative.

Sulawesi yellows

A leaf yellowing disease of coconuts was reported from central Sulawesi in 1989. It has an unknown aetiology and is most prevalent in the areas of Donggal (Kecamatan Ampibabo) and Poso (Kecamatan Tojo). In these districts, ~175,000 ha of coconuts are grown on smallholder farms (Simatupang, 1999).

Typically, symptoms are first evident as a yellowing at the tips of leaflets on older leaves that progresses to the lamina. Yellowing progresses to involve successively younger leaves, but not to the spear, which remains green. Foliar discoloration may be accompanied by leaf necrosis. As new leaves are produced, they do not expand fully and remain small, imparting a stunted appearance to the crown. Nut production declines progressively. Nuts are small in size or misshapen. The syndrome affects both tall and hybrid varieties equally, but appears to be non-lethal (Simatupang, 1999). Its cause is not known.

Tatipaka disease

The disease first appeared in the Tatipaka village of Razole Taluk in the East Godavari District of Andhra Pradesh, India, after the cyclone in 1949, and has since been identified in the Srikakulam and Nellore Districts.

Symptoms

Tatipaka affects palms between 20 and 60 years of age. The onset of symptoms is characterized by sudden, profuse growth. Leaf production increases slightly, as does yield.

The crown assumes a rosette appearance as leaves lengthen and become dark green. As this phase recedes, successive leaves are reduced in both size and number and become light green with chlorotic spots before slowly yellowing. In some cases, leaflets adhere together, resulting in a fasciated appearance. Expanded leaves often exhibit a bow-like appearance due to bending of the rachis; which may also twist. Senescent leaves remain attached and are found hanging below the crown instead of falling to the ground. Spathes are very small and contain few rachillae, and inflorescence production is progressively reduced. Fruit bunches contain both normal and atrophied nuts, or may be barren. Nuts become rounded with a soft mesocarp, whereas atrophied nuts consist only of husk. At an advanced stage, palms display extensive root rot and a reduced capacity for root regeneration. Tapering of the trunk and bending into an 'S' shape is common. Palms shed yellowed leaves in slow succession and may eventually die (Rethinam *et al*., 1989).

Causal agent

Ultrastructural examination of new roots, beneath the meristem, petioles of developing leaves and rachilla of new inflorescences has revealed the presence of phytoplasmas in phloem sieve tubes of some affected palms. Transmission electron microscopy observations have been corroborated by light microscopy using Dienes's stain, or fluorescent microscopy employing aniline blue as fluorochrome. Temporary remission of symptoms on diseased palms after treatment with tetracycline hydrochloride further supports the concept of a phytoplasma aetiology for the disease.

Management

Resistance has been noted in the dwarf cultivar 'Gangabondam', which is a good parent for the production of hybrids. Attempts to control the disease by foliar and soil application of various chemicals, hormones and nutrients were not successful. The primary control strategy is to arrest the disease's spread by identifying and roguing diseased palms. Eradication of 87% of the affected palms has been completed, and monitoring is continuing (Rajamannar, 1994).

Acknowledgements

Florida Agricultural Experiment Station Journal Series No. R-08777.

References

Abdullah, F. (2000) Spatial and sequential mapping of the incidence of basal stem rot of oil palms (*Elaeis guineensis*) on a former coconut (*Cocos nucifera*) plantation. In: Flood, J., Bridge, P.D. and Holderness, M. (eds) Ganoderma *Diseases of Perennial Crops*. CAB International, Wallingford, UK, pp. 183–194.

Agnihothrudu, V. and Barua, G.C.S. (1965) Notes on fungi from North-East India. XVIII. The occurrence of *Pestalotia palmarum* Cooke (sensu Guba) and the *Rynchosphaeria* perfect stage in Assam. *Journal of the Indian Botanical Society* 44, 290–292.

Aisagbonhi, C.I. and Kolade, K.O. (1994) A survey of insects associated with coconut palms in NIFOR Benin with emphasis on possible vectors of bronze leaf wilt ('Awka wilt') disease. *Principes* 38, 95–99.

Ashburner, G.R., Córdova, I.I., Oropeza, C., Illingworth, R. and Harrison, N.A. (1996) First report of coconut lethal yellowing disease in Honduras. *Plant Disease* 80, 960.

Been, B.O. (1995) Integrated pest management for the control of lethal yellowing, quarantine, cultural practices and optimal use of hybrids. In: Oropeza, C., Howard, F.W. and Ashburner, G.R. (eds) *Lethal Yellowing, Research and Practical Aspects*. Kluwer Academic Publishers, Dordrecht, The Netherlands, pp. 101–109.

Bhaskaran, R. (2000) Management of basal stem rot disease of coconut caused by *Ganoderma lucidum*. In: Flood, J., Bridge, P.D. and Holderness, M. (eds) Ganoderma *Diseases of Perennial Crops*. CAB International, Wallingford, UK, pp. 121–128.

Bhaskaran, R. and Ramanathan, T. (1984) Occurrence and spread of Thanjavur wilt disease of coconut. *Indian Coconut Journal* 15, 1–3.

Bigornia, A.E. (1977) Evaluation and trends of researches on the coconut cadang-cadang disease. *Philippine Journal of Coconut Studies* 2, 5–34.

Blair, G.P. (1964) Red ring disease of the coconut palm. *Journal of the Agricultural Society of Trinidad and Tobago* 64, 31–49.

Boccardo, G. (1985) Viroid aetiology of Tinangaja and its relationship with cadang-cadang disease of coconut. In: Maramorosch, K. and McKelvey, J.J. (eds) *Subviral Pathogens of Plants and Animals: Viroids and Prions*. Academic Press, Orlando, Florida, pp. 75–99.

Broschat, T.K., Harrison, N.A. and Donselman, H. (2002) Losses to lethal yellowing cast doubt on coconut cultivar resistance. *Palms* 46, 185–189.

Bryce, G. (1921) Report of the work of botanical and mycological division. In: *Report of the Ministry of Agriculture 1920*. Ceylon Department of Agriculture, pp. 13–15.

Butler, E.J. (1906) Some diseases of palms. *Agricultural Journal of India* 1, 299–310.

Calvez, C., Renard, J.L. and Marty, G. (1980) Tolerance of the hybrid coconut Local × Rennell to New Hebrides disease. *Oléagineux* 35, 443–451.

Cook, A.A. (1971) *Diseases of Tropical and Subtropical Fruits and Nuts*. Hafner Press, New York.

Cordova, I., Oropeza, C., Almeyda, H. and Harrison, N.A. (2000) First report of a phytoplasma-associated leaf yellowing syndrome of palma jipi plants in southern Mexico. *Plant Disease* 84, 807.

Davis, T.A. (1962) Coconut withering diseases of Ceylon. *Coconut Research Institute of Sri Lanka Bulletin No. 22*. Coconut Research Institute, Lunuwila, Sri Lanka.

Dean, C.G. (1979) Red ring disease of *Cocos nucifera* L. caused by *Rhadinaphelenchus cocophilus* (Cobb, 1919) Goodey, 1960. An annotated bibliography and review. *Technical Communication No. 47*. Commonwealth Institute of Helminthology, St Albans, UK.

Desmier de Chenon, R. (1984) Research on the genus *Lincus* Stal, Hemiptera Pentatomidae Discocephalinae, and its possible role in the transmission of the marchitez of oil-palm and hart-rot of coconut. *Oléagineux* 39, 1–4.

Dollet, M. (1984) Plant diseases caused by flagellated protozoa (*Phytomonas*). *Annual Review of Phytopathology* 22, 115–132.

Dollet, M., Giannotti, J., Renard, J.L. and Ghosch, S.K. (1977) Etude d'un jaunissement létal des coctiers au Cameroun: la maladie de Kribi. Observations d'organismes de type mycoplasmes. *Oléagineux* 32, 317–322.

Eden-Green, S.J. (1997a) An updated survey of coconut diseases of uncertain aetiology. In: Eden-Green, S.J. and Ofori, F. (eds) *Proceedings of an International Workshop on Lethal Yellowing-like Diseases of Coconut*, Elmina, Ghana, November 1995. Natural Resources Institute, Chatham, UK, pp. 77–84.

Eden-Green, S.J. (1997b) History, distribution and present status of lethal yellowing-like disease of palms. In: Eden-Green, S.J. and Ofori, F. (eds) *Proceedings of an International Workshop on Lethal Yellowing-like Diseases of Coconut*, Elmina Ghana, November, 1995. Natural Resources Institute, Chatham, UK, pp. 9–25.

Ekanayake, U.B.M. (1963) Leaf scorch disease of coconut – a preliminary note. *Ceylon Coconut Review* 3, 81.

Elliott, M.L. and Broschat, T.K. (2001) Observations and pathogenicity experiments on *Ganoderma zonatum* in Florida. *Palms* 45, 62–72.

Erwin, D.C. and Ribiero, O.K. (1996) *Phytophthora Diseases Worldwide*. APS Press, St Paul, Minnesota.

Fernando, L.C.P. (1999) Decline disorders of coconut in Sri Lanka. In: Allorerung, D., Harries, H. C., Jones, P. and Warokka, J.S. (eds) *Proceedings of the International Workshop on Lethal Diseases of Coconut Caused by Phytoplasmas in South East Asia*. APCC Publishers, Jakarta, Indonesia, pp. 59–66.

Frison, E.A., Putter, C.A.J. and Diekmann, M. (1993) *FAO/IBPGR Technical Guidelines for the Safe Movement of Germplasm*. Food and Agriculture Organization of the United Nations, Rome/ International Board for Plant Genetic Resources, Rome.

Gerber, K. and Giblin-Davis, R.M. (1990) Association of the red ring nematode and other nematode species with the palm weevil, *Rhynchophorus palmarum*. *Journal of Nematology* 22, 143–149.

Gerber, K., Giblin-Davis, R.M., Griffith, R., Escobar-Goyes, J. and D'Ascoli Cartaya, A. (1989) Morphometric comparisons of geographic and host isolates of the red ring nematode, *Rhadinaphelenchus cocophilus*. *Nematropica* 19, 151–159.

Giblin-Davis, R.M. (1993) Interaction of nematodes with insects. In: Khan, M.W. (ed.) *Nematode Interactions*, Chapman & Hall, London, pp. 302–334.

Giblin-Davis, R.M., Gerber, K. and Griffith, R. (1989) *In vivo* and *in vitro* culture of the red ring nematode, *Rhadinaphelenchus cocophilus*. *Nematropica* 19, 135–142.

Giblin-Davis, R.M., Oehlschlager, A.C., Perez, A., Gries, G., Gries, R., Weissling, T.J., Chinchilla, C.M., Peña, J.E., Hallet, R.H., Pierce, H.D., Jr and Gonzalez, L.M. (1996) Chemical and behavioral ecology of palm weevils (Circulionidae:Rhynchophorinae). *Florida Entomologist* 79, 153–167.

Goberdhan, L. (1964) Observations on coconut palms artificially infected by the nematode *Rhadinaphelenchus cocophilus* (Cobb, 1919) Goodey, 1960. *Journal of Helminthology* 38, 25–30.

Griffiths, R. (1987) Red ring disease of coconut palm. *Plant Disease* 71, 193–196.

Griffiths, R. and Koshy, P.K. (1990) Nematode parasites of coconut and other palms. In: Luc, M., Sikora, R.A. and Bridge, J. (eds) *Plant Parasitic Nematodes in Subtropical and Tropical Agriculture.* CAB International, Wallingford, UK, pp. 363–386.

Guerrini, F., Segur, C., Gargani, D., Tibayrenc, M. and Dollet, M. (1992) An isozyme analysis of the genus *Phytomonas*: genetics, taxonomic and epidemiologic significance. *Journal of Protozoology* 39, 516–521.

Hanold, D. and Randles, J.W. (1997) *Report on ACIAR-Funded Research on Viroids and Viruses of Coconut Palm and Other Tropical Monocotyledons 1985–1993*. ACIAR Monograph 45, Canberra, Australia. http://www.aciar.gov.au/publications/monographs/45/index.htm

Harries, H.C. (1978) Lethal yellowing disease of coconuts in global perspective. *Philippine Journal of Coconut Studies* 3, 1–4.

Harries, H.C. (1992) Biogeography of the coconut *Cocos nucifera* L. *Principes* 36, 155–162.

Harries, H.C. (1995) The genetics of durable resistance to lethal yellowing disease. In: Oropeza, C. Howard, F.W. and Ashburner, G.R. (eds) *Lethal Yellowing, Research and Practical Aspects.* Kluwer Academic Publishers, Dordrecht, The Netherlands, pp. 149–171.

Harries, H.C. (2001) Coconut varieties and lethal yellowing: a regional perspective for the Americas. *Palms* 45, 148–150.

Harrison, N.A. and Oropeza, C. (1997) Recent studies on detection of lethal yellowing disease phytoplasmas in the Americas. In: Eden-Green, S.J. and Ofori, F. (eds) *Proceedings of an International Workshop on Lethal Yellowing-like Diseases of Coconut*, Elmina, Ghana, November, 1995. Natural Resources Institute, Chatham, UK, pp. 221–234.

Harrison, N.A., Bourne, C.M., Cox, R.L., Tsai, J.H. and Richardson, P.A. (1992) DNA probes for detection of the mycoplasmalike organisms associated with lethal yellowing disease of palms in Florida. *Phytopathology* 82, 216–224.

Harrison, N.A., Richardson, P.A., Kramer, J.B. and Tsai, J.H. (1994) Detection of the mycoplasma-like organism associated with lethal yellowing disease of palms in Florida by polymerase chain reaction. *Plant Pathology* 43, 998–1008.

Harrison, N.A., Womack, M. and Carpio, M.L. (2002) Detection and characterization of a lethal yellowing (16SrIV) group phytoplasma in Canary Island date palms affected by lethal decline in Texas. *Plant Disease* 86, 676–681.

Haseloff, J., Mohamed, N.A. and Symons, R.H. (1982) Viroid RNAs of cadang-cadang disease of coconuts. *Nature* 299, 316–321.

Hodgson, R.A.J. and Randles, J.W. (1998) Specific identification of coconut tinangaja viroid for differential field diagnosis of viroids in coconut palm. *Phytopathology* 88, 774–781.

Hodgson, R.A.J. and Randles, J.W. (1999) Detection of coconut cadang-cadang viroid-like sequences. In: Oropeza, C., Verdeil, J.L., Ashburner, G.R., Cardena, R. and Santamaría, J.M. (eds) *Current Advances in Coconut Biotechnology*. Kluwer Academic Publishers, Dordrecht, The Netherlands, pp. 227–246.

Holliday, P. (1980) *Fungus Diseases of Tropical Crops*. Cambridge University Press, Cambridge.

Hoof, H.A. and Seinhorst, J.W. (1962) *Rhadinaphelenchus cocophilus* associated with little leaf of coconut and oil palm. *Tijdschrift over Plantenziekten* 68, 251–256.

Howard, F.W. (1983) World distribution and possible geographic origin of palm lethal yellowing disease and its vectors. *FAO Plant Protection Bulletin* 31, 101–112.

Howard, F.W. and McCoy, R.C. (1980) Reduction in spread of mycoplasma-like organism associated lethal decline of the palm *Veitchia merrillii* by use of insecticides. *Journal of Economic Entomology* 73, 268–270.

Howard, F.W., Norris, R.C. and Thomas, D.L. (1983) Evidence of transmission of palm lethal yellowing agent by a planthopper, *Myndus crudus* (Homoptera, Cixiidae). *Tropical Agriculture, Trinidad* 60, 168–171.

Howard, F.W., Williams, D.S. and Norris, R.C. (1984) Insect transmission of lethal yellowing to young palms. *International Journal of Entomology* 26, 331–338.

Imperial, J.S., Bautista, R.M. and Randles, J.W. (1985) Transmission of cadang-cadang viroid to six species of palms by inoculation with nucleic acid extracts. *Plant Pathology* 34, 391–401.

Jackson, G.V.H. and Firman, I.D. (1982) Seedborne marasmioid fungi on coconut. *Plant Pathology* 31, 187–188.

Jackson, G.V.H. and McKenzie, E.H.C. (1988) *Marasmiellus cocophilus* on coconuts in Solomon Islands. *Plant Protection Bulletin, India* 36, 91–97.

Jayasekara, C., Ramasinghe, C.S. and Silva, I.R.S. (1990) *Report of the Coconut Research Institute for 1990.* Lunuwila, Sri Lanka, pp. 121–122.

Jones, P., Cronje, C.P.R. and Warokka, J.S. (1999) Investigations into coconut diseases of uncertain etiology in Indonesia. *Proceedings of the 5th International Conference on Plant Pathology in the Tropics.* Kuala Lumpur, Malaysia, March 15–18, 1999, pp. 39–42.

Joseph, T. and Radha, K. (1975) Role of *Phytophthora palmivora* in bud rot of coconut. *Plant Disease Reporter* 59, 1014–1017.

Julia, J.F. (1982) *Myndus taffini* (Homoptera: Cixiidae), vector of foliar decay of coconuts in Vanuatu. *Oléagineux* 37, 409–414.

Kastelein, P. and Parsadi, M. (1986) Phytomonas flagellates (Trypanosomatidae) in the phloem of a diseased *Bentinckia nicobarica* (Kurz) Becc. and *Roystonea regia* (H.B.K.) Cookpalm. *Surinaamse-Landbouw* 34, 1–3, 8–14.

Keese, P., Osorio-Keese, M.E. and Symons, R.H. (1988) Coconut Tinangaja viroid: sequence homology with coconut cadang-cadang viroid and other potato spindle tuber viroid related RNAs. *Virology* 162, 508–510.

Kirthisnghe, J.K.F (1966) Leaf scorch decline of coconut palm in Ceylon. *Sequence of Events, Experimental Work etc. in Chronological Order up to December 1966.* Coconut Research Institute, Lunuwila, Sri Lanka.

Koshy, P.K. (1999) Root wilt and Tatipaka diseases of coconut in India. In: Allorerung, D., Harries, H.C., Jones, P. and Warokka, J.S. (eds). *Proceedings of the International Workshop on Lethal Diseases of Coconut Caused by Phytoplasmas in South East Asia.* APCC Publishers, Jakarta, Indonesia, pp. 27–35, 55–58.

Kullaya, A., Mpunami, A., Harries, H.C., Schuiling, M. and Kaiza, D.A. (1997) Lethal disease resistance screening in Tanzania and prospects for rehabilitation. In: Eden-Green, S.J. and Ofori, F. (eds) *Proceedings of an International Workshop on Lethal Yellowing-Like Diseases of Coconut*, Elmina Ghana, November, 1995. Natural Resources Institute, Chatham, UK, pp. 163–172.

Louise, C., Dollet, M. and Mariau, D. (1986) Research into hartrot of the coconut, a disease caused by Phytomonas (Trypanosomatidae), and into its vector *Lincus* sp. (Pentatomidae) in Guiana. *Oléagineux* 41, 437–449.

Mahindapala, R. and Chandrasena, A.M. (1975) Yield and drupe characteristics of coconut palms affected by leaf scorch decline. *Ceylon Coconut Quarterly* 26, 73–76.

Marche, S., Roth, C., Manohar, S.K., Dollet, M. and Baltz, T. (1993) RNA virus-like particles in pathogenic plant trypanosomatids. *Molecular and Biochemical Parasitology* 57, 261–267.

Marche, S., Roth, C., Phillipe, H., Dollet, M. and Baltz, T. (1995) Characterization and detection of plant trypanosomatids by sequence analysis of the small subunit ribosomal RNA gene. *Molecular and Chemical Parasitology* 71, 15–26.

McCoy, R.E., Carroll, V.J., Poucher, C.P. and Gwin, G.H. (1976) Field control of coconut lethal yellowing with oxytetracycline hydrochloride. *Phytopathology* 66, 1148–1150.

McCoy, R.E., Howard, F.W., Tsai, J.H., Donselman, H.M., Thomas, D.L., Basham, H.G., Atilano, R.A., Eskafi, F.M., Britt, L. and Collins, M.E. (1983) Lethal yellowing of palms. *University of Florida Agricultural Experiment Station Bulletin No. 834.* University of Florida, Gainesville, Florida.

Menara, A., Dollet, M., Gargani, D. and Louise, C. (1988) Culture *in vitro* sur cellules d'invertébrés de *Phytomonas* sp. (Trypanosomatidae) associés au Hartrot, maladie du cocotier. *Compte Rendu de l'Academie des Science, Paris, Serie III* 307, 597–602.

Menon, K.V.P. and Pandalai, K.M. (1960) *The Coconut Palm.* Indian Central Coconut Committee, Ernakulam. India.

Miller, R.N.G., Holderness, M., Bridge, P.D., Chung, G.F. and Zarakia, M.H. (1999) Genetic diversity of *Ganoderma* in oil palm plantings. *Plant Pathology* 48, 595–603.

Mordue, J.E.M. and Holliday, P. (1971) Pestilotiopsis palmarum. *CMI Descriptions of Pathogenic Fungi and Bacteria No. 319.* Commonwealth Mycological Institute, Kew, UK.

Mohamed, N.A., Bautista, R.M., Buenaflo, G.G. and Imperial, J.S. (1985) Purification and infectivity of the coconut cadang-cadang viroid. *Phytopathology* 75, 79–84.

Moncalvo, J.M. (2000) Systematics of *Ganoderma*. In: Flood, J., Bridge, P.D. and Holderness, M. (eds) *Ganoderma Diseases of Perennial Crops*. CAB International, Wallingford, UK, pp. 23–45.

Mpunami, A., Tymon, A., Jones, P. and Dickinson, M.J. (1999) Genetic diversity in the coconut lethal yellowing disease phytoplasmas of East Africa. *Plant Pathology* 48, 109–114.

Mpunami, A., Tymon, A., Jones, P. and Dickinson, M.J. (2000) Identification of potential vectors of the coconut lethal disease phytoplasma. *Plant Pathology* 49, 355–361.

Muller, E., Ahomadegbe, J.C., Coulaud, D., Gargani, D., Fernandez-Becerra, C. and Dollet, M. (1995) Variability in kinetoplast DNA from plant trypanosomatids responsible for Hartrot and Marchitez diseases. *Phytopathology* 85, 942–947.

Nair, M.K., Nambiar, K.K.N., Koshy, P.K. and Jayasankar, N.P. (1991) *Coconut (Root) Wilt Disease*. Central Plantation Crops Research Institute, Kasaragod, Kerala, India.

Nowell, W. (1919) Red ring or root disease of coconut palms. *West Indian Bulletin* 17, 189–192.

Nutman, F.J. and Roberts, P.M. (1955) Lethal yellowing, the 'unknown disease' of coconut palms in Jamaica. *Empire Journal of Agriculture* 23, 257–267.

Oehlschlager, A.C., MacDonald, R.S., Chinchilla, C.M. and Patschke, S.N. (1995) Influence of pheromone-based mass-trapping system on the distribution of *Rhynchophorus palmarum* (Coleoptera: Circulionidae) in oil palm. *Environmental Entomology* 24, 1005–1012.

Oropeza, C. and Zizumbo, D. (1997) A history of lethal yellowing in Mexico. In: Eden-Green, S.J. and Ofori, F. (eds) *Proceedings of an International Workshop on Lethal Yellowing-like Diseases of Coconut*, Elmina, Ghana, November, 1995. Natural Resources Institute, Chatham, UK, pp. 69–76.

Park, M. (1927) On the occurrence of coconut *Rhizoctonia bataticola* (Taub.) Butler. *Tropical Agriculturalist* 60, 7–8.

Parthasarathy, M.V., van Slobbe, W.G. and Soudant, C. (1976) Trypanosomatid flagellate in the phloem of diseased palms. *Science* 192, 1346–1348.

Peries, O.S. (1968) Studies on leaf scorch decline of coconut palm. *Ceylon Coconut Quarterly* 19, 109–115.

Peries, O.S. (1974) Ganoderma basal stem rot of coconut: a new record of the disease in Sri Lanka. *Plant Disease Reporter* 58, 293–295.

Persley, G.J. (1992) *Replanting the Tree of Life*. CAB International, Wallingford, UK.

Plavsic-Banjac, B., Hunt, P. and Maramorosch, K. (1972) Mycoplasma-like bodies associated with lethal yellowing disease of coconut palms. *Phytopathology* 62, 298–299.

Ploetz, R., Harrison, N. and Jones, P. (1999) Diseases of coconut (*Cocos nucifera* L.). In: *Common Names of Plant Diseases*, http://www.aps.org

Price, W.C. (1971) Cadang-cadang of coconut – a review. *Plant Science* 3, 1–13.

Purseglove, J.W. (1972) *Tropical Crops: Monocotyledons*. Longman Group, Harlow, UK. .

Rajagopal, V., Sasikala, M., Amma, B.S.K., Chempakam, B. and Rawther, T.S.S. (1988) Early diagnostic techniques on the root (wilt) disease of coconut in India. *Philippine Journal of Coconut Studies* 13, 31–35.

Rajamannar, M. (1994) Tatipaka disease of coconut. In: Chadha, K.L. and Rehinam, P. (eds) *Advances in Horticulture, Volume 10, Plantation and Spices Crops Part 2*. Vedam eBooks P Ltd, New Delhi, India, pp. 921–937.

Rajapakse, C.N.K. and Fernando, L.C.P. (1997) Leaf scorch decline of coconut in Sri Lanka: a review. In: Eden-Green, S.J. and Ofori, F. (eds) *Proceedings of an International Workshop on Lethal Yellowing-like Diseases of Coconut*, Elmina, Ghana, November, 1995. Natural Resources Institute, Chatham, UK, pp. 97–106.

Randles, J.W. (1975) Association of two ribonucleic acids species with cadang-cadang disease of coconut palm. *Phytopathology* 65, 163–167.

Randles, J.W. and Hanold, D. (1989) Coconut foliar decay virus particles are 20-nm icosahedra. *Intervirology* 30, 177–180.

Randles, J.W., Hanold, D. and Julia, J.F. (1987) Small circular single-stranded DNA associated with foliar decay disease of coconut palm in Vanuatu. *Journal of General Virology* 68, 273–280.

Randles, J.W., Miller, D.C., Morin, J.P., Rhode, W. and Hanold, D. (1992) Localisation of coconut foliar decay virus in coconut palm. *Annals of Applied Biology* 121, 601–617.

Randles, J.W., Boccardo, G., Retuerma, M.L. and Rillo, E.P. (1997) Transmission of the RNA species associated with cadang-cadang of coconut palm, and the insensitivity of the disease to antibiotics. *Phytopathology* 67, 1211–1216.

Randles, J.W., Hanold, D., Pacumbaba, F.P. and Rodriguez, M.J.B. (1997) Cadang-cadang disease of coconut palm – an overview. In: Hanold, D. and Randles, J.W. (eds) *Report on ACIAR-funded Research on Viroids and Viruses of Coconut Palm and Other Tropical monocotyledons 1985–1993.* ACIAR Monograph 45 (electronically published).

Randles, J.W., Weffels, E., Hanold, D., Miller, D.C., Morin, J.P. and Rhode, W. (1999) Detection and diagnosis of coconut foliar decay disease. In: Oropeza, C., Verdeil, J.L., Ashburner, G.R., Cardena, R. and Santamaría, J.M. (eds) *Current Advances in Coconut Biotechnology.* Kluwer Academic Publishers, Dordrecht, The Netherlands, pp. 247–258.

Rethinam, P., Rajamannar, M. and Narasimhachari, C.L. (1989) Tatipaka disease of coconut in Andhra Pradesh. *Indian Coconut Journal* 20, 1–4.

Rodriguez, M.J.B. (1993) Molecular variation in coconut cadang-cadang viroid (CCCVd). PhD thesis, University of Adelaide, Adelaide, South Australia.

Rodriguez, M.J.B. and Randles, J.W. (1993) Coconut cadang-cadang viroid (CCCVd) mutants associated with severe disease vary in both pathogenicity domain and the central conserved region. *Nucleic Acids Research* 212, 2771.

Rohde, W., Randles, J.W., Langridge, P. and Hanold, D. (1990) Nucleotide sequence of a circular single-stranded DNA associated with coconut foliar decay virus. *Virology* 176, 648–651.

Ryvarden, L. (2000) Studies in neotropical polypores. 2: A preliminary key to neotropical species of *Ganoderma* with a laccate pileus. *Mycologia* 92, 180–191

Sasikala, M., Mathen, K., Govindankutty, M.P., Solomon, J.J. and Geetha, L. (1988) Transmission of mycoplasma-like organisms from *Cocos nucifera* with root (wilt) to *Catharanthus roseus* by *Cassytha filiformis. Netherlands Journal of Plant Pathology* 94, 191–194.

Schuiling, M., Mpunami, A., Kaiza, D.A. and Harries, H.C. (1992) Lethal disease of coconut palm in Tanzania. 3. Low resistance of imported germplasm. *Oléagineux* 47, 693–697.

Seo, G.-S. and Kirk, P.M. (2000) *Ganodermataceae*: nomenclature and classification. In: Flood, J., Bridge, P. and Holderness, M. (eds) Ganoderma *Diseases of Perennial Crops.* CAB International, Wallingford, UK, pp. 3–22.

Simatupang, O. (1999) The history and status of yellow disease of coconut palm in Central Sulawesi. In: Allorerung, D., Harries, H.C., Jones, P. and Warokka, J.S. (eds) *Proceedings of the International Workshop on Lethal Diseases of Coconut Caused by Phytoplasmas in South Esat Asia.* APCC Publishers, Jakarta, Indonesia, pp. 51–54.

Simone, G.W. (1994a) Ganoderma butt rot. In: Ploetz, R.C., Zentmyer, G.A., Nishijima, W.T., Rohrbach, K.G. and Ohr, H.D. (eds) *Compendium of Tropical Fruit Diseases.* APS Press, St Paul, Minnesota.

Simone, G.W. (1994b) Stem bleeding. In: Ploetz, R.C., Zentmyer, G.A., Nishijima, W.T., Rohrbach, K.G. and Ohr, H.D. (eds) *Compendium of Tropical Fruit Diseases.* APS Press, St Paul, Minnesota.

Sitepu, D. (1983) Coconut wilt in Natuna islands of Indonesia. In: Singh, K.G. (ed.) *Exotic Plant Quarantine Pests and Procedures for Introduction of Plant Materials.* Asian Plant Quarantine Centre and Training Institute, Serdang, Malaysia.

Solomon, J.J. (1994) Root (wilt) disease of coconut. In: Chadha, K.L. and Rehinam, P. (eds) *Advances in Horticulture Volume 10. Plantation and Spices Crops part 2.* Vedam eBooks P Ltd, New Delhi, India, pp. 883–898.

Solomon, J.J. (1997) Current status of research on root (wilt) disease of coconut in India. In: Eden-Green, S.J. and Ofori, F. (eds) *Proceedings of an International Workshop on Lethal Yellowing-like Diseases of Coconut*, Elmina, Ghana, November, 1995. Natural Resources Institute, Chatham, UK, pp. 85–96.

Solomon, J.J., Govindankutty, M.P. and Nienhaus, F. (1987) Association of mycoplasma-like organisms with coconut root (wilt) disease in India. *Zeitschrift fur Pflanzenkrankheiten und Pflanzenschutz* 90, 295–297.

Stamps, D.J. (1985) Phytophthora arecae. *CMI Descriptions of Pathogenic Fungi and Bacteria No. 833*, Commonwealth Mycological Institute, Kew, UK.

Steyaert, R.L. (1975a) Ganoderma boninense. *CMI Descriptions of Pathogenic Fungi and Bacteria. No. 444*, Commonwealth Mycological Institute, Kew, UK.

Steyaert, R.L. (1975b) Ganoderma zonatum. *CMI Descriptions of Pathogenic Fungi and Bacteria No. 448*, Commonwealth Mycological Institute, Kew, UK.

Thevenin, J.M., Motulo, H.F.J., Kharie, S. and Renard, J.L. (1995) Lutte chimique contre la pourriture du coeur a Phytophthora du cocotier en Indonesie. *Plantations Recherche Developpement* 2, 41–50.

Thomas, D.L (1979) Mycoplasma-like bodies associated with lethal declines of palms in Florida. *Phytopathology* 69, 928–934.

Thomas, D.L. and Norris, R.C. (1980) The use of electron microscopy for lethal yellowing diagnosis. *Proceedings of the Florida State Horticultural Society* 93, 196–199.

Tymon, A.M., Jones, P. and Harrison, N.A. (1997) Detection and differentiation of African coconut phytoplasmas: RFLP analysis of PCR-amplified 16S rDNA and DNA hybridization. *Annals of Applied Biology* 131, 91–102.

Tymon, A.M., Jones, P. and Harrison, N.A. (1998) Phylogenetic relationships of coconut phytoplasmas and the development of specific oligonucleotide PCR primers. *Annals of Applied Biology* 132, 437–452.

Uchida, J.Y., Aragaki, M., Ooka, J.J. and Nagata, N.M. (1992) Phytophthora fruit and heart rots of coconut in Hawaii. *Plant Disease* 76, 925–927.

Vakili, N.G. (1963) A leaf spotting disease of *Musa* seedlings incited by *Pestilotia palmarum*. *Plant Disease Reporter* 47, 644–646.

Venkatarayan, S.V. (1936) The biology of *Ganoderma lucidum* on areca and coconut palms. *Phytopathology* 26, 153–175.

Waters, H. (1978) A wilt disease of coconuts from Trinidad associated with *Phytomonas* sp., a sieve tube-restricted protozoan flagellate. *Annals of Applied Biology* 90, 293–302.

Waters, H. and Hunt, P. (1980) The *in vivo* three-dimensional form of a plant mycoplasma-like organism by the analysis of serial ultrathin sections. *Journal of General Microbiology* 116, 111–131.

Warokka, J.S. (1999) The impact and etiology of coconut wilt diseases in Indonesia. In: Allorerung, D., Harries, H.C., Jones, P. and Warokka, J.S. (eds) *Proceedings of the International Workshop on Lethal Diseases of Coconut Caused by Phytoplasmas in South Esat Asia*. APCC Publishers, Jakarta, Indonesia, pp. 27–35.

Zelazny, B. and Niven, B.S. (1980) Duration of the stages of cadang-cadang diseases of coconut palm. *Plant Disease* 64, 841–842.

Zelazny, B. and Pacumbaba, E. (1982) Incidence of cadang-cadang of coconut palm in the Philippines. *Plant Disease* 66, 547–549.

Zelazny, B., Randles, J.W., Boccardo, G. and Imperial, J.S. (1982) The viroid nature of the cadang-cadang disease of coconut palm. *Scientia Filipina* 2, 46–53.

9 Diseases of Date

R.C. Ploetz[1], Howard D. Ohr[2], John B. Carpenter[3] and Yaacov Pinkas[4]
[1] *University of Florida, Tropical Research and Education Center, Homestead, Florida, USA;*
[2] *University of California, Plant Pathology Department, Riverside, California, USA;*
[3] *(deceased) USDA Date and Citrus Experiment Station, Indio, California, USA;*
[4] *(deceased) ARO, The Volcani Center, Bet-Dagan, Israel*

Introduction

Date, *Phoenix dactylifera*, is an ancient crop that may have been cultivated as early as Neolithic times (Oudejans, 1976; Diamond, 1998). Its earliest record is from 4500-year-old pits of the fruit that were found in Egypt, and domestication is thought to have occurred simultaneously in several different locations between the Indus River and Atlantic Ocean ~4000 years ago. Throughout this region, dates have long been a staple food of the native peoples.

The date palm grows to 30 m in height, and has pinnately compound leaves, 4–7 m long, and basal suckers (Purseglove, 1975; Oudejans, 1976). It is dioecious and pollinated by wind and insects. Since seedling progeny are very heterozygous, most plantings are established with suckers. The ratio of female:male trees in commercial plantations is usually 50:1, and females are pollinated artificially by hand to maximize fruit set. The fruit are drupes that are borne in large infructescences, and individual trees may produce 30–100 kg of fruit year^{-1}.

Superior individuals have been selected for many years, and hundreds, or even thousands, of cultivars now exist. Deliberate attempts to improve this crop by breeding are relatively recent events (Oudejans, 1976).

World production of dates in 2000 exceeded 5.2 Mt (Anonymous, 2001). With few exceptions, most notably Oman and Sudan, major production is confined to the subtropical deserts of North Africa and the Near East (Table 9.1). Although the date palm is hardy and widely adaptable, it fruits poorly, or not at all, in humid environments. Optimum fruit production requires abundant irrigation water, 5–7 months of intense heat (means of 30°C), low humidity and no rain between pollination and harvest.

Table 9.1. The world's major producers of date.[a]

Country	Production (1000 t)
Iran	930
Egypt	890
Saudi Arabia	650
Pakistan	580
Iraq	540
Algeria	430
United Arab Emirates	318
Sudan	176
Oman	135
Libya	132

[a]Figures are from the online database of the Food and Agricultural Organization of the United Nations, and are in 1000s of metric tonnes (Anonymous, 2001).

In the Middle East, >800 uses have been documented for date palm (Purseglove, 1975). In addition to its primary role as a fruit crop, it is used for timber, thatch and to make palm wine. It is also used widely as an ornamental, including locations where the climate is too cold or wet to allow fruit to mature.

Date Diseases

There are relatively few major diseases of date, due probably to the arid conditions under which the crop is usually grown. Still, diseases can have a major impact on fruit production and the health and longevity of orchards. For example, Bayoud disease has destroyed two-thirds of the date palms in Morocco, and Khamedj can dramatically reduce the amount of fruit that is harvested in Libya and Iraq. Diseases that are caused by phytoplasmas are less widely spread, but are potentially serious due to their lethal impact. This chapter focuses primarily on the most significant diseases of date in commercial production. Less important diseases are listed in Table 9.2.

Bayoud (Fusarium wilt)

Bayoud is a lethal disease. It is also known as Fusarium wilt, but since this name can also refer to two other date palm diseases, only bayoud will be used for the present disease, in this chapter. Although bayoud currently is restricted to Algeria and Morocco, it jeopardizes date production worldwide.

The term bayoud originates from the Arabic word for white, 'abiadh', which refers to the white coloration of affected fronds (Vanachter, 1989). The disease was recognized before 1870 in the Drâa Valley of southern Morocco, and within a few years was reported in western and eastern portions of the country. By 1898, it had spread to western Algeria along trade routes between the two countries, and by 1988 20 million date palms had been killed by the disease (Vanachter, 1989; Freeman and Maymon, 2000). Bayoud is now found throughout Morocco, and has spread to western and central portions of Algeria.

Two other diseases can be confused with bayoud. Fusarium wilt of Canary Island date palm, *Phoenix canariensis*, is widespread. To date, it has been reported in Australia, France, Italy, Japan, Morocco, Tenerife and the USA (California, Florida and Nevada) (Summerell *et al.*, 2001). Since the pathogen that causes this disease, *Fusarium oxysporum* f. sp. *canariensis*, also affects *P. dactylifera* (Feather *et al.*, 1989; Ohr, 1991), it may be responsible for reports of bayoud from countries other than Algeria and Morocco (see Cook, 1975).

Table 9.2. Minor diseases and disorders of date palm.[a]

Disease	Cause	Disease	Cause
Al-Wijm	Unknown	Nematode damage	*Meloidogyne arenaria*
Barhee disorder	Unknown		*Meloidogyne hapla*
Belaat	*Phytophthora* sp.		*Meloidogyne incognita*
Black leaf spot	*Chaetosphaeropsis* sp.		*Meloidogyne javanica*
Blacknose	Physiological disorder of fruit		*Pratylenchus penetrans*
Black scald	Physical disorder	Omphalia root rot	*Omphalia tralucida*
Crosscut	Anatomical defect, fruitstock		*Omphalia pigmentata*
Diplodia disease	*Diplodia phoenicum*	Pestalotia leaf spot	*Pestalotiopsis palmarum*
	Diplodia theobromae	Rapid decline	Unknown
Faround	Unknown	Taches brunes	*Mycosphaerella tassiana* (anamorph:
Inflorescence rot	*Diplodia* spp.	(brown leaf spot)	*Cladosporium herbarum*)
	Fusarium spp.	Terminal bud rot	*Ceratocystis paradoxa* (anamorph:
	Thielaviopsis spp.		*Chalara paradoxa*)
Internal browning	Unknown		

[a]Source: Ohr and Carpenter (1993).

Another wilt disease of date palm was reported recently in Saudi Arabia (Abdalla *et al.*, 2000). *Fusarium proliferatum* was recovered from palms with wilt and dieback symptoms that were similar to those caused by bayoud. Isolates of the fungus produced severe wilt symptoms on seedlings of the 'Succary' cultivar. They also caused a fruit rot, which triggered additional concern since all isolates that were tested produced an array of mycotoxins. Although an isolate of *F. solani* that was recovered during these studies was also highly virulent on date palm seedlings, this species was uncommon and, therefore, deemed a minor problem. At this time, *F. proliferatum* and *F. solani* have not been reported as pathogens of date palm outside Saudi Arabia.

Symptoms

Symptoms of bayoud appear first on recently matured leaves (Cook, 1975; Carpenter and Elmer, 1978). Leaflets become chlorotic at the base of one side of the leaf and die progressively up the rachis to the leaf apex in a unilateral fashion (Plate 68). Necrosis then proceeds down the opposite side of the leaf until it is killed. At the same time, the adaxial side of the rachis becomes brown and its surface depressed. These external symptoms may take from a few days to several weeks to develop. Alternatively, a brownish lesion may appear on the abaxial side of the rachis and extend to the tip of the leaf, which then whitens and dies. Death of leaflets then progresses downward to the leaf base, and they may also become brown or red, depending on the cultivar that is affected. Those who are experienced with the disease can recognize a slight discoloration in recently matured leaves 1 or 2 months before the appearance of more conspicuous symptoms. In all of the above cases, the vascular system of the rachis becomes reddish brown (Fig. 9.1).

As the disease progresses, entire whorls of leaves in the canopy wither and collapse (Plate 69). New leaves are not formed, and the terminal bud and entire palm eventually die. In cross-sections of affected palms, the vascular systems are reddish brown. From the first appearance of symptoms to the death of a palm may be a month or as long as 10 years or more.

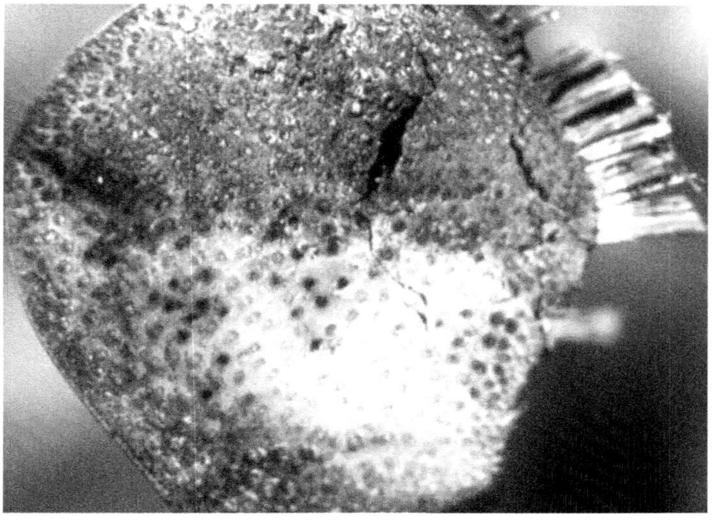

Fig. 9.1. Internal discoloration of the vascular system of the rachis of a date palm that is affected by bayoud (photo: J. Carpenter).

Causal agent

F. oxysporum f. sp. *albedinis* causes bayoud (Cook, 1975; Carpenter and Elmer, 1978). It is soilborne and causes a typical vascular wilt (see sections on *F. oxysporum* in Chapter 1 and on Panama disease in Chapter 4).

Tantaoui *et al.* (1996) assessed the relatedness of 42 isolates of the pathogen from Morocco and two from Algeria. Vegetative compatibility, restriction fragment length polymorphisms (RFLPs) of mitochondrial DNA and random amplified polymorphic DNAs (RAPDs) all demonstrated that the fungus was incredibly uniform. No RFLP or RAPD polymorphisms were observed, and all strains belonged to a single vegetative compatibility group (VCG), 0170.

To examine the pathogen's population structure further, the DNA transposable element *Fot*1 was used (Tantaoui *et al.*, 1996). Repetitive DNA patterns were produced when *Eco*RI-digested DNA of the isolates was probed with *Fot*1, and 23 distinct hybridization patterns were recognized. Four patterns accounted for >50% of the isolates, one was found twice, and 18 were represented by a single isolate. The pattern produced by the Algerian isolates was also found where the disease was first reported, the Drâa Valley. Cluster analysis grouped most of the strains at a genetic distance of 0.11, indicating a close genetic relationship. Thus, *F. oxysporum* f. sp. *albedinis* is comprised of a single clonal lineage that probably originated in Morocco and spread to Algeria.

Recently, Freeman and Maymon (2000) developed polymerase chain reaction (PCR) primers that amplified a specific 400 bp band for isolates of *F. oxysporum* f. sp. *albedinis*, but not other members of the species, including *F. oxysporum* f. sp. *canariensis*. It will be utilized by the Plant Protection Services Department of the Israel Ministry of Agriculture in efforts to exclude the pathogen from Israel.

Epidemiology

F. oxysporum f. sp. *albedinis* infects roots with pneumathodes, rootlets that originate in the pericycle and pierce the hypodermis (Belarbi-Halli and Mangenot, 1986). As few as four or five of the 1000 roots on a large palm need to be infected in order for systemic colonization of the xylem to occur.

The pathogen can be spread by water, wind and soil or in infected tissues of date, henna and lucerne (Carpenter and Elmer, 1978). The latter plants, which often are grown in date palm groves, are symptomless hosts. Bayoud becomes epidemic in groves with susceptible cultivars and abundant surface irrigation. Movement from oasis to oasis is thought to occur via infected suckers of the host and parts of the palm that are used to weave baskets, saddle pads or other products.

Management

Exclusion of the pathogen from non-infested areas is most important. The above PCR protocols could be used to ensure that foreign germplasm is free of the pathogen before it is introduced. Whenever possible, introductions should be made as micropropagated plantlets.

Once the disease is present, few options are available. Reducing the amount of water that a tree receives impedes the development of bayoud, but it also limits fruit production and is used seldom. Disease development is also retarded in saline soils, and this fact is used to manage the disease in many areas (Amir *et al.*, 1996). In Algeria, an integrated approach was suggested that used these measures and resistant cultivars (Brac de la Perriere *et al.*, 1995).

Unfortunately, most desirable cultivars are susceptible, and those that are tolerant usually produce inferior fruit (Saaidi *et al.*, 1981). Cook (1975) listed six tolerant cultivars, but only 'Takerboucht' and 'Bou Ijjou' produced acceptable fruit, and they were of a poorer quality than those of the susceptible 'Deglet Noor' and 'Medjool'. It is hoped that resistant genotypes eventually will be bred that produce desirable fruit (Sedra and Besri, 1994).

Black scorch, inflorescence blight, bud rot, heart rot and trunk rot

Several different diseases of date palm are incited by *Chalara paradoxa* (Streets, 1933; Cook, 1975; Carpenter and Elmer, 1978).

They are usually of minor economic importance, and include black scorch, inflorescence blight, bud rot, heart rot and trunk rot. The diseases have been reported in North Africa, India and the USA. Another disease, le couer que penche (bending head), is associated with the black scorch pathogen and *Diplodia theobromae*.

Symptoms

Dark brown to black lesions, in which abundant chlamydospores of the pathogen are produced, develop on all organs (Carpenter and Elmer, 1978). Black scorch lesions are irregular and occur along the leaf petiole. Affected leaves are reduced in length and the pinnae (leaflets) often are severely distorted (Fig. 9.2).

Bud rot symptoms include internal and external lesions down the base of the youngest leaves into the bud region (Fig. 9.3). Buds often are killed or severely damaged, and emerging fronds may have a 'bitten-leaf' appearance, reduced leaflets and black, necrotic tips (Fig. 9.4). Some palms recover, probably by development of lateral buds, and exhibit a characteristic bend where they are infected. This is the medjnoon or fool-disease symptom.

Fig. 9.3. Bud rot symptoms on date palm caused by *Chalara paradoxa*. Note the 'bitten-leaf' appearance of the youngest leaves and absence of a central whorl of new leaves (photo: J. Carpenter).

Fig. 9.2. Distortion and reduced length of pinnae (leaflets) on a date palm leaf affected by black scorch (photo: J. Carpenter).

Inflorescence rot develops before the spathe opens. Lesions develop on the young fruit stalks and spathes, and fruit strands and flowers can become partially or totally blackened (Fig. 9.5).

Trunk rot begins as a soft, yellow decay that darkens with age. Eventually, a reddish brown liquid exudes from lesions to discolour large patches of the trunk. Lesions may coalesce to cause a heart rot, and trunks ultimately may become hollow. Defoliation and death follow.

Causal agent

The teleomorph of *Chalara paradoxa*, *Ceratocystis paradoxa* is uncommon on date palm (Klotz and Fawcett, 1932; Chase and Broschat, 1991). The pathogen is described in Chapter 1.

Fig. 9.4. Date palm (foreground) with severe black scorch and bud rot symptoms caused by *Chalara paradoxa*. Note the sparseness of canopy compared with healthy trees in the background, and the bitten, newest leaves (photo: J. Carpenter).

Fig. 9.5. Inflorescence rot caused by *Chalara paradoxa*. Dark, blackish lesions develop on the young fruit stalks, spathes, fruit strands and flowers (photo: J. Carpenter).

Epidemiology

All parts of the date palm are susceptible (Carpenter and Elmer, 1978). Infection occurs in young, tender leaf tissues, wounds and growth cracks. Because of the sporadic occurrence of these diseases, little is known of their epidemiology.

Management

Resistance occurs in some commercial cultivars and could be used to breed resistant cultivars (Carpenter and Elmer, 1978). Black scorch has been observed on at least 17 cultivars, with 'Thoory' being the most susceptible. 'Deglet Noor' appears to be less susceptible than most cultivars and 'Tazizoot' was not affected.

The disease can be partially controlled by removing and burning all infected leaves and inflorescences. Applying a powdered Bordeaux mix in the spring to the apical bud and where inflorescences emerge has also been suggested. Minor trunk infections can be excised and treated with benomyl, but trees with major infections or heart rot should be removed.

Fruit rots

Preharvest fruit rots are common in all date-growing areas (Fawcett and Klotz, 1932; Carpenter and Elmer, 1978; Djerbi, 1988). Pathogen-incited fruit rot is part of a broader phenomenon of fruit spoilage that is caused by various biotic and abiotic factors (Turrell *et al.*, 1940). Since some of these factors interact, it often is difficult to estimate the damage that is caused by only microbes. Due to environmental differences, damage varies seasonally and geographically. In spite of the control measures that were practised in California, annual losses of 5% occurred. In severe cases, 25% of the fruit of 'Deglet Noor' and 'Medjool' are discarded in Israel due to fungal spoilage, and similar levels are reported from other date-growing countries.

Symptoms

Side-spot decay appears as a translucent or opaque-brown stain that becomes darkened and circular (Carpenter and Elmer, 1978). It begins 1–2 cm before the advancing margin of ripening tissue and, depending on the stage of maturity, may develop anywhere between the tip and the calyx end of the fruit.

Areas affected by calyx-end rot are soft and irregular (Fig. 9.6) (Carpenter and Elmer, 1978). *Aspergillus niger* often penetrates the seed cavity through the calyx, filling it with a black, powdery mass of spores. This is especially common on 'Deglet Noor' and 'Medjool'. Internal contamination often is obscure, and it is not uncommon to find infected fruit packed with first quality fruit.

Causal agents

Species of *Alternaria*, *Helminthosporium* and *Macrosporium* cause side-spot decays, and *A. niger*, *A. niger* var. *phoenicis* and *Citromyces ramusus* cause calyx-end rot (Fawcett and Klotz, 1932; Rieuf, 1963; Carpenter and Elmer, 1978; Djerbi, 1988). *A. niger* is described in Chapter 1. Although wounding enhances infection by these fungi, it is not required. Fruit spoilage can also be initiated by wound parasites. Twenty species, including filamentous fungi, yeasts and bacteria, have been isolated from affected fruit, and 12 of these are in the genus *Aspergillus*.

Epidemiology

Fruit rots are most common during periods of rain or high humidity and the khalal and rutab stages of fruit development (Carpenter and Elmer, 1978). During the early (kimri) and late (tamar) stages, non-wounded fruit are relatively resistant to fungal invasion. Standing water, weeds or intercrops that increase humidity in the grove favour rot development. High soil water content or rain that causes the fruit surface to crack also facilitate infection. Under high humidity, fruit rot caused by *A. niger* develops without wounds. Spores of these pathogens are abundant in date groves, and large quantities can be recovered from the surface of healthy fruits and covering nets.

Management

An integrated approach to reducing fruit rot combines: (i) insertion of wire rings into fruit bunches prior to the khalal stage to increase ventilation and decrease humidity; (ii) paper wraps to protect bunches from rain; and (iii) dusting with fungicides (Darley and Wilbur, 1955). Although a mixture of 5% ferbam, 5% malathion, 50% sulphur (active against fungi, insects and mites, respectively) and 40% talc was used for many years in the USA, ferbam is no longer registered for use on dates in the USA and there are no other fungicides that are labelled for this purpose there.

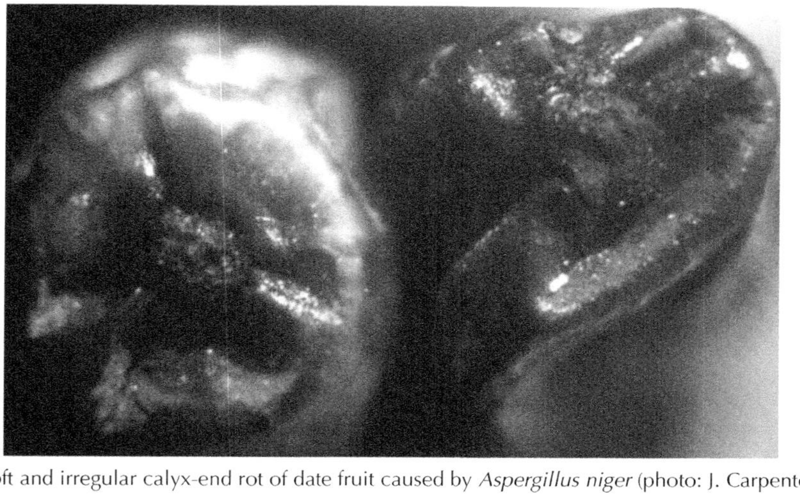

Fig. 9.6. Soft and irregular calyx-end rot of date fruit caused by *Aspergillus niger* (photo: J. Carpenter).

Graphiola leaf spot

Graphiola leaf spot is the most widespread disease of date palm and probably occurs wherever it is cultivated under humid conditions (Laville, 1973; Carpenter and Elmer, 1978). It is a common disease in the Punjab of India, the humid coastal areas of North Africa and the Near East. It causes premature death of infected leaves, which may result in reduced fruit yields.

Symptoms

An early symptom is the appearance of small spots on both sides of the pinnae and on the rachis (Carpenter and Elmer, 1978; Chase and Broschat, 1991). As the disease progresses, abundant small, black sori are produced on both sides of the pinnae and on the rachis. They are most abundant on apical portions of the pinnae, and increase in number as leaves age. Sori produce masses of yellow basidiospores, and deflate to become rough black craters after spore dispersal is completed (Fig. 9.7).

Causal agent

Graphiola phoenicis was described as the cause of this disease in 1823 (Chase and Broschat, 1991). It is an unusual fungus that is related to true smuts in the *Ustilaginales* (Cole, 1983). Its sori are 1–3 mm in dia, and consist of two layers: a dark, hard, outer peridium that is persistent; and a thin, hyaline peridium that degenerates after the spores mature (Sinha *et al.*, 1970; Chase and Broschat, 1991). Basidiospores are produced in fertile areas of the sori and are interspersed with groups of sterile filaments (Fig. 9.7). They are spherical to ellipsoidal, 3–6 μm in diameter, and have a thick, smooth, hyaline wall.

Epidemiology

Basidiospores germinate by either budding or producing septate hyphae. Infection is through stomata, and mycelium grows inter- and intracellularly for 10–11 months before sori are produced. At maturity, columns of fertile hyphae extend from the sori and appear yellowish due to sporulation. After basidiospores are disseminated, only the rough black craters of the sori remain. Chlorophyll content is reduced in moderately and severely affected cultivars, and the lifespan of severely infected leaves is shortened from the normal 6–8 years to 3 years (Sinha *et al.*, 1970).

Management

Leaf pruning is the most common measure for combating this problem (Carpenter and Elmer,

Fig. 9.7. Black, globose sorus of *Graphiola phoenicis* from which columns of fertile hyphae have emerged that are yellowish due to their production of basidiospores (right) and a sorus that has collapsed after spore dispersal has been completed (left) (photo: S. Koike).

1978). However, removal of too many leaves reduces the vigour of trees and fruit yields. In India, application of copper-containing compounds at 15 day intervals beginning with the first signs of sporulation controlled the disease in three to four sprays (Chase and Broschat, 1991). Resistance is found in several commercial cultivars (Sinha *et al.*, 1970).

Khamedj

Khamedj, or inflorescence rot, is present in North Africa, the Mediterranean, the Arabian peninsula and the Near East (Carpenter and Elmer, 1978). It is the most important disease of date in Tunisia, and is the only date disease of economic importance in the Tigris and Euphrates Valleys (Cook, 1975). In Iraq, important outbreaks are sporadic and occur only after prolonged cold, humid weather in the winter.

Although khamedj is usually a minor disease, it can cause substantial losses. On severely affected palms, 30–40 kg of fruit can be lost annually, and in some areas 40% of the crop is lost. The incidence of symptomatic trees in affected areas usually ranges from 5 to 65%.

Symptoms

Khamedj appears in the late winter or early spring as spathes begin to emerge (Carpenter and Elmer, 1978). In the early stages, it is difficult to distinguish affected from healthy spathes. Brownish or rusty areas develop on the non-opened spathe after the pathogen has already invaded the floral tissues. Lesions may be confluent and are most common near the top of the spathe, which, at the time of infection, is soft and still hidden in the leaf base. The internal face of the spathe under the lesions is yellow and translucent and may have brown dots corresponding to points of contact with diseased flowers. The pathogen attacks flowers and strands and may move on to the stalk of the inflorescence. Spathes, severely damaged when young, may remain closed; however, the spathe usually splits and reveals partial to near complete involvement of the flowers and strands. Typically, some palms develop symptoms every year, whereas others in the same planting develop symptoms only occasionally, even under favourable conditions.

Causal agent

Mauginiella scaettae is the major cause of khamedj, with minor inflorescence rots being caused by *Ceratocystis paradoxa* and species of *Fusarium* (Michael and Sabet, 1970; Cook, 1975; Carpenter and Elmer, 1978). Sporulation of *M. scaettae* is essentially the same in culture and on infected palm tissues. The fungus develops a white to cream-coloured mycelial mat that becomes covered with powdery conidia. The conidial chains fragment into units of one to several cells that are 10–50 μm long and 5–10 μm wide. One- and two-celled units predominate. The optimum temperature for germination is 21°C and for growth is ~20°C. The fungus is sensitive to temperatures above 25°C, and growth is arrested at 30–35°C.

Epidemiology

Khamedj is most severe in areas with excessive or prolonged winters and spring rains, on neglected palms in marginal areas of oases, or in waterlogged soil, salty depressions, or low-lying lands (Carpenter and Elmer, 1978). Male palms frequently are grown in these marginal areas as communal property pollen sources. They may be heavily infected because they receive little attention and the old inflorescences are not removed regularly. Pollination has been implicated in the movement of *M. scaettae* from diseased to healthy trees.

Khamedj does not spread rapidly. It is presumed that *M. scaettae* survives in old tissues as mycelium (Michael and Sabet, 1970). Since spores are short lived, they are not considered to be important for the persistence of the pathogen. The fungus is capable of direct penetration and is an aggressive parasite. Infected leaf bases are especially important because they press tightly against the highly susceptible emerging spathes.

Management

A good sanitation programme is essential to control Khamedj (Michael and Sabet, 1970; Carpenter and Elmer, 1978). All diseased parts of the inflorescence should be collected and burned immediately after harvest. In addition, old flower stalks, spathes, old leaves and leaf bases should be removed.

Fungicides should be applied immediately after harvest to the old flowering area of diseased palms, and in early spring on the new flowering area just before the spathes emerge. A number of fungicides appear effective, including Bordeaux mix.

The 'Hamraia', 'Tafezouine' and 'Takermest' cultivars are resistant, 'Hillawi' and 'Zahidi' are somewhat resistant, and 'Khadrawy' and 'Sayer' are highly susceptible.

Phytoplasma-associated diseases

At least three diseases of date palm are associated with phytoplasmas (formerly called mycoplasma-like organisms or MLOs). Although none of the diseases are major factors in date production and only one, lethal yellowing, is widespread, they are included in this chapter because they kill their host.

Lethal yellowing occurs in the northern Caribbean (Hispaniola and Jamaica) and areas that border the region in Belize, Guatemala, Honduras, Mexico and the USA (Florida and Texas). It affects 38 species of palms (Harrison *et al.*, 1995). Date palm is highly susceptible, and from 1978 to 1980 the disease killed 50% of the date palms in the lower Rio Grande Valley in Texas (McCoy *et al.*, 1980). Although reports of lethal yellowing on date palm have been limited to Florida and Texas (McCoy *et al.*, 1980; Howard and Harrison, 2001), it is probable that it would be affected elsewhere if it were planted where the disease is found.

Two diseases were reported recently in Sudan. Slow decline, which is also known as El Arkish, occurs commonly along the Nile river between Dongola and Merowe-Karema (Cronjé *et al.*, 2000b). Annually, it kills ~6% of the mature date palms in the area. White tip dieback is scattered throughout northern Sudan, but its relative impact in the area was not reported (Cronjé *et al.*, 2000a).

Symptoms

All three disorders cause a general chlorosis and necrosis of the canopy, but differences exist in syndrome development, the age of palms that are affected and the length of time that elapses before palms die.

Lethal yellowing initially causes death of adventitious roots at the base of trees and necrosis of the newest whorl of leaves in the crown (McCoy *et al.*, 1980). These leaves turn a dull green and then brown as they die. Initial symptoms in other leaves are a browning of the pinnae margins that later turn a dusty grey colour (Chase and Broschat, 1991). Fruit drop and necrosis of the inflorescence occur on mature palms and, as the apical bud dies, it can develop a soft, putrid decay. Eventually, all leaves in the canopy lose turgor and wilt along the trunk, and the entire crown falls from the palm. Death occurs within 4 months. Ultimately, all specimens in a given area are affected, but those that are mature are the first to die.

Slow decline kills date palms 12–24 months after symptoms first appear (Cronjé *et al.*, 2000b). Symptoms include chlorosis of the oldest leaves in the canopy that progresses to involve younger leaves. Leaves turn white to a light brown and eventually die, and a smelly decay can develop in the apical bud. Leaves then fall, to leave an erect dry tuft of young leaves at the trunk apex that may also break off to leave only the trunk standing. Alternatively, white, dry spear leaves appear before foliar chlorosis develops. Suckers on affected palms exhibit the same symptoms as larger plants. Unlike lethal yellowing, slow decline apparently does not cause fruit drop.

White tip dieback affects immature palms, 5–8 years old, which die within 6–12 months (Cronjé *et al.*, 2000a). The emerging spear leaf becomes severely chlorotic, as do the tips of pinnae in older leaves. White and necrotic streaks extend along the length of the leaf midrib. The crown does not become chlorotic, but changes quickly from green to dry white.

Causal agents

Since they have not been cultured, it has not been possible to complete Koch's postulates for these microbes, or to assign them definitive names. However, their intimate association with diseased, but not healthy, palms strongly suggests that they are responsible for these diseases.

Based on sequences of 16S/23S rDNA, the lethal yellowing phytoplasma is distinct from those that are associated with the Sudanese date palm diseases (N. Harrison, personal communication). The sequences of the slow decline and white tip agents were 99% homologous (GenBank accession numbers, AF268000 and AF100411, respectively) and, in turn, were both 99% homologous with sequences of Bermuda grass white leaf phytoplasmas from Southeast Asia and Sudan (GenBank accession numbers Y14645 and AF100412) (Cronjé *et al.*, 2000a,b). Preliminary results suggest that these agents are closely related to the sugarcane whiteleaf phytoplasma that recently was implicated in lethal declines of coconut palm in Southeast Asia (see Chapter 7).

In contrast, the lethal yellowing agent is distinct from, but related to, phytoplasmas from coconut palms that are affected by lethal decline diseases in East and West Africa (Harrison *et al.*, 1994; Tymon *et al.*, 1997). These agents are associated with similar disease syndromes in coconut palm in each area, but have not been reported on date palm in Africa.

Epidemiology

The epidemiology of these diseases is poorly understood. Lethal yellowing exhibits a jump-spread type distribution that is characteristic for diseases with flying vectors (McCoy *et al.*, 1980). Affected palms occur in discrete, but diffuse foci. Spread does not always occur from tree to adjacent tree, and several trees may be skipped before another in a focus is affected. Although specific work has not been conducted for lethal yellowing on date palm, it has been shown that the planthopper, *Myndus crudus*, is responsible for spreading the disease among other palm species (Howard and Harrison, 2001).

Next to nothing is known about the epidemiology of slow decline and white-tip dieback. Although losses from slow decline are known to occur steadily over time, the distribution and spread of the disease were not mentioned (Cronjé *et al.*, 2000b). White-tip dieback was reported to occur in scattered foci, but its spread was not discussed (Cronjé *et al.*, 2000a).

Management

No control measures were reported for slow decline and white-tip dieback (Cronjé *et al.*, 2000a, b). Although specific measures have not been reported for lethal yellowing on date palm, some work has been conducted on coconut and other ornamental palms (Chase and Broschat, 1991). Insecticidal control of *M. crudus* reduces the disease's spread, but the degree of control is insufficient to warrant the repeated applications that would be needed in both ornamental landscapes and palm plantations. Trunk injections of tetracycline antibiotics are effective, but they are needed on either a preventative basis or once fruit and flower symptoms appear; they are largely ineffective after foliar symptoms develop. Also, since 4-month application intervals are required, trees ultimately accrue very significant damage to their vascular systems (monocots do not regenerate vascular elements). Tolerance to lethal yellowing among different date palm cultivars has not been reported.

References

Abdalla, M.Y., Al-Rokibah, A., Moretti, A. and Mulé, G. (2000) Pathogenicity of toxigenic *Fusarium proliferatum* from date palm in Saudi Arabia. *Plant Disease* 84, 321–324.

Amir, H., Amir, A. and Riba, A. (1996) The role of microflora in the resistance to vascular fusarium wilt induced by soil salinity in palm groves. *Soil Biology and Biochemistry* 28, 113–122.

Anonymous (2001) http://www.fao.org/default.htm

Belarbi-Halli, R. and Mangenot, F. (1986) Bayoud disease of date palm: ultrastructure of root infection through pneumathodes. *Canadian Journal of Botany* 64, 1703–1711.

Brac de la Perriere, R.A., Amir, H. and Bounaga, N. (1995) Prospects for integrated control of 'bayoud' (Fusarium wilt of the date palm) in Algerian plantations. *Crop Protection* 14, 227–235.

Carpenter, J.B. and Elmer, H.S. (1978) *Pests and Diseases of the Date Palm.* US Department of Agriculture Handbook 527, Washington, DC.

Chase, A.R. and Broschat, T.K. (eds) (1991) *Diseases and Disorders of Ornamental Palms.* APS Press, St Paul, Minnesota.

Cole, G.T. (1983) *Graphiola phoenicis*: a taxonomic enigma. *Mycologia* 75, 93–116.

Cook, A.A. (1975) *Diseases of Tropical and Subtropical Fruits and Nuts.* Hafner Press.

Cronjé, P., Dabek, A.J., Jones, P. and Tymon, A.M. (2000a) First report of a phytoplasma associated with a disease of date palms in North Africa. *New Disease Reports.* http://www.bspp.org.uk/ndr/2000/2000–4.htm

Cronjé, P., Dabek, A.J., Jones, P. and Tymon, A.M. (2000b) Slow decline a new disease of mature date palms in North Africa associated with a phytoplasma. *New Disease Reports.* http://www.bspp.org.uk/ndr/2000/2000–7.htm

Darley, E.F. and Wilbur, W.D. (1955) Results of experiments on control of fruit spoilage of Deglet Noor and Saidy dates in California, 1935–1954. *Annals of the Date Growers' Institute* 32, 14–15.

Diamond, J. (1998) *Guns, Germs and Steel. A Short History of Everybody for the Last 13,000 Years.* Vintage.

Djerbi, M. (1988) *Les Maladies du Palmier Dattier.* FAO/ PNUD/ RAB/ 84/018. FAO, Rome.

Fawcett, H.S. and Klotz, L.J. (1932) *Diseases of the Date Palm,* Phoenix dactylifera. California Agricultural Experiment Station Bulletin 522.

Feather, T.V., Ohr, H.D., Munnecke, D.E. and Carpenter, J.B. (1989) The occurrence of *Fusarium oxysporum* on *Phoenix canariensis,* a potential danger to date production in California. *Plant Disease* 73, 78–80.

Freeman, S. and Maymon, M. (2000) Reliable detection of the fungal pathogen *Fusarium oxysporum* f. sp. *albedinis,* causal agent of bayoud disease of date palm, using molecular techniques. *Phytoparasitica* 28, 1–8.

Harrison, N.A., Richardson, P.A., Jones, P., Tymon, A.M., Eden-Green, S.J. and Mpunami, A.A. (1994) Comparative investigation of MLOs associated with Caribbean and African coconut lethal decline diseases by DNA hybridisation and PCR assays. *Plant Disease* 78, 507–511.

Harrison, N.A., Richardson, P.A. and Tsai, J.H. (1995) Detection and diagnosis of lethal yellowing: Conventional and molecular techniques. In: Oropeza, C., Howard, F.W. and Ashburner, G.R. (eds) *Lethal Yellowing: Research and Practical Aspects.* Kluwer Academic Publishers, Dordrecht, The Netherlands, pp. 79–92.

Howard, F.W. and Harrison, N.A. (2001) *Lethal Yellowing of Palms.* Palm Beach County Cooperative Extension, University of Florida.

Klotz, L.J. and Fawcett, H.S. (1932) Black scorch of the date palm caused by *Thielaviopsis paradoxa. Journal of Agricultural Research* 44, 155–166.

Laville, E. (1973) Les maladies du dattier. In: Munier, P. (ed.) *Le Palmier-dattier.* GP. Maisonneuve and Larose, Paris, pp. 95–108.

McCoy, R.E., Miller, M.E., Thomas, D.L. and Amador, J. (1980) Lethal decline of Phoenix palms in Texas associated with mycoplasma-like organisms. *Plant Disease* 64, 1038–1040.

Michael, I.F. and Sabet, K.A. (1970) Biology and control of *Mauginiella scaettae* Cav., the pathogen of khamedj disease in the United Arab Republic. *Date Growers' Institute Report* 47, 5–8.

Ohr, H.D. (1991) Fusarium wilt. In: Chase, A.R. and Broschat, T.K. (eds) *Diseases and Disorders of Ornamental Palms.* APS Press, St Paul, Minnesota, pp. 11–12.

Ohr, H.D. and Carpenter, J.B. (1993) Diseases of date palm (*Phoenix dactylifera* L.). http://www.scisoc.org/resource/common/names/datepalm.htm

Oudejans, J.H.M. (1976) Date palm. In: Simmonds, N.W. (ed.) *Evolution of Crop Plants.* Longman, London, pp. 229–231.

Purseglove, J.W. (1975) *Tropical Crops. Monocotyledons,* Vols 1 and 2 combined. Longman, London.

Rieuf, P. (1963) Contribution a la lutude du charbon de la datte *Aspergillus phoenicis* (Cda.) Thom. *Al Awamia* 6, 1–16.

Saaidi, M., Toutain, G., Bannerot, H. and Louvet, J. (1981) Selecting date palm (*Phoenix dactylifera* L.) for resistance to Bayoud Disease, *Fusarium oxysporum albedinis. Fruits* 36, 241–249.

Sedra, M.H. and Besri, M. (1994) Evaluation of date palm resistance to bayoud disease due to *Fusarium oxysporum* f. sp. *albedinis.* Development of an efficient method for discriminating *in vivo* plantlets acclimatised in the glasshouse. *Agronomie* 14, 467–472.

Sinha, M.K., Singh, R. and Jeyarajan, R. (1970) Graphiola leaf spot on date palm (*Phoenix dactylifera*): susceptibility of date varieties and effect on chlorophyll content. *Plant Disease Reporter* 54, 617–619.

Streets, R.B. (1933) Heart rot of the date palm. *Arizona Agricultural Experiment Station Technical Bulletin* 48, 443–469.

Summerell, B.A., Kistler, H.C. and Gunn, L.V. (2001) Fusarium wilt of *Phoenix canariensis* caused by *Fusarium oxysporum* f. sp. *canariensis*. In: Summerell, B.A., Leslie, J.F., Backhouse, D., Bryden, W.L. and Burgess, L.W. (eds) Fusarium. *Paul E. Nelson Memorial Symposium*. APS Press, St Paul, Minnesota, pp. 271–294.

Tantaoui, A., Ouinten, M., Geiger, J.P. and Fernandez, D. (1996) Characterization of a single clonal lineage of *Fusarium oxysporum* f. sp. *albedinis* causing Bayoud disease of date palm in Morocco. *Phytopathology* 86, 787–792.

Turrell, F.W., Sinclair, W.B. and Bliss, D.E. (1940) Structure and chemical factors in relation to fungus spoilage of dates. *Annals of the Date Growers' Institute* 17, 5–11.

Tymon, A.M., Jones, P. and Harrison, N.A. (1997) Detection and differentiation of African coconut phytoplasmas: RFLP analysis of PCR-amplified 16S rDNA and DNA hybridisation. *Annals of Applied Biology* 131, 91–102.

Vanachter, A. (1989) Strategies for the control of Bayoud disease of the date palm caused by *Fusarium oxysporum* f. sp. *albedinis*. In: Tjamos, E.C. and Beckman, C.H. (eds) *Vascular Wilt Diseases of Plants: Basic Studies and Control*. Springer-Verlag, Berlin, pp. 501–513.

10 Diseases of Durian

T.-K. Lim[1] and S. Sangchote[2]
[1]*Biosecurity Australia, Department of Agriculture, Fisheries and Forestry Australia, Canberra, Australia;* [2]*Department of Plant Pathology, University of Kasetsart, Bangkok, Thailand*

Introduction

Durian, *Durio zibethinus* (family: *Bombacaceae*), is one of 12 edible species in this genus. It is native to the Malay Archipelago, and thrives naturally in Borneo, Malaysia and Sumatra.

Durian is a tall tree growing to a height of 20 m. Vegetatively propagated trees are shorter than those established from seed, but are more precocious and bear within 5–7 versus 10–12 years. The flowers are hermaphroditic and are borne in corymbose inflorescences. They are large, odoriferous and creamy white. The stamens are borne in five bundles and the central style has a sticky stigma. The fruit is a large loculicidal capsule, ovoid or oval to globose in shape, spiny, greenish brown to golden brown, and weighs 1–4 kg. It splits into five segments when ripe and has 1–7 fleshy arils in each segment. They are cream, white or yellow, envelop a large ellipsoidal seed, and have a sulphurous odour that many find disagreeable. Despite this smell, durian is one the world's most esteemed fruits.

Durian requires hot (mean range of 22–33°C), humid, tropical environments with rainfall of 2–3 m distributed evenly throughout the year. The paramount environmental constraints are the durations of absolute minimum temperature and relative humidity. In its native habitat, durian prospers on well-drained, deep, fertile, loamy soil, rich in nutrients and organic matter. It is extremely sensitive to drought stress, and variability in drought susceptibility occurs among cultivars (Lim, 1990). Durian is a shallow-rooted crop with 60% of the total root length confined to within 60 cm of the canopy edge and 0–30 cm of the soil surface.

The leading producers of durian are, in descending order, Thailand, Malaysia and Indonesia. Each of these countries has >200 recognized commercial cultivars. Popular cultivars in Thailand include 'Chanee', 'Gaan Yaow', 'Gradumtong', 'Monthong', 'Nok Yib' and 'Puang Mani'. In Malaysia, the renowned cultivars are 'Ang Bak', 'Durian Paduka', 'D2', 'D16', 'D24', 'D96', 'D168', 'Hor Lor', 'Lempur Emas', 'MDUR 78', 'MDUR 79', 'MDUR 88' and 'Tawa'. In Indonesia, 'Parung', 'Perwira', 'Petruk', 'Si Dodol', 'Si Japang', 'Sitokong', 'Sunan' and 'Tembaga' predominate. Other countries where production is mainly for domestic consumption include Australia, Brunei, Kampuchea (Cambodia), Laos, Mynamar (Burma), the Philippines and Vietnam.

Foliar and Fruit Diseases

Algal diseases

Several green algae occur on durian, but only *Cephaleuros virescens*, the cause of algal leaf spot, is of great significance. It affects the foliage of many agriculturally important fruit trees, plantation crops and ornamentals (Singh, 1980). On durian, it debilitates and affects the vitality of trees by growing on, and reducing the photosynthetic area of, leaves. It also invades cortical tissue and cracks the bark of twigs and tender branches. Damage is caused by its growth and expansion inside the host tissue rather than from parasitism. The disease is more prevalent when trees are neglected or environmentally stressed (Lim, 1990).

Symptoms

Algal leaf spot produces orange, rust coloured velutinous spots on the upper surface of leaves, twigs and branches. Such spots may merge to form larger patches that eventually turn greyish green. When the algal growth is scraped away, a thin greyish white to dark coloured necrotic crust is exposed.

Infestations of *Trentepohlia* spp. form 3–4 mm thick, orange coloured mats that envelope the sun-exposed bark of the trunk and branches, and greenish white mats on the sun-shadow side. The thick, felt-like incrustations block lenticels on the bark, and interfere with the ramiflorous flowering of durian. Other epiphytic algae form thin, superficial, greenish white layers or mats on the bark and foliage.

Causal agents

C. virescens is in the family *Trentepholiaceae* (division: *Chlorophyta*). It is described in Chapter 1.

Trentepohlia aurea, T. arborueum and *T. monile* commonly infest the trunks and branches of old durian trees (Lim and Kamaruzaman, 1989b). They are also in the family *Trentepholiaceae*. Other species of green epiphytic algae in the genera *Trebouxia, Pleurococcus* and *Phycopeltis* are found on the foliage, trunks and branches of durian.

Epidemiology

Infestations of green algae are most serious on neglected trees that are unthrifty and in poor vigour. Rainy periods and humid conditions are conducive to the spread and proliferation of *C. virescens* and *Trentepohlia* spp. since algal fragments, zoospores and detached sporangia are dispersed via rain, wind and watersplash.

Management

Trees should be maintained in a vigorous state by proper fertilization, irrigation and control of weeds underneath trees. Removal of infested shoots and leaves is also helpful. Stubborn infestations can be controlled with an algicide or copper containing fungicide. In Thailand, sprays of copper oxychloride 85% WP 50 g 20 l^{-1} are recommended when >30% of the leaves are affected by algal leaf spot in August/September (Disthaporn *et al.*, 1994).

Anthracnose

Anthracnose is an important postharvest fruit disease (see below), but is less important on other organs of the host. It is common on leaves of seedlings and mature trees, and causes a dieback of twigs (Lim, 1980).

Symptoms usually start at the tip and, less commonly, margins of leaves as discrete regular necrotic spots or irregular necrotic patches. These spread down the leaf blade and develop grey–brown centres with concentric rings of tiny spots (acervuli of the causal fungus) (Fig. 10.1). Under humid conditions, acervuli form and become covered with salmon coloured spores. Young leaves are often distorted, and premature abortion of leaves may occur. Twig dieback occurs as a progressive dieback starting from the tip downwards.

Anthracnose is caused by *Colletotrichum gloeosporioides* (teleomorph: *Glomerella cingulata*), which is described in Chapter 1. It is spread by conidia that are produced on necrotic lesions and are moved via wind and rainsplash. Wet weather is conducive to the spread and establishment of the disease. Mancozeb, copper-based and other fungicides are effective, especially during the nursery stage. Keeping trees in good vigour, and removing diseased twigs and shoots are also helpful (Lim, 1990).

Fig. 10.1. Foliar symptoms of anthracnose on durian (photo: T.K. Lim).

Phomopsis leaf spot

Phomopsis leaf spot is a minor disease on durian seedlings and neglected trees. Necrotic, brown circular spots, ~1 mm in diameter with dark margins and yellow halos, form on leaves. Pycnidia of the pathogen appear as black, pin-sized dots in the centre of lesions. On seedlings, affected leaves abort, leaving bare shoots that are prone to sun scorching and dieback caused by *Diplodia theobromae* and *C. gloeosporioides*.

Phomopsis durionis produces ellipsoid to oval, hyaline α-conidia and longer, filiform and bent β-stylospores in dark brown to black, ostiolate pycnidia. The fungus also causes a postharvest fruit rot on durian (Fig. 10.2).

The disease is most common under warm, humid conditions on stressed trees. On seedlings, overcrowding and excessive watering are conducive to its establishment and spread. Mancozeb controls premature defoliation of nursery trees.

Pink disease

This destructive disease attacks woody shoots and small branches to cause wilting and dieback of pockets of associated foliage. The disease is prevalent in cloudy, heavy rainfall areas (e.g. Malaysia, northern Queensland and Thailand).

Pink disease is caused by the fungus, *Erythricium salmonicolor* (anamorph: *Necator decretus*), and is described in Chapter 1. It produces pinkish white mycelial threads that envelop branches and shoots. As infected bark dies, the mycelia mature to become a pink crust. The crust develops irregular cracks as the weather dries, and the foliage subtended by the infected bark wilts and dies.

The disease is spread by rainsplash and wind, hence the Malay colloquial name for the disease, 'Cendawan Angin', which means wind fungus. The pathogen has a wide host range including economic crops and fruit trees such as carambola, cocoa, guava, jackfruit, mango, rubber and species of *Aleurites*, *Ficus*, *Gardenia*, *Jasminum*, *Malus*, *Pittosporum*, *Pyrus* and *Severina*.

Wide tree spacing and well-pruned and ventilated canopies are helpful. Dead, diseased branches should also be removed and burned, and the cut surface should be protected with a fungicide. Topical sprays or brush-on applications of triadimefon, tridemorph, oxycarboxin, flusilazol and copper-based fungicides afford some control.

Postharvest fruit rots

Many microbes are associated with postharvest fruit rots, most of which are fungi. Some are primary invaders, whilst others are secondary, infecting through wounds and cracks caused by insects, injuries and natural desiccation of the fruit surface.

Symptoms

It is very difficult to determine the cause of the different fruit rots based on visual symptoms. Unless the pathogen sporulates on the fruit, it may be necessary to isolate it before diagnosis is possible. In general, these diseases cause irregular necrotic patches in varying shades of brown that, with time, become dark brown to black.

Symptoms of Phytophthora fruit rot initially are small, brown lesions. They develop when harvested fruit begin to ripen, but intact, immature fruits are also affected on the tree. The lesions are soft and extend into the aril and seed, and whole fruits can be rotted within a few days (Plate 70) (Lim and Chan, 1986a). Anthracnose produces small, dark brown spots between the spines of fruit, mostly while in the grove. These spots coalesce to produce larger lesions in which orange masses of conidia are produced. Fusarium rot produces soft, water-soaked, brownish lesions that are rapidly covered with whitish, salmon-coloured mycelia. Diplodia fruit rot produces small, dark brown lesions on ripening fruit. They expand rapidly, and masses of dark greenish mycelia and dusty dark conidia are produced on the necrotic area. Lesions extend to and rot the aril. Phomopsis fruit rot causes small, dark brown, round to oval-shaped lesions on ripening fruit. They expand rapidly, and small, grey masses of mycelia are produced on the lesions (Fig. 10.2). The tips of the fruit spine are killed. Rhizopus and Mucor rot produce large, soft necrotic patches on the fruit during storage. Under humid conditions, sparse cottony, grey, mycelial strands are produced over the lesions. *Phyllosticta* sp. and *Curvularia eragrostidis* cause firm, superficial lesions.

Fig. 10.2. Symptoms of Phomopsis fruit rot on durian. Note the sporulation of the causal fungus, *Phomopsis durionis*, on the fruit surface (photo: T.K. Lim).

Causal agents

The most important and common postharvest rots are caused by *Phytophthora palmivora* (Phytophthora fruit rot), *C. gloeosporioides* (anthracnose), *Phomopsis durionis* (Phomopsis fruit rot), *Diplodia theobromae* (Diplodia fruit rot) and *Fusarium solani* (Fusarium fruit rot). Other less important postharvest pathogens include: *Erwinia* sp., *Rhizopus stolonifer* (Rhizopus rot), *Mucor* sp. (Mucor rot), *Phyllosticta* sp. and *Curvularia eragrostidis* (teleomorph: *Cochliobolus eragrastidis*) (Lim, 1990; Sangchote *et al.*, 1996).

Epidemiology

The epidemiology of *P. palmivora*-induced diseases is elaborated under soilborne diseases.

Conidia of *C. gloeosporioides* are spread by rain and wind. The fungus infects immature and unripened fruit quiescently, and symptoms develop once ripening begins. The pathogen also causes leaf anthracnose and leaf blight.

P. durionis produces conidia in pycnidia on infected lesions that are spread by rain and wind. Lesions develop on fruit as they ripen after harvest.

D. theobromae has a wide host range. It also infects fruits of rambutan, mangosteen and mango, and can survive in plant debris and soil (Sangchote, 1989). Young durian fruits, which are thinned and drop on the ground, are a good substrate for this fungus (Sangchote *et al.*, 1996).

Species of *Rhizopus* and *Mucor* are secondary invaders that affect bruised fruit. *F. solani* is both a common soilborne and storage fungus, whereas *Phyllosticta* sp. and *C. eragrostidis* are common field fungi.

Management

Fruit should neither come in contact with soil and plant debris, nor be damaged or bruised during harvest and handling. To control anthracnose and Phomopsis and Diplodia fruit rots, fruits are treated in Thailand within 6 h of harvest by dipping in 500 μg of carbendazim or thiabendazole l^{-1} for 2–3 min (Sangchote *et al.*, 1996). Also, pruning of diseased plant parts and spraying fruits with carbendazim is recommended prior to harvest, especially after rain.

Proper orchard care for managing Phytophthora fruit rot is discussed below under soilborne diseases. Also, postharvest treatment with 2000 μg of fosetyl-Al l^{-1} for 1 min has been recommended (Disthaporn *et al.*, 1994).

Rhizoctonia leaf blight

This is a destructive and common disease that affects the foliage of durian seedlings and trees. In the nursery, mortality can reach 50% if the disease is left unchecked (Lim *et al.*, 1987). It is economically important in Malaysia and Thailand where it is rife in durian orchards.

Causal agent

The pathogen is a member of anastomosis group (AG)-1 of *Rhizoctonia solani* (teleomorph: *Thanatephorus cucumeris*) (Lim *et al.*, 1987; Lim, 1990). It has multinucleate hyphae (3–16 (8)) that branch at right angles, and sclerotia that turn from white to various shades of brown when mature.

Symptoms

Symptoms start as water-soaked spots on leaves (Fig. 10.3) that coalesce to form larger, irregular, water-soaked patches that dry into light brown necrotic lesions. Affected leaves are curled, and woody twigs and branches appear shrivelled, desiccated and necrotic. Under humid conditions, light brown tufts of fungal mycelia and sclerotial bodies form on affected plant parts. Diseased leaves abscise, and twigs and branches die back. In severe cases, portions of the canopy defoliate and appear bald.

Epidemiology

Heavy rain hastens the spread and enhances the severity of the disease. Infection of foliage originates from mycelia and sclerotia, contact with diseased plant parts, and the

Fig. 10.3. Symptoms of Rhizoctonia leaf spot on durian (photo: T.K. Lim).

agency of insects and rainsplash. Affected leaves fall to the ground where the fungus survives in the soil and leaf litter as sclerotia and mycelial strands.

Management

In the nursery, pasteurized soil should be used, and overcrowding and excessive watering avoided. The disease can be managed effectively with several applications of any of the following as foliar sprays or soil drenches: pencycuron, benomyl, carbendazim, flutolanil, thiophanate methyl and triadimefon (Lim, 1990). When 10% defoliation is reached in Thailand, canopy sprays of 12 g of carbendazim 60% WP 20 l^{-1} or 50 g of copper oxychloride 85% WP 20 l^{-1} are recommended (Disthaporn *et al.*, 1994). Affected shoots and foliage should be removed and burned. Weed control under trees helps reduce inoculum.

Sclerotium fruit rot

This disease occurs on fallen, ripe durian fruit in Malaysia (Lim and Kamaruzaman, 1989a). It is not a problem when fruit are picked when mature but not ripe, as is done in Thailand, but can be after ripe fruit are left on the ground for a couple of days. Sclerotium fruit rot can extend into the aril and seeds and makes the fruit unpalatable.

Sclerotium rolfsii (teleomorph: *Athelia rolfsii*) forms a thick, white mycelial mat over the fruit and occasionally forms sclerotial bodies. It can also cause seedling blight in the nursery. It is a facultative parasite that survives in the soil as thick mycelial strands and sclerotia. Sclerotia turn from white to brown to dark brown, and have a thick rind, distinct medulla and cortex. Clamp connections are common on its 4–7 μm wide primary hyphae.

S. rolfsii is found in a diverse range of soils in the tropics and subtropics. The fungus has been reported to cause root and stem diseases on ~500 plant species in 100 families. It survives saprotrophically on decaying plant debris and organic matter, and thrives in warm, humid environments.

Weeds should be controlled under trees, especially during the harvest period. Fallen fruit should be harvested daily. Raised canvas or polyethylene platforms or strong netting beneath trees can be used to break the fall of fruits and prevent their contact with soil.

Sooty mould and black mildew

The sooty mould and black mildew fungi are common in all durian-growing areas. The sooty mould fungi are superficial, epiphytic saprotrophs with dark coloured hyphae and fruiting bodies. They are deuteromycetes and ascomycetes. In contrast, the black mildew fungus is an epiphytic parasitic ascomycete that has dark-coloured hyphae and produces a range of absorptive structures that penetrate host tissue.

Sooty mould and black mildew debilitate durian trees by covering leaf surfaces, thereby decreasing photosynthesis. They increase the humidity within tree canopies, and are most problematic in stressed, neglected trees that are infested with lac insects, mealy bugs, scales, planthoppers and other sucking insects. As a parasite, black mildew obtains some of its nutrients from the host plant tissues it infects.

Symptoms

Sooty mould and black mildew occur on both surfaces of durian leaves forming discrete, velutinous or powdery, black spots, black spongy pellicles or thin, black, effuse films. On twigs and leaf petioles, they form a hard lumpy crust, and on fruit they form a spongy crust on the surface.

Causal agents

Sooty mould and black mildew have been studied extensively on durian only in Malaysia. *Meliola durionis*, the only black mildew species known on durian, was reported by Turner (1971) in Sarawak and Johnston (1960) in peninsular Malaysia. The same authors recorded the sooty mould species, *Capnodium moniliforme*. Lim (1989) later described seven additional sooty mould species in Malaysia: three dematiaceous hyphomycetes, *Polychaeton* sp., *Leptoxyphium* sp. and *Tripospermum* sp.; and four ascomycetes, *Scorias spongiosa*, *Phragmocapnias betle*, *Trichomerium grandisporum* and *Trichopeltheca asiatica*.

Epidemiology

Sooty mould thrives on the honeydew of lac insects, mealybugs, scale insects, planthoppers and flatids. Their various asexual spores, ascospores and mycelial fragments are dispersed by insects, rainsplash and wind. Warm, humid conditions and the presence of these insects enhance their proliferation.

Management

Controlling the insects that produce honeydew helps control these problems, and reduced humidity and increased air circulation in orchards via wider tree spacing, good weed control and pruning of intertwining branches are also helpful.

Velvet felt

The velvet felt fungus, *Septobasidium* sp., occurs in Malaysia (Lim, 1990) and Thailand (Chandrasikul, 1962). Infestations often are mistaken for pink disease. However, unlike *E. salmonicolor*, *Septobasidium* sp. is not parasitic on durian and grows only superficially around twigs and branches. It can be distinguished by its soft, velvet, violet or violet–grey superficial growth on twigs and branches. The advancing margin of the fungus is white and finely fan shaped, and the incrustation is removed easily.

Septobasidium sp. is a basidiomycete. It does not penetrate the underlying host tissues, and is in fact a beneficial fungus that parasitizes certain scale insects. The fungus produces haustoria that extract nutrients from the cadavers of scale insects found in the bottom layer of the fungal stroma. It is rampant in heavy rainfall areas. It has a complicated relationship with scale insects for whom it offers protection and insulation against predators and from which it derives nutrients. It need not be controlled on durian.

Soilborne Diseases

Diseases caused by *Phytophthora palmivora*

Phytophthora palmivora causes the most devastating, widespread and common diseases of durian in Southeast Asia and Australia. It causes root rot and patch cankers on the trunk and stem that ultimately are lethal. It also blights leaves, and causes twig dieback and pre- and postharvest fruit rots. On seedlings, losses from dieback and foliar blight often exceed 50% (Chan and Lim, 1987).

Symptoms

Patch canker causes necrosis on the bark that is accompanied by discoloration and exudation of a reddish brown, gummy, resinous substance. The underlying cortical tissues and wood appear dull and reddish brown with dark claret patches (Plate 71). Cankers can girdle branches and cause branch dieback, whereas trunk cankers and extensive root rot can kill trees. Severe defoliation and unthriftiness precedes tree death. Water-soaked spots on leaves darken and coalesce into larger necrotic patches, and affected leaves abscise. Seedling dieback can result from root or stem infections.

Causal agent

P. palmivora is a ubiquitous, soilborne oomycete. It was first reported to cause patch cankers on durian in Malaysia by Thompson (1934) and in Thailand by Bhavakul and Jaengsri (1969).

The hyphae of *P. palmivora* are coenocytic and diploid. All isolates that have been examined from durian are of the A1 mating type (Suzui *et al.*, 1978; Lim and Chan, 1986a). Thus, even though mating and oospore formation have not been investigated under natural conditions, the absence of the A2 mating type on durian suggests that the sexual stage of the pathogen may not be common. Other features of the pathogen are described in Chapter 1.

Epidemiology

The pathogen persists in soil in infected roots and as chlamydospores. Sporangia, zoospores and chlamydospores are dispersed by wind, rainsplash, rain, insects, snails and planting material, especially if seeds from infected fruits are used (Lim, 1990). Infected fallen fruits are also important sources of inoculum. Low-lying, waterlogged areas or heavy soils are conducive to the development of root and stem rot. In Thailand, fruit rot is rife during the rainy season in May–June.

Isolates of *P. palmivora* from durian did not affect cacao, jackfruit, mandarin orange, passion fruit, pulasan, rambutan and tangelo, but were highly pathogenic to durian and moderately pathogenic to papaya (Tai, 1971; Lim and Chan, 1986a; Chan and Lim, 1987). Five phenotypes were revealed in isozyme analyses of isolates from durian (Mchau and Coffey, 1994).

Management

Many systemic fungicides exhibit prophylactic and therapeutic properties against *P. palmivora*, including metalaxyl, fosetyl-Al, ofurace, milfuram and cyprofuram (Lim, 1990). In Thailand, once a single fruit on a tree is affected, the entire tree is sprayed with 50 g 80% WP fosetyl-Al 20 l^{-1}. For patch canker control, trunks and branches are painted with red lime or metalaxyl. Affected trees are also trunk-injected with fosetyl-Al. Preliminary trials in Vietnam indicated that disease incidence and severity were reduced following trunk injections of phosphorous acid, the active breakdown product of fosetyl-Al (Thanh, 1998).

Variation in the reaction of commercial clones to *P. palmivora* has been noted in Australia, Malaysia and Thailand (Tai, 1971, 1973; Pongpisutta and Sangchote, 1994; Lim, 1997). The Malaysian Agricultural Research and Development Institute (MARDI) recently released three hybrids between the Malaysian selections 'D10' and 'D24' that possess improved resistance. A large, potential source of disease resistance exists in allied *Durio* species and *Neesia*, *Coelostegia* and *Kostermansia*, related genera that evolved in damp, low-lying environments (Lim, 1997).

Alternatively, root rot-resistant rootstocks could be developed (Lim, 1997). The seedling rootstocks that are widely used with clonal scions of durian cultivars are heterogeneous and give rise to variable planting material. Genetically uniform rootstocks could be produced by marcottage or from cuttings, and would be valuable tools in durian production if they were from graft-compatible, root rot-resistant selections. Early fruiting, high yield potential, excellent fruit quality and habitat adaptability are additional, potentially selectable traits.

Awarun (1994) reported that two isolates of *Trichoderma harzianum* protected durian against *P. palmivora* in Thailand. Lim and Chan (1986b) indicated that sporangia and chlamydospores of *P. palmivora* were parasitized by *Gliocladium roseum* that was isolated from durian field soil. In Thailand, Disthaporn *et al.* (1994) recommended using *T. harzianum* inoculum in sorghum:rice bran plus compost for application to soil around affected trees, and another antagonistic fungus, *Chaetomium globosum*, has also been used (S. Vichitrananda, Thailand, 1997, personal communication). Adjusting the pH of acidic soils to 6.5 with lime promotes the growth and development of antagonists.

Other management tools include: the use of multiclonal plantings; the use of pasteurized potting mix fortified with antagonists; planting on raised mounds; the use of mulches and well-composted, organic manures; proper fertilization and irrigation practices; and removal and burning of diseased fruit and branches.

Pythium root rot

Pythium root rot was first recorded on durian trees in Singapore by Thompson (1938). The disease also affects and usually kills seedlings. The foliage of affected seedlings becomes flaccid and, eventually, brown. The cortical tissues and main vascular bundles of the primary roots are extensively decayed. On trees, infection usually begins on young rootlets and spreads to laterals before extending towards the bole of the tree. The aboveground symptom is dieback of branches in a portion of the tree that can be confused with damage caused by *P. palmivora*. Thus, isolation of the pathogen is needed to confirm this disease.

The ubiquitous soilborne oomycete, *Pythium vexans*, causes Pythium root rot. It is homothallic and readily produces oospores. Sporangia are subglobose, ovoid to slightly pyriform, and produce zoospores in an extruded external sac (Lim, 1990). It survives saprophytically or parasitically in the soil. Oospores are the primary survival structures, and infection occurs primarily via zoospores. The fungus has optimum and maximum temperatures for growth of, 30 and 35°C respectively. It thrives and spreads under wet conditions.

Durian should not be planted in areas that are waterlogged or prone to flooding. Raised mounds, good drainage, mulches, and sound fertilizer and irrigation management should be utilized. Metalaxyl drenches of the soil can also be used. Use of pasteurized potting mix amended with species of *Trichoderma* will help to control the disease in the nursery.

White root disease

White root disease is common on land that previously was planted to rubber or cassava. The disease can be devastating during the first 6 years of tree establishment, and infected trees will die if the disease is not arrested at an early stage of development (Lim, 1990).

White root disease causes wilting of foliage, yellow and brown discoloration of leaves, shrivelling of leaves and, ultimately, death of the tree. Belowground signs of the disease are thick, white rhizomorph threads that envelop roots. On dead tree stumps, the pathogen produces leathery, hard, orange–yellow brackets (basidiocarps) at the tree collar.

Rigidoporus lignosus causes white root disease. It is a basidiomycete with a wide host range that includes forest species, rubber, cassava and other fruit crops. It is described in Chapter 1. The disease is spread by windblown and insect-vectored basidiospores that are produced in abundance on the lower, porate surface of basidiocarps. Spread also occurs when roots come in contact with affected roots or decaying tree stumps and root fragments.

Complete clearing and removal of debris during land preparation can prevent the disease. All stumps, roots and debris of the previous crop must be removed and burned. Dead trees should be felled, all roots and stumps excavated and burned, and the planting hole should be lined with sulphur or drenched with a fungicide. Trenches should be dug around affected trees to isolate them from healthy trees, and the trench lined with sulphur. During early disease development, affected roots should be treated by removing the adhering soil and painting them with an appropriate fungicide, such as quintozene, tridemorph, penconazole or triadimenol (Lim *et al.*, 1990).

Another control option consists of establishing leguminous cover crops that foster the development and growth of antagonists, such as species of *Trichoderma* and *Gliocladium*. Lim and Teh (1990) reported that *T. harzianum*, *T. hamatum* and *T. koningii* were antagonistic to *R. lignosus*.

References

Awarun, S. (1994) Selection and application of antagonistic microorganisms to control root and stem rot of durian caused by *Phytophthora palmivora* (Butl.) Butl. MSc thesis, Department of Plant Pathology, Kasetsart University, Bangkok (in Thai).

Bhavakul, K. and Jaengsri, V. (1969) Root rot of durian. In: *Plant Diseases Control*. Agricultural Science Society, Bangkok, pp. 60–61.
Chan, L.G. and Lim, T.K. (1987) Control of *Phytophthora palmivora* on cacao and durian seedlings. *Journal of Plant Protection in the Tropics* 4(1), 9–13.
Chandrasikul, A. (1962) *A Preliminary Host List of Plant Disease in Thailand*. Technical Bulletin No. 6, Department of Agriculture, Bangkok, Thailand.
Disthaporn, S., Uttayopas, W., Chantarapannik, S., Kraturuek, C., Namroengsri, W. and Palakul, S. (1994) *Study on Integrated Pest Management in Durian, 1991–1994 Annual Report*. Department of Agricultural Extension and Department of Agriculture, Thai–German Project IPM in Selected Fruit Crops.
Johnston, A. (1960) A supplement to a host list of plant disease in Malaya. *Mycological Paper No. 77*. Commonwealth Mycological Institute, Kew, UK.
Lim, T.K. (1980) Anthracnose and related problems in some local fruit trees. *National Fruit Seminar*, Serdang, Malaysia, November 5–7, 1980, Preprint No. 15.
Lim, T.K. (1989) Studies of some lesser known mycoflora of durian: sooty mould and black mildew. *Pertanika* 12, 159–166.
Lim, T.K. (1990) *Durian Diseases and Disorders*. Tropical Press, Kuala Lumpur.
Lim, T.K. (1997) Durian – sources of resistance to *Phytophthora palmivora*. In: Johnson, G.I., Highley, E. and Joyce, D.C. (eds) *Disease Resistance in Fruits*. Proceedings of an International Workshop held in Chiang Mai, Thailand, May 18–21, 1997, ACIAR Proceedings No. 80, pp. 217–222.
Lim, T.K. and Chan, L.G. (1986a) Fruit rot of durian caused by *Phytophthora palmivora*. *Pertanika* 9, 269–276.
Lim, T.K. and Chan, L.G. (1986b) Parasitism of *Phytophthora palmivora* by *Gliocladium roseum*. *Zeitschrift fur Pflanzenkranheiten und Pflanzenschutz* 93, 509–514.
Lim, T.K. and Kamaruzaman, S. (1989a) A rot of detached durian fruits caused by *Sclerotium rolfsii*. *Pertanika* 12, 11–14.
Lim, T.K. and Kamaruzaman, S. (1989b) Occurrence of the green alga, *Trentepholia* on the trunk and branches of durian. *Planter, Kuala Lumpur* 65, 328–333.
Lim, T.K. and Teh, B.K. (1990) Antagonism, *in vitro* of *Trichoderma* species against several basidiomycetous soil-borne pathogens and *Sclerotium rolfsii*. *Zeitschrift fur Pflanzenkranheiten und Pflanzenschutz* 97, 33–41.
Lim, T.K., Ng, C.C. and Chin, C.L. (1987) Etiology and control of durian foliar blight caused by *Rhizoctonia solani*. *Annals of Applied Biology* 110, 301–307.
Lim, T.K., Hamm, R.T. and Mohamad, R.B. (1990) Persistency and volatile behaviour of selected chemicals in treated soil against three basidiomycetous root disease pathogens. *Tropical Pest Management* 36, 23–26.
Mchau, G.R.A. and Coffey, M.D. (1994) Isozyme diversity in *Phytophthora palmivora*: evidence for a south-east Asian centre of origin. *Mycological Research* 98, 1035–1043.
Pongpisutta, R. and Sangchote, S. (1994) *Phytophthora* fruit rot of durian (*Durio zibethinus*) In: Champs, B.R., Highley, E. and Johnson, G.I. (eds) *Post-harvest Handling of Tropical Fruits*. Proceedings of an International Conference held at Chiang Mai, Thailand, July 19–23, 1993. ACIAR Proceedings No. 50, pp. 460–461.
Sangchote, S. (1989) *Botryodiplodia* stem end rot of mango and its control. *Acta Horticulturae* 291, 296–303.
Sangchote, S., Pongpisutta, R. and Bunjoedchoedchu, R. (1996) Diseases of durian fruits after harvest. *Proceedings of the 34th Kasetsart University Annual Conference*, Kasetsart University, Bangkok, pp. 148–152 (abstract in English).
Singh, K.G. (1980) *A Check List of Host and Disease in Malaysia*. Bulletin No. 154. Ministry of Agriculture, Malaysia.
Subhadrabandhu, S. and Ketsa, S. (2001) *Durian. King of Tropical Fruit*. CAB International, Wallingford, UK.
Suzui, T., Kueprakone, U. and Kamhangridthirong, T. (1978) Mating types of *Phytophthora palmivora*, *P. nicotianae* var. *parasitica* and *P. botryosa* in Thailand. *Transactions of the Mycological Society of Japan* 19, 261–267.
Tai, L.H. (1971) Studies on *Phytophthora palmivora*, the causal organism of patch canker disease in durian. *Malayan Agriculture Journal* 48, 1–9.
Tai, L.H. (1973) Susceptibility of durian clones to patch canker. *Mardi Research Bulletin* 1, 5–9.
Thanh, H.V. (1998) Current status of *Phytophthora* disease of durian in the lowland of Vietnam. In: *Management of Phytophthora Diseases in Durian Workshop No. 1*, University of Melbourne, December 6–12, 1998. ACIAR Project PHT 95/134.

Thompson, A. (1934) A disease of durian tree. *Malayan Agriculture Journal* 22, 369–371.

Thompson, A. (1938) A root disease of the durian tree caused by *Pythium complectans* Braun. *Malayan Agriculture Journal* 26, 460–464.

Turner, G.J. (1971) Fungi and plant disease in Sarawak. *Phytopathological Paper No. 12*, Commonwealth Mycological Institute, Kew, UK.

11 Diseases of Fig

Themis J. Michailides
University of California, Kearney Agricultural Center, Parlier, California, USA

Introduction

The common edible fig, *Ficus carica*, probably originated in southern Arabia since the caprifig, ancestor of the edible fig, is still found there growing wild (Solms-Laubach, 1885). The species name refers to Caria, an ancient region of Asia Minor noted for its figs. Fig spread through the Old World in ancient times, and was first introduced into the New World by Spanish and Portuguese missionaries (Cuba and Peru).

Origin and Taxonomy of Important Species

F. carica is a member of the mulberry (*Moraceae*) family. It contains 60 genera and possibly >2000 species of trees, shrubs, vines and herbs. Common edible figs and their pollinating counterpart, caprifigs, are the only members of the subgenus *Eusyce* that are cultivated for their fruit.

The edible 'fruit' of *F. carica* is unusual. It is a morphologically unique syconium that is comprised almost entirely of vegetative peduncular tissue. The true fruits are tiny, pedicellate drupelets, which are found within the syconium (Fig. 11.1). *F. carica* is a gynodioecious species with two distinct forms: the monoecious, non-edible caprifig, which serves as a pollenizer, and the pistillate edible fig. Pollination is achieved by the female fig wasp, *Blastophaga psenes* (Fig. 11.2), which develops in the syconium of the caprifig. The symbiotic relationship between fig and fig wasps is an example of co-evolution between plant and insect (Galil and Neeman, 1977; Beck and Lord, 1988a,b; Brostein, 1988). Other related forms that are described as distinct species include *F. geraniifolia*, *F. palmata*, *F. serrata* (the Indian form of *F. carica*), *F. pseudo-carica* and several others that are used in breeding programmes.

History

The literature on figs is voluminous and dates from ancient times (Condit and Enderud, 1956). Figs have been prized for both medicinal and dietary value. The early Greeks considered it an honour to bestow either the foliage or the fruit, and winning athletes in the original Olympic games were crowned with fig wreaths and given figs to eat. Figs were so highly valued that their export from ancient Greece was forbidden. Athenians who informed authorities about those who illegally exported figs from Attica were named 'sycophants', a practice from which the word took its modern meaning (*sykon* in Greek means fig fruit).

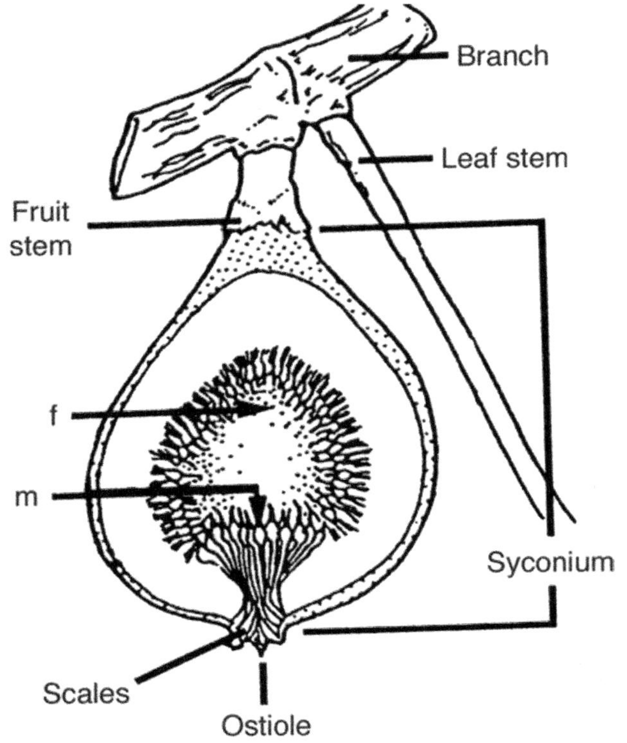

Fig. 11.1. A fig syconium in cross-section depicting the location of female (f) and male (m) inflorescences (diagram: T.J. Michailides).

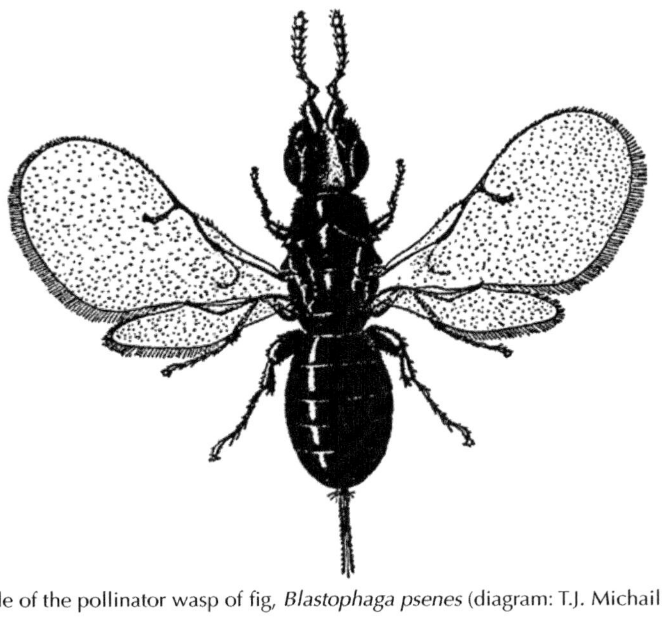

Fig. 11.2. A female of the pollinator wasp of fig, *Blastophaga psenes* (diagram: T.J. Michailides).

Attributes and Present Status

Fig fruit are enjoyed both fresh in areas where they are grown, and dried throughout the world. They recently have been described as the fruit with the highest content of fibre (Duxbury, 1986; Bamford, 1990) and minerals and polyphenols (Vinson, 1999).

In 1999, world production was almost 1.2 Mt (FAO, 2000). With few exceptions, the leading producers either were from the Mediterranean region or were areas that had Mediterranean climates (Table 11.1). Currently, Turkey is the leading country in world fig production.

Caprifigs

Although the various crops of the fig and caprifig trees are given different names in different countries, the convention in this chapter is to refer to the edible, pistillate, cultivated members of *F. carica* as 'figs' and the inedible, monoecious forms that serve as pollinators as 'caprifigs'. Caprifigs are essential for pollination of the edible fruit of the Smyrna-type cultivars. Their syconia contain both female and male flowers (hermaphrodites). The fig wasp transfers pollen from the caprifig to pistillate flowers of the edible fig, a process known to fig growers as caprification.

Table 11.1. Leading producers of fig, *Ficus carica*, in 1999.[a]

Country	Production (t)
Turkey	260,000
Egypt	216,500
Greece	80,000
Iran	78,555
Morocco	70,000
India	60,000
Spain	55,000
USA	46,000
Algeria	45,000
Syria	45,000

[a]Source: FAO (2000). Production figures are in metric tonnes.

Caprifig trees produce three to four crops of syconia annually (Fig. 11.3) (Brostein and Patel, 1992). The winter crop (mamme) is initiated in the autumn and matures in late winter, and the spring crop (profichi), initiated in late winter, matures in mid-spring. The profichi syconia contain the most pollen and are used for the pollination of the edible Smyrna-type figs.

Edible figs

Edible fig cultivars are divided into three horticultural categories, Smyrna, San Pedro and Common, which are distinguished by their cultural and pollination requirements. There are >600 named cultivars, many of which are synonymous (Condit, 1955). They are produced clonally to maintain their desirable characters.

Smyrna-type figs are grown mostly in Algeria, California, Greece, Portugal and Turkey, and usually require pollination in order to set fruit. Thus, they usually do not produce a first, breba crop in summer when pollen is not available. However, the following cultivars can set a significant amount of breba crop: 'Bardajik' (Turkey), 'Calimyrna' ('Lob Injir') (California), 'Cheker Injir' (Turkey), 'Kassaba' (Turkey), 'Tameriout' (Algeria) and 'Taranimt' (Kabylia). In California, the main cultivar, 'Sari Lop' (renamed 'Calimyrna' = California plus Smyrna), produces fair breba seedless crops, but requires pollination to set a main crop.

The San Pedro-type figs initiate moderate to large breba crops and an adequate main crop (which requires pollination) on the same branch in the same season. Cultivars of it include 'Banqhette' (Morocco), 'Dauphine' (France), 'Gentile' (England), 'King' and 'Lampeira' (Italy, Portugal) and 'San Pedro' (Spain) (Condit, 1955). 'King', which is of minor importance, is the only San Pedro-type fig used in the California fig industry (Ferguson *et al.*, 1990).

Most common fig cultivars produce a moderate to large main crop without caprification, and are thus parthenocarpic. However, climatic differences among locations can markedly alter the expression of partheno-

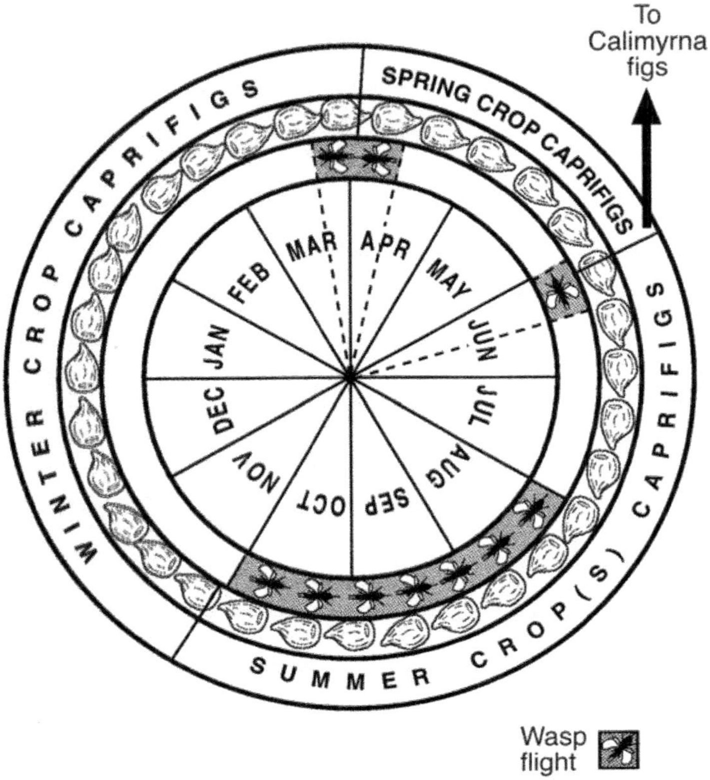

Fig. 11.3. A diagram of *Blastophaga psenes* flights and caprifig crops during a calendar year under California conditions. The fig wasp has three major flights in late March to early April, early June, and from August to October. Caprifig trees produce at least three major crops per year in winter (mamme), spring (profichi) and summer (mammoni). The spring crop caprifigs are collected by growers, and hung on trees in Calimyrna orchards (diagram: T.J. Michailides).

carpy in common figs. Common cultivars include: 'Adriatic' (Italy), 'Beall' (California), 'Black Spanish' (Spain), 'Blanche' (France), 'Bourjassotte Grise' (UK), 'Brown Turkey' (Turkey), 'Celeste' (Malta), 'Dottato' (Italy), 'Mission' (Spain) and 'Troiano' (Italy) (Condit and Swingle, 1947). Condit (1955) described 470 major cultivars in this class.

Fig fruit are marketed either dry or fresh. Figs for drying are allowed to ripen fully and partially dry on the tree. Partially dried figs then fall to the ground where they continue to dry and are later collected by hand and dried under the sun. Since commercial figs are harvested every 2 weeks, disease and insect infestation is common due to their extended contact with the soil. Grade standards vary by the producing country. Grade defects include insect infestation; smut, mould and endosepsis; sour rot; and filthy or worthless fruit.

Fig Diseases

Fig spoilage was mentioned in the Old Testament of the Bible (de Lagarde, 1881), and continues to have a major impact on the production of this fruit. Many diseases, some of which can be quite destructive, affect the fig plant and fruit. Detailed information on 18 of the most significant diseases and disorders are found below. Fungi cause most of these problems, but two are caused by bacteria, one by viruses, and another is physiological. Table

11.2 lists relatively minor problems of this crop that could not be covered in detail.

Anthracnose

This disease was described in Germany and Louisiana (Stevens and Hall, 1909; Edgerton, 1911). Both foliage and fruit are affected. Leaf lesions have a dark brown margin, and affected leaves often turn brown, dry at the edges, and may abscise. Occasionally, spots and lesions can be found on leaf petioles and blades. Symptoms on fruit range from localized, sunken lesions to a general rot of the entire fruit. Pink, slimy masses of conidia are produced in acervuli on lesions, and affected fruit may remain on the trees as dried 'mummies'.

Anthracnose is caused by *Colletotrichum caricae* (teleomorph: *Glomerella cingulata*). Conidia are hyaline and generally cylindrical or slightly elliptical. Setae, long, black, hair-like structures, are present in acervuli. Acervuli consist of a few layers of pseudoparenchymatous tissue, upon which are borne small, parallel conidiophores. Conidia are borne singly on the tips of the conidiophores. The pathogen is described in detail in Chapter 1.

Mummified fruit are a source of conidia for new infections; they are dispersed in rain and irrigation water. Removal of sources of inoculum, such as infected fruit fallen on the ground or hanging on the tree (Matz, 1918) and pruning all the dead limbs and twigs, may help to reduce the disease (Edgerton, 1911). 'De Constantine' and 'Celeste' are resistant.

Armillaria root rot

Armillaria root rot is the most important of several different root diseases in California that are caused by basidiomycetes (Adaskaveg and Ogawa, 1990). Leaves on affected trees are discoloured, severely wilted and abscise prematurely (Thomas *et al.*, 1948).

Table 11.2. Miscellaneous diseases and disorders of fig, *Ficus carica*.

Disease	Cause
Alternaria internal rot	*Alternaria alternata*, *Curvularia* spp., *Dendryphiella vinosa*, *Epicoccum purpurescans*, *Stemphylium botryosum* and *Ulocladium atrum*
Alternaria leaf spot	*Alternaria alternata* and *Alternaria* spp.
Branch wilt	*Hendersonula toruloidea*
Brown leaf spot	*Phyllosticta sycophilla*
Canker	*Tubercularia fici*
Cephalosporium leaf spot	*Cephalosporium fici*
Dagger nematode	*Xiphinema index*
Frost dieback	Abiotic disorder
Immature fruit drop	Abiotic disorder
Leaf blight	*Rhizoctonia solani* AG 1-IB
Leaf spot	*Cercospora fici*
Lesion nematode	*Pratylenchus vulnus*
Limb blight	*Erythricium salmonicolor* (synonym: *Corticium salmonicolor*)
Macrophoma canker and fruit rot	*Macrophoma fici*
Ormathodium spot	*Ormathodium fici*
Reniform nematode	*Rotylenchulus macrodoratus*
Root rots and wood decay	*Inonotus cuticularis*, *I. rickii*, *Pleurotus ostreatus* and *Schizophyllum commune*
Sclerotinia shoot blight	*Sclerotinia sclerotiorum*
Sclerotium blight	*Sclerotium rolfsii*
Stem gall and canker	*Nectriella pironii*
Sunburn	Abiotic disorder
Thread blight	*Corticium* (*Ceratobasidium*?) *stevensii*

These symptoms may be confused with those caused by asphyxiation, drought, insects and Rosellinia root rot. Death of trees usually follows periods of excessive wind, drought or heavy fruit loads (La Massèse *et al.*, 1984).

The causal agent, *Armillaria mellea*, is described in Chapter 1. Its wide host range and ability to survive as a saprophyte make it a difficult pathogen to control. If roots of previously grown plants exhibit signs of *A. mellea* during the preparation of new plantations, two fumigants, carbon bisulphide and methyl bromide (most effective), can be used.

Aspergillus mould

Although Aspergillus mould occurs in other countries, the only published work comes from California where the disease is considered the second most common fruit decayer (Doster *et al.*, 1996). In its early stages, small, brown, soft spots with white mycelia develop on the interior flesh of green fruit, usually at the eye end of the figs. In contrast to smut, water-soaked areas do not appear at the lower part of the fig around the eye. When the fruit is ripe, the pathogens produce abundant condiophores and conidia in the fruit cavity (Plate 72A and B).

Twenty-three species of *Aspergillus* have been reported on fig in California, only a few of which produce an *Eurotium* teleomorph (Doster *et al.*, 1996). Among these, *E. amstelodami* (anamorph: *A. amstelodami*) may be the most common. Other species include: *A. flavus* (Fig. 11.4), *A. parasiticus* (Fig. 11.5) and *A. tamarii* in Section *Flavi*, and *A. alliaceus, A.*

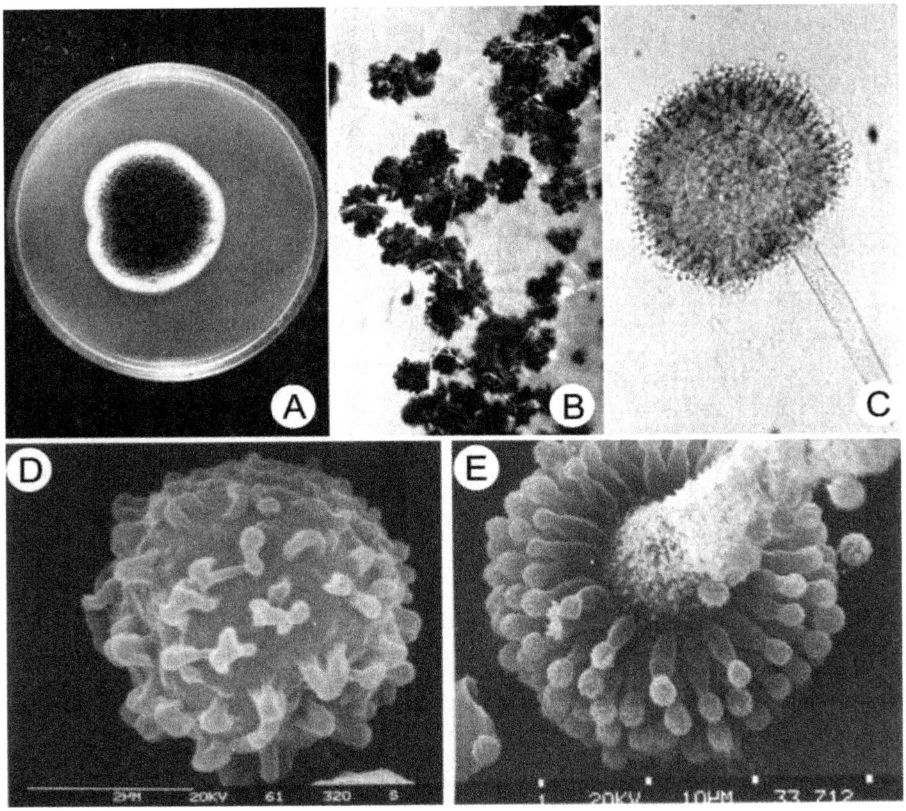

Fig. 11.4. (A) Culture of *Aspergillus flavus* on Czapek agar after 1 week at 25°C, (B) conidial heads (note splitting), (C) conidial head, (D) SEM of a conidium and (E) SEM of a conidial head of the fungus (from CMI descriptions nos 91 and 1251).

Fig. 11.5. SEM of (A) a conidial head and (B) conidium of *Aspergillus parasiticus* (from CMI description no. 994).

melleus, *A. ochraceus* and *A. sclerotiorum* in Section *Circumdati*. Although only a small percentage of figs are affected by Aspergillus mould, it should be considered a serious problem since some of the agents produce carcinogenic aflatoxins (*A. flavus* and *A. parasiticus*) or ochratoxins (*A. alliaceus*, *A. melleus*, *A. ochraceus* and *A. sclerotiorum*). The conidia of *A. flavus* are pale green and finely roughened, those of *A. parasiticus* are dark green and very rough, and those of *A. ochraceus* are 2.5–3.0 µm, finely rough-walled and ochre–yellow to olivaceous.

Figs become more susceptible as they mature (Doster and Michailides, 1997). Wounding does not increase infection or aflatoxin contamination in very immature or very mature figs, and insect damage does not predispose fruit to infection. However, insects could play an important role in transmitting spores of these fungi. Reducing dust in orchards may reduce the colonization of figs by these pathogens.

Bacterial canker

Bacterial canker is the only bacterial disease that has been reported on figs. In California, it has only been found on limbs that are >4 years old, but in Italy it causes dark lesions on leaves and elongated lesions and wilting on new shoots. Symptoms on limbs start as wet spots, which dry later, causing cracks 1.25–2.5 cm in length. These small surface cracks may cover much larger cankers (to 15 cm) on the cambium (Hansen, 1948). Cankers can girdle and kill limbs.

Bacterial canker is caused by *Pseudomonas fici*. The bacterium is Gram negative, 0.5–0.6 × 1.5–2.6 µm, motile with four to five flagella, and has optimum and maximum temperatures for growth of 15 and 35–37°C respectively. Colonies on nutrient agar are white to creamy, and on King's B medium produce a fluorescent pigment.

Infected shoots should be pruned (Pilgrim, 1950a). A 2% Bordeaux mixture should be applied when buds are opening, followed by a 0.75% mixture after they are open, a 1% mixture at the end of April, and a 1.5% solution in August. In California, the Adriatics and caprifigs are more susceptible than Kadotas, Calimyrnas and Missions (Pilgrim, 1951).

Botrytis limb blight (dieback) and fruit rot

First reported by Prunet (1903), the disease is now widely recognized (Massee, 1911; Condit and Stevens, 1919; Manziano *et al.*, 1990). It can blight a significant number of shoots in trees during wet and cool springs.

Symptoms

Initial infections start at the base of young shoots that wilt, bend and develop light green to brown leaves. Blighted shoots and fruits of both fig and caprifig can be covered with abundant buff sporulation of the pathogen (Figs 11.6A and B). Under extreme wet conditions, infected midribs and the surrounding areas of leaves turn black (Manziano *et al.*, 1990).

Causal agent

The disease is caused by *Botrytis cinerea* (teleomorph: *Botryotinia fuckeliana*), a cosmopolitan ascomycete on numerous crops. It is described in Chapter 1.

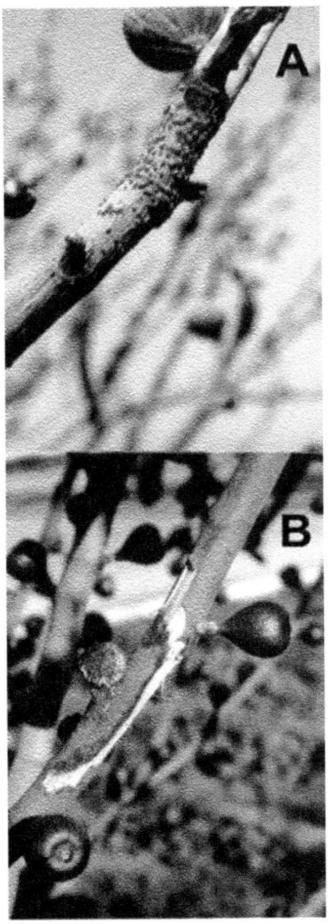

Fig. 11.6. (A) Sporulation of *Botrytis cinerea* on a fig shoot affected by Botrytis blight (dieback) and (B) a canker initiated from a fruit affected by Botrytis rot (photos: T.J. Michailides).

Isolates from fig orchards produce sparse to abundant conidia and sclerotia. Sclerotia are small, hard, black survival structures and often are observed firmly attached to the outside of decayed fig fruits (Fig. 11.6B). Carpogenic germination to produce apothecia can occur after a dormant period of 2–6 months, but has not been reported in fig orchards.

Epidemiology

B. cinerea can infect through injuries on fruit, and leaf and fruit scars to cause cankers below and above the entry point (Fig. 11.6B). Infected shoots blight, resulting in fruit blight. Conidia from blighted shoots and fruits are disseminated by air and initiate new infections during cool, rainy weather on susceptible tissues (leaf or bud scars) or weakened shoots and fig fruits. English (1962) showed that winter injury to fruits or branches followed by infection by *B. cinerea* caused shoot dieback.

This disease is more common on caprifigs than on 'Calimyrna' figs. Since the pathogen sporulates on caprifigs as late as July, these shoots probably constitute an important inoculum source (Ricci, 1972). Conidia contaminate the mamme crop of caprifigs and, unless these are treated before storage, a significant percentage decay. Although control of this disease in figs has not been studied, dipping the mamme crop in iprodione (Rovral 50W) before cold storage can significantly reduce the disease.

Endosepsis

Endosepsis occurs in California, Greece, Turkey and possibly other areas. Endosepsis is also called 'brown rot', 'eye-end rot', 'pink rot' and 'soft rot', terms that describe various phases and symptoms of the disease (Caldis, 1927). In California, endosepsis, smut and souring diseases can cause combined losses of up to 50% (Hansen, 1928). Handpicking and sorting of fruit for later treatment also raise the costs of production in some districts.

Symptoms

Any cultivar that requires pollination by the fig wasp can be affected. Green fruit exhibit brown or rust coloured areas on flower stigmas, flower bases or the entire flower. Upon ripening, the brown streaks become yellow–brown spots that involve a number of flowers and can develop on any part of the pulp, but usually are found near the eye. No external symptoms are evident at this stage. When fruit soften, the skin appears water soaked, usually around the eye in a circular spot and extending down the sides to the neck. The water-soaked areas gradually turn pink or purple. In other cases, only a small water-soaked ring appears around the eye, and a drop of dense, clear or amber syrup is exuded. This is common in 'Calimyrna', but not caprifig, fruits. The pulp becomes amber and is considerably darker than the beige colour of healthy fruit (Fig. 11.7). Affected 'Calimyrna' fruit will dry and appear healthy if the pulp is not greatly affected. However, even slightly affected pulp has off flavours. A 'slip-skin' condition can also develop whereby, the fruit epidermis is removed easily.

Causal agents

The most common endosepsis agent is *Fusarium lactis* (O'Donnell *et al.*, 1998), and less common are *F. solani* and *F. dimerum* (Michailides *et al.*, 1987; Wilson and Ogawa, 1979).

F. lactis produces microconidia in chains on polyphialides and few, small falcate macroconidia (Gerlach and Nirenberg, 1982). Some of the microconidia are pyriform. Colonies are white or light pink, with aerial hyphae. Unless indicated otherwise, the discussions below pertain to this fungus.

Epidemiology

The pathogen overwinters in mummified mammoni (summer crop) caprifigs and in mamme (winter crop) caprifigs until April of the following year (Plate 73). Propagules of the pathogen can also be found on the surface of caprifig and 'Calimyrna' trees (Michailides *et al.*, 1987, 1996). Conidia of the fungus are introduced from these locations into caprifig and 'Calimyrna' fruit by female fig wasps as they enter to lay eggs. Since the pathogen sporulates at approximately the same time that adult female wasps emerge, the wasps become contaminated with conidia and transfer them to healthy, developing fruit of the following caprifig crop, or from profichi caprifigs to 'Calimyrna' fruit during the caprification process (Caldis, 1927). Parthenocarpic cultivars are not infected by *F. lactis* unless the wasps have entered them accidentally. The pathogen first invades the dead stigmas, but eventually fills the interior of the fruit as it begins to ripen.

Endosepsis decreases with distance from an inoculum source, decreasing faster to the south than in other directions (Michailides and Morgan, 1998). Although endosepsis can complete as many cycles per year as its vector in caprifigs (polycyclic disease), there is no secondary spread in 'Calimyrna' and 'Smyrna' figs (Michailides and Morgan, 1998).

Management

In California, control efforts are directed towards reducing the proximity of 'Calimyrna' to caprifig trees and producing clean female fig wasps. Caprifigs are now planted in separate orchards, usually far

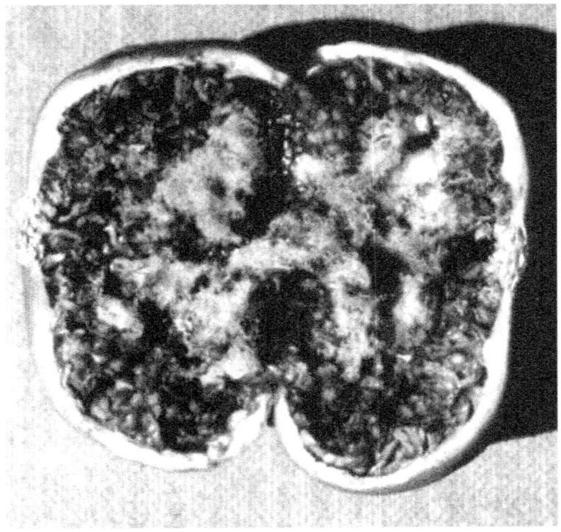

Fig. 11.7. Endosepsis on a 'Calimyrna' fig (photo: T.J. Michailides).

from 'Calimyrna' crops. Excess caprifig trees and those in abandoned orchards are destroyed (Smith and Hansen, 1935). Dipping mammes in a solution of benomyl, chlorothalonil, DCNA and potassium sorbate lowers the incidence of *F. lactis*, as well as *Rhizopus* and *Alternaria* (Obenauf *et al.*, 1982).

Good sanitation practices are even more beneficial than chemical treatments (Michailides *et al.*, 1996). Discoloured mammes should be discarded, since wasps from these fruits usually are contaminated with the pathogen (Michailides and Ogawa, 1989). By treating disease-free mamme fruit, endosepsis is reduced in the following profichi fruit. Profichi figs that show external symptoms of discoloration should be discarded before hanging them in 'Smyrna' fig trees for pollination (Fig. 11.8). In California, caprifig growers test profichi for infection by *Fusarium* spp. (Michailides *et al.*, 1994). If the caprifig orchard is close to the edible fig orchard, surplus profichi should be destroyed before they are mature (Smith and Hansen 1935). Avoiding over-caprification is another way to reduce endosepsis in 'Smyrna' type figs (Michailides and Morgan, 1994).

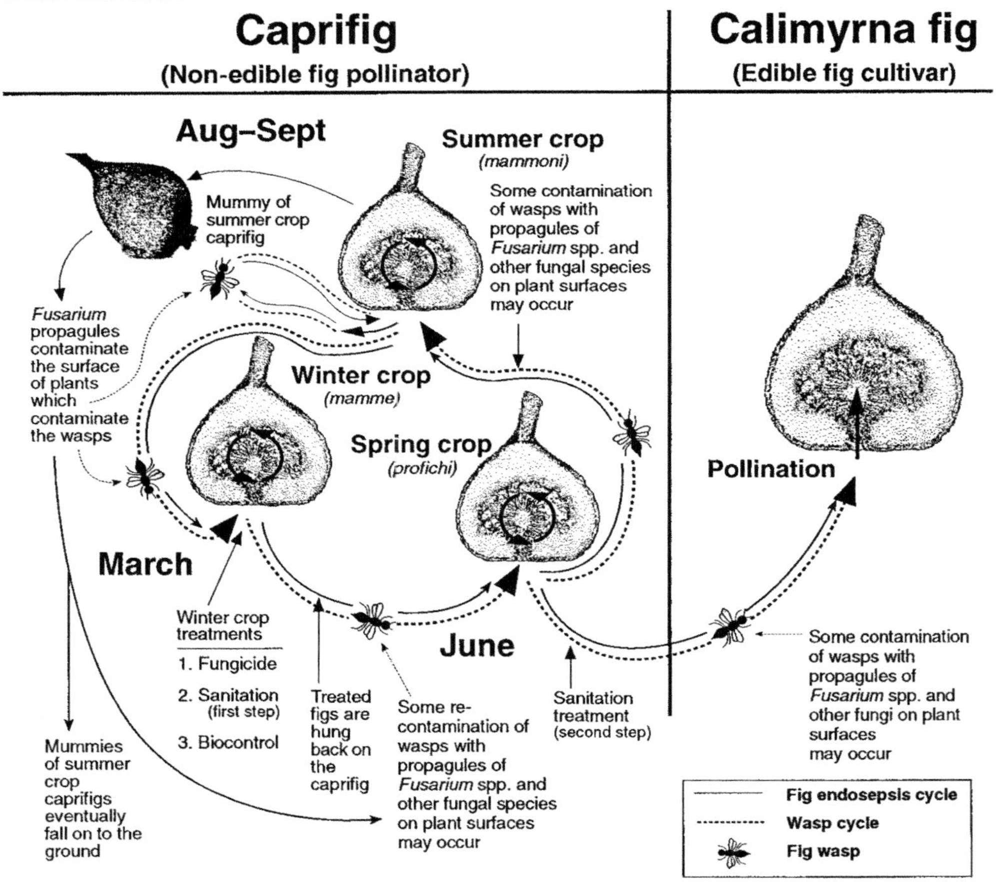

Fig. 11.8. Diagram depicting pollination in non-edible caprifigs and edible 'Calimyrna' figs, the life cycle of fig wasp, *Blastophaga psenes*, and the disease cycle of fig endosepsis caused by *Fusarium lactis* and other *Fusarium* spp. Depending on weather conditions, one to two additional wasp and disease cycles can be completed during the summer and early autumn in California. The circular arrows inside the caprifigs indicate successful development and emergence of adult female wasps through the fig ostiole.

Fig foot rot

Fig foot rot was first reported in Brazil, and in 1982 in Japan (Kato *et al.*, 1982). Initial symptoms are wilting twig tips followed by leaf fall. Branches dry and entire trees may die. Longitudinal sections of the trunk show necrosis in cortical tissues and the cambium.

The disease is caused by *Ceratocystis fimbriata* (anamorph: *Chalara* sp.). Its perithecia are superficial or immersed in the substrate, globose, 90–210 μm in diameter, with a neck 29–715 μm in length, 25–37 μm at the base and 15–25 μm wide at the tip. The number of ostiolar hyphae ranges from eight to 14, with a mean width of 2–3 μm. The ascospores are unicellular, hyaline, hat shaped and 2.7 × 6.4 μm (Valarini and Tokeshi, 1980). The fungus produces macroconidia and endoconidia.

Valarini and Tokeshi (1980) assumed that the beetle *Xyleborus ferrugineus* was a vector of the pathogen based on its frequent association with diseased plants. In the field, 'Branco' and 'Portugues' were resistant when they were used as rootstocks for 'Roxo de Valinhos' (Valarini and Tokeshi, 1980).

Fig mosaic

Fig mosaic has been reported in the British Isles, California, Greece, Italy, Libya, New South Wales and Yugoslavia. The first complete description of the disease was given by Condit and Horne (1933). In California, it is thought to have originated on 'Samson' caprifigs that probably were introduced from Asia Minor in 1882. The disease then spread through cuttings from this cultivar (Condit and Horne, 1933, 1941). It is also common on edible figs.

Symptoms

On leaves, mosaic spots are distinctive, contrasting with the normal green colour of the foliage (Fig. 11.9). The margins of the yellow spots blend gradually through a light yellow colour into the dark green of healthy tissue. Mosaic spots or lesions may be scattered uniformly over the surface of the leaves or may appear as irregular patches of light green diffused widely throughout the leaf blade. Later in the season, a rust-coloured band develops along the border of the mosaic spots, appar-

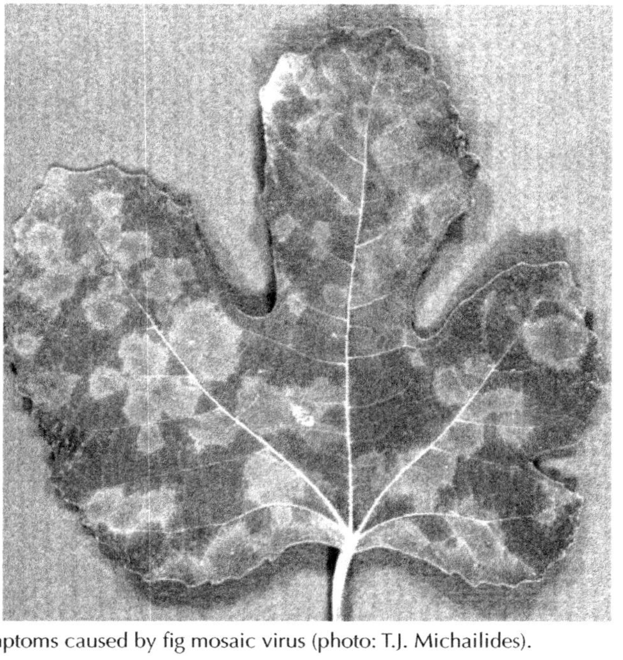

Fig. 11.9. Foliar symptoms caused by fig mosaic virus (photo: T.J. Michailides).

ently caused by the death of epidermal or subepidermal cells. Malformed leaves, the size and shape of which vary greatly, also develop.

Mosaic spots on fruits are similar to those on leaves. Premature fruit drop is sometimes associated with the presence of fruit or leaf symptoms on certain cultivars (Condit and Horne, 1943).

Causal agent

Fig mosaic virus causes fig mosaic. To date, it has not been characterized or isolated. Although Quacquarelli (1971) isolated *Sowbane mosaic virus* from figs with mosaic symptoms, it does not cause fig mosaic (Plavsic and Milicic, 1980).

Symptoms of fig mosaic are not transmissible mechanically. They are associated with polymorphic particles that are present only in symptomatic plants, and are either spherical and 120–160 nm in diameter or elongated and 200 nm in length (Plavsic and Milicic, 1980).

Epidemiology

Fig mosaic is not seedborne, but can be transmitted by grafting and females (but not males) of the eriophyid mite, *Aceria ficus* (Flock and Wallace, 1955; Smith, 1972). A single female is capable of transmission, and minimal acquisition and infection feeding periods are <15 min. For optimal transmission, a feeding period of 24 h on the source and test plants is necessary (Smith, 1972). Symptoms are produced <10 days after feeding (Jeppson *et al.*, 1975).

Management

'Black Mission' is the most seriously damaged cultivar, and 'Kadota' and 'Calimyrna' are the least affected. 'Black Bursa' is affected in Turkey. *F. palmata*, or progeny with *F. palmata* as the male parent, appear to be immune (Condit and Horne, 1933).

Care must be taken to ensure that stocks are not propagated from mosaic-affected trees (Khalil, 1982). Apical meristem culture and thermotherapy of alternating light (16 h 5000 lux at 37°C) and dark (8 h at 34°C) are effective (Gella *et al.*, 1997).

Fruit splitting

This is a severe problem in some years, especially in 'Calimyrna' and 'Adriatic' figs. Splitting is the result of sudden changes in the internal fruit pressure brought on by either cool temperatures and high humidity as the fruit matures or showers during the fruit-ripening period (Obenauf *et al.*, 1978). Splitting of 'Calimyrna' also can result from over-caprification (too many seeds are developed).

Splitting begins in the eye of fruit, and ruins the fruit, since the pulp is exposed to insect and fungus attack. Over-caprification can be prevented by reducing the number of caprifigs or by isolating caprifig plantings. Insect control may prevent further spoilage of split fruits.

Phomopsis canker

Phomopsis canker, which is also known as fig canker, occurs on all commercial cultivars of fig in California. 'Kadota' is seriously affected, 'Calimyrna' suffers minor damage, and 'Mission' and 'Adriatic' are rarely attacked (English, 1951). The disease has also been reported in Italy (Manziano *et al.*, 1990).

Symptoms

Dead bark and wood develop on pruning wounds or on injuries caused by frost or sunburn. The bark in the older portions of cankers becomes bleached, cracked and sunken (Wilson and Ogawa, 1979). Pycnidia of the pathogen develop on the outer bark layers. Cirrhi of conidia exude from pycnidia under humid conditions. Elongation of elliptical cankers results in yearly zonations, with the fungus sporulating more profusely in the outermost zone. Enlarged cankers girdle and kill branches on which the withered foliage remains.

Causal agent

The disease is caused by *Phomopsis cinerascens* (teleomorph: *Diaporthe cinerascens*). Its pycnidia are black, 250–500 μm, globose and immersed in the bark from which their ostioles emerge. They produce two types of pycnidiospores on host bark: α-conidia, which are elliptic–fusoid and 6–9 × 2–2.5 μm; and β-conidia, which are filiform, 20–25 × 1 μm, mostly hooked and on short pedicels.

Epidemiology

P. cinerascens is a wound parasite. The frequent association of Phomopsis canker with pruning wounds explains the disease's importance on 'Kadota', a cultivar that is pruned heavily (Hansen, 1949a, b), and infection through leaf scars probably occurs. The same fungus causes a canker on weeping fig, *F. benjamina*, and a twig blight on the same species indoors (Hampson, 1981; Anderson and Hartman, 1983).

The fungus survives in cankers and on pruned branches that are left on the ground. During rainy weather, the spores are exuded from pycnidia and washed down along the branches and trunk of the tree. They germinate and start new cankers wherever they come in contact with an unhealed wound. Pruning tools contaminated with pycnidiospores can disseminate the fungus.

Canker extension on 'Kadota' trees is restricted between April and October, apparently due to active growth of the host. However, from November to February, they are highly susceptible (English, 1952).

Management

Although removing and burning infected branches from young trees is beneficial, older trees with a high incidence of cankers may need to be cut back to the trunk and treated with a protective fungicide (English, 1953). Young, non-bearing trees should be pruned late in the dormant season, and pruning tools that are used on affected trees should be disinfested with dilute sodium hypochlorite (2500 μm ml^{-1}) (Pilgrim, 1950b; English, 1953). Because vigorous trees appear to be least susceptible, cultural practices that are designed to keep the trees in good health should be followed.

Phytophthora fruit rot

Phytophthora fruit rot is caused by *Phytophthora palmivora*, which is described in Chapter 1. It was first reported in Japan, and is also known in Florida, India, New South Wales and Taiwan (Nisikado *et al.*, 1941; El-Gholl and Alfieri, 1984).

The fungus infects both wounded and non-wounded green fruits, creating water-soaked lesions that are covered with white fluffy mycelium of the pathogen. The disease is more common in the lower part of the tree, but fruits and limbs in higher parts of the trees are also affected. In Japan, the pathogen caused a white powdery rot on fig fruits, leaves, young buds and young stems (Katsura and Yamamoto, 1969). Isolates of *P. palmivora* from 'White Genoa' figs caused rot on apple, Japanese pear, persimmon, aubergine, tomato and potato (Nisikado *et al.*, 1941).

Root-knot nematodes

Root-knot nematodes are probably the most important and widespread parasites of figs (McBeth, 1949). Although the exact species of *Meloidogyne* that are involved have not been reported, *M. arenaria* has been isolated from roots of ornamental figs (Davide, 1979). Trees in sandy soil are very susceptible (Matz, 1918).

Root-knot nematodes cause characteristic knots or galls on fig roots. The adult female is pearl white, pyriform and ~2 mm long. Egg masses as large as her body are often found near or attached to her body, but they may also be found outside the gall. In the juvenile stages, this nematode is long and slender. The adult males retain their vermiform shape while the adult females become pyriform as they reach maturity. The females usually develop just beneath the surface of the root and, as a result of their growth and secretions, induce the formation of galls. Heavily infected roots may actually die. As a result, affected plants have reduced absorption and transport of water and nutrients, and exhibit stunting, chlorosis and reduced fruit production (McBeth, 1949).

To date, no research has been done to explore resistance to nematodes in *Ficus* spp.

Fumigation before the establishment of a fig nursery can limit the dissemination of nematodes, but in orchards is very expensive and may need to be limited only to the planting rows (McKenry and Thomason, 1975; Carles, 1985).

Rosellinia root rot

Rosellinia root rot was recorded in figs grown the Aegean plains in Turkey (Igriboz, 1940; Karaca, 1979). It made the Turkish fig industry move gradually to the more hilly areas of the Aegean Region. Rosellinia root rot is a more common and devastating disease of fig than Armillaria root rot (La Massèse *et al.*, 1984).

Armillaria root rot and Rosellinia root rot can be distinguished reliably only on the basis of underground signs. *Rosellinia necatrix* (anamorph: *Dematophora necatrix*) is described in Chapter 1. It produces cottony white mycelia on root surfaces that are significantly different from the rhizomorphs and mycelial plaques formed by *A. mellea*. As an ascomycete, *R. necatrix* also does not produce basidiomes.

Controlling Rosellinia root rot is extremely difficult. Fumigants, such as allyl bromide, ammonium hydrosulphide, carbon bisulphide, carbon tetrachloride, chloroform, chloropicrin, ethylene dibromide and formalin pentachlorethane have all failed to control the disease. Removal of all infected roots from planting sites is also futile since the fungus can survive in small pieces of dead roots.

Rust

Fig rust is widely distributed on *F. carica*, other species of *Ficus*, *Morus alba*, *M. nigra* and species of *Broussonetia* and *Maclura*. The disease has been reported in Bermuda, Egypt, India, Italy, New Zealand, the USA (Arkansas and Florida), Venezuela and Yugoslavia (Matz, 1918; Dale, 1958).

Symptoms

Angular leaf spots are pale and reddish brown due to the formation of uredinia. Severe cases on fig result in premature defoliation, leaving only immature fruit that may drop (McKenzie, 1986).

Causal agent

Rust is caused by *Cerotelium fici* (Laundon and Rainbow, 1971). Pycnia and aecia of this fungus are unknown. Uredinia are hypophyllous and scattered on the fruit or sometimes grouped around the edge of leaves or in dew drop runs on lamina (McKenzie, 1986). Urediniospores are globose, obovoid, ellipsoidal, or angular, 19–30 $\times$ 15–23 μm and sparsely echinulate. Telia have been reported only twice (Laundon and Rainbow, 1971). Teliospores are hyaline, smooth, barrel shaped, oblong or ovoid, 14–22 $\times$ 10–13 μm, form in loose chains and tend to separate.

Epidemiology

No information is available on physiological specialization in *C. fici*, and its life cycle is incompletely known (Laundon and Rainbow, 1971). In the southern USA, it is believed to overwinter as urediniospores on fallen leaves that infect young leaves in the spring (Krezdorn and Adriance, 1981).

Management

Fallen leaves should be removed from the orchard and destroyed. Fig rust can be controlled with sulphur dusts, 0.05% Vitavax (Sinhri and Mishra, 1984), or Bordeaux mixture when applied to young leaves (Laundon and Rainbow, 1971). Other fungicides, such as zineb and maneb, are also effective if applied before the appearance of disease (Assawah *et al.*, 1966).

Smut

Smut is one of the most studied diseases of figs. It is not a true smut, but refers to the black powdery conidia of the causal fungi that form on affected figs.

Symptoms

In its early stages, small, brown, soft spots develop on the interior flesh of green fruit while they are still on the tree (Phillips *et al.*, 1925). Decay usually begins at the eye end and progresses as water-soaked areas around the eye. When fruit are ripe, the fungus produces abundant black masses of conidia in the fruit cavity (Fig. 11.10).

Causal agents

The disease is caused by species of *Aspergillus* in Section *Nigri*. Initially, it was thought that only one species, *A. niger*, was involved; it is described in Chapter 1. However, Doster *et al.* (1996) showed that five distinct, black-spored taxa are involved in California: *A. niger* var. *niger*, *A. niger* var. *awamori*, *A. japonicus* var. *aculeatus*, *A. japonicus* var. *japonicus* and *A. carbonarius*. *A. niger* vars *niger* and *awamori* were responsible for 93% of the affected 'Calimyrna' figs in 1993 and >99% of the 'Conadria' figs in 1994 (Doster *et al.*, 1996).

Epidemiology

Hodgson (1918) first proposed that insects were responsible for transmitting smut. Transmission tests showed that the dried fruit beetle, *Carpophilus* spp., and the vinegar fly, *Drosophila melanogaster*, were able to transmit spores of the fungus to edible figs (Phillips *et al.*, 1925). More recently, Michailides *et al.* (1991) showed that beetles infested with *A. niger* and caged with healthy 'Calimyrna' figs did not increase levels of smut relative to fruit that were inoculated with *A. niger* alone, but that dusting figs with soil powder and *A. niger* did.

Thrips were also implicated in carrying fig spoilage fungi including *A. niger* (Hansen, 1929; Smith and Hansen, 1931; Davey and Smith, 1933; Baker, 1939). The fig mite, although present in essentially all ripening figs, has not been implicated as a carrier of any fruit pathogens (Hansen and Davey, 1932). In the Smyrna fig district of Turkey, Hagan (1929) proposed that an ant, *Formicomus ionicus*, which entered ripening fruit was a prolific disseminator of smut spores and a greater threat in Turkey than *Carpophilus* spp.

Management

Fruit of the 'Black Mission' and 'Kadota' cultivars are affected less than those of 'Adriatic' and 'Calimyrna' (Phillips *et al.*, 1925). Removing old fruit culls and refuse on which dried fruit beetles and vinegar flies might persist and preventing dust in the orchard may help to reduce decays caused by these fungi.

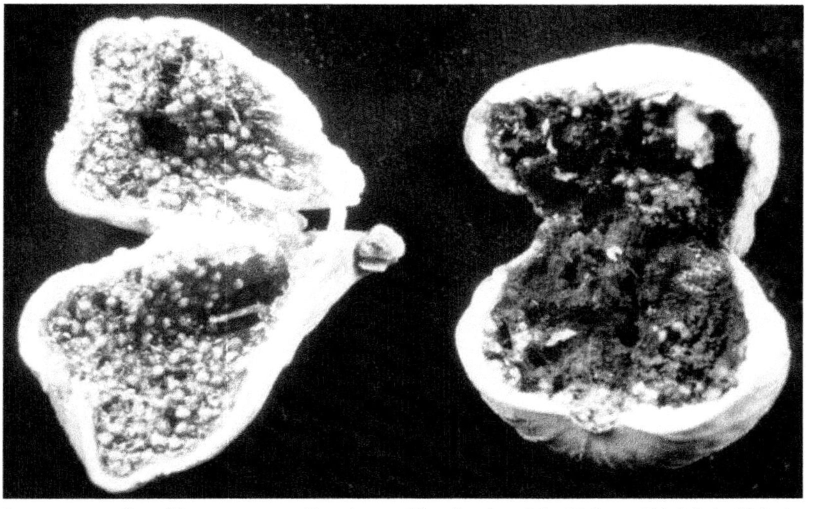

Fig. 11.10. Smyrna type fig with smut caused by *Aspergillus* Section *Nigri* (photo: T.J. Michailides).

Soft rot

The disease was first described in Louisiana (Edgerton, 1911). It also occurs in California, Italy, Texas and Turkey (Taubenhaus, 1936; Manziano *et al.*, 1990). Mature fruit become mushy, are eventually covered by mycelium and sporangia of the pathogen, and ultimately fall from the tree.

Soft rot is caused by *Rhizopus stolonifer* (Fig. 5.3) and *Mucor* spp. It is found regularly in the cavity of 'Calimyrna' and 'Conadria' figs (T.J. Michailides and M.A. Doster, unpublished data). A similar soft rot of fig fruit is caused by *Choanophora cucurbitarum* (Taubenhaus, 1936).

Soft rot is very common when fruits are ripening during rainy weather in the summer. Rains during ripening result in cracks near the eye. Airborne sporangiospores of the causal fungi are abundant and can infect fruit through these cracks. Spores are also disseminated by the dripping juice of affected fruit, rainwater, wind and insects (Edgerton, 1911).

Nearly all large, soft-skinned cultivars are very susceptible to this decay, while smaller figs, such as 'Celeste' and 'Reine Blanche', are less so. During conducive conditions, little can be done to control this disease. Edgerton (1911) suggested that figs should be harvested regularly and that ants be kept out of trees.

Souring

Fermentative spoilage or 'souring' of figs is a common problem. It occurs while fruit are still on the tree.

Symptoms

Symptoms are most distinct on fruit of parthenocarpic cultivars, such as 'Adriatic', that have not been caprified (Caldis, 1930). In caprified figs, the symptoms of souring may be confused with those of endosepsis. The symptoms are noticeable only when the fruit ripens and the eye is wide open.

Deterioration of any kind does not begin before the eye opens, and figs with closed eyes have sterile pulp (Phillips *et al.*, 1925). Affected figs develop a pink colour and later become water soaked. A pink syrupy liquid exudes through the eye, dropping on to the leaves or congealing at the eye. The gas bubbles and characteristic alcohol odour that develop indicate fermentation of the pulp. In later stages, the pulp disintegrates and usually is covered by a white scum. Affected fruits lose firmness, sag, turn black, shrivel and either drop or hang on the twig. A dead spot or 'eye canker' often develops where the peduncle is attached to the tree.

Causal agents

Souring is caused by yeasts and bacteria (Mrak *et al.*, 1942). From affected fruit, Miller and Phaff (1962) recovered species of yeast in *Saccharomyces* and *Pichia*, and of the apiculate yeasts *Hanseniaspora kloeckera*, *Candida* and *Torulopsis*. Approximately the same yeast flora are found on the exterior and in the intestinal tract of a beetle, *Carpophilus hemipterus*, which feeds on soured figs. Because Koch's postulates were not performed with each of these yeasts, it is not known whether they all cause souring.

Epidemiology

Insects transmit the organisms that cause souring (Howard, 1933). Beetles and fruit fly species that have been implicated as vectors apparently cannot enter the fig until its ostiole opens upon ripening, and yeasts that cause fermentation were not present in experimental figs until after the ostioles had opened. Periodic swarms of the cotton leaf worm moth, *Alabama argillacea*, also transmit the disease during feeding (Hull, 1929).

Management

Exclusion of insects from trees almost completely eliminated souring (Phillips *et al.*, 1925), although it did not prevent smut and mould (Davey and Smith, 1933). Souring was also reduced when figs were protected either by bagging or by closing the ostiole with tanglefoot (Howard, 1933).

Surface mould or contact spot

Surface moulds or contact spots are caused by *Alternaria alternata*, *Aspergillus niger* and *Cladosporium herbarum*. Other fungi that are involved include *A. flavus*, *A. parasiticus* and occasionally other species of *Aspergillus* and *Penicillium* (T.J. Michailides, unpublished data). Some of these diseases were first reported in California (Smith and Hansen, 1931). All fig cultivars are susceptible, but Alternaria surface mould is most severe on 'Kadota' figs. Alternaria surface rot is a limiting factor in the production of fresh fruit or fruit for canning (Bewaji *et al.*, 1977).

Symptoms

Surface moulds occur on both green and ripe fruit, developing mainly on the surface of fruit that are in contact with other fruit (Fig. 11.11). Symptoms differ according to the causal agent. For *C. herbarum*, initial dark olive green specks develop and enlarge to become depressed and yellowish olive in colour. The first symptoms of Alternaria fruit rot are water-soaked areas, which soon are covered with olivaceous conidia of the fungus. The first symptoms of Aspergillus infections are water-soaked areas, which soon are covered with white, then yellow, and finally black conidiophores of *A. niger* or yellow-greenish to yellow-dark greenish conidiophores of *A. flavus* and *A. parasiticus*.

Causal agents

A. alternata produces conidiophores on the surface of infected areas, bearing long, sometimes branched, chains. *A. niger* produces conidiophores bearing heads of conidia in chains on the surface of infected areas. Both are described in Chapter 1. *A. flavus*, *A. ochraceus* and *A. parasiticus* are described under Aspergillus mould in this chapter. Conidiophores of *C. herbarum* are erect, arising in tufts from the epidermal cells. It is described in Chapter 6 under Cladosporium spot.

Epidemiology

C. herbarum, *A. alternata* and *Aspergillus* spp. are common in the air, on plant surfaces and in soil (Ingol, 1971), usually developing on dying or dead tissues of plants. On figs, *C. herbarum* and *A. alternata* can be found on tissues of green fruit damaged by frost or sunburn and on wounded green or mature fruit. While *C. herbarum* predominates on green fruit, *A. alternata* is a serious problem on ripe fruit (English, 1954). If rains prevail during harvest, Alternaria rot becomes a problem in 'Calimyrna' fruit, usually developing on the surface where figs touch.

Fig. 11.11. Surface or contact spot on 'Calimyrna' figs (photo: T.J. Michailides).

An important source of inoculum is the picking box. Wooden boxes become contaminated with fungi during the season, and 14 different species were isolated from the inner surfaces of picking boxes, with *C. herbarum* and *A. alternata* being most prevalent (Harvey, 1956). Wooden boxes have been replaced with plastic ones that are easier to disinfest and keep clean.

Management

Sprays with protectants, such as zinc coposil and Dithane Z78 (zinc ethylene bis-dithio-carbamate), in the first week of September resulted in reduced disease caused by *C. herbarum* and *A. alternata* (English, 1954). A combination of benomyl, potassium sorbate, DCNA (Botran 75W) and chlorothalonil appeared to be very promising in reducing Alternaria rot incidence in profichi caprifigs (Obenauf *et al.*, 1982). Maneb is registered for spraying 'Kadota' figs to control Alternaria rot in California (Dibble *et al.*, 1972). On fresh-market fruit, preharvest and postharvest sprays with chlorothalonil reduced Alternaria surface rot more effectively than benomyl or maneb (Bewaji and English, 1976; Bewaji *et al.*, 1977).

Picking the fruit before it becomes over-ripe reduced Alternaria rot, and a modified atmosphere of ≥23% CO_2 during storage and transit was as effective as immediate storage at 0°C (Brooks and McColloch, 1938). To prevent contamination of harvested fruits, picking boxes should be kept clean and washed regularly.

Acknowledgements

I thank M.A. Doster and D.P. Morgan for their technical assistance in the studies on endosepsis and smut, J. Doyle and L. Ferguson for useful discussions on fig, and the California Fig Instutute (Fresno, California) for financial support of research on diseases of figs.

References

Adaskaveg, J.E. and Ogawa, J.M. (1990) Wood decay pathology of fruit and nut trees in California. *Plant Disease* 74, 341–352.

Anderson, R.G. and Hartman, J.R. (1983) Phomopsis twig blight on weeping fig indoors: a case study. *Foliage Digest* 6(1), 5–7.

Assawah, M.W., Amr, K.M. and Boulos, Z.Y. (1966) Studies on determining the suitable period and intervals for controlling the rust of fig trees. *Alexandria Journal of Agricultural Research* 13, 339–352.

Bamford, R. (1990) Use of figs and fig products in bakery foods. *AIB Research Department Technical Bulletin* 12(10), 1–6.

Baker, E.W. (1939) The fig mite and other mites of the fig tree. *California Department of Agriculture Bulletin* 28, 266–275.

Beck, N.G. and Lord, E.M. (1988a) Breeding systems in *Ficus carica*. The common fig. I. Floral diversity. *American Journal of Botany* 75, 1904–1912.

Beck, N.G. and Lord, E.M. (1988b) Breeding systems in *Ficus carica*. The common fig. II. Pollination events. *American Journal of Botany* 75, 1912–1922.

Bewaji, O. and English, H. (1976) Control of Alternaria surface rot of Kadota figs (abstract). *American Phytopathological Society Proceedings* 3, 278.

Bewaji, O., English, H. and Schick, F.J. (1977) Control of Alternaria surface rot of Kadota figs. *Plant Disease Reporter* 61, 351–355.

Brooks, C. and McColloch, L.P. (1938) Spotting of figs on the market. *Journal of Agricultural Research* 56, 473–488.

Brostein, J.L. (1988) Mutualism, antagonism, and the fig–pollinator interactions. *Ecology* 69, 1298–1302.

Brostein, J.L. and Patel, A. (1992) Temperature-sensitive development: consequences for local persistence of two subtropical fig wasp species. *American Midlands Naturalist* 128, 397–403.

Caldis, P.D. (1927) Etiology and transmission of endosepsis (internal rot) of the fruit of the fig. *Hilgardia* 2, 287–328.

Caldis, P.D. (1930) Souring of figs by yeasts and the transmission of the disease by insects. *Journal of Agricultural Research* 40, 1031–1051.
Carles, L. (1985) Lefiguier (2e partie) principal maladies. *Arboriculture Fruitiere* 32(375), 56–58.
Condit, I.J. (1955) Fig varieties: a monograph. *Hilgardia* 11, 323–538.
Condit, I.J. and Enderud, J. (1956) A bibliography of the fig. *Hilgardia* 25, 1–663.
Condit, I.J. and Horne, W.T. (1933) A mosaic of the fig in California. *Phytopathology* 23, 887–896.
Condit, I.J. and Horne, W.T. (1941) Further notes on fig mosaic. *Phytopathology* 31, 561–563.
Condit, I.J. and Horne, W.T. (1943) Mosaic spots of fig fruits. *Phytopathology* 33, 719–723.
Condit, I.J. and Stevens, H.L. (1919) 'Dieback' of the fig in California. *Fig and Olive Journal* 4, 11–12.
Condit, I.J. and Swingle, W.T. (1947) *The Fig*. Chronica Botanica Company, Waltham, Massachusetts.
Dale, J.L. (1958) Two rusts previously unreported from Arkansas. *Plant Disease Reporter* 42, 402.
Davey, A.E. and Smith, R.E. (1933) The epidemiology of fig spoilage. *Hilgardia* 7, 523–551.
Davide, R. (1979) Reactions of different crops to infections by *Meloidogyne arenaria* isolated from fig and the influence of temperature on development of the nematode. *Plant Disease Reporter* 63, 207–211.
de Lagarde, P. (1881) Ueben die semitischen Namen des Feigenbaumes und der Feige K. Gesell. der Wissenschaften zu Göttingen Nach 1881, 368–396.
Dibble, I.E., English, W.H., Hart, W.H., Lownsberry, B.F., Moller, W.W. and Stafford, E.M. (1972) *Pest and Disease Control Program for Figs*. California Agricultural Experiment Station, Division of Agricultural Science, University of California.
Doster, M.A. and Michailides, T.J. (1997) Susceptibility of maturing Calimyrna figs to decay by aflatoxin-producing fungi in California. *Acta Horticulturae* 480, 187–191.
Doster, M.A., Michailides, T.J. and Morgan, D.P. (1996) *Aspergillus* species and mycotoxins in figs from California orchards. *Plant Disease* 80, 484–489.
Duxbury, D.D. (1986) Ingredients spotlight: dried figs contain more dietary fiber than most fruits and cereals. *Food Processing* (May). California Fig Institute, Fresno, California.
Edgerton, C.W. (1911) Diseases of the fig tree and fruit. *Louisiana Agricultural Experiment Station Bulletin* 126.
El-Gholl, N.E. and Alfieri, S.A. Jr (1984) Fruit rot of fig caused by *Phytophthora palmivora*. *Proceedings of the Florida State Horticultural Society* 97, 327–328.
English, H. (1951) Phomopsis canker: a progress report. *5th Annual Research Conference of the California Fig Institute Proceedings*, pp. 45–58.
English, H. (1952) Phomopsis canker of figs (abstract). *Phytopathology* 42, 513.
English, H. (1953) Further work on the control of Phomopsis canker. *7th Annual Research Conference of the California Fig Institute Proceedings*, pp. 12–15.
English, H.W. (1954) Further experiments on the control of surface mold rot of Kadota figs. *8th Annual Research Conference of the California Fig Institute Proceedings*, pp. 16–20.
English, H. (1962) Canker ant dieback disorders of fig trees. *16th Annual Research Conference of the California Fig Institute Proceedings*, pp. 13–15.
FAO (2000) FAOSTAT online database at: http://www.fao.org/default.htm
Ferguson, L., Michailides, T.J. and Shorey, H.H. (1990) The California fig industry. *Horticultural Reviews* 12, 409–490.
Flock, R.A. and Wallace, J.M. (1955) Transmission of fig mosaic by the eriophyid mite *Aceria ficus*. *Phytopathology* 45, 52–54.
Galil, J. and Neeman, G. (1977) Pollen transfer and pollination in the common fig (*Ficus carica* L.). *New Phytologist* 79, 163–171.
Gella, R., Marin, J.A., Lopez Corrales, M. and Toribio, F. (1997) Elimination of fig mosaic from fig shoot-tip cultures by thermotherapy. *Acta Horticulturae* 480, 173–177.
Gerlach, W. and Nirenberg, H. (1982) *The Genus* Fusarium *– a Pictorial Atlas*. Paul Parey. Berlin.
Hagan, H.R. (1929) The fig-insect situation in the Smyrna fig district. *Journal of Economic Entomology* 22, 900–909.
Hampson, M.C. (1981) Phomopsis canker on weeping fig in Newfoundland. *Canadian Plant Disease Survey* 81(1), 3–5.
Hansen, H.N. (1928) Endosepsis and its control in caprifigs. *Phytopathology* 18, 931–938.
Hansen, H.N. (1929) Thrips as carriers of fig-decaying organisms. *Science* 69, 356–357.
Hansen, H.N. (1948) A canker disease of figs (abstract). *Phytopathology* 38, 314–315.
Hansen, H.N. (1949a) Phomopsis canker. *3rd Annual Research Conference of the California Fig Institute Proceedings*, pp. 18–19.

Hansen, H.N. (1949b) Phomopsis canker of fig. *California Agriculture* 3(11), 13–14.
Hansen, H.N. and Davey, A.E. (1932) Transmission of smut and molds in figs. *Phytopathology* 22, 247–252.
Harvey, J.M. (1956) Post-harvest decay studies with fresh figs. *10th Annual Research Conference of the California Fig Institute Proceedings*, pp. 27–28.
Hodgson, R.W. (1918) Black smut of figs. *California State Department of Agriculture Monograph Bulletin 7*, 188–189.
Howard, B.J. (1933) The influence of insects in the souring of figs. *Journal of Economic Entomology* 28, 917–918.
Hull, F.M. (1929) Some possible means of control of the damage caused by the cotton leaf worm moth to the fig. *Journal of Economic Entomology* 22, 792–798.
Igriboz, N. (1940) Incir Hastaliklari, Ziraat vekaleti Neriyati. Kultur Basimevi, Izmir.
Ingol, C.T. (1971) *Fungal Spores: Their Liberation and Dispersal*. Oxford University Press, London.
Jeppson, L.R., Keifer, H.H. and Baker, E.W. (1975) *Mites Injurious to Economic Plants*. ARS, USDA, University of California Press, Berkeley.
Karaca, I. (1979) *Sistematik Bitki Hastaliklari (Ascomycetes Cilt. 3.2.) Baski*. Ege Universitesi, Ziraat Fakultesi Yayinlari, No. 143.
Kato, K., Hirota, K. and Miyagawa, T. (1982) A new disease, Ceratocystis canker of fig caused by *Ceratocystis fimbriata* Ellis & Halst. (in Japanese) *Nihon Shokubutsu Boeki Kyokai* 36, 55–59.
Katsura, K. and Yamamoto, K. (1969) Two species of *Phytophthora* on fig (*Ficus carica* L.). *Scientific Reports, Kyoto University of Agriculture* 21, 24–31.
Khalil, J.A. (1982) Occurrence of fig mosaic disease on fig trees in Wadi El-Hai agricultural project-Libya. *Libyan Journal of Agriculture* 11, 207–208.
Krezdorn, A.H. and Adriance, G.W. (1981) Fig growing in the south. *United States Department of Agriculture Handbook 196*.
La Massèse, C.S., Deportes, L., Mercier, S. and Roger, J.-P. (1984) Les principaux ennemis du figuier: le nematodes et les maladies. *Phytoma* 364, 39–41.
Laundon, G.F. and Rainbow, A.F. (1971) Cerotelium fici. *CMI. Descriptions of Pathogenic Fungi and Bacteria* No. 281. Commonwealth Mycological Institute, Kew, UK.
Manziano, F., Scalione, M., Nanni, B., Mangione, M.T. and Astore, A. (1990) Le principali malattie crittogamiche del fico nel Cilento centrale. *Agricultura e Ricerca* 112–113, 113–122.
Massee, G. (1911) A fig disease. *Garden Chronicles* 28, 5 (July 7).
Matz, J. (1918) Some diseases of the fig. *Florida Agricultural Experiment Station Bulletin 149*.
McBeth, C.W. (1949) Nematodes affecting figs. *3rd Annual Research Conference of the California Fig Institute Proceedings*, pp. 16–17.
McKenry, M.V. and Thomason, I.J. (1975) The dosage-response of *Meloidogyne* infected *Ficus* roots to *cis*-1, 3-dichloropropene nematicides. In: *Proceedings of the 3rd International Congress of Plant*, Moscow, USSR, August 21–27, 1975, pp. 507–513.
McKenzie, E.H.C. (1986) New plant disease record in New Zealand: fig rust (*Cerotelium fici*) on *Ficus carica*. *New Zealand Journal of Agricultural Research* 29, 707–710.
Michailides, T.J. and Morgan, D.P. (1994) Dynamics of *Blastophaga psenes* populations, availability of caprifigs and incidence of fig endosepsis caused by *Fusarium moniliforme*. *Phytopathology* 84, 1254–1263.
Michailides, T.J. and Morgan, D.P. (1998) Spread of endosepsis in Calimyrna fig orchards. *Phytopathology* 88, 637–647.
Michailides, T.J. and Ogawa, J.M. (1989) Investigations on the correlation of fig endosepsis on Calimyrna fig with caprifig infestations by *Fusarium moniliforme*. *Fig Research Report*. Crop Year 1988. University of California, Kearney Agricultural Center, pp. 1–34.
Michailides, T.J., Ogawa, J.M. and Ferguson, L. (1987) Investigations on the correlation of fig endosepsis on Calimyrna fig with caprifig infestations by *Fusarium moniliforme*. *Fig Research Report*. Crop Year 1987. University of California, pp. 1–25.
Michailides, T.J., Morgan, D.P., Belinger, C. and Wang, J. (1991) Fig smut in California commercial orchards as affected by different cultural practices. *Annual Research Report Crop Year 1990*. California Fig Institute, pp. 1–13.
Michailides, T.J., Morgan, D.P. and Klamm, R. (1994) Comparison of three methods for determining fig endosepsis caused by *Fusarium moniliforme* and other molds in caprifigs and Calimyrna figs. *Plant Disease* 78, 44–50.
Michailides, T.J., Morgan, D.P. and Subbarao, K.V. (1996) Fig endosepsis. An old disease still a dilemma for California growers. *Plant Disease* 80, 828–841.

Miller, M.W. and Phaff, H.J. (1962) Successive microbial populations in Calimyrna figs. *Applied Microbiology* 10, 394–400.

Mrak, E.M., Phaff, H.J., Vaughn, R.H. and Hansen, H.N. (1942) Yeasts occurring in souring figs. *Journal of Bacteriology* 44, 441–450.

Nisikado, Y., Hirata, K. and Kimura, K. (1941) On a *Phytophthora* rot of fig. *Ber. Ohara Institute* 8, 427–442.

Obenauf, G.L., Gerdts, M., Leavitt, G. and Crane, J. (1978) *Commercial Dried Fig Production in California.* Division of Agricultural Science. University of California, Leaflet 21051.

Obenauf, G.L., Ogawa, J.M., Lee, K. and Frate, C.A. (1982) Fungicide control of molds that attack caprifigs. *Plant Disease* 66, 566–567.

O'Donnell, K., Cigelnik, E. and Nirenberg, H.I. (1998) Molecular systematics and phylogeography of the *Gibberella fujikuroi* species complex. *Mycologia* 90, 465–493.

Phillips, E.H., Smith, E.H. and Smith, R.E. (1925) Fig smut. *University of California Agricultural Experiment Station Bulletin 387.*

Pilgrim, A.J. (1950a) Bacterial canker of Adriatics. *4th Annual Research Conference of the California Fig Institute Proceedings*, pp. 32–33.

Pilgrim, A.J. (1950b) Phomopsis canker of Kadotas. *4th Annual Research Conference of the California Fig Institute Proceedings*, pp. 30–32.

Pilgrim, A.J. (1951) Bacterial canker of figs. *5th Annual Research Conference of the California Fig Institute Proceedings*, pp. 48–51.

Plavsic, B. and Milicic, D. (1980) Intracellular changes in trees infected with fig mosaic. *Acta Horticulturae* 110, 281–286.

Prunet, A. (1903) A disease of the branches of fig. *Compte Rendu de l'Academie des Sciences, Paris* 136, 395–397.

Quacquarelli, A. (1971) Il mosaico del fico e il virus latente del *Chenopodium*. *Phytopatologia Mediterranea* 10, 283–286.

Ricci, P. (1972) Observations surla pourriture de figues fraîches après la recolte. *Annals of Phytopathology* 4, 109–117.

Sinhri, A. and Mishra, A. (1984) Efficacy of different fungicides against fig rust (*Cerotelium fici*). *Indian Journal of Mycology and Plant Pathology* 8, 38.

Smith, K.M. (1972) *A Textbook of Plant Virus Diseases*, 3rd edn. Academic Press, New York.

Smith, R.E. and Hansen, H.N. (1931) Fruit spoilage diseases of figs. *University of California Agricultural Experiment Station Bulletin No. 500.*

Smith, R.E. and Hansen, H.N. (1935) Directions for control of endosepsis in figs. 1934–1935. *University of California Agricultural Experiment Station.*

Solms-Laubach, H.G. (1885) Die Geschlechterdifferenzierung bei den Feigenbaumen (Comparison of *F. carica* with other species). *Botanische Ztg*. 43, 514–522.

Stevens, F.L. and Hall, J.G. (1909) Eine neue Feigen-Anthraknose. *Zeitschrift für Pflanzenkrankheiten* 19, 65–68.

Taubenhaus, J.J. (1936) A fig decay. *Texas Agricultural Experiment Station Annual Report* 49, 113.

Thomas, H.E., Roberts, C. and Amstutz, A. (1948) Rootstock susceptibility to *Armillaria mellea*. *Phytopathology* 38, 152–154.

Valarini, P.J. and Tokeshi, H. (1980) *Ceratocystis fimbriata*: agente causal da 'seca da figueira' e seu controle. *Summa Phytopathologia* 8, 102–106.

Vinson, J.A. (1999) The functional food properties of fig. *Cereal Foods World* 44(2), 82–87.

Wilson, E.E. and Ogawa, J.M. (1979) *Fungal, Bacterial and Certain Nonparasitic Diseases of Fruit and Nut Crops in California*. Agricultural Science Publications, Division of Agricultural Sciences, University of California, Berkeley, Publication 4090.

12 Diseases of Guava

T.-K. Lim[1] and B.Q. Manicom[2]
[1] *Biosecurity Australia, Department of Agriculture, Fisheries and Forestry Australia, Canberra, Australia;* [2] *Institute for Tropical and Subtropical Crops, Nelspruit, Republic of South Africa*

Introduction

Guava, *Psidium guajava* (family: *Myrtaceae*), is indigenous to tropical America. It is widely distributed in the tropics and subtropics where it has become naturalized in many areas, in some of which it is deemed a noxious weed. Although the tree can thrive under neglect, yields are low, and fruit flies and other insect pests often infest fruits. The fruit contains high levels of vitamin C (approx. 200 mg 100 g^{-1}) and is utilized fresh or for processing. Leading producers are Brazil, India and Mexico. Other important producing regions include Cuba, Dominican Republic, Egypt, Hawaii, Jamaica, Kenya, Malaysia, the Philippines, Puerto Rico, Taiwan, Thailand and Venezuela.

Guava seedlings produce fruit within 2–3 years, and trees grow to a height of 3–10 m, branching close to the ground. Its shallow root system often gives rise to ground suckers. It has distinct opposite, decussate leaves borne on a four-angled ridged, pubescent stem. The flowers are white, solitary or in two- to three-flowered cymes, hermaphroditic and have numerous stamens. The fruit is a berry, obovate to globose or oblong to pyriform in shape, and green to pale green or yellow when mature ripe. The flesh contains abundant seeds and is white, reddish pink or yellow usually with a strong pleasant aroma when ripe. Parthenocarpic seedless fruits from triploid cultivars are formed without fertilization. Some well-known, red-fleshed cultivars are: 'Branca' (Brazil); 'Basateen Alsohbia' (Egypt); 'Beaumont', 'Hong Kong Pink' and 'Kua Hua Ula' (Hawaii); and 'Fan Retief' (South Africa). Popular white-fleshed cultivars are: 'Allahabad Safeda', 'Allahabad Seedless' and 'Lucknow No. 4' (India); and 'Crystal Seedless', 'Glom Sali', 'Kampuchea', 'seedless Domron' and 'Taiwan Pear' (Thailand and Malaysia).

Guava fruits all year round in warm climates. Its cropping cycle can be regulated by irrigation, fertilization, pruning, deflowering, fruit removal, and by defoliation with sprays of urea and ethepon. It is cross-pollinated by insects, although self-pollination occurs in isolated plantings.

Foliar Diseases

Bacterial disease

This destructive disease has only been reported from Brazil where it is capable of killing trees (Tokeshi *et al.*, 1980). Symptoms consist of leaf roll, wilt, yellowing and reddening of veins and adjacent areas, followed by collapse of vascular tissues. Leaves, new shoots, flowers and fruit eventually die, culminating in plant death. Twigs and branches are also affected.

A bacterium, *Erwinia psidii* sp. nov., causes this disease (Rodriguez *et al.*, 1987). Affected plant parts and trees should be removed.

Rust

This is an extremely destructive disease that affects guava and other members of the *Myrtaceae*, such as pimento (*Pimenta officinalis*), *Eugenia* spp., *Eucalyptus* spp. and *Melaleuca* spp. (Anonymous, 1985). The disease was first recorded on guava in Brazil in 1884 (Anonymous, 1985). Its present distribution is limited to Central and South America and the Caribbean. Recently it was reported on melaleuca in Florida (Rayachhetry *et al.*, 1997). There is an unconfirmed report of rust on *Eucalyptus* spp. in India (Anonymous, 1985).

Symptoms

Conspicuous orange to reddish pustules of the causal fungus are produced on foliage, young shoots, flowers and fruit. Distortion, severe defoliation, growth reduction and mortality occur when the disease is severe.

Causal agent

Puccinia psidii produces pale yellow, amphigenous uredia, 0.1–0.5 mm in diameter, in groups on brownish or blackish spots up to 5 mm in diameter (Laundon and Waterson, 1965). Uredospores are ellipsoid to obovoid, 21–26 × 16–19 μm, with finely echinulate cell walls that are 1.5–2.5 μm thick (Fig. 12.1). Teliospores are ellipsoid to cylindrical, rounded above, slightly constricted at the septum, and 30–48 × 19–22 μm. Cell walls are buff, smooth, 1–1.5 μm thick at the side and 2–4 μm thick above, and pedicels are fragile and often deciduous.

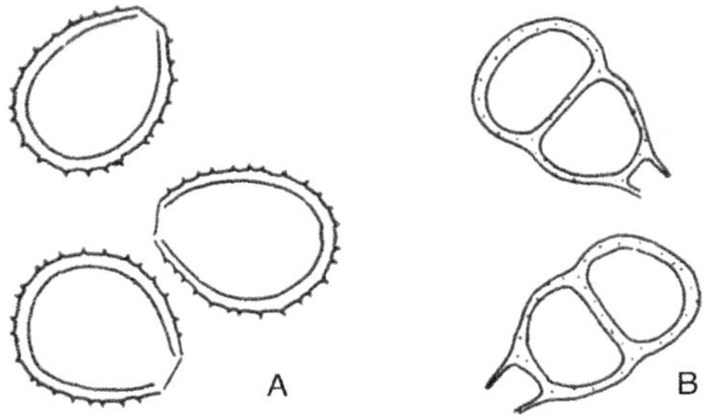

Fig. 12.1. (A) Urediospores and (B) teliospores of *Puccinia psidii* (from CMI description no. 56).

Epidemiology

de Toledo Piza and Ribeiro (1988) reported that uredospore germination occurs between 18 and 22°C, with 18°C and 8 h of darkness being optimum. In Bahia, severe disease on leaves and twigs of eucalyptus was correlated with a relative humidity of >90% and temperatures between 15 and 25°C (Carvalho *et al.*, 1994). Isolates from guava, eucalyptus and *Syzygium jambos* were capable of cross-infection (de Castro *et al.*, 1983).

Management

Ruiz *et al.* (1987) reported that bitertanol, captafol, chlorothalonil, mancozeb, copper oxychloride, oxycarboxin, triadimenol, triforine and propiconazole gave complete control on seedlings of *Eucalyptus grandis*, but that captafol and propiconazole were phytotoxic. Ruiz *et al.* (1991) also reported that chlorothalonil, copper oxychloride, mancozeb, oxycarboxin and triforine gave up to 10 days protective control before inoculation. Triadimenol, triforine and oxycarboxin exhibited therapeutic effects.

In areas where the disease is not found, strict quarantine measures must be observed to avoid its introduction. In Australia, the importation of plants of susceptible species from guava rust areas is permitted only under strict conditions that include post-entry quarantine and disease screening for a minimum of 12 months after arrival (Anonymous, 1985).

Fruit Diseases

Anthracnose

Anthracnose is the most common pre- and postharvest fruit disease in all guava-growing countries. It can cause considerable postharvest losses.

Symptoms

Anthracnose is first evident on mature fruits on the tree. Symptoms consist of distinctive sunken, dark coloured, necrotic lesions that become covered with pinkish spore masses of the pathogen under humid conditions (Plate 74). The small sunken spots often coalesce to form larger patches, and the disease extends into the flesh.

Causal agents

The most common anthracnose pathogen is *Colletotrichum gloeosporioides* (teleomorph: *Glomerella cingulata*), although *C. acutatum* (teleomorph: *Glomerella acutata*) is responsible in Australia (Cooke, 1982). Both are described in Chapter 1.

In Puerto Rico, Liu (1972) isolated two strains of *C. gloeosporioides* from diseased guava fruit that differed in cultural appearance, conidial size and physiology, one growing best between 24 and 28°C and the other between 28 and 32°C. When crossed, they produced perithecia on potato dextrose agar (PDA) at 24–28°C. The teleomorph is rare in the field. Attributes of *C. acutatum* and whether isolates of it from guava produce perithecia in the laboratory or field have not been reported.

Epidemiology

Conidia are the most important means by which the pathogen and disease are spread in the field. They are produced on necrotic fruit lesions, dead twigs, leaves and other host tissues, and spread via rainsplash. Although symptoms may develop shortly after infection, latent infections that remain quiescent for months are common. Disease spread is greatest during wet conditions. On fruit, lesions develop at any stage of development but expand most rapidly at 30°C. Epidemics can develop during prolonged warm, wet weather, and disease development on non-injured fruit is low (34%) compared with those injured by punctures (100%) or sand (70%) (Pandey *et al.*, 1997).

Management

Resistant cultivars provide the best option for controlling this disease. In Hissar, India, susceptibility to anthracnose varied among fruit from various crosses. Resistant hybrids were obtained from 'Allahabad Safeda' and 'Banarsri Surkha' (Naresh *et al.*, 1987). A few hybrids from crosses with 'Apple Colour' were resistant. In another study, Sharma (1981) reported that anthracnose development was delayed up to 4 days in 'Apple Colour', 'Lucknow 49', 'Dharwar', 'Chakaiya' and a hybrid between 'Banarsri Surkha' and 'Apple Colour'.

The use of benomyl and carbendazim in the field and as postharvest treatments in combination with hot water can reduce postharvest anthracnose. Growth of, and acervulus formation by, *C. gloeosporioides* was strongly inhibited by benomyl, thiabendazole and thiophanate methyl at 5–50 $\mu g\ l^{-1}$ (Butt *et al.*, 1995). Vitigran blue inhibited acervulus development at 50 $\mu g\ l^{-1}$, and boric acid inhibited colony growth more than borax or bleaching powder.

Pestalotiopsis fruit canker

Pestalotiopsis fruit canker is also known as fruit necrosis (Montiel, 1997), fruit canker and fruit scabby canker. It occurs in Australia, Burma, Ecuador, India, Malaysia, Mozambique, Nigeria, Puerto Rico, Venezuela and Zambia, and attacks fruit at all stages of development (Mordue, 1969a). Losses occur in the field as well as during postharvest storage.

Distinctive dark brown to black, raised spots develop on fruit that may coalesce. The necrotic epidermis tears open to form a small raised crater with an elevated margin and sunken centrum, the corky scab symptom. As the fruit expands, cracks emanate from the lesions.

The disease is caused by *Pestalotiopsis psidii*, a weak parasite that occurs in the woody tissues of twigs as an endophyte. It is an opportunist that invades fruits through insect injuries (Verma and Sharma, 1976; Lim and Khoo, 1990). Characteristic conidia with three apical appendages are produced in acervuli on leaves and fruit (Fig. 12.2).

Cankers are associated with punctures caused by insect feeding. In Malaysia, the cacao mirid bug, *Helopeltis theobromae*, is involved (Lim and Khoo, 1990), whereas in Australia, it is fruit spotting bugs, *Amblypelta* spp. (S. Smith, Darwin, 1997, personal communication). Dense plantings and canopies are prone to greater attack by such insects and, therefore, greater development of scabby cankers.

Applications of insecticides to reduce feeding by the above insects are helpful. Pruning trees and wider spacing can suppress mirid damage. Also, natural enemies, such as *Oecophylla smaragdina*, can reduce populations of these insects (Lim and Khoo, 1990).

Phytophthora fruit rot

Phytophthora fruit rot has been reported in Cuba (Ariosa, 1982), Hawaii (Ko *et al.*, 1982), India (Sohi and Sridhar, 1971; Singh *et al.*, 1976, 1978; Gupta *et al.*, 1977; Sharma *et al.*, 1978; Gupta, 1988; Mathur *et al.*, 1992) and Malaysia (Lim and Chin, 1987). Lim and Chin (1987) also reported a Phytophthora foliar blight on guava seedlings and trees.

Symptoms

Symptoms begin as water-soaked areas on the fruit. The lesions turn greyish, and under humid conditions become covered by fine fluffy, cottony masses of whitish mycelia. The disease is common on young and mature green fruit, and ripe yellow fruit are less susceptible (Ko *et al.*, 1982). Affected leaves appear wet and blighted, and young shoots on seedlings may die back (Lim and Chin, 1987).

Causal agents

At least two taxa are involved: the A2 mating type of *Phytophthora citricola* (Hawaii) and *P. nicotianae* (Cuba, India and Malaysia) (Singh *et al.*, 1978; Ariosa, 1982; Ko *et al.*, 1982; Lim and Chin, 1987). Only the A1 mating type of *P. nicotianae* was reported in Malaysia. Both species can be isolated from soil using selective media or guava fruit as baits (Singh *et al.*, 1978). Both species are described in Chapter 1.

Epidemiology

P. nicotianae produces thick-walled chlamydospores that persist in the soil as the primary source of inoculum. The fungus can survive for 12–16 months in the soil and on dead host tissues (Prasad and Mehrotra, 1985). Infected leaves and twigs serve as secondary sources of inoculum for fruit infection.

The disease is favoured by relative humidity >90%. Mathur *et al.* (1992) reported that rain for at least 3–4 days per week with maximum temperatures of 28–30°C and minimum temperatures of 21–22°C are conducive to the development and spread of the disease.

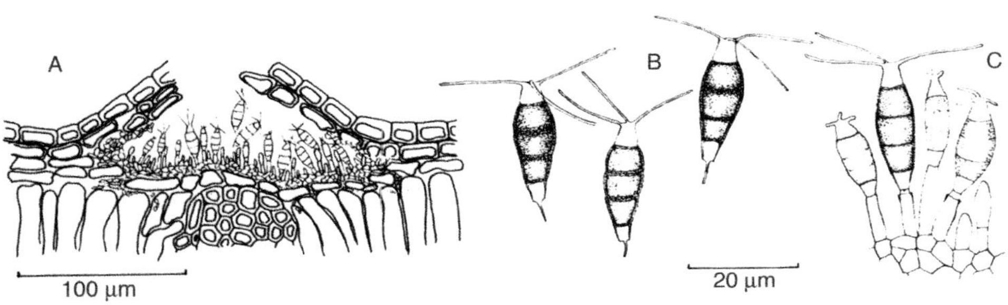

Fig. 12.2. (A) Acervulus, (B) mature conidia and (C) immature conidia and conidiogenous cells of *Pestalotiopsis psidii* (from CMI description no. 515).

Management

The disease can be controlled with sprays of fosetyl-Al, metalaxyl, milfuram, cymoxanil or their combinations with mancozeb and etridiazole (Lim and Khoo, 1990). Control was also obtained with zineb and aureofungin (Sohi and Sridhar, 1971), and captafol and zineb (Gupta, 1988). Copper fungicides were reported to be phytotoxic (Sohi and Sridhar, 1971; Lim and Khoo, 1990).

Singh *et al.* (1976) wrote that susceptibility ranged from 5 to 20% on 'Allahabad Safeda', 'Apple Guava', and red and pink fleshed cultivars. Gupta *et al.* (1977) reported that 'Lucknow 49', 'Banasri Surkha', 'Allahabad Safeda' and 'Mishiri' were highly susceptible, 'Telshidar' moderately susceptible, and 'Chitidar' and 'Apple Guava' were relatively resistant.

Incidence of the disease on guava seedlings in the nursery can be reduced via the use of pasteurized soil and by avoiding soil splash during irrigation.

Postharvest fruit rots

Many postharvest fruit rots begin as preharvest rots on the tree. Some, such as anthracnose, Phytophthora fruit rot, scabby canker and stylar end rot, even occur on immature green fruits. However, most postharvest fruit rots occur on mature or ripening fruits prior to harvest, at harvest, and in transit and storage. Depending on the associated fungi, severe postharvest losses can result.

Symptoms

Generally, there are two types of postharvest fruit rots. Dry firm rots produce necrotic lesions that usually are superficial and shallow, or are depressed on the fruit surface and either restricted (spots) or extensive (patches). In contrast, extensive and invasive rots start as small water-soaked spots that expand and extend to the mesocarp and deep into the seed cavity. The rots can also be distinguished by the colour of the lesion and the presence of fungal mycelia, fruiting bodies and spores. The most important fruit rots are detailed below.

Cylindrocladium fruit rot causes orange–brown lesions on near ripe or ripe fruit. The rot is firm and not sunken. Under humid conditions, lesions are covered with fine, lanose mycelia with light brown centres and white peripheries. Guignardia fruit rot also causes firm, dry lesions on near ripe fruit. They are sunken and dark brown to black. During moist weather, rough pycnidial spots appear on the lesion and grey masses of spores exude from the pycnidia. The fungus causes a greenish black necrosis of the mesocarp.

Lasiodiplodia fruit rot occurs on near ripe and ripe fruits in the field and in postharvest storage. It is characterized by brown discoloration that spreads rapidly over the fruit. The rot turns dark brown and necrotic and becomes covered by grey mycelium of the pathogen during wet weather. Pin-sized pycnidia are produced on affected areas.

Mucor and Rhizopus fruit rot produce water-soaked lesions (Plate 75). Botryosphaeria brown fruit rot also begins as water-soaked, soft lesions. They develop on near ripe fruit and turn brown, necrotic and ultimately produce black, erumpent, globose pycnidia in stroma that appear as tiny pinheads. The rot imparts a purplish tinge to the mesocarp.

Causal agents

Diverse microorganisms cause postharvest rots of guava fruit (Table 12.1). Some are field pathogens while others are strictly postharvest.

Epidemiology

Many postharvest rots commence in the field from direct infection of fruit on the tree and by secondary infection through wounds, bruises and injuries that are caused by insects. Such rots undergo further development in transit and storage and are aggravated by improper temperature and relative humidity conditions in storage. Invasion by storage fungi is also enhanced by improper storage conditions and by poor phytosanitation.

Table 12.1. Microorganisms associated with pre- and postharvest fruit rots of guava.

Microorganism	Type of rot	Pre-harvest	Post-harvest	Location (reference)
Acremonium terricola			+	India (Madhukar and Reddy, 1989)
Aspergillus niger var. *awamori*	Soft		+	India (Lal *et al.*, 1980b)
Aspergillus fumigatus	Dry		+	Nigeria (Adisa, 1985)
Aspergillus niger	Soft		+	India (Utikar *et al.*, 1986), Malaysia (Lim and Khoo, 1990), Nigeria (Adisa, 1985)
Diplodia theobromae	Soft	+	+	India (Majumdar and Pathak, 1989), Malaysia (Lim and Khoo, 1990), Nigeria (Adisa, 1985)
Botryosphaeria ribes	Soft		+	India (Majumdar, 1985)
Botryosphaeria sp. (resembles *B. ribes*)	Soft	+	+	Malaysia (Lim, 1987)
Ceratocystis adiposa			+	India (Badyal and Sumbali, 1992)
Choanephora cucurbitarum	Soft		+	Nigeria (Adisa, 1985)
Cladosporium cladosporioides			+	India (Madhukar and Reddy, 1989)
Cladosporium oxysporum			+	India (Majumdar, 1985)
Cladosporium sp.	Dry		+	Nigeria (Adisa, 1985)
Cochliobolus spicifer			+	India (Majumdar, 1985)
Colletotrichum gloeosporioides		+	+	India (Majumdar and Pathak, 1989), Malaysia (Lim and Khoo, 1990), Nigeria (Adisa, 1985)
Colletotrichum sp.			+	India (Utikar *et al.*, 1986)
Curvularia tuberculata			+	India (Kapoor, 1983)
Cylindrocladiella parva	Firm	+	+	Malaysia (Lim and Khoo, 1990)
Cylindrocladium scoparium			+	India (Badyal and Sumbali, 1992), Malaysia (Sepiah, 1990)
Dichotomyces cejpii			+	India (Saxena and Saksena, 1983)
Diplodia theobromae	Dry		+	India (Rajagopalan and Wilson, 1972)
Dothiorella sp. (teleomorph: *Botryosphaeria berengiana*)	Brown	+	+	Hawaii (Kunimoto and Ko, 1984)
Drechslera halodes			+	India (Madhukar and Reddy, 1989)
Drechslera (*Cochliobolus*) *hawaiinensis*			+	India (Madhukar and Reddy, 1989)
Erwinia sp.	Soft		+	Nigeria (Adisa, 1985)
Fusarium equisetti	Dry		+	Nigeria (Adisa, 1985)
Fusarium decemcellulare			+	India (Majumdar, 1985)
Fusarium oxysporum	Dry		+	Nigeria (Adisa,1985)
Fusarium semitectum			+	India (Majumdar, 1985)
Fusarium solani	Soft		+	India (Chakrabati, 1983); Majumdar and Pathak, 1989)
Geotrichum candidum	Sour		+	India (Shankhapal and Hatwalne, 1976)
Gilbertella persicaria			+	India (Majumdar, 1985)
Guignardia musae	Dry	+	+	Hawaii (Ko and Kunimoto, 1980)
Guignardia psidii (anamorph: *Phyllosticta psidiicola*)	Dry	+	+	India (Ullasa and Rawal, 1984), Malaysia (Lim and Khoo, 1990)
Hormoconis sp.			+	India (Majumdar, 1985)
Macrophoma allahabadensis			+	India (Srivastava and Tandon, 1969)
Macrophoma sp.	Dry	+	+	Venezuela (Quintero and Urdaneta, 1997)
Macrophomina phaseolina	Dry	+	+	Pakistan (Jan and Shah, 1991)

Continued

Table 12.1. *Continued.*

Microorganism	Type of rot	Pre-harvest	Post-harvest	Location (reference)
Macrophomina sp.	Dry		+	Venezuela (Diaz and Rondon, 1971)
Mucor hiemalis	Soft		+	Hawaii (Kunimoto *et al.*, 1977; Ito *et al.*, 1979)
Mucor sp.	Soft		+	Malaysia (Lim and Khoo, 1990)
Penicillium chrysogenum			+	India (Badyal and Sumbali, 1992)
Penicillium multicolor	Dry		+	Nigeria (Adisa, 1985)
Penicillium sp.	Dry		+	Nigeria (Adisa, 1985)
Pestalotia olivacea	Dry, firm		+	India (Dhingra and Mehrotra, 1980)
Pestalotiopsis palmarum			+	India (Majumdar, 1985)
Pestalotiopsis psidii	Dry, firm	+	+	India (Utikar *et al.*, 1986; Patel and Patel, 1988; Majumdar and Pathak, 1989), Malaysia (Lim and Khoo, 1990)
Pestalotiopsis versicolor	Dry, firm		+	India (Majumdar and Pathak, 1989)
Phoma psidii		+	+	India (Singh and Bhargawa, 1977)
Phomopsis destructum		+	+	India (Rao *et al.*, 1976)
Phomopsis psidii		+	+	India (Srivastava and Tandon, 1969; Utikar *et al.*, 1986; Rawal and Ullasa, 1988); Malaysia (Lim and Razak, 1986);
Phytophthora citricola	Firm–soft	+		Hawaii (Ko *et al.*, 1982)
Phytophthora nicotianae	Firm–soft	+	+	India (Majumdar and Pathak, 1989), Malaysia (Lim and Chin, 1987)
Rhizoctonia solani	Soft		+	Nigeria (Adisa, 1985)
Rhizopus arrhizus	Soft		+	India (Majumdar, 1985; Badyal and Sumbali, 1992), Nigeria (Adisa, 1985)
Rhizopus microsporus	Soft		+	India (Majumdar, 1985)
Rhizopus stolonifer	Soft	+	+	Hawaii (Ooka, 1980), Malaysia (Lim and Khoo, 1990), Nigeria (Adisa, 1985)
Sclerotium rolfsii		+	+	India (Ullasa and Rawal, 1985)
Syncephalastrum racemosum			+	India (Utikar *et al.*, 1986)
Chalara paradoxa (teleomorph: *Ceratocystis paradoxa*)	Soft		+	India (Lal *et al.*, 1980a)
Thievalia microspora			+	India (Majumdar, 1985)
Thievalia terricola			+	India (Madhukar and Reddy, 1989)

Management

The incidence of preharvest fruit rot is reduced with sound orchard management (e.g. correct planting densities and adequate pruning, fertilization, irrigation, weed control and sanitation) in concert with fruit bagging, monitoring insects that cause injuries, and the use of natural enemies and appropriate chemicals. Sound orchard management also mitigates postharvest losses, as do avoiding injury during harvest, postharvest fungicide treatment alone or in combination with hot water treatment, and proper cartons, temperatures (10°C) and relative humidity during storage.

Soilborne Diseases

Damping-off

Pre- and post-emergence damping-off is a common problem. In Malaysia, seedlings are affected 2–8 weeks after germination (Lim and Khoo, 1990), and in India the disease is common on seedlings grown in non-sterilized field soils (Gupta, 1979). Damping-off is also important on grafted and marcotted plants.

Symptoms

Pre-emergence damping-off is recognized by rotting and death of the seed or seedling before it emerges from the soil. After emergence, leaves and stems of seedlings wilt and stems collapse above the soil line and fall over, leading to the death of seedlings within 2 days. Under moist and humid conditions, strands of tough, coarse mycelia with bits of soil and numerous dark brown tiny, round sclerotia dangle from affected parts of the seedling.

Causal agents

Pre- and post-emergence damping-off is caused by *Rhizoctonia solani* (teleomorph: *Thanatephorus cucumeris*) in India (Gupta, 1979), whereas both *R. solani* and *Sclerotium rolfsii* (teleomorph: *Athelia rolfsii*) cause post-emergence damping-off in Malaysia (Lim and Khoo, 1990).

R. solani has wide hyphae, 5–12 μm in diameter, that branch at right angles (Fig. 12.3) (Mordue, 1969b). It produces barrel-shaped monilioid cells, has prominent septal pores in multinucleate hyphae (usually 4–8 per cell), and produces sclerotia, <1 mm in diameter, that are not differentiated into a medulla and rind. The teleomorph is relatively uncommon.

In contrast, *S. rolfsii* produces clamp connections and hard brown sclerotia that are differentiated into a rind and medulla. The teleomorph is only produced occasionally.

Epidemiology

R. solani and *S. rolfsii* persist in soil on plant debris, decaying organic matter and roots of weeds as sclerotia or hyphal strands. They are more prevalent in acidic soils, have wide host ranges and are disseminated via rain-splashed sclerotia.

Management

Potting mixes should be sterilized or pasteurized with methyl bromide, dazomet or soil solarization. Treating seeds with Bavistin and quintozene at 3 and 5 g kg^{-1} seed gave good control of pre- and post-emergence damping-off caused by *R. solani* (Gupta,

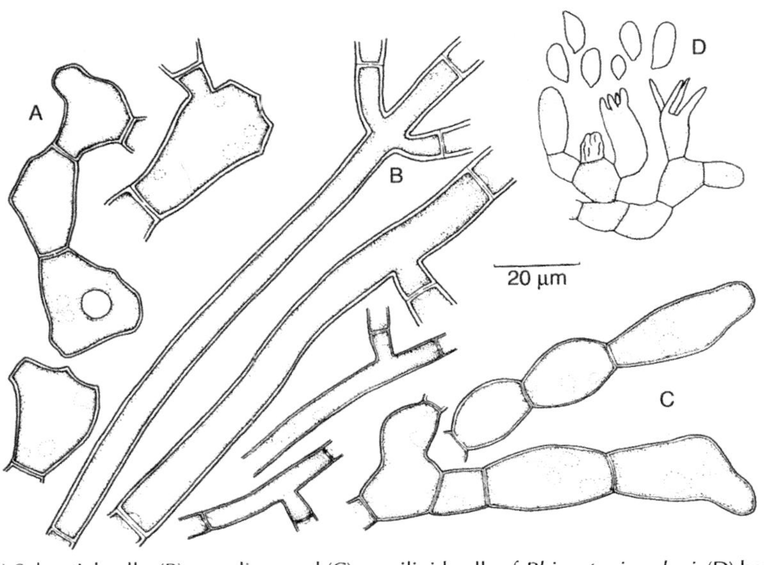

Fig. 12.3. (A) Sclerotial cells, (B) mycelium and (C) monilioid cells of *Rhizoctonia solani*, (D) basidia and basidiospores of its teleomorph, *Thanatephorus cucumeris* (from CMI description no. 406).

1979). Infested soil can be treated with drenches of PCNB, quintozene, benomyl, carboxin, pencycuron, propiconazole, toclofos methyl and flutolanil (Lim and Khoo, 1990). Lim and Teh (1990) reported that *Trichoderma harzianum, T. hamatum* and *T. koningii* were antagonistic to *S. rolfsii in vitro*, and suggested that these fungi could be mixed into pasteurized soil that was used to produce guava seedlings.

Guava wilt disease

Guava wilt disease was first reported in Taiwan by Kurosawa (1926). A wilt of guava with almost identical symptoms appeared in South Africa in 1980 and was found in the Johor Province of Malaysia in 1995 (Schoeman, 1997). It is probable that the diseases in these countries are caused by the same, or a closely related, fungus. Although there is no record of planting material being moved among these countries, we assume that guava wilt in Malaysia, South Africa and Taiwan is the same disease.

In South Africa, the disease first appeared in southeastern Mpumalanga Province. Within 10 years, it had spread throughout the guava-growing areas of this and the Northern Province, reducing the area planted to guava from ~700 ha to 100 ha. The plantings in these areas are almost entirely of a clonally propagated pink cultivar, 'Fan Retief'. To date, quarantine measures implemented in 1985 have prevented spread of the disease to the Western Cape.

In Taiwan, the disease is present in all guava-producing areas, where it reduces the life expectancy of orchards from ≥25 to 10–12 years. Either the disease is less aggressive in Taiwan than in South Africa, or the popular cultivars that are grown there are more resistant. Infection appears to occur mainly through pruning wounds in Taiwan, whereas it is more of a soilborne disease in South Africa (Grech, 1986)

In Malaysia, the disease affected 42% of a commercial planting of 270 ha of the clonally propagated pink cultivar 'Beaumont' (Schoeman, 1997). The extent of the disease in small plantings is unknown.

A different wilt disease caused losses of 30% in India (Pandy and Dwivedi, 1985). In this chapter, only the disease in Malaysia, South Africa and Taiwan will be referred to as guava wilt.

Symptoms

The rate of symptom development varies. In the fast syndrome, wilting first appears on leaves at the tips of branches in the upper canopy. Within 2–4 weeks, all leaves have wilted and dried, imparting a scorched appearance to the tree. Development of fruit is arrested and they mummify on the tree. Blisters, which contain masses of white to salmon pink spores, develop in the bark of dead wood. The fungus is isolated easily from blisters or by incubating sections of wood or main roots in a humid environment. Fast wilting can also occur in sectors, giving rise to trees with dead and healthy branches. In time, the tree dies.

When the disease progresses slowly, the entire tree usually is affected (sectors are uncommon). Shoot growth may stop and the subterminal leaves become distorted. Leaves turn yellow and then assume various shades of red and brown, often in spots, resembling what occurs during normal senescence. Defoliation occurs from the base of the shoots upward. There is no dramatic wilting, but over a few months the tree defoliates and dies. This occurs fastest during the summer.

In India, symptoms commence with chlorosis and desiccation of leaves at the tips of branches and gradually extend downward. Death of the tree may take 3–4 years. Sectorial symptoms are common but, in contrast to guava wilt, vascular discoloration occurs. The disease is soilborne.

Causal agents

The taxonomy of the fungus that causes guava wilt is confused. In Taiwan, the causal fungus was identified as *Myxosporium psidii* (Kurosawa, 1926). Although this genus is no longer valid, the pathogen appears to be identical to that found in Malaysia and South Africa.

Workers in Taiwan and South Africa reported that the fungus had a paecilomycetous conidial stage (Grech, 1986). In South Africa, it has gone through several name changes. First identified as *Paecilomyces* (Manicom, 1980), it was later changed to *Septofusidium* (Grech, 1982, 1985), *Nalanthamala madreeya* (Centraalbureau voor Schimmelcultures, unpublished, 1983), *Gliocladium* sp. (Grech, 1984), *Gliocladium roseum* (Commonwealth Mycological Institute, unpublished, 1984), *Hyalosporium psidicola* (unpublished, 1986) and *Acremonium diospyri* (Benade *et al.*, 1991). Most recently, Dr Gams (CBS, The Netherlands) tentatively identified it as *Penicillium vermoesini*, although he was of the opinion that it is neither a *Penicillium* nor a *Gliocladium* species (Schoeman *et al.*, 1997).

In contrast, the disease in India is caused by *Fusarium oxysporum* f. sp. *psidii* (Pandey and Dwivedi, 1985). A less common wilt in this country is caused by *Macrophomina phaseolina* (Dwivedi, 1990).

Epidemiology

Cardinal temperatures for the guava wilt pathogen are 10, 35 and 30°C. Guava is most susceptible during the warm summer months, and death of inoculated plants was significantly lower at 20–24°C than at 24–28 or 28–32°C (Schoeman, 1995).

The fungus moves in the xylem, favouring that most recently formed, and the cambium. It can be isolated from all symptomatic woody parts, and in sectorial infections can also be recovered from asymptomatic branches.

Spread via infested pruning shears occurs in Taiwan. Experimentally, severe disease developed after shoots were removed from seedlings or nursery trees and sprayed with conidia of the fungus, and after mycelium was introduced into roots or the trunk (Leu and Kao, 1979). No disease developed when conidia were applied to the soil in the absence of wounding.

In contrast, inoculation after wounding in South Africa did not result in extensive damage in potted plants and caused only minor branch dieback in the field (Schoeman, 1996). However, sectorial symptoms developed when the main roots or base of the stem were inoculated in the field. Wounding enhanced disease development, and trees developed severe symptoms in 3 months and died after 6 months. In square plantings, the disease spread in rows where furrow irrigation was used and radially where fields were under sprinkler irrigation. Otherwise, spread was a function of distance from diseased trees.

In summary, pruning wounds are regarded as a prime infection site in Taiwan, whereas roots and the stem at ground level are primary infection courts in South Africa. Furthermore, spread appears to occur through root grafts in South Africa.

In India, seedlings planted in soil infested with *F. oxysporum* f. sp. *psidii* wilted and died within 60 days (Pandey and Dwivedi, 1985).

Management

Effective chemical control measures are not available for guava wilt. Spread can be slowed by roguing affected plants (Leu *et al.*, 1979). The fungus does not survive for more than a few months in soil, but can survive in root pieces for more than a year. Thus, it is important to remove as much root debris from the soil as possible before replanting.

In South Africa, strip fumigation of soil did not isolate diseased areas (Grech, 1984). Although treatment of pruning wounds with benomyl and copper oxychloride is advocated, it is rarely effective. The sterilization of picking crates is also recommended to ensure that inoculum is not moved between farms, but its effectiveness has not been demonstrated (Schoeman, 1996).

In Malaysia, disease reductions were obtained by eliminating the use of chicken manure, which burned and predisposed roots to infection, and by adding inoculum of arbuscular mycorrhizae to soil before planting (C.T. Ho, Malaysia, 1997, personal communication).

In South Africa, 'Dimple' is three times and 'Pai-pa' twice as resistant as 'Fan Retief' (Schoeman, 1996). Two resistant rootstocks, TS-G1 and TS-G2, were selected from seedlings of 'Fan Retief', and are being used to re-establish the South African industry (Vos *et al.*, 2002).

Nematodes

Nematode species in 16 genera have been reported on guava. However, those in *Meloidogyne* are most problematic and widespread.

Symptoms

Aboveground symptoms include chlorosis, and reduced yield, growth and leaf size. Below ground, these pathogens reduce fine root densities and severely distort roots; *Meloidogyne* spp. produce small to large multiple galls.

Causal agents

The root knot nematodes, *Meloidogyne* spp., are the most widely studied on guava. *M. incognita* has been reported in Brazil, Cuba, Malaysia and Venezuela (Fernandez and Silveira, 1975; Razak and Lim, 1987; Crozzoli *et al.*, 1991). Races 1 and 2 have been detected using various host differentials (de Moura and de Moura, 1989; Crozzoli and Casassa, 1998). *M. arenaria* has been reported in Cuba and Venezuela (Fernandez and Silveira, 1975; Crozzoli *et al.*, 1991), *M. javanica* in Mexico (Carillo *et al.*, 1990) and *M. mayaguensis* in South Africa (Willers, 1997).

Epidemiology

Fernandez *et al.* (1986) described the biology of *M. incognita* and *M. arenaria* on guava in Cuba. The larvae penetrate roots within 24 h of hatching, and the life cycle lasts 26–30 days. For hatching, the optimum pH is 6–7, and the optimum temperature is 27–30°C for *M. arenaria* and 25–30°C for *M. incognita*. Light textured soils favour the development of root knot nematodes, and they are spread among farms via planting material and infested soil (Razak and Lim, 1987).

Management

Exclusion is the best control method. This can be achieved at the nursery stage by using non-infested planting materials and disinfested soil (Lim and Khoo, 1990). Fernandez *et al.* (1987) reported that 50 g of 98% methyl bromide m^{-2} and 100 g of dazomet m^{-2} controlled *Meloidogyne* spp. for 5 months and resulted in 100% healthy seedlings.

If seedlings are infested, they should be treated with suitable nematicides prior to planting in the field. As obligate sedentary endoparasites, root knot nematodes are well protected in host root tissues during most of their life cycle. Thus, they are extremely difficult to eradicate once they are in an orchard.

In Malaysia, fenamiphos was effective for 3 months and no detectable residues of the chemical were found on fruit samples 1 week to 2 months after soil application (Lim and Khoo, 1990). Carbofuran is also effective. In Venezuela, ethoprophos reduced nematode populations and no residues were detected in fruit after a second application after 4 months; ethoprophos was more effective than fenamiphos and carbofuran at similar rates (Casassa *et al.*, 1996). In Mexico, satisfactory control of *M. javanica* was obtained with carbofuran and ethoprophos (Carillo *et al.*, 1990). In South Africa, cadusafos controlled *M. mayaguensis* (Willers, 1997). Finally, DBCP and metamsodium reduced populations of *Helicotylenchus* spp., but other species were unaffected or increased in number (Rodriguez-Fuentes and Landa-Bolanos, 1977).

In Cuba, Fernandez and Silveira (1975) reported that *Psidium friedrichsthalianum* was more resistant than *P. molle* and *P. guajava* to *M. incognita* and *M. arenaria*; it was recommended as a rootstock.

References

Adisa, V.A. (1985) Fruit rot diseases of guava in Nigeria. *Indian Phytopathology* 38, 427–430.

Anonymous (1985) Guava rust. *Australia Commonwealth Department Primary Industry Plant Quarantine Leaflet No. 45.*

Ariosa, T.I.M.D. (1982) Una neuva enfermedad de la guayabe (*Psidium guajava* L.) en la provincia Sancti Spiritus. *Centro-agricola* 9, 3–7 (in Spanish).

Badyal, K. and Sumbali, G. (1992) Some new diseases of guava fruits. *Indian Phytopathology* 45, 277–278.

Benade, E., Kemp, G.H.J., Wingfield, M.J. and Kock, J.F.L. (1991) Comparison of *Acremonium diospyri* with the guava wilt pathogen in South Africa. *Phytophylactica* 23, 98.

Butt, A.A., Nasir, M.A. and Bajwa, M.N. (1995) *In vitro* evaluation of different chemicals against *Gloeosporium psidii*, the cause of anthracnose of guava. *Pakistan Journal of Phytopathology* 7, 92–93.

Carillo, R.J., Carrillo, F.C. and Dominguez-Alvarez, J.L. (1990) Nematods asociados al cultivo de la guayabe (*Psidium guajava* L.) y control quimico en el Canon de Juchipila, Zac., Mexico. *Revista Chapingo* 15, 67–68, 94–97 (in Spanish).

Carvalho, A. de O. de, Alfenas, A.C., Maffia, L.A. and Carmo, M.G.F. do (1994) Avaliacao do progresso de ferrugem (*Puccinia psidii*) em brotacoes de *Eucalpytus cloeziana* no sudeste da Bahia. *Revista Arvore* 18, 265–274 (in Portuguese).

Casassa, P.A.M., Matheus, C.J.M., Crozzoli, P.R. and Casanova, A. (1996) Control quimico de *Meloidogyne* spp. em el cultivo del guayabo (*Psidium guajava* L.) en el Minicipio Mara del Estado Zulia, Venezuela. *Revista de la Facultad de Agronomia, Universidad del Zulia* 13, 303–312 (in Spanish).

Chakrabarti, N. (1983) A note on postharvest rot of guava caused by *Fusarium solani*. *Indian Phytopathology* 36, 556.

Cooke, A. (1982) Guava. In: Persley, D. (ed.) *Diseases of Fruit Crops*. Department of Primary Industries, Queensland, p. 59.

Crozzoli, R. and Casassa, A.M. (1998) Especies y razas de *Meloidogyne* en el cultivo del guayabo en Venezuela. *Revista de la Facultad de Agronomia, Universidad del Zulia* 15, 107–108 (in Spanish).

Crozzoli, P.R., Cassa, P.M., Rivas, G.D. and Matheus, C.J. (1991) Nematodos fitoparasitos asociados al cultivo del guayaba en el estado Zulia, Venezuela. *Fitopatologia Venezolana* 4, 2–6 (in Spanish).

de Castro, H.A., Krugner, T.L., Ideriha, C.H.F., Capello, M.S.C. and Marchi, A.B. (1983) Inoculacao cruzada de *Eucalyptus*, goiaba (*Psidium guajava*) e Jambeiro (*Syzygium jambos*) com *Puccinia psidii*. *Fitopatalogia Brasileira* 8, 491–497 (in Portuguese).

de Moura, R.M. and de Moura, A.M. (1989) Meloidogyninose da goiabeira: doenca de alta severidade no estado de Pernambuco, Brasil. *Nematologia Brasileira* 13, 13–19 (in Portuguese).

de Toledo Piza, S.M. and Ribeiro, I.J.A. (1988) Influencia da luz e da temperatura na germinacao de uredosporos de *Puccinia psidii*. *Bragantia* 47, 75–78 (in Portuguese).

Dhingra, R. and Mehrotra, R.S. (1980) A few unrecorded postharvest diseases of fruits and vegetables. *Indian Phytopathology* 33, 475–476.

Diaz, P.C. and Rondon, A. (1971) Un tipo de *Macrophomina* patogeno en frutos de guayaba. *Agronomia Tropical* 21, 111–118.

Dwivedi, S.K. (1990) Guava wilt incited by *Macrophomina phaseolina*. *National Academy of Science Letters* 13, 301–303.

Fernandez, D. and Silveira, M. (1975) El *Psidium friedrichsthalianum* como patron para guayabo, resistente a los nematodos del genero *Meloidogyne*. *Revista de Agricultura Cuba* 3, 80–85 (in Spanish).

Fernandez, E., Valdes, S. and Carrasco, J. (1986) Estudio del la biologia de especies de *Meloidogyne* que atacan el cultivo de la guayabe (*Psidium guajava*). *Cienca y Tecnica en la Agricultura, Proteccion de Plantas* 9, 27–41 (in Spanish).

Fernandez, E., Miralles, L., Navarro, A. and Sanchez, Y.M. (1987) Efecto de diferentes dosis de bromuro de metilo sobre *Meloidogyne* spp. en suelo para viveros de guayabo (*Psidium guajava*). *Cienca y Tecnica en la Agricultura, Proteccion de Plantas* 10, 59–71 (in Spanish).

Grech, N.M. (1982) Disease management progress report. *CSFRI, Nelspruit*.

Grech, N.M. (1984) Disease management progress report. *CSFRI, Nelspruit*.

Grech, N.M. (1985) First report of guava rapid death syndrome caused by *Septofusidium* sp. in South Africa. *Plant Disease* 69, 726.

Grech, N.M. (1986) Study visit on guava wilting disease in the Republic of China. *CSFRI Report, Nelspruit*.

Gupta, J.H. (1979) Control of damping-off of guava by seed treatment with systemic and non-systemic fungicides. *Progressive Horticulture* 10, 53–55.

Gupta, J.H. (1988) Fungicidal control of fruit rot of guava caused by *Phytophthora parasitica* Dastur. *Progressive Horticulture* 20, 331–332.

Gupta, P.C., Madaan, R.L. and Suhag, L.S. (1977) Varietal reaction of guava fruits to *Phytophthora nicotianae* var. *parasitica*. *Indian Journal of Mycology and Plant Pathology* 7, 177.

Ito, P.J., Kunimoto, R. and Ko, W.H. (1979) Transmission of *Mucor* fruit rot of guava fruits by three species of fruit flies. *Tropical Agriculture* 56, 49–52.

Jan, H. and Shah, S.F.A. (1991) First report of *Macrophomina phaesolina* (Tassi) Goid. on guava in North West Frontier Province, Pakistan. *Pakistan Journal of Phytopathology* 3, 67.

Kapoor, I.J. (1983) Pathological and biochemical studies on *Curvularia* rot of guava. *Zeitschrift für Pflankrankheiten und Pflanzenschutz* 90, 591–598.

Ko, W.H. and Kunimoto, R.K. (1980) Guava fruit firm rot induced by bruising. *HortScience* 15, 722–723.

Ko, W.H., Kinimoto, R.K. and Nishijima, W.T. (1982) Fruit rot of guava caused by *Phytophthora citricola*. *Plant Disease* 66, 854–855.

Kunimoto, R.K., Ito, P.J. and Ko, W.H. (1977) *Mucor* rot of guava fruits caused by *Mucor hiemalis*. *Tropical Agriculture* 54, 185–187.

Kunimoto, R.K and Ko, W.H. (1984) Brown rot of guava fruit in Hawaii caused by *Botryopshaeria berengiana*. *Plant Disease* 68, 918.

Kurosawa, E. (1926) Guava Tachigare Byd. *Report of the Taiwan Museum* 83, 47–61 (in Japanese).

Lal, B., Rai, R.N., Arya, A. and Tewari, D.K. (1980a) A new rot of *Psidium guajava* L. *National Academy of Science Letters* 3, 361–362.

Lal, B., Rai, R.N., Arya, A. and Tewari, D.K. (1980b) A new soft rot of guava. *National Academy of Science Letters* 3, 259–260.

Laundon, G.F. and Waterson, J.M. (1965) Puccinia psidii. CMI Descriptions of Pathogenic Fungi and Bacteria No. 56. Commonwealth Mycological Institute, Kew, UK.

Leu, L.S. and Kao, C.W. (1979) Artificial inoculation of guava with *Myxosporium psidii*. *Plant Disease Reporter* 63, 1077–1079.

Leu, L.S., Kao, C.W., Wang, C.C., Liang, W.J. and Hsieh, S.P.Y. (1979) Myxosporium wilt of guava and its control. *Plant Disease Reporter* 63, 1075–1077.

Lim, W.H. (1987) *Botryosphaeria* rot of ripe guava fruits. *MAPPS Newsletter* 11(4), 55–56.

Lim, T.K. and Chin, C.L. (1987) Foliar blight of guava seedlings caused by *Phytophthora nicotianae* var. *nicotianae*. *Fitopatologia Brasileira* 12, 251–254.

Lim, T.K. and Khoo, K.C. (1990) *Guava in Malaysia Production, Pests and Diseases*. Tropical Press, Kuala Lumpur, Malaysia.

Lim, T.K. and Razak, A.R. (1986) Studies on a *Phomopsis* rot of bagged guava (*Psidium guajava*) fruits in Malaysia. *Fitopatologia Brasileira* 11, 227–236.

Lim, T.K. and Teh, B.K. (1990) Antagonism, *in vitro* of *Trichoderma* species against several basidiomycetous soil-borne pathogens and *Sclerotium rolfsii*. *Zeitschrift für Pflanzenkranheiten und Pflanzenschutz* 97, 33–41.

Liu, L.J. (1972) Identification and occurrence of perfect stage and cultural and morphological variants of *Colletotrichum gloeosporioides* from guava in Puerto Rico. *Journal of Agriculture of the University of Puerto Rico* 56, 171–180.

Madhukar, J. and Reddy, S.M. (1989) Hitherto unrecorded postharvest diseases of guava. *Indian Phytopathology* 42, 479.

Majumdar, V.L. (1985) Some fungi hitherto unrecorded on guava (*Psidium guajava* L.) fruits. *Indian Botanical Reporter* 4, 195.

Majumdar, V.L. and Pathak, V.N. (1989) Incidence of major postharvest diseases of guava fruits in Jaipur markets. *Indian Phytopathology* 42, 469.

Manicom, B.Q. (1980) Disease management progress report. *CSFRI, Nelspruit*.

Mathur, S., Bhatnagar, M.K. and Mathur, K. (1992) Occurrence and epidemiology of fruit rot of guava caused by *Phytophthora nicotianae* var. *parasitica*. *Indian Phytopathology* 45, 217–220.

Montiel, C.A. (1997) *Pestalotiopsis psidii* (Pat.) Mordue causante de necrosis de fructos de guayabo (*Psidium guajava* L.) en plantaciones de loss municiplios Baralt y Mara del estado Zulia. *Revista de la Facultad de Agronomia, Universidad del Zulia* 14, 341–347 (in Spanish).

Mordue, J.E.M. (1969a) Pestalotiopsis psidii. *CMI Descriptions of Pathogenic Fungi and Bacteria No. 515*. Commonwealth Mycological Institute, Kew, UK.

Mordue, J.E.M. (1969b) Thanatephorus cucumeris. *CMI Descriptions of Pathogenic Fungi and Bacteria No. 406*. Commonwealth Mycological Institute, Kew, UK.

Naresh, M., Madaan, R.L. and Daulta, B.S. (1987) Evaluation of guava hybrids to anthracnose fruit rot under field conditions. *Haryana Journal of Horticultural Sciences* 16, 82–85.

Ooka, J.J. (1980) Guava fruit rot caused by *Rhizopus stolonifer* in Hawaii. *Plant Disease* 64, 412–413.

Pandey, R.R. and Dwivedi, R.S. (1985) *Fusarium oxysporum* f. sp. *psidii* as a pathogen causing wilt of guava in Varanasi district, India. *Phytopathologiese Zeitschrift* 114, 243–248.

Pandey, R.R., Arora, D.K. and Dubey, R.C. (1997) Effect of environmental conditions and inoculum density on infection of guava fruits by *Colletotrichum gloeosporioides*. *Mycopathologia* 137, 165–172.

Patel, G.S. and Patel, R.B. (1988) Market diseases of guava and their control. *Indian Journal of Mycology and Plant Pathology* 18, 202–203.

Prasad, G. and Mehrotra, R.S. (1985) Persistence of *Phytophthora nicotianae* var. *parasitica* in artificially infested soil at different inoculum levels in dead host tissues. *Indian Journal of Mycology and Plant Pathology* 15, 279–282.

Quintero, E. and Urdaneta, L. (1997) Evaluacion *in vitro* de fungicidas para el control del hongo *Macrophoma* sp. agente causal de la pudricion apical del fructo del guayabo (*Psidium guajava* L.). *Revista de la Facultad de Agronomia, Universidad del Zulia* 14, 233–244 (in Spanish).

Rajagopalan, B. and Wilson, K.T. (1972) *Diplodia* dry rot of guava fruit. *Agriculture Research Journal of Kerala* 10, 194–195.

Rao, D.P.C., Agrawal, S.C. and Saksena, S.B. (1976) *Phomopsis destructum* on *Psidium guaiava* fruits in India. *Mycologia* 68(5), 1132–1134.

Rawal, R.D. and Ullasa, B.A. (1988) Management of fruit diseases of guava (*Psidium guajava*) through fungicidal sprays. *Indian Journal of Agricultural Science* 58(12), 950–952.

Rayachhetry, M.B., Elliot, M.L. and Van, T.K. (1997) Natural epiphytotic of the rust *Puccinia psidii* on *Melaleuca quinquenervia* in Florida. *Plant Disease* 81, 831.

Razak, A.R. and Lim, T.K.(1987) Occurrence of the root knot nematode *Meloidogyne incognita* on guava in Malaysia. *Pertanika* 10, 265–270.

Rodriguez, N.J., Robbs, C.F. and Yamashiro, T. (1987) A bacterial disease of guava (*Psidium guajava*) caused by *Erwinia psidii* sp. nov. *Fitopatologia Brasileira* 12, 345–350.

Rodriguez-Fuentes, M.E. and Landa-Bolanos, J. (1977) Desinfeccion quimmica del seulo para semilleros de guayabo (*Psidium guajava* L.) contra nemtatodos fitoparasitos. *Centro Agricola* 4, 57–77 (in Spanish).

Ruiz, R.A.R., Alfenas, A.C., Ferreira, F.A. and Zambolim, L. (1987) Fungicidas protetoes e sistemicos para o controle da ferrugem do eucalipto causada por *Puccinia psidii*. *Revista-Avore* 11, 56–59 (in Portuguese).

Ruiz, R.A.R., Alfenas, A.C. and Demuner, N.L. (1991) Efficiencia de fungicidas para o controle da ferrugem (*Puccinia psidii*) em goiabeira (*Psidium guajava*). *Summa-Phytopathologica* 17, 147–153 (in Portuguese).

Saxena, A.K. and Saksena, S.B. (1983) A new fruit rot disease of guava from India. *Indian Phytopathology* 36, 170–172.

Schoeman, M.H. (1995) Effect of temperature on guava disease development. *ITSC Information Bulletin*, August, 5–6.

Schoeman, M.H. (1996) Epidemiology and control of guava wilt disease. *ITSC Research Report, Nelspruit.*

Schoeman, M.H. (1997) Verslag oor 'n besoek aan Malaysië om die voorkoms van koejawelverwelksiekte daar te ondersoek. *ITSC Report, Nelspruit.*

Schoeman, M.H., Benade, E. and Wingfield, M.J. (1997) The symptoms and cause of guava wilt in South Africa. *Journal of Phytopathology* 145, 37–41.

Sepiah, M. (1990) New storage disease of guava fruit caused by *Cylindrocladium scoparium*. *Plant Disease* 74, 253.

Shankhapal, K.V. and Hatwalne, V.G. (1976) Sour rot of guava in India. *Current Science* 45, 565–566.

Sharma, B.S., Champawat, R.S. and Bhanagar, G.C. (1978) Effect of chemicals and hot water treatments on *Phytophthora* fruit rot of guava. *Indian Journal of Mycology and Plant Pathology* 8, 40.

Sharma, S.K. (1981) Control of anthracnose of guava fruits caused by *Glomerella cingulata* (Stonem.) Spauld. and Schrenk. Thesis/Abstract 7, 157.

Singh, G., Chohan, J.S. and Mann, S.K. (1976) Fruit rot of guava – a new disease problem in the Punjab. *Indian Journal of Mycology and Plant Pathology* 6, 77

Singh, K.B., Prasad, G., Bhargava, K.S. and Mehrotra, R.S. (1978) Studies on the occurrence of fruit rot of guava due to *Phytophthora nicotianae* var. *parasitica*. *Indian Phytopathology* 31, 263.

Sohi, H.S. and Sridhar, T.S. (1971) Controlling fruit rot of guava. *Indian Horticulture* 16, 9–10.

Srivastava, M.P. and Tandon, R.N. (1969) Post-harvest rots of guava in India. *Plant Disease Reporter* 53, 206–208.

Tokeshi, H., Valdenbenito, R.M. and Dias, A.S. (1980) Ocorrencia de bacteriose de goiabeira no Estado de Sao Paulo. *Summa-Phytopathologica* 6, 85–87 (in Portuguese).

Ullasa, B.A. and Rawal, R.D. (1984) *Guignardia* fruit rot of guava – a new disease from Bangalore. *Current Science* 53, 435–436.

Ullasa, B.A. and Rawal, R.D. (1985) A new fruit rot of guava caused by *Sclerotium rolfsii*. *Current Science India* 54, 470–471.

Utikar, P.G., Shinde, P.A. and Soawane, C.S. (1986) Influence of temperature and incubation period on fruit for initiation and development by postharvest fungi of guava. *Current Research Reporter, Mahatma Phule Agricultural University* 2, 209–211.

Verma, B.R. and Sharma, S.L. (1976) Seasonal variation in symptoms caused by *Pestalotia psidii* on guava fruits. *Indian Journal of Mycology and Plant Pathology* 6, 97–98.

Vos, J.E., Schoeman, M.H., Berjak, P., Watt, M.P., Toerien, A.J. and Verhoyen, M.N.J. (2000). In: *In vitro selection and commercial release of Guava wilt resistant rootstock.* XXV International Congress, Part 3: Culture techniques with special emphasis on environmental implications, Brussels, Belgium, August 2–7, 1998. *Acta Horticulturae* 513, 69–79.

Willers, P. (1997) First record of *Meloidogyne mayaguensis* Rammah and Hirschmann, 1988. Heteroderidae on commercial crops in the Mpumalanga province, South Africa. *Inligtingsbulletin Insituut vir Tropiese en Subtropiese - Gewasse no. 294*, 19–20.

13 Diseases of Kiwifruit

B.A. Latorre[1] and H.A. Pak[2]

[1]*Facultad de Agronomía e Ingeniería Forestal, Pontificia Universidad Católica de Chile, Santiago, Chile;* [2]*The Horticulture and Food Research Institute of New Zealand Ltd, Mount Albert Research Centre, Auckland, New Zealand*

Introduction

The genus *Actinidia* originated in Asia. It contains >60 species in four sections, *Leiocarpae, Maculatae, Strigosae* and *Stellatae,* all of which are functionally dioecious, perennial, deciduous vines. Most species are endemic to the temperate and humid forests of Central China (Ferguson, 1990; Zuccherelli, 1994).

The kiwifruit was first included in the variable species *A. chinensis* that later was divided into two varieties, *chinensis* and *deliciosa,* by Chevalier (1941). Liang and Ferguson (1984) raised these varieties to specific status, retaining *A. chinensis* for the variant with smooth-skinned, hairless fruit, chromosome number, $2n = 58$, and A. *deliciosa* for the variant with fruit with stiff hairs and $2n = 6x = 174$. Since then, ploidy races within *A. chinensis* have been recognized: $2n = 58$ and $2n = 4x = 116$. Three varieties have been described within *A. deliciosa*: vars *deliciosa, chlorocarpa* and *longipila.* The vast majority of cultivated kiwifruit throughout the world are of *A. deliciosa* var. *deliciosa* (Ferguson, 1984, 1990).

Kiwifruit is one of the most recently developed fruit crops in temperate regions. Known as 'mihoutao' in China, it was introduced to Europe, New Zealand and the USA early in the 20th century. It originally was named Chinese gooseberry, probably in New Zealand. Commercial planting began in New Zealand during the early 1930s, and in 1959, when fruit were first exported to the USA, it was renamed kiwifruit. The major producing countries are listed in Table 13.1.

Worldwide commercial production relies heavily on a single cultivar, 'Hayward'. Fruit quality, yield, storage characteristics and a high vitamin C content make 'Hayward' a very desirable cultivar. It is a late flowering cultivar requiring 600–850 h of chilling (<7°C).

Table 13.1. Estimated annual production of kiwifruit in major producing countries.

Country	Area (ha)	Production (1000 t)
Australia	450	5.5
Brazil	900	3
Chile	7,694	120
France	4,000	50
Greece	4,500	40
Iran	2,000	55
Italy	18,740	250
Japan	3,740	43.9
Korea	1,500	13
New Zealand	10,430	253
Portugal	1,100	10.5
Spain	1,500	10
USA	1,880	28

Source: XVI Conference, International Kiwifruit Organisation (IKO), Talca, Chile, September 1998.

Staminate flowers have imperfect ovaries with poorly developed styles. Pistillate plants appear perfect but their stamens produce non-viable pollen, necessitating interplanting with male cultivars (e.g. 'Matua' and 'Tomuri'). Flowers occur singly or in inflorescences that arise from axils of the basal nodes on the current season's shoots on 1-year-old canes. They are cup shaped, with 3–7 brown sepals, 5–7 white petals, and numerous stamens and styles.

Fruit of 'Hayward' are brown, hairy, ovoid berries, 55–70 × 40–50 mm, with green flesh. Pollination and subsequent seed formation strongly influence fruit growth, with large numbers of small seeds being produced. The number of seeds a fruit contains largely determines its size. Kiwifruit have excellent storage qualities, weigh 80–120 g, and have a vitamin C content of ~0.85 mg g^{-1} fresh weight, which is higher than that of citrus fruit.

Kiwifruit are climacteric, but generally after they have reached eating firmness. Respiration rates at maturity range from 20 to 30 mg CO_2 kg^{-1} h^{-1} at 20°C (Beveer and Hopkirk, 1990). Fruit are sensitive to ethylene, with concentrations as low as 0.05 $\mu l\ l^{-1}$ being sufficient to induce fruit softening (Arpaia *et al.*, 1984). Storage at 0°C maintains fruit firmness and delays ripening, allowing a storage life of 4–6 months (McDonald, 1990).

A major factor in the successful commercialization of kiwifruit has been the relative paucity of important diseases and pests. Pests and diseases not found in cultivated kiwifruit, such as a rust, caused by *Pucciniastrum actinidiae*, and a powdery mildew, caused by *Uncinula actinidiae*, occur in its native range. Strict quarantine measures should be adopted in other production areas to avoid introducing these exotic threats.

Reliance of a worldwide industry on a single cultivar carries inherent risks with respect to new diseases. Breeding programmes to develop new cultivars rely primarily on the wide range of *Actinidia* germplasm in China and neighbouring countries. Several exotic species of *Actinidia* have potential as fruit crops, particularly in tropical areas where 'Hayward' cannot be cultivated. Research programmes to develop new cultivars and to improve 'Hayward' are underway in Europe and New Zealand, and a new, yellow-fleshed cultivar of *A. chinensis*, 'Hort16A', is already in production in New Zealand.

Diseases and Disorders

The major disease problems are Botrytis stem-end rot, Phytophthora root rot, Pseudomonas blossom blight and damage caused by the root-knot nematode. Neither viruses nor viroids have been reported on kiwifruit. Kiwifruit is highly susceptible to iron deficiency (Fig. 13.1) in high pH soils (>7.2). Similarly, it is highly intolerant of flooding (Robertson, 1982; Save and Serrano, 1986). Wind damage is the most important abiotic disorder and wind protection is needed to ensure high yields.

Abiotic Disorders

Frost damage

The susceptibility of kiwifruit vines to frost damage is dependent on the stage of crop development (Davison, 1990). In mid-winter, severe damage can result when temperatures drop below −9°C. Plants exposed to severe frosts during winter usually collapse during the spring or summer. This is the result of extensive necrosis beneath the cortex that partially or completely girdles the base of the trunk (Plate 76). Although this syndrome can be confused with Phytophthora root rot, new shoots will develop from the roots that remain healthy.

Sun scald

Fruit are very sensitive to sun scald, particularly when well-shaded fruit are exposed directly to sun after summer pruning (Sale and Lyford, 1990). Over-exposure of fruit to sun induces a dark brown skin that significantly reduces fruit quality. Leaves directly exposed to the sun often develop a yellow patch that becomes necrotic and turns brown. Defoliation may also occur.

Fig. 13.1. Chlorosis due to iron deficiency in 1-year-old kiwifruit (photo: B.A. Latorre).

Wind damage

Wind is considered the most limiting factor for kiwifruit production (Sale and Lyford, 1990). Strong winds can significantly reduce yields due to breakage of replacement canes and loss of flowers. Wind also reduces pollination activity by bees, resulting in poor fruit set and reduced fruit size, and may also cause leaf abrasion, defoliation and friction rubs on fruit that are known as 'proximity marks'. Wind shelters (e.g. living trees or artificial shelter cloth) are recommended to prevent wind damage (McAneney *et al.*, 1984).

Major Diseases

Bleeding canker

Symptoms are wilting, cane blight and red, rusty cankers on canes, cordons and trunks that usually exude a red discharge. The bacteria *Pseudomonas syringae* pv. *syringae* and *P. syringae* pv. *actinidia* have been described as the causal agents of this syndrome in, respectively, California and Italy (Opgenorth *et al.*, 1983; Scortichini and Margarita, 1989), and Japan (Takikawa *et al.*, 1989) and Italy (Scortichini, 1994).

Blossom blight

Blossom blight is the major bacterial disease of kiwifruit. It was identified in New Zealand in 1973 and was found subsequently in Italy, France, Japan and the USA (Wilkie *et al.*, 1973; Luisetti and Gaignard, 1987; Conn *et al.*, 1993). A recent report indicates that the pathogen was probably introduced from China to New Zealand on vegetative materials of *A. chinensis* (Hu *et al.*, 1999).

Symptoms

Severe infections may result in rotting of the entire bud, while less severely affected buds may be symptomless until they open (Young *et al.*, 1988). In pistillate flowers, the sterile staminate organs (anthers and filaments) are the tissues that are affected most often. The anthers of lightly affected flowers are orange–brown in colour, while severe infection results in a chocolate brown coloured rot, which completely destroys the male parts. Affected flowers may abscise, or when fruit set occurs the fruit may be deformed or subsequently drop from the vine. Symptoms on leaves start as small spots, 1–2 mm in diameter, with yellow haloes and dark, necrotic centres. These lesions may expand to form large, irregular patches of necrotic tissue.

Causal agents

The causal agent is an unusual fluorescent pseudomonad (Hu *et al.*, 1999). It was identified originally as *Pseudomonas viridiflava* based on the LOPAT determinative scheme (Lelliot *et al.*, 1966; Wilkie *et al.*, 1973; Bradbury, 1986). A re-examination of the bacterium with DNA hybridization and phenotypic characters indicated that it is actually a member of the *P. savastanoi* genomic species (Young *et al.*, 1997). The authors provided a series of phenotypic tests that enable the kiwifruit pathogen and *P. viridiflava* to be distinguished.

Epidemiology

The infection process is not well understood. However, the consistent isolation of the bacterium from apparently healthy buds, flowers and leaves suggests that it is an epiphyte (Young, 1988; Balestra and Varvaro, 1997). Infection may develop after prolonged cool, wet periods (Young, 1988). The disease has been shown to be polycyclic, with secondary spread occurring in unopened buds (Everett and Henshall, 1994).

Management

Disease incidence varies widely between districts and seasons. In New Zealand, incidence of the disease is usually low, rarely exceeding 15% of the buds (Pennycook and Triggs, 1991). Winter applications of copper compounds may be beneficial (Young, 1988; Balestra and Varvaro, 1997).

Botrytis stem-end rot

Botrytis stem-end rot, which is also known as grey mould, is the most economically important disease of kiwifruit. It occurs in all the major producing countries (Opgenorth, 1983; Sommer *et al.*, 1983; Bisiach *et al.*, 1984; Pennycook, 1985; Ieki, 1993). In addition to direct losses resulting from rotten fruit, indirect costs are incurred when severely affected lines of fruit need to be repacked. Losses of >30% have been reported in New Zealand, but 0.2–2% are typical (Michailides and Elmer, 2000). Annual monetary losses of US$8 million have been recorded in New Zealand.

Symptoms

Symptoms first appear in the tissue surrounding the picking scar, after 4–6 weeks in cold storage (0°C). Affected flesh is conspicuously darker, and has a glassy, water-soaked appearance (Pennycook, 1985). The rot advances to the distal end of the fruit, with a characteristic margin between healthy and affected tissue (Plate 77). Fluffy, white to grey mycelium and black sclerotia may appear on the surface of the fruit. Normally, primary rots occur within 3 months of cold storage, and later 'nesting' may develop when mycelium spreads to healthy fruit. Rots may also occur on the side of fruit when they are infected via lateral wounds (Pennycook, 1985; Pinilla *et al.*, 1994).

Causal agent

Botrytis cinerea (teleomorph: *Botryotinia fuckeliana*) commonly produces asexual spores, macroconidia and microconidia, and sclerotia. Macroconidia are abundant under cool and humid conditions, and are most important for disease dispersal. Microconidia play a role as spermatia in sexual reproduction. Ascospores are produced in apothecia that arise from sclerotia, but they are rarely observed in nature (Faretra *et al.*, 1988). The fungus is described in Chapter 1.

Epidemiology

B. cinerea is widely distributed and affects >235 host plants. It overwinters saprophytically and as sclerotia. Fruits are infected during the harvesting and packing process after the picking wound, which results from snapping the fruit from its pedicel, is contaminated by conidia (Pennycook, 1985).

Environmental conditions during the growing season influence disease incidence in storage. High incidences are most likely to occur following a cool (15–23°C) and wet growing season, conditions that favour a buildup of inoculum in the canopy. Rainfall prior to harvest can increase the incidence of stem-end rots in storage (Brook, 1992). Latent infections arise from colonization of floral senescent tissues (sepals, stamens and receptacles), resulting in rots that originate at the distal end of fruit (Opgenorth, 1983; Michailides and Morgan, 1996).

Infected fruit produce ethylene, which can trigger ripening within affected trays (Niklis *et al.*, 1997). High rates of ethylene (780 $\mu l\ h^{-1}\ g^{-1}$ mycelial dry weight) are also produced by *B. cinerea*, which may be important in disease development in cold storage (Qadir *et al.*, 1997).

Management

Several methods have been developed to predict disease incidence in cold storage based on its relationship to contamination and colonization of fruit by the pathogen in the field (Michailides and Morgan, 1996; Elmer *et al.*, 1997). These provide the industry with tools for improved management of the disease in the cold storage and marketing chain.

Postharvest environments have a major impact upon disease development. Fruit is stored at 0°C to reduce the rate of fruit ripening and increase storage life. Low-temperature storage also inhibits the development of Botrytis stem-end rot. Controlled atmosphere (CA) storage is often used to extend the storage life of fruit by maintaining fruit firmness. The effect of CA storage on the incidence of Botrytis stem-end rot is unclear. Although Brigati and Pratella (1991) reported an increase in stem-end rot in CA compared with fruit stored in air, Manning and Lallu (1997) showed that increased disease in CA storage was due mostly to fungi other than *B. cinerea*.

Dicarboximide fungicides (e.g. iprodione, procymidone and vinclozolin), applied either at flowering or immediately before harvest, have been the standard control measures in most countries. However, the development of dicarboximide-resistant strains limits their effectiveness (Manning and Pak, 1995). Newer fungicides, such as the anilino pyrimidines (e.g. cyprodinil, mepanipyrim and pyrimethanil), phenylpyrroles (e.g. fludioxonil) and hydroxyanilides (e.g. fenhexamide) may prove useful on kiwifruit (Lyr, 1995). Antagonistic microorganisms are also being investigated as potential biological control agents, but further research is needed before they are used extensively (Walter *et al.*, 1996).

Suppression of Botrytis stem-end rot can be achieved by delaying cold storage of the fruit for several days after harvest ('curing') (Pennycook and Manning, 1992). Although temperature and relative humidity are critical factors, there is no agreement on the optimum conditions for curing (Ippolitto *et al.*, 1994; Pinilla *et al.*, 1994; Zoffoli and Latorre, 1995; Bautista-Baños *et al.*, 1997). Curing is thought to increase the resistance of fruit to infection through changes in the activity of enzymes, especially chitinases (McLeod and Poole, 1994; Wurms *et al.*, 1997).

Crown gall

Although widespread in agricultural soils throughout the world, crown gall only occasionally occurs on kiwifruit, mostly in nurseries. Biovars 2 and 3 of *Agrobacterium tumefaciens* are responsible (Sawada and Ieki, 1992). Sanitation to reduce injuries to the roots or crown, particularly in the nursery, lowers the risk of infection. Biological treatments are available, and *A. radiobacter* strain K84 was effective against biovar 2 when applied as a soil drench or root dip before planting.

Phytophthora root and crown rot

Phytophthora root and crown rot are important kiwifruit diseases, particularly in highly saturated soils. Single or small groups of diseased vines occur in low-lying areas of orchards. The diseases have been reported in Australia, Chile, France, Italy, New Zealand, Spain and the USA.

Symptoms

Foliar symptoms consist of small, chlorotic leaves and sparse foliage (Fig. 13.2). Bronze and necrotic leaves appear as the disease progresses, leading to partial defoliation of vines, exposing fruit to sunburn. Budbreak may be delayed and shoot growth of affected vines is significantly reduced. Eventually, shoot dieback, collapse and death of the entire vine may occur.

Roots are partially or completely affected by a brown to reddish rot (Fig. 13.3). Necrotic, firm lesions appear on feeder roots, and the outer root tissues usually slough off. Soft and watery roots, with a dark brown, black or even bluish discoloration, are often caused by microorganisms other than species of *Phytophthora* (Erwin and Ribeiro, 1996). Brown to reddish decay may also develop under the bark at the soil line. Eventually, crown rot girdles the trunk and the vines collapse and die.

Causal agents

Phytophthora cactorum, P. cinnamomi, P. citricola, P. citrophthora, P. cryptogea, P. drechsleri, P. gonapodyides, P. lateralis and *P. megasperma* have been identified on kiwifruit. Differences in geographical distribution (Table 13.2) and virulence of the various species have been reported. However, the most virulent species appear to be *P. cinnamomi* and *P. cryptogea* (Conn *et al.*, 1991; Latorre *et al.*, 1991; Stewart and McCorrison, 1991; Erwin and Ribeiro, 1996).

P. cinnamomi, P. citricola and *P. citrophthora* are described in Chapter 1. *P. cryptogea* has a wide host range of plants in 23 families (Erwin and Ribeiro, 1996). Its sporangia are non-papillate, non-caducous, and oval to obpyriform with rounded bases (Fig. 13.4). They are 20–93 (37–40) μm in length and 15–51 (23) μm in width, with a length:breadth ratio of 1.45:1–1.57:1. Hyphal swellings are abundant in aqueous culture, form in chains or clusters and average 11 μm in diameter. Chlamydospores do not form on kiwifruit. *P. cryptogea* is heterothallic and requires pairs of opposite mating types to form oogonia and oospores. Antheridia are amphigynous, spherical and

Fig. 13.2. Sparse canopy and small leaves on a kiwifruit vine that is affected by Phytophthora root rot. Note the larger leaves on the unaffected vine on the left (photo: B.A. Latorre).

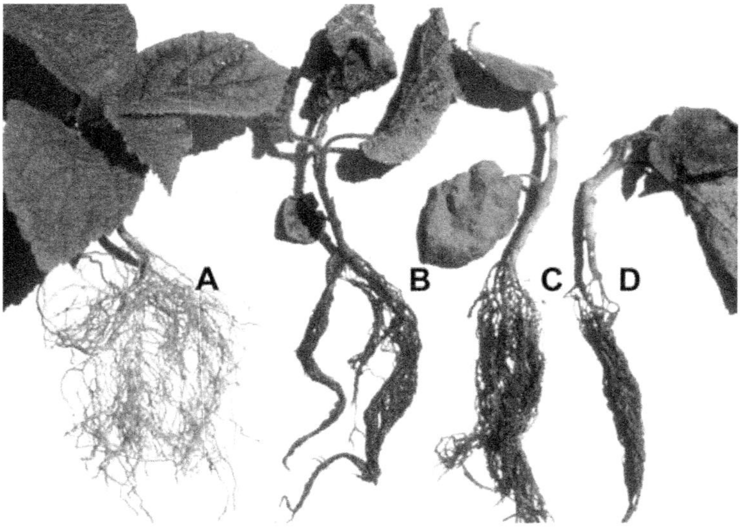

Fig. 13.3. Kiwifruit seedlings that were treated with (A) water, (B) an isolate of *Phytophthora citrophthora*, and (C) and (D) different isolates of *P. cryptogea* (photo: B.A. Latorre).

Table 13.2. Cardinal temperatures for growth and geographical distribution of the major species of *Phytophthora* that are associated with kiwifruit.

	Growth temperature (°C)		
Species	Range	Optimum	Geographical distribution
P. cactorum	2–31	25	New Zealand, USA
P. cinnamomi	5–34	24–28	New Zealand, USA
P. citricola	3–31	25–28	New Zealand
P. citrophthora	<5–33	24–28	Chile, USA
P. cryptogea	<1–<35	22–25	Chile, France, New Zealand, USA
P. drechsleri	>5–37.5	28–31	USA
P. gonapodyides	<5–30	25	New Zealand
P. lateralis	3–<26	20	New Zealand
P. megasperma	5–30	20–27	New Zealand, USA

Source: Erwin and Ribeiro (1996).

average 10 μm in diameter. Oogonia have tapered bases and are 28–40 (30) μm in diameter. Oospores are plerotic, 20–32 (26) μm in diameter, and thick walled (3.5 μm).

Epidemiology

The *Phytophthora* species that affect kiwifruit do the greatest damage in heavy, cool, poorly drained soils. Most survive as chlamydospores or oospores.

Soil saturation for >48 h predisposes plants to infection (Erwin and Ribeiro, 1996). When free water is available, polycyclic epidemics can be initiated from very small, often non-detectable levels of inoculum. Cool, wet conditions favour the production and release of zoospores that swim towards roots, encyst and germinate. It is generally assumed that infection occurs through either intact or wounded roots during spring and autumn when cool temperatures are coupled with

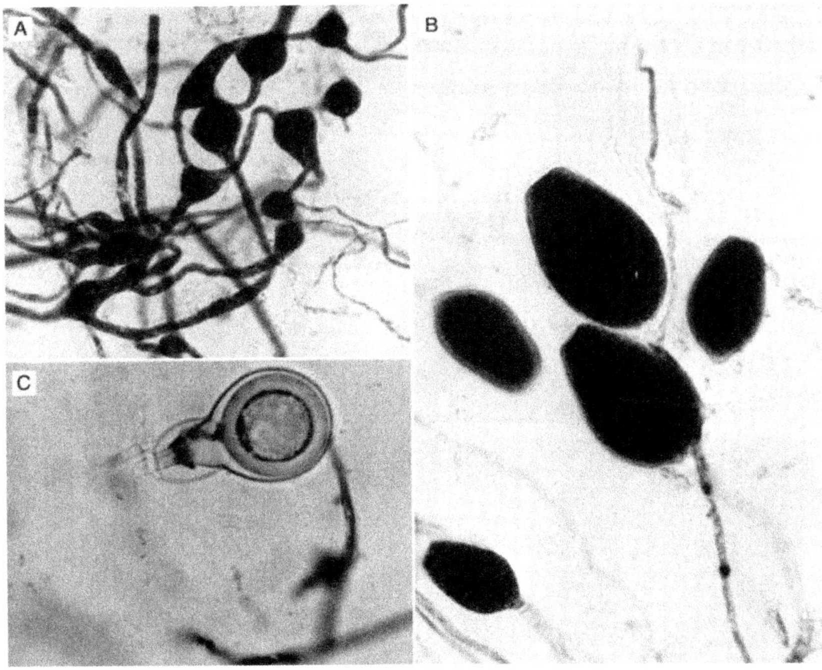

Fig. 13.4. (A) Hyphal swellings, (B) sporangia and (C) an oogonium of *P. cryptogea* with an amphigynous antheridium and oospore (from CMI description no. 592).

periods of soil saturation. Dissemination occurs through surface drainage, rainsplash or irrigation. Long-distance dissemination may occur through transport of infested tools, machinery and infected plants.

Management

Resistant rootstocks are not available. Sanitation and cultural measures, such as avoiding sites with heavy soils, high water tables and hardpans, and selecting those with good drainage, are useful. Before planting, fields should be graded and sloped to remove low areas where water may accumulate during irrigation. Planting on raised beds (>50 cm) and using herbicides rather than manual weeding also assist disease control (Robertson, 1982; Erwin and Ribeiro, 1996).

Several pesticides control Phytophthora root rots (Zaviezo *et al.*, 1993; Lyr, 1995). Ethyl phosphonates (e.g. fosetyl-Al and its breakdown product phosphorous acid) and phenylamides (e.g. metalaxyl, ofurace and oxidaxyl) are effective (Fig. 13.5). Fosetyl-Al and metalaxyl have protective and curative action. Since both are water soluble and efficiently absorbed through the roots, soil drench or granular soil applications are commonly used. Fosetyl-Al is also phloem mobile; it can be applied effectively to foliage.

Ripe rot

Ripe rot is a complex of diseases that generally affects fruit after prolonged periods of cold storage or during ripening at ambient temperatures. Ripe rot reduces shelf life, and deteriorates the quality of, and imparts undesirable odours and flavours to, the flesh. The disease has been reported in Italy, Japan, New Zealand and the USA (Sommer and Behara, 1975; Hawthorne *et al.*, 1982; Sommer *et al.*, 1983; Pennycook, 1985; Pennycook and Samuels, 1985; Testoni *et al.*, 1997).

Fig. 13.5. Effectiveness of metalaxyl applied as soil drench on 1-year-old kiwifruit vines inoculated with *Phytophthora citrophthora*. Non-inoculated, non-treated control plants are on the right; inoculated, non-treated plants are in the middle row; and inoculated plants treated with metalaxyl are on the left (photo: B.A. Latorre).

Symptoms

First symptoms normally appear as pitting on the fruit surface after 4 months in cold storage. A decay eventually develops in the flesh beneath the pits. Lesions caused by *Cryptosporiopsis* spp. are circular, tan or fawn coloured spots, 4–8 mm in diameter, with a shallow layer of yellow tissue beneath the spot that remains attached to the skin when it is removed (Pennycook, 1985). Lesions caused by *Phialophora* spp. are dark brown, round to oval spots, 2–15 mm in diameter, that extend 1–5 mm into the flesh as a light brown and spongy tissue (Testoni *et al*., 1997). Individual fruit affected by either *Cryptosporiopsis* spp. or *Phialophora* spp. may have numerous lesions. Those caused by *Botryosphaeria* spp. are oval, pale brown, up to 30 mm long, penetrate deeply into the fruit, and commonly occur as a single lesion on a fruit (Pennycook and Samuels, 1985). The skin peels back easily, exposing white, water-soaked tissue surrounded by a narrow, green margin.

Causal agents

The most frequent causes of ripe rots are *Botryosphaeria dothidea* (anamorph: *Fusicoccum aesculi*), *Cryptosporiopsis* spp., *Phialophora* spp. and *Diaporthe actinidiae* (anamorph: *Phomopsis* sp.). *B. dothidea* is described in Chapter 1.

Epidemiology

The fungi that cause ripe rots are all components of the orchard mycoflora. Consequently, primary inoculum probably comes from infected wood on the vine or from surrounding hosts that are used as shelterbelts. Infection takes place in the orchard during the growing season, and remains quiescent until the ripening process triggers further development. Susceptibility increases as fruits ripen, but it is unclear whether ripe rot enhances the fruit ripening process.

Management

Since the major sources of inoculum are unclear, it is difficult to target control measures, such as the application of protectant fungicides. The major emphasis should be on orchard sanitation with removal of dead wood from the canopy, and either removal or mulching of kiwifruit prunings. Summer pruning to maintain an open canopy should also help to reduce inoculum.

Maintenance of good cold storage conditions to maximize fruit quality and decrease the rate of ripening will decrease the incidence of ripe rots. A wider range of fungi can be isolated from fruit that ripen at ambient temperatures than in cold storage. The trend towards reduced fungicide usage in kiwifruit orchards may increase the importance of ripe rots in the future since these fungicides provide some level of control.

Root-knot nematodes

Root-knot nematodes are the most important plant parasitic nematodes that affect kiwifruit. They disrupt water and nutrient uptake by the roots, thus reducing photosynthesis, growth and yields.

Symptoms

Swellings and small white to creamy galls appear on the roots. Severely affected plants have fewer feeder roots than healthy plants.

Causal agents

Meloidogyne arenaria, *M. hapla*, *M. incognita* and *M. javanica* affect kiwifruit (Vovlas and Roca, 1976; González, 1987; Cohn and Duncan, 1990; Pinochet *et al.*, 1990). The geographical distributions of these species vary (Table 13.3). The wide distribution of *M. hapla* possibly is because it is well adapted to the cooler regions where most kiwifruit are grown. The species are identified primarily with morphological features, but, recently, molecular identification methods have been developed, such as protein profiles, and isozyme and random amplified polymorphic DNA (RAPD) analyses (Hartman and Sasser, 1985; Cenis, 1993; Philippi *et al.*, 1996).

Table 13.3. Geographical distribution of the major species of root-knot nematodes, *Meloidogyne* spp., that are associated with kiwifruit.

Species	Countries
M. arenaria	Australia, Brazil, Chile, Italy
M. hapla	Australia, Chile, France, New Zealand, Spain
M. incognita	Australia, Chile, USA
M. javanica	Australia, Chile, Italy

Data from: Vovlas and Roca (1976), González (1987), Cohn and Duncan (1990) and Pinochet *et al.* (1990).

Epidemiology

The population densities of nematodes that kiwifruit can tolerate is species dependent. Densities of *M. hapla* as high as 1–5 $\times$ 10^4 nematodes ml^{-1} can be tolerated, whereas densities of *M. incognita* higher than 1 nematode ml^{-1} may significantly reduce growth (Di Vito *et al.*, 1988; Philippi and Budge, 1992).

Root galls are inhabited by one or more females, which lay eggs in a protective gelatinous matrix. The first stage juveniles develop inside eggs, and the second stage hatches from eggs, migrates and penetrates the root cortex. It establishes in the vascular system prior to developing into a sedentary endoparasite. *M. hapla* is able to survive and hatch at lower soil temperatures (<10°C) than other species (Van Gundy, 1985).

In response to infection, multinucleate giant cells are produced in the host plant root tissue. Three additional moults occur before the juveniles develop into pyriform adult females. Males are vermiform and rare. The life cycle takes 25–30 days to complete under favourable conditions, and several generations may occur each year. Dissemination occurs through irrigation, movement of infected plants and use of contaminated tools or machinery.

Management

If possible, only root-knot-free planting sites should be used. If infestation levels are high, soil should be fumigated prior to planting with methyl bromide, dazomet or vapam-sodium. Similarly, root immersion in 0.1% phenamiphos or 1% methomyl for 1 h has been suggested prior to planting (Grandison, 1983). On established orchards, applications of aldicarb, carbofuran, oxamyl or phenamiphos to soil can reduce populations significantly.

Sclerotinia field rot

Sclerotinia field rot occurs during the growing season and is the only fungal disease of kiwifruit that affects fruit while they are still on the vine (Brook, 1990). Infection of pistillate buds, flowers and fruit leads to direct crop losses, while scarring of the fruit (Plate 78) results in crop losses either during the season by fruit thinning or after harvest, by down-grading or rejection of fruit for cosmetic reasons. The disease has been reported in Italy, Japan and New Zealand.

Symptoms

Generally the first symptoms appear as a blight of the staminate flowers, since they open before the pistillate flowers (Pennycook, 1985). Whole flower clusters on staminate vines may develop a soft, wet rot, with the affected tissues turning brown. Petals of staminate and pistillate flowers are very susceptible, and leaves and developing fruitlets can become infected when they touch infected petals. Pistillate flowers may rot rapidly, soften, turn brown, dry out and eventually drop from the vine. Under humid conditions, fluffy, white mycelium and black sclerotia may be visible.

Fruit rot appears within 2 months of fruit set. It usually originates from floral tissues that were colonized by the fungus and remained attached to the fruit. Initially, lesions are small, soft and pale green, but under favourable conditions expand to form a watery, whitish lesion, which can penetrate deep in the flesh. Unfavourable weather conditions may arrest disease development, leaving the fruits scarred.

Causal agent

Sclerotinia sclerotiorum is a widespread pathogen with a broad host range. It produces white, fluffy mycelium and numerous black sclerotia that normally are formed within internal cavities of infected tissues, but can also form on the fruit surface (Plate 78). Apothecia are trumpet shaped, off-white to fawnish brown with a 3–10 mm diameter disc that is borne on a 3–30 mm long stipe (Fig. 13.6) (Kohn, 1979). Ascospores are released from the fertile, upper surface of the apothecium in response to changes in humidity or pressure.

Epidemiology

The disease cycle starts with the emergence of apothecia beneath vines in the spring and early summer that is determined largely by soil temperature and moisture (Willets and Wong, 1980). A single sclerotium is capable of producing several apothecia. Ascospores are forcibly discharged into the air and germinate on and infect senescing floral material. Apothecium production is optimum between 10 and 20°C, and no apothecia are produced below 10°C (Willets and Wong, 1980). Ascospores may be released continually over several days, and may survive 3–5 days depending on the relative humidity. Secondary spread of the disease within the canopy results solely from the growth of the pathogen's mycelium from diseased to healthy tissue. Infected tissues can develop sclerotia on the vine or ground where they are then available to produce apothecia the following season.

Management

Control relies mainly on fungicides. An application of a benzimidazole (e.g. benomyl or carbendazim) at early flowering can be effective in reducing infection, as can applications of dicarboximide fungicides (e.g. iprodione, procymidone and vinclozolin) after flowering. These protectant fungicides need to be applied prior to the occurrence of wet weather that favours infection. New methods are being evaluated for the control of Sclerotinia field rot on kiwifruit based on the use of elicitors to induce resistance within the host plant (Reglinski *et al.*, 1997).

Minor Diseases

Fruit rots

A wide range of fungi has been isolated from fruit that have been kept in cold storage for

Fig. 13.6. A sclerotium of *Sclerotinia sclerotiorum* bearing apothecia. The scale is in 1 mm increments (photo: H.A. Pak).

extended periods or ripened under ambient conditions. These include: *Alternaria alternata, Aspergillus niger, Botryosphaeria parva, Diaporthe perniciosa, Fusarium avenaceum, F. acuminatum, Glomerella cingulata, G. acutata, Mucor piriformis, Penicillium expansum, Phoma exigua, P. macrostoma, P. nigricans, Rhizopus stolonifer* and *Trichoderma harzianum* (Sommer *et al.*, 1983; Morales and Ulloa, 1985; Manning and Lallu, 1997; Guerber and Correll, 2001).

Leaf spots

Several fungi have been isolated from necrotic leaf lesions including: *A. alternata, B. parva, Cladosporium* spp., *Colletotrichum acutatum, F. acuminatum, Glomerella cingulata, Phoma exigua* and *Phomopsis* sp. (Hawthorne *et al.*, 1982). While most of these are considered weak, opportunistic pathogens (Hawthorne and Otto, 1986), leaf infections can be important sources of inoculum for ripe rots.

Minor nematodes

Nematodes in the genera *Criconemoides, Helicotylenchus, Hemicycliophora, Heterodera, Paratrichodorus, Paratylenchus, Pratylenchus, Rotylenchus, Tylenchus, Trichodorus, Tylenchorhynchus* and *Xiphinema* have been found on roots of kiwifruit (González, 1987; Watson *et al.*, 1991). Unlike root-knot nematodes, these nematodes have not been shown to cause economically significant losses of kiwifruit.

Root rots

Two basidiomycetes cause root rots that lead to the death of kiwifruit vines within 2 years (Brook, 1990). *Armillaria mellea* has been reported in North America (Smith, 1971) and Italy (Ciccarese *et al.*, 1992), and *A. novae-zelandiae* in New Zealand (Brook, 1990). Initial symptoms include leaf yellowing and a reduction in vine vigour and fruit size. A wet, pulpy rot develops on the roots, and the presence of white mycelium under the bark, rhizomorphs growing from diseased roots, and a strong mushroom-like smell are characteristic signs and symptoms of the disease. Rhizomorphs, growing either from diseased wood in the soil or from felled shelterbelt trees that are infected after basidiospores of the pathogen colonize stumps (Horner, 1991), are primary sources of inoculum. *A. mellea* is described in more detail in Chapter 1.

Rosellinia necatrix (anamorph: *Dematophora necatrix*) causes root rot on kiwifruit, particularly where orchards were established where susceptible crops were grown previously (Ciccarese *et al.*, 1992; Ieki, 1993). It is described in Chapter 1. *Macrophomina phaseolina, Cylindrocladium crotalariae* (teleomorph: *Calonectria crotalaria*) and *Sclerotium rolfsii* (teleomorph: *Athelia rolfsii*) have also been identified as causes of root and crown rots (Krausz and Caldwell, 1987; González *et al.*, 1988; Raabe, 1988).

Silver leaf

Chondrostereum purpureum causes silver leaf on several host plants, and is a minor problem on kiwifruit in Chile (Alvarez *et al.*, 1991).

References

Alvarez, M., Elorriaga, A. and Pinilla, B. (1991) Determinación de plateado en kiwi. *Revista Frutícola* (Chile) 12, 10–12.

Arpaia, M.I., Mitchel, F.G., Kader, A.A. and Mayer, G. (1984) Effects of delays in establishing controlled atmospheres on kiwifruits. Softening during and following storage. *Journal of the American Society of Horticultural Science* 109, 584–587.

Balestra, G.M. and Varvaro, L. (1997) Epiphytic survival and control of *Pseudomonas viridiflava* on *Actinidia deliciosa*. *Acta Horticulturae* 444, 745–749.

Bautista-Baños, S., Long, P.G. and Ganesh, S. (1997) Curing of kiwifruit for control of postharvest infection by *Botrytis cinerea*. *Postharvest Biology and Technology* 12, 137–145.

Beveer, D.J. and Hopkirk, G. (1990) Fruit development and fruit physiology. In: Warrington, I.J. and Weston, G.C. (eds) *Kiwifruit: Science and Management*. Ray Richards Publisher, Auckland, New Zealand, pp. 97–126.

Bisiach, M., Minervini, G. and Vercesi, A. (1984) Biological and epidemiological aspects of the kiwifruit (*Actinidia chinensis* Planchon) rot, caused by *Botrytis cinerea* Pers. *Rivista Patologia Vegetale S. IV* 20, 38–55.

Bradbury, J.F. (1986) *Guide to Plant Pathogenic Bacteria*. International Mycological Institute, Kew, UK.

Brigati, S. and Pratella, G.C. (1991) Effetto indotto dalla CO_2, dalla refrigerazione e dai trattamenti di campo sulla *B. cinerea* nell'actinidia frigoconservata. *Informatore Fitopatologico* 12, 44–46.

Brook, P.J. (1990) Diseases of kiwifruit. In: Warrington, I.J. and Weston, G.C. (eds) *Kiwifruit: Science and Management*. Ray Richards Publisher, Auckland, New Zealand, pp. 420–428.

Brook, P.J. (1992) Botrytis stem-end rot and other storage diseases of kiwifruits – a review. *Acta Horticulturae* 297, 545–550.

Ciccarese, F., Frisullo, S. and Amenduni, M. (1992) Osservazioni sui marciumi del colletto dell'actinidia nell'Italia meridionale. *Informatore Fitopatologico* 42, 57–58.

Cenis, J.L. (1993) Identification of four major *Meloidogyne* spp. by random amplified polymorphic DNA (RAPD-PCR). *Phytopathology* 83, 76–78.

Chevalier, A. (1941) Un *Actinidia* à fruits comestibles intéressant pour la France (*A. chinensis* Planch.var. *deliciosa* Chev.). *Revue de Botanique Appliquée et d'Agriculture Tropicale* 21, 240–244.

Cohn, E. and Duncan, L.W. (1990) Nematode parasites of subtropical and tropical fruit trees. In: Luc, M., Sikora, R.A. and Bridge, J. (eds) *Plant Parasitic Nematodes in Subtropical and Tropical Agriculture*. CAB International, Wallingford, UK, pp. 347–362.

Conn, K.E., Gubler, W.D., Mircetich, S.M. and Hasey, J.K. (1991) Pathogenicity and relative virulence of nine *Phytophthora* spp. from kiwifruit. *Phytopathology* 81, 974–979.

Conn, K.E., Gubler, W.D. and Hasey, J.K. (1993) Bacterial blight of kiwifruit in California. *Plant Disease* 77, 228–230.

Davison, R.M. (1990) The physiology of the kiwifruit vine. In: Warrington, I.J. and Weston, G.C. (eds) *Kiwifruit: Science and Management*. Ray Richards Publisher. Auckland, New Zealand, pp. 127–154.

Di Vito, M., Vovlas, N. and Simeone, A.M. (1988) Effect of root-knot nematode *Meloidogyne incognita* on the growth of kiwi (*Actinidia deliciosa*) in pots. *Advances in Horticultural Science* 2, 109–112.

Elmer, P.A.G., Whelan, H.G., Boyd-Wilson, K.S.H. and Pyke, N.B. (1997) Relationship between *Botrytis cinerea* inoculum in kiwifruit vines, contamination of the fruit surface at harvest and stem end rot in cool storage. *Acta Horticulturae* 444, 713–717.

Erwin, D.C. and Ribeiro, O.K. (1996) *Phytophthora Diseases Worldwide*. APS Press, St Paul, Minnesota.

Everett, K.R. and Henshall, W.R. (1994) Epidemiology and population ecology of kiwifruit blossom blight. *Plant Pathology* 43, 824–830.

Faretra, F., Antonacci, E. and Pollastro, D. (1988) Sexual behaviour and mating system of *Botryotinia fuckeliana*, teleomorph of *Botrytis cinerea*. *Journal of General Microbiology* 134, 2543–2550.

Ferguson, A.R. (1984) Kiwifruit: a botanical review. *Horticultural Reviews* 6, 1–64.

Ferguson, A.R. (1990) The genus *Actinidia*. In: Warrington, I.J. and Weston, G.C. (eds) *Kiwifruit: Science and Management*. Ray Richards Publisher, Auckland, New Zealand, pp. 15–35.

González, G., Santelices, C. and Sanfuentes, E. (1988) Pudrición carbonosa de raíces de kiwi. *Agro-Ciencia* (Chile) 4, 159–160.

González, H. (1987) Situación nematológica del kiwi en Chile. *Revista Frutícola* (Chile) 8, 33–36.

Grandison, G.S. (1983) Root-knot nematode control on kiwifruit (*Actinidia chinensis*) by chemical bare-root dip. *Plant Disease* 67, 899–900.

Guerber, J.C. and Correll, J.C. (2001) Characterization of *Glomerella acutata*, the teleomorph of *Colletotrichum acutatum*. *Mycologia* 93, 216–229.

Hartman, K.M. and Sasser, J.N. (1985) Identification of *Meloidogyne* species on the basis of differential host test and perineal-pattern morphology. In: Sasser, J.N. and Carter, C.C. (eds) *An Advanced Treatise on Meloidogyne*, Vol. II, *Methodology*. North Carolina State University, Raleigh, North Carolina, pp. 60–70.

Hawthorne, B.T. and Otto, C. (1986) Pathogenicity of fungi associated with leaf spots of kiwifruit. *New Zealand Journal of Agricultural Research* 29, 533–538.

Hawthorne, B.T., Rees-George, J. and Samuels, G.J. (1982) Fungi associated with leaf spots and postharvest fruit rots of kiwifruit (*Actinidia chinensis*) in New Zealand. *New Zealand Journal of Botany* 20, 143–150.

Horner, I.J. (1991) Epidemiology of *Armillaria* root-rot of kiwifruit. *Acta Horticulturae* 297, 573–578.

Hu, F.-P., Young, J.M. and Jones, D.S. (1999) Evidence that bacterial blight of kiwifruit, caused by *Pseudomonas* sp., was introduced into New Zealand from China. *Journal of Phytopathology* 147, 89–97.

Ieki, H. (1993) Kiwifruit diseases in Japan. *Japan Pesticide Information* 61, 11–13.

Ippolito, A., Lattanzio, V., Nigro, F., di Venere. D., Lima, G., Castellano, M.A. and Salerno, M. (1994) Improvement of kiwifruit resistance to *Botrytis* storage rot by curing. *Phytopathologia Mediterranea* 33, 132–136.

Kohn, L.M. (1979) A monographic revision of the genus *Sclerotinia*. *Mycotaxon* 9, 365–444.

Krausz, J.P. and Caldwell, J.D. (1987) *Cylindrocladium* root rot of kiwifruit. *Plant Disease* 71, 374–375.

Latorre, B.A., Alvarez, C. and Ribeiro, O.K. (1991) *Phytophthora* root rot of kiwifruit in Chile. *Plant Disease* 75, 949–952.

Lelliot, R.A., Billing, E. and Hayward, A.C. (1966) A determinative scheme for the fluorescent plant pathogenic pseudomonads. *Journal of Applied Bacteriology* 29, 470–489.

Liang, C.-F. and Ferguson, A.R. (1984) Emendation of the Latin name of *Actinidia chinensis* Planch. *Guihaia* 5, 71–72.

Luisetti, J. and Gaignard, J.L. (1987) Deux maladies bactériennes du kiwi en France. *Phytoma* 391, 42–45.

Lyr, H. (ed.) (1995) *Modern Selective Fungicides*, 2nd edn. Gustav Fischer Verlag, New York.

Manning, M.A. and Lallu, N. (1997) Fungal diseases of kiwifruit stored in controlled atmosphere conditions in New Zealand. *Acta Horticulturae* 444, 725–731.

Manning, M.A. and Pak, H.A. (1995) *Botrytis* storage rot of kiwifruit: efficacy of pre-harvest sprays in orchards with dicarboximide-resistant *Botrytis* populations. *Proceedings of the 48th New Zealand Plant Protection Conference*, 1995, pp. 22–26.

McAneney, K.J., Judd, M.J. and Trought, M.C.T. (1984) Wind damage to kiwifruit (*Actinidia chinensis* Planch) in relation to windbreak performance. *New Zealand Journal of Agricultural Research* 27, 255–263.

McDonald, B. (1990) Precooling, storage and transport of kiwifruit. In: Warrington, I.J. and Weston, G.C. (eds) *Kiwifruit: Science and Management*. Ray Richards Publisher, Auckland, New Zealand, pp. 429–459.

McLeod, L.C. and Poole, P.R. (1994) Changes in enzymic activities after harvest and in the early stages of *Botrytis cinerea* infection of kiwifruit. *Journal of the Science of Food and Agriculture* 64, 95–100.

Michailides, T.J. and Elmer, P.A.G. (2000) Botrytis gray mold of kiwifruit caused by *Botrytis cinerea* in the United States and New Zealand. *Plant Disease* 84, 208–223.

Michailides, T.J. and Morgan, D.P. (1996) Using incidence of *Botrytis cinerea* in kiwifruit sepals and receptacles to predict gray mold decay in storage. *Plant Disease* 80, 248–254.

Morales, A. and Ulloa, A. (1985) Kiwi: pudriciones fungosas en postcosecha. *ACONEX* (Chile) 11, 35–37.

Niklis, N., Sfakiotakis, E. and Thanassoulopoulos, C.C. (1997) Ethylene production by *Botrytis cinerea*, kiwifruit and Botrytis rotted kiwifruit under storage temperatures. *Acta Horticulturae* 444, 733–737.

Opgenorth, D.C. (1983) Storage rot of California-grown kiwifruit. *Plant Disease* 67, 382–383.

Opgenorth, D.C., Lai, M., Sorrell, M. and White, J.B. (1983) *Pseudomonas* canker of kiwifruit. *Plant Disease* 67, 1283–1284.

Pennycook, S.R. (1985) Fungal fruit rots of *Actinidia deliciosa* (kiwifruit). *New Zealand Journal of Experimental Agriculture* 13, 289–299.

Pennycook, S.R. and Manning, M.A. (1992) Picking wound curing to reduce Botrytis storage rot of kiwifruit. *New Zealand Journal of Crop and Horticultural Science* 20, 357–360.

Pennycook, S.R. and Samuels, G.J. (1985) *Botryosphaeria* and *Fusicoccum* species associated with ripe fruit rot of *Actinidia deliciosa* (kiwifruit) in New Zealand. *Mycotaxon* 24, 445–458.

Pennycook, S.R. and Triggs, C.M. (1991) Bacterial blossom blight of kiwifruit – a 5-year survey. *Acta Horticulturae* 297, 559–565.

Philippi, I. and Budge, A. (1992) Efectos de *Meloidogyne hapla* en plantas jóvenes de kiwi. *Nematropica* 22, 47–54.

Philippi, I., Latorre, B.A., Perez, G.F. and Castillo, L. (1996) Identificación de los nematodos del nudo (*Meloidogyne* spp.) del kiwi por análisis de isoenzimas, en Chile. *Fitopatología* 31, 6–101.

Pinilla, B.C., Alvarez, M.A. and García, M.A. (1994) Pudrición penduncular de post-cosecha causada por *Botrytis cinerea* en kiwi. *Revista Frutícola* (Chile) 15, 63–66.

Pinochet, J., Verdejo, S. and Soler, A. (1990) Observations on the seasonal fluctuation of *Meloidogyne hapla* on kiwi (*Actinidia deliciosa*) in Spain. *Nematrópica* 20, 31–37.

Qadir, A., Hewett, E.W. and Long, P. (1997) Ethylene production by *Botrytis cinerea*. *Postharvest Biology and Technology* 11, 85–91.

Raabe, R.D. (1988) *Sclerotium rolfsii* crown and root rot of kiwi (*Actinidia chinensis*). *Plant Disease* 72, 1077.

Reglinski, T., Poole, P.R., Whitaker, G. and Hoyte, S.M. (1997) Induced resistance against *Sclerotinia sclerotiorum* in kiwifruit leaves. *Plant Pathology* 46, 716–721.

Robertson, G.I. (1982) Kiwifruit can tolerate *Phytophthora* but not wet feet. *The Orchardist of New Zealand* 55, 148–151.

Sale, P.R. and Lyford, P.B. (1990) Cultural, management and harvesting practices for kiwifruit in New Zealand. In: Warrington, I.J. and Weston, G.C. (eds) *Kiwifruit: Science and Management*. Ray Richards Publisher, Auckland, New Zealand, pp. 247–296.

Save, R. and Serrano, L. (1986) Some physiological and growth responses of kiwi fruit (*Actinidia chinensis*) to flooding. *Physiologia Plantarum* 66, 75–78.

Sawada, H. and Ieki, H. (1992) Crown gall of kiwi caused by *Agrobacterium tumefaciens* in Japan. *Plant Disease* 76, 212.

Scortichini, M. (1994) Occurrence of *Pseudomonas syringae* pv. *actinidiae* on kiwifruit in Italy. *Plant Pathology* 43, 1035–1038.

Scortichini, M. and Margarita, L. (1989) *Pseudomonas syringae* pv. *syringae* van Hall agente causale di seccumi rameali su actinidia in Italia. *Informatore Fitopatologico* 10, 49–52.

Smith, R.L. (1971) Chinese gooseberry, a new host for *Armillaria mellea*. *Plant Disease Reporter* 55, 1099–1100.

Sommer, N.F. and Behara, L. (1975) *Diaporthe actinidiae*, a new species causing stem-end rot of Chinese gooseberry fruits. *Mycologia* 67, 650–653.

Sommer, N.F., Fortlage, R.J. and Edwards, D.C. (1983) Minimizing postharvest diseases of kiwifruit. *California Agriculture* 37(1–2), 1618.

Stewart, A. and McCorrison, A.M. (1991) Pathogenicity and relative virulence of seven *Phytophthora* species on kiwifruit. *New Zealand Journal of Crop and Horticultural Science* 19, 73–76.

Takikawa, Y., Serizawa, S., Ichikawa, T., Tsuyumu, S. and Goto, M. (1989) *Pseudomonas syringae* pv. *actinidiae* pv. nov.: the causal bacterium of canker of kiwifruit in Japan. *Annals of the Phytopathological Society of Japan* 55, 437–444.

Testoni, A., Grassi, B., Quaroni, S., Saracchi, M. and Sardi, P. (1997) Pitting on kiwifruit in storage caused by *Phialophora* sp. *Acta Horticulturae* 444, 751–756.

Van Gundy, S.D. (1985) Ecology of *Meloidogyne* spp.: emphasis on environmental factors affecting survival and pathogenicity. In: Sasser, J.N. and Carter, C.C. (eds) *Advance Treatise on Meloidogyne*, Vol. 1, *Biology and Control*. Graphics North Carolina State University, Raleigh, North Carolina, pp. 177–182.

Vovlas, N. and Roca, F. (1976) *Meloidogyne hapla* su *Actinidia chinensis* in Italia. *Nematologia Mediterranea* 4, 115–116.

Walter, M., Boyd Wilson, K.S.H., Elmer, P.A.G. and Köhl, J. (1996) Selection of antagonistic saprophytes for suppression of *Botrytis cinerea* sporulation on kiwifruit tissues. *XIth International Botrytis Symposium. Programme and Book of Abstracts*. Wageningen, The Netherlands, p. 89.

Watson, R.N., Wilson, E.A. and Marsden, R.S. (1991) Distribution of plant-parasitic nematodes in the rhizosphere of kiwifruit. *Acta Horticulturae* 297, 537–543.

Wilkie, J.P., Dye, D.W. and Watson, D.R. (1973) Further hosts of *Pseudomonas viridiflava*. *New Zealand Journal of Agricultural Research* 16, 315–323.

Willets, H.J. and Wong, J.A.-L. (1980) The biology of *Sclerotinia sclerotiorum*, *S. trifoliorum*, and *S. minor* with emphasis on specific nomenclature. *Botanical Review* 46, 101–165.

Wurms, K.V., Sharrock, K.R., Long, P.G., Greenwood, D.R. and Ganesh, S. (1997) Responses of chitinases in kiwifruit to curing and to long-term storage. *New Zealand Journal of Crop and Horticultural Science* 25, 213–220.

Young, J.M. (1988) Bacterial blight of kiwifruit in New Zealand. *OEPP/EPPO Bulletin* 18, 131–140.

Young, J.M., Cheesmur, G.J., Welham, F.V. and Henshall, W.R. (1988) Bacterial blight of kiwifruit. *Annals of Applied Biology* 112, 91–105.

Young, J.M., Gardan, L., Ren, X.-Z. and Hu, F.-P. (1997) Genomic and phenotypic charcterization of the bacterium causing blight of kiwifruit in New Zealand. *Plant Pathology* 46, 857–864.

Zaviezo, T., Latorre, B.A. and Torres, R. (1993) Effectiveness of three phenylamide fungicides against *Phytophthora cryptogea* isolated from kiwi and their mobility in the soil. *Plant Disease* 77, 1239–1243.

Zoffoli, J.P. and Latorre, B.A. (1995) Efecto de la temperatura y tiempo de curado en el control de *Botrytis cinerea* en frutos de kiwi. *Simiente* (Chile) 65, 22.

Zuccherelli, G. (1994) *L'Actinidia e i Nuovi Kiwi*. Edagricole, Edizioni Agricole, Bologna, Italy.

14 Diseases of Longan, Lychee and Rambutan

L.M. Coates[1], S. Sangchote[2], G.I. Johnson[3] and C. Sittigul[4]
[1]*Department of Primary Industries, Indooroopilly, Queensland, Australia;* [2]*Department of Plant Pathology, Kasetsart University, Bangkok, Thailand;* [3]*Australian Centre for International Agricultural Research, Canberra, Australia;* [4]*Department of Plant Pathology, Chiang Mai University, Chiang Mai, Thailand*

Introduction

Longan, *Dimocarpus longan*, lychee, *Litchi chinensis*, and rambutan, *Nephelium lappaceum*, are members of the *Sapindaceae*, a pantropical family of trees, shrubs and climbers. There are ~131 genera in the family that include lesser-known fruit crops such as the pulasan, *Nephelium mutabile*, and akee, *Blighia sapida*. A characteristic of the family is the presence in leaves and fruit skins of saponins, which can act as anti-feedants and molluscicides (Duke, 1990), and may confer disease resistance (Osbourne, 1996). The latter trait may lower buildup of intra-tree inoculum of fruit pathogens such as *Colletotrichum gloeosporioides* and *Diplodia theobromae*.

Each fruit is a camara, with only one mericarp, developing from a three-carpellate gynoecium by abortion, and a fleshy aril (Spjut, 1994). Fruit are borne on panicles that generally project clear of the foliage, and hang down as they approach maturity. Lychee and longan have a rough, leathery rind with numerous microscopic crevices; the skin is more verrucose in lychee. In lychee, high levels of anthocyanins develop in the skin, resulting in pink or red fruit. Longan skin becomes light brown as fruit mature. Rambutan fruit skin is fleshy, crisp and pliable, rather than leathery, and has numerous prominent spinterns that are red or yellow on mature fruit. Upon maturity, the skin of each of these fruits separates easily from the fleshy aril that surrounds a single seed. The aril is watery, translucent, clear to milky white and delicately scented.

The lychee is native to Malaysia, northern Vietnam and southern China. Principal producers of lychee are China, India, Taiwan and Thailand, with smaller industries in Australia, Israel, Madagascar, South Africa and Vietnam (Singh, 1998). The longan is native to the mountainous areas from China to Myanmar, and principal producers are China, Thailand and Vietnam. The rambutan is native to Southeast Asia, and the greatest production occurs in Indonesia, Malaysia and Thailand (Arora, 1998). Commercial production of all three fruits has increased in the last 20 years, but there remains considerable potential for improvement in genotype, agronomy and postharvest technologies.

Commercial nursery stock is produced by marcotting (lychee, longan), approach grafting (longan) or budding on to seedlings (rambutan). Orchard trees generally are pruned to limit size and optimize branch spacing. Saponins in lychee, longan and rambutan may be responsible for the relative paucity of major diseases that affect these crops. In general, leaf spots and flower blights are not serious problems. Root rots and diebacks are not widespread, and when they do occur, may be exacerbated by water stress. When stored at ambient temperatures, fruit of each species deteriorate quickly due to a combination of skin desiccation, browning and decay.

Soilborne Diseases

Armillaria root rot

Armillaria root rot of lychee was first reported by Cohen (1955) in Florida. In Florida, it is considered to be the most serious disease of lychee because it can result in tree death (McMillan, 1994b). The disease is also becoming increasingly important in certain lychee-growing areas in South Africa, where it has caused serious tree losses in a small number of orchards (Darvas, 1992; Manicom, 1995). In Australia, Armillaria root rot occasionally causes death or slow tree decline (Menzel *et al.*, 1988).

Symptoms

Affected lychee trees are non-vigorous and produce no new growth (Darvas, 1992). Foliage may turn yellow and fall prematurely (McMillan, 1994b) or become necrotic while remaining attached to the tree (Manicom, 1995). Branch and eventually tree death then occur. The disease is distinguished from other forms of dieback by the presence of white to light tan sheets of mycelium between the bark and the wood of the root crown and large roots (McMillan, 1994b). The underlying wood becomes discoloured, and clumps of honey coloured mushrooms occasionally form at the base of infected trees, particularly during cool, rainy weather.

Causal agents

Two basidiomycetes, *Armillaria mellea* and *A. socialis*, cause this disease in South Africa (Darvas, 1992) and Florida (Cohen, 1955), respectively. *A. mellea* is sometimes referred to as the honey mushroom, honey agaric, oak root fungus or shoestring fungus, and is described in Chapter 1 (Shaw and Kile, 1991). The fungi form a number of morphological structures including basidiomes (mushrooms), basidiospores, mycelia, pseudosclerotial tissue and rhizomorphs (compacted masses of hyphal strands that are covered by a cortex).

Epidemiology

Infection results when lychee roots contact infected plants or rhizomorphs of the fungus in the soil. Infected tree stumps and large roots of dead hosts are common sources of inoculum, and the fungi may persist as saprophytes for 10 years or more (Cohen, 1955; Darvas, 1992). Following penetration of a healthy lychee root, the fungi grow up the root along the cambial layer, eventually girdling the root crown and causing tree death (Cohen, 1955). Although clumps of honey coloured basidiomes are produced occasionally at the base of a dying tree, they have no known role in the disease cycle (Shaw and Kile, 1991).

In South Africa, it has been suggested that lychee trees planted in poorly drained soils or trees weakened by nematodes may be more susceptible to *A. mellea* (Darvas, 1992).

Management

See Chapter 1.

Longan decline

Longan decline is a serious disease in Thailand, affecting up to 41 and 33% of the longan trees in the Chiang Mai and Lam Phun provinces, respectively (Visitpanich *et al.*, 1999).

Symptoms

Affected trees are non-vigorous: 1- to 2-year-old plants are stunted and have small leaves, whereas those over 5 years of age produce new shoots rarely. As a result, affected trees have thin canopies with many bare twigs and branches (Plate 79). They break easily at the root collar during windy weather, and produce reduced numbers of brittle adventitious roots. In waterlogged soils, root, and ultimately tree, mortality may occur (Visitpanich *et al.*, 1999).

Affected trees continue to flower and set small, low quality fruit. Trees also become more susceptible to attack by insect pests and to surface colonization by lichens and algae.

Cause and epidemiology

The cause of longan decline has not been clearly established, although the syndrome appears to be associated with high numbers of the nematode *Rotylenchulus reniformis* in soil beneath affected trees (Visitpanich *et al.*, 1999). High groundwater in lowland orchards may also be a contributing factor, since improving drainage reduces the severity of symptoms in these locations. However, the syndrome also occurs in upland regions where, conversely, a lack of water during the dry season exacerbates the problem.

Management

In Thailand, it was found that improving drainage and amending soil with cow manure and granular fertilizers in lowland orchards helped alleviate symptoms. In upland orchards, additional irrigation during the dry season and application of soil amendments were partially effective (Visitpanich *et al.*, 1999). In more recent studies, it was found that applications of chicken manure plus urea fertilizer increased leaf size in decline-affected trees, and dramatically reduced numbers of *R. reniformis* in pot trials (C. Sittigul, Thailand, 2001, personal communication).

Marcot death

Sudden death of lychee nursery stock or field transplants raised from marcots can be serious. The problem has been recorded in Australia and associated with marcots that were taken from trees with branch cankers. Affected lychee transplants wilt and die suddenly. In some cases, leaves may also drop from the trees, and roots are necrotic.

Fusarium spp. have been recovered from affected transplants and unthrifty marcots. Symptom expression may be more serious following water stress during early establishment. Infection appears to occur during the growth of marcots on tree branches, whereas dieback symptoms develop after transplanting.

Marcots should be made on vigorous, canker-free lychee trees, and excessive numbers should not be made on a given tree. Incorporation of benzimidazole fungicides into the growth medium that is used to produce marcots appears to be beneficial. Transplants should be kept well watered until they are established.

Nematode damage

In South Africa, nematodes have been associated with a severe dieback and decline syndrome of lychee (Milne *et al.*, 1971). Aboveground symptoms include many bare twigs and branches, leaf chlorosis and tip burn, poor flowering, excessive fruit drop and erratic leaf flushes. Affected roots become stunted and darken. The feeder root mass is reduced, leading to decreased water and nutrient uptake and, ultimately, dieback of aboveground portions of the plant.

Large numbers of the ring nematode, *Hemicriconemoides mangiferae*, and dagger nematode, *Xiphinema brevicolle*, are associated with these symptoms. *H. mangiferae* penetrates the cortex of young lychee roots, whereas *X. brevicolle* feeds more superficially (Milne *et al.*, 1971).

Pre-plant soil fumigation and the application of nematicides after planting have shown considerable promise for the control of these nematodes (Milne *et al.*, 1971; Cohn and Duncan, 1990). Because these nematodes survive in air-layer soil, the practice of using orchard soil for marcotting is not recommended (Milne *et al.*, 1971).

Lychee decline

Recent surveys in northern Vietnam have highlighted the widespread occurrence of a lychee decline syndrome (Trung, 1999; Trung *et al.*, 1999). Symptoms include leaf fall, dull leaves and branch dieback, i.e. those that are associated with symptoms that are produced by *Phytophthora* spp. on other hosts. Species of *Fusarium*, *Pythium*, *Cylindrocladium*, *Scopulariopsis* and *Rhizoctonia* that have been isolated are thought to be secondary colonizers (Trung *et al.*, 1999). The decline appears to be favoured by poor drainage, deep planting and inadequate nutrition, and has been reduced by the application of phosphonates (Trung, 1999).

Foliar, Floral, Stem and Preharvest Fruit Diseases

Algal spot

Algal spot, which is also known as algal rust or red rust, is a common, yet easily controlled, disease of many fruit crops in the humid subtropics and tropics. It is particularly prevalent in wet production areas and during periods of frequent rain. The disease affects lychee (Mishra *et al.*, 1974; Menzel *et al.*, 1988; McMillan, 1994b), longan (Visarathanonth, 1999a) and rambutan (Visarathanonth, 1999b).

Circular, slightly raised patches of velvety, greenish grey growth of the causal alga, *Cephaleuros virescens*, appear on the surface of leaves. These patches become light brown and eventually reddish brown when the alga produces sporangia. On lychee, this disease has not been observed on twigs and branches in Florida (McMillan, 1994b), whereas it affects both leaves and young stems in India (Gupta, 1992). On rambutan, it affects mainly mature leaves but may also occur on stems and branches. A leaf blight results on rambutan when spots on this organ coalesce. *C. virescens* is described in Chapter 1.

Menzel *et al.* (1988) reported that the cultivars 'Sewy Tung' and 'Haak Yip' are very susceptible. Algal spot is controlled readily by a combination of copper-based preparations and cultural practices (Bhavakul, 1977; Menzel *et al.*, 1988; Gupta, 1992; Tindall, 1994).

Corky bark

Corky bark is a recently recognized disease of lychee in South Florida (R.C. Ploetz, unpublished results). In some ways, it resembles the gall and scaly bark problem on mango (see Chapter 15). Small to large irregular patches of bark become raised, cracked and rough (Fig. 14.1). The bark surface in these areas is friable and can be rubbed off easily. As these patches enlarge and become more frequent, branch death occurs. Eventually, large portions of the canopy can be killed (Fig. 14.2).

Fusarium decemcellulare, the same fungus that causes gall and scaly bark on mango, has been recovered from affected trees (R.C. Ploetz, unpublished results). It is a weak pathogen on mango and requires wounding to infect and cause symptoms on this host. An armoured scale, *Andaspis punicae*, has also been associated with corky bark lesions on lychee (J.E. Peña, personal communication). Although one could speculate that the scale wounds or vectors the fungus on lychee, this relationship, as well as Koch's postulates for *F. decemcellulare* on lychee, have not been proven.

To date, no pesticides have been identified to control this problem. At this time, the following measures are suggested: (i) removal and destruction of affected branches and trees in the orchard; (ii) disinfestation of pruning equipment to ensure that the pathogen is not spread during pruning operations; and (iii) use of healthy planting material in new orchards (in no instance should air layers be taken from affected trees).

Downy blight

Downy blight is an important disease of lychee fruit and flowers in China (Chi *et al.*, 1984), Taiwan (Kao and Leu, 1980; Ann and Ko, 1984), Thailand (C. Sittigul, Thailand, 2000, personal communication) and Vietnam (Vien *et al.*, 2001).

Fig. 14.1. Symptoms of corky bark on 'Mauritius' lychee. Note the large irregular patches of raised bark along the stem (photo: R.C. Ploetz).

Fig. 14.2. Large portions of a lychee canopy that have been killed by corky bark. This damage can be highly localized or affect the entire canopy (photo: R.C. Ploetz).

Symptoms

The disease affects both young and ripe fruit as well as leaves, flowers and pedicels of lychee (Ann and Ko, 1984; Chi *et al.*, 1984). Symptoms appear as a brown blight of affected organs, and during wet conditions lesions become covered with whitish masses of hyphae, sporangia and sporangiophores (Fig. 14.3). Infected fruit may fall prematurely.

Causal agent

Downy blight is caused by the chromistan *Peronophythora litchii* (Chen, 1961; Ko *et al.*, 1978). The facultative necrotroph produces colourless, aseptate mycelium that is 4–6 μm wide and branched irregularly at right or acute angles (Fig. 14.4) (Hall, 1989). Thin-walled hyphal swellings are 30–35 μm in diameter, and sporangia are ovoid to obovoid, occasionally irregular, 28–33 × 19–22 μm, with distinct, flattened apical papilla. Sporangiophores are arborescent, one- to five-celled, 440–1325 × 4–6 μm and dichotomously branched in the upper one-fifth to one-third. Antheridia are paragynous, ellipsoidal and 9–13 μm, and oogonia are spherical to ellipsoidal, colourless and 28–33 × 24–30 μm. Oospores are aplerotic, colourless and 22–26 μm in diameter.

Epidemiology

Fruit isolates of *P. litchii* from Taiwan caused symptoms on both wounded and non-wounded mature fruit, suggesting that the disease could develop in the field without mechanical injury (Chen *et al.*, 1998). The disease spreads rapidly during extended periods of rain and causes extensive fruit drop in Taiwan during May and June, which coincides with the beginning of the rainy season and fruit ripening (Kao and Leu, 1980). In Vietnam, the disease has been most serious during periods of unusually cool, wet weather (Vien *et al.*, 2001).

Fig. 14.3. Lychee fruit affected by downy blight. Note the white masses of hyphae, sporangia and sporangiophores of the pathogen, *Peronophythora litchii*, covering portions of the fruit surface (photo: Pipob Lumyong).

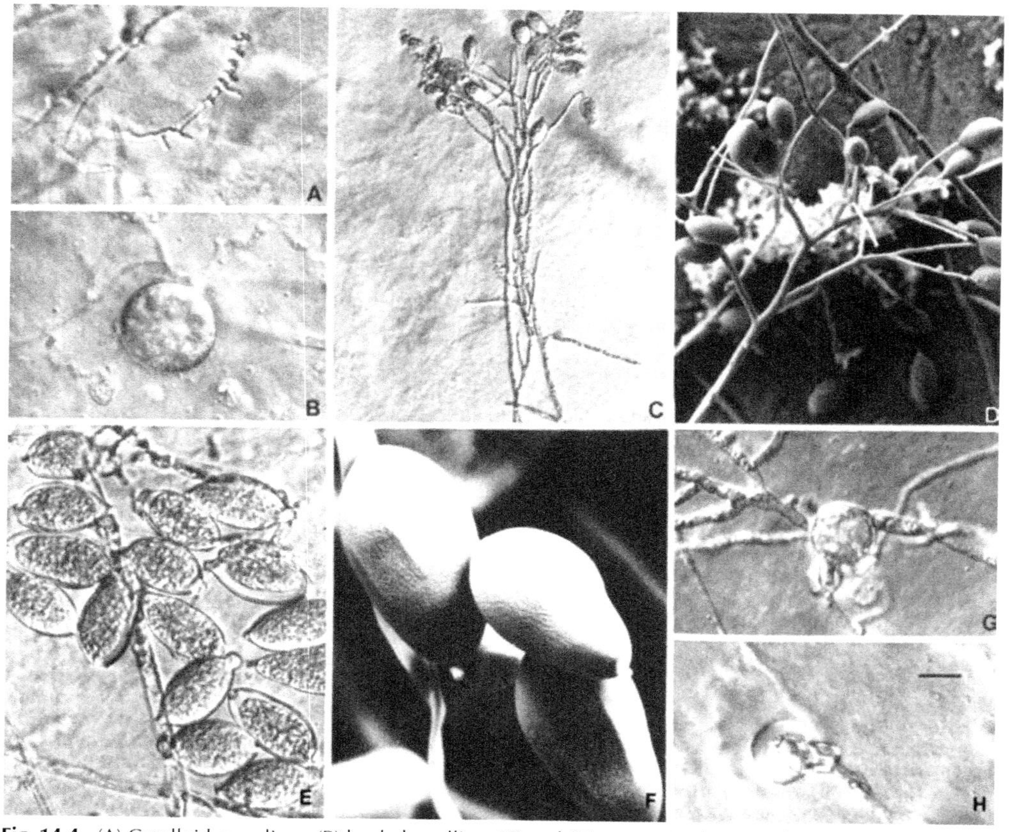

Fig. 14.4. (A) Coralloid mycelium, (B) hyphal swelling, (C) and (D) sporangiophores and sporangia, (E) and (F) sporangia, (G) oospore and (H) attached paragynous antheridium of *Peronophythora litchii*. Bar = 40 μm for (B), (E), (G) and (H); 24 μm for (D); 15 μm for (A) and (C); and 6 μm for (F) (from CMI description no. 974).

Management

To reduce inoculum, dead branches should be pruned after harvest, and canopies should be sprayed with copper oxychloride during winter (Li, 1997). Spraying the ground surface twice with a copper sulphate preparation followed by the application of lime has also been recommended. Metalaxyl, fosetyl-Al and mancozeb have been effective when applied at flower initiation, fruitlet and pre-ripe stages (Li, 1997; Ou *et al.*, 1999).

Pepper spot

In Australia, pepper spot has increased in prevalence in recent years (Cooke and Coates, 2002). In 1993, <5% of the lychee growers in Australia reported pepper spot in their orchards, but by 1999 this figure had increased to 43% (Drew and Drew, 1999). Pepper spot only causes superficial blemishes, but affected fruit can be downgraded and unmarketable. Symptoms are slightly raised, pinhead-sized dark spots that develop mostly on the stem end and shoulders of fruit (Fig. 14.5) (Bagshaw *et al.*, 1995). In severe cases, spots coalesce and cause the entire fruit to darken. Spots also appear on leaves and petioles.

Pepper spot is spread by waterborne conidia of the pathogen, *Colletotrichum gloeosporioides*. Inoculum for early infections of fruit is thought to come from infected leaves (Yip, 1997). Warm, wet conditions favour the development and spread of pepper spot, and the disease is usually most prevalent on the lower branches of trees (Yip, 1997). 'Kwai May Pink' is most susceptible (Drew, 1999).

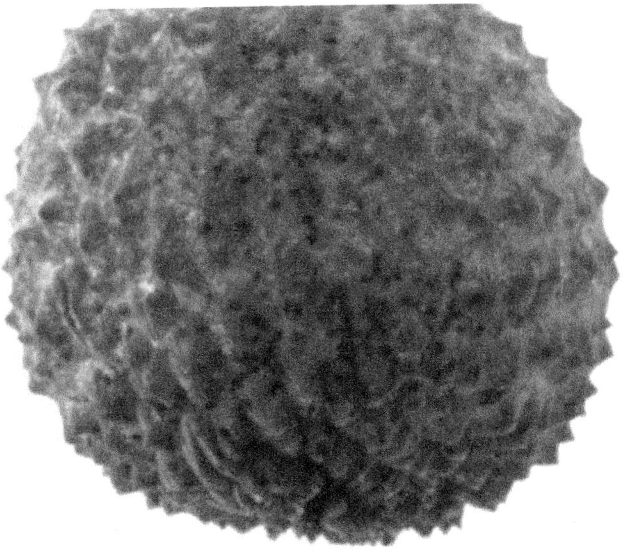

Fig. 14.5. Slightly raised, small dark lesions of pepper spot on a lychee fruit (photo: A.W. Cooke).

Several fungicides currently are being evaluated for control of pepper spot (Drew, 1999). In preliminary trials, a combination of mancozeb and prochloraz significantly reduced disease severity (L.M. Coates, Australia, 1997, unpublished results).

Phytophthora-incited diseases

Where fruit are induced to set during the off-season in Thailand, Phytophthora foliage blight and fruit rot are two serious diseases of longan (Visitpanich *et al.*, 2000). They are particularly important during cool weather after 2–3 days of rainfall (Bhavakul *et al.*, 1998). Phytophthora fruit rot of rambutan is widespread in most production areas (Tindall, 1994).

Symptoms

Young shoots, panicles and fruit of longan are affected most often. Symptoms include a dark necrosis of young shoots, a brown blight on panicles, flower drop, irregular brown lesions on fruit and premature fruit drop. During rainy weather, fruit are cracked and lesions are covered with white sporangia and sporangiophores of the pathogen.

On rambutan, black spots develop on the skin of fruit, often causing the fruit to wither and fall. Young fruit are most susceptible, particularly during the early part of the rainy season. The decay penetrates the aril, and affected fruit smell putrid (Tindall, 1994).

Causal agents and epidemiology

Phytophthora palmivora affects longan, whereas *P. nicotianae* and *P. botryosa* affect rambutan (Tindall, 1994). On both hosts, disease development is favoured by moist conditions, since water is needed for sporangium production as well as zoospore release and movement. *P. palmivora* and *P. nicotianae* are described in Chapter 1.

Management

Infected fruit and plant material should be removed from affected orchards. During disease-conducive periods, fungicide application may be necessary. Tindall (1994) reported that Phytophthora fruit rot on rambutan may be reduced with mancozeb and copper-based fungicides.

Pink disease

Pink disease affects a wide range of economically important tree crops, including rambutan, particularly in humid, tropical areas. The disease is most severe during wet weather in rambutan trees with a dense canopy. Symptoms begin as white patches of mycelial growth, mainly on twigs and young branches, that later become light pink. Affected bark may crack, and severely infected branches die. The causal agent, *Erythricium salmonicolor* (anamorph: *Necator decretus*), is described in Chapter 1.

Although there is no information available on the life cycle of this fungus on rambutan, on other crops both the teleomorph and anamorph are formed on host tissue during wet weather (Lim, 1994). Cultural practices, that reduce moisture and inoculum in the canopy, and fungicides are helpful (Bhavakul, 1977).

Powdery mildew

Powdery mildew is the most serious disease of rambutan in the field, and occurs in most production areas. The disease affects mainly young and actively growing leaves, inflorescences and fruit. On very young fruit, white powdery areas of mycelial growth appear on the surface, later becoming brown. Fruit more than 1 cm in diameter are usually stunted and covered by white mycelium on the spinterns and skin between them (Plate 80) (Tindall, 1994). The spinterns are stunted and later become brown.

Powdery mildew is caused by *Oidium nephelii*. Conidia of the fungus are spread by wind. All stages of growth can be infected, but young leaves, fruit and inflorescences are particularly susceptible. Cultivars with sweet fruit reportedly are more susceptible than those with acidic fruit (Tindall, 1994).

Good orchard hygiene is an important strategy in the control of powdery mildew (Tindall, 1994). In particular, canopy ventilation should be improved by pruning, and weeds should be controlled to remove possible alternative host plants. If applied in the early stages of disease development, certain fungicides are effective (Bhavakul, 1977; Tindall, 1994).

Rambutan seedling disease

This disease can cause complete defoliation and death of rambutan seedlings in Brunei and Malaysia (Peregrine *et al.*, 1990). Symptoms first appear as small light brown necrotic spots on the leaves of seedlings, that enlarge and become tan or light brown with age. Older lesions may be slightly zonate. When lesions coalesce, a 70–80% reduction in leaf area can occur (Peregrine *et al.*, 1990). There have been no reports of this disease on mature rambutan trees in the field.

This disease is caused by the fungus *Pseudocercospora nephelii*. Good control can be achieved with mancozeb and benomyl at recommended commercial rates. Also beneficial are cultural measures such as sowing seed thinly and at a uniform spacing, removing polythene covers immediately after seed germination, and avoiding over-watering (Peregrine *et al.*, 1990).

Sooty mould

Sooty mould can affect both fruit and leaves of longan and rambutan that are damaged by mealy bugs, scale insects and red mites (Tindall, 1994). Although damage is only superficial, affected fruit are downgraded due to their unsightly appearance. Dense, black mats of mycelium form on the surface of leaves and fruit (Fig. 14.6).

The fungus, *Meliola nephelii* var. *singalensis*, causes sooty mould on rambutan, whereas an unidentified species of *Meliola* is found on longan. Both fungi feed on honeydew that is excreted by sucking insects and mites. On longan, the longan wax scale, *Ceroplastes pseudoceriferus*, and soft scale, *Drepanococcus chiton*, are involved. The fungi colonize the surface of, but do not parasitize, host tissue.

Effective control of sooty mould usually can be achieved by reducing infestations of scale and other insects with insecticides (e.g. carbaryl and mineral oil). If necessary, fungicides such as copper oxychloride or

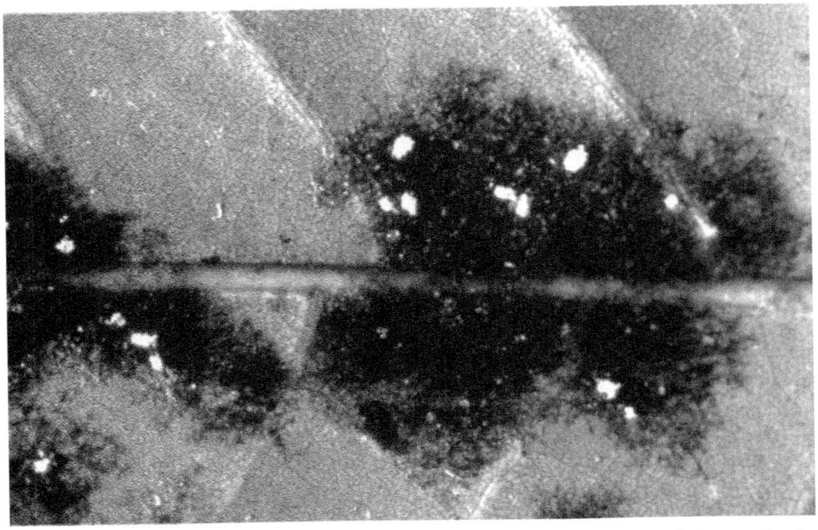

Fig. 14.6. Symptoms of sooty mould on the underside of a longan leaf. The white scale insect that is dispersed along the leaf is one of the sucking insects that produce honeydew on which the sooty mould fungus grows (photo: C. Sittigul).

benomyl can also be applied (Tindall, 1994).

Witches' broom

Witches' broom is the most serious preharvest disease of longan. It has been reported in China, Hong Kong and Thailand (So and Zee, 1972; Menzel *et al.*, 1990; Sarindu, 1993).

Symptoms

Infected trees have abnormally crowded panicles that lose their flowers prematurely, resulting in the characteristic 'broom-like' appearance of inflorescences (Plate 81) (So and Zee, 1972; Menzel *et al.*, 1990). The disease also affects leaves. Young leaves on infected shoots may appear stunted, discoloured and deformed (Visitpanich *et al.*, 1996; Zhang and Zhang, 1999). Older leaves may yellow and brown along veins and form blisters (Menzel *et al.*, 1990). Witches' broom can develop whenever the tree is producing new shoots.

In Thailand, it has been reported that an erinium consisting of fine light green hairs is apparent on both sides of affected leaves (Visitpanich *et al.*, 1996). Four-legged mites, *Aceria dimocarpi*, reside inside the erinium mass.

Causal agents

A causal agent for witches' broom has not been clearly established. In Thailand, Visitpanich *et al.* (1999) observed phytoplasmas in longan seedlings with symptoms of the disease. However, Sdoodee *et al.* (1999) were unable to confirm the presence of phytoplasmas in affected longan tissue using the polymerase chain reaction (PCR), even though the isolated DNA suggested the presence of a prokaryote.

In China, viral particles have been observed in affected, but not in healthy, trees (Chen *et al.*, 2000). Similar viral particles were observed in the salivary glands of insects that vectored the disease (Chen *et al.*, 1992).

Epidemiology

Different cultivars vary in their susceptibility to this disease (So and Zee, 1972). In Thailand, 'Beaw Kiew', 'Deang Klom' and 'Ma Teen Klong' develop severe symptoms, whereas others, such as 'Heaw' and 'Daw', are affected mildly (Visitpanich *et al.*, 1996).

Aceria dimocarpi transmits the phytoplasma that has been implicated in Thailand (Visitpanich *et al.*, 1996). In contrast, the virus in China can be transmitted by a variety of methods including the lychee stinkbug (*Tessaratoma papillosa*), the longan psylla (*Cornegenapsylla sinica*), dodder (*Cuscuta campestris*), seed and budwood (Zhang and Zhang, 1999; Chen *et al.*, 2000).

Management

Control of insect and mite vectors is an important strategy in the control of the disease (Visitpanich *et al.*, 1996; Zhang and Zhang, 1999; Chen *et al.*, 2000). Other strategies include the use of resistant cultivars, careful selection of propagating material, adoption of cultural practices that increase tree vigour and removal of infected branches, panicles and seedlings (Chen *et al.*, 2000). Strict quarantine inspection of all longan material that is received from the affected areas is essential.

Minor Foliar, Floral, Stem and Preharvest Fruit Diseases

Several foliar diseases of minor economic importance have been reported on lychee in Florida, including leaf necrosis (caused by *Colletotrichum gloeosporioides*), Gloeosporium leaf blight (*Gloeosporium* sp.), Phyllosticta leaf spot (*Phyllosticta* sp.) and a leaf spot caused by *Phomopsis* sp. (Alfieri *et al.*, 1994). A leaf blight and dieback caused by *Pestalotiopsis mangiferae* has been reported in India (Kang and Singh, 1991). Other dieback diseases due to *Diplodia* sp., *Phomopsis* sp. and *Spheropsis* sp. have been reported in Florida (Alfieri *et al.*, 1994). In Australia, a 'slow decline' and 'sudden death' syndrome have been recorded, but no causal agent has been identified (Menzel *et al.*, 1988). In China, witches' broom of lychee has been recorded (Chen *et al.*, 1996). Symptoms are similar to those that are observed for longan witches' broom.

On rambutan, vein necrosis is caused by the bacterium *Xanthomonas nepheliae*, and a leaf spot is caused by *Phomopsis* sp. (Sirithorn, 1985; Tindall, 1994). Velvet blight, caused by *Septobasidium bogoriense*, causes twig and stem damage (Peregrine and Ahmad, 1982). Brown panicle, caused by species of *Botrytis* and *Cladosporium*, is a minor disease of rambutan flowers. Symptoms appear as a grey mould covering florets that eventually turn brown (Tindall, 1994).

Postharvest Fruit Diseases

Anthracnose

Anthracnose is an important postharvest disease of lychee worldwide, particularly in high rainfall regions. In Florida, anthracnose is the major factor limiting lychee production (McMillan, 1994a), and in Australia the disease causes significant fruit losses (Johnson, 1989). It is also a serious pre- and postharvest disease of rambutan that affects leaves, flowers and fruit in high rainfall growing regions. Anthracnose is a minor disease on longan foliage (Fig. 14.7).

Symptoms

Anthracnose can develop on immature lychee fruit and cause premature abscission (Nakasone and Paull, 1998), but more commonly affects harvested fruit as well as ripening fruit in the field (McMillan, 1994b). Symptoms appear as circular, dark brown to black lesions on the fruit rind. If humid conditions occur during postharvest storage, white mycelial growth and salmon-coloured conidial masses of the causal fungus may cover the surface of lesions (Plate 82).

On rambutan, anthracnose affects both harvested fruit and ripening fruit in the field. Symptoms are similar to those described for lychee.

Circular, necrotic brown spots also develop on the foliage of these crops. Alahakoon and Brown (1994) reported that a marginal necrosis is observed commonly on young leaves of both seedlings and mature trees of rambutan in Sri Lanka.

Causal agents

Colletotrichum gloeosporioides causes anthracnose on lychee and rambutan, although on

Fig. 14.7. Anthracnose symptoms on longan leaves (photo: C. Sittigul).

lychee *C. acutatum* plays a minor role in Australia (G.I. Johnson, Australia, 1990, personal communication). The *Glomerella* teleomorphs of these fungi have not been reported on these hosts. *Colletotrichum* sp. causes the disease on longan. *C. gloeosporioides* and *C. acutatum* are described in Chapter 1.

Epidemiology

Infected leaves and twigs of lychee have been reported as year-round sources of inoculum for the disease in Florida (McMillan, 1994a). Conidia are spread by watersplash, and free moisture is required for infection to take place. Infections of lychee and rambutan fruit occur in the orchard, but they usually remain quiescent until fruit ripen or are harvested (Visarathanonth and Ilag, 1987). High storage temperatures favour disease development.

Management

Anthracnose is best controlled by a combination of orchard hygiene, fungicide application and postharvest temperature management. Increasing tree ventilation and reducing inoculum sources in the orchard by pruning out dead wood, twigs and leaves are beneficial. Although direct penetration of the pericarp is probably the major mode of infection, controlling insect pests to reduce alternative entry sites is also helpful.

In Florida, preharvest applications of benomyl plus mancozeb gave excellent control of anthracnose in lychee field trials (McMillan, 1994a). Good results were also obtained with tebuconazole and mancozeb (Crane *et al.*, 1997). On rambutan fruit, applications of mancozeb, zineb and benomyl from flowering to harvest provide good control (Visarathanonth and Ilag, 1987; Tindall, 1994).

Postharvest applications of fungicides such as benomyl (Scott *et al.*, 1982; Huang and Scott, 1985; Visarathanonth and Ilag, 1987) and prochloraz (Brown *et al.*, 1984) have been reported for the control of anthracnose and other fruit rots of lychee and rambutan, but increasing restrictions on the use of postharvest fungicides may limit this practice in the future. Sulphur dioxide fumigation or slow release pads have also proven effective (Tongdee, 1986; Johnson, 1990). When fumigation is followed by acid dipping to restore skin colour, blue mould, caused by *Penicillium* spp., can cause serious damage (Zauberman *et al.*, 1991; Coates *et al.*, 1993).

Cool storage of lychee fruit at 5°C is an important strategy for anthracnose suppression. Disease develops quickly, however, when cool-stored fruit are returned to warmer temperatures. Postharvest storage for rambutan fruit should be at 10°C (Visarathanonth and Ilag, 1987; Tindall, 1994).

Gliocephalotrichum fruit rot

Gliocephalotrichum fruit rot is an important postharvest disease of rambutan fruit. In a survey conducted in Bangkok in 1984, 10% of the fruit that were sampled in the marketplace were affected by this disease (Visarathanonth and Ilag, 1987). This disease also occurs in the Philippines (Pordesimo and Luna-Ilag, 1982) and Sri Lanka (Sivakumar *et al.*, 1997).

Symptoms, which develop during ripening and after harvest, arise from quiescent field infections (Visarathanonth and Ilag, 1987). They initially appear as light brown, water-soaked lesions on the rind and in the pulp of fruit (Visarathanonth and Ilag, 1987). These lesions become dark brown as they enlarge. Under humid conditions, severely infected fruit are covered with light yellow, cottony mycelium of the causal fungi (Plate 83) (Visarathanonth and Ilag, 1987; Farungsang *et al.*, 1994). In Thailand, the disease is caused by *Gliocephalotrichum bulbilium*, and in Sri Lanka by *G. microchlamydosporum* (Sivakumar *et al.*, 1997).

Fungicide applications, both before and after harvest, are useful (Saenyoung and Visarathanonth, 1985; Farungsang *et al.*, 1994). Carbendazim, in particular, has been shown to be an effective fungicide for general control of rambutan fruit rots. Low storage temperatures (10°C) suppress disease development (O'Hare *et al.*, 1994).

Greeneria fruit rot

Greeneria fruit rot, caused by *Greeneria* sp., is a major postharvest disease of rambutan fruit in Thailand. From 1990 to 1993, this disease accounted for ~50% of all postharvest diseases of rambutan in Thailand (Farungsang *et al.*, 1998).

Symptoms appear as dark brown lesions on the rind of fruit. The pathogen is readily isolated from surface-sterilized rind and spintern tissue of immature rambutan fruit (Sangchote *et al.*, 1998), suggesting that symptoms that develop after harvest probably arise from quiescent field infections.

Fruit should be stored at 10°C to reduce rot development. Other control measures commonly used on rambutan for fruit rots (e.g. hot water and fungicides) have proven to be ineffective against this disease (Farungsang *et al.*, 1998). Biological control is one strategy that is being investigated for the control of this disease.

Pestalotiopsis fruit rot

Pestalotiopsis fruit rot is one of the most serious postharvest diseases of rambutan fruit in Thailand (Sangchote *et al.*, 1998). Dark brown lesions develop on the rind. Under humid conditions, a whitish mycelial growth of the causal fungus, *Pestalotiopsis* sp., may develop on the surface of lesions. Harvested fruit should be stored at 10°C.

Stem-end rot

Depending on the season, geographic location and causal fungus, stem-end rot can cause significant losses of lychee, rambutan and longan fruit.

Symptoms

The disease first appears as a brown discoloration of the rind at the stem end of fruit (Plate 84). Lesions expand rapidly, particularly those that are caused by *Diplodia theobromae*. It is difficult to distinguish stem-end rot caused by *D. theobromae*, *Phomopsis* sp. and *Botryosphaeria* spp. on the basis of symptoms. Thus, isolation of the pathogens is usually required for correct diagnosis.

Causal agents

D. theobromae, *Phomopsis* sp. and anamorphs of *Botryosphaeria* spp. are the most common causes of stem-end rot. *Colletotrichum* sp. has also been reported as a causal agent of this disease in lychee (Johnson *et al.*, 1998).

Epidemiology

The process by which these fungi infect lychee, rambutan and longan fruit has not been established clearly. In rambutan, it has been suggested that *D. theobromae* is a weak pathogen requiring wounds for entry (Visarathanonth and Ilag, 1987). More recent research, however, supports the theory that symptoms arise primarily from quiescent infections in the skin and stem end of fruit (Johnson, 1995; Sangchote *et al.*, 1998). *Phomopsis* sp., *Botryosphaeria* spp. and *Colletotrichum* sp. have also been isolated as endophytes from lychee, rambutan and longan stem tissue (Johnson *et al.*, 1998), suggesting another possible mode of infection.

Management

Inoculum of these fungi can be reduced in the orchard by pruning out dead leaves and twigs in tree canopies. Pruning also increases ventilation, making conditions less favourable for the pathogens. After harvest, fruit should be cool-stored to suppress disease development, and fungicide application may give some control.

Other postharvest fruit diseases

A wide range of other fungi are reported to cause postharvest disease in lychee, longan and rambutan, but in most cases do not cause serious losses if fruit are handled correctly and marketed promptly.

Causal agents

In lychee, these include *Alternaria* sp., *Aspergillus* sp., *Cladosporium* sp., *Curvularia* sp., *Cylindrocarpon tonkinense*, *Fusarium* sp., *Geotrichum candidum*, *Geotrichum ludwigii*, *Mycosphaerella* sp., *Penicillium lilacinum*, *Pestalotiopsis* sp., *Phoma epicoccina*, *Stemphylium* sp. and *Trichothecium* sp. (Prasad and Bilgrami, 1973; Tandon and Tandon, 1975; Snowdon, 1990; Coates *et al.*, 1994; Kooariyakul and Sardsud, 1998; Tsai and Hsieh, 1998). Several other fungi, including many yeasts, have been isolated from diseased lychee fruit (Roth, 1963; Scott *et al.*, 1982), but causal relationships have not been determined. Yeasts rapidly invade the flesh of lychee fruit via insect wounds in the pericarp (Roth, 1963; Fitzell and Coates, 1995), resulting in fermentation of the aril.

Fungi commonly isolated from diseased rambutan fruit during postharvest storage include *Aspergillus* sp., *Fusarium* sp., *Penicillium* sp., *Phoma* sp. and *Rhizopus stolonifer* (Visarathanonth and Ilag, 1987; O'Hare *et al.*, 1994). Lam (1982) isolated the yeast *Candida* sp. from fermented rambutan fruit in Malaysia.

Fungi commonly associated with postharvest decay of longan include *D. theobromae*, *Pestalotiopsis* sp., *Cladosporium* sp. and *Fusarium* sp. *Aspergillus niger* is isolated occasionally (Sardsud *et al.*, 1994), and yeasts are also associated with decay (Tongdee, 1997).

Epidemiology

Infection of lychee, longan and rambutan fruit by most pathogens (excluding yeasts) is restricted largely to the fruit rind (pericarp), resulting in browning of the rind and saprophytic growth on the fruit surface. Under conditions of high moisture and humidity, fruiting bodies may form on the surface of lesions, aiding in pathogen identification.

Management

Postharvest decay of lychee, longan and rambutan fruit can be reduced significantly by cool storage (Visarathanonth and Ilag, 1987; Fitzell and Coates, 1995; Tongdee, 1997). Postharvest fungicide treatment, such as a hot benomyl dip, may give partial or good control of some pathogens but not others (Scott *et al.*, 1982; Huang and Scott, 1985). However, in many countries, the use of benomyl or certain other fungicides after harvest is not permitted. Fumigation with sulphur dioxide (SO_2) can give very effective control of lychee fruit rots (Lonsdale and Kremer-Köhne, 1991; Coates *et al.*, 1993; Duvenhage, 1993), although treated fruit can become more susceptible to surface colonization by *Penicillium* spp. (Zauberman *et al.*, 1991; Coates *et al.*, 1993). Fumigation with SO_2 in association with pre-cooling (usually hydro-cooling) and cool storage has been very effective for the control of decay and browning in longan fruit exported from Thailand (Tongdee, 1994, 1997). Several types of SO_2 application systems are available, but the method currently in use in Thailand is a high concentration, short duration system where SO_2 is generated by burning sulphur (Tongdee, 1997). Careful control of fumigation is required to provide a sufficient dose to control fungal growth, but not high enough to cause off-flavours in the aril and rind damage. Fumigated fruit can also become more susceptible to surface colonization by *Penicillium* sp. during subsequent storage and handling (Tongdee, 1994).

Controlling insect pests in the orchard can help reduce skin injuries and, therefore, potential infection sites for many pathogens including yeasts (Roth, 1963; Fitzell and Coates, 1995). Increasing tree ventilation through pruning and removing inoculum sources such as dead twigs and leaves will help reduce the incidence of disease initiated in the field.

Acknowledgements

The authors thank Ken Pegg for reviewing the chapter, and Tony Cooke for preparation of the colour plates.

References

Alahakoon, P.W. and Brown, A.E. (1994) Host range of *Colletotrichum gloeosporioides* on tropical fruit crops in Sri Lanka. *International Journal of Pest Management* 40(1), 23–26.

Alfieri, S.A. Jr, Langdon, K.R., Kimbrough, J.W., El-Gholl, N.E. and Wehlburg, C. (1994) *Diseases and Disorders of Plants in Florida*. Bulletin No. 14. Division of Plant Industry, Gainesville, Florida.

Ann, P.J. and Ko, W.H. (1984) Blossom blight of litchi in Taiwan caused by *Peronophythora litchii. Plant Disease* 68, 826.

Arora, R.K. (1998) Genetic resources of native tropical fruits in Asia. In: Arora, R.K. and Ramanatha Rao, V. (eds) *Tropical Fruits in Asia, Diversity, Maintenance, Conservation and Use. Proceedings of the IPGRI-ICAR-UTFANET Regional Training Course on the Conservation and Use of Germplasm of Tropical Fruits in Asia*, Bangalore, India, May 18–31, 1997, pp. 42–53.

Bagshaw, J.S., Underhill, S.J.R. and Fitzell, R.D. (1995) Lychees – disorders and injuries. In: Coates, L., Cooke, T., Persley, D., Beattie, B., Wade, N. and Ridgway, R. (eds) *Postharvest Diseases of Horticultural Crops*, Vol. 2, *Tropical Fruit*. Department of Primary Industries, Queensland, pp. 43–44.

Bhavakul, K. (1977) Diseases of rambutan. *Kasikorn* 50, 3–7.

Bhavakul, K., Tospol, M., Rakvitayasart, V. and Suwanketnikom, S. (1998) Studies on Phytophthora leaf blight of longan: symptoms, causal organism and chemical control. *Proceedings of a Seminar and Workshop on Longan Producing Technology*, Chiang Mai Phucome Hotel, Chiang Mai, Thailand, September 14–15, 1998, pp. 62–73.

Brown, B.I., Scott, K.J. and Mayer, D.G. (1984) Control of fruit rots of guava, lychee and custard apple by prochloraz dips. *Singapore Journal of Primary Industries*, 12, 40–49.

Chen, C.C. (1961) A species of *Peronophythora* gen. nov. parasitic on litchi fruit in Taiwan. *Special Publication College of Agriculture National Taiwan University* 10, 1–37.

Chen, J.Y., Xu, C.F., Li, K.B. and Xia, Y.H. (1992) On transmission of longan witches' broom disease by insect vectors. *Acta Phytopathologica Sinica* 22, 245–249.

Chen, J.Y., Li, K.B., Chen, J.Y. and Fan, G.C. (1996) A preliminary study on litchi witches' broom and its relation to longan witches' broom. *Acta Phytopathologica Sinica* 26(4), 331–335.

Chen, J.Y., Chen, J.Y. and Xu, X. (2000) Advances in the research of longan witches' broom disease. *First International Symposium on Litchi and Longan*, Program and Abstracts, Guangzhou, China, June 19–23, 2000, p. 74.

Chen, L.C., Lai, S.C., Lee, C.C., Chung, Y.W. and Ann, P.J. (1998) Effect of environmental factors on mycelial growth of *Peronophythora litchii. Plant Pathology Bulletin* 7(4), 189–200.

Chi, P.K., Pang, S.P. and Liu, R. (1984) On downy blight of *Litchi chinensis* Sonn. 1. The pathogen and its infection process. *Acta Phytopathologica Sinica* 14, 113–119.

Coates, L.M., Doogan, V.J. and Gardiner, S. (1993) Postharvest disease control in lychees using sulphur dioxide treatment. *Australasian Postharvest Conference Proceedings*, Gatton, Australia, September, 1993, pp. 73–76.

Coates, L.M., Johnson, G.I., Sardsud, U. and Cooke, A.W. (1994) Postharvest diseases of lychee in Australia, and their control. In: Johnson, G.I. and Highley, E. (eds) *Development of Postharvest Handling Technology for Tropical Tree Fruits*, Thailand, July 16–18, 1992. ACIAR Proceedings No. 58, pp. 68–69.

Cohen, M. (1955) Clitocybe rot of lychee trees. *Proceedings of the Florida State Horticultural Society* 68, 329–332.

Cohn, E. and Duncan, L.W. (1990) Nematode parasites of subtropical and tropical fruit trees. In: Luc, M., Sikora, R.A. and Bridge, J. (eds) *Plant Parasitic Nematodes in Subtropical and Tropical Agriculture*. CAB International, Wallingford, UK, pp. 347–362.

Cooke, A.W. and Coates, L.M. (2002) Pepper spot: a preharvest disease of lychee caused by *Colletotrichum gloeosporiodes*. *Australasian Plant Pathology* 31, 303–304.

Crane, J.H., Sanford, R.E. and McMillan, R.T., Jr (1997) Control of lychee anthracnose by foliar applications of tebuconazole, mancozeb, and copper hydroxide on 'Mauritius' lychee fruit under South Florida conditions. *Proceedings of the Florida State Horticultural Society* 110, 149–152.

Darvas, J.M. (1992) Armillaria rot of litchi trees (*Litchi chinensis*) in South Africa. *South African Litchi Growers' Association Yearbook* 4, 2–4.

Drew, H. (1999) Pepper spot – a new disease affecting lychee in Australia. *Proceedings of the Fifth National Lychee Conference*, Sunshine Coast, Queensland, September 13–15, 1999, pp. 21–23.

Drew, H. and Drew, J. (1999) Lychee pepper spot update – November 1999. *Proceedings of the Fifth National Lychee Conference*, Sunshine Coast, Queensland, September 13–15, 1999, p. 24.

Duke, S.O. (1990) Natural pesticides from plants. In: Janick, J. and Simon, J.E. (eds) *Advances in New Crops*. Timber Press, Portland, Oregon, pp. 511–517.

Duvenhage, J.A. (1993) Control of postharvest decay and browning of litchi fruit by sodium metabisulphite and low pH dips. *South African Litchi Growers' Association Yearbook* 5, 31–21.

Farungsang, U., Sangchote, S. and Farungsang, N. (1994) Rambutan postharvest diseases in Thailand. In: Johnson, G.I. and Highley, E. (eds) *Development of Postharvest Handling Technology for Tropical Tree Fruits*. Proceedings of a Workshop held in Bangkok, Thailand, July 16–18, 1992. ACIAR Proceedings No. 58, pp. 51–59.

Farungsang, N., Farungsang, U. and Sangchote, S. (1998) Biological control of postharvest fruit rot (*Greeneria* sp.) in rambutan with phylloplane yeasts. In: Coates, L.M., Hofman, P.J. and Johnson, G.I. (eds) *Disease Control and Storage Life Extension in Fruit*. Proceedings of an International Workshop held at Chiang Mai, Thailand, May 22–23, 1997. ACIAR Proceedings No. 81, pp. 113–119.

Fitzell, R.D. and Coates, L.M. (1995) Lychees – diseases. In: Coates, L., Cooke, T., Persley, D., Beattie, B., Wade, N. and Ridgway, R. (eds) *Postharvest Diseases of Horticultural Crops*, Vol. 2, *Tropical Fruit*. Department of Primary Industries, Queensland, pp. 41–42.

Gupta, J.H. (1992) Chemical control of algal rust of litchi caused by *Cephaleuros virescens* Kunze. *Progressive Horticulture* 24(1–2), 109–110.

Hall, G. (1989) Peronophythora litchi. *CMI Descriptions of Pathogenic Fungi and Bacteria No. 974*. Commonwealth Mycological Institute, Kew, UK.

Huang, P.Y. and Scott, K.J. (1985) Control of rotting and browning of litchi fruit after harvest at ambient temperatures in China. *Tropical Agriculture* 62, 2–4.

Johnson, G.I. (1989) Lychee disease control. *Proceedings of the Second National Lychee Seminar*, Cairns, Australia, September 21–23, 1989, pp. 90–93.

Johnson, G.I. (1990) Postharvest disease assessment. Rating systems. In: Beattie, B.B. (ed.) *Managing Postharvest Horticulture in Australasia, Proceedings of the Australasian Postharvest Conference 1989*. AIAS Occasional Publication No. 46, pp. 107–117.

Johnson, G.I. (1995) Rambutans – diseases. In: Coates, L., Cooke, T., Persley, D., Beattie, B., Wade, N. and Ridgway, R. (eds) *Postharvest Diseases of Horticultural Crops*, Vol. 2, *Tropical Fruit*. Department of Primary Industries, Queensland, p. 84.

Johnson, G.I., Joyce, D.C. and Gosbee, M.J. (1998) *Botryosphaeria* (anamorphs *Fusicoccum* and *Dothiorella*), *Diaporthe* (anamorphs *Phomopsis* spp.) and *Lasiodiplodia*: infection and defence. In: Johnson, G.I., Highley, E. and Joyce, D.C. (eds) *Disease Resistance in Fruit*. Proceedings of an International Workshop held at Chiang Mai, Thailand, May 18–21, 1997. ACIAR Proceedings No. 80, pp. 46–52.

Kao, C.W. and Leu, L.S. (1980) Sporangium germination of *Peronophythora litchii*, the causal organism of litchi downy blight. *Mycologia* 72, 737–748.
Kang, M.S. and Singh, I. (1991) Die-back and leaf blight of litchi incited by *Pestalotiopsis mangiferae* (Henn.) Steyaert. *Plant Disease Research* 6, 103–104.
Ko, W.H., Chang, H.S., Su, H.J., Chen, C.C. and Leu, L.S. (1978) Peronophythoraceae, a new family of Peronosporales. *Mycologia* 70, 380–384.
Kooariyakul, S. and Sardsud, V. (1998) Bagging of lychee fruit to reduce postharvest disease. In: Coates, L.M., Hofman, P.J. and Johnson, G.I. (eds) *Disease Control and Storage Life Extension in Fruit*. Proceedings of an International Workshop held at Chiang Mai, Thailand, May 22–23, 1997. ACIAR Proceedings No. 81, pp. 92–100.
Lam, P.F. (1982) Malaysian summary of research report on mango and rambutan project. *Proceedings of the Workshop on Mango and Rambutan*, University of the Philippines at Los Baños, College, Laguna, Philippines, April 18–25, 1982, pp. 21–25.
Li, J. (1997) Diseases and pests and their control. In: Zhang, Z. (ed.) *Litchi Pictorial Narration of Cultivation*. Pomology Research Institute, Guangdong Academy of Agricultural Sciences, Guangzhou, China.
Lim, T.K. (1994) Pink disease. In: Ploetz, R.C., Zentmyer, G.A., Nishijima, W.T., Rohrbach, K.G. and Ohr, H.D. (eds) *Compendium of Tropical Fruit Diseases*. APS Press, St Paul, Minnesota, pp. 37–38.
Lonsdale, J.H. and Kremer-Köhne, S. (1991) Maintaining market quality of fresh litchis during storage, Part 2: control of postharvest decay. *South African Litchi Growers' Association Yearbook* 3, 18–20.
Manicom, B.Q. (1995) Litchi dieback. *South African Litchi Growers' Association Yearbook* 7, 3–4.
Menzel, C.M., Watson, B.J. and Simpson, D.R. (1988) The lychee in Australia. *Queensland Agricultural Journal* January–February 1988, pp. 19–26.
Menzel, C.M., Watson, B.J. and Simpson, D.R. (1990) Longan. In: Bose, T.K. and Mitra, S.K. (eds) *Fruits: Tropical and Subtropical*. Naya Prokash, Calcutta, pp. 521–545.
Milne, D.L., de Villiers, E.A. and Holtzhausen, L.C. (1971) Litchi tree decline caused by nematodes. *Phytophylactica* 2, 37–44.
Mishra, B., Prakash, O. and Misra, A.P. (1974) A serious leaf spot disease of litchi caused by *Cephaleuros virescens*. *Indian Journal of Mycology and Plant Pathology* 3(2), 219–220.
McMillan, R.T. Jr (1994a) Epidemiology and control of anthracnose of lychee. *Proceedings of the Florida State Horticultural Society* 107, 345–346.
McMillan, R.T. Jr (1994b) Diseases of *Litchi chinensis* in South Florida. *Proceedings of the Florida State Horticultural Society* 107, 360–362.
Nakasone, H.Y. and Paull, R.E. (1998) *Tropical Fruits*. Crop Production Science in Horticulture Series No. 7. CAB International, Wallingford, UK.
O'Hare, T.J., Prasad, A. and Cooke, A.W. (1994) Low temperature and controlled atmosphere storage of rambutan. *Postharvest Biology and Technology* 4, 147–157.
Osbourne, A. (1996) Saponins and plant defense – a soap story. *Trends in Plant Science* 1, 4–9.
Ou, Z.J., Deng, W.S. and Wu, C.T. (1999) Experiments to control litchi downy mildew disease using 80% mancozeb wettable powder. *China Fruits* 3, 32.
Peregrine, W.T.H. and Kassim bin Ahmad (1982) Brunei – a first annotated list of plant diseases and associated organisms. *Phytopathological Papers* 27, 1–87.
Peregrine, W.T.H, Kassim bin Ahmad, Ahmad bin Abas and Sutton, B.C. (1990) A serious disease of seedling rambutan caused by *Pseudocercospora nephelii* sp. nov. *Plant Pathology* 39, 197–201.
Pordesimo, A.N. and Luna-Ilag, L. (1982) Postharvest diseases of mango and rambutan in the Philippines. *Proceedings of the Workshop on Mango and Rambutan*, University of the Philippines at Los Baños, College, Laguna, Philippines, April 18–25, 1982, pp. 211–232.
Prasad, S.S. and Bilgrami, R.S. (1973) Investigations on diseases of 'litchi' – III. Fruit rots and their control by postharvest treatments. *Indian Phytopathology* 26, 523–527.
Roth, G. (1963) Post harvest decay of litchi fruit (*Litchi chinensis* Sonn.). *Citrus and Subtropical Fruit Research Institute, Nelspruit Technical Communications* 11, 1–16.
Saenyoung, S. and Visarathanonth, N. (1985) Effects of four fungicides for control of fruit rots of rambutan. *Proceedings of the 23rd National Conference on Agriculture and Biological Sciences* (Poster session), Kasetsart University, Bangkok, pp. 371–379.
Sangchote, S., Farungsang, U. and Farungsang, N. (1998) Pre- and postharvest infection of rambutan by pathogens and effects of postharvest treatments. In: Coates, L.M., Hofman, P.J. and Johnson, G.I. (eds) *Disease Control and Storage Life Extension in Fruit*. Proceedings of an International Workshop held at Chiang Mai, Thailand May 22–23, 1997. ACIAR Proceedings No. 81, pp. 87–91.

Sardsud, V., Sardsud, U., Sittigul, C. and Chaiwangsri, T. (1994) Effects of postfumigation washing treatments and storage temperature on disease development in fresh longan. In: Johnson, G.I. and Highley, E. (eds) *Development of Postharvest Handling Technology for Tropical Tree Fruits*. Proceedings of a Workshop held in Bangkok, Thailand, July 16–18, 1992. ACIAR Proceedings No. 58, pp. 77–79.

Sarindu, N. (1993) Researches on plant mycoplasma diseases in Thailand. In: *Plant Pathology and Microbiology*. Department of Agriculture, Bangkok, Thailand, pp. 123–143.

Scott, K.J., Brown, B.I., Chaplin, G.R., Wilcox, M.E. and Bain, J.M. (1982) The control of rotting and browning of litchi fruit by hot benomyl and plastic film. *Scientia Horticulturae* 16, 253–262.

Sdoodee, R., Schneider, B., Padovan, A.C. and Gibb, K.S. (1999) Detection and genetic relatedness of phytoplasmas associated with plant diseases in Thailand. *Journal of Biochemistry, Molecular Biology and Biophysics* 3, 133–140.

Shaw, G.C. and Kile, G.A. (1991) *Armillaria Root Disease*. Agricultural Handbook No. 691, Forest Service, US Department of Agriculture, Washington, DC.

Singh H.P. (1998) Genetic diversity, breeding and utilisation of the genepool of litchi. In: Arora, R.K. and Ramanatha Rao, V. (eds) *Tropical Fruits in Asia, Diversity, Maintenance, Conservation and Use. Proceedings of the IPGRI-ICAR-UTFANET Regional Training Course on the Conservation and Use of Germplasm of Tropical Fruits in Asia*, Bangalore, India, May 18–31, 1997, pp. 185–195.

Sirithorn, P. (1985) Morphological, biochemical and pathologic properties of the bacterium causing vein necrosis of rambutan (*Nephelium lappaceum* L.). University of the Philippines at Los Baños, College, Laguna, Philippines.

Sivakumar, D., Wijeratnam, R.S.W., Wijesundera, R.L.C. and Abeysekera, M. (1997) Post-harvest diseases of rambutan (*Nephelium lappaceum*) in the Western Province. *Journal of the National Science Council of Sri Lanka* 25, 225–229.

Snowdon, A.L. (1990) *A Colour Atlas of Post-Harvest Diseases and Disorders of Fruits and Vegetables, Vol. 1: General Introduction and Fruits*. Wolfe Scientific, London.

So, V. and Zee, S.Y. (1972) A new virus of longan (*Euphoria longana* Lam.) in Hong Kong. *PANS* 18, 283–285.

Spjut, R.W. (1994) A systematic treatment of fruit types. *Memoirs of the New York Botanical Garden* 70, 112–115.

Tandon, J.M.P. and Tandon, R.N. (1975) Rot of fruits of litchi (*Litchi chinensis*) in marketing processes. *Indian Phytopathology* 28, 530–531.

Tindall, H.D. (1994) Rambutan cultivation. In: *FAO Plant Production and Protection Paper No. 121*, pp. 135–141.

Tongdee, S.C. (1986) Sulfur dioxide fumigation in the storage of fresh litchi. *TISTR Research News*, March 1986.

Tongdee, S.C. (1994) Sulphur dioxide fumigation in postharvest handling of fresh longan and lychee for export. In: Champ, B.R., Highley, E. and Johnson, G.I. (eds) *Postharvest Handling of Tropical Fruits*. Proceedings of an international conference held at Chiang Mai, Thailand, July 19–23, 1993. ACIAR Proceedings No. 50, 186–195.

Tongdee, S.C. (1997) Longan. In: Mitra, S.K. (ed.) *Postharvest Physiology and Storage of Tropical and Subtropical Fruits*. CAB International, Wallingford, UK, pp. 335–345.

Trung, H.M. (1999) Lychee production in Vietnam – prospects and problems. *Proceedings of the Fifth National Lychee Conference*, Sunshine Coast, Queensland, September 13–15, 1999, pp. 83–87.

Trung, H.M., Vien, N.V., Benyon, F.H.L., Nguyen, H.T., Summerell, B.A. and Burgess, L.W. (1999) Lychee decline in Northern Vietnam. *Conference Handbook, Australasian Plant Pathology Society, 12th Biennial Conference*, Canberra, Australia, September 27–30, 1999, p. 52.

Tsai, J.N. and Hsieh, W.H. (1998) Occurrence of litchi sour rot and characteristics of the pathogens *Geotrichum candidum* and *G. ludwigii*. *Plant Pathology Bulletin* 7, 10–18.

Vien, N.V., Benyon, F.H.L., Trung, H.M., Summerell, B.A., Van, N.K. and Burgess, L.W. (2001) First record of *Peronophythora litchii* on litchi fruit in Vietnam. *Australasian Plant Pathology* 30, 287–288.

Visarathanonth, N. (1999a) *Diseases of Fruits*. J. Film Process Co. Ltd, Bangkok, Thailand.

Visarathanonth, N. (1999b) *Diseases of Rambutan*. A. Plus Tree Media Co. Ltd, Bangkok, Thailand.

Visarathanonth, N. and Ilag, L.L. (1987) Postharvest diseases of rambutan. In: Lam, P.F. and Kosiyachinda, S. (eds) *Rambutan – Fruit Development, Postharvest Physiology and Marketing in ASEAN*. ASEAN Food Handling Bureau, Kuala Lumpur, pp. 51–57.

Visitpanich, J., Sittigul, C. and Sardsud, V. (1996) Longan leaf curl symptoms in Chiang Mai and Lam Phun. *Journal of Agriculture* 12(3), 203–218.

Visitpanich, J., Sittigul, C., Sardsud, V., Chanbang, Y., Chansri, P. and Aksorntong, P. (1999) Determination of the causal agents of decline, witches' broom and sudden death symptoms of longan and their control. *Final Report, Thailand Research Fund Project*, Department of Plant Pathology, Chiang Mai University, Chiang Mai, Thailand.

Visitpanich, J., Sittigul, C. and Chanbang, Y. (2000) Longan leaf blight and fruit drop. *House Agricultural Magazine* 24(1), 144–148.
Yip, H.Y. (1997) Lychee fruit pepper spot. *Living Lychee* 11, 6.
Zauberman, G., Ronen, R., Akerman, M., Weksler, A., Rot, I. and Fuchs, Y. (1991) Postharvest retention of the red colour of litchi fruit pericarp. *Scientia Horticulturae* 47, 89–97.
Zhang, Q. and Zhang, Q. (1999) Investigation of the occurrence of longan witches broom disease and its control. *South China Fruits* 28, 1, 24.

15 Diseases of Mango

R.C. Ploetz
University of Florida, Tropical Research and Education Center, Homestead, Florida, USA

Introduction

Mango, *Mangifera indica* (family: *Anacardiaceae*), is the world's fifth most important fruit crop (FAO, 2000). Global production in 2000 was 24.8 million tonnes (Mt), 12 Mt of which were harvested in India. Other major producers included, in descending order, China, Mexico, Thailand, Pakistan, Philippines, Indonesia, Nigeria and Brazil. The overwhelming majority of these fruit are consumed in the producing countries. In 1999, only 2% of the total harvest was exported as fresh fruit (Mexico was the leading exporter with 204,002 t), and <0.1% was exported as juice or pulp.

M. indica is one of 69 species in the genus, at least 27 of which produce edible fruit (Mukherjee, 1997). Although the species name suggests that it originated in India, it is now thought to have evolved in a large area that includes present day Bangladesh, northeastern India and northwestern Mynamar (Burma). Mango is highly adaptable, and is grown between ~35° N and S latitudes in a wide range of soils and environmental conditions.

Within the species, two types are recognized. Monoembryonic mangoes originated in subtropical India and produce a seed with a single zygotic embryo. In contrast, polyembryonic mangoes arose in Southeast Asia and produce a seed with several embryos, one of which is zygotic and the others which are usually nucellar. Nucellar embryos of the polyembryonic 'Turpentine' and '13–1' are used for rootstocks in Florida and Israel, and those of 'Kensington Pride' are used to propagate this cultivar in Australia (Crane *et al.*, 1997).

Hundreds of mango cultivars exist. Although superior types have been selected for several centuries in India, organized introductions of parents and the evaluation of their progeny has occurred only during the last century (Knight, 1997). In 1889, grafted cultivars were first introduced from India to Florida. During the ensuing decades, superior individuals from several generations of open-pollinated progeny were selected and brought into production. The Florida cultivars possess desirable traits such as uniform bearing, reddish skin coloration and tolerance to anthracnose. Many of these cultivars, most notably 'Tommy Atkins', are now used worldwide.

The introduction of superior parents and evaluation of their seedling progeny now occurs in several countries (Knight, 1997). In virtually all of these programmes, the male, pollen parent is not known. Mango has a long juvenile period, and fruit retention after pollination is very low. Most parents are heterogeneous and only one zygotic embryo is produced per fruit. Thus, evaluation of large seedling populations is a major undertaking. Classical breeding has produced some useful hybrids, but open pollinated germplasm is far more important (Iyer and Degani, 1997).

The Major Diseases

Diseases affect every organ of this important tree (Ploetz and Prakash, 1997). They damage seedlings and grafted plants in the nursery, decrease fruit set and retention, and are important pre- and postharvest problems on fruit. In many areas, they are the most important constraint to fruit production. Although most of these diseases are caused by fungi, an alga and bacterium are also important pathogens. No mango disease is known to be caused by phytoplasma, protozoa, viroid or virus agents.

Algal leaf spot (red rust)

Algal leaf spot, which is also known as red rust, is a common problem in the tropics and subtropics (Joubert and Rijkenberg, 1971). On mango, the disease is usually serious only in poorly managed orchards (Lim and Khoo, 1985). In these situations, mites, insects and other foliar diseases can increase the severity of the disease.

Algal leaf spot is caused by the green alga, *Cephaleuros virescens* (Lim and Khoo, 1985). It is described in Chapter 1, as are the symptoms, epidemiology and management of the disease.

Alternaria rot (black spot)

Alternaria rot, which is also known as black spot, causes postharvest fruit rot in Australia, Egypt, India, Israel and South Africa. Most commercial cultivars are susceptible. The pathogen also affects leaves and panicles; the latter problem is covered below under Blossom blight.

Symptoms

On leaves, symptoms are present throughout the year as round, black spots that are 1–3 mm in diameter and most noticeable on the underside of leaves (Prusky, 1994). Similar lesions develop around lenticels on fruit (Fig. 15.1). They usually start near the stem end, but can expand and merge to cover much of the fruit surface. Affected areas do not soften or penetrate more than 1–2 mm into the flesh until late in symptom development. Lesion centres are depressed and develop olive–brown conidia of the pathogen under moist conditions. Symptoms of Alternaria rot are more restricted, and affected tissue is darker and harder than that affected by anthracnose (Plate 85).

Causal agent

Alternaria alternata is described in Chapter 1 (Prusky *et al.*, 1983; Cronje *et al.*, 1990). The fungus is widespread and has a large host range (Neergaard, 1945; Domsch *et al.*, 1980). Diseases that it causes on mango are generally significant only in arid environments. For example, in Israel, it is the most important fruit disease (Prusky, 1994).

Epidemiology

Infected leaves, twigs and inflorescences are significant sources of inoculum for fruit infection, as are leaves on the orchard floor. Conidia of *A. alternata* are moved by air currents and in dew runoff (Prusky, 1994).

Fruit infection has been related to the length of time over 350 h that relative humidity is >80%. These infections develop after ripening begins (Prusky *et al.*, 1983). Fungitoxic levels of two resorcinol compounds have been associated with infection latency (Droby *et al.*, 1986, 1987). They are present in the peel prior to ripening, and as ripening begins their concentrations decline and postharvest decay begins.

Management

'Keitt' is more resistant than 'Haden' and 'Tommy Atkins' (Droby *et al.*, 1986). Losses can be minimized by regular fungicide applications in the field, postharvest fungicide treatment and enhanced ripening with ethylene (Prusky, 1994). Three applications of the fungicide maneb beginning 2 weeks after

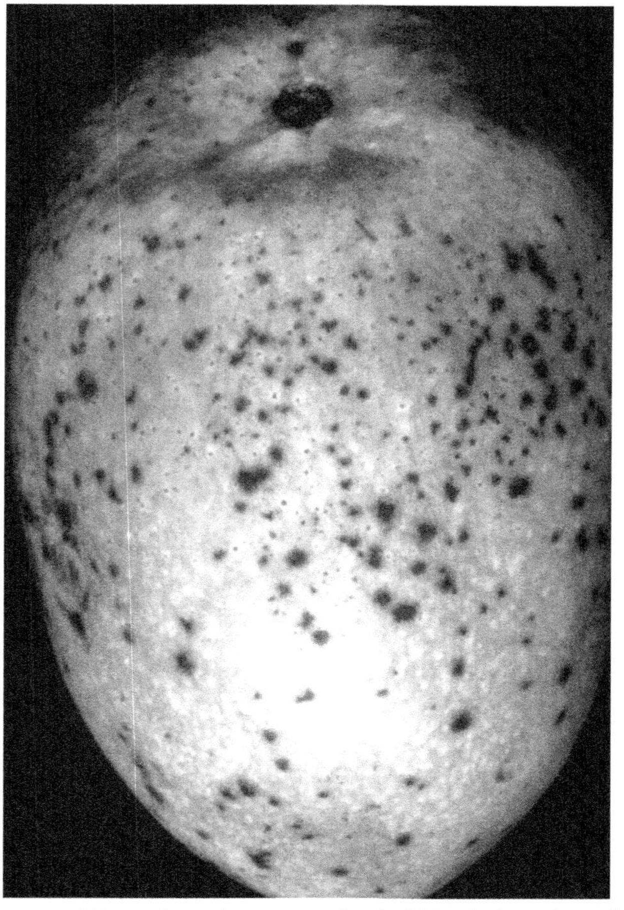

Fig. 15.1. Initial symptoms of Alternaria rot that are associated with lenticels on the surface of a mango fruit (photo: D. Prusky).

fruit set were most effective before harvest, and efficacy decreased if treatment was delayed. Postharvest treatment with prochloraz was as effective as three preharvest treatments. A brush and hot water (50–60°C) treatment was shown to be as effective as postharvest treatment with prochloraz 3 weeks after storage (Prusky *et al.*, 1997).

Anthracnose

Anthracnose is the most important disease of mango in all but arid production areas (Cook, 1975; Lim and Khoo, 1985; Dodd *et al.*, 1997; Ploetz and Prakash, 1997). Because it causes unsightly blemishes on fruit, it is a major pre- and postharvest problem. Anthracnose can also damage foliage, and under crowded and moist conditions causes serious problems in nurseries and young orchards (Bose *et al.*, 1973). The damage it causes on panicles is covered later in this chapter under Blossom blight.

Symptoms

New leaves that emerge during rainy periods are most susceptible (Fitzell and Peak, 1984; Jeffries *et al.*, 1990). Leaf lesions begin as small, dark brown spots that are surrounded by chlorotic haloes, have irregular margins and are not limited by leaf veins. Leaf lesions usually remain small, but under warm and humid conditions enlarge and coalesce to form irregular patches, ≥ 1 cm in diameter (Plate 86). Centres of old lesions deteriorate and fall from the leaf, resulting in a perforated, tattered appearance.

Anthracnose also causes twig dieback, although it is a less common cause of this symptom than other fungi (Lim and Khoo, 1985; Ploetz *et al.*, 1996a). In severe cases, elongated, blackened lesions form on twigs that dry from the tip.

Small fruit can develop minute brown spots and abort if infected early in their development. After fruit exceed 4–5 cm in diameter, abortion is less common and infections tend to stop development once an appressorium is formed. Further development occurs after fruit mature and begin to ripen. In these cases, irregular, dark brown to black lesions develop that are somewhat depressed and can crack the fruit surface (Fig. 15.2). Under humid conditions, large areas may be involved, and orange to pinkish masses of conidia are formed on the decaying surface. Lesions can form anywhere on fruit, but linear smears that radiate from the stem end to the apex are common (Plate 87). Lesions on fruit initially are superficial, and penetrate deeper than 5 mm into the mesocarp only during the final stages of development.

Causal agents

Anthracnose is caused by three closely related taxa of fungi. In most regions, *Colletotrichum gloeosporioides* (teleomorph: *Glomerella cingulata*) is the responsible agent (Dodd *et al.*, 1997). In Australia, a variant of this fungus that produces smaller conidia, *C. gloeosporioides* var. *minor*, is involved, and in Australia and India, *C. acutatum* (teleomorph: *Glomerella acutata*) plays a minor role (Fitzell, 1979; Prakash, 1990). *C. gloeosporioides* and *C. acutatum* are described in Chapter 1.

Gantotti and Davis (1993) reported that isolates of *C. gloeosporioides* from different mango organs produced different pectin-degrading enzymes. Ten different pectic zymogram groups (PZGs) were identified, based on the array of enzymes that the iso-

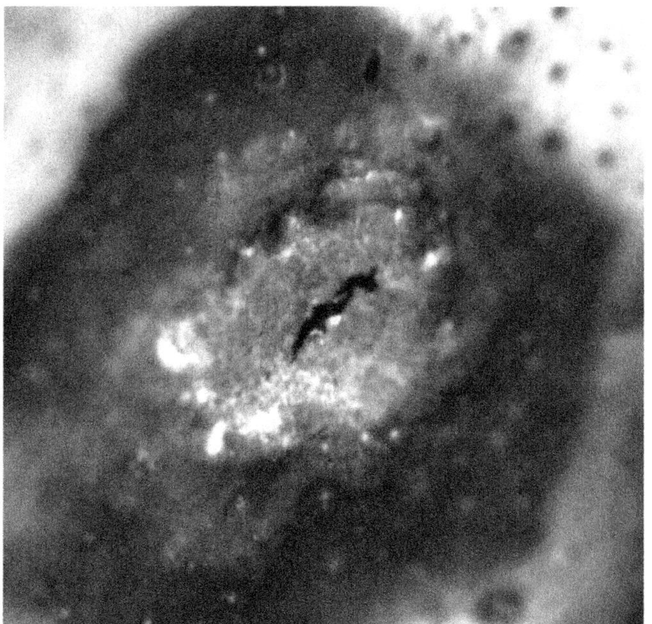

Fig. 15.2. An anthracnose lesion on the surface of mango fruit that has cracked and become covered with mycelia and conidia of the causal fungus, *Colletotrichum gloeosporioides* (photo: R.C. Ploetz).

lates produced. The authors indicated that the fungus on mango was genetically diverse, and suggested that isolates from some of the PZGs were adapted to different organs. In contrast, Alahakoon *et al.* (1994b) indicated that >80% of the isolates that were isolated from mango leaves, inflorescences and fruit in Sri Lanka were identical, based on restriction fragment length polymorphisms (RFLPs) of genomic DNA.

Related work indicated that isolates from mango may comprise a pathogenically and genetically distinct population of *C. gloeosporioides* (Hodson *et al.*, 1993; Alahakoon *et al.*, 1994a; Hayden *et al.*, 1994). Although most of the isolates that were tested were collected in Australia and Sri Lanka, isolates from other countries were also represented. Analyses of RFLPs and random amplified polymorphic DNAs (RAPDs) indicated that isolates from mango were genetically uniform and differed from those that were recovered from avocado, carambola and papaya. Isolates that were genetically similar to those from the latter fruits were found infrequently on mango. The mango isolates were not found on the other crops and usually were virulent only on mango. Alahakoon *et al.* (1994b) suggested that the mango isolates comprised a population of *C. gloeosporioides* that was disseminated throughout the world from a single source, perhaps as an endophyte.

Epidemiology

Fitzell and Peak (1984) determined that conidia were the most important type of inoculum in Australia. They were produced on branch terminals, mummified inflorescences, flower bracts and leaves. New leaf flushes were the most significant source of inoculum, a finding that was corroborated by Dodd *et al.* (1991) in the Philippines. Optimum production of conidia occurred between 25 and 30°C when free moisture was available, but also formed at relative humidities of 95–97%. The pathogen's teleomorph played no apparent role in the spread of the disease (Fitzell and Peak, 1984).

Conidia usually are dispersed by rainsplash, and infection requires free moisture (Jeffries *et al.*, 1990). In water, conidia form germtubes within 6–8 h, and appressoria within 10–12 h. Germ tubes are usually <20 μm long and terminate in appressoria. As they age, appressoria become melanized. This attribute strengthens the appressorium and facilitates penetration of the cuticle by infection pegs that it produces. Infection is direct.

Management

Although some mango cultivars are moderately tolerant, none are sufficiently resistant to be produced without fungicides in humid areas (Dodd *et al.*, 1997). Several systemic and non-systemic fungicides are effective, and local authorities should be consulted regarding their registration and use. Care should be exercized when one of these, chlorothalonil, is used since it is phytotoxic when applied to fruit that are larger than a golf ball (Fig. 15.3).

Fungicide application focuses on reducing damage to fruit and inflorescences. Since infected foliage and branches are important sources of inoculum, fruit set and anthracnose control on fruit are enhanced if disease control is exercized prior to flowering (Jeffries *et al.*, 1990). Trees may also require protection in crowded nurseries that receive overhead irrigation.

Dodd *et al.* (1991) developed a model that was based on key criteria for infection. It was used to reduce the number of applications of a systemic fungicide to control anthracnose on fruit.

Although fruit-to-fruit spread of anthracnose after harvest is unlikely, postharvest control of latent infections often is needed, particularly if fruit are stored or shipped (Dodd *et al.*, 1997). Hot water treatments and fungicide dips are most common. Cultivars vary in their tolerance to hot water, and treatment temperatures should never exceed 55°C for 5 min. When hot-water treatment is combined with fungicides (e.g. benomyl or imazalil), water temperatures should be reduced to 52 or 53°C. Cold fungicide dips are less effective. Gamma-ray irradiation has been tested, but is not very effective at the doses that fruit of most cultivars tolerate (Johnson *et al.*, 1997).

Fig. 15.3. Cracks and fissures on the surface of 'Keitt' mango fruit caused by applications of chlorothalonil. Although this fungicide is effective against blossom blight and anthracnose, it should not be applied when fruit are larger than a golf ball (photo: R.C. Ploetz).

Bacterial black spot (black canker)

Bacterial black spot is a major constraint in many production areas, and fruit losses of >50% can occur on the most susceptible cultivars. It can be the most important disease where fungal-induced diseases are well controlled (Gagnevin and Pruvost, 2001). In India, the disease is called bacterial canker due to the cankers it causes on stems of susceptible cultivars (Prakash *et al.*, 1994). Bacterial black spot has been identified in Australia, the Comoros, India, Japan, Kenya, Malaysia, Mauritius, New Caledonia, Pakistan, the Philippines, Réunion, Rodrigues, South Africa, Taiwan, Thailand and the United Arab Emirates (Fukuda *et al.*, 1990; Pruvost *et al.*, 1992; Prakash *et al.*, 1994; Gagnevin and Pruvost, 1995, 2001; Kishun, 1995). The legitimacy of the only report from the western hemisphere (Brazil) has been questioned (Gagnevin and Pruvost, 2001).

Symptoms

Mango leaves, stems and fruit are affected (Manicom and Pruvost, 1994; Gagnevin and Pruvost, 2001). On leaves, water-soaked spots are initially 1–3 mm in diameter. As they enlarge, they become raised, black and angular, are limited by veins and surrounded by chlorotic haloes (Fig. 15.4A). These lesions are larger and more conspicuously raised than those caused by bacterial spot (Fig. 15.4B). Lesions can merge to form large necrotic patches, and bacteria may ooze from lesions during wet conditions. Old lesions dry out, turn white or grey, and crack. Defoliation occurs in severe cases. Anthracnose lesions are not raised or as black and angular as those that are caused by bacterial black spot.

Fruit lesions start as water-soaked haloes around lenticels and wounds. They become raised, blacken, crack along the long axis, and range from 1 to 15 mm in diameter. They are irregular in shape, can extend 8–15 mm

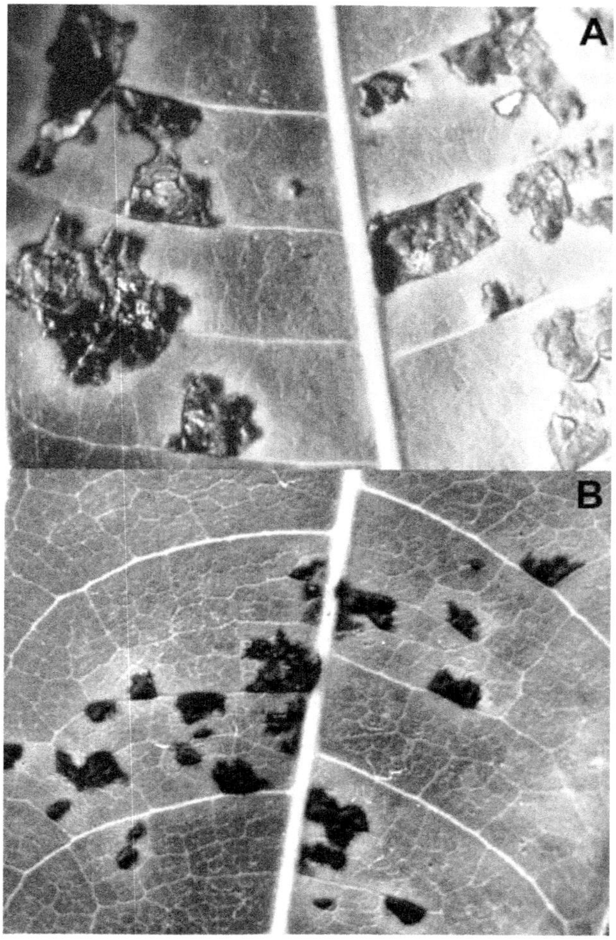

Fig. 15.4. (A) Symptoms of bacterial black spot on the undersurface of a mango leaf, caused by *Xanthomonas* sp. pv. *mangiferaeindicae* (photo: O. Pruvost) and (B) symptoms of bacterial spot on the undersurface of a mango leaf, caused by a yellow pigmented xanthomonad (photo: R.C. Ploetz). Bacterial black spot lesions are larger and more raised than those of bacterial spot.

into the mesocarp, and usually produce a gummy exudate (Plate 88). Young fruit abscise when they or panicles are affected.

On branches, bacterial black spot lesions are dark and cracked along the long axis (Plate 89). They develop only on highly susceptible cultivars and often are associated with wounds.

Causal agent

Xanthomonas campestris pv. *mangiferaeindicae* is the most commonly used epithet for this pathogen. However, Gagnevin and Pruvost (2001) noted that the pathogen had not been included in recent taxonomic studies. Until it was examined with current methods, they felt that the provisional name *X.* sp. pv. *mangiferaeindicae* was more appropriate.

Early reports of *Pseudomonas mangiferae-indicae* and *Erwinia mangiferae* as incitants were probably in error (Patel *et al.*, 1948; Steyn *et al.*, 1974). The bacterium's placement in *Pseudomonas* was probably due to it producing non-pigmented colonies in culture, and Cook (1975) indicated that *E. mangiferae* was a saprophyte that reached high populations in old lesions. *Pseudomonas syringae* pv. *syringae* causes a different disease of mango, apical necrosis (Cazorla *et al.*, 1998).

The pathogen has one flagellum, is Gram-negative, rod-shaped and 0.4–0.5 × 1.0–1.5 μm (Manicom and Wallis, 1984). It is oxidase negative and does not reduce nitrates to nitrites. It cannot use asparagine as a sole carbon and nitrogen source, but is able to hydrolyse starch, esculin, gelatin and casein. On artificial media, colonies are cream coloured. (The latter trait is atypical for xanthomonads, which usually are yellow in culture.) Yellow-pigmented xanthomonads have been recovered from mango leaves in Brazil, Réunion, South Africa and the USA (Florida). They are weak pathogens (see Fig. 15.4B) that do not affect fruit and differ in several criteria from *bona fide* strains of the pathogen (M.J. Davis, personal communication). Since these strains can confuse diagnosis, reliable detection of the disease relies on pathogenicity tests and identification of the pathogen. In the latter regard, a semi-selective medium, mono- and polyclonal antibodies, and RFLP analysis of the *hrp* gene cluster have been used (Gagnevin and Pruvost, 2001).

Epidemiology

The pathogen is disseminated by wind-driven rain (Gagnevin and Pruvost, 2001). Long-distance dissemination occurs via infected propagation material and, less frequently, in infected fruit pits. True seed are not infected, but surface contamination is known. Insects may play a role in dissemination, but these interactions are little studied.

The pathogen is an epiphytic colonist of leaves (Manicom, 1986; Pruvost *et al.*, 1990), buds (Pruvost *et al.*, 1993) and fruit (Pruvost and Luisetti, 1991). Infection occurs through wounds and, less often, stomata on old leaves (young leaves are thought to be resistant due to their non-functional stomata) (Gagnevin and Pruvost, 2001). High relative humidity (>90%) and moderate temperatures (25–30°C) favour disease development (Kishun and Sohi, 1983; Pruvost and Luisetti, 1991), and disease severity on leaves and fruit is correlated (Manicom, 1986; Pruvost *et al.*, 1990).

Other plants in the *Anacardiaceae* can be infected experimentally, including cashew, imra, mombin and pepper tree (Pruvost *et al.*, 1992; Kishun, 1995). However, strains from these hosts appear to pose little threat to mango (Gagnevin and Pruvost, 2001).

Management

Resistance to bacterial black spot varies greatly among mango cultivars, and resistant cultivars should be used where disease pressure is high (Manicom and Pruvost, 1994). During rainy weather, applications of copper-based bactericides are recommended on susceptible cultivars (Pruvost *et al.*, 1989). Schedules for their application focus on protecting fruit.

When new orchards are established, pathogen-free planting material should be utilized. Since the pathogen moves only short distances in wind-blown aerosols (usually within orchards), the long-distance spread of the pathogen depends almost entirely upon the movement of infected plants (Manicom and Pruvost, 1994). Windbreaks should be used to reduce wounding, and infected twigs should be pruned to reduce inoculum in the canopy.

Black mildew and sooty moulds

Although black mildew and sooty mould are minor problems in well-maintained orchards, the layers of hyphae that these fungi form may be thick enough to block sunlight and reduce photosynthesis. These blemishes also reduce the aesthetic quality and marketability of fruit.

Symptoms

The black mildew and sooty mould fungi produce dark coloured growths on the surfaces of stems, leaves and fruit of mango (Lim and Khoo, 1985). These range from diffuse webs to thick, felty layers of dark hyphae. The variation in symptoms is probably caused by the different species of fungi that are involved.

Causal agents

Black mildew and sooty mould are similar in appearance, but their respective causal agents are distinct (Lim and Khoo, 1985). The black mildew agent, *Meliola mangiferae*, is an obligate parasite. Members of the order *Melioliales* produce dark mycelium with two types of hyphopodia (Alexopoulos *et al.*, 1996). Capitate hyphopodia are lobed appressoria from which infection haustoria are formed, whereas mucronate hyphopodia function as conidiogenous cells. Recent molecular work places these fungi in the pyrenomycetes.

In contrast, the fungi that cause sooty moulds are diverse saprophytes that require honeydew (insect excreta) to colonize plant surfaces. In Malaysia, Lim and Khoo (1985) listed genera in the coelomycetes (*Polychaeton*), hyphomycetes (*Tripospermum*) and loculoascomycetes (*Antennulariella*, *Chaetothyrium*, *Limacinula* and *Scorias*), whereas the reported agents in India were hyphomycetes (*Leptoxyphium*, *Microxyphium* and *Tripospermum*) and loculoascomycetes (*Capnodium*) (Butler and Bisby, 1931; Prakash, 1988). Identifying the specific sooty mould agents on affected tissues can be difficult since these fungi may not sporulate and are often found together on a given leaf surface.

Epidemiology and management

Sooty moulds develop on honeydew that is produced on plant surfaces by aphids, mealybugs, scales and other sucking insects. These problems dissipate when the associated insects are controlled with oils or insecticides (Lim and Khoo, 1985).

Blossom blight

Blossom blight can reduce fruit set and production considerably since affected flowers die and large areas of the panicle may be affected. When this disease was controlled with fungicides in the Philippines, a 55–80% increase in fruit set occurred (Pordesimo, 1982).

Symptoms

Blossom blight starts as a wilt of the affected part of the inflorescence that is often curved (Fig. 15.5). The peduncle blackens and dies back from the tip. Internally, discoloration advances beyond the surface lesion. Large black lesions can develop lower on the peduncle, and once it is girdled will kill the apex. Symptoms that were caused by

Fig. 15.5. Symptoms of blossom blight on panicles of mango. Note the blighted, curved terminals and almost complete lack of fruit set (photo: D. Benscher).

Alternaria alternata and *Colletotrichum gloeosporioides* in South Africa were small, mainly superficial black spots, 1–2 × 2–5 mm, on the peduncle (Darvas, 1993; Lonsdale and Kotzé, 1993). Rather than blossom blight, Lonsdale and Kotzé (1993) indicated that these pathogens caused blossom spot.

Causal agents

The cause of blossom blight is somewhat confused and in need of additional research. *C. gloeosporioides* and *C. gloeosporioides* var. *minor* (in Australia) have been reported most frequently as the responsible fungi (Fitzell *et al.*, 1984; Jeffries *et al.*, 1990), although *A. alternata* has also been reported to attack panicles and reduce fruit set (Cronje *et al.*, 1990). Powdery mildew, caused by *Oidium mangiferae*, also damages panicles, but its symptoms are distinct from blossom blight; it is covered separately in this chapter, as are the descriptions of *A. alternata* (Alternaria rot) and *C. gloeosporioides* (Anthracnose).

Recent studies in South Africa indicate that two species of *Fusicoccum* may be important causes of blossom blight. Lonsdale and Kotzé (1993) reported that *F. parvum* (listed as *Dothiorella mangiferae*) caused extensive, systemic damage, and Darvas (1993) indicated that *F. aesculi* (reported as *D. dominicana*) was the only fungus that caused typical symptoms of blossom blight. Descriptions of these fungi appear in Chapters 3 and 1, respectively.

Although *F. aesculi* and *F. parvum* have not been reported as blossom blight agents in other countries, they are associated with mango decline and stem-end rot in several different areas (see below). Work is needed to determine the extent of *Fusicoccum*-incited panicle disease, and what agents are the most important blossom blight pathogens.

Epidemiology

Little is known about the epidemiology of *Fusicoccum*-incited panicle disease. Studies on the stem-end rot diseases have shown that the causal fungi are endophytes. The roles that internal and external sources of inoculum play in the development of panicle disease are not known.

Fitzell *et al.* (1984) investigated environmental conditions that were conducive to infection by *C. gloeosporioides* var. *minor*. They indicated that temperature and free moisture were important determinants of infection, and developed a model that was used to schedule fungicide application. The model reduced the number of applications that were needed to control blossom blight.

Management

Blossom blight must be controlled in order for optimum fruit set and development to occur. Once flowering begins, early and frequent fungicide applications are often needed, especially in areas with high rainfall. Presumably, systemic fungicides would be needed to control *Fusicoccum*-incited disease.

Galls and scaly bark

Gall and scaly bark disorders of mango are known in several producing regions. These diseases are usually minor problems, but can be serious. Although the symptoms vary, the disorders are discussed here under the same heading due to their gross similarities and their probable common cause.

Symptoms

Bark scaling develops as deep cracks along the entire rootstock portion of the plant (Fig. 15.6), and cracks may penetrate the phloem and become necrotic (Prakash and Srivastava, 1987). These symptoms resemble those of a scaly bark disorder, 'cuarteado', in Colombia (Cook, 1975). In Hawaii and Oahu, similar symptoms developed on mango seedlings (Cook *et al.*, 1971). The bark from the soil line to the first branches was rough and scaly, and xylem pegs, 5 mm long, were evident when the bark was removed around leaf scars and secondary branches.

In Mexico, a disorder known as 'nanahuate', 'bolas' or 'buba of mango', causes galls, 5–10 cm in diameter, that resemble cauliflower, are initially light green, but become dark brown as they die (Angulo and Villapudua, 1982). The galls remain attached

Fig. 15.6. Bark scaling on the rootstock portion of a mango tree in Israel (photo: R.C. Ploetz).

to trees for many years, and severely affected branches die. Similar symptoms have been noted in Florida, and are associated with pruning injuries (Fig. 15.7). Larger galls have also been noted in Puerto Rico and Florida in the USDA-ARS mango germplasm collection in Miami (Ploetz *et al.*, 1996b; R. Rodriguez, personal communication). The latter galls have been present for >20 years, and range up to 45 cm in diameter (Fig. 15.8). They have rough, scaly exteriors, and usually are found on the main trunk or scaffold limbs of affected trees. Cracks may penetrate the phloem and become necrotic, but the branch death that is associated with galls in Mexico and India has not been observed.

Causal agent

Fusarium decemcellulare causes the above diseases in Florida, Mexico and Venezuela (Malaguti and Reyes, 1964; Angulo and Villapudua, 1982; Ploetz *et al.*, 1996b). Colonies on potato dextrose agar (PDA) are dark carmine red on the underside. The fungus produces microconidia in false heads or chains on branched and non-branched monophialides (Fig. 15.9). Very large macroconidia (55–92 × 5.5–7 μm) are produced in slimy yellow sporodochia, ~1 mm in diameter. The fungus's teleomorph, *Albonectria rigidiuscula* (Rossman *et al.*, 1998), has not been observed on mango.

F. decemcellulare causes corky bark, gall, canker and dieback diseases on diverse woody hosts in the subtropics and tropics (Holliday, 1980; Farr *et al.*, 1989; Alfieri *et al.*, 1994). It causes an important disease of cacao, the cushion gall disease, as well as a stem gall on loquat, and scaly bark of pongam. It has been associated with the corky bark disease of lychee in Florida (see Chapter 14). Host specialization in the fungus has not been reported.

Although *F. decemcellulare* has not been reported to cause gall and scaly bark disorders of mango in other areas, neither have other agents. In the early 1990s, the possible role of *Agrobacterium tumifaciens* was examined in Miami. The bacterium could not be recovered from affected tissues (R. McGuire, personal communication). In Hawaii, microorganisms were not recovered from affected plants, and mechanical transmission tests to common virus indicator plants were negative. Cook (1975) compared the symptoms in Hawaii with those of citrus exocortis, a disease that was thought to be caused by a virus at the time, but is now known to be caused by a viroid. Since viroids were first recognized as plant pathogens at about the same time, Cook *et al.* (1971) did not examine tissue for viroids.

Epidemiology

Isolates of *F. decemcellulare* from mango are only mildly aggressive (Ploetz *et al.*, 1996b). Experimentally, they require wounding in order to infect mango. A recent outbreak of scaly bark in a commercial mango orchard in Florida was associated with pruning wounds (Fig. 15.7).

In the cushion gall disease on cacao, *F. decemcellulare* interacts with several different pest and pathogenic agents, most

Fig. 15.7. Gall, caused by *Fusarium decemcellulare*, that has formed at the site of a previous pruning wound on mango (photo: R.C. Ploetz).

notably capsid bugs (Holliday, 1980). In the corky bark disease of lychee, an armoured scale is associated with the disease (J.E. Peña, personal communication). These insects apparently facilitate infection and disseminate the pathogen. Insect–*F. decemcellulare* interactions have not been investigated on mango.

Management

No pesticides have been identified to control this problem. Measures that should be helpful include the removal and destruction of affected branches and trees in the orchard, disinfestation of pruning equipment to ensure that the pathogen is not spread during pruning operations, and the use of healthy planting material in new orchards.

Internal breakdown

The fruit of many mango cultivars are susceptible to physiological disorders that are referred to, generically, as internal breakdown (Schaffer, 1994). They result from the breakdown of the mesocarp in various locations of the fruit. Three general types of damage are recognized, 'stem-end cavity', 'jelly seed' and 'soft nose'.

These disorders can lead to severe postharvest losses. For example, 70% of the fruit of 'Tommy Atkins' was lost due to internal breakdown in some production areas in Guatemala in 1990 and Nicaragua in 2002 (Schaffer, 1994; R.C. Ploetz, personal observations). Losses in South Florida usually range from 5 to 20%, but can exceed 50% when fruit ripen on the tree.

Fig. 15.8. A large gall on mango cv. 'Langra' in the USDA mango germplasm collection in Miami, USA, caused by *Fusarium decemcellulare*. The machete provides scale (photo: R.C. Ploetz).

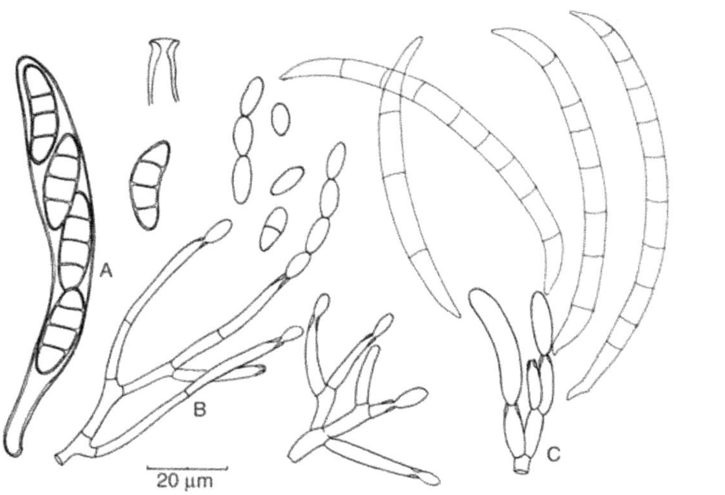

Fig. 15.9. (A) Ascus and ascospores of *Albonectria rigidiuscula*, and (B) microconidia and conidiophores, and (C) macroconidia and conidiophores of its anamorph, *Fusarium decemcellulare* (from CMI description no. 21).

Symptoms

The mesocarp loses physical integrity and dissolves or becomes softened and gelatinous. These problems are most pronounced in ripe fruit (Raymond *et al.*, 1998).

Stem-end cavity begins 8 weeks after fruit set, and eventually becomes an open cavity in which a physical connection between the pedicle and seed is lost (Plate 90) (Raymond *et al.*, 1998). The affected area may be darkened, due possibly to the accumulation of tannins or a phytotoxic buildup of resin (Raymond *et al.*, 1998).

Jelly seed is a general term for the watery breakdown of the mesocarp adjacent to the seed (Fig. 15.10) (Schaffer, 1994). Affected tissue appears to be over-ripe, and the symptoms also begin ~8 weeks after fruit set. Eventually, large areas around the seed can be affected (Raymond *et al.*, 1998). Although these areas usually are water soaked, they can also be spongy and greyish black, not unlike tissue in the stem-end cavity disorder.

Soft nose refers to water soaking of the mesocarp as occurs in the jelly seed disorder (Schaffer, 1994; Raymond *et al.*, 1998). It almost always develops at the distal end of fruit, but can eventually affect larger areas and become darkened (Fig. 15.11).

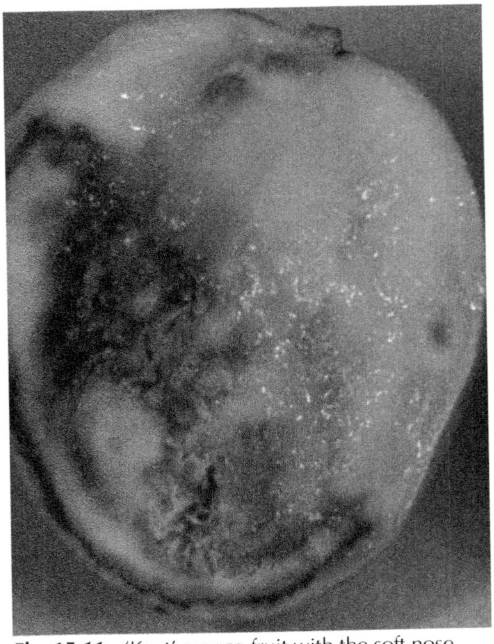

Fig. 15.11. 'Kent' mango fruit with the soft nose disorder (photo: C.W. Campbell).

During a recent episode of internal breakdown on 'Tommy Atkins' in Nicaragua, the overwhelming symptom was of 'pepita negra' (black seed). Externally, the first

Fig. 15.10. Mango fruit with the jelly seed disorder (photo: C.W. Campbell).

symptom of pepita negra is a small black dot at the sinus or beak that internally involves a blackened mesocarp and associated seed (Fig. 15.12). Eventually, classic soft nose symptoms develop on affected fruit. Why symptoms of pepita negra begin in the sinus region, and whether soft nose also begins development in this region are not known. Soft nose and pepita negra always develop on mature fruit.

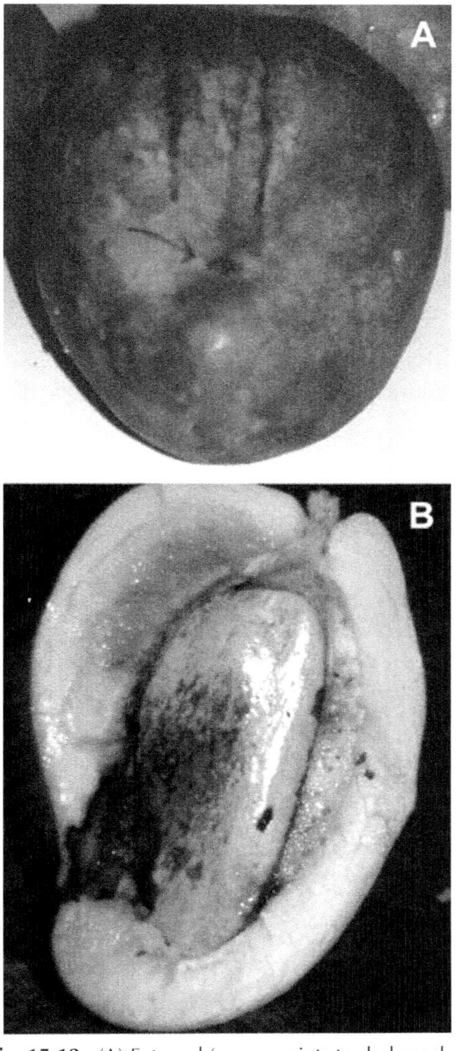

Fig. 15.12. (A) External (arrow points to darkened sinus) and (B) internal symptoms of pepita negra on a 'Tommy Atkins' mango fruit in Nicaragua (photo: R.C. Ploetz).

Cause

The cause of these disorders is somewhat controversial, but several lines of evidence suggest that insufficient levels of calcium in the mesocarp are responsible. That calcium is essential for the integrity and postharvest quality of several different fruits, including mango, is well established (Simmons *et al.*, 1998). Calcium plays a crucial role in preserving cell membrane integrity and strengthening cell walls, and calcium deficiencies in fruit are often associated with increased cell permeability and disintegration.

Low levels of calcium in either mango foliage or fruit have been correlated with internal breakdown (Young, 1957; Young and Miner, 1960; Burdon *et al.*, 1991; Tarmizi *et al.*, 1993). In addition, increased nitrogen fertilization, which would dilute calcium in tissues due to a stimulation of vegetative growth, has also been correlated with an increased incidence of this problem.

Recently, Cracknell (personal communication) completed compelling studies on Tenerife. He applied different levels of calcium and nitrogen to 'Tommy Atkins' to induce different concentrations of these elements in fruit. Breakdown severity was correlated with mesocarp concentrations of nitrogen (positive, $R^2 = 0.95$), calcium (negative, $r^2 = 0.73$) and nitrogen:calcium ratios ($R^2 = 0.93$). When plants received 0.72 mM nitrogen and either 0.62 or 2.5 mM calcium, 75% or more of the fruit were edible (breakdown ratings of 2 or lower). However, when plants received 5.36 mM nitrogen and either 0.62 or 2.5 mM calcium, only 2 and 15% of the fruit, respectively, were edible.

Management

Cultivars with an Indian pedigree are most susceptible to internal breakdown, and those with a fibrous mesocarp are least susceptible (Malo and Campbell, 1978; Schaffer, 1994). The incidence of the disorders is influenced by season and geographic location, and a higher incidence has been reported in sandy versus calcareous soils. On acidic, sandy soils, the disorder has been reduced in fruit of 'Keitt' with soil applications of $CaCO_3$ or foliar sprays of $Ca(NO_3)_2$ (Young and Miner, 1960). Since high rates of nitrogen fertilization, particularly in the NH_4 form, have increased the prevalence of this problem, excessive levels of this nutrient should not be used.

Malformation

One of the most serious diseases of mango is malformation (Prakash and Srivastava, 1987; Kumar and Beniwal, 1992; Ploetz, 2001). The term 'malformation' refers to the abnormal growth of inflorescences that occurs on affected trees. Since affected inflorescences do not set fruit, the disease can greatly reduce fruit production. Estimated losses range as high as 80% (Ginai, 1965). Vegetative shoots are also affected, and the names 'bunchy top' and 'witches' broom' are used to describe the distortion of these organs in some areas.

Malformation was first reported in India in 1891 (Watt, 1891), and is now recognized in Brazil, Central America, Egypt, Florida (USA), Israel, Malaysia, Mexico, Pakistan, South Africa, Sudan, Swaziland and Uganda (Crookes and Rijkenberg, 1985; Lim and Khoo, 1985; Kumar and Beniwal, 1992; Ploetz and Prakash, 1997). The disease's greatest impact is in arid production areas.

Symptoms

Young trees in nurseries are most vulnerable to vegetative malformation (Kumar and Beniwal, 1992). Shoots from apical or axillary buds are misshapen and have dramatically shortened internodes (Plate 91). Leaves are dwarfed, and are narrow, brittle and bend back towards the supporting stem. Shoots do not expand fully, resulting in a tightly bunched appearance of these portions of the plant. If all buds on a plant are affected, it remains stunted.

Primary and secondary axes on inflorescences are shortened, thickened and highly branched (Fig. 15.13) (Kumar and Beniwal, 1992; Ploetz and Prakash, 1997). Malformed panicles produce up to three times the normal number of flowers, and these are usually

Fig. 15.13. Symptoms of floral malformation of mango. Note the shortened internodes, large number and size of flowers, and leaves within the inflorescence (phyllody) (photo: R.C. Ploetz).

larger than normal size (Fig. 15.14). Malformation increases the number of male flowers in an inflorescence, and hermaphroditic flowers that are produced are either sterile or, if fertilized, eventually abort. Affected blossoms often produce vegetative structures. Malformed panicles retain a robust, healthy appearance long after non-malformed panicles have set fruit and shed non-fertilized flowers.

Causal agents

The cause of malformation has been controversial for many years. Among the suggested causes have been mites (Narasimhan, 1954), nutritional problems (Prasad *et al.*, 1965), hormonal imbalances (Singh and Dhillon, 1989), viruses (Kauser, 1959), phytoplasmas (Kumar and Beniwal, 1992) and unknown factors (Kumar and Beniwal, 1992).

Convincing evidence that a fungus causes malformation has been in the literature for decades. In India, Summanwar *et al.* (1966) and Varma *et al.* (1974) were the first to report that *Fusarium moniliforme* (recognized later as *F. subglutinans*) caused, respectively, the floral and vegetative forms of the disease. Since the first reports from India, this fungus has been shown to cause malformation in Egypt, Florida, Israel and South Africa (Ibrahim *et al.*, 1975; Manicom, 1989; Ploetz and Gregory, 1993; Freeman *et al.*, 1999).

Recently, Steenkamp *et al.* (1999, 2000) examined phylogenetic relationships in this pathogen with β-tubulin and histone H3 gene sequences. They indicated that a group of isolates from Florida, India, Israel and South Africa were closely related. They were described later as members of a new species, *F. mangiferae* (Britz *et al.*, 2002). The authors indicated that *F. mangiferae* is conspecific with isolates of '*F. subglutinans*' that previously had been shown to cause mango malformation worldwide. In the same studies, another group of malformation isolates from South Africa was shown to be phylogenetically distinct from the above group. It was described as a second, new species, *F. sterilihyphosum* (Britz *et al.*, 2002). *F. sterilihyphosum* recently has been recovered from malformed mango panicles in Brazil (Zheng and Ploetz, 2002). Although pathogenicity tests have not been performed with isolates of *F. sterilihyphosum*, their recovery from only malformed mango trees suggests that this species also

Fig. 15.14. A comparison of healthy and malformed mango inflorescences (photo: R.C. Ploetz).

causes malformation. *F. mangiferae* and *F. sterilihyphosum* are members of the *Gibberella fujikuroi* species complex, but do not form a *G. fujikuroi* teleomorph (Leslie, 1995; Steenkamp *et al.*, 2000; Ploetz *et al.*, 2002).

Radial growth of *F. mangiferae* on PDA averages 3.4 mm day^{-1} at 25°C (Britz *et al.*, 2002). Cultures produce aerial, white mycelium, and may be rosy buff or dark purple in reverse. Discrete or confluent sporodochia are cream to orange. Conidiophores are sympodially branched and bear mono- and polyphialides. Microconidia are usually oval, single-celled and 4.3–14.4 (9.0) × 1.7–3.3 (2.4) μm. Macroconidia are long and slender, usually four- to six-celled and 43.1–61.4 (51.8) × 1.9–3.4 (2.3) μm. Chlamydospores are not produced.

On PDA, *F. sterilihyphosum* grows 4.8 mm day^{-1} at 25°C (Britz *et al.*, 2002). Aerial mycelium is nearly white and the undersides of colonies range from straw to greyish rose to light purple. Sporodochia are cream to orange, but rarely present. Conidiophores are sympodially branched and bear mono- and polyphialides. Microconidia are usually oval, single-celled and 4.5–14.2 (8.8) × 1.6–3.5 (2.6) μm. Macroconidia have a slightly beaked apical cell, a footlike basal cell, and are usually four- to six-celled and 28.4–47.1 (37.1) × 2.4–4.1 (3.2) μm. Sterile, coiled hyphae are formed (Fig. 15.15), but chlamydospores are not.

Two other species of *Fusarium* have been associated with malformation. *Fusarium oxysporum* was reported to cause malformation in Egypt, Mexico and India (Bhatnagar and Beniwal, 1977; Kumar and Beniwal, 1992; Ochoa *et al.*, 1994; El Khoreiby, 1997). Its causal role has not been clearly demonstrated (Ploetz and Prakash, 1997). The D mating population of *Gibberella fujikuroi* (anamorph: *F. proliferatum*) was recovered from malformed mango trees in Malaysia (Leslie, 1995), but pathogenicity tests with the latter isolates were not conducted.

Epidemiology

The epidemiology of malformation is poorly understood. Long-distance spread of the disease relies on the movement of infected nursery stock, but the means by which within-tree and tree-to-tree spread is accomplished is not known.

Malformation moves slowly in affected orchards (Kumar and Beniwal, 1992). Macro- and microconidia are probable infective propagules since they form on dead, malformed tissues. However, their sensitivity to UV irradiation (Pinkas, personal communication) and the prevalence of malformation in arid countries which have high levels of sunlight confuse this issue. In a study in India, conidia of *F. subglutinans* (probably *F. mangiferae*) were not detected in rotary traps that were placed in an affected mango orchard (Prakash and Srivastava, 1987). Thus, aerial dissemination of conidia of this pathogen may be uncommon.

Artificial inoculation experiments suggest that wounding is required for natural infection. Manicom (1989) was unable to reproduce symptoms unless tissue was wounded prior to inoculation, and artificial induction of the disease by others relied on wounding before inoculation (Summanwar *et al.*, 1966; Ploetz and Gregory, 1993).

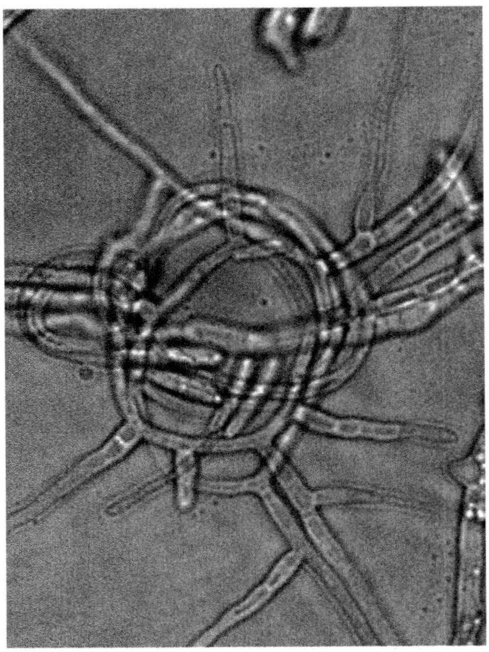

Fig. 15.15. Coiled, sterile hyphae of *Fusarium sterilihyphosum* (photo: Q. Zheng).

Circumstantial evidence suggests that the mango bud mite, *Aceria mangiferae*, is an important factor in the natural development of malformation. It feeds on epidermal cells of floral and vegetative buds of mango, and is often observed in high numbers on malformed trees. Although it has been shown to carry *F. subglutinans* (*F. mangiferae*?) on its body, the pathogenicity of these isolates was not determined (Abdel-Sattar, 1973; Manicom, 1989). Still, it is clear that contaminated mites could play a role in disseminating the fungus and enabling it to infect its mango host (Crookes and Rijkenberg, 1985). Once relationships among *E. mangiferae*, the pathogen and host are understood better, they should provide useful insights into the epidemiology and management of this disease.

F. mangiferae has a highly localized distribution within malformed mango trees (Ploetz, 1994). It predominates in symptomatic tissues, and is found far less frequently, or not all, in asymptomatic organs. This distribution suggests that vegetative and floral buds of mango are the primary sites of infection, and that systemic colonization of older, subtending tissues does not occur. The levels of colonization that occur in affected, but not healthy, tissues appear to indicate that infection thresholds must be reached before symptoms are produced.

Work in South Africa provided good evidence that hormonal imbalances in mango are responsible for the symptoms of malformation (Nicholson and van Staden, 1988; van Staden and Nicholson, 1989; van Staden *et al.*, 1989). Three cytokinins, *trans*-zeatin, dihydrozeatin and ribosyldihydrozeatin, were found in healthy, but not malformed, flowers. In contrast, *iso*-pentenyladenine was detected in malformed, but not healthy, flowers. In culture, the fungus produced *iso*-pentenyladenine derivatives, as well as glucosyl-*O*-zeatin and ribosylzeatin that occur at elevated levels in affected blossoms. The authors suggested that the pathogen produces *iso*-pentenyladenine derivatives in the host that are biologically active and are not found in healthy mango, and also produces glucosyl-*O*-zeatin and ribosylzeatin in the host, resulting in higher than normal levels of these hormones. Furthermore, the fungus may block production of dihydrozeatin-like compounds in affected tissues. In total, these events would alter the cytokinin:auxin ratio to the extent that the morphology of infected tissues would be altered.

Management

Few approaches to manage malformation are effective (Ploetz and Prakash, 1997). New plantings should be established with pathogen-free nursery stock. Scion material should never be taken from an affected orchard, and affected plants that are observed in the nursery should be removed and destroyed. Nurseries should also not be established in orchards, especially when they are affected by malformation (this practice is common in Egypt and India, two of the most severely affected countries). A new polymerase chain reaction (PCR)-based method for identifying *F. mangiferae* could be useful in preventing the introduction of this pathogen with new germplasm (Zheng and Ploetz, 2002).

Once the disease is found in an orchard, control is possible, but more difficult. Removing symptomatic tissues from trees is usually effective, presumably because it reduces the levels of inoculum in an orchard (Ploetz, 2001). Conventionally, affected terminals and the three subtending nodes are cut from trees, removed from the field and burned. If these measures are practised for 2 or 3 consecutive years, the disease can be reduced to insignificant levels. Thereafter, the disease can be kept in check by removing symptomatic tissues every other year.

Mango decline

Mango decline is a general term that describes several different diseases including blight, canker, gummosis, twig blight, tip dieback and stem bleeding. These diseases are widespread, and can be very destructive (for a thorough review of these diseases, refer to Ploetz and Prakash, 1997). Because these diseases have comparable symptoms and related causal agents, they are treated under the same heading in this chapter.

Symptoms

These problems are not well understood. Symptoms include: marginal necrosis of leaf lamina (Fig. 15.16); defoliation (Fig. 15.16); foliar symptoms of nutritional deficiencies, particularly of iron, manganese and zinc (Fig. 15.16); dieback of small branches from the apex (Fig. 15.16); production of clear or cloudy discharge from terminal buds or cracks in branches and trunks (Plate 92); vascular discoloration; and root degeneration (Ploetz and Prakash, 1997).

Causal agents

Ascomycetes and coelomycetes are associated most often with mango decline (Ploetz and Prakash, 1997), although other biotic or abiotic factors have also been suggested (McSorley *et al.*, 1980; Kadman and Gazit, 1984; Schaffer *et al.*, 1988). The fungal agents are often the same endophytes that cause stem-end rots of mango fruit.

Johnson (1994b) and Johnson *et al.* (1991b, 1992) have shown that several fungi are responsible for stem-end rot, and that host and environmental factors influence the impact of different species. Salient features of these fungi that are listed below are from Johnson (1994b).

Some of the agents have *Botryosphaeria* teleomorphs and *Diplodia* or *Fusicoccum* anamorphs. *D. theobromae* (teleomorph: *B. rhodina*) and *F. aesculi* (teleomorph: *B. dothidea*) are described in Chapter 1. *D. theobromae* causes wilting, defoliation, leaf scorch, gummosis, vascular discoloration and cankers in El Salvador, Egypt, Florida, India, Indonesia, Malaysia and Puerto Rico (Muller, 1940; Das Gupta and Zacchariah, 1945; Alvarez-García and López-García, 1971; Acuña and Waite, 1977; Lim and Khoo, 1985; Ploetz *et al.*, 1996a). Orchard trees and grafted seedlings in the nursery are affected. *F. aesculi* causes twig and branch dieback in Australia, Florida and South Africa (Johnson *et al.*, 1991a; Darvas, 1993; Ploetz *et al.*, 1996a). The same fungus is an important stem-end rot pathogen in Australia.

F. mangiferum causes sudden wilting of shoots and large branches, leaf scorching, trunk cankers and gummosis in Niger (Reckhaus and Adamou, 1987). On oatmeal agar (OA), it produces grey, felted mycelium, and on PDA, dark grey mycelium with a 'pepper-spot' pattern of pycnidial initials. Conidia are shorter and broader than those produced by *F. aesculi*, and usually are single-celled, hyaline, ellipsoid to ovoid and 9–18 × 4–6 μm (Fig. 15.17); microconidia, 2.5 × 4.5 μm, sometimes are observed. Mycelium is composed of branched, brown hyphae that eventually break up into one- or two-celled thallospores (Punithalingham and Waterson, 1970). A *Botryosphaeria* teleomorph develops occasionally on OA.

In Florida, *Alternaria alternata*, *Colletotrichum gloeosporioides*, *Diplodia theobromae*, *F. aesculi* and two *Phomopsis* spp. caused one or more of the above symptoms

Fig. 15.16. In the foreground, tip dieback of a mango shoot, and in the background zinc-deficient shoot tips (lower) and marginal leaf necrosis (upper) on a tree affected by mango decline. The tree in a desert production area in Israel was water stressed, one of the factors that predisposes mango to this syndrome (photo: R.C. Ploetz).

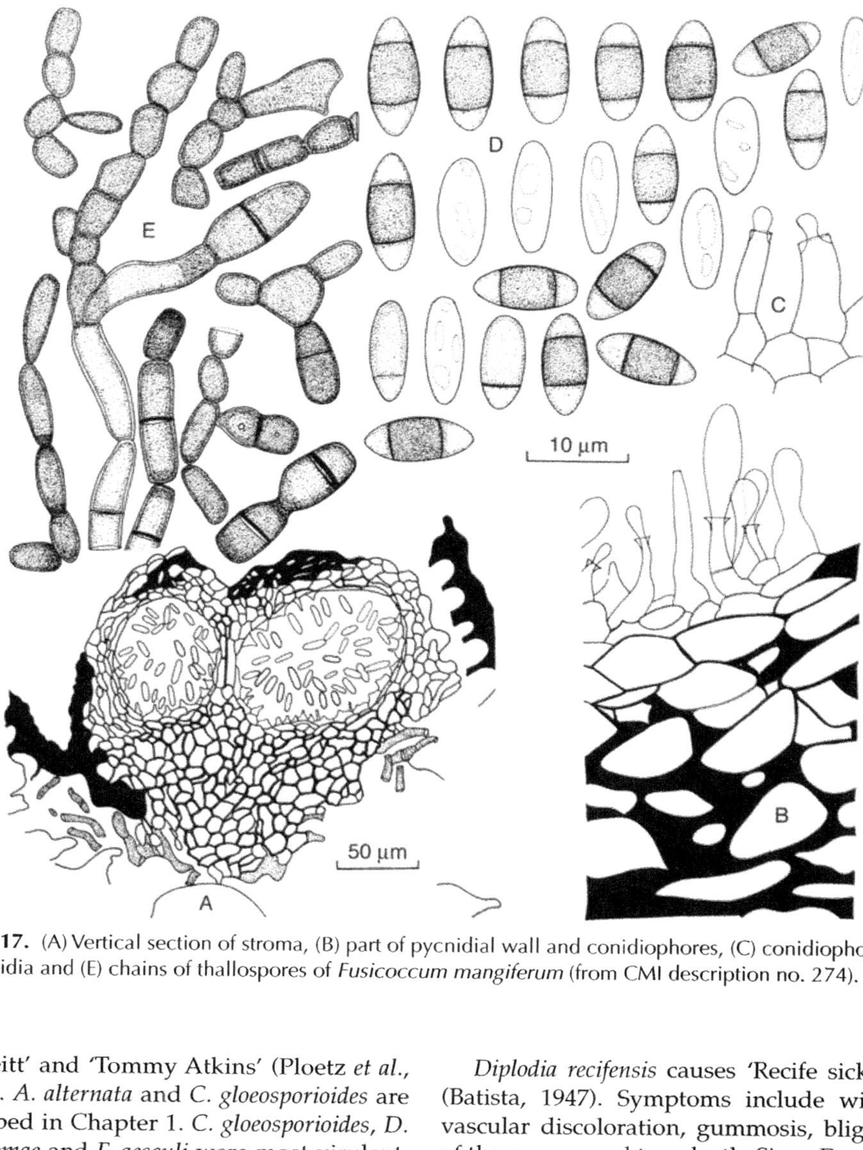

Fig. 15.17. (A) Vertical section of stroma, (B) part of pycnidial wall and conidiophores, (C) conidiophores, (D) conidia and (E) chains of thallospores of *Fusicoccum mangiferum* (from CMI description no. 274).

on 'Keitt' and 'Tommy Atkins' (Ploetz *et al.*, 1996a). *A. alternata* and *C. gloeosporioides* are described in Chapter 1. *C. gloeosporioides, D. theobromae* and *F. aesculi* were most virulent, and caused significant necrosis, gummosis and vascular discoloration. Bud necrosis, tip dieback and gummosis symptoms that they caused could not be distinguished unless *C. gloeosporioides* sporulated on inoculated branches.

In Brazil, two species of fungi cause decline disorders. Although the associated symptoms are similar to those that are described above, trees are killed more quickly and frequently than elsewhere.

Diplodia recifensis causes 'Recife sickness' (Batista, 1947). Symptoms include wilting, vascular discoloration, gummosis, blighting of the canopy and tree death. Since *D. recifensis* caused these symptoms only if trees were severely wounded, it was considered a weak pathogen.

An equally destructive disease, called 'seca', 'murcha' or mango blight, is caused by *Ceratocystis fimbriata* (anamorph: *Chalara* sp.) (Viegas, 1960; de Toledo Piza, 1966; Ribiero, 1980). Symptoms are identical to those of Recife sickness, and it is found in many of the same areas. Inoculations with *C. fimbriata* reproduced these symptoms and verified the tolerance of some cultivars to the disease (Viegas, 1960; Ribiero, 1980). *C. fimbriata* produces brown to black globose perithecia, 140–220 μm in diameter, with long (up to 900 μm), slightly tapered necks (Fig. 15.18) (Morgan-Jones, 1967). Ascospores are elliptical, hyaline, non-septate, smooth, 4.5–8 × 2.5–5.5 μm, and have a gelatinous sheath that forms a hat-shaped brim. Conidiophores are slender, septate, hyaline to pale brown, and up to 160 μm long. Conidia are cylindrical, hyaline, smooth and 11–25 (15) × 4–5.5 μm. Chlamydospores are terminal in chains, obovate to oval, thick walled, brown and 9–18 × 6–13 μm.

Epidemiology

The above fungi are usually weak pathogens. Some are endophytes, and infection by all is facilitated by wounding. Damage that the fungi cause is greatest in trees that have been weakened by different factors, including water stress; extreme summer and winter temperatures; sun scorch; tar or tanglefoot; high humidity in propagation nurseries; hardpan soils; and nutritional deficiencies, particularly of manganese, iron and zinc (Muller, 1940; Das Gupta and Zacchariah, 1945; Alvarez-García and López-García, 1971; Acuña and Waite, 1977; Schaffer *et al.*, 1988; Ploetz *et al.*, 1996a).

Scolytid beetles are thought to play key roles in the development of Recife sickness and seca in Brazil (Batista, 1947; Viegas, 1960; de Toledo Piza, 1966; Ribiero, 1980). The insects produce galleries in the cambium of affected trees. They feed primarily on fungi, but may act as wounding agents that facilitate infection by, and transmission of, these

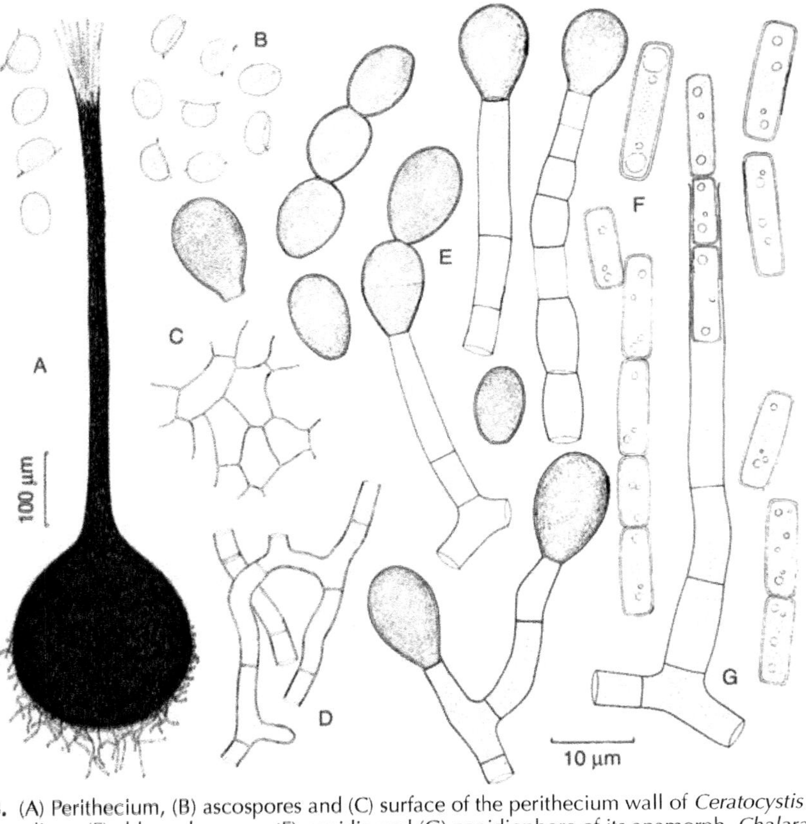

Fig. 15.18. (A) Perithecium, (B) ascospores and (C) surface of the perithecium wall of *Ceratocystis fimbriata*, and (D) mycelium, (E) chlamydospores, (F) conidia and (G) conidiophore of its anamorph, *Chalara* sp. (from CMI description no. 141).

pathogens. Ribiero (1980) reported that *Hypocryphalus mangiferae* was the primary species involved in disseminating *C. fimbriata* since it was the only scolytid found on healthy, as well as diseased, trees.

Management

Managing the decline diseases is difficult since many facets of their epidemiology are not well understood. Reducing the impact of predisposing factors that can be managed, such as drought stress, should be beneficial, but apparently has not been studied. Presently, no protocols exist for either identifying or producing pathogen-free germplasm that could be used to establish decline-free orchards. The internal location and diversity of causal agents also decrease the opportunity for controlling these diseases with fungicides (Peterson *et al.*, 1991). Broad-spectrum, systemic fungicides might be effective, but have not been widely tested. Injecting fungicides, as is done to control Dutch elm disease, might also be effective, but has not been tested. In India, dieback was controlled by removing affected portions of the canopy and treating the wounded areas with a 5:5:50 Bordeaux mix (Prakash and Raoof, 1989).

***Phytophthora*-incited diseases**

Phytophthora palmivora causes diseases of mango in several different areas. It caused wilt, crown rot, root rot and the death of nursery trees in Arizona, the Philippines and Thailand (Kueprakone *et al.*, 1986; Matheron and Matejka, 1988; Tsao *et al.*, 1994). Gumming and conspicuous bark lesions develop aboveground on these plants, whereas root and crown rots are evident at or below the ground level (Fig. 15.19). Crowded conditions, in addition to excessive irrigation and rainfall, exacerbate these diseases. Presumably, sanitation, the use of less crowded conditions and reduced frequency of irrigation would help to control these problems.

Damage has also been recorded on the trunks of field-grown, mature trees in Côte d'Ivoire (Lourd and Keuli, 1975), and on fruit in Malaysia and West Africa (Turner, 1960; Chee, 1969). Non-confirmed reports of damage on mature trees come from Colima and Yucatan, Mexico and Esquintla, Guatemala (C.W. Campbell and R. Campbell, personal communications). Mortality of trees is not observed, but substantial stem cracking and bleeding do occur. The specific symptoms which occur on fruit have not been recorded.

P. palmivora is described in Chapter 1. Considering its cosmopolitan distribution and wide host range, as well as the diverse symptoms it causes on mango, it is surprising that these diseases have not been reported more widely. Failures to use selective media or specialized techniques that are required to isolate members of this genus may be responsible (Tsao, 1990).

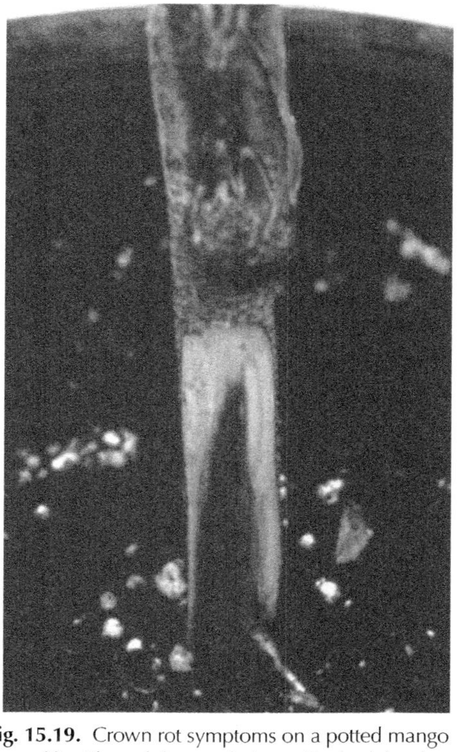

Fig. 15.19. Crown rot symptoms on a potted mango caused by *Phytophthora palmivora*. The bark has been removed from the lower stem to expose the interior of the damaged crown (photo: M.E. Matheron).

Pink disease

Pink disease affects many economically important woody plants in the humid tropics, and is one of the most destructive diseases of mango in these regions (Holliday, 1980). The disease is also known as cobweb, rubellosis and thread blight (Prakash and Srivistava, 1987). It has been studied most thoroughly on rubber, a very susceptible and important host in Southeast Asia (Rao, 1975).

On mango, pink disease can significantly reduce the vigour of trees and fruit yields (Lim and Khoo, 1985). Trees with dense canopies are most susceptible. In Malaysia, the disease is very common in inland areas with heavy rainfall.

Symptoms

Silky, white mycelial threads develop at the forks of branches or twigs (Lim and Khoo, 1985). Under moist conditions, threads coalesce to form a thin, rough pink incrustation on the bark. This stage takes up to several months to develop and coincides with the penetration of bark and internal colonization of the tree. Affected bark often cracks. As the fungus kills the vascular and cambial areas beneath the bark, branches above the colonized areas die, resulting in a sparse, patchy canopy. Basidiospores of the pathogen are formed in a pinkish white hymenial layer on the pink crust. Orange pustules, the pathogen's sporodochia, may also develop on infected bark.

Causal agent

The causal organism is a basidiomycete, *Erythricium salmonicolor* (anamorph: *Necator decretus*) (Holliday, 1980; Lim and Khoo, 1985). It has a wide host range and is described in Chapter 1.

Epidemiology and management

The anamorph and teleomorph of the pathogen are formed during wet conditions and are dispersed by rainsplash and wind. Cultural practices and fungicides are helpful. See Chapter 1 for details.

Powdery mildew

Powdery mildew is a widespread and important disease of leaves, panicles and fruit of mango. The disease can reduce yields by as much as 90%, due mainly to the effect the disease has on fruit set and development (Schoeman *et al.*, 1995).

Symptoms

Mango cultivars vary considerably in their resistance to powdery mildew (Palti *et al.*, 1974). On the most susceptible cultivars, all foliar, floral and fruit parts of the plant are affected (Fig. 15.20). Since affected panicles may set few or no fruit, and because infected small fruit may abort, it is on these portions of the plant that the disease is most important.

Leaves may also suffer considerable damage, and young leaves are most susceptible. A white, powdery coating, the conidia, forms on the underside, both sides or the top side of leaves, depending on the cultivar that is

Fig. 15.20. Powdery mildew on panicles of a resistant (left, 'Glenn') and susceptible (right, 'Nam Doc Mai') mango cultivar (photo: G.I. Johnson).

affected (Fig. 15.21). When damage occurs on the underside of leaves, often it is restricted to the midrib, and in all cases the leaf becomes distorted. These areas become purplish and eventually necrotic. Mature leaves are not susceptible.

On affected panicles, the powdery growth can cover all tissues, resulting eventually in a brown, shrivelled necrosis. Fruit which have set on such panicles may be shed. Skin cracking and corky tissue occurs on fruit that are infected when they are about the size of a pea, and young fruit abscise (Fig. 15.22). As they enlarge past the size of a marble, they are no longer susceptible.

Causal agent

Berthet (1914) first described *Oidium mangiferae* in Brazil, and it is now recognized in most of the important mango-producing regions in the world (Palti *et al.*, 1974). The fungus affects only mango (Prakash and Srivistava, 1987).

Fig. 15.21. Mango leaves that are affected by powdery mildew. Note the superficial coverage of mycelium of the causal fungus, *Oidium mangiferae*, on the leaves (photo: G.I. Johnson).

Conidium and haustorium traits indicate that *O. mangiferae* belongs to the *Erysiphe polygoni* group (Johnson, 1994a). Although the pathogen originally was classified as *E. cichoracearum* by Wagle (1928), Uppal *et al.* (1941) indicated that it produced saccate and lobed appressoria, which are not characteristic of *E. cichoracearum*.

The pathogen has no known teleomorph, a common trait for fungi that cause powdery mildews in the tropics (Holliday, 1980). Conidia of *O. mangiferae* are single-celled, hyaline, elliptical to barrel shaped, and 33–43 × 18–28 μm (Uppal *et al.*, 1941; Palti *et al.*, 1974). They are produced in very large numbers on host surfaces, and impart a powdery appearance to affected tissues that is responsible for the disease's common name. Conidiophores are of the pseudoidium type, with 2–4 septa and a straight basal cell (Boesewinkel, 1980). Germ tubes terminate in appressoria that in turn produce globular haustoria in the host's epidermal cells.

Epidemiology

Powdery mildew is most severe during cool, dry weather (Gupta, 1989). Conidia are wind disseminated and are released on a diurnal basis (Schoeman *et al.*, 1995). Peak spore release, between 1100 and 1600 h, was positively correlated with temperature and negatively correlated with humidity, vapour pressure deficit and leaf wetness (all significant at $P < 0.01$). The cardinal temperatures for conidium germination are 9, 32 and 23°C (Palti *et al.*, 1974). Conidia germinate at a very wide range of relative humidity (as low as 20%).

Infection can take place within 5–7 h, and conidia are produced within 5 days of infection. Disease development occurs between 10 and 31°C, and 60 and 90% relative humidity.

Management

Schoeman *et al.* (1995) noted that symptoms of powdery mildew first appeared 2–3 weeks after 20% of the inflorescences were at the red coloured to red open stage of panicle development. On this basis, they recom-

Fig. 15.22. Powdery mildew on small mango fruit. Note the cracking and conspicuous sporulation of *Oidium mangiferae* on the fruit surface. Once fruit become larger than a pea, they are no longer susceptible (photo: G.I. Johnson).

mended that the first fungicide application should occur when panicles begin to change colour. Assuming an effective period of 3 weeks for a given application, they concluded that applications should continue every 3 weeks after the initial application and until panicle susceptibility decreased at the end of fruit set.

Powdery mildew is easily controlled with sulphur, although its use may burn flowers and young fruit during warm, sunny conditions (Johnson, 1994a). Several newer fungicides are effective, but must be used at the onset of symptom development in order to be effective. Work in Israel demonstrated the efficacy of a sterol-inhibiting fungicide, diniconazole, a strobilurin fungicide, Kresoxym-methyl, and a phosphate fertilizer, KH_2PO_4, over several years (Reuveni *et al.*, 1998). When the fungicides were used either in alternation or in tank mixes with the fertilizer, enhanced disease control was realized.

Root rot and damping-off

Pythium vexans has been reported to cause root rot and wilt of seedlings (Lim and Khoo, 1985). In Malaysia, this pathogen caused seedling losses of up to 30% in nurseries. Symptoms included wilting of foliage which initially becomes pale green, but later develops necrotic patches. Roots develop a wet, blackened necrosis which begins in fine roots and progresses to larger roots and the root collar. Death of seedlings often occurs. In addition to overcrowding and excessive moisture, Lim and Khoo (1985) indicated that the use of polybags were also factors in the development of this problem.

Prakash and Singh (1980) reported that the basidiomycete *Rhizoctonia solani* (teleomorph: *Thantephorus cucumeris*) caused root and damping-off of seedlings in India. It is described in Chapter 12. Affected tissues become soft, dark brown or black, and seedlings ultimately may become completely girdled and collapse. Mycelia and sclerotia of the pathogen are formed conspicuously on affected tissues.

Scab

The disease was first recognized in Cuba and Florida in the 1940s. It is now widespread in the western hemisphere, and recently was reported in Australia (Cook, 1975; Alcorn *et al.*, 1999). It is caused by *Denticularia mangiferae*.

Scab is most important in nurseries since young host tissues are most susceptible, and because the moist environments which are often found there aid infection (Ruehle and Ledin, 1955). Lesions, usually first observed on the underside of leaves, are dark brown to black, and 1–2 mm in diameter. They may enlarge to 5 mm in diameter and become light grey with narrow, dark borders. Infected foliage develops a distorted appearance. Greyish blotches are produced on stems.

Although Bitancourt and Jenkins did not prove an associated teleomorph, *Elsinoë mangiferae*, was related to *D. mangiferae* (Bitancourt and Jenkins, 1943; Alcorn *et al.*, 1999), it is described on the assumption that they are related. *E. mangiferae* produces brownish ascocarps, 30–48 × 80–160 μm, in the host epidermis. Globular asci, 10–15 μm in width, are dispersed in ascocarps, and contain 1–8 hyaline ascospores. Ascospores are 4–6 × 10–13 μm, four-celled and constricted at the median septum; the subapical cell is longitudinally septate. Conidiophores of *F. mangiferae* are erect, sinuous, 2.5–3.5 × 12–35 μm, and wider at the base. Conidia are brown, one- or two-celled, and 2–4 × 6–29 μm.

Host tissues are most susceptible when young, and generally decrease in susceptibility as they mature. Although rainy weather promotes sporulation of the fungus, specific information is not available on the epidemiology of scab. Whether or not conidia and ascospores are infectious is not known.

Stem-end rot

Johnson (1994b) wrote a very thorough description of stem-end rot and its diverse causal agents. Interested readers are referred to this publication for more information than is provided in this chapter.

Stem-end rot is a postharvest disease that increases in importance as orchards become older and when preharvest fungicide programmes reduce the incidence and severity of anthracnose. Losses can increase during prolonged storage at low temperatures, or when fruit ripen at >28°C.

Symptoms

Variable symptoms are caused by different fungi at the stem end upon ripening. *D. theobromae, Fusicoccum aesculi* (Plate 93) and *F. mangiferum* (Plate 94) cause diffuse, water-soaked symptoms that spread from the stem end and darken. Necrosis is subcuticular and may affect the entire mesocarp within a week at 25°C. Subsequently, mycelium of the causal fungi may develop on, and tan liquid may drain from, the fruit exterior. *Phomopsis mangiferae* produces dark lesions at the stem end that penetrate the mesocarp, but do not spread as quickly as those caused by the above fungi (Plate 95). These lesions may resemble stem-end anthracnose caused by *Colletotrichum gloeosporioides*, but the latter lesions usually penetrate the mesocarp no deeper than 10–20 mm and are covered with salmon pink spore masses of the pathogen (Fig. 15.2). *Cytosphaera mangiferae* causes slow-spreading, tan lesions at the stem end similar to the symptoms produced by *Aspergillus niger*. *C. mangiferae* produces characteristic conidiomata around the stem end. *Pestalotiopsis mangiferae* causes tan lesions that develop slowly and eventually are covered with acervuli.

Causal agents

Several fungi cause stem-end rot (Johnson, 1994b). *D. theobromae* is most important in hot regions, whereas *F. aesculi* predominates in cooler areas. Less frequent causes of stem-end rot are *Colletotrichum gloeosporioides*, *F.*

mangiferum, *Phomopsis mangiferae*, *Pestalotiopsis mangiferae*, *Cytosphaera mangiferae* and two *Dothiorella* taxa (*Dothiorella* 'long' and *D. aromatica*) that were placed recently in *Fusicoccum* (Slippers *et al.*, 2001). Diagnostic features of *C. gloeosporioides*, *Diplodia theobromae* and *F. aesculi* appear in Chapter 1, and those of *F. mangiferum* are found in this chapter under 'Mango decline'.

Phomopsis mangiferae produces dark brown, subglobose, usually solitary conidiomata that are 1.2 mm wide and 0.75 mm high (Punithalingham, 1993) (Fig. 15.23). They are uni- or multilocular and have no ostiole. Conidiophores are hyaline filiform, simple or branched, septate and 15–75 μm long. Conidia exude white and occasionally yellow tendrils. α-Conidia are hyaline, single-celled, 6–9 × 2–2.5 μm, and fusoid with a large guttule at each end.

Pestalotiopsis mangiferae produces acervuli on fruit, twigs and leaves that are up to 300 μm in diameter, amphigenous, scattered, globose to lenticular, or ellipsoidal and subepidermal (Fig. 15.24) (Mordue, 1980). At maturity, acervuli rupture the epidermis with a pore through which conidia exude. Conidiophores are hyaline, cylindrical, septate, branched and 5–15 × 1.5–4 μm, and conidiogenous cells are holoblastic, indeterminate and smooth. Conidia are fusiform, straight or slightly curved, slightly constricted at the septa, 16–24 (19.8) μm long × 4–7.5 (5.6) μm wide. They have three olivaceous brown median cells and a hyaline apical cell that terminates in 1–4 (3) × 3–25 (9.8) μm appendages with obtuse ends.

Cytosphaera mangiferae produces colonies on PDA with light grey to buff mycelium, and dark, immersed crenate margins. Pycnidia are produced in concentric bands. Conidia are hyaline, oblong to ovate, aseptate, very thick walled and 16–26 × 10–13 μm (Fig. 15.25).

Epidemiology

The causal fungi are endophytes in mango stem tissue (Johnson *et al.*, 1992). They infect the inflorescence early in its development and reach the fruit pedicel several weeks after flowering begins. They then remain quiescent until fruit ripen.

When they occur simultaneously in fruit, *D. theobromae* predominates over *F. aesculi* at high temperatures, whereas *F. aesculi* predominates at ≤ 25°C. Although these fungi generally out-compete other stem-end

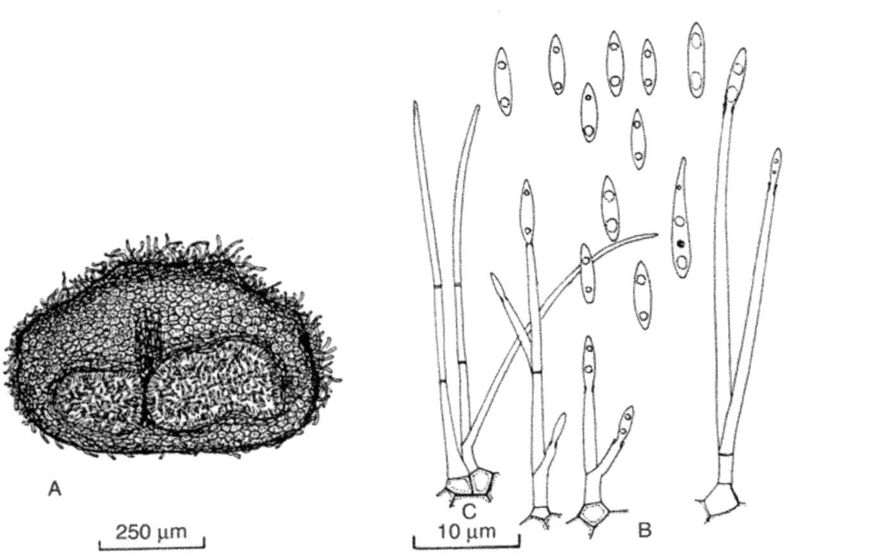

Fig. 15.23. (A) Vertical section of a locular conidiomata, (B) conidiophores, conidiogenous cells and α-conidia, and (C) paraphyses of *Phomopsis mangiferae* (from CMI description no. 1168).

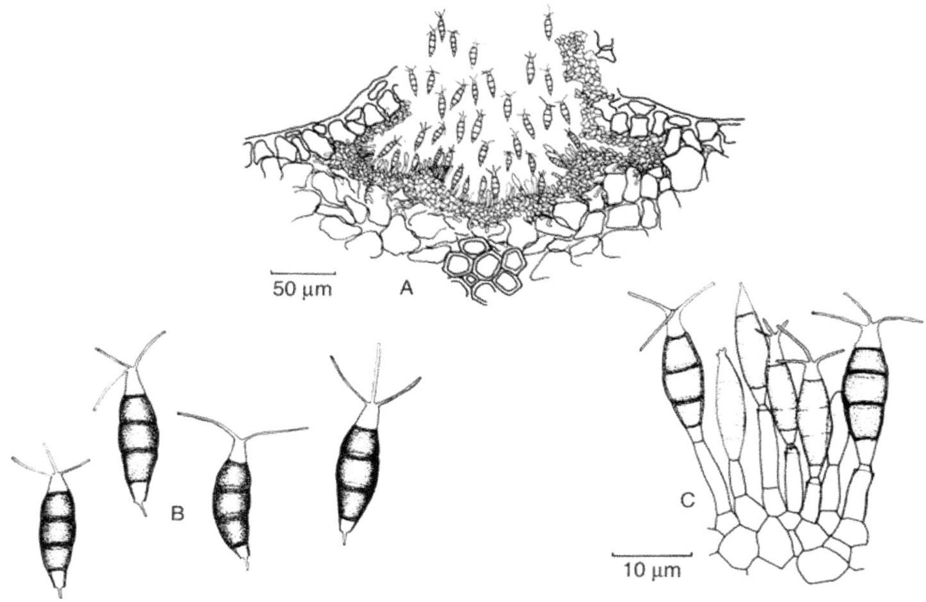

Fig. 15.24. (A) Vertical section of an acervulus, (B) mature conidia and (C) conidiogenous cells and developing conidia of *Pestalotiopsis mangiferae* (from CMI description no. 676).

colonists, *C. gloeosporioides* can repress up to 25% of the *F. aesculi* infections at 13–18°C. Symptoms appear 3–7 days after harvest at 25°C, but can be delayed for an additional 7–12 days at 13°C. Ethylene treatment before storage at 13°C can promote stem-end rot development.

Fruit can also be infected by *D. theobromae* at harvest, particularly when the stem end is inserted in soil to remove sap. These infections develop earlier than endophytic infections. Fruit-to-fruit spread of these pathogens can also occur after harvest.

Management

A dip in hot (52°C) benomyl for 5 min is effective, but not allowed in many areas (Johnson, 1994b). Hot water (55°C) alone or vapour heat treatment are less effective control measures. When fruit are placed in cool storage for long periods, they must be treated with a fungicide. Preharvest applications of copper oxychloride and postharvest application of prochloraz may also reduce stem-end rot, but the level of control will be lower than that obtained with hot benomyl. Fruit should not be placed in soil for sap removal.

Verticillium wilt

Verticillium wilt of mango was first reported in Florida (Marlatt *et al.*, 1970). Although the cause of the disease was reported originally to be *Verticillium albo-atrum*, this was before *V. dahliae* was widely recognized as a distinct species. Since the causal fungus described by Marlatt *et al.* (1970) formed microsclerotia, it is clear that *V. dahliae* was responsible. *V. dahliae* is described in Chapter 3.

Symptoms of the disease resemble those caused on avocado (see Chapter 3). They include a 'firing' and necrosis of leaves, usually in only a portion of the canopy. The unilateral development of the disease often does not progress to other portions of the tree and, in time, trees may recover. Killed leaves usually remain attached to the tree, and the xylem of affected branches is brown. Internally, the vascular system is brown (Fig. 15.26).

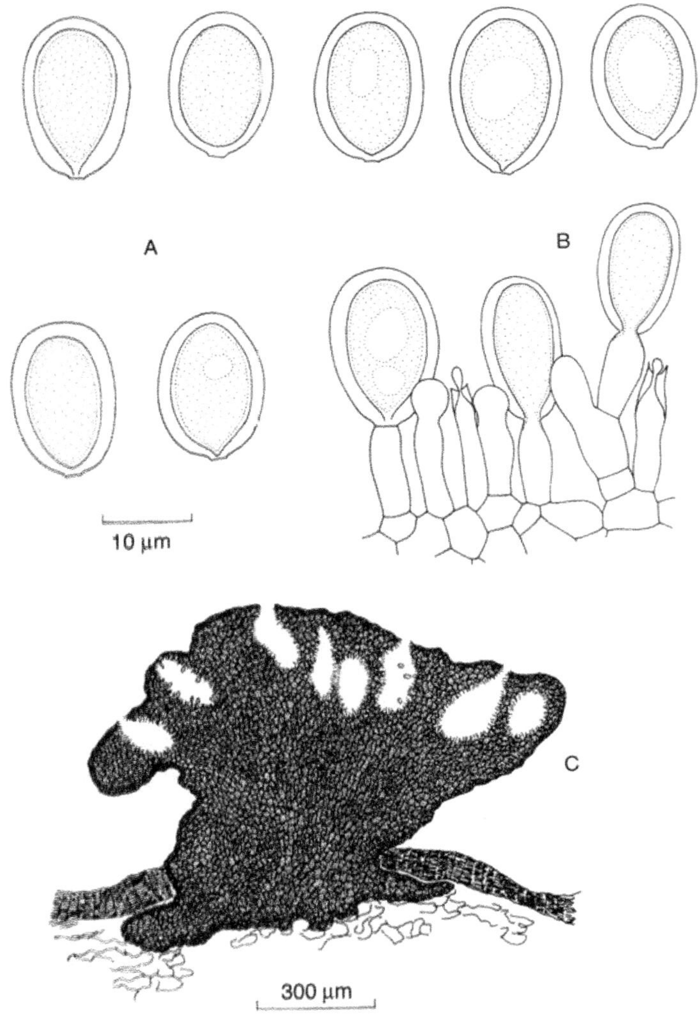

Fig. 15.25. (A) Thick-walled conidia, (B) conidia and conidiophores, and (C) pycnidium of *Cytosphaera mangiferae* (from Sutton, 1980).

Verticillium wilt is relatively uncommon and is usually found on land on which susceptible vegetable crops, such as aubergine and tomato, were grown recently (Pohronezny and Marlatt, 1982). In Egypt, almost all trees that were interplanted with another suscept, potato, were affected by the disease (R.C. Ploetz, unpublished observations). New mango orchards should not be planted on such sites. Alternative control measures for Verticillium wilt are not available. Fumigation with methyl bromide or other fumigants would not be practical for large sites, and fungicides which are effective against this disease do not exist.

White root disease

Rigidoporus lignosus, the basidiomycete that causes white root disease, is a common soil inhabitant in the humid tropics of Africa and Asia (Holliday, 1980). It has also been reported in the western hemisphere, but

Fig. 15.26. Discoloration of the vascular system in a mango branch caused by Verticillium wilt (photo: R.C. Ploetz).

the identity of the fungus in the latter regions is unclear. *R. lignosus* has a large host range on woody perennials that contains many important crops, including rubber, the host on which the pathogen was first reported. The importance of rubber and the very significant losses witnessed by rubber producers in the eastern hemisphere have resulted in a considerable body of research being generated on this pathogen (Nandris *et al.*, 1987). *R. lignosus* is described in Chapter 1.

The most effective means for controlling white root rot rely on eliminating or avoiding colonized woody debris when new orchards are established (Lim and Khoo, 1985). Unfortunately, this is impractical in many situations. Alternative measures include: (i) treating planting holes with elemental sulphur to promote the growth of antagonistic microorganisms; (ii) treating stumps left after clearing operations with chemicals which discourage their colonization by basidiospores; and (iii) establishment of leguminous cover crops which promote growth of the pathogen and the eventual exhaustion of its energy reserves.

Acknowledgements

I thank Carl Campbell, Dov Prusky and Greg Johnson for reviewing all or part of this chapter, Andy Cracknell for sharing unpublished results on internal breakdown, and Pedro Crous, Bernard Slippers and Glen Stanosz for reprints and information on the taxonomy of *Botryosphaeria*.

References

Abdel-Sattar, M.M.A. (1973) Histopathology of mango malformation. MSc thesis. Al-Azhar University, Cairo.

Acuña, H.E. and Waite, B.H. (1977) La muerte regresiva del mango (*Mangifera indica* L.) en El Salvador. *Proceedings of the American Society of Horticultural Sciences, Tropical Region* 21, 15–16.

Alahakoon, P.W., Brown, A.E. and Sreenivasaprasad, S. (1994a) Cross-infection potential of genetic groups of *Colletotrichum gloeosporioides* on tropical fruits. *Physiological and Molecular Plant Pathology* 44, 93–103.

Alahakoon, P.W., Brown, A.E. and Sreenivasaprasad, S. (1994b) Genetic characterization of *Colletotrichum gloeosporioides* isolates obtained from mango. *International Journal of Pest Management* 40, 225–229.

Alcorn, J.L., Grice, K.R.E. and Peterson, R.A. (1999) Mango scab in Australia caused by *Denticularia mangiferae* (Bitanc. & Jenkins) comb. nov. *Australasian Plant Pathology* 28, 115–119.

Alexopoulos, C.J., Mims, C.W. and Blackwell, M. (1996) *Introductory Mycology*, 4th edn. John Wiley & Sons, New York.

Alfieri, S.A. Jr, Langdon, K.R., Kimbrough, J.W., El-Gholl, N.E. and Wehlburg, C. (1994) *Diseases and Disorders of Plants in Florida.* Bulletin No. 14, Florida Department of Agricultural and Consumer Services.

Alvarez-García, L.A. and López-Gracía, J. (1971) Gummosis, dieback, and fruit rot disease of mango (*Mangifera indica* L.) caused by *Physalospora rhodina* (B. & C.) Cke. in Puerto Rico. *Journal of Agriculture University of Puerto Rico* 55, 435–450.
Angulo, S.M. and Villapudua, J.R. (1982) Buba of mango (*Mangifera indica* L.) in the state of Sinaloa, Mexico. *Phytopathology* 72, 171 (abstract).
Batista, A.C. (1947) Mal do Recife (grave doenca da Manggueira). Unpublished thesis, Pernambuco College of Agriculture (abstracted in *Review of Applied Mycology* 27, 77–78, 1948).
Berthet, J.A. (1914) Molestia da mangueira. *Bolm Agriculturae. Sao Paolo* 15, 818–819.
Bhatnagar, S.S. and Beniwal, S.P.S. (1977) Involvement of *Fusarium oxysporum* in causation of mango malformation. *Plant Disease Reporter* 61, 894–898.
Bitancourt, A.A. and Jenkins, A.E. (1943) Scab of mango caused by *Elsinoë*. *Phytopathology* 33, 1 (abstract).
Boesewinkel, H.J. (1980) The identity of mango mildew, *Oidium mangiferae*. *Phytopathologische Zeitschrift* 99, 126–130.
Bose, S.K., Sindhan, G.S. and Pandey, B.H. (1973) Studies on dieback disease of mango in the Tarai region of Kuaon. *Progressive Horticulture* 70, 557–584.
Britz, H., Steenkamp, E.T., Coutinho, T.A., Wingfield, B.D., Marasas, W.F.O. and Wingfield, M.J. (2002) Two new species of *Fusarium* section *Liseola* associated with mango malformation. *Mycologia* 94, 722–730.
Burdon, J.N., Moore, K.G. and Wainright, H. (1991) Mineral distribution in mango fruit susceptible to the physiological disorder 'soft nose'. *Scientia Horticulturae* 48, 329–336.
Butler, E.J. and Bisby, G.R. (1931) The fungi of India. *Scientific Monograph No. 1*. The Imperial Council of Agricultural Research in India, Calcutta.
Cazorla, F.M., Tores, J.A., Olalla, I., Perez-Garcia, A., Farre, J.M. and de Vincente, A. (1998) Bacterial apical necrosis of mango in southern Spain: a disease caused by *Pseudomonas syringae*. *Phytopathology* 88, 614–620.
Chee, K.H. (1969) Hosts of *Phytophthora palmivora*. *Review of Applied Mycology* 48, 337–344.
Cook, A.A. (1975) *Diseases of Tropical and Subtropical Fruits and Nuts*. Hafner Press, New York.
Cook, A.A., Milbrath, G.M. and Hamilton, R.A. (1971) Woody gall and scaly bark of *Mangifera indica* in Hawaii. *Phytopathology* 61, 1320 (abstract).
Crane, J.H., Bally, I.S.E., Mosqueda-Vazquez, R.V. and Tomer, E. (1997) Crop production. In: Litz, R.E. (ed.) *The Mango: Botany, Production and Uses*. CAB International, Wallingford, UK, pp. 203–256.
Crookes, C.A. and Rijkenberg, F.H.J. (1985) A literature review of the distribution, symptomatology, cause and control of mango blossom malformation. *South African Mango Grower's Association Research Report* 5, 15–24.
Cronje, C., Wehnwe, F.C. and Kotzé, J.M. (1990) *Alternaria alternata* as a lesion pathogen of mango inflorescences in South Africa. *Phytophylactica* 22, 117–118.
Darvas, J.M. (1993) *Dothiorella dominicana*, an important mango pathogen in South Africa. *Acta Horticulturae* 341, 321–328.
Das Gupta, S.N. and Zacchariah, H.T. (1945) Studies in the diseases of *Mangifera indica*. Part V. On the die-back of mango trees. *Journal of the Indian Botanical Society* 24, 101–108 (abstracted in the *Review of Applied Mycology* 25, 267–268, 1946).
de Toleda Piza, C. (ed.) (1966) Anais do Simposio Sobre a Seca da Mangueira (abstracted in *Review of Applied Mycology* 46, 378, 1967).
Dodd, J.C., Estrada, A.B., Matcham, J., Jeffries, P. and Jeger, M.J. (1991) The effect of climatic factors on *Colletotrichum gloeosporioides*, causal agent of mango anthracnose, in the Philippines. *Plant Pathology* 40, 568–575.
Dodd, J.C., Prusky, D. and Jeffries, P. (1997) Fruit diseases. In: Litz, R.E. (ed.) *The Mango: Botany, Production and Uses*. CAB International, Wallingford, UK, pp. 257–280.
Domsch, K.H., Gams, W. and Anderson, T.H. (1980) *Compendium of Soil Fungi*. Academic Press, New York.
Droby, S., Prusky, D., Jacoby, B. and Goldman, A. (1986) Presence of antifungal compounds in the peel of mango fruits and their relation to latent infections by *Alternaria alternata*. *Physiological and Molecular Plant Pathology* 29, 173–183.
Droby, S., Prusky, D., Jacoby, B. and Goldman, A. (1987) Induction of antifungal resorcinols in mesocarp of unripe mango fruits and its relation to latent infection by *Alternaria alternata*. *Physiological and Molecular Plant Pathology* 30, 285–292.
El Khoreiby, A. (1997) Mango culture in Egypt: an overview. In: Crane, J.H. (ed.) *Mango Subsector Horticultural Assessment Report*. UF/IFAS-ATUT-RONCO project, Appendix E.
FAO (2000) FAOSTAT online database at: http://www.fao.org/default.htm

Farr, D.F., Bills, G.F., Chamuris, G.P. and Rossman, A.Y. (1989) *Fungi on Plant and Plant Products in the United States*. APS Press, St Paul, Minnesota.
Fitzell, R.D. (1979) *Colletotrichum acutatum* as a cause of anthracnose of mango in New South Wales. *Plant Disease Reporter* 63, 1067–1070.
Fitzell, R.D. and Peak, C.M. (1984) The epidemiology of anthracnose disease of mango: inoculum sources, spore production and dispersal. *Annals of Applied Biology* 104, 53–59.
Fitzell, R.D., Peak, C.M. and Darnell, R.E. (1984) A model for estimating infection levels of anthracnose disease of mango. *Annals of Applied Biology* 104, 451–458.
Freeman, S., Maimon, M. and Pinkas, Y. (1999) Use of GUS transformants of *Fusarium subglutinans* for determining etiology of mango malformation disease. *Phytopathology* 89, 456–461.
Fukuda, T., Uehara, K., Azegami, K., Tabei, H. and Nishiyama, K. (1990) Bacterial canker of mango in Japan caused by *Xanthomonas campestris* pv. *mangiferaeindicae*. *Annals of the Phytopathological Society of Japan* 56, 474–480.
Gagnevin, L. and Pruvost, O. (1995) Assessment of genomic variability in *Xanthomonas campestris* pv. *mangiferaeindicae*. *Phytopathology* 85, 1163 (abstract).
Gagnevin, L. and Pruvost, O. (2001) Epidemiology and control of mango bacterial black spot. *Plant Disease* 85, 928–935.
Gantotti, B.V. and Davis, M.J. (1993) Pectic zymogram analysis for characterizing genetic diversity of the mango anthracnose pathogen. *Acta Horticulturae* 341, 353–359.
Ginai, M.A. (1965) Malformation of mango inflorescence (West Pakistan). *Journal of Agricultural Research* 3, 248–251.
Gupta, J.H. (1989) Perpetuation and epidemiology of powdery mildew of mango. *Acta Horticulturae* 231, 528–533.
Hayden, H.L., Pegg, K.G., Aitken E.A.B. and Irwin, J.A.G. (1994) Genetic relationships as assessed by molecular markers and cross-infection among strains of *Colletotrichum gloeosporioides*. *Australian Journal of Botany* 42, 9–18.
Hodson, A., Mill, P.R. and Brown, A.E. (1993) Ribosomal and mitochondrial DNA polymorphisms in *Colletotrichum gloeosporioides* isolated from tropical fruits. *Mycological Research* 97, 329–335.
Holliday, P. (1980) *Fungus Diseases of Tropical Crops*. Cambridge University Press, Cambridge.
Ibrahim, A.N., Satour, M.M., El-Tobshy, Z.M. and Abdel Sattar, M.A. (1975) Pathological and histological note on mango malformation in Egypt. *Current Science* 44, 443–444.
Iyer, C.P.A. and Degani, C. (1997) Classical breeding and genetics. In: Litz, R.E. (ed.) *The Mango: Botany, Production and Uses*. CAB International, Wallingford, UK, pp. 49–68.
Jeffries, P., Dodd, J.C., Jeger, M.J. and Plumbley, R.A. (1990) The biology and control of *Colletotrichum* species on tropical fruit crops. *Plant Pathology* 39, 343–366.
Johnson, G.I. (1994a) Powdery mildew. In: Ploetz, R.C., Zentmyer, G.A., Nishijima, W., Rohrbach, K. and Ohr, H.D. (eds) *Compendium of Tropical Fruit Diseases*. APS Press, St Paul, Minnesota, pp. 38–39.
Johnson, G.I. (1994b) Stem-end rot. In: Ploetz, R.C., Zentmyer, G.A., Nishijima, W., Rohrbach, K. and Ohr, H.D. (eds) *Compendium of Tropical Fruit Diseases*. APS Press, St Paul, Minnesota, pp. 39–41.
Johnson, G.I., Cooke, A.W., Mead, A.J. and Wells, I.A. (1991a) Stem end rot of mango in Australia: causes and control. *Acta Horticulturae* 219, 288–295.
Johnson, G.I., Mead, A.J., Cooke, A.W. and Dean, J.R. (1991b) Mango stem end rot pathogens – infection levels between flowering and harvest. *Annals of Applied Biology* 119, 465–473.
Johnson, G.I., Mead, A.J., Cooke, A.W. and Dean, J.R. (1992) Mango stem end rot pathogens – fruit infection by endophytic colonisation of the inflorescence and pedicel. *Annals of Applied Biology* 120, 225–234.
Johnson, G.I., Sharp, J.L., Milne, D.L. and Oosthuyse, S.A. (1997) Postharvest technology and quarantine treatments. In: Litz, R.E. (ed.) *The Mango: Botany, Production and Uses*. CAB International, Wallingford, UK, pp. 447–508.
Joubert, J.J. and Rijkenberg, F.H.J. (1971) Parasitic green algae. *Annual Review of Phytopathology* 9, 45–64.
Kadman, A. and Gazit, S. (1984) The problem of iron deficiency in mango trees and experiments to cure it in Israel. *Journal of Plant Nutrition* 7, 282–290.
Kauser, A.G. (1959) Malformation of inflorescence in mango. *Punjab Fruit Journal* 22, 19–21.
Kishun, R. (1995) Detection and management of *Xanthomonas campestris* pv. *mangiferaeindicae*. In: Varma, J.P., Verma, A. and Kumar, D. (eds) *Detection of Plant Pathogens and Their Management*. Angkar Publishers, New Delhi, pp. 173–182.
Kishun, R. and Sohi, H.S. (1983) Bacterial canker in mangoes. *Indian Farmer's Digest* 14, 21–23.

Knight, R.J. Jr (1997) Important mango cultivars and their descriptors. In: Litz, R.E. (ed.) *The Mango: Botany, Production and Uses*. CAB International, Wallingford, UK, pp. 545–565.
Kueprakone, U., Saengkong, S., Pienpuck, K. and Choobumroong, W. (1986) *Phytophthora palmivora* (Butl.) Butl., the causal organism of black rot of mango seedlings. *Thai Agricultural Research Journal* 4, 67–73 (in Thai with an English abstract).
Kumar, J. and Beniwal, S.P.S. (1992) Mango malformation. In: Kumar, J., Chaube, H.S., Singh, U.S. and Mukhopadhyay, A.N. (eds) *Plant Diseases of International Importance. Diseases of Fruit Crops*, Vol. III. Prentice Hall, Englewood Cliffs, New Jersey, pp. 357–393.
Leslie, J.F. (1995) *Gibberella fujikuroi*: available populations and variable traits. *Canadian Journal of Botany* 73 (Suppl. 1), S282–S291.
Lim, T.K. and Khoo, K.C. (1985) *Diseases and Disorders of Mango in Malaysia*. Tropical Press, Kuala Lumpur.
Lonsdale, J.H. and Kotzé, J.M. (1993) Chemical control of mango blossom diseases and the effect on fruit set and yield. *Plant Disease* 77, 558–562.
Lourd, M. and Keuli, S.D. (1975) Note sur un chancre à *Phytophthora* du manguier en Côte d'Ivoire. *Fruits* 30, 541–544.
Malaguti, G. and de Reyes, C. (1964) A gall disease of cacao and mango in Venezuela caused by *Calonectria rigidiuscula*. *Phytopathology* 54, 499 (abstract).
Malo, S.E. and Campbell, C.W. (1978) Studies on mango fruit breakdown in Florida. *Proceedings of the American Society of Horticultural Sciences, Tropical Region* 22, 1–15.
Manicom, B.Q. (1986) Factors affecting bacterial black spot of mangoes caused by *Xanthomonas campestris* pv. *mangiferaeindicae*. *Annals of Applied Biology* 109, 129–135.
Manicom, B.Q. (1989) Blossom malformation of mango. *South African Mango Grower's Association Yearbook* 10, 11–12.
Manicom, B.Q. and Pruvost, O.P. (1994) Bacterial black spot. In: Ploetz, R.C., Zentmyer, G.A., Nishijima, W.T., Rohrbach, K.G. and Ohr, H.D. (eds) *Compendium of Tropical Fruit Diseases*. APS Press, St Paul, Minnesota, pp. 41–42.
Manicom, B.Q. and Wallis, F.M. (1984) Further characterization of *Xanthomonas campestris* pv. *mangiferaeindicae*. *International Journal of Systematic Bacteriology* 34, 77–79.
Marlatt, R.B., Knight, R.J. Jr and Goldweber, S. (1970) *Verticillium* wilt of mango (*Mangifera indica*) in Florida. *Plant Disease Reporter* 54, 569–571.
Matheron, M.E. and Matejka, J.C. (1988) Phytophthora crown and root rot on nursery-grown mango trees delivered to Arizona. *Phytopathology* 78, 1572 (abstract).
McSorley, R., Parrado, J.L. and Goldweber, S. (1980) Observations on a mango decline in South Florida. *Proceedings of the Florida State Horticultural Society* 93, 132–133.
Mordue, J.E.M. (1980) Pestalotiopsis mangiferae. *CMI Descriptions of Pathogenic Fungi and Bacteria No. 676*. Commonwealth Mycological Institute, Kew, UK.
Morgan-Jones, G. (1967) Ceratocystis fimbriata. *CMI Descriptions of Pathogenic Fungi and Bacteria No. 141*. Commonwealth Mycological Institute, Kew, UK.
Mukherjee, S.K. (1997) Introduction: botany and importance. In: Litz, R.E. (ed.) *The Mango: Botany, Production and Uses*. CAB International, Wallingford, UK, pp. 1–19.
Muller, H.R.A. (1940) Overzicht van de belangrijkste Mangga-ziekten in Nederlandsch Indie. (English abstract in the *Review of Applied Mycology* 19, 355, 1940.)
Nandris, D., Nicole, M. and Geiger, J.P. (1987) Root rot diseases of rubber trees. *Plant Disease* 71, 298–306.
Narasimhan, M.J. (1954) Malformation of panicles in mango incited by spp. of *Eriophyes*. *Current Science* 23, 297–298.
Neergaard, P. (1945) *Danish Species of* Alternaria *and* Stemphylium. *Taxonomy, Parasiticism, Economical Significance*. Einar Munksgaard, Copenhagen.
Nicholson, R.I.D. and van Staden, J. (1988) Cytokinins and mango flower malformation. I. Tentative identification of the complement in healthy and malformed inflorescences. *Journal of Plant Physiology* 132, 720–724.
Ochoa, R., Aguilar, H. and Vargas, C. (1994) *Phytophagous Mites of Central America: an Illustrated Guide*. CATIE, Turrialba, Costa Rica.
Palti, J., Pinkas, Y. and Chorin, M. (1974) Powdery mildew of mango. *Plant Disease Reporter* 58, 45–49.
Patel, M.K., Kulkarni, Y.S. and Moniz, L. (1948) *Pseudomonas mangiferae-indicae*, pathogenic on mango. *Indian Phytopathology* 1, 147–152.
Peterson, R.A., Johnson, G.I., Schipke, L.G. and Cooke, A.W. (1991) Chemical control of stem end rot. *Acta Horticulturae* 291, 304–307.

Ploetz, R.C. (1994) Distribution and prevalence of *Fusarium subglutinans* in mango trees affected by malformation. *Canadian Journal of Botany* 72, 7–9.

Ploetz, R.C. (2001) Malformation: a unique and important disease of mango, *Mangifera indica* L. In: Summerell, B.A., Leslie, J.F., Backhouse, D., Bryden, W.L. and Burgess, L. (eds) Fusarium: *Paul E. Nelson Memorial Symposium*. APS Press, St Paul, Minnesota.

Ploetz, R.C. and Gregory, N. (1993) Mango malformation in Florida: distribution of *Fusarium subglutinans* in affected trees, and relationships within and among strains from different orchards. *Acta Horticulturae* 341, 388–394.

Ploetz, R.C. and Prakash, O. (1997) Foliar, floral and soilborne diseases. In: Litz, R.E. (ed.) *The Mango: Botany, Production and Uses*. CAB International, Wallingford, UK, pp. 281–326.

Ploetz, R.C., Benscher, D., Vázquez, A., Colls, A., Nagel, J. and Schaffer, B. (1996a) A re-evaluation of mango decline in Florida. *Plant Disease* 80, 664–668.

Ploetz, R., Vázquez, A. and Benscher, D. (1996b) First report of *Fusarium decemcellulare* as a pathogen of mango in the United States. *Plant Disease* 80, 1207.

Ploetz, R., Zheng, Q., Vázquez, A. and Sattar, M.A.A. (2002) Current status and impact of mango malformation in Egypt. *International Journal of Pest Management* 48, 279–285.

Pohronezney, K. and Marlatt, R.B. (1982) Some common diseases of mango in Florida. *Plant Pathology Fact Sheet PP-23*. University of Florida, Gainesville, Florida.

Pordesimo, A.N. (1982) Development of programmed spray application for the control of anthracnose. In: *PCARRD Research Highlights 1982*. Los Baños, PCARRD, pp. 38–39.

Prakash, O. (1988) Sooty mould disease of mango and its control. *International Journal of Tropical Plant Disease* 9, 277–280.

Prakash, O. (1990) Report of the plant pathologist. *Annual Report CIHNP (CMRS), Lucknow for the Year 1989–1990*.

Prakash, O. and Raoof, M.A. (1989) Die back disease of mango (*Mangifera indica*), its distribution, incidence, cause and management. *Fitopatologia Brasiliensis* 14, 207–215.

Prakash, O. and Singh, U.N. (1980) Root rot and damping off of mango seedlings caused by *Rhizoctonia solani*. *Indian Journal of Mycology and Plant Pathology* 10, 69.

Prakash, O. and Srivastava, K.C. (1987) *Mango Diseases and their Management – a World Review*. Tommorow's Printer, New Delhi.

Prakash, O., Misra, A.K. and Raoof, M.A. (1994) Studies on mango bacterial canker disease. *Biological Memoirs* 20, 95–107.

Prasad, A., Singh, H. and Shukla, T.N. (1965) Present status of mango malformation disease. *Indian Journal of Horticulture* 22, 254–265.

Prusky, D. (1994) Alternaria rot (black spot). In: Ploetz, R.C., Zentmyer, G.A., Nishijima, W., Rohrbach, K. and Ohr, H.D. (eds) *Compendium of Tropical Fruit Diseases*. APS Press, St Paul, Minnesota, pp. 34–35.

Prusky, D., Fuch, Y. and Yanko, U. (1983) Assessment of latent infections as a basis for control of postharvest disease of mango. *Plant Disease* 67, 816–818.

Prusky, D., Falick, E., Kobiler, I., Fuchs, Y., Zauberman, G., Pesis, E., Roth, I., Weksler, A., Akerman, M., Ykutiely, O., Waisblum, A., Keinan, A. and Ofek, G. (1997) Hot water brush: a new method for the control of post harvest disease caused by Alternaria rot in mango fruits. *Acta Horticulturae* 455, 780–785.

Pruvost, O. and Luisetti, J. (1991) Effect of time of inoculation with *Xanthomonas campestris* pv. *mangiferaeindicae* on mango fruits susceptibility. Epiphytic survival of *X. c.* pv. *mangiferaeindicae* on mango fruits in relation to disease development. *Journal of Phytopathology* 133, 139–151.

Pruvost, O., Couteau, A. and Luisetti, J. (1989) Efficacité de différentes formulations chemiques pour lutter contre la maladie des taches noires de la mangue (*Xanthomonas campestris* pv. *mangiferaeindicae*). *Fruits* 44, 343–350.

Pruvost, O., Couteau, A. and Luisetti, J. (1990) Development of bacterial black spot of mangoes and epiphytic populations of the pathogen (*Xanthomonas campestris* pv. *mangiferaeindicae*) under natural conditions in Réunion Island. *Fruits* 45, 125–140.

Pruvost, O., Couteau, A. and Luisetti, J. (1992) Pepper tree (*Schinus terebenthifolius* Radii), a new host plant for *Xanthomonas campestris* pv. *mangiferaeindicae*. *Journal of Phytopathology* 135, 289–298.

Pruvost, O., Couteau, A., Verniere and Luisetti, J. (1993) Epiphytic survival of *Xanthomonas campestris* pv. *mangiferaeindicae* on mango buds. *Acta Horticulturae* 341, 337–344.

Punithalingham, E. (1993) Phomopsis mangiferae. *CMI Descriptions of Pathogenic Fungi and Bacteria No. 1168*. Commonwealth Mycological Institute, Kew, UK.

Punithalingham, E. and Waterson, J.M. (1970) Hendersonula toruloidea. *CMI Descriptions of Pathogenic Fungi and Bacteria No. 274*. Commonwealth Mycological Institute, Kew, UK.

Rao, B.S. (1975) *Maladies of Hevea in Malaysia*. Rubber Research Institute of Malaysia. Kuala Lumpur.

Raymond, L., Schaffer, B., Brecht, J.K. and Crane, J.H. (1998) Internal breakdown in mango fruit: symptomatology and histology of jelly seed, soft nose and stem-end cavity. *Postharvest Biology and Technology* 13, 59–70.

Reckhaus, P. and Adamou, I. (1987) Hendersonula dieback of mango in Niger. *Plant Disease* 71, 1045.

Reuveni, M., Harpaz, M. and Reuveni, R. (1998) Integrated control of powdery mildew on field-grown mango trees by foliar sprays of mono-potassium phosphate fertilizer, sterol inhibiting fungicides and the strobilurin Kresoxym-methyl. *European Journal of Plant Pathology* 104, 853–860.

Ribiero, I.J.A. (1980) Seca de mangueira. Agentes causais e estudio da molesta. In: *Anais do I Simpósio Brasiliero Sobre a Cultura de Mangueira*. Sociedad Brasileira de Fruticultura, Jaboticobal, November 24–28, 1980, pp. 123–130.

Rossman, A.Y., Samuels, G.J., Rogerson, C.T. and Lowen, R. (1998) Genera of Bionectriaceae, Hypocreaceae and Nectriaceae (Hypocreales, Ascomycetes). *Studies in Mycology* 42, 1–248.

Ruehle, G.D. and Ledin, R.B. (1955) *Mango Growing in Florida*. Agricultural Experiment Station Bulletin 574. University of Florida, Gainesville, Florida.

Schaffer, B. (1994) Decline. In: Ploetz, R.C., Zentmyer, G.A., Nishijima, W., Rohrbach, K. and Ohr, H.D. (eds) *Compendium of Tropical Fruit Diseases*. APS Press, St Paul, Minnesota, pp. 43.

Schaffer, B., Larson, K.D., Snyder, G.H. and Sanchez, C.A. (1988) Identification of mineral deficiencies asscociated with mango decline by DRIS. *HortScience* 23, 617–619.

Schoeman, M.H., Manicom, B.Q. and Wingfield, M.J. (1995) Epidemiology of powdery mildew on mango blossoms. *Plant Disease* 79, 524–528.

Simmons, S.L., Hofman, P.J., Whiley, A.W. and Hetherington, S.E. (1998) Effects of leaf:fruit ratios on fruit growth, mineral concentration and quality of mango (*Mangifera indica* L. cv. Kensington Pride). *Journal of Horticultural Science and Biotechnology* 73, 367–374.

Singh, Z. and Dhillon, B.S. (1989) Hormonal changes associated with vegetative malformation of mango (*Mangifera indica* L.). *Journal of Phytopathology* 125, 193–197.

Slippers, B., Johnson, G.I., Cooke, A.W., Crous, P.W., Coutinho, T.A., Wingfield, B.D. and Wingfield, M.J. (2001) Taxonomy of *Botryosphaeria* spp. causing stem end rot of mango. In: *Proceedings of the 13th Biennial Australasian Plant Pathology Conference*. Cairns, Australia, September 24–27, 2001.

Steenkamp, E.T., Wingfield, B.D., Coutinho, T.A., Wingfield, M.J. and Marasas, W.F.O. (1999) Differentiation of *Fusarium subglutinans* f. sp. *pini* by histone gene sequence data. *Applied and Environmental Microbiology* 65, 3401–3406.

Steenkamp, E., Britz, H., Coutinho, T., Wingfield, B., Marasas, W. and Wingfield, M. (2000) Molecular characterization of *Fusarium subglutinans* associated with mango malformation. *Molecular Plant Pathology* 1, 187–193.

Steyn, P.L., Viljoen, N.M. and Kotzé, J.M. (1974) The causal organism of bacterial black spot of mangoes. *Phytopathology* 64, 1400–1404.

Summanwar, A.S., Raychaudhuri, S.P. and Phatak, S.C. (1966) Association of the fungus *Fusarium moniliforme* Sheld. with the malformation in mango (*Mangifera indica* L.). *Indian Phytopathology* 19, 227–228.

Sutton, B.C. (1980) *The Coelomycetes*. Commonwealth Mycological Institute, Kew, UK.

Tarmizi, S.A., Malik, T.M.T.A., Pauziah, M. and Zahrah, T. (1993) Incidence of insidious fruit rot as related to mineral nutrients in Harumanis mangoes. *MARDI Research Journal* 21, 43–49.

Tsao, P.H. (1990) Why many phytophthora root rots and crown rots of tree and horticultural crops remain undetected. *OEPP/EPPO Bulletin* 20, 11–17.

Tsao, P.H., Luzaran, P.B., de los Santos, A.B., Portales, L.A., Gochango, A.M. and Gruber, L.C. (1994) Phytophthora crown and root rot of mango detected in Philippine nurseries. *Plant Disease* 78, 100.

Turner, P.D. (1960) *Annual Report of the West African Research Institute 1958–1959* (as cited in Chee, 1969).

Uppal, B.N., Patel, M.K. and Kamat, M.N. (1941) Powdery mildew of mango. *Journal of the University of Bombay* 9, 12–16.

van Staden, J. and Nicholson, R.I.D. (1989) Cytokinins and mango flower malformation. II. The cytokinin complement produced by *Fusarium moniliforme* and the ability of the fungus to incorporate ($8–^{14}C$) adenine into cytokinins. *Physiological and Molecular Plant Pathology* 35, 423–431.

van Staden, J., Bayley, A.D. and Macrae, S. (1989) Cytokinins and mango flower malformation. III. The metabolism of (^{3}H) *iso*-pentenyladenine and ($8–^{14}C$) zeatin by *Fusarium moniliforme*. *Physiological and Molecular Plant Pathology* 35, 433–438.

Varma, A., Lele, V.C., Raychauduri, S.P., Ram, A. and Sang, A. (1974) Mango malformation: a fungal disease. *Phytopathologische Zeitschrift* 79, 254–257.

Viegas, A.P. (1960) Mango blight. *Bragantia* 19, 163–182 (abstracted in *Review of Applied Mycology* 42, 696, 1963).

Wagle, P.V. (1928) Studies in the shedding of mango flowers and fruits. *Part I, Memoirs of the Department of Agriculture of India, Botanical Sciences* 8, 219–249.

Watt, G. (1891) *A Dictionary of Economic Products of India*. Government Printing Press, Calcutta.

Young, T.W. (1957) 'Soft-nose', a physiological disorder in mango fruits. *Proceedings of the Florida State Horticultural Society* 70, 280–283.

Young, T.W. and Miner, J.T. (1960) Response of 'Kent' mango to nitrogen fertilization. *Proceedings of the Florida State Horticultural Society* 73, 334–336.

Zheng, Q. and Ploetz, R. (2002) Genetic diversity in, and development of a PCR assay for identifying, the mango malformation pathogen. *Plant Pathology* 51, 208–216.

16 Diseases of Mangosteen

T.-K. Lim[1] and S. Sangchote[2]
[1]*Biosecurity Australia, Department of Agriculture, Fisheries and Forestry Australia, Canberra, Australia;* [2]*Department of Plant Pathology, University of Kasetsart, Bangkok, Thailand*

Introduction

The species name of mangosteen, *Garcinia mangostana* (family: *Clusiaceae* or, alternatively, *Guttiferae*), is derived from the Malay word 'manggis', and the genus is named after the French botanist, Laurent Garcinia (1683–1751). The family contains ~35 genera and 800 species. More than 40 of the 400 species in the genus yield edible fruits, but mangosteen is the most renowned. Its extreme popularity and delicious taste has earned it the title the 'Queen of Tropical Fruits'.

Mangosteen originated in the Malay Archipelago and thrives in truly tropical conditions (Verheij and Coronel, 1992). It is polyploid ($2n = 96$), whereas other *Garcinia* species are diploid ($2n = 48$). Mangosteen is found only in cultivation and only female trees are known. The tree is evergreen and grows to a height of 10–25 m with a dense canopy of large, broad, dark green, opposite leaves that are ovoid to oblong, 15–25 × 7–13 cm and have short petioles. The fruit is a round berry, 5–8 cm in diameter, with a pericarp that changes from green to dark purple as it matures. The soft, white flesh inside is arranged in segments. Mangosteen is genetically uniform and its polyembryonic seeds are produced apomictically and the fruit are parthenocarpic.

Mangosteen is grown mainly in Southeast Asia, and the main commercial countries are Indonesia, Malaysia and Thailand (Verheij and Coronel, 1992). Annual production of the fruit probably does not exceed 150,000 t. It grows best in warm, humid areas, with average temperatures of 30°C, and those below 20°C and above 38°C reduce growth. It thrives in shade on well-drained, slightly acidic soil that is rich in organic matter. It requires plenty of water and shade during the first 3–4 years of establishment. Seedlings have a juvenile period of 10–12 years. Grafted plants bear after ~5–7 years, but exhibit slower growth and lower yields.

Diseases

Compared with other tropical fruits, the pests and diseases that attack mangosteen have been studied and documented less extensively.

Many of the pathogens that attack mangosteen are widespread and attack other tropical trees. For example, some of the soilborne basidiomycetes that affect mangosteen are also found on other tropical trees. Many of these have similar survival strategies and occur in disease complexes (e.g. the root rot complex on rubber) that have similar epidemiologies. Other basidiomycetes that cause stem and trunk diseases, such as pink disease and white thread blight, are also rife on other tropical tree species. The ubiquitous algal leaf spot has a very extensive host range and is also widespread. Fruit diseases on mangosteen are similar to those on other tropical fruit species, the most noteworthy being anthracnose, Diplodia, Pestalotiopsis and Phomopsis fruit rots. In contrast, sooty mould and black mildew on mangosteen may involve different species in different areas.

Foliar Diseases

Algal leaf spot

This disease is widespread on many tropical fruit including mangosteen. It can be serious on unthrifty trees in poorly maintained orchards.

The causal alga, *Cephaleuros virescens*, forms conspicuous, orange to rust coloured velutinous spots of varying diameter on the foliage. Often the spots coalesce to form larger irregular spots. Damage is brought about by parasitism and growth of the alga beneath the cuticle causing degeneration and discoloration of the host cells. The epidemiology and management of the disease is discussed in Chapter 1.

Horse-hair blight

Turner (1971) reported this disease in Sarawak where it causes dieback of shoots and branches. It is not as common as white thread blight on mangosteen, but the disease is rampant on cocoa in Malaysia.

Diagnostic features of the disease include the presence of thick, black, horse-hair-like, rhizomorphic strands that are attached to and dangle among dead foliage and shoots. Threads of the pathogen, *Marasmiellus equicrinus*, also encircle branches and shoots. It produces small, white basidiocarps, and also attacks bamboo, chempedak, citrus, cocoa, langsat, mango, pineapple, rambutan, rubber and tea (Singh, 1980).

The epidemiology and management of the disease are similar to those for white thread blight.

Pestalotiopsis leaf blight, stem canker and fruit rot

Pestalotiopsis leaf blight has been reported in Thailand (Sonthirat *et al.*, 1994), Malaysia (Johnston, 1960; Williams and Liu, 1976) and North Queensland (L. Vawdrey, Queensland Department of Primary Industries, 1998). In the same areas, the pathogen causes pre- and postharvest fruit rots and, in North Queensland, it also causes a stem canker and dieback. Although the pathogen is common on mangosteen, these diseases are not major problems.

Symptoms

On leaves, lesions commence as small brown spots. They expand gradually and become larger and straw coloured with a dark brown margin. Young infected leaves may be distorted and become blighted. Symptoms develop as slightly depressed, brown lesions.

Stem canker symptoms include branch splitting, gummosis, blistering of the bark and dieback. The symptoms become obvious a few days after a storm (L. Vawdrey, Queensland Department of Primary Industries, 1998).

Diseased fruits become hard, and infected areas turn light pink. Black, pin-sized acervuli of the pathogen form in lesions.

Causal agents

In Thailand, the pathogen is *Pestalotiopsis flagisettula*, whereas the species that is responsible in Malaysia and Queensland has not been identified.

Acervuli are globose to lenticular and rupture the epidermis by a pore, which becomes wide and irregular. Conidia emerge in a black column and gradually become effuse over the fruit surface. Conidia are fusiform to slightly clavate, five-celled and average 6.4×19.3 μm. The three median cells are olivaceous, while the apical and basal cells are hyaline. The apical cell has three hyaline, apical appendages (Sangchote and Pongpisuta, 1995).

Epidemiology

Symptom expression is greatest in stressed trees. The fungus exists as an endophyte in woody branches and survives as a saprophyte in plant trash and in the soil. Conidia are spread by rain (Sangchote and Pongpisuta, 1995).

In Queensland, leaf blight and stem canker occur on trees that have been predisposed by sun scalding. In Malaysia and Australia, the fungus also infects mangosteen fruit, usually through injuries caused by feeding activities of insects such as fruit spotting, red-banded and mirid bugs.

Management

Leaf blight and stem canker can be managed with light pruning, burning diseased shoots and leaves, and fungicides. Applying copper fungicides, mancozeb or iprodione at leaf flush will also help to reduce inoculum. To avoid sun scalding, windbreaks and plant shades, especially during the early stages of establishment, should be provided. Controlling insects that feed on the fruit also helps lower the incidence of fruit rot in the field.

Pink disease

This disease has been recorded in Sabah (Williams and Liu, 1976), North Queensland (L. Vawdrey, Queensland, 1998, personal communication), and probably occurs in Indonesia (Semangoen, 1971). It can be damaging, but is usually sporadic and not considered economically important.

Symptoms

The disease manifests itself as pinkish white, mycelial threads that envelop the branches and shoots. The foliage above the zone of infection dries up and dies. Pink incrustations develop as the weather begins to dry; these represent a more mature stage of the pathogen.

Causal agent

Erythricium salmonicolor (anamorph: *Necator decretus*) causes pink disease. It is described in Chapter 1. Pink disease was attributed to a different pathogen, *Pellicularia koleroga*, in Indonesia (Semangoen, 1971), although the principal author has observed the same symptoms on mangosteen in Indonesia that are caused by *E. salmonicolor* elsewhere.

Epidemiology and management

Refer to Chapter 1.

Sooty mould and black mildew

Sooty mould and black mildew are common on the foliage of mangosteen trees in Southeast Asia. The causal fungi subsist on honeydew produced by insects, and produce velutinous, black sooty films or layers on mangosteen leaves and petioles. In Sarawak, Turner (1971) reported two sooty mould species, *Brooksia tropicalis* and *Grallomyces portoricensis*. Two black mildew species, *Meliola garcinae* and an unidentified *Meliola* sp., were recorded in Peninsular Malaysia (Johnston, 1960) and in Sarawak (Turner, 1971).

These problems have a minor effect on productivity, and specific control measures usually are not required. Since leaves in the interior of the canopy are affected more frequently, pruning internal portions of the canopy to improve ventilation and penetration of insecticides can be used for severe infestations. Applications of copper sprays are also effective.

Thread blight (white thread blight disease)

Thread blight is also known as white thread blight disease. It was first reported on mangosteen in Sarawak by Turner (1971).

Symptoms

White, thread-like mycelial signs of the pathogen develop on the underside of leaves and on twigs and branches. On leaves, the mycelial threads fan out to form a silvery white network. The fungal strands turn opaque white or creamy in colour. Foliage above the affected areas is killed and becomes yellow, and finally brown. Dead leaves abscise but remain attached to the canopy by fungal strands.

Causal agent

The pathogen is a basidiomycete, *Marasmiellus scandens*. In Sabah, white thread blight was associated with *Rhizomorpha* sp. (Williams and Liu, 1976).

During wet weather, the fungus forms small, gilled basidiocarps on infested plant debris that are white, resupinate and shell like, with a pileus. Basidiospores are hyaline, tiny, and are formed in the hymenial lining of the gills.

Epidemiology

Basidiospores are disseminated via wind, rainsplash and insects, and mycelium spreads via contact with foliage. Warm, shady, humid or wet conditions are conducive to disease spread and establishment. The fungus is very common in the tropics on bamboo, bougainvillea, carambola, cinnamon, cocao, coffee, durian, jackfruit, Java apple, leguminous shrubs, litchi, mango, black pepper, rambutan, rubber, sapodilla, soursop and water apple (Singh, 1980).

Management

Disease incidence can be reduced by wider tree spacing, frequent corrective pruning and application of any of the following fungicides: benodanil, flusilazol, oxycarboxin, propiconazole, triadimefon and tridemorph (T.-K. Lim, unpublished data).

Fruit Diseases

Anthracnose

Anthracnose was reported on fruit in Thailand (Dhirabhava *et al.*, 1980; Sangchote and Pongpisuta, 1995) and on seedlings in Sri Lanka (Alahakoon and Brown, 1994).

Symptoms

Lesions on fruit are hard and light brown; pin-sized black spots, acervuli of the causal fungus, appear in zonate rings on the necrotic tissues.

Causal agent

Colletotrichum gloeosporioides (teleomorph: *Glomerella cingulata*) is described in Chapter 1.

Epidemiology

The fungus produces conidia in acervuli on diseased lesions. These then spread via rainsplash to infect leaves and fruits. Although symptoms may develop shortly after infection, latent infections are common and may remain quiescent on fruit for months (Sangchote and Pongpisuta, 1995).

C. gloeosporioides has a wide host range, but there is great genetic and pathogenic variation among isolates from different hosts. In Sri Lanka, three isolates from mangosteen were found to be highly pathogenic on leaves of durian, guava, mango and rambutan, whereas isolates from these hosts and pini jambu were essentially non-pathogenic on mangosteen (Alahakoon *et al.*, 1994).

Management

Fruit should not be bruised or injured at or after harvest, since injuries provide avenues for secondary infection. Spraying fruits with carbendazim, iprodione or mancozeb after rain and 2 weeks before harvest is recommended in Thailand (Sangchote and Pongpisuta, 1995).

Diplodia fruit rot

Johnston (1960) first reported Diplodia fruit rot on mangosteen in Malaysia. This disease is the most important postharvest disease of mangosteen in Thailand (Dhirabhava *et al.*, 1980; Sangchote and Pongpisuta, 1995). The causal fungus was also associated with a disorder of the trunk and collar of mangosteen trees in Sabah (Williams and Liu, 1976).

Symptoms

Diseased fruit becomes hard. Diseased areas appear grey to black, and are covered with fluffy mycelia and abundant pycnidia (Plate 96). Pycnidia are immersed in the epidermis, and later become erumpent.

Causal agent

The causal fungus is *Diplodia theobromae*. It produces ostiolate, erumpent pycnidia in lesions from which conidia are extruded in a black mass, and is described in Chapter 1.

Epidemiology

D. theobromae is a common wound and secondary pathogen. It can also exist as a saprophyte and, with species of *Pestalotiopsis*, *Dothiorella* and *Phomopsis*, is also a common endophyte on mature woody branches of trees (Johnson *et al.*, 1994). *D. theobromae* is common in warm climates and has a high optimum temperature for growth, 30°C. It has been found on a wide range of hosts, and sporulates readily on host tissue.

Management

To reduce development of the disease, fruit should not be injured or bruised, and trees should be kept in a healthy state.

Phomopsis fruit rot

This is a common postharvest disease of mangosteen in Thailand (Sangchote and Pongpisuta, 1995). Diseased fruit become hard. Affected areas are light brown in colour, and black pin-sized dots, pycnidia, are formed as lesions age.

The disease is caused by *Phomopsis* sp. The fungus produces pycnidia that are immersed in the rind in necrotic areas and later break through the host epidermis to produce a white stream of conidia. The conidia are hyaline and ellipsoid. The fungus can live as an endophyte in twigs and branches, or as a saprophyte in plant debris. Conidia, the infective propagules, are dispersed by rainsplash. Bruised or injured fruit are infected most often.

Measures that are used against anthracnose are also effective against Phomopsis fruit rot.

Gliocephalotrichum fruit rot

This postharvest fruit rot was reported from Thailand (Sangchote and Pongpisuta, 1995). Epidermal tissues of affected fruit become swollen and light pink.

This fruit rot is caused by *Gliocephalotrichum bulbilium*. The fungus produces conidia and conidiophores in the subepidermis of the fruit wall that emerge through the epidermis. Conidia are produced in clusters at the end of the conidiophore, and are uniseptate and cylindrical to oblong. Like other postharvest rot pathogens, the fungus can affect other fruit crops such as rambutan (Farungsang *et al.*, 1994).

Soilborne Diseases

Brown root disease

This destructive disease affects mangosteen in Peninsular Malaysia (Johnston, 1960). It occurs sporadically on mangosteen that is planted on former jungle or rubber land.

Symptoms

The disease causes leaf discoloration, wilting of foliage and ultimately death of trees. More diagnostic, however, are brown, crusty, mycelial mats of the pathogen that grow on roots and trunk bases. Often, soil particles adhere to the brown crust.

Causal agent

Phellinus noxius affects tropical trees in >50 genera, and ranges throughout tropical Africa, Australia and Southeast Asia (Holliday, 1980). The fungus produces a distinctive woody, tomentose basidiocarp with zonate, brown and white margins at the base of tree stumps. The porate undersurface is pale coloured, and basidiospores are produced in the fertile hymenial lining that lines the pores. The pathogen is described in Chapter 5.

Epidemiology

The fungus persists in the soil on decaying wood and tree stumps. The disease spreads mainly by root contact with infested wood or tree stumps via thick rhizomorphs.

Management

Strategies that are described for white root disease in Chapter 10 are effective. These include: isolation trenches to inhibit spread from diseased trees; injection of root collars with 10% tridemorph in a bitumen emulsion; treating tree stumps with creosote; and using legume cover crops to deplete the pathogen food base and augment populations of antagonistic organisms.

In India, Jacob *et al.* (1991) reported that *Trichoderma harzianum* improved seedling growth and prevented infection by *P. noxius* when added to the soil. The following approaches that were used by Barrance (1989) on rubber in Vanuatu are amenable for brown and red root control on mangosteen: use of the trap blocking plants *Plectranthus amboinicus*, *Coleus scutellaroides* and *Zingiber* spp.; use of trap crops, such as cocoa, to identify inoculum sources; and proper site selection and land preparation.

Red root

This disease occurs sporadically on mangosteen that is planted on former jungle or rubber land in Peninsular Malaysia (Johnston, 1960). Infected trees ultimately are killed.

Aboveground symptoms are similar to those described for brown root disease. The presence of red rhizomorphic incrustations on the roots and the presence of distinctive hard, woody, dark brown basidiocarps distinguishes red root from brown root disease. The advancing margin of the pathogen's rhizomorph is white and fan shaped, and fuses to form an encrusting skin which becomes red–brown with age.

Ganoderma philippi causes red root. It produces very variable, but more often thin and flat basidiocarps (Fig. 16.1) (Steyaert, 1975). Their upper surfaces are army to bone brown, either with concentric grooves or knobbles, and those with grooves sometimes with silver speckles. The pore surface is usually cinnamon buff, up to 150–160 mm in diameter, and up to 40 mm thick at the base. Basidiospore production often is sparse and may be absent. They are ovoid, 6–8.5 (7.5) × 4–8 (5.2) μm, and chamois coloured.

G. philippi survives in the soil on decaying wood and stumps. It spreads via its rhizomorph from diseased roots or infected stumps to healthy roots, and by airborne basidiospores.

Control methods are similar to those described for brown root disease. Chemicals that are effective include tridemorph and drazoxolon.

Stem Disease

Zignoella stem canker

This disease has only been reported in Malaysia (Yaacob and Tindall, 1995).

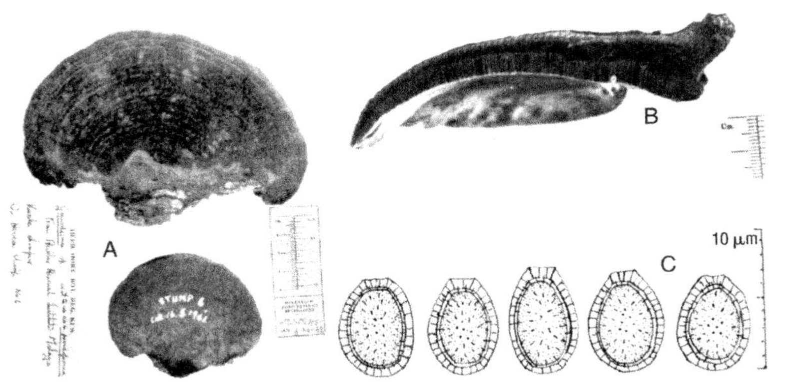

Fig. 16.1. (A) Upper and lower surface of basidiocarp, (B) section of basidiocarp and (C) basidiospores of *Ganoderma philippi* (from CMI description no. 446).

Symptoms

On stems and branches, tuberous, cankerous growths can be seen. This is followed by a dieback of foliage and branches. In serious cases, it can result in the death of the tree.

Causal agents

In Malaysia, the causal agent was reported to be *Zignoella garcineae* (Yaacob and Tindall, 1995). However, the pathogenicity of this fungus has not been studied. The genus *Zignoella* was placed into synonymy with the genus *Chaetosphaeria* by Booth (1957, 1958). *Chaetosphaeria* was originally described for a hyaline-spored species with smooth ascomata (Tulasne and Tulasne, 1863), but when the genus was taken up by Saccardo (1883) it was conceived as having smooth or tomentose ascomata and hyaline or brown ascospores. Species originally described under *Zignoella* have hyaline, multiseptate spores. *Chaetosphaeria* occurs mainly on decaying twigs or wood.

Epidemiology

Little is known about the epidemiology of the disease.

Management

Removal of infected branches and trees are recommended to curb the spread of this disease (Yaacob and Tindall, 1995).

References

Alahakoon, P.W. and Brown, A.E. (1994) Host range of *Colletotrichum gloeosporioides* on tropical fruits in Sri Lanka. *International Journal of Pest Management* 40 (1), 23–26.

Alahakoon, P.W., Brown A.E. and Sreenivasaprasad, S. (1994) Cross-infection potential of genetic groups of *Colletotrichum gloeosporioides* on tropical fruits. *Physiological and Molecular Plant Pathology* 44 (2), 93–103.

Barrance, A.J. (1989) *Phellinus noxius* in Vanuatu – management considerations. *Forest Research Report Vanuatu No. 1A-89.*

Booth, C. (1957) Studies of Pyrenomycetes. I and II. *Mycological Paper* 68, 1–27.

Booth, C. (1958) The genera *Chaetosphaeria* and *Thaxteria* in Britain. *Naturalist (Hull)*, pp. 83–90.

Dhirabhava, V., Akaravesapong, P., Sukcharoen, T. and Anotharom, S. (1980) Postharvest diseases and storage techniques of mangosteen fruits. In: *Annual Report No.1, Division of Plant Pathology and Microbiology, Department of Agriculture, Bangkok*, pp. 712–718 (in Thai).

Farungsang, U., Sangchote, S. and Farungsang, N. (1994) Rambutan postharvest diseases in Thailand. In: Johnson, G.I. and Highly, E. (eds) *Development of Postharvest Handling Technology for Tropical Tree Fruits*. ACIAR Proceedings No. 58, pp. 51–59.

Holliday, P. (1980) *Fungus Diseases of Tropical Crops*. Cambridge University Press, Cambridge.

Jacob, C.K., Annajutty, J., Jayaratnam, K. and Joseph, A. (1991) Effect of antagonists on *Phellinus noxius* causing brown root disease of *Hevea*. *Indian Journal of Natural Rubber Research* 4(2), 142–145.

Johnson, G.I., Mead, A.J., Cooke, A.W. and Wells, I.A. (1994) Stem-end rot diseases of tropical fruit – mode of infection in mango, and prospects for control. In: Johnson, G.I. and Highley, E. (eds) *Development of Postharvest Handling Technology for Tropical Fruits*. ACIAR Proceedings No. 58, pp. 70–71.

Johnston, A. (1960) *A Supplement to a Host List of Plant Diseases in Malaya. Mycological Paper No. 77,* Commonwealth Mycological Institute, Kew, UK.

Saccardo, P.A. (1883) Sylloge Fungorum Omnium Hucusque Cognitorum Digessit P.A. Saccardo, Patavii (Padua) 2, pp. 882.

Sangchote, S. and Pongpisuta, R. (1995) *Fruit Rots of Mangosteen and Their Control*. ACIAR Project No. 9313, Annual Report 1995, Department of Plant Pathology, Kasetsart University, Bangkok.

Semangoen, H. (1971) *Penyakit-Penyakit Tanaman Pertanian di Indonesia*. Universitas Gadja Mada, Yogyarkarta, Indonesia (in Indonesian).

Singh, K.G. (1980) *A Check List of Host and Disease in Malaysia Bulletin No. 154*. Ministry of Agriculture, Malaysia.

Steyaert, R.L. (1975) *CMI Descriptions of Pathogenic Fungi and Bacteria No. 446*. Commonwealth Mycological Institute, Kew, UK.

Sonthirat, P., Pitakpaivan, P., Khamhanggridthirong, T., Choobamrong and Kueprakone, U. (1994) *Host Index of Plant Diseases in Thailand*, 3rd edn. Department of Agriculture, Bangkok, Thailand.

Tulasne, L.R. and Tulasne, C. (1863) *Selecta Fungorum Carpologia*, Vol. II. Paris. (English translation, W.B. Grove, 1931. Oxford University Press.)

Turner, G.J. (1971) *Plant Disease in Sarawak. Phytopathological Paper No. 13*. Commonwealth Mycological Institute, Kew, UK.

Verheij, E.W.M. and Coronel, R.E. (eds) (1992) *Plant Resources of South-East Asia. No 2. Edible Fruits and Nuts*. Prosea Foundation, Bogor, Indonesia.

Williams, T.H. and Liu, P.S.W. (1976) *A Host List of Plant Diseases in Sabah, Malaysia. Phytopathological Paper 19*. Commonwealth Mycological Institute, Kew, UK.

Yaacob, O. and Tindall, H.D. (1995) *Mangosteen Cultivation*. FAO Plant Production and Protection Paper 129.

17 Diseases of Papaya

Denis M. Persley[1] and Randy C. Ploetz[2]
[1]*Agency for Food and Fibre Sciences, Queensland Department of Primary Industries, Indooroopilly, Queensland, Australia;* [2]*University of Florida, Tropical Research and Education Center, Homestead, Florida, USA*

Introduction

Papaya or papaw, *Carica papaya*, is a widely grown and important perennial fruit crop throughout the lowland tropical and subtropical regions of the world. Its ease of propagation from seed, popularity and versatility, and relatively short period from planting to production have made it an important food in many countries (Manshardt, 1992). The fruit is used mainly as a fresh fruit, but can be frozen, pickled, dried and also eaten green as a vegetable. The fruit is a good source of vitamins A and B, whereas the enzyme papain, obtained from green fruit, has diverse uses in the pharmaceutical and food industries.

Origin and Distribution

The papaya is the most economically important member of the *Caricaceae*, a dicotyledonous family containing small, latex-producing trees and shrubs. All have terminal clusters of leaves and most are dioecious (Purseglove, 1968). The *Caricaceae* belong to a group of plant families that are characterized by the production of glucosinalates, with the clade demonstrating a common phylogenetic origin with the *Brassicaceae* (Bremer *et al.*, 1998). There currently are five genera within the family: *Carica*, *Cylicomorpha*, *Horovitzia*, *Jacaratia* and *Jarilla* (Badillo, 1993). The former contains 22 species that are distributed from Argentina and Chile to southern Mexico (Manshardt, 1992). Recently, Badillo (2000) recommended dividing *Carica* into two genera: *Carica*, containing *C. papaya* only; and *Vasconcellea*, containing 21 species. This revision is based on morphological differences and molecular analysis that has shown that *C. papaya* is only distantly related to other members of *Carica* (Jobin-Décor *et al.*, 1997; Aradhya *et al.*, 1999; Kim *et al.*, 2002; Van Droogenbroeck *et al.*, 2002). *C. papaya* is a polygamous, diploid species with a small genome (372 Mbp) and nine pairs of chromosomes (Storey, 1976; Arumuganathan and Earle, 1991; Ming *et al.*, 2001).

Papaya is thought to have originated on the Caribbean coast of Central America (Purseglove, 1968; Manshardt, 1992). The Spaniards took papaya from Central America to Southeast Asia in the 16th century and from there it spread to most tropical and subtropical countries by the 18th century (Purseglove, 1968; Storey, 1976). Papaya is now an important tropical fruit crop worldwide with an annual production in 2001 of 5.47 Mt (FAO, 2001). Brazil and Mexico are major producers, with production for domestic and export consumption occurring throughout Southeast Asia, China, Australia, South America, South Africa and the USA (Hawaii and Florida) (Table 17.1).

Table 17.1. Papaya production statistics in 2001.[a]

Country	Hectares	Metric tonnes (× 1000)	Yield (kg ha^{-1} × 1000)
Brazil	40,300	1450	3598
Nigeria	90,000	748	831
India	57,000	644	1130
Mexico	17,977	613	3409
Indonesia	34,890	470	1347
Congo	13,000	213	1638
Peru	13,500	174	1286
Taiwan	3,400	135	3970
Thailand	9,800	119	1214
Colombia	4,600	114	2467
Venezuela	7,000	105	1500
Ecuador	5,250	101	1924
Philippines	6,500	77	1191
Yemen	4,106	68	1656
Malaysia	5,600	60	1071
Bangladesh	5,700	41	719
Cuba	2,300	40	1739
Costa Rica	750	35	4693
Mozambique	3,400	31	912
USA	770	25	3240
Guatemala	700	25	3571

[a]Data are from FAO, http://www.fao.org/default.htm

Botany

Papaya is a rapidly growing perennial herbaceous plant, 2–10 m in height, usually unbranched with hollow stems and petioles, and latex vessels (laticifers) in all organs and tissues. Leaves are clustered near the apex of the trunk and are palmately lobed with 25–100 cm long petioles (Purseglove, 1968).

Flowers are borne in modified cymose inflorescences that appear in the axils of leaves. The type of inflorescence that is produced depends on the sex of the tree. Papaya trees are classified as male, female or hermaphrodite (Nakasone and Paull, 1998). Male trees produce long, pendulous, many flowered cymose inflorescences. The flower is unisexual and lacks a functional pistil. Female trees have inflorescences with few flowers that have large pistils without stamens. Hermaphrodite trees normally produce bisexual flowers but are sexually variable. Warm night temperatures and water stress induce the production of male flowers, whereas cool temperatures transform the stamens into fleshy carpel-like structures. Selection seems to have favoured hermaphroditism since perfect flowered types are common among domesticated *C. papaya*, but naturally occurring populations are uniformly dioecious with male and female flowers on separate individuals (Manshardt, 1992).

Sex inheritance is controlled by a single locus with multiple alleles (Storey, 1976). Staminate plants and hermaphrodites are heterozygous, pistillate plants are homozygous for the recessive allele, and the homozygous condition involving either dominant allele is lethal (Manshardt, 1992). The existence of different sex types has important horticultural consequences. Hermaphrodite papaya can be inbred to produce gynodioecious cultivars (plants bearing female and hermaphrodite flowers on separate individuals) that have stable horticultural characteristics from generation to generation. Examples are the Hawaiian 'Solo' and the Malaysian 'Eksotica' types (Manshardt, 1992). Pollination in gynodioecious cultivars is efficient, with every plant in a plantation being a potential fruit producer. In the dioecious cultivars, outcrossing results in considerable and rapid genetic drift causing variation in horticultural characteristics unless the parental lines can be maintained (Aquilizan, 1987) or the cultivar is cloned by micropropagation (Drew, 1988). Gynodioecious cultivars are best suited to tropical regions with mild climates and little seasonal variation, since sex expression and fruit development in hermaphrodites are greatly affected by environmental conditions (Manshardt, 1992). Dioecious cultivars tolerate seasonal fluctuations better and are favoured in subtropical regions.

The fruit is a large fleshy berry weighing from a few hundred grams to 9 kg (Purseglove, 1968). It is covered with a smooth, thin, green skin that turns yellow or orange when mature. The flesh is thick, succulent and easily bruised, and varies in colour from yellow to orange to red. Fruit contain hundreds of black, wrinkled seeds that are attached to the wall of the ovary and enclosed in a gelatinous sarcotesta (Purseglove, 1968).

Ecological Requirements and Production

Papaya is a tropical plant and is very sensitive to frost. Optimum temperatures for growth are between 21 and 33°C, with growth and production being severely affected below 12–14°C (Nakasone and Paull, 1998). Plantations should be situated in warm sheltered sites with windbreaks to maximize yield and fruit quality. Papaya can be grown on a variety of soil types. However, good soil drainage is essential, since poor drainage results in major waterlogging and root rot problems (Nakasone and Paull, 1998). A minimum monthly precipitation of ~100 mm sustains good growth without supplementary irrigation. Mulching of plants with coarse grass hay promoted earlier flowering and increased fruit set, yield and weight in the subtropics (Elder *et al.*, 2000). A soil pH of 5.5–6.5 is optimum, and crops require a balanced supply of nitrogen, phosphorus, potassium, calcium and magnesium (Ross and Chay-Prove, 2000). An adequate supply of minor elements such as boron and zinc is also essential. Application rates of fertilizers are best determined by soil and leaf analysis (Ross and Chay-Prove, 2000).

Desirable characteristics of papaya cultivars and the objectives in breeding programmes are outlined by Nakasone and Paull (1998). Commercial cultivars may be inbred gynodioecious lines (Hawaiian 'Solo' lines), out-crossing dioecious populations, F_1 hybrids or clones such as 'Hortus Gold'.

Diseases of Papaya

Cultivated papaya has a narrow genetic base that may be partly responsible for the susceptibility of papaya to a wide range of diseases (Kim *et al.*, 2002). The following review is limited to the major diseases of this crop. Minor diseases of papaya are listed in Table 17.2.

Diseases that are Caused by Bacteria

Bacterial canker and decline

Bacterial canker is an important disease in the Caribbean. The disease was described originally as St Croix papaya decline, after the island of St Croix on which it was first reported (Webb, 1985). A disease with similar symptoms, decline, has been reported from the Mariana Islands in the Pacific (Trujillo and Schroth, 1982).

SYMPTOMS Angular, water-soaked lesions develop on leaves. They coalesce and spread along the veins, resulting in large areas of necrotic tissue. Lesions cincture the petiole, causing leaves, particularly at the top of the canopy, to wilt and hang downwards. Water-soaked lesions occur on the stem and spread to form cankers, which eventually girdle the stem and collapse plants (Webb, 1985). Small, water-soaked lesions develop on green fruit, developing into firm, depressed lesions. These symptoms are very similar to those of decline in the Northern Mariana Islands (Trujillo and Schroth, 1982).

Table 17.2. Minor diseases and disorders of papaya.

Disease/disorder	Cause
Algal leaf spot	*Cephaleuros virescens*
Angular leaf spot	*Leveillula taurica*
Blossom spot	*Choanephora cucurbitarum*
Bacterial leaf spot	*Pseudomonas carica-papayae*
Bacterial wilt	*Ralstonia solanacearum*
Black rot	*Erwinia cypripedii*
Feather leaf	Unknown virus
Foot rot	*Pythium aphanidermatum*
	Pythium ultimum
Fruit rot	*Monilia* sp.
Fruit spot	*Cercospora mamaonis*
Greasy spot	*Corynespora cassiicola*
Leaf spot	*Alternaria* sp.
	Asperisporium caricae
	Cercospora mamaonis
	Cercospora papayae
	Choanephora cucurbitarum
	Curvularia carica-papayae
	Gloeosporium sp.
	Mycosphaerella caricae (anamorph: *Phoma caricae-papayae*)
	Phyllosticta sp.
Nivum Haamir dieback	Unknown cause
Petiole spot	*Didymella* sp.
Sclerotium blight	*Athelia rolfsii* (anamorph: *Sclerotium rolfsii*)
Seedling blight	*Colletotrichum gloeosporioides*
Stem-end rot	*Alternaria alternata*
	Colletotrichum gloeosporioides
	Fusarium sp.
	Diplodia theobromae
	Mycosphaerella caricae (anamorph: *Phoma caricae-papayae*)
	Phomopsis sp.
	Rhizopus stolonifer
Stemphylium fruit spot	*Stemphylium lycopersici*
Stem rot	*Haematonectria haematococca* (anamorph: *Fusarium solani*)
	Fusarium sp.
	Phytophthora palmivora
	Pythium aphanidermatum
	Pythium ultimum
Target spot	*Phyllosticta caricae-papayae*
Terminal necrosis and wilt	*Tobacco ringspot virus*
Verticillium wilt	*Verticillium dahliae*
Yellow strap leaf	Production of toxins by *Aspergillus wentii* in the rhizosphere

Information from Nishijima (http://www.apsnet.org/online/common/names/papaya.asp) and Conover (1979).

CAUSAL AGENT The diseases in the Caribbean and Mariana Islands are caused by bacteria in the genus *Erwinia* (Trujillo and Schroth, 1982; Webb, 1985). Neither organism has been identified to the species level. However, the two differ in several physiological and biochemical characteristics, the most notable of which is the absence of flagella in the agent from St Croix (Webb, 1985). *E. stewartii* is the only other member of the genus with this characteristic.

The pathogen in the Mariana Islands may be a member of the *E. chrysanthemi* group (Trujillo and Schroth, 1982).

EPIDEMIOLOGY The bacterium causing canker in the Caribbean is disseminated by rain, but free moisture does not increase disease severity or pathogen survival on leaf surfaces (Webb, 1985). It did not survive for longer than 2 weeks in soil, but survived indefinitely in leaf lesions and cankers on infected trees, and as an epiphyte on the leaves of several non-host species (Webb, 1985). Unlike decline in the Mariana Islands, there is no evidence for spread by the African snail or insects.

Internal yellowing

This disease occurs in Hawaii and affects the internal flesh of ripening fruit (Nishijima *et al.*, 1987). The disease was first seen in fruit following hot water treatments to control fruit fly.

SYMPTOMS Ripe or over-ripe fruit are most severely affected. The flesh develops a yellow discoloration, with the affected areas being soft with diffuse, spreading margins and an offensive, rotting odour (Plate 97). Symptoms develop in areas around the seed cavity, usually near the calyx and middle sections of the fruit, but not on the fruit exterior (Nishijima *et al.*, 1987). The symptoms differ from purple stain fruit rot, caused by *Erwinia herbicola*, which discolours internal vascular and parenchyma tissue of ripe fruit purple (Nelson and Alvarez, 1980).

CAUSAL AGENT Internal yellowing is caused by the bacterium *Enterobacter cloacae*, a member of the *Enterobacteriaceae*. It is a facultatively anaerobic, Gram-negative rod that is $0.3–0.6 \times 0.8–2.0$ μm, has peritrichous flagella and is oxidase negative and catalase positive (Nishijima *et al.*, 1987). *E. cloacae* grows well on standard bacteriological media on which yellow pigment and purple stain are not produced. Colonies are creamy tan on yeast extract–dextrose–calcium carbonate agar, dark pink to burgundy with translucent margins on tetrazolium chloride agar, and orange on Miller–Schroth medium (Nishijima, 1994). Nishijima (1994) reported a range of biochemical reactions for the pathogen.

EPIDEMIOLOGY The epidemiology of the disease is not fully understood. The bacterium has been isolated from papaya flowers, homogenates of papaya seeds, and the crop and midgut of the oriental fruit fly, *Dacus dorsalis* (Nishijima *et al.*, 1987). *E. cloacae* is thought to be transmitted to papaya flowers by fruit flies and other insects, and remains quiescent until symptoms develop in fully ripe fruit.

MANAGEMENT Hot water treatment of fruit reduced the incidence of internal yellowing (Nishijima *et al.*, 1987). In the absence of this treatment, the incidence of the disease in Hawaii was as high as 43% (Nishijima, 1994).

Mushy canker

This disease occurs in the Northern Mariana Islands and is caused by an *Erwinia* taxon that is distinct from that that causes decline in the same area (Trujillo and Schroth, 1982). The disease is characterized by blackish, water-soaked, mushy cankers near or in the leaf axils of the fleshy upper portion of the stem. It is more prevalent after wind and rain that is associated with severe storms (Trujillo and Schroth, 1982).

Papaya bunchy top

Papaya bunchy top (PBT) is a serious disease that often limits commercial production throughout the American tropics (Davis, 1994). It was first reported from Puerto Rico in 1931 and currently occurs throughout much of the Caribbean region and in Central and South America (Davis *et al.*, 1996).

SYMPTOMS The initial symptoms include diffuse chlorosis in young leaves with a reduction in the growth of leaves and petioles. Small, discrete water-soaked spots develop on affected petioles and stems, later developing into irregular blotches ~1–2 mm in diameter. Petioles are rigid, almost horizontal and shortened. Leaf blades are thickened, stiff and cupped downwards, with marginal and interveinal chlorosis and necrosis (Fig. 17.1). Apical growth ceases, which, with the shortened internodes, gives plants a bunched appearance (Cook, 1972). Chronically affected plants often have only a tuft of small leaves at the apex. Flowering and fruit set seldom occur in affected plants (Davis, 1994). Although the absence of latex flow from fresh puncture wounds in leaves, stems and fruit was reported to be unreliable for diagnosing PBT (Webb and Davis, 1987), it is now considered a valid test for the disease (M.J. Davis, personal communication, 2002).

CAUSAL AGENT Virus and phytoplasma aetiologies had been suggested for this disease in the past (Storey and Halliwell, 1969; Cook, 1972). Recent molecular studies have disproved the phytoplasma hypothesis and consistently have associated an obligate, laticifer-inhabiting bacterium in the genus *Rickettsia* with PBT (Davis *et al.*, 1996, 1998). The bacterium was detected in diseased plants with primers that amplified a 705 bp fragment of a putative succinate dehydrogenase (SdhA) gene of the organism, whereas no amplification was detected in extracts from healthy papaya plants. The bacterium was also detected in adults of a known vector of bunchy top, *Empoasca papayae* (Davis *et al.*, 1998).

Fig. 17.1. Apex of a papaya plant in the advanced stages of bunchy top (photo: M. Ferwerda-Licha).

EPIDEMIOLOGY Bunchy top is transmitted by two leafhoppers, *E. papayae* and *E. stevensi*, and distribution of the disease largely coincides with their distribution (Adsuar, 1946; Haque and Parasram, 1973). The former is the principal vector, and is the only leafhopper known to breed on papaya, whereas *E. stevensi* is important in areas such as Trinidad where *E. papayae* does not occur (Davis, 1994). Single insects of both species can transmit the agent, with symptoms appearing 30–45 days after inoculation (Davis, 1994). The level of tolerance in a cultivar affects the rate and severity of symptom expression.

MANAGEMENT Tolerant cultivars are of some value in areas where disease pressure is low (Davis, 1994). Control has been achieved by applying persistent insecticides for vector control and topping affected plants to allow development of healthy axillary shoots (Cook, 1972). Antibiotic therapies through drenching and root dip treatments have resulted in remission of symptoms (Storey and Halliwell, 1969).

Diseases that are Caused by Fungi and Fungus-like Agents

Alternaria fruit spot

Alternaria fruit spot can be a major fruit disease where papaya is grown in dry areas, such as Israel and on the island of Maui (Alvarez *et al.*, 1977; Snowdon, 1990). It is usually not important unless fruit are kept in prolonged cold storage (e.g. 14 days at 10°C) (Alvarez and Nishijima, 1987). Lesions are circular to ellipsoidal, depressed, and eventually blacken as the pathogen sporulates (Plate 98). Minimal rotting of the underlying flesh occurs. Ultimately, lesions can coalesce to cover the entire fruit surface.

The disease is caused by *Alternaria alternata*, which is described in Chapter 1. It produces large numbers of conidia on fruit and senescing leaf petioles.

Chlorothalonil or mancozeb sprayed biweekly reduced the disease by ~50%, but did not provide sufficient control for export (Alvarez *et al.*, 1977). Postharvest, hot water dips (20 min at 48°C) also reduce the disease.

Anthracnose

This disease is an important postharvest disease of papaya in most production areas, particularly when fruit are transported to distant markets (Alvarez and Nishijima, 1987).

SYMPTOMS Small, water-soaked spots appear on the fruit surface as ripening commences. As infection advances, circular sunken lesions with translucent light brown margins form (Alvarez and Nishijima, 1987). Light orange to pink conidial masses cover the central portion of the lesions and frequently occur in a concentric ring pattern (Plate 99) (Dickman, 1994). The internal tissue is firm with a greyish white discoloration, which later turns brown (Alvarez and Nishijima, 1987).

A second type of symptom appears as irregular to circular, sharply defined spots, 1–10 mm in diameter (Fig. 17.2). These 'chocolate spots' are slightly depressed and reddish brown (Dickman, 1994). The spots enlarge rapidly as fruit ripen to form characteristic, sunken lesions.

CAUSAL AGENT Anthracnose is caused by *Glomerella cingulata* (anamorph: *Colletotrichum gloeosporioides*), which is described in Chapter 1.

EPIDEMIOLOGY The primary inoculum occurs on senescing petioles and leaves, and is disseminated by wind or rain. The fungus infects intact, unwounded, immature green fruit in the field. Spores germinate, form appressoria on the fruit surface, and penetrate the cuticle directly with the aid of enzymes and mechanical pressure (Dickman and Alvarez, 1983; Latunde-Dada, 2001). The subcuticular hyphae remain latent until fruit ripening commences.

High temperatures and relative humidity favour disease development. Spores require free moisture for germination, with little germination occurring below 97% relative humidity (Dickman, 1994). Spores are released from acervuli when moisture is abundant, and rainsplash is an important means of dispersal. The fungus is not active during hot, dry weather (Dickman, 1994).

MANAGEMENT Anthracnose is managed by applying fungicides to fruit in the plantation, particularly when weather conditions favour disease development. Postharvest fungicide and hot water treatments also provide control.

Lines and cultivars highly susceptible to anthracnose should be avoided, and plantations are best established in warm, sheltered locations.

Asperisporium black spot

This disease occurs in Australia, Africa, Central and South America, India and the USA (Ellis and Holliday, 1972; Peterson *et al.*, 1993).

SYMPTOMS Leafspots develop on older leaves and are 1–3 mm in diameter, circular and initially water soaked, later becoming necrotic and brown in colour. The centres of lesions become bleached with age (Fig. 17.3). Dark brown to black conidial masses form

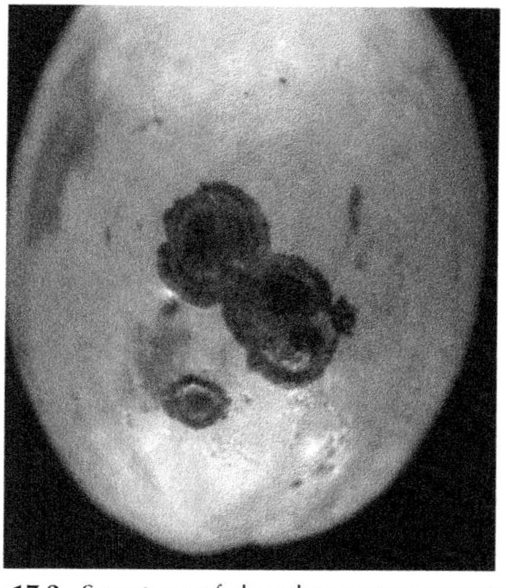

Fig. 17.2. Symptoms of chocolate spot on papaya (photo: W.T. Nishijima).

on the undersurface of leaves. Severely affected leaves curl, turn brown and die, resulting in extensive defoliation. Slightly raised lesions, 2–6 mm in diameter, develop on trunks. Lesions on fruit are slightly sunken, circular, 2–6 mm in diameter and brown to black. The tissue beneath these lesions becomes corky, but internal decay does not occur (Peterson *et al.*, 1993).

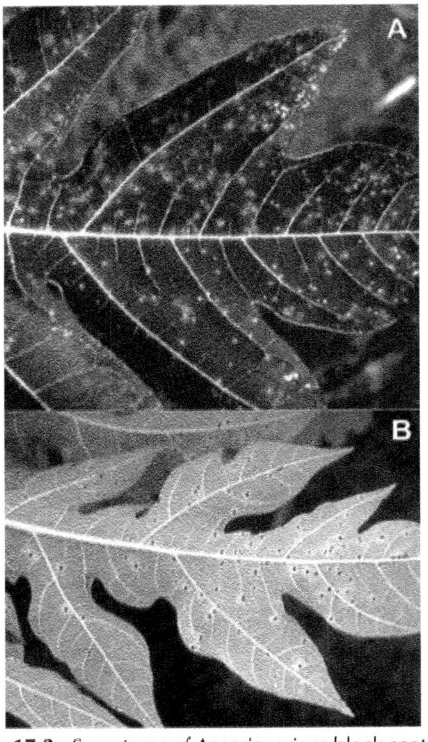

Fig. 17.3. Symptoms of Asperisporium black spot symptoms on the (A) upper and (B) lower surface of a papaya leaf (photos: A.W. Cooke).

CAUSAL AGENT Black spot is caused by the fungus *Asperisporium caricae*. Conidiophores are closely packed and cover the surface of the stromata (Ellis and Holliday, 1972). They are unbranched or occasionally branched, straight or flexuous, hyaline to olivaceous brown, and smooth with several prominent conidial scars at the apex (Fig. 17.4). Conidiophores are up to 45 μm in length and 6–9 μm wide. Conidia are solitary, 14–26 × 7–10 μm, dry, ellipsoidal, pyriform or clavate, almost always one septate, hyaline to pale brown and distinctly verrucose (Ellis and Holliday, 1972). The fungus appears to be a specialized biotroph since only limited growth occurred on three of 29 culture media that were tested by Chambers and Rijkenberg (1987).

EPIDEMIOLOGY Inoculum from infected papaya plants is dispersed by wind and wind-driven rain. Optimum germination of conidia occurs between 18 and 20°C, and does not occur above 30°C (Peterson and Grice, 1994). Cooler temperatures and intermittent moisture, such as dew and rain showers, favour disease development, whereas hot, dry conditions are inhibitory.

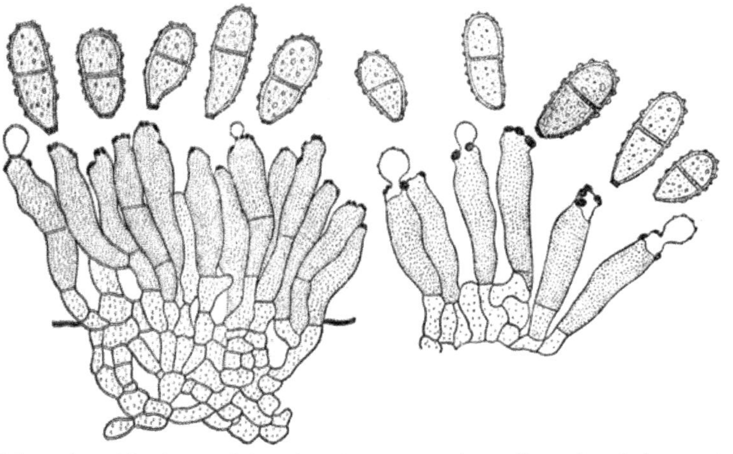
Fig. 17.4. Conidia and conidiophores of *Asperisporium caricae* (from Ellis and Holliday, 1972).

MANAGEMENT Severe outbreaks usually occur in areas where plantation hygiene and management have been neglected. Control involves removing severely diseased leaves and fruit, combined with regular and thorough application of fungicides, particularly to the undersides of young leaves (Ross and Chay-Prove, 2000).

Black (dry) rot

This disease occurs in most countries where papaya is grown. The disease can occur as a fruit rot, stem rot or as spots on leaves, flowers and young fruit.

SYMPTOMS Very young fruit may develop a black, sunken rot originating at the stem end or point of contact with a dead leaf (Simmonds, 1965). These fruit usually wither and fall.

On ripening fruit, small, brown, slightly sunken, water-soaked lesions develop, later becoming black and up to 4 cm in diameter (Plate 100). The margins of the lesions are light brown and translucent, and their surface dries and wrinkles with age, and are covered with hyphae and pycnidia of the pathogen (Simmonds, 1965; Nishijima, 1994a). Under humid conditions, light coloured tendrils of pycnidiospores ooze from pycnidia. The infected tissue is dry and firm.

The fungus also causes a stem-end rot on ripening fruit. Initial infection is through broken peduncles and causes slight browning and latex discharge. As the disease develops, the infected area becomes wrinkled, dry and black (Nishijima, 1994a).

The fungus is able to invade the upper regions of the stem through wounds or from diseased fruit, causing an extensive dark, sunken rot and eventual death of the upper portion of the trunk (Simmonds, 1965).

CAUSAL AGENT Black rot is caused by *Mycosphaerella caricae* (anamorph: *Phoma caricae-papayae*). Perithecia are 100–180 × 70–200 μm, dark brown to black, and flask shaped to oval (Fig. 17.5) (Sivanesan, 1990). Asci are 29–53 × 7–13 μm, and ascospores are 8–15 × 3–5 μm, straight to slightly curved, hyaline, septate and constricted in the middle. Conidia are hyaline, 7–12 × 2.5–3.5 μm, have one or two cells, are short cylindrical or slightly reniform, and are produced in pycnidia (Sivanesan, 1990).

EPIDEMIOLOGY The fungus colonizes senescing leaves and petioles, producing abun-

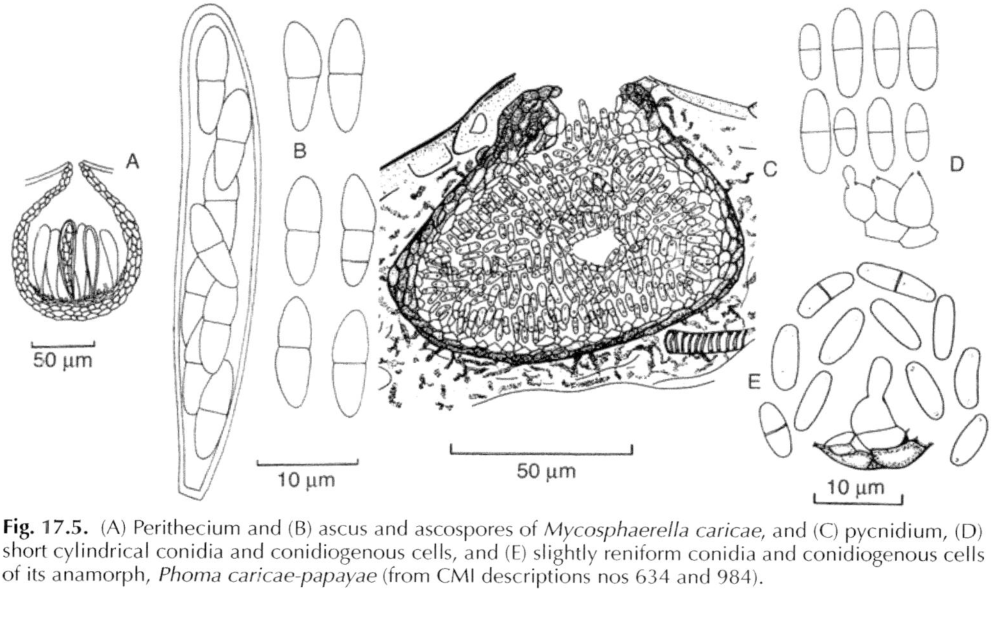

Fig. 17.5. (A) Perithecium and (B) ascus and ascospores of *Mycosphaerella caricae*, and (C) pycnidium, (D) short cylindrical conidia and conidiogenous cells, and (E) slightly reniform conidia and conidiogenous cells of its anamorph, *Phoma caricae-papayae* (from CMI descriptions nos 634 and 984).

dant ascospores and conidia, both of which are infective (Chau and Alvarez, 1979). Ascospores are discharged within an hour after relative humidity reaches 100%. Conidia and ascospores are deposited on the surface of fruit during rain but require wounds to infect (Nishijima, 1994a). Papaya latex is a suitable medium for ascospore germination and fungal development. Wounds created during and after harvest are colonized quickly if suitable conditions exist. Papaya is susceptible to chilling injury at temperatures below 7°C, and normal storage and transport temperatures provide a suitable environment for spore germination and disease development (Nishijima, 1994a).

MANAGEMENT Good plantation hygiene and regular fungicide applications reduce inoculum levels in plantations. Postharvest hot water and fungicide treatments provide good control (Snowdon, 1990; Nishijima, 1994a).

Brown spot

Brown spot, which is also known as Corynespora leaf spot, occurs in most countries where papaya is grown (Ellis and Holliday, 1971). The disease can cause premature defoliation and reduced yields and fruit quality.

SYMPTOMS Small, light brown, almost circular lesions develop on the lower leaves, gradually spreading to younger leaves (Fig. 17.6A). Fully developed lesions have brown centres 2 mm in diameter surrounded by a prominent, 4–8 mm diameter, yellow halo (Peterson *et al.*, 1993). On close examination, lesions have faint, concentric rings (Pernezny and Litz, 1993). The fungus sporulates in lesions on both surfaces of the leaf, but is more abundant on the lower. Under favourable conditions, the lesions may coalesce causing extensive areas of necrosis and premature death (Simmonds, 1965). Long, elliptical lesions coated by dark masses of conidia develop on petioles (Fig. 17.6B). Small, dark, sunken lesions may also develop on the surface of fruit (Fig. 17.6C) (Simmonds, 1965; Peterson *et al.*, 1993).

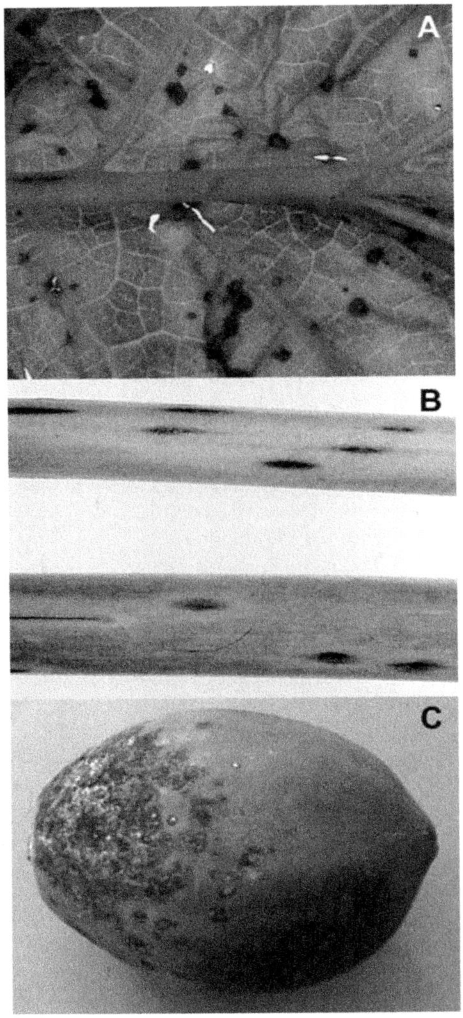

Fig. 17.6. Symptoms of brown spot on a (A) leaf, (B) petioles and (C) fruit of papaya (photos: A.W. Cooke).

CAUSAL AGENT *Corynespora cassiicola* causes brown spot, and papaya is one of its many hosts (Holliday, 1980).

Fungal mycelium is mostly immersed. Conidiophores are erect and usually simple, straight or slightly flexuous, pale to mid-brown, smooth and septate with up to nine successive cylindrical proliferations (Fig. 17.7) (Ellis and Holliday, 1971). Conidiophores are 110–850 μm long and 4–11 μm wide. Conidia are solitary or catenate, very variable in shape and usually

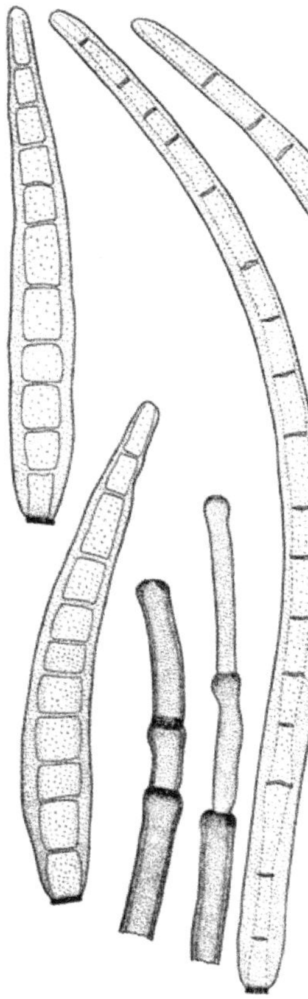

Fig. 17.7. Conidia and conidiophores of *Corynespora cassiicola* (from Ellis and Holliday, 1971).

brown, 40–220 μm in length, and 9–22 μm wide at the broadest part.

EPIDEMIOLOGY Conidia are dispersed by air, particularly during wet, windy weather. Disease development is favoured by hot, humid weather (Peterson *et al.*, 1993). The fungus has a wide host range including rubber, tomato, cucurbits and legumes, with little evidence of host specialization (Ellis and Holliday, 1971). The fungus can be seedborne and can survive in host debris for up to 2 years (Ellis and Holliday, 1971).

MANAGEMENT Regular applications of fungicides, for example dithiocarbamates or copper, provide good control of the disease (Pernezny and Litz, 1993; Ross and Chay-Prove, 2000).

Cercospora black spot

This disease occurs in most countries where papaya is grown and is most common in poorly maintained plantations.

Leaf spots are irregularly shaped, greyish white and 1–5 mm in diameter. Severe leaf spotting can result in leaf yellowing and defoliation (Nishijima, 1994b). Fruit symptoms begin as tiny black spots, which enlarge to ~3 mm in diameter. They are indistinct on green fruit but become prominent as fruit ripen. Although the spots cause little damage, they detract from the market acceptability of fruit (Nishijima, 1994b).

Cercospora papayae causes black spot. Hyaline, multiseptate conidia (20–75 × 3–5 μm) are produced on brown, multiseptate, non-branched, 50–200 × 3–6 μm conidiophores (Fig. 17.8) (Ellis, 1976). Stromata are usually produced on the upper leaf surface (Nishijima, 1994b).

Primary inoculum is produced on papaya leaves. Conidia are dispersed by wind, and disease development is favoured by warm, wet weather. The disease is seldom a problem in well-managed plantations where

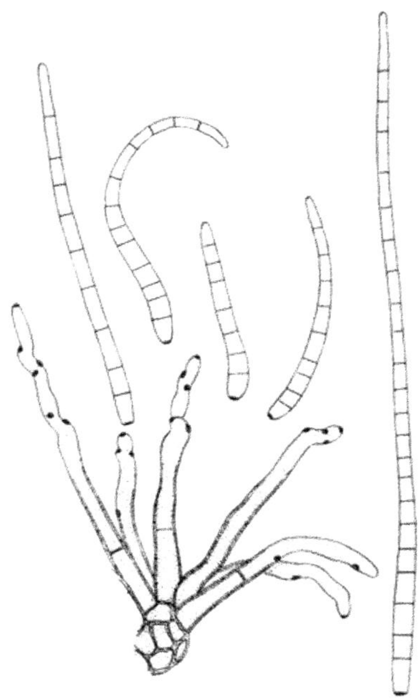

Fig. 17.8. Conidia and conidiophores of *Cercospora papayae* (from Ellis, 1976).

fungicides are applied to manage several diseases, including black spot.

Collar rot

This disease is most important in the Puna district on the island of Hawaii where annual rainfall is 366 cm year^{-1} (Laemmlen and Aragaki, 1971; Nishijima and Aragaki, 1973). It is a sporadic problem in replanted fields where drainage is poor, and is not found in other locations in the islands where lower amounts of rainfall occur. The causal agent, *Calonectria ilicicola* (anamorph: *Cylindrocladium parasiticum*), has also been recovered from papaya in Queensland, Australia, and another species, *Cy. gracile*, has been reported on papaya in Brazil (Crous, 2002).

SYMPTOMS The root collar is discoloured initially and then water soaked (Laemmlen and Aragaki, 1971; Nishijima and Aragaki, 1973). As decay of the stem becomes more conspicuous, extensive deterioration of the roots develops in concert with chlorosis, wilting and stunting of the canopy (Figs 17.9 and 17.10). Ultimately, the collar and entire root system are rotted and covered with orange to red perithecia of the pathogen. Single plants in planting holes may remain healthy while others are affected.

Fig. 17.9. Symptoms of collar rot of papaya. The rotted tissues are covered with the orange–red perithecia of the causal fungus, *Calonectria ilicicola* (photo: W.T. Nishijima).

CAUSAL AGENT The former name of *Ca. ilicicola*, *Ca. crotalariae* (anamorph: *Cy. crotalariae*), was changed when it was determined that the former name had priority and that a Latin description for the latter species was never published (Crous, 2002). Its orange to red perithecia are subglobose to ovoid, 300–550 μm high and 280–400 μm wide. Asci are clavate, 90–140 × 12–19 μm, taper to a long, thin stalk and contain eight hyaline ascospores (Fig. 17.11). They are fusoid with rounded ends, straight to slightly curved, two- to four-celled, and 30–65 × 4–7 μm. The anamorph produces hyaline, cylindrical, two- to four-celled, 45–90 × 4–7 μm macroconidia. Macroconidiophores consist of a septate 50–100 × 5–6 μm stipe, penicillate arrangement of fertile branches, a septate, 120–240 × 3–4 μm stipe extension and a sphaeropedunculate, 6–12 μm in diameter vesicle. The pathogen produces thick-walled, brown microsclerotia.

EPIDEMIOLOGY Collar rot was a problem primarily on papaya seedlings in fields that were recently cleared of native ohia forests, but became sporadic after this practice was discontinued (Nishijima and Aragaki, 1973). It now occurs mainly in replanted fields following heavy rainfall in areas where drainage is poor.

The pathogen has a wide host range that includes groundnut, lucerne, anthurium, eucalyptus and koa. Conidia and ascospores are responsible for long-distance spread, and microsclerotia for long-term survival (Hwang and Ko, 1976). The fungus can survive for >3 years in the absence of host plants.

Fig. 17.10. Wilting and lodging of a papaya plant caused by collar rot (photo: W.T. Nishijima).

MANAGEMENT The following cultivars are listed in descending order of their resistance to collar rot: 'Kapoho Solo', 'Line 8', 'Waimanalo' and 'Sunrise Solo' (Nishijima and Aragaki, 1973). Although most broad-spectrum fungicides are effective, they are seldom used due to the disease's scattered and sporadic occurrence. Spot treatments with fungicides or fumigants and replanting with healthy seedlings have been effective.

Fusarium fruit rot

This disease has been reported from Hawaii, India, Israel and the Philippines (Snowdon, 1990; Nishijima, 1994c). Lesions are up to 15 mm in diameter, depressed and usually covered by mycelia and conidia of the causal fungi (Fig. 17.12). They develop on the fruit surface and stem end.

Several species of *Fusarium* cause papaya fruit rot (Nishijima, 1994c). *F. solani* is most common and widespread in Hawaii, India and the Philippines, and is described in Chapter 18. Its teleomorph, *Haematonectria haematococca*, has not been reported on papaya. In Mexico, *F. solani* causes a seedling collar rot. During wet weather, it also infects young papaya fruit by entering the seed cavity through the blossom end and spreading within the fruit, ultimately causing the fruit to abscise. *F. moniliforme* and *F. equiseti* have been reported to cause fruit rots in India (Saxena and Sharma, 1981).

F. solani requires a predisposing stress or injury to establish on fruit. It is often a secondary invader of lesions caused by *Colletotrichum gloeosporioides* and other fungi. In Hawaii, Fusarium fruit rot is most common during rainy conditions. Preventive field sprays and hot water dips are effective.

Lasiodiplodia fruit and stem-end rot

Lasiodiplodia fruit and stem-end rot is a relatively minor problem but can be important under certain conditions (Hunter and Buddenhagen, 1972). Its common name refers to the former name of its causal agent, *Lasiodiplodia theobromae*, which is now referred to as *Diplodia theobromae*; it is described in Chapter 1. The disease has been

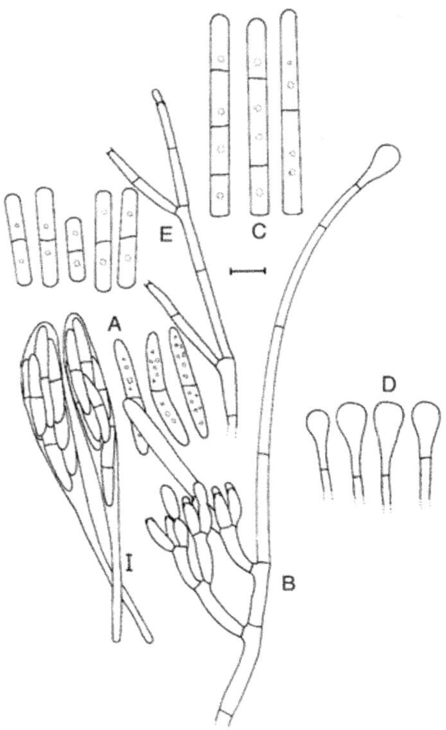

Fig. 17.11. (A) Asci and ascospores of *Calonectria ilicicola*, and (B) macroconidiophore, (C) macroconidia, (D) macrovesicles and (E) microconidiophore and microconidia of its anamorph, *Cylindrocladium parasiticum*. Bars = 10 μm (from Crous, 2002).

reported only in Hawaii and India, but non-identified species in the genus cause papaya stem cankers in Brazil and Mexico (Gupta and Nema, 1979; Saldana *et al.*, 1985; Alvarez and Nishijima, 1987). *D. theobromae* causes numerous diseases throughout the tropics and subtropics. Although postharvest stem-end rot and surface rot of papaya fruit are most common, young, developing fruit in the field can also be affected.

The overall symptoms of fruit and stem-end rot resemble those caused by black rot, but can be distinguished by its wider, more extensive, translucent lesion margin. *D. theobromae* is fast growing and can rot the entire fruit. Grey mycelium develops over the affected area and later turns black as masses of pycnidia form (Plate 101). Flesh becomes bluish black in the soft, water-soaked areas. Air pockets often form in affected tissue, presumably caused by the flesh shrinking, and later become filled with grey mycelium. Sporulating lesions are black and have a rough surface caused by erumpent, confluent pycnidia in stromata.

Stem-end rots caused by *D. theobromae* have been most severe on fruit that are vapour heated for fruit fly quarantine purposes (Alvarez and Nishijima, 1987). No specific control measures have been developed, but field sprays with protective fungicides should reduce field inoculum levels and disease incidence (Alvarez *et al.*, 1977). Hot water, postharvest dips (48°C for 20 min) are also effective.

Phytophthora fruit, root and stem rot

This disease can cause significant losses of fruit and plants during rainy periods (Snowdon, 1990). Damage is especially severe after hurricanes (Hamill, 1987). The disease is found worldwide and can develop whenever favourable environmental conditions occur.

SYMPTOMS Ripe fruits are most susceptible because concentrations of the proteolytic enzyme papain decline during ripening (Hine *et al.*, 1965a). Immature fruit that are infected exhibit water-soaked lesions that ooze latex (Plate 102). If favourable conditions prevail, the disease continues to develop and fruit wither. Mature fruit become covered with off-white mycelium of the pathogen that produces masses of sporangia (Hunter and Buddenhagen, 1969). On the ground, such fruit are important reservoirs of inoculum for root infection. During periods of heavy rainfall, entire columns of fruit can be lost. In addition, apical portions of the fruit-bearing stem can develop water-soaked lesions in association with leaf scars during rainy periods that can weaken and collapse the upper stem.

Seedlings are especially vulnerable to root rot, and plants generally become more resistant as they age (Erwin and Ribiero, 1996). However, in saturated soils, plants of all ages can be affected. As the disease progresses, the entire root system darkens and disintegrates

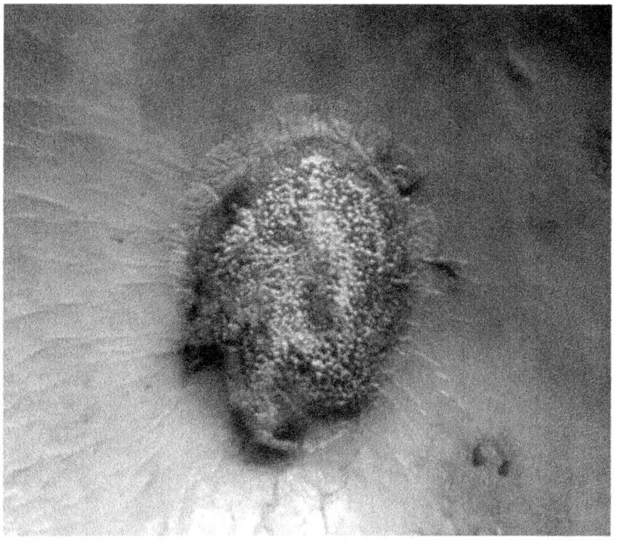

Fig. 17.12. Circular, depressed lesion on papaya fruit caused by Fusarium fruit rot. Note sporulation of the causal fungus (photo: W.T. Nishijima).

(Fig. 17.13), and plants are stunted with chlorotic, wilted leaves (Fig. 17.14). Before the plant dies, only a few small leaves may remain at the apex. Root rot development stops when soils dry, but the root systems of affected plants may be sufficiently damaged to result in lodging if fruit set is high.

CAUSAL AGENTS *Phytophthora palmivora* is the most important causal agent (Erwin and Ribiero, 1996). It previously was misidentified as *P. parasitica* (*P. nicotianae*) (Parris, 1942; Hunter and Buddenhagen, 1969). *P. cinnamomi* has been reported to cause root rot in Peru (Bazán de Segura, 1951), and *P. capsici* has also

Fig. 17.13. Root rot on papaya plants caused by (from the left) *Phytophthora palmivora*, *P. nicotianae* and *Pythium* sp. (photo: A.W. Cooke).

Fig. 17.14. Wilting and chlorosis of papaya plants caused by root rot (photo: A.W. Cooke).

been reported to cause papaya fruit rot in Hawaii (Ko, 1994). The latter isolates were described recently as a new species, *P. tropicalis* (Aragaki and Uchida, 2001). *P. palmivora* and *P. cinnamomi* are described in Chapter 1, and *P. tropicalis* is described in Chapter 2.

EPIDEMIOLOGY *P. palmivora* survives in soil primarily as chlamydospores (Hunter and Buddenhagen, 1969). Oospores that can survive for long periods are uncommon since both mating types of this heterothallic pathogen seldom occur in the same location. Sporangia and zoospores are sensitive to dry conditions and survive poorly.

The cardinal temperatures for *P. palmivora* are 12, 30 and 36°C, and production of sporangia is greatest at 25°C. Disease development is most severe when excessive soil moisture or rainfall and temperatures between 20 and 30°C are present.

Saturated soils enable zoospores to swim towards infection sites on the root surface and may also increase the susceptibility of the host. Wind-blown rain is essential for primary infections and epidemic development of fruit rot (Hunter and Kunimoto, 1974). Rainfall disperses sporangia from infected fruit and soil and, in concert with wind, can move inoculum considerable distances within and among plantings. High numbers of sporangia are produced on the surface of affected fruit, and they are prime causes of disease outbreaks (Hunter and Buddenhagen, 1969). Since wounding is required for stem infections, they often develop after hurricanes (Hamill, 1987). Chlamydospores formed in fallen fruit survive in soil and serve as the main source of inoculum for infection of roots of papaya seedlings in subsequent plantings.

MANAGEMENT Cultural and chemical measures can be used to manage these diseases (Erwin and Ribiero, 1996). Plantings should be established in soils in which papaya and other hosts of *P. palmivora* have not been grown. Nursery plants should be grown in steamed or fumigated soil, and care should be taken to ensure that they remain free of the pathogen before they are moved to the field. In infested soils, drainage within orchards should be maximized and affected fruit on plants and on the ground should be removed. In northern Australia, establishing plants on raised mounds and amending soil with sawdust plus urea significantly reduced root rot (Vawdrey *et al.*, 2002). Where volcanic soils are infested with the pathogen in Hawaii, a virgin soil technique, in which clean soil is used to fill 30-cm-diameter planting holes to a depth of 10 cm and a height of 4 cm, has been effective since plants are generally resistant to the pathogen by the time they extend into infested soil (Ko, 1987).

Preventive fungicides in the field and with hot water dips after harvest can control Phytophthora fruit rot of papaya (Aragaki *et al.*, 1981; Alvarez and Nelson, 1982). In all but the wettest areas, 'Sunrise Solo' is resistant to anthracnose and chocolate spot, but is highly susceptible to Phytophthora blight (Alvarez and Nishijima, 1987).

Powdery mildew

Powdery mildew of papaya, first described in 1898 from Brazil, occurs in most countries where papaya is grown. Plants of all ages are affected, with young plants being particularly susceptible.

SYMPTOMS The disease mainly occurs on the younger parts of the plant. The young crown leaves develop yellowish green patches, producing a mosaic-like pattern. In the early stages, the lower leaf surfaces are speckled with small, water-soaked dots that later coalesce to produce larger, water-soaked areas. Under humid conditions, these areas become covered with a white, powdery mycelium that consists of short chains of oval conidia standing erect from the mycelium (Fig. 17.15A). The mycelium is often abundant adjacent to leaf veins and also develops on petioles, flower pedicels and young fruit. As the disease progresses, the yellow patches may turn necrotic, giving the leaf a scorched appearance (Simmonds, 1965).

Circular, white areas of mycelium occur on the surface of young fruit and may coalesce to cover much of the fruit surface. As fruit develop, the mycelium disappears but leaves light grey, scarred areas on the surface; the growth of underlying areas is checked, causing malformation of the mature fruit (Fig. 17.15B). This scarring remains evident on ripe fruit, causing downgrading at harvest (Simmonds, 1965; Peterson *et al.*, 1993).

CAUSAL ORGANISMS The fungus *Sphaerotheca caricae-papayae* (anamorph: *Oidium caricae*) causes powdery mildew, although the disease has been ascribed to *Sphaerotheca humuli* in Australia (Clare, 1964; Simmonds, 1965; Farr *et al.*, 1989). The mycelium of *O. caricae* is hyaline and septate, with haustoria developing in the epidermal cells of the host. Conidia are hyaline and granular and borne in chains of at least 3–5 conidia.

Another form of powdery mildew, often termed angular leaf spot, occurs mainly on the older leaves and is caused by *Oidiopsis taurica* (Simmonds, 1965).

EPIDEMIOLOGY Papaya is the only known host of *S. caricae-papayae*. The fungus is dispersed by wind-borne conidia, and disease development is usually favoured by low light levels, high humidity, moderate temperatures and moderate rainfall (Rawal, 1987; Ooka, 1994). In Australia, however, powdery mildew is most prevalent during the cooler, usually dry winter months and low to moderate humidity (Simmonds, 1965).

MANAGEMENT Fungicides containing sulphur are effective, although some can cause phytotoxicity when applied during hot weather. Several systemic fungicides provide good control and are recommended for severe outbreaks, particularly on young plants.

Root rot and damping-off

Damping-off kills seedlings in the nursery and root rot can kill plants in the field. These problems can be serious, especially when pythiaceous pathogens are involved in wet areas and when sequential crops of papaya are grown in the same soil (Figs 17.13 and 17.14). General aboveground symptoms of root rot include foliar chlorosis, a reduction in leaf size and number, and wilting.

Several different agents are involved. The most important of these is *Phytophthora palmivora*. Other agents include *Glomerella cingulata* (anamorph: *Colletotrichum gloeosporioides*), *Fusarium* spp., *Ph. nicotianae*, *Pythium aphanidermatum*, *Py. ultimum*, *Pythium* sp., and anastomosis group (AG) 4 of *Thanatephorus cucumeris* (anamorph:

Fig. 17.15. Symptoms of powdery mildew on a papaya (A) leaf and (B) fruit caused by *Sphaerotheca humuli* in Australia. Note that signs of the fungus have largely disappeared from the fruit to leave a scarred surface. Powdery mildew is caused by *Sphaerotheca caricae-papayae* elsewhere (photos: A.W. Cooke).

Rhizoctonia solani) (Da Costa, 1944; Trujillo and Hine, 1965; Yamamoto and Aragaki, 1982; Nishijima, 1999). *G. cingulata*, *Ph. palmivora* and *Ph. nicotianae* are described in Chapter 1, and *T. cucumeris* in Chapter 12.

Soil or potting mix in the nursery must be pathogen free. Production fields should be well drained and either not have been previously used for papaya production or fumigated prior to use. Where *Ph. palmivora* is the primary concern, Ko's (1987) virgin soil technique can be used.

Soft rot

This disease, which is also called transit rot or Rhizopus soft rot, occurs throughout the world on a wide range of fruit and vegetable crops. As the name transit rot implies, the disease is important during the storage, transport and marketing of produce.

SYMPTOMS Affected fruit first show circular, water-soaked areas and the skin is easily displaced from the flesh. The disease develops very rapidly into a soft, wet rot covered by a cottony white mycelium in which small white, then black, sporangia are produced (Plate 103). The mycelium extends to the surface of healthy parts of the fruit and sometimes to the surface of the packing container. Soft rotting is caused when enzymes produced by the pathogen break down and dissolve cell walls, causing collapse and leakage of cell contents, and a soft decay. The released juices often have a fermented or acidic odour.

CAUSAL AGENT *Rhizopus stolonifer* causes soft rot. The fungus produces white, cottony then brownish black colonies on potato dextrose agar (PDA) (Sarbhoy, 1966). The fungus spreads rapidly by a mass of stolons fixed at various points to the substrate by rhizoids. A description of the fungus can be found in Chapter 5.

EPIDEMIOLOGY The fungus is common on dead and decaying plant material. Spores are abundant in the atmosphere and invade mature fruit, usually through wounds. Infection and disease development are favoured by warm, moist conditions. Spore germination can occur in condensate formed on fruit after removal from cold rooms. Once established, soft rot can spread rapidly from fruit to fruit, and an entire carton of fruit can be rotted within a few days (Alvarez and Nishijima, 1987; Persley, 1993).

MANAGEMENT Effective management involves strict hygiene to reduce inoculum, careful handling of fruit to minimize wounding, and cool storage. Postharvest fungicide treatments also provide control.

Wet fruit rot

This disease affects fruit and senescing petioles (Nishijima, 1994d). Its incidence is sporadic, but damage can be severe.

SYMPTOMS Affected tissues become wrinkled, translucent and light green to yellow (Fig. 17.16). A band of water-soaked tissue rapidly advances from the infection site to the seed cavity, and the affected portion often can be lifted free from the rest of the fruit. Pycnidia usually form on the fruit surface of advanced infections (Alvarez and Nishijima, 1987). Affected tissue is soft, mushy and wet, but, unlike tissues affected by Rhizopus soft rot, seldom produces exudates (Nishijima, 1994d). Affected areas become covered with a white to grey mycelium under very humid conditions.

Fig. 17.16. Phomopsis wet rot on a papaya fruit (photo: W.T. Nishijima).

CAUSAL AGENT Wet fruit rot is caused by *Phomopsis caricae-papayae* in Australia and India, and by *Phomopsis* sp. elsewhere (Simmonds, 1965; Snowdon, 1990). *P. caricae-papayae* produces α-conidia that are hyaline, fusiform, single-celled, 5–7 × 2–2.5 μm, and with a guttulate at either end (Fig. 17.17) (Punithalingham, 1985). β-Conidia are not known from host tissue but on agar are 20–27 × 0.5 μm and curved at the apex. α-Conidia of *Phomopsis* sp. are similar, but larger (6.4–8 × 2.7–3.1 μm), and β-conidia (13.7–20 × 1–1.8 μm) are formed on the host.

EPIDEMIOLOGY The pathogen sporulates abundantly on senescing petioles on papaya plants. Conidia are discharged during rainy periods and deposited on fruit surfaces. The conidia are unable to penetrate intact skin and can only initiate infection through

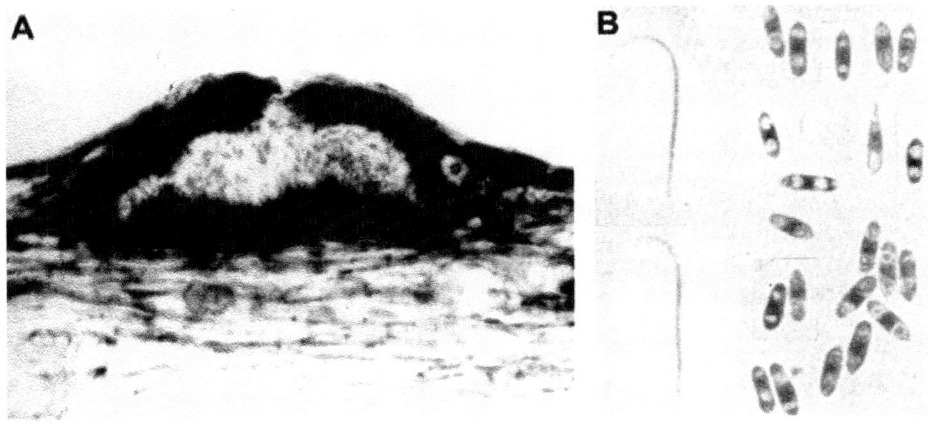

Fig. 17.17. (A) Partially erumpent pycnidial conidioma and (B) β- and α-conidia of *Phomopsis caricae-papayae* (from CMI description No. 827).

wounds such as the broken pedicel at harvest or punctures or abrasions on the fruit surfaces, including those caused by other fungal pathogens (Nishijima, 1994d). Wet rot usually develops only on fully ripened fruit.

MANAGEMENT Good plantation hygiene, including the removal of dead leaves and petioles, combined with a regular fungicide programme reduces inoculum in plantations. Careful harvesting to minimize damage to fruit, and postharvest hot water and fungicide treatments minimize losses from wet rot.

Diseases that are Caused by Nematodes

Reniform nematodes

Reniform nematodes are the most damaging on papaya (McSorley, 1981; Cohn and Duncan, 1990). The adult females of these pests are sedentary and embedded in the root cortex. 'Reniform' refers to their kidney-shaped posterior that protrudes from the root surface.

Rotylenchulus reniformis is the predominant nematode on papaya, and another species, *R. parvus*, has been reported in Kenya (McSorley, 1981; Cohn and Duncan, 1990). They cause leaf chlorosis, stunting and wilting aboveground that cannot be distinguished from damage caused by many other soilborne pathogens. Fruit are smaller than normal and may have a bland taste. Belowground symptoms are subtle but, with a hand lens, egg masses of the nematode that resemble grains of sand are evident on the root surface.

Juveniles initially are <0.5 mm long. After females mature, they penetrate the root cortex and become sedentary (Sivakumar and Seshadri, 1972). Feeding that occurs near the phloem induces giant cells that are sinks for nutrients. Feeding also occurs in the root cortex, damaging the root and providing sites of entry for fungi. The nematode's life cycle is completed in 25 days.

Reniform nematodes spread via cultivation and moving water. They have a wide host range, and are particularly important when papaya is planted after pineapple (Hine *et al.*, 1965b). Infested sites and infected seedlings should be avoided. Preplant soil fumigation reduces nematode populations and increases crop vigour and yields (Lange, 1960; Ayala *et al.*, 1971). Unfortunately, fumigants and effective nematicides are either phytotoxic, no longer available or are losing registration.

Root-knot nematodes

Root-knot nematodes may cause severe galling of roots, stunting of plants, and death of seedlings in sandy soils (Hine *et al.*, 1965; McSorley, 1981; Cohn and Duncan, 1990). Aboveground symptoms are similar to those described for the reniform nematode. Papaya roots attacked by root-knot nematode show varying degrees of galling depending on the severity of attack (Fig. 17.18). Since the female and her egg mass are usually completely embedded in the root, roots must be dissected in order to make a positive identification. When root terminals are infected, no further elongation occurs.

Two species predominate on papaya, *Meloidogyne incognita* and *M. javanica*, and, due to their adaptation to cooler temperatures two others, *M. arenaria* and *M. hapla*, are involved at high elevations (Cohn and Duncan, 1990). Root penetration by these sedentary pests usually occurs near the tip. As the female juvenile feeds in the root stele, giant cells are formed, resulting in the characteristic, globular galls for which these nematodes are named. The sedentary female moults several times until she is pear shaped. Although single nematodes cause little or no root swelling, multiple infections in an area can result in large galls. A single female can produce 350 eggs, and as many as 14–17 generations year^{-1} occur in the subtropics and tropics.

Pre-plant fumigation is effective (Lange, 1960; Ayala *et al.*, 1971). However, heavily galled roots must decompose to the extent that penetration of the fumigant is possible.

Diseases that are Caused by Phytoplasmas

Diseases caused by phytoplasmas (previously mycoplasma-like organisms or MLOs) cause serious crop losses in papaya. The most important is papaya dieback, a problem that has caused sporadic, but significant mortality in Australia for many years (Glennie and Chapman, 1976).

Papaya dieback

Papaya dieback has been recognized in Australia since 1922 and causes annual losses of between 5 and 100% (Glennie and Chapman, 1976). Dieback is the major constraint to successful papaya production in Queensland. The disease does not appear to occur elsewhere, although a decline disease of papaya in the Jordan Valley with similar symptoms was associated inconsistently with a mollicute (Franck and Bar-Joseph, 1992).

SYMPTOMS The initial symptom is a bunched appearance of the inner crown

Fig. 17.18. Extensive galling on a papaya root system caused by root-knot nematodes (photo: K. Pernezny).

leaves that become slightly chlorotic. This bunching is due to a shortening of the petioles and an inward curving of the lamina. One or more of the young leaves begin to shrivel and die, with the necrosis extending down the petiole to the top of the stem (Fig. 17.19A). A dark, water-soaked spot develops at this point, later turning brown and drying out to form a hard, superficial scab. The first one or two leaves to wilt usually develop areas with a water-soaked, etched pattern. The larger crown leaves become chlorotic and then necrotic, and the entire crown can die within 1–4 weeks, with the stem gradually dying back from the top. As the disease progresses, the older leaves yellow but remain attached to form a fringe around the dead top (Fig. 17.19B) (Simmonds, 1965). Fruit on affected trees become flabby, drop off and often develop severe fungal spotting. The phloem laticifers die and there is a marked reduction in latex flow (Simmonds, 1965; Harding and Teakle, 1988; Siddique *et al.*, 1998). Young plants invariably die, but older plants often recover if the affected stem is excised to allow side shoots to develop (Simmonds, 1965; Elder *et al.*, 2002).

CAUSAL AGENT Several causes of the disease were suggested (Glennie and Chapman 1976), but were disproven after rigorous examination (Harding *et al.*, 1991; Harding and Teakle, 1993; Aleemullah and Walsh, 1996). A major advance occurred when phytoplasmas were detected consistently in affected plants by the polymerase chain reaction (PCR) amplification of the 16S rRNA and 16S–23S rRNA spacer region (Gibb *et al.*, 1996; Liu *et al.*, 1996). The papaw dieback phytoplasma is a member of the European stolbur group within the aster yellows strain cluster (Gibb *et al.*, 1996). It is very closely related to the

Fig. 17.19. Papaya plants in the (A) mid and (B) terminal stages of dieback (photos: A.W. Cooke).

phytoplasmas that are associated with Australian grapevine yellows and *Phormium* yellow leaf disease in New Zealand. On the basis of phylogenetic analyses of the 16S rRNA genes and 16S–23S spacer regions, it was proposed that these phytoplasmas be designated *'Candidatus* Phytoplasma australiense' (Liefting *et al.*, 1998; White *et al.*, 1998). These phytoplasmas can be differentiated further by analysis of their *tuf* gene sequences (Schneider *et al.*, 1997). The dieback phytoplasma is distinct from the one that is associated with yellow crinkle and mosaic diseases of papaya in Australia, *'Candidatus* Phytoplasma australasia' (White *et al.*, 1998). A model for the pathogenesis of *'Candidatus* Phytoplasma australiense' was proposed by Guthrie *et al.* (2001).

EPIDEMIOLOGY Although dieback occurs throughout the year, there is usually a marked peak in incidence during spring (October–December) and autumn (March–May) (Glennie and Chapman, 1976; Elder *et al.*, 2002). Vector species have not been identified for this disease, but it is likely that sap-sucking leafhoppers or planthoppers are involved since they transmit other phytoplasmas and because papaya plants remain disease free when grown inside insect-proof netting in plots that are severely affected by dieback (Elder *et al.*, 2002).

It is thought that the phytoplasma resides in alternative hosts in bushland and is moved into plantations by leafhoppers that arrive via weather events. Major epidemics occur in years when dry conditions in late winter and early spring favour the immigration of insects from unattractive dry vegetation (Elder *et al.*, 2002). The insects do not remain on papaya plants for long since it is not a favoured host.

MANAGEMENT Effective management is difficult because of the rapid onset and spread of dieback. Ratooning plants by cutting them ~75 cm above ground level as soon as symptoms develop results in healthy side shoots and can save many plants in a plantation. These side shoots generally are free of detectable phytoplasmas although plants are susceptible to further dieback episodes (Guthrie *et al.*, 1998). Other measures include protecting plants with insect netting and the application of systemic insecticides to protect plants during times of expected dieback activity. Control through plant transformation warrants investigation using, for example, antibody fragments specific to the dieback phytoplasma.

Yellow crinkle and mosaic

Sporadic outbreaks of mosaic and, particularly, yellow crinkle can cause serious losses in plantations in Australia. In northern Australia, these diseases are more important than dieback (Padovan and Gibb, 2001). Yellow crinkle and mosaic are covered here together, as current evidence indicates that both are associated with the same phytoplasma (Gibb *et al.*, 1996; White *et al.*, 1998; De La Rue *et al.*, 1999).

SYMPTOMS The first symptom of yellow crinkle disease is a pronounced yellowing of leaves about halfway up the canopy, giving a yellow ring appearance to plants (Plate 104). The chlorosis is accompanied by a downward bending of the petioles. The crown leaves develop thin, translucent areas along the margin and between the main veins. These areas become necrotic and tattered as growth proceeds, giving a ragged appearance to the leaf (Simmonds, 1965). The crown leaves eventually become claw like in appearance (Fig. 17.20), and the older leaves dry and fall, leaving a bare stem with a few stunted leaves at the top. Plants may remain in this condition for many months, making no growth and eventually dying. Floral parts develop phyllody and proliferation of axillary leaves, which are strap like and thickened.

Plants affected by mosaic develop several symptoms, although not all may be present at any one time. Affected plants develop stunted, chlorotic young leaves with translucent areas around the margins that may wither to give a ragged effect. The petioles and upper portions of the stem develop narrow, water-soaked streaks, and latex flow is much reduced or absent. Affected plants

Fig. 17.20. Apex of a papaya plant severely affected by yellow crinkle (photo: A.W. Cooke).

often are stunted, producing multiple side shoots which often then develop typical mosaic symptoms. The younger fruit on plants that have been affected for some time often show characteristic light green areas, sharply delimited from the darker green normal areas and extending out from the stem end. Latex is absent from the light areas but occurs in normal quantities outside these areas (Simmonds, 1965).

If often is difficult to differentiate yellow crinkle from mosaic disease, but the water-soaked streaks on the petioles and upper portion of the stem and the reduced or absent latex flow are characteristics of mosaic. A distinguishing feature of yellow crinkle is phyllody, which has not been observed on plants with mosaic disease (Peterson *et al.*, 1993).

CAUSAL AGENT Although viruses were once considered to be the causal agents (Simmonds, 1965), phytoplasmas were detected in the phloem of plants affected by yellow crinkle (Gowanlock *et al.*, 1976). Transmission studies using dodder demonstrated that the agent is very similar or identical to that causing tomato big bud (TBB) disease (Greber, 1966), which is now known to be caused by a phytoplasma (Davis *et al.*, 1997). A consistent association between phytoplasma and yellow crinkle and mosaic has been demonstrated using PCR and phytoplasma-specific primers (Gibb *et al.*, 1996; Liu *et al.*, 1996). Based on DNA sequence analyses, the yellow crinkle- and mosaic-associated phytoplasmas are indistinguishable and have been included in the taxon '*Candidatus* Phytoplasma australasia' (White, 1998), which is distinct from '*Candidatus* Phytoplasma australiense' the agent that is associated with dieback disease (Gibb *et al.*, 1998).

EPIDEMIOLOGY It is probable that the phytoplasma associated with yellow crinkle and mosaic is transmitted by phloem-feeding leafhoppers or planthoppers since both diseases do not develop under insect proof netting (Elder *et al.*, 2002), and the tomato big bud phytoplasma is transmitted by a leafhopper, *Orosius argentatus* (Hill, 1943).

In a 3-year study in northern Australia, Padovan and Gibb (2001) reported that five strains of phytoplasma were found in plants with yellow crinkle, and that the TBB and sweet potato little leaf (SPLL) strains of phytoplasma were found in 94% of these plants. There was a significant, but not complete, correlation between phyllody and TBB and virescence and SPLL, and mixed infections were not detected. The phytoplasmas were rarely detected in leafhoppers and other plant species in the area.

Outbreaks of yellow crinkle usually occur following hot, dry weather that causes vectors to migrate from dry vegetation into papaya plantations (Peterson *et al.*, 1993; Elder *et al.*, 2002). No consistent pattern was observed for mosaic (Elder *et al.*, 2002).

MANAGEMENT Ratooning diseased plants is of little benefit as the new shoots almost always develop disease symptoms. Thus, affected plants should be removed from plantations (Simmonds, 1965). Various forms of netting should protect plants from potential vectors.

Diseases that are Caused by Viruses

Viruses cause increasing losses in many countries. In particular *Papaya ringspot virus* (PRSV) is a major pathogen in most production areas.

Leaf curl disease

Leaf curl was first reported from papaya in India in the 1930s (Nariani, 1956). Its incidence has increased in recent years (Saxena *et al.*, 1998a), which is almost certainly due to the wide distribution and activity of the whitefly vector, *Bemisia tabaci*.

SYMPTOMS The disease is characterized by downward curling and cupping of leaves, with vein clearing and thickening. Enations develop in the form of frills on green veins. Affected leaves become brittle and leathery, and petioles twist in a ziz-zag manner (Nariani *et al.*, 1956; Saxena *et al.*, 1998a). Affected plants often fail to flower, and the few fruits that are set are distorted and often fall prematurely. In advanced stages, defoliation occurs and growth ceases (Nariani, 1956).

CAUSAL AGENT *Papaya leaf curl virus* is a species in the genus *Begomovirus* in the family *Geminiviridae* (Saxena *et al.*, 1998b; van Regenmortel *et al.*, 2000). Begomoviruses have a genome consisting of two components, each of which are 2.5–2.8 kb in size and required for infectivity. The nucleotide sequences of the two DNAs are very different, except for a 200 nucleotide non-coding region that is almost identical (Hull, 2002).

The virus is transmitted by *B. tabaci* in a persistent, circulative manner (Nariani, 1956; Hull, 2002). The virus is not transmitted mechanically but has been transmitted via the whitefly vector to tobacco and tomato (Nariani, 1956).

Meleira disease

'Meleira' or sticky disease is a major disease of papaya in several states of Brazil (Kitajima *et al.*, 1993; Barbosa *et al.*, 1999). The disease is characterized by an intense exudation of watery latex from fruit and the leaves of young plants (Rezende and Costa, 1993). Affected fruit becomes stained as the latex darkens by oxidation. The aetiology of the disease has not been fully resolved but appears to be associated with the presence of isometric, 50 nm in diameter, virions in the laticifers (Kitajima *et al.*, 1993). Double-stranded RNA with an estimated size of 6×10^6 kDa has been isolated from the roots, flowers, leaves and young fruit of infected plants (Rezende and Costa, 1993; Barbosa *et al.*, 1999).

Papaya droopy necrosis and papaya apical necrosis

These diseases cause similar symptoms on papaya and both are associated with infection by rhabdoviruses (Zettler and Wan, 1994). Papaya droopy necrosis (PDN) has been reported from Florida (Wan and Conover, 1981) and papaya apical necrosis (PAN) occurs in Venezuela (Lastra and Quintero, 1981). In Venezuela, apical necrosis virtually eliminated papaya production in the state of Zulia in 1979, whereas droopy necrosis destroyed all experimental fields of papaya at the University of Florida in Homestead where old and new plantings overlapped (Zettler and Wan, 1994).

SYMPTOMS The symptoms of both diseases are more severe during cool winter months. The initial symptom of droopy necrosis is a drooping and recurvature of the crown leaves, sometimes associated with marginal necrosis. The apical leaves are pale yellow, do not expand normally and are sharply recurved. Petioles are shortened and the crown is rounded, giving plants a distinct bunchy appearance. During the winter months, leaves abscise, the stem tip becomes necrotic and the plant eventually dies (Wan and Conover, 1981).

Plants affected by apical necrosis become chlorotic with a rapid wilting and decline of the apex, followed by death (Lastra and Quintero, 1981).

CAUSAL AGENTS Both diseases are associated with rhabdoviruses in the nuclei of vascular parenchyma cells (Zettler and Wan, 1994). The sizes of the virions associated with both diseases are similar: PDN virions are 87–98 × 180–254 nm and those of PAN are 80–84 × 210–230 nm (Lastra and Quintero, 1981; Wan and Conover, 1981). Neither virus is mechanically transmissible.

EPIDEMIOLOGY AND MANAGEMENT The vector of PDN has not been identified. Symptoms of PAN developed in eight of ten seedling papaya plants exposed to *Empoasca papayae* leafhoppers that had fed on virus-infected plants; however, rhabdovirus particles were not detected in these plants by electron microscopy (Lastra and Quintero, 1981).

The incidence of PDN in Florida was reduced considerably by avoiding overlapping papaya crops and by prompt removal of infected plants (Zettler and Wan, 1994).

Papaya leaf distortion mosaic

Papaya leaf distortion mosaic is the most prevalent virus disease of papaya in Japan and is also present in Saipan and Taiwan (Maoka *et al.*, 1995; Chen *et al.*, 2002). The disease was first reported from Okinawa and originally was thought to be papaya ringspot (Kawano and Yonaha, 1992).

SYMPTOMS The symptoms on papaya vary according to virus strain, cultivar, age of plants and environment. The crowns of infected plants have a rosette appearance with small leaves and short petioles. Leaves develop a mosaic pattern, yellow mottling and become distorted (Kawano and Yonaha, 1992). Dark green or water-soaked streaks and spots develop on the stem and petioles. Stem diameter is often less than half that of healthy plants (Kawano and Yonaha, 1992). Chlorotic spots with green centres develop on young fruit and frequently become swollen and surrounded by a dark green or brown ring. Symptoms are more severe during cool weather and when plants are stressed, for example when growing in dry, infertile soils.

The rosettes of leaves and slender petioles on the crowns of infected plants and the swellings around the ringspots on fruit help distinguish this disease from papaya ringspot.

CAUSAL AGENT The disease is caused by *Papaya leaf distortion mosaic virus* (PLDMV) in the family *Potyviridae*. It is a distinct species in the family with a coat protein amino acid sequence that is 49–59% similar to that of PRSV and other potyviruses (Maoka *et al.*, 1996).

PLDMV can be detected, and distinguished from PRSV, by enzyme-linked immunosorbent assay (ELISA) and reverse transcription (RT)–PCR (Kawano and Yonaha, 1992; Chen *et al.*, 2002; Maoka *et al.*, 2002).

EPIDEMIOLOGY AND MANAGEMENT The virus is spread by aphids in a non-persistent manner. There is no evidence for seed transmission. Papaya is the only known natural host, although several cucurbit species were infected after manual inoculation (Kawano and Yonaha, 1992).

The potential for rapid spread of the virus by aphids has made control difficult, and no resistant cultivars have been reported (Kawano and Yonaha, 1992). Unfortunately, PRSV-resistant transgenic lines in Taiwan are affected by PLDMV (Chen *et al.*, 2002)

Papaya lethal yellowing disease

A disease causing chlorosis, mottling, leaf curl, leaf abscission and rapid decline of papaya plants, particularly those in the

Solo group, occurs in Pernambuco and neighbouring states of Brazil (Loreto *et al.*, 1983; Silva *et al.*, 1997). Disease incidence has reached 40% in plantations. The virus causing the disease, *Papaya lethal yellowing virus* (PLYV), has isometric virions ~30 nm in diameter and reaches a high concentration in infected plants. The virions have an RNA of ~4.8 kb and a coat protein with an estimated molecular weight of 36 kDa (Silva *et al.*, 1997). The virus is readily detected in papaya leaves by ELISA (Silva *et al.*, 1997). The amino acid sequence of the coat protein of PLYV is 58.3–83.3% similar to that of isolates of *Tomato bushy stunt virus*, the type species of the *Tombusvirus* genus (Silva *et al.*, 1997).

The virus is transmitted mechanically to papaya, the only known host. It has also been isolated from the testa of seeds collected from fruits from infected plants and can survive in soils and irrigation water (Comarco *et al.*, 1998). Although the field distribution of the disease suggests an insect vector, none has been identified (Silva *et al.*, 1997).

Papaya mild yellowing disease

A virus causing mild interveinal chlorosis in both young and mature leaves of papaya has been described from Venezuela and named *Papaya mild yellowing virus* (PMYV) (Marys *et al.*, 1995). The supercoiled filamentous particles with lengths of 400–700 nm resemble the morphology of tenuiviruses, although a definitive link to this genus has not been established (Marys *et al.*, 1995). The virus can be transmitted mechanically to papaya and to several cucurbit species in which it causes mild mottling and vein clearing (Marys *et al.*, 1995). In a recent survey, PMYV was the most common virus found in papaya in Venezuela, often occurring as mixed infections with PRSV (Marys *et al.*, 2000).

Papaya mosaic

Papaya mosaic was first described by Conover (1964). The disease occurs in several countries including the USA (Florida), Mexico and some South American countries (Cook, 1972; Brunt *et al.*, 1996). It is usually of minor importance.

SYMPTOMS Infected plants develop leaf mottling, which is most evident in young leaves. In younger plants, vein clearing occurs in new growth that later becomes rugose and reduced in size. Unlike papaya ringspot, symptoms of papaya mosaic do not usually occur on stems, petioles and fruits (Cook, 1972; McMillan *et al.*, 1993).

CAUSAL AGENT Papaya mosaic is caused by *Papaya mosaic virus* (PapMV), a member of the *Potexvirus* genus. It has also been called *Papaya mild mosaic virus*. It has filamentous virions, 470–580 nm long and 13 nm wide, and a single-stranded positive sense RNA genome of 6.65 kb (Brunt *et al.*, 2000). The virus can be identified by virion morphology, serology and molecular tests (e.g. PCR) (Abou Haidar and Gellatly, 1999).

EPIDEMIOLOGY The natural host range of the papaya-infecting strain of PapMV appears to be restricted to papaya and cucurbits (Purcifull and Hiebert, 1971; Noa-Carrazana and Silva-Rosales, 2001). PapMV is readily sap transmissible and can be spread via plant-to-plant contact. Insect vectors do not spread the virus and there are no reports of seed transmission (Brunt *et al.*, 1996). Experimental hosts include several *Carica* species, *Citrullus vulgaris*, *Cucumis melo*, *Sesamum indicum* and *Zinnia elegans* (Cook, 1972). *Gomphrena globosa* develops local lesions on inoculated leaves and is a useful diagnostic host (Purcifull and Hiebert, 1971).

MANAGEMENT Since the virus is easily spread by contact, care should be taken to prevent transmission in the nursery and field with contaminated implements and handling of plants.

Papaya ringspot

Papaya ringspot is a widespread and destructive disease in most countries where papaya is grown. The disease, caused by *Papaya ringspot virus*-type P (PRSV-P), severely limits production in many countries of Southeast Asia (the Philippines, Taiwan, Thailand, Vietnam), Africa, India, South America (Brazil, Venezuela), the Caribbean islands, Mexico and the USA (Hawaii, Florida, Texas) (Purcifull *et al.*, 1984; Opina, 1986; Yeh *et al.*, 1988; Teliz *et al.*, 1991; Rezende and Costa, 1993; Gonsalves, 1994; Maoka *et al.*, 1995; Jain *et al.*, 1998; Marys *et al.*, 2000; Ndunguru and Rajabu, 2002). The disease has a restricted distribution in Queensland, Australia (Thomas and Dodman, 1993) and has not been recorded from South Africa and several island nations of the South Pacific (Kiritani and Su Hong, 1999).

Papaya ringspot was first described by Jensen (1949), and in earlier literature was sometimes called papaya distortion ringspot and papaya mosaic. The latter should not be confused with papaya mosaic that is caused by *Papaya mosaic virus*.

SYMPTOMS Plants of all ages are susceptible, and symptoms are generally more severe during cooler weather. The disease derives its name from the characteristic dark green sunken rings that develop on fruit of affected plants (Fig. 17.21) (Jensen, 1949). These rings often persist as dark orange to brown mark-

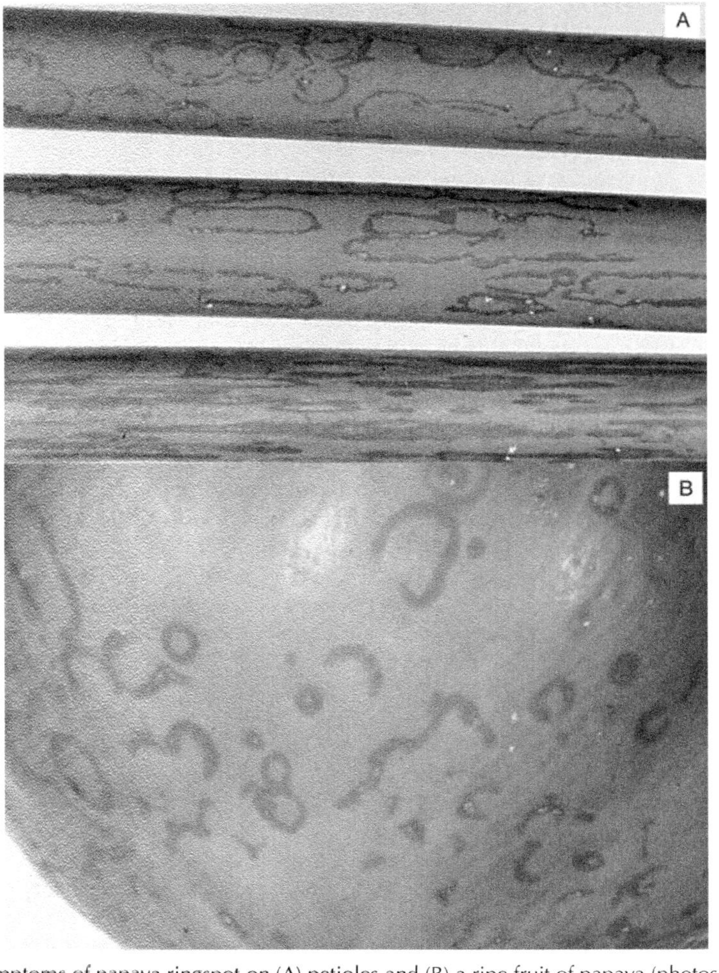

Fig. 17.21. Symptoms of papaya ringspot on (A) petioles and (B) a ripe fruit of papaya (photos: A.W. Cooke).

ings as fruit mature. Dark green, water-soaked streaks develop on petioles and stems. Mottle and mosaic patterns of varying severity develop on leaves that often have a ruffled appearance. One or more leaf lobes may become severely distorted and reduced in size, giving a shoestring appearance (Plate 105). Affected plants become stunted and fruit set is markedly reduced or absent. Fruit from affected plants have poor flavour, a leathery appearance and are predisposed to fungal fruit rots (Cook, 1972; Peterson *et al.*, 1993; Gonsalves, 1994).

CAUSAL AGENT PRSV is a species in the genus *Potyvirus* in the family *Potyviridae* (van Regenmortel *et al.*, 2000). The virus occurs as two closely related types that are serologically indistinguishable, but can be separated by host range (Purcifull *et al.*, 1984). PRSV-P affects papaya and cucurbits and is the causal agent of papaya ringspot disease. PRSV-W affects cucurbits but not papaya. The latter type previously was called *Watermelon mosaic virus*-1 (WMV-1) and causes important diseases of cucurbits worldwide (Purcifull *et al.*, 1984). PRSV is transmitted in a non-persistent manner by many aphid species to a narrow range of hosts (Conover, 1964; Teliz *et al.*, 1991). It is not regarded as being seed transmitted in either papaya or cucurbits (Purcifull *et al.*, 1984), although there is one report of seed transmission of PRSV-P in seed of a local papaya cultivar in the Philippines (Bayot *et al.*, 1990).

The virus has flexuous, filamentous virions that are 780 × 12 nm. PRSV has a single type of coat protein of 36 kDa and induces cylindrical (pinwheel) and amorphous inclusions in the cytoplasm of host cells (Gonsalves and Ishii, 1980; Purcifull *et al.*, 1984). Virions have a single-stranded, positive sense monopartite RNA genome, which is 3′ polyadenylated and has a virus protein, VPg, of ~24 kDa covalently linked to the 5′ terminus (Shukla *et al.*, 1994). The genome is monocistronic and acts as mRNA. It is translated as a single polyprotein of ~340–368 kDa which is cleaved post-translationally by proteolytic processing to produce nine individual protein products, two of which, VPg and coat protein (CP), are present in virions.

The complete sequence of the 10,366 nucleotides of the PRSV genome has been determined (Yeh *et al.*, 1992; Wang and Yeh, 1997). The virus is similar in organization to other potyviruses, but has a very long P1 coding region. The nucleotide sequences of coat protein and 3′-untranslated regions indicate that PRSV-P and -W are closely related (Quemada *et al.*, 1990; Bateson and Dale, 1992; Wang and Yeh, 1992). Further studies (Bateson *et al.*, 1994; Wang *et al.*, 1994; Jain *et al.*, 1998; Silva-Rosales *et al.*, 2000; Bateson *et al.*, 2002) have shown that nucleotide and amino acid sequence divergence of up to 14% exists in the coat protein region between isolates obtained from different geographic areas, and that this variation is more closely related to geographic distribution than to strain type. For example, Australian PRSV-P and -W isolates had almost identical coat protein sequences and were more closely related to each other than to isolates from Southeast Asia, leading Bateson *et al.* (1994) to postulate that Australian PRSV-P was derived from Australian PRSV-W.

Virions of PRSV can be detected by direct electron microscopy of sap. The virus can be identified by serological assays (Gonsalves and Ishii, 1980), and by PCR using either virus-specific or degenerate potyvirus group primers (Bateson *et al.*, 1994). Determination of virus type (P or W) requires inoculation of differential host species.

EPIDEMIOLOGY The spread of papaya ringspot is similar to that of many diseases that are caused by non-persistent, aphid-borne potyviruses (Magdalita *et al.*, 1989; Mora-Aguilera *et al.*, 1992; Nelson, 1995). Papaya is the major primary and secondary source of inoculum, and rapid secondary spread can occur, with plantations being totally affected in <6 months after planting (Gonsalves, 1994). This is most probable when young plantations are close to infected plants and when populations of winged aphids are high (Gonsalves, 1994). Although cultivated cucurbits are natural hosts of at least some isolates of PRSV-P (Persley, 1998), they appear to have only a minor role in the epidemiology of the virus in papaya

(Gonsalves, 1998). Transmission of PRSV-P is almost entirely due to transitory aphid populations, as papaya is not a preferred host for aphids and colonies are very rarely found on plants (Mora-Aguilera *et al.*, 1992).

MANAGEMENT Diverse measures have been used to manage papaya ringspot.

Quarantines and rouging. These methods can delay the spread of the disease but do not usually provide long-term control. In Taiwan, quarantine measures were not successful in preventing virus movement from the west coast into new production areas on the east coast (Yeh *et al.*, 1988). PRSV-P was present on the island of Hawaii for >30 years before it finally reached the major production area of Puna in 1992, subsequently devastating commercial production, despite vigorous attempts at eradication and containment (Gonsalves, 1998).

A quarantine area was established in Queensland, Australia, following the detection of PRSV-P in the southeast region in 1991 (Thomas and Dodman, 1993). Restrictions on the movement of papaya and cucurbit plants out of the zone and eradication of severely affected plantations so far has prevented movement of the virus into the major production areas of north and central Queensland (Lines *et al.*, 2002).

Cultural practices. Several cultural practices have proven useful in slowing epidemics and reducing crop damage. Establishing plantations with seedling plants free of PRSV-P is essential, and new plantings should be situated as far as possible from affected plantations (Gonsalves, 1994). Plantations can be surrounded by non-host crops or interplanted with other tree crops. These measures help dilute the amount of virus inoculum reaching papaya, as aphids probe first on non-host species and lose their ability to transmit the virus in the process. Covering crops with aphid-proof netting in Taiwan enabled economic production for at least one season (Yeh *et al.*, 1988).

Cross-protection. Cross-protection is the use of a mild strain of a virus to protect plants against the effects of infection by a severe strain of the same virus (Gonsalves and Garnsey, 1989). Considerable effort has been devoted to developing the concept as an effective and economical control for papaya ringspot (Yeh and Gonsalves, 1994). Mild strains of PRSV-P were developed by nitrous acid mutagenesis from the PRSV-HA strain from Hawaii (Yeh and Gonsalves, 1984), and protection against severe strains was demonstrated in field experiments in Taiwan and Hawaii (Yeh *et al.*, 1988; Yeh and Gonsalves, 1994). Seedling plants manually inoculated with the mild strain of PRSV were established on up to 500 ha of commercial papaya in Taiwan during the 1980s (Yeh *et al.*, 1988) and on the island of Oahu in Hawaii in the 1990s (Yeh and Gonsalves, 1994). Several factors have limited the continued commercial application of cross-protection, including: specificity of the strains that have been used; the limited protection observed on some cultivars, particularly during cool weather; the possible reversion of strains to become virulent; and the costs and high level of management that is required (Yeh and Gonsalves, 1994; Gonsalves, 1998).

Tolerant and resistant cultivars. Resistance has not been found in *Carica papaya* (Mekako and Nakasone, 1975). However, *C. cauliflora*, *C. quercifolia*, *C. pubescens* and *C. stipulata* are highly resistant (Conover, 1964; Horovitz and Jimenez, 1967) and hybridization with *C. papaya* has been achieved with the aid of embryo culture (Horovitz and Jimenez, 1967; Khuspe *et al.*, 1980; Manshardt and Wenslaff, 1989; Magdalita *et al.*, 1998).

Most of the work has involved *C. cauliflora*, which generally has produced hybrids with poor vigour and no useful progeny, with the exception of a single fertile cross reported in India (Khuspe *et al.*, 1980). More success has been achieved with hybrids between papaya and *C. quercifolia* and *C. pubescens*, although overcoming post-zygotic barriers to develop fertile, virus-resistant plants remains a major challenge (Manshardt and Drew, 1998).

The difficulties encountered with hybridization have been at least partly explained by recent work that has shown *C. papaya* to be only distantly related to other *Carica* species (Aradhya *et al.*, 1999; Van Droogenbroeck *et al.*, 2002). A revision of

Carica has placed *C. papaya* in its own genus, *Carica*, and the other species in *Vasconcellea* (Badillo, 2000). These data indicate that intergeneric rather than interspecific hybridization is being attempted and that *C. cauliflora* is among the species most distantly related to *C. papaya* (Kim *et al.*, 2002)

Although resistant papaya germplasm is not available, several tolerant cultivars have been developed. Plants of tolerant cultivars are infected by PRSV-P but generally develop milder leaf symptoms and are able to set reasonable quantities of marketable fruit (Escudero *et al.*, 1994; Gonsalves, 1994). Of these, the most widely used is 'Cariflora', a dioecious cultivar developed in Florida (Conover and Litz, 1978). 'Cariflora' was derived from several dioecious seed line accessions obtained from a commercial plantation in South Florida. Plants were selected in the field following manual inoculation with PRSV-P, and the most tolerant were used in sib-mating and recurrent selection over 4 years, followed by two seasons of polycrossing and recurrent selection (Conover and Litz, 1978; Conover *et al.*, 1986).

Tolerance in 'Cariflora' is polygenic and quantitatively inherited (Conover and Litz, 1978). The cultivar is suited to South Florida and the lowland Caribbean and has also been used as a source of tolerance in breeding programmes in Taiwan to produce Tainung lines, e.g. 'Tainung No. 5' (Yeh and Gonsalves, 1994).

Genetically engineered resistance against papaya ringspot was achieved by Fitch *et al.* (1992). They transformed the cultivars 'Kapoho' and 'Sunset' with the coat protein-coding region of a mild Hawaiian strain of PRSV-P by bombarding immature zygotic embryos with microprojectiles. The transgenic lines were resistant to Hawaiian PRSV-P in the glasshouse and field (Fitch *et al.*, 1992; Lius *et al.*, 1997). They were used to develop the cultivars 'Sun Up', derived from line 55–1 which is homozygous for a single insert of the coat protein gene (Tennant *et al.*, 2001), and 'Rainbow', a hybrid of 'Sun Up' and 'Kapoho' that is hemizygous for the coat protein gene (Gonsalves, 1998). Both cultivars have received regulatory approval for commercial use and currently are being grown in the Puna district of Hawaii. They are credited with saving the papaya industry in this important production area (Ferreira *et al.*, 2002).

The transgenic lines that were highly resistant to Hawaiian isolates of PRSV-P showed varying degrees of susceptibility when challenged with isolates from different countries (Tennant *et al.*, 1994, 2001). The degree of susceptibility was related largely to the level of sequence homology between the coat protein transgene and the coat protein-coding region of the challenge isolate (Chiang *et al.*, 2001; Tennant *et al.*, 2001).

The work pioneered in Hawaii has now been followed successfully in several countries using local isolates of PRSV-P (Gonsalves, 1998; Yeh *et al.*, 1998; Ying *et al.*, 1999; Lines *et al.*, 2002). Several studies (Chiang *et al.*, 2001; Lines *et al.*, 2002) have shown that the transgenic protection against PRSV-P is RNA mediated, with the underlying mechanism being post-transcriptional gene silencing (Baulcombe, 1999).

Tomato spotted wilt

Tomato spotted wilt virus (TSWV), the type species of the genus *Tospovirus* in family *Bunyaviridae*, has caused severe decline in papaya plants in Hawaii (Gonsalves and Trujillo, 1986). The virus caused spotting, chlorosis and necrosis of apical leaves, bending of the apical shoot and water-soaked lesions on the stem and petioles. Fruit on infected trees were deformed, with ripe fruit developing dark green rings in a yellow background.

TSWV has a wide host range. It is transmitted in a persistent, circulative manner by several species of thrips, with the western flower thrips, *Frankliniella occidentalis*, being a widespread and efficient vector (Bautista *et al.*, 1995; Mumford *et al.*, 1996). Serious outbreaks of tomato spotted wilt in papaya plantations in Hawaii were associated with high incidences of TSWV in a common weed species, *Emila fosbergii* (Gonsalves and Trujillo, 1986).

Disorders of Papaya

Bumpy fruit

Bumpy fruit is a common disorder of papaya fruit that is caused by boron deficiency (Wang and Ko, 1975; Nishina, 1991). The fruit surface develops bumps and exudes latex and as a result, there are localized deficient areas that stop growing. As the fruit enlarges, non-deficient areas continue to grow, giving the fruit a lumpy appearance (Fig. 17.22). Internally, seeds either abort or are poorly developed, and vascular bundles are darkened. Severely deficient plants may have a dwarfed, bunched appearance.

The normal and deficient ranges for boron in papaya have been reported to be either between 20 and 50 p.p.m. and below 16 p.p.m. (Weir *et al.*, 1995) or $\geq$ 25 p.p.m. and $\leq$ 20 p.p.m. (Nishina, 1991). Applications of 0.25% borax ($Na_2B_4O_7{\cdot}10H_2O$) to foliage and 1–3 kg boron ha^{-1} to the soil have alleviated the problem in Hawaii (Nishina, 1991).

Carpellody (cat-face)

Carpellody of fruit, which is also known as cat-face, develops on hermaphroditic plants that are exposed to low night temperatures, and high moisture and nitrogen levels. Under these conditions, the stamens of some genotypes develop into carpel-like structures. Fruit are oblong, rather than pyriform, and either look like female fruit or have pronounced longitudinal ridges and seams that make them unmarketable (Fig. 17.23). Since this trait is genetically controlled, selection of seed from unaffected

Fig. 17.23. A column of papaya fruit with the carpellody disorder (photo: W.T. Nishijima).

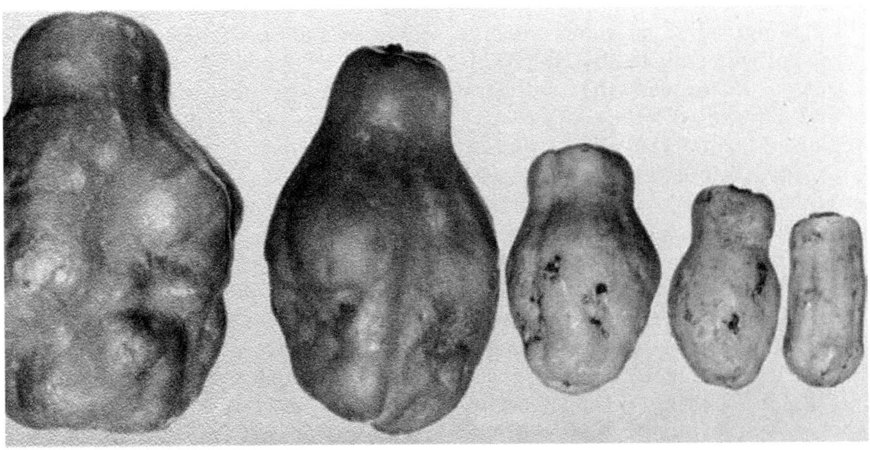

Fig. 17.22. Papaya fruit exhibiting various degrees of severity of the bumpy fruit disorder, caused by boron deficiency (photo: W.T. Nishijima).

plants is helpful. The incidence of carpellody on Solo cultivars is low.

Freckles

Freckles is a common problem on commercial cultivars (Hine *et al.*, 1965). Symptoms begin as pinpoint spots on immature fruit that increase to 4–13 mm in diameter, become brown and reticulated, and can have a water-soaked margin as fruit enlarge (Fig. 17.24). Large spots may have grey centres. Freckles are superficial and do not affect the flesh. They generally develop on exposed surfaces on the exterior of the fruit column, usually during sunnier times of the year.

Although pathogens have not been associated with the disorder, enclosing fruit in pollination bags after they are set controls the problem.

A severe type of freckles, in which high numbers of freckles coalesce and russet the fruit surface, develops on 'Sunrise Solo' during dry periods in Brazil, Fiji and Hawaii.

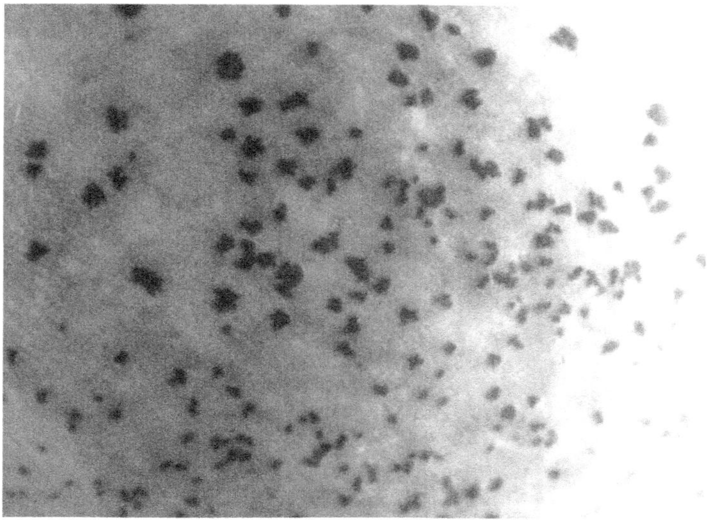

Fig. 17.24. Symptoms of the freckle disorder on the surface of a papaya fruit (photo: W.T. Nishijima).

References

Abou Haidar, M.G. and Gellatly, D. (1999) Potexviruses. In: Granoff, A. and Webster, R.G. (eds) *Encyclopedia of Virology*, 2nd edn. Academic Press, San Diego, pp. 1364–1368.

Adsuar, J. (1946) Transmission of papaya bunchy top by leafhoppers of the genus *Empoasca*. *Science* 103, 316.

Aleemullah, M. and Walsh, K.B. (1996) Australian papaya dieback: evidence against the calcium deficiency hypothesis and observation on the significance of laticifer autofluorescence. *Australian Journal of Agricultural Research* 47, 371–385.

Alvarez, A.M. and Nelson, M.G. (1982) Control of *Phytophthora palmivora* in papaya orchards with weekly sprays of chlorothalonil. *Plant Disease* 66, 37–39.

Alvarez, A.M. and Nishijima, W.T. (1987) Postharvest diseases of papaya. *Plant Disease* 71, 681–686.

Alvarez, A.M., Hylin, J.W. and Ogata, J.N. (1977) Postharvest diseases of papaya reduced by biweekly orchard sprays. *Plant Disease Reporter* 61, 731–735.

Aquilizan, F.A. (1987) Breeding systems for fixing stable papaya inbred lines and breeding potential for hybrid variety production. In: *The Breeding of Horticultural Crops*. Proceedings of an International Symposium, Taiwan, December 1986. FFTC Book Series No. 35, pp. 101–106.

Aradhya, M.K., Manshardt, R.M., Zee, F. and Morden, C.W. (1999) A phylogenetic analysis of the genus *Carica* L. (*Caricaceae*) based on restriction fragment length variation in a cpDNA intergenic spacer region. *Genetic Resources and Crop Evolution* 46, 579–586.

Aragaki, M. and Uchida, J.Y. (2001) Morphological distinctions between *Phytophthora capsici* and *P. tropicalis* sp. nov. *Mycologia* 93, 137–145.

Aragaki, M., Kimoto, W.S. and Uchida, J.Y. (1981) Limitations of hot water treatment in the control of *Phytophthora* fruit rot of papaya. *Plant Disease* 65, 744–745.

Arumuganathan, K. and Earle, E.D. (1991) Nuclear DNA content of some important plant species. *Plant Molecular Biology Reporter* 9, 208–218.

Ayala, A., Acosta, N. and Adsuar, J.A. (1971) A preliminary report on the response of *Carica papaya* to foliar applications of two systemic nematicides. *Nematropica* 1, 10 (abstract).

Badillo, V. (1993) *Caricaceae, Segundo Esquema*. Publicada por la Asociacion de professors, Alcance 43, Universidad Central de Venezuela, Maracay.

Badillo, V.M. (2000) *Carica* L. vs *Vasconcella* St Hil. (Caricaceae): con la rehabilitacion de este ultimo. *Ernstia* 10, 74–79.

Barbosa, C. de J., Meissner Filho, P.E., Habibe, T.C., Vilarinhos, A.B. and Matrangolo, W.J.R. (1999) Distribution of virus replicative forms in papaya affected by the stick disease. *Revista Brasileira de Fruticultura* 21, 356–358.

Bateson, M.F. and Dale, J.L. (1992) The nucleotide sequence of the coat protein gene and the 3′ untranslated region of papaya ringspot virus type W (Aust). *Archives of Virology* 123, 101–109.

Bateson, M.F., Henderson, J., Chaleeprom, W., Gibbs, A. and Dale, J.L. (1994) Papaya ringspot potyvirus: isolate variability and origin of PRSV type P (Australia). *Journal of General Virology* 75, 3547–3553.

Bateson, M.F., Lines, R.E., Revill, P., Chaleeprom, W., Cuong, W., Ha., Gibbs, A.J. and Dale, J.L. (2002) On the evolution and molecular epidemiology of papaya ringspot virus. *Journal of General Virology* 83, 2575–2585.

Baulcombe, D.C. (1999) Viruses and gene silencing in plants. *Archives of Virology* 15 (Supplement), 189–201.

Bautista, R.C., Mau, R.F.L., Cho, J.J. and Custer, P.M. (1995) Potential of tomato spotted wilt tospovirus plant hosts in Hawaii as virus reservoirs for transmission by *Frankliniella occidentalis*. *Phytopathology* 85, 953–958.

Bayot, R.G., Villegas, V.N., Magdalita, P.M., Jovellana, M.D., Espino, T.M. and Exconde, S.B. (1990) Seed transmissibility of papaya ringspot virus. *Philippine Journal of Crop Science* 15, 107–111.

Bazán de Segura, C. (1951) Plant diseases new to Peru. *Plant Disease Reporter* 35, 465–466.

Bremer, K., Chase, M.W. and Stevens, P.F. (1998) An ordinal classification for the families of flowering plants. *Annals of the Missouri Botanical Garden* 85, 531–553.

Brunt, A.A., Crabtree, K., Dallwitz, M.J., Gibbs, A.J. and Watson, L. (1996) *Viruses of Plants*. Descriptions and lists from the VIDE database. CAB International, Wallingford, UK.

Brunt, A.A., Foster, G.D., Martelli, G.P. and Zavriev, S.K. (2000) Potexvirus. In: van Regenmortel, M.H.V., Fauquet, C.M.F. and Bishop, D.H.L. (eds) *Virus Taxonomy. Seventh Report of the International committee on Taxonomy of Viruses*. Academic Press, pp. 975–981.

Chambers, K.R. and Rijkenberg, F.H.J. (1987) Culture of *Asperisporium caricae*, the papaya black spot organism. *Phytophylactica* 19, 113.

Chau, K.F. and Alvarez, A.M. (1979) Role of *Mycosphaerella* ascospores in stem-end rot of papaya fruit. *Phytopathology* 69, 500–503.

Chen, L.F., Baul, H.J. and Yeh, S.D. (2002) Identification of viruses capable of breaking transgenic resistance of papaya conferred by the coat protein gene. Proceedings International Symposium on Tropical and Sub Tropical Fruits. *Acta Horticulturae* 575, 465–474.

Chiang, C.H., Wang, J.J., Jan, F.J., Yeh, S.D. and Gonsalves, D. (2001) Comparative reactions of recombinant papaya ringspot viruses with chimeric coat protein genes and wild-type viruses on coat protein transgenic papaya. *Journal of General Virology* 82, 2827–2836.

Clare, B.G. (1964) *Erysiphacae of South-eastern Queensland*. University of Queensland Papers, Department of Botany, Vol. IV, 10, pp. 112–144.

Cohn, E. and Duncan, L.W. (1990) Nematode parasites of subtropical and tropical fruit trees. In: Luc, M., Sikora, R.A. and Bridge, J. (eds) *Plant Parasitic Nematodes in Subtropical and Tropical Agriculture*. CAB International, Wallingford, UK, pp. 347–362.

Comarco, R.F., Lima, J.A. and Pio-Ribeiro, G. (1998) Transmission and presence in the soil of papaya lethal yellowing virus. *Fitopatologia Brasileira* 23, 453–458.

Conover, R.A. (1964) Mild mosaic and faint mottle ringspot, two papaya virus diseases of minor importance in Florida. *Proceedings of the Florida State Horticultural Society* 77, 444–448.

Conover, R.A. (1979) Yellow strap leaf, a new disease of Florida papayas. *Proceedings of the Florida State Horticultural Society* 92, 276–277.

Conover, R.A. and Litz, R.E. (1978) Progress in breeding papayas with tolerance to papaya ringspot virus. *Proceedings of the Florida State Horticultural Society* 91, 182–184.

Conover, R.A., Litz, R.E. and Malo, S.E. (1986) *'Cariflora'*– a papaya ringspot virus tolerant papaya for south Florida and the Caribbean. *HortScience* 21, 1072.

Cook, A.A. (1972) *Virus Diseases of Papaya*. Florida Agricultural Experiment Station Bulletin 750 (technical).

Crous, P.W. (2002) *Taxonomy and Pathology of* Cylindrocladium *(*Calonectria*) and Allied Genera*. APS Press, St Paul, Minnesota.

Da Costa, E.W.B. (1944) Diseases of the pawpaw. *Queensland Agricultural Journal*. 1 May, pp. 282–293.

Davis, M.J. (1994) Bunchy top. In: Ploetz, R.C., Zentmyer, G.A., Nishijima, W.T., Rohrbach, K.G. and Ohr, H.D. (eds) *Compendium of Tropical Fruit Diseases*. APS Press, St Paul, Minnesota. pp. 69–70.

Davis, M.J., Kramer, J.B., Ferwerda, F.H. and Brunner, B.R. (1996) Association of a bacterium and not a phytoplasma with papaya bunchy top disease. *Phytopathology* 86, 102–109.

Davis, M.J., Ying, Z., Brunner, B., Pantoja, A. and Ferwerda, F.H. (1998) Rickettsial relative associated with papaya bunchy top disease. *Current Microbiology* 36, 80–84.

Davis, R.I., Schneider, B. and Gibb, K.S. (1997) Detection and differentiation of phytoplasmas in Australia. *Australian Journal of Agricultural Research* 48, 535–544.

De La Rue, S.J., Schneider, B. and Gibb, K.S. (1999) Genetic variability in phytoplasmas associated with papaya yellow crinkle and papaya mosaic diseases in Queensland and the Northern Territory. *Australasian Plant Pathology* 28, 108–114.

Dickman, M.B. (1994) Anthracnose. In: Ploetz, R.C., Zentmyer, G.A., Nishijima, W.T., Rohrbach, K.G. and Ohr, H.D. (eds) *Compendium of Tropical Fruit Diseases*. APS Press, St Paul, Minnesota. p. 58.

Dickman, M.B. and Alvarez, A.M. (1983) Latent infection of papaya caused by *Colletotrichum gloeosporioides*. *Plant Disease* 67, 748–750.

Drew, R.A. (1988) Rapid clonal propagation of papaya *in vitro* from mature field-grown trees. *HortScience* 23, 609–611.

Elder, R.J., Macleod, W.N., Reid, D.J. and Gillespie, R.L. (2000) Growth and yield of three hybrid papayas (*Carica papaya*) under mulched and bare ground conditions. *Australian Journal of Experimental Agriculture* 40, 747–754.

Elder, R.J., Milne, J.R., Reid, D.J., Guthrie, J.N. and Persley, D.M. (2002) Temporal incidence of three phytoplasma-associated diseases of *Carica papaya* and their potential hemipteran vectors in central and south-east Queensland. *Australasian Plant Pathology* 31, 165–176.

Ellis, M.B. (1976) *Cercospora papayae*. In: Ellis, M.B. (ed.) *More Dematiaceous Hyphomycetes*. Commonwealth Mycological Institute, Kew, UK, p. 247.

Ellis, M.B. and Holliday, P. (1971) Corynespora cassiicola. *CMI Descriptions of Pathogenic Fungi and Bacteria No. 303*. Commonwealth Mycological Institute, Kew, UK.

Ellis, M.B. and Holliday, P. (1972) Asperisporium caricae. *CMI Descriptions of Pathogenic Fungi and Bacteria No. 347*. Commonwealth Mycological Institute, Kew, UK.

Erwin, D.C. and Ribiero, O.K. (1996) *Phytophthora Diseases Worldwide*. APS Press, St Paul, Minnesota.

Escudero, J., Acosta, A., Ramirez, L.V., Caloni, I. and Ruiz Sifre, G. (1994) Yield of three papaya genotypes and their tolerance to papaya ringspot virus in Puerto Rico. *Journal of Agriculture of the University of Puerto Rico* 78, 111–121.

FAO (2001) Online database of the Food and Agricultural Organization of the United Nations. (http://apps.fao.org/default.htm)

Farr, D.F., Bills, G.F., Chamuris, G.P. and Rossman, A.Y. (1989) *Fungi on Plants and Plant Products in the United States*. APS Press, St Paul, Minnesota.

Ferreira, S.A., Pitz, K.Y., Manshardt, R., Zee, F., Fitch, M. and Gonsalves, D. (2002) Virus coat protein transgenic papaya provides practical control of papaya ringspot virus in Hawaii. *Plant Disease* 86, 101–105.

Fitch, M., Manshardt, R., Gonsalves, D., Slightom, J.L. and Sanford, J.C. (1992) Virus resistant papaya plants derived from tissues bombarded with the coat protein gene of papaya ringspot virus. *Biotechnology* 10, 1466–1472.

Franck, A. and Bar-Joseph, M. (1992) Use of netting and whitewash spray to protect papaya plants against Nivun Haamir (NH) – dieback disease. *Crop Protection* 11, 525–528.

Gibb, K.S., Persley, D.M., Schneider, B. and Thomas, J.E. (1996) Phytoplasmas associated with papaya diseases in Australia. *Plant Disease* 80, 174–178.

Gibb, K.S., Schneider, B. and Padovan, A.C. (1998) Differential detection and genetic relatedness of phytoplasma in papaya. *Plant Pathology* 47, 325–332.

Glennie, J.D. and Chapman, K.R. (1976) A review of dieback – a disorder of the papaw (*Carica papaya* L.) in Queensland. *Queensland Journal of Agriculture and Animal Sciences* 33, 177–188.

Gonsalves, D. (1994) Papaya ringspot. In: Ploetz, R.C., Zentmyer, G.A., Nishijima, W.T., Rohrbach, K.G. and Ohr, H.D. (eds) *Compendium of Tropical Fruit Diseases*. APS Press, St Paul, Minnesota, pp. 67–68.

Gonsalves, D. (1998) Control of papaya ringspot virus in papaya: a case study. *Annual Review of Phytopathology* 36, 415–437.

Gonsalves, D. and Garnsey, S.M. (1989) Cross protection techniques for control of virus diseases in the tropics. *Plant Disease* 73, 592–597.

Gonsalves, D. and Ishii, M. (1980) Purification and serology of papaya ringspot virus. *Phytopathology* 70, 1028–1032.

Gonsalves, D. and Trujillo, E.E. (1986) Tomato spotted wilt virus in papaya and detection of the virus by ELISA. *Plant Disease* 70, 501–506.

Gowanlock, D.H., Greber, R.S., Behncken, G.M. and Finlay, J. (1976) Electron microscopy of mycoplasma-like bodies in several Queensland crop species. *Australian Plant Pathology Society Newsletter* 5, Abstract 223.

Greber, R.S. (1966) Identification of the virus causing papaw yellow crinkle with tomato big bud virus by transmission tests. *Queensland Journal of Agricultural and Animal Sciences* 23, 147–153.

Gupta, O. and Nema, K.G. (1979) Effect of different temperature and relative humidity on the development of fruit rots of papaya caused by *Botryodiplodia theobromae* and *Colletotrichum papayae*. *Indian Phytopathology* 32, 106–107.

Guthrie, J.N., White, D.T., Walsh, K.B. and Scott, P.T. (1998) Epidemiology of phytoplasma-associated papaya diseases in Queensland, Australia. *Plant Disease* 82, 1107–1111.

Guthrie, J.N., Walsh, K.B., Scott, P.T. and Rasmussen, T.S. (2001) The phytopathology of Australian papaya dieback: a proposed role for the phytoplasma. *Physiological and Molecular Plant Pathology* 58, 23–30.

Hamill, S.D. (1987) Fruit rot of papaw caused by *Phytophthora palmivora* in Queensland. *Australasian Plant Pathology* 16, 22.

Haque, S.Q. and Parasram, S. (1973) *Empoasca stevensi*, a new vector of bunchy top disease of papaya. *Plant Disease Reporter* 57, 412–413.

Harding, R.M. and Teakle, D.S. (1988) Autofluorescence of necrotic phloem cells, and a reduced latex flow: new symptoms for papaw dieback disease in Australia. *Australasian Journal of Agricultural Research* 39, 857–862.

Harding, R.M. and Teakle, D.S. (1993) Failure of five viruses to cause typical Australian papaw dieback disease. *Australian Plant Pathology* 22, 62–67.

Harding, R.M., Teakle, D.S. and Dale, J.L. (1991) Double-stranded RNA in *Carica papaya* is not associated with dieback disease and is unlikely to be of viral origin. *Australian Journal of Agricultural Research* 42, 1179–1186.

Hill, A.V. (1943) Insect transmission and host plants of virescence (big bud of tomato). *Journal of the Council for Scientific and Industrial Research* 16, 85–90.

Hine, R.B., Aragaki, M. and Tokunaga, J. (1965a) Enzymatic inactivation of the papaya blight fungus, *Phytophthora palmivora*, by papain and other proteolytic enzymes. *Phytopathology* 55, 1223–1226.

Hine, R.B., Holtzmann, O.V. and Raabe, R.D. (1965b) *Diseases of Papaya (Carica papaya L.) in Hawaii*. Hawaiian Agricultural Experiment Station Bulletin 136, University of Hawaii.

Holliday, P. (1980) *Fungus Diseases of Tropical Crops*. Cambridge University Press, Cambridge, pp. 114–115.

Horovitz, S. and Jimenez, H. (1967) Cruzamientos interspecificos y in tergenericos in *Caricaceae* y sus implicaciones fitotecnicas. *Agronomie Tropicale (Maracay)* 17, 323–343.

Hull, R. (2002) *Matthews' Plant Virology*, 4th edn. Academic Press.

Hunter, J.E. and Buddenhagen, I.W. (1969) Field biology and control of *Phytophthora parasitica* on papaya (*Carica papaya*) in Hawaii. *Annals of Applied Biology* 63, 55–60.

Hunter, J.E. and Buddenhagen, I.W. (1972) Incidence, epidemiology and control of fruit diseases of papaya in Hawaii. *Tropical Agriculture (Trinidad)* 49, 61–71.

Hunter, J.E. and Kunimoto, R.K. (1974) Dispersal of *Phytophthora palmivora* sporangia by wind-blown rain. *Phytopathology* 64, 202–206.

Hwang, S.C. and Ko, W.H. (1976) Biology of conidia, ascospores, and microsclerotia of *Calonectria crotalariae* in soil. *Phytopathology* 66, 51–54.

Jain, R.K., Pappu, H.R., Pappu, S.S., Varma, A. and Ram, R.D. (1998) Molecular characterisation of papaya ringspot potyvirus isolates from India. *Annals of Applied Biology* 132, 413–425.

Jensen, D.D. (1949) Papaya virus diseases with special reference to papaya ringspot. *Phytopathology* 39, 191–211.

Jobin-Décor, M.P., Graham, G.C., Henry, R.J. and Drew, R.A (1997) RAPD and isozyme analysis of genetic relationship between *Carica papaya* and wild relatives. *Genetic Resources and Crop Evolution* 44, 471–477.

Kawano, S. and Yonaka, T. (1992) *The Occurrence of Papaya Leaf Distortion Mosaic Virus in Okinawa*.

Technical Bulletin FFTC 132, 13–23. Food and Fertilizer Technology Centre for the Asian and Pacific Region, Taipei.

Khuspe, S., Hendre, R., Mascarenhas, A., Jugannathan, V., Thombre, M. and Joshi, A. (1980) Utilization of tissue culture to isolate interspecific hybrids in *Carica*. In: *National Symposium on Plant Tissue Culture, Genetic Manipulation and Somatic Hybridization of Plant Cells*. Bhabha Atomic Research Center, Bombay, India.

Kim, M.S., Moore, P.H., Zee, F., Fitch, M.M.M., Steiger, D.L., Manshardt, R.M., Paull, R.E., Sekioka, T. and Ming, R. (2002) Genetic diversity in *Carica papaya* as revealed by AFLP markers. *Genome* 45, 503–512.

Kiritani, K. and Su Hong, J. (1999) Papaya ringspot, banana bunchy top and citrus greening in the Asia and Pacific region: occurrence and control strategy. *Japan Agricultural Research Quarterly* 33, 23–30.

Kitajima, E.W., Rodrigues, C.H., Silveira, J.S., Alves, F., Ventura, J.A., Aragao, F.T.L. and Oliveira, L.H.R. (1993) Association of isometric virus-like particles restricted to the laticifers, with 'meleira' ('sticky disease') of papaya (*Carica papaya*). *Fitopatologia Brasileira* 18, 118–122.

Ko, W.H. (1987) Biological control of Phytophthora root rot of papaya with virgin soil. *Plant Disease* 66, 446–448.

Ko, W.H. (1994) Phytophthora fruit rot and root rot. In: Ploetz, R.C., Zentmyer, G.A., Nishijima, W.T., Rohrbach, K.G. and Ohr, H.D. (eds) *Compendium of Tropical Fruit Diseases*. APS Press, St Paul, Minnesota, pp. 61–62.

Laemmlen, F.F. and Aragaki, M. (1971) Collar rot of papaya caused by *Calonectria* sp. *Plant Disease Reporter* 55, 743–745.

Lange, A.H. (1960) The effect of fumigation on the papaya replant problem in two Hawaiian soils. *Proceedings of the American Horticultural Society* 75, 305–312.

Lastra, R. and Quintero, E. (1981) Papaya apical necrosis, a new disease associated with a rhabdovirus. *Plant Disease* 65, 439–440.

Latunde-Dada, A.O. (2001) *Colletotrichum*: tales of forcible entry, stealth, transient confinement and breakout. *Molecular Plant Pathology* 2, 187–198.

Liefting, L.W., Padovan, A.C., Gibb, K.S., Beever, R.E., Anderson, M.T., Beck, D.L. and Forster, R.S. (1998) '*Candidatus* Phytoplasma australiense' is the phytoplasma associated with Australian grapevine yellows, papaya dieback and *Phormium* yellow leaf diseases. *European Journal of Plant Pathology* 104, 619–623.

Lines, R.E., Persley, D.M., Dale, J.L., Drew, R.A. and Bateson, M.F. (2002) Genetically engineered immunity to papaya ringspot virus in Australian papaya cultivars. *Molecular Breeding* 10, 119–129.

Liu, B., White, D.T., Walsh, K. and Scott, P.T. (1996) Detection of phytoplasmas in dieback, yellow crinkle, and mosaic diseases of papaya using polymerase chain reaction techniques. *Australian Journal of Agricultural Research* 47, 387–394.

Lius, S., Manshardt, R.M., Fitch, M.M.M., Slightom, J.L., Sanford, J.C. and Gonsalves, D. (1997) Pathogen-derived resistance provides papaya with effective protection against papaya ringspot virus. *Molecular Breeding* 3, 161–168.

Loreto, T.J.T., Vital, A.F., Rezende, J.A.M., Vega, J. and Costa, A.S. (1983) Ocorrencia de um amarelo letal do mamoeiro Solo, no estado de Pernambuco. *O Biologico* 49, 275–279.

Magdalita, P.M., Opina, O.S., Espino, R.R.C. and Villegas, V.N. (1989) Epidemiology of papaya ringspot in the Philippines. *Philippine Phytopathology* 25, 1–11.

Magdalita, P.M., Drew, R.A., Godwin, I.D. and Adkins, S.W. (1998) An efficient interspecific hybridization protocol for *Carica papaya* L × *C. cauliflora* Jacq. *Australian Journal of Experimental Agriculture* 38, 523–530.

Manshardt, R.M. (1992) Papaya. In: Hammerschlag, F.A. and Litz, R.E. (eds) *Biotechnology of Perennial Fruit Crops*. CAB International, Wallingford, UK, pp. 489–511.

Manshardt, R.M. and Drew, R.A. (1998) *Biotechnology of Papaya*. Proceedings International Symposium in Biotechnology of Tropical and Subtropical Species. *Acta Horticulturae* 461, 65–73.

Manshardt, R.M. and Wenslaff, T.F. (1989) Interspecific hybridisation of papaya with other *Carica* species. *Journal American Society of Horticultural Science* 114, 689–694.

Maoka, T., Kawano, S. and Usugi, T. (1995) Occurrence of the P-strain of papaya ringspot virus in Japan. *Annals of the Phytopathological Society of Japan* 61, 34–37.

Maoka, T., Kasriwazaki, S., Isuda, S., Usugi, T. and Hibino, H. (1996) Nucleotide sequence of the capsid protein gene of papaya leaf-distortion mosaic potyvirus. *Archives of Virology* 141, 197–204.

Maoka, T., Tsuda, S., Usugi, T., Noda, C. and Iwasaki, M. (2002) Changes in serological reactivity of papaya leaf distortion mosaic virus caused by papain in *Carica papaya* L. and its detection using antipain or papain. *Journal of General Plant Pathology* 68, 89–93.

Marys, E., Carballo, O. and Izaguirre-Mayoral, M.L. (1995) Properties of a previously undescribed supercoiled filamentous virus infecting papaya in Venezuela. *Archives of Virology* 140, 891–898.
Marys, E.E., Carballo, O. and Izaguirre-Mayoral, M.L. (2000) Occurrence and relative incidence of viruses infecting papaya in Venezuela. *Annals of Applied Biology* 136, 121–124.
McMillan, R.J., Tennant, P. and Gonsalves, D. (1993) Reoccurrence of papaya mosaic potexvirus in Florida. *Proceedings of the Florida State Horticultural Society* 106, 146–147.
McSorley, R. (1981) *Plant Parasitic Nematodes Associated with Tropical and Subtropical Fruit.* Florida Agricultural Experiment Station Bulletin 823, University of Florida.
Mekako, H.U. and Nakasone, H.Y. (1975) Interspecific hybridisation among six *Carica* species. *Journal of the American Society for Horticultural Science* 100, 237–242.
Ming, R., Moore, P.H., Zee, F., Abbey, C.A., Ma, H. and Paterson, A.H. (2001) Construction and characterization of a papaya BAC library as a foundation for molecular dissection of a tree-fruit genome. *Theoretical and Applied Genetics* 102, 892–899.
Mora-Aguilera, G., Teliz, D., Campbell, C.L. and Catarino, A.V. (1992) Temporal and spatial development of papaya ringspot in Veracruz, Mexico. *Journal of Phytopathology* 136, 27–36.
Mumford, R.A., Barker, I. and Wood, K.R. (1996) The biology of the *Tospoviruses*. *Annals of Applied Biology* 128, 159–183.
Nakasone, H.Y. and Paull, R.E. (1998) Papaya. In: *Tropical Fruits.* Crop Production Science in Horticulture Series No. 7. CAB International, Wallingford, UK, pp. 239–269.
Nariani, T.K. (1956) Leaf curl of papaya. *Indian Phytopathology* 9, 151–155.
Ndunguru, J. and Rajabu, C.A. (2002) Papaya ringspot virus disease in the Lake Victoria basin. *Tropical Science* 42, 11–16.
Nelson, M.N. and Alvarez, A.M. (1980) Purple stain of *Carica papaya*. *Plant Disease* 64, 93–95.
Nelson, S.C. (1995) Spatiotemporal distance class analysis of plant disease epidemics. *Phytopathology* 85, 37–43.
Nishijima, K.A. (1994) Internal yellowing. In: Ploetz, R.C., Zentmyer, G.A., Nishijima, W.T., Rohrbach, K.G. and Ohr, H.D. (eds) *Compendium of Tropical Fruit Diseases.* APS Press, St Paul, Minnesota, p. 65.
Nishijima, K.A., Couey, H.M. and Alvarez, A.M. (1987) Internal yellowing, a bacterial disease of papaya fruits caused by *Enterobacter cloacae*. *Plant Disease* 71, 1029–1034.
Nishijima, W.T. (1994a) Dry rot. In: Ploetz, R.C., Zentmyer, G.A., Nishijima, W.T., Rohrbach, K.G. and Ohr, H.D. (eds) *Compendium of Tropical Fruit Diseases.* APS Press, St Paul, Minnesota, p. 59.
Nishijima, W.T. (1994b) Cercospora black spot. In: Ploetz, R.C., Zentmyer, G.A., Nishijima, W.T., Rohrbach, K.G. and Ohr, H.D. (eds) *Compendium of Tropical Fruit Diseases.* APS Press, St Paul, Minnesota, p. 59.
Nishijima, W.T. (1994c) Fusarium fruit rot. In: Ploetz, R.C., Zentmyer, G.A., Nishijima, W.T., Rohrbach, K.G. and Ohr, H.D. (eds) *Compendium of Tropical Fruit Diseases.* APS Press, St Paul, Minnesota, p. 60.
Nishijima, W.T. (1994d) Wet fruit rot. In: Ploetz, R.C., Zentmyer, G.A., Nishijima, W.T., Rohrbach, K.G. and Ohr, H.D. (eds) *Compendium of Tropical Fruit Diseases.* APS Press, St Paul, Minnesota, p. 64.
Nishijima, W.T. (1999) Common names of plant diseases. Diseases of papaya (*Carica papaya* L.). http://www.apsnet.org/online/common/names/papaya.asp
Nishijima, W.T. and Aragaki, M. (1973) Pathogenicity and further characterization of *Calonectria crotalariae* causing collar rot of papaya. *Phytopathology* 63, 553–558.
Nishina, M.S. (1991) *Bumpy Fruit of Papaya as Related to Boron Deficiency.* Hawaii Institute of Tropical Agriculture and Human Resources, Commodity Fact Sheet PA-4 (B), University of Hawaii.
Noa-Carrazana, J.C. and Silva-Rosales, L. (2001) First report of a Mexican isolate of *Papaya mosaic virus* in papaya (*Carica papaya*) and pumpkin (*Cucurbita pepo*). *Plant Disease* 85, 558.
Ooka, J.J. (1994) Powdery mildew. In: Ploetz, R.C., Zentmyer, G.A., Nishijima, W.T., Rohrbach, K.G. and Ohr, H.D. (eds) *Compendium of Tropical Fruit Diseases.* APS Press, St Paul, Minnesota, pp. 62–63.
Opina, O.S. (1986) Studies on a new virus disease of papaya in the Philippines. In: *Plant Virus Diseases of Horticultural Crops in the Tropics and Subtropics.* FFTC Book Series No. 33, pp 158–167.
Padovan, A.C. and Gibb, K.S. (2001) Epidemiology of phytoplasma diseases in papaya in northern Australia. *Journal of Phytopathology* 149, 649–658.
Parris, G.K. (1942) *Phytophthora parasitica* on papaya (*Carica papaya*) in Hawaii. *Phytopathology* 32, 314–320.
Pernezny, K. and Litz, R.E. (1993) *Some Common Diseases of Papaya in Florida.* University of Florida Cooperative Extension Service PP35.
Persley, D.M. (1993) Common diseases in perennial fruit crops. In: Persley, D.M. (ed.) *Diseases of Fruit Crops.* Department of Primary Industries, Queensland, Australia, pp. 9–11.
Persley, D.M. (1998) Identification, epidemiology and control of papaya ringspot virus, a recently recorded virus of papaya (*Carica papaya*) in Australia. Masters thesis, Queensland University of Technology.

Peterson, R. and Grice, K. (1994) Papaw black spot. *Queensland Fruit and Vegetable News*, April 21, p. 11.
Peterson, R.A., Coates, L.M. and Persley, D.M. (1993) Papaw diseases. In: Persley, D.M. (ed.) *Diseases of Fruit Crops*. Department of Primary Industries, Queensland, Australia, pp. 70–76.
Punithalingham, E. (1985) Phomopsis caricae-papayae. *CMI Descriptions of Pathogenic Fungi and Bacteria No. 827*. Commonwealth Mycological Institute, Kew, UK.
Purcifull, D.E. and Hiebert, E. (1971) Papaya mosaic virus. *CMI/AAB Descriptions of Plant Viruses No. 56*. Commonwealth Mycological Institute, Kew, UK.
Purcifull, D., Edwardson, J., Hiebert, E. and Gonsalves, D. (1984) Papaya ringspot virus. *CMI/AAB Descriptions of Plant Viruses No. 292*. Commonwealth Mycological Institute, Kew, UK.
Purseglove, J.W. (1968) *Tropical Crops: Dicotyledons*. Longman, London.
Quemada, H., L'Hostis, B., Gonsalves, D., Reardon, T.M., Heinrikson, R., Hiebert, E.L., Sieu, L.C. and Slightom, J.L. (1990) The nucleotide sequences of the 3′-terminal regions of papaya ringspot virus strains W and P. *Journal of General Virology* 71, 203–210.
Rawal, R.D. (1987) Influences of some weather factors on the development of powdery mildew in papaya. *Plant Disease Research* 2, 97–99.
Rezende, J.A.M. and Costa, A.S. (1993) Papaya diseases caused by virus and mycoplasma. *Summa Phytopathogicia* 19, 73–79.
Ross, P. and Chay-Prove, P.M. (2000) *Papaw Information Kit*. Agrilink Series, Queensland Horticulture Institute, Department of Primary Industries, Queensland, Australia.
Saldana, M.I., Marquez, M. and Ruiz, P. (1985) Identificacion de enfermedades fungosas del cultivo de la papaya (*Carica papaya* L.) en el estado de Tabasco. *Revista Mexicana de Fitopatologia* 3, 17–19.
Sarbhoy, A.K. (1966) Rhizopus stolonifer. *CMI Descriptions of Pathogenic Fungi and Bacteria No. 110*. Commonwealth Mycological Institute, Kew, UK.
Saxena, R.M. and Sharma, K.D. (1981) Deterioration of papaya fruits by fungi. *Agricultural Science Digest–India*. 1, 140–142.
Saxena, S., Hallan, V., Singh, B.P. and Lane, P.V. (1998a) Leaf curl disease of *Carica papaya* from India may be caused by a bipartite geminivirus. *Plant Disease* 82, 126.
Saxena, S., Hallan, V., Singh, B.P. and Sane, P.V. (1998b) Nucleotide sequence and intergeminiviral homologies of the DNA-A of papaya leaf curl geminivirus from India. *Biochemistry and Molecular Biology International* 45, 101–113.
Schneider, B., Gibb, K.S. and Seemuller, E. (1997) Sequence and RFLP analysis of the gene coding for the elongation factor Tu of several phytoplasma strains for differentiation and classification of phytoplasmas. *Microbiology* 143, 3381–3389.
Shukla, D.D., Ward, C.M. and Brunt, A.A. (1994) *The Potyviridae*. CAB International, Wallingford, UK.
Siddique, A.B., Guthrie, J.N., Walsh, K.B., White, D.T. and Scott, P.T. (1998) Histopathology and within-plant distribution of phytoplasma associated with Australian papaya dieback. *Plant Disease* 82, 1112–1120.
Silva, A.M.R., Kitajima, E.W., Sousa, M.V. and Resend, R.O. (1997) Papaya lethal yellowing virus: a possible member of the *Tombusvirus* genus. *Fitopatologia Brasileira* 22, 529–534.
Silva-Rosales, L., Becerra-Lear, N., Ruiz-Castro, S., Teliz-Ortiz, D. and Noa-Carrazone, J.C. (2000) Coat protein sequence comparisons of three Mexican isolates of papaya ringspot virus with other geographical isolates reveal a close relationship to American and Australian isolates. *Archives of Virology* 145, 835–843.
Simmonds, J.H. (1965) Papaw diseases. *Queensland Agricultural Journal* 91, 666–667.
Sivakumar, C.V. and Seshadri, A.R. (1972) Histopathology of infection by the reniform nematode, *Rotylenchulus reniformis* Linford and Oliveira, 1940 on castor, papaya and tomato. *Indian Journal of Nematology* 2, 173–181.
Sivanesan, A. (1990) Mycosphaerella caricae. *CMI Descriptions of Pathogenic Fungi and Bacteria No. 984*. Commonwealth Mycological Institute, Kew, UK.
Snowdon, A.L. (1990) *A Color Atlas of Postharvest Diseases and Disorders of Fruits and Vegetables*, Vol. 1, *General Introduction and Fruits*. Wolfe Scientific Publishers, London.
Storey, G.F. and Halliwell, R.S. (1969) Association of a mycoplasma-like organism with bunchy top disease of papaya. *Phytopathology* 59, 1336–1337.
Storey, W.B. (1976) Papaya. In: Simmonds N.W. (ed.) *Evolution of Crop Plants*. Longman, London, pp. 21–24.
Teliz, D., Mora, G., Nieto, D., Gonsalves, D., Garcia, E., Matheis, L. and Oolia, C. (1991) Papaya ringspot virus in Mexico. *Revista Mexicana de Fitopatologia* 9, 64–68.
Tennant, P., Gonsalves, C., Ling, K.-S., Fitch, M., Manshardt, R., Slightom, J.L. and Gonsalves, D. (1994) Differential protection against papaya ringspot virus isolates in coat protein gene transgenic papaya and classically cross-protected papaya. *Phytopathology* 84, 1359–1366.

Tennant, P., Fermin, G., Fitch, M., Manshardt, R., Slightom, J. and Gonsalves, D. (2001) Papaya ringspot virus resistance of transgenic Rainbow and Sunup is affected by gene dosages, plant development and coat protein homology. *European Journal of Plant Pathology* 107, 645–653.

Thomas, J.E. and Dodman, R.L. (1993) The first record of papaya ringspot virus-type P from Australia. *Australasian Plant Pathology* 22, 2–7.

Trujillo, E.E. and Hine, R.B. (1965) The role of papaya residues in papaya root rot caused by *Pythium aphanidermatum* and *Phytophthora parasitica*. *Phytopathology* 55, 1293–1298.

Trujillo, E.E. and Schroth, M.N. (1982) Two diseases of papaya trees caused by *Erwinia* species in the northern Mariana Islands. *Plant Disease* 66, 116–120.

Van Droogenbroeck, B., Breyne, P., Goetghebeur, P., Romeijn-Peeters, E., Kyndt, T. and Gheysen, G. (2002) AFLP analysis of genetic relationships among papaya and its wild relatives (*Caricaceae*) from Ecuador. *Theoretical and Applied Genetics* 105, 289–297.

van Regemortel, M.H.V., Fauquet, C.M. and Bishop, D.H.L. (2000) *Virus Taxonomy – Classification and Taxonomy of Viruses.* Seventh Report of the International Committee on Taxonomy of Viruses, Academic Press.

Vawdrey, L.L., Martin, T.M. and De Faveri, J. (2002) The potential of organic and inorganic soil amendments, and a biological control agent (*Trichoderma* sp.) for the management of Phytophthora root rot of pawpaw in far northern Queensland. *Australasian Plant Pathology* 31, 391–399.

Wan, S.H. and Conover, R.A. (1981) A rhabdovirus associated with a new disease of Florida papayas. *Proceedings of the Florida State Horticulture Society* 94, 318–321.

Wang, C.H. and Yeh, S.D. (1992) Nucleotide sequence comparison of the 3′ primed-terminal regions of severe, mild and non-papaya infecting strains of papaya ringspot virus. *Archives of Virology* 127, 345–354.

Wang, C.H. and Yeh, S.D. (1997) Divergence and conservation of the genomic RNAs of Taiwan and Hawaii strains of papaya ringspot virus. *Archives of Virology* 142, 285.

Wang, C.H., Bau, H.J. and Yeh, S.D. (1994) Comparison of the nuclear inclusion to protein and coat protein genes of five papaya ringspot virus strains distinct in geographic origins and pathogenicity. *Phytopathology* 84, 1205–1210.

Wang, D.N. and Ko, W.H. (1975) Relationship between deformed fruit diseases of papaya and boron deficiency. *Phytopathology* 65, 445–447.

Webb, R.R. (1985) Epidemiology and control of bacterial canker of papaya caused by an *Erwinia* sp. on St. Croix, US Virgin Islands. *Plant Disease* 69, 305–309.

Webb, R.R. and Davis, M.J. (1987) Unreliability of latex-flow test for diagnosis of bunchy top of papaya covered by a mycoplasma-like organism. *Plant Disease* 71, 192.

Weir, R.G., Cresswell, G.C. and Loebel, M.R. (1995) *Plant Nutrient Disorders 2. Tropical Fruit and Nut Crops.* Inkata Press, Melbourne.

White, D.T., Blackall, L.L., Scott, P.T. and Walsh, K.B. (1998) Phylogenetic positions of phytoplasmas associated with dieback, yellow crinkle and mosaic diseases in papaya, and their proposed inclusion in '*Candidatus* Phytoplasma australiense' and a new taxon, '*Candidatus* Papaya australasia'. *International Journal of Systemic Bacteriology* 48, 941–951.

Yamamoto, D.T. and Aragaki, M. (1982) Pathogenicity of *Rhizoctonia* isolates to papaya in Hawaii. *Plant Disease* 66, 1136–1137.

Yeh, S.D. and Gonsalves, D. (1994) Practices and perspective of control of papaya ringspot virus by cross protection. In: Harris, K.F. (ed.) *Advances in Disease Vector Research 10*. Springer-Verlag, New York, pp. 237–257.

Yeh, S.D., Bau, H.J., Cheng, Y.H., Yu, T.A. and Yang, J.S. (1998) Greenhouse and field evaluations of coat protein transgenic papaya resistant to papaya ringspot virus. *Acta Horticulturae* 461, 321–328.

Yeh, S.D., Gonsalves, D., Wang, H.L., Namba, R. and Chiu, R.J. (1988) Control of papaya ringspot virus by cross protection. *Plant Disease* 72, 375–380.

Yeh, S.D., Jan, F.J., Chiang, C.H., Doong, T.J., Chen, M.C., Chung, P.H. and Bau, H.J. (1992) Complete nucleotide sequence and genetic organization of papaya ringspot virus RNA. *Journal of General Virology* 73, 2531–2541.

Ying, Z., Yu, X. and Davis, M.J. (1999) New method for obtaining transgenic papaya plants by Agrobacterium-mediated transformation of somatic embryos. *Proceedings of the Florida State Horticultural Society* 112, 201–205.

Zettler, F.W. and Wan, S.H. (1994) Papaya droopy necrosis and papaya apical necrosis. In: Ploetz, R.C., Zentmyer, G.A., Nishijima, W.T., Rohrbach, K.G. and Ohr, H.D. (eds) *Compendium of Tropical Fruit Diseases*, APS Press, St Paul, Minnesota, pp. 66–67.

18 Diseases of Passion Fruit

Barry Manicom[1], Carlos Ruggiero[2], Randy C. Ploetz[3] and Antonio de Goes[2]

[1]*Institute of Tropical and Subtropical Crops, Nelspruit, Republic of South Africa;* [2]*UNESP, Campus Jaboticabal, Jaboticabal, Brazil;* [3]*University of Florida, Tropical Research and Education Center, Homestead, Florida, USA*

Introduction

Passion flowers were valued ornamentals in Europe as early as 1610, and created a sensation when the first specimens arrived from the New World (Killip, 1938). Jacomo Bosio, a monk in Rome, saw in the flower structure symbols of the passion of Christ and called it 'Flos passionis'. From this association came the plant's common and scientific names.

The leaves of some species are used as a substitute for tea, and the leaves and roots of others have medicinal properties. However, fruit are the most significant economic product of members of the *Passifloraceae*. The aromatic pulp that surrounds the seed of several species is used in sherbets, jams and jellies. The juice is a natural concentrate and will impart its distinctive flavour to beverage and dessert items even when highly diluted (Chan, 1980). Purees and punches, in which its juice is added to those of orange, banana, papaya and guava, are popular tropical drinks.

Origins, Taxonomy and Traits

The *Passifloraceae* contains ~500 species and 12 genera (Purseglove, 1968). They are perennial herbaceous and woody plants that usually have a vine-like growth habit. The most important genus, *Passiflora*, contains ~400 species; all but three of these are native to the New World. Fifty or 60 species in the genus produce edible fruit, but relatively few of these are palatable, and fewer still are significant as fruit crops (Martin and Nakasone, 1970).

Three major factors determine the usefulness of a species of *Passiflora* in a given area: (i) adaptation to temperature and elevation; (ii) self-incompatibility; and (iii) reaction to important diseases (Purseglove, 1968). Several species that produce desirable fruit do so only at high elevations that have cool night temperatures, whereas other species that do well in the humid lowland environments of the tropics do not tolerate the former conditions. Self-incompatible types must be cross-pollinated, either by hand or by insects (usually bees), in order for fruit to set. This can be a serious constraint in commercial production. Diseases can seriously reduce the productivity of all of the species that are important commercially. Due to their impact, most plantations around the world are productive for 5 or less years (Rínderman and Cruz, 1997).

The Important Species

P. edulis is the most common and important species. It originated in southern Brazil, but was widely distributed in the tropics and subtropics during the 19th century. Two forms, the purple, *edulis*, and yellow, *flavicarpa*, are recognized. Although the purple form has superior flavour, it can be grown commercially only in the subtropics or at high elevations in the tropics (up to 2000 m). The yellow form does better in hot, lowland environments, produces more and larger fruit (5–7 versus 4–5 cm in diameter), and tolerates Phytophthora blight, nematodes and brown spot (Winks *et al.*, 1988). It also has resistance to Fusarium wilt, and is used as a rootstock for f. *edulis* in affected areas (Cox and Kiely, 1961).

Hybrids between f. *edulis* and f. *flavicarpa* have been developed in Australia, Hawaii, Israel, South Africa and the USA (Florida), and are commercially important in the lowland environments that are inhospitable to f. *edulis* (Knight, 1980; Winks *et al.*, 1988). They are usually high yielding, and tolerate nematodes, Fusarium wilt and woodiness. These hybrids must be propagated clonally since seedling progeny segregate. For this reason, rooted cuttings are used, but they also perpetuate diseases such as Haematonectria canker and some that are caused by viruses.

P. quadrangularis, the giant granadilla, is a native of tropical South America. It was first cultivated in the 18th century, and is now widely distributed (Purseglove, 1968). It is adapted to hot, humid environments, and does not do well at the elevations that are required by *P. edulis* f. *edulis*. In addition to the common beverage and dessert uses for its pulp and juice, the green fruit are boiled and used as a vegetable, and the fleshy, tuberous root is used as a substitute for yams.

P. ligularis, the sweet granadilla, is distributed from central Mexico to Venezuela, south-central Peru and western Bolivia, and is generally not utilized outside this region (Martin and Nakasone, 1970). It does poorly at low elevations and can be grown as high as 3000 m. It has become naturalized at high elevations in Hawaii.

P. mollissima, the curuba, tacso, tacsonia or banana passion fruit, originated in the Andes and is adapted to high elevations (2000–3000 m) and cool temperatures. Known as banana poka in the Hawaiian Islands, where it was introduced as an ornamental in 1921, it has become a naturalized, aggressive weed on Hawaii, Kauai and Maui. Biological control programmes in the islands recently have employed the Septoria blotch pathogen (Trujillo *et al.*, 1994; Norman and Trujillo, 1995). Another disease, brown spot, limits commercial production of the species.

P. foetida, the stinking passion flower, is from the West Indies and South America, and produces an edible, but rarely used fruit. It has been introduced to many tropical countries where it has become a naturalized weed. It has been grown as a cover crop in Malaysia and East Africa, and has been used to control erosion.

Production and Importance

Worldwide production figures for passion fruit are not available. However, statistics from different producing regions clearly indicate that it is an important crop. Abeysinghe (1973) listed 26 areas in which passion fruit was produced commercially, and Winks *et al.* (1988) estimated that 80–90% of all production occurred in Australia, Brazil, Fiji, Hawaii, Kenya, Papua New Guinea, South Africa, Sri Lanka, Taiwan and Venezuela. Latin America is the most important producing region, and Brazil, Colombia, Ecuador and Peru supply the world export market for juice. Given the crop's popularity and widespread cultivation, worldwide production probably approaches 1 Mt per annum.

The area under production fluctuates dramatically. The crop has a short cycle (fruit are produced 6–9 months after planting and full production is reached after 18 months), but the economic life of a plantation is only 3–5 years. Because growers can rapidly enter or leave the market in response to market conditions, the area under production is price sensitive.

PASSION FRUIT DISEASES THAT ARE CAUSED BY BACTERIA

Bacteria cause diverse symptoms on passion fruit. *Agrobacterium tumefaciens* causes tumours; *Erwinia carotovora* ssp. *carotovora* causes a soft rot; *Ralstonia solanacearum* causes a vascular wilt; and *Pseudomonas syringae* pv. *syringae*, *P. syringae* pv. *passiflorae* and *P. viridiflava* cause leaf spots (Bradbury, 1986). Premature plant death, a disease whose aetiology is still uncertain, has been associated with a bacterium (Malavolta, 1998).

Bacterial spot

The most important bacterial disease is bacterial spot. Bacterial spot has been recorded in Australia and Colombia, and is especially important in Brazil (Malavolta, 1998). It is the most important bacterial disease of passion fruit due to the high susceptibility of economically important cultivars, the high level of damage the disease causes and the difficulty with which it is controlled.

Symptoms

Small, well defined, generally circular spots that reach 1 cm in diameter under favourable conditions are the most common symptoms on leaves. They may occur on any part of the leaf, but are most common along the edges (Fig. 18.1). They are translucent, dark green, anasarcous, encircled by a chlorotic halo, and have an oily aspect. As the disease progresses, it may cause defoliation. Leaf infections may become systemic and affect branches, which suffer progressive drying and develop longitudinal grooves and darkened vascular bundles (Fig. 18.2). Necrotic, circular or irregular lesions are formed on fruit. These are dark or brownish green and oily with well-defined edges. They form a hard crust that may cover several originally separate lesions. These spots penetrate to the pulp, making the fruit unmarketable. The disease may also occur on petals and flowers of *P. alata* (J. Rodrigues Neto, Campinas, 1999, personal communication).

Causal agent

Xanthomonas campestris pv. *passiflorae* is rod shaped, 0.50 × 1.50 μm, Gram-negative, aerobic and has a polar flagellum (Bradbury, 1986). In culture, there is a strain of the pathogen that produces yellow pigments that are typical of xanthomonads and another that is colourless. Its optimum temperature for growth is 27°C.

Fig. 18.1. Bacterial spot lesions, caused by *Xanthomonas campestris* pv. *passiflorae*, on the underside of a passion fruit leaf (photo: V.A. Malavolta).

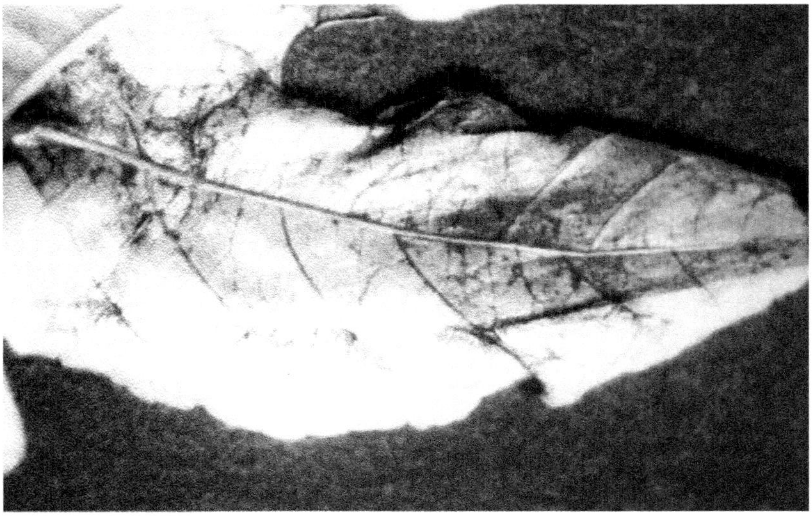

Fig. 18.2. Symptoms of systemic infection of a passion fruit leaf, caused by *Xanthomonas campestris* pv. *passiflorae* (photo: V.A. Malavolta).

Biochemical, serological and pathogenic variability is recognized in the taxon, and different strategies are required to manage disease caused by different strains. This behaviour has important practical significance when genetic resistance to this disease is sought.

Epidemiology

P. alata, *P. amethystina*, *P. serrato-digitata* and *P. edulis* ff. *edulis* and *flavicarpa* are affected (L.O.S. Beriam, Campinas, 1999, personal communication).

Penetration by *X. campestris* pv. *passiflorae* occurs most frequently via stomata and hydrathodes. Injury probably also contributes to the infection process. Infection is favoured by high relative humidity, a water film on the leaf surface and frequent rainfall. Local dissemination of the pathogen is enhanced by wind-blown rain and irrigation, and by workers handling wet plants, whereas long-distance dispersal occurs on seedlings and, according to Dias (1990), externally and internally on seeds. Insect dissemination has not been demonstrated.

The bacterium survives better in waste from contaminated crops than it does in contaminated seeds. Often, *X. campestris* pv. *passiflorae* occurs simultaneously with *Colletotrichum gloeosporioides* in leaves and fruit and *Fusarium* sp. in stems and roots. These interactions remain little studied.

Management

Seeds and seedlings should be from healthy plants and, if possible, should be obtained from disease-free areas. Alternatively, seeds should be treated at 50°C for 15–30 min (Dias, 1990). Other complementary measures that should be adopted include: planting in areas that have not had the disease for the preceding 2 years; use of wind breaks; avoiding work on plants when they are wet; disinfesting pruning tools and hands; and using fertilizers judiciously, especially with respect to nitrogen.

Chemical control is based on the use of mixtures of cupric and carbamate fungicides, or products that contain streptomycin or oxytetracycline. These measures have shown variable effectiveness that may be due to crop management, the quality and frequency of applications, the level of infection and susceptibility of the host plant, and virulence of the pathogen. Currently available products rarely afford satisfactory control of systemic infections under favourable environmental conditions.

Genetic resistance would be a most effective and desirable alternative for the control of bacterial spot. Thus, much effort has been channelled towards the search for resistant materials that possess good agronomic and commercial characteristics (Malavolta, 1998). Among the species that have shown resistance are *P. caerulea*, *P. cincinata*, *P. foetida*, *P. gilberti*, *P. mollissima* and two non-identified species, Introduction I 48669 ('curuba de la Sierra de Santa Marta') and 'maracujá-de-cobra'.

PASSION FRUIT DISEASES THAT ARE CAUSED BY FUNGI AND FUNGUS-LIKE ORGANISMS

Aboveground Diseases

The most important fungal diseases of the aerial portions of the plant are anthracnose and brown spot. They affect leaves and cause pre- and postharvest fruit spoilage in most producing countries. Several other diseases vary in importance in different countries. These include Diplodia rot, Phomopsis rot, stem-end rot and soft rot. Less important postharvest diseases are Fusarium rot, Penicillium rot and Septoria rot.

Anthracnose and fruit canker

Anthracnose and fruit canker probably occur wherever this crop is grown in humid environments (Cedeño *et al.*, 1993; Lutchmeah, 1993; de Goes, 1998; Wolcan and Larran, 2000). *P. alata*, *P. edulis* and *P. quadrangularis* may suffer extensive defoliation, and affected fruit command a lower market price and have greatly reduced postharvest shelf lives.

SYMPTOMS All aerial organs of the plant are attacked (Persley, 1993; de Goes, 1998). Spots, initially 2–3 mm in diameter and oily in appearance, are produced on the leaf. They become dark brown, round or irregularly shaped and >1 cm in diameter. The centres of spots become brittle and may break apart. Under high temperature and humidity, lesions also develop on petioles. As foliar lesions coalesce, large areas of the leaf die, resulting, eventually, in abscission. Dark brown spots, 4–6 mm in diameter, are produced on the branches and tendrils, eventually turning into cankers. Severe lesions can cause the death of shoots and a partial blighting of the plant (Fig. 18.3).

Affected flowers abort, and immature fruit abscise. Lesions on fruit initially are superficial and light brown, and later become sunken and greyish to dark brown (Plate 106). They may be larger than 1 cm in diameter and may reach interior portions of the fruit. As fruit mature, the spots enlarge and become oily or light tan. The fruit skin becomes papery and acervuli are formed on lesions here and on leaves. Under high humidity, masses of red and orange spores form in acervuli.

Dieback, characterized by reduced elongation of shoots, shortened internodes and an eventual wilting and death of these structures, has been described in several countries. However, these symptoms are associated normally with other causes such as water stress, nutritional problems or a poor root system. Although *C. gloeosporioides* was isolated consistently from plants with dieback in Suriname, these symptoms were associated

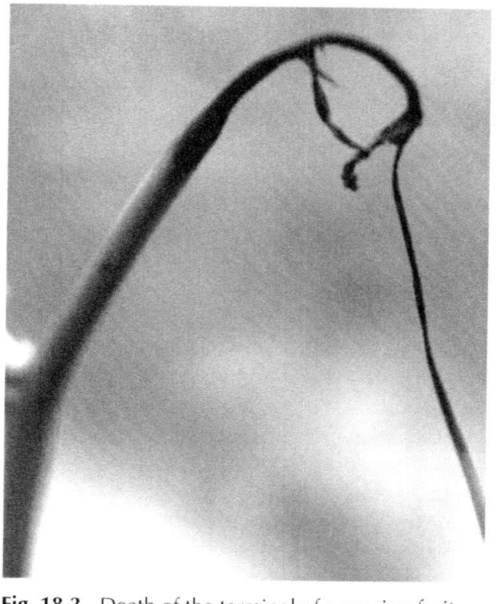

Fig. 18.3. Death of the terminal of a passion fruit vine caused by anthracnose (photo: R.C. Ploetz).

with root disease caused by *Fusarium solani* (Power and Verhoff, 1984). In Brazil, dieback symptoms normally are associated with factors that cause plant stress, which may include a simultaneous infection by other pathogens, particularly the bacterial spot agent, *X. campestris* pv. *passiflorae* (de Goes, 1998).

CAUSAL AGENT *C. gloeosporioides* is described in Chapter 1. Its teleomorph, *Glomerella cingulata*, has not been observed on passion fruit in nature, but isolates from *P. edulis* have produced it in culture (Wolcan and Larran, 2000). Unlike strains of *C. gloeosporioides* from other hosts, there does not appear to be physiological specialization among those recovered from passion fruit (Dodd *et al.*, 1992).

EPIDEMIOLOGY Conidia are produced in lesions and dead and senescent tissues of the plant (Dodd *et al.*, 1992). Within vines and plantings, the pathogen is spread mainly by rain or irrigation water, whereas long-distance dissemination relies on infected seeds, seedlings and cuttings. High temperature and humidity, reduced ventilation and light, and host injury encourage disease development. Maximum germination of conidia occurred between 30 and 33°C in the dark, and was accelerated between 22 and 25°C in the presence of light (Francisco Neto *et al.*, 1994).

Host injury increases infection, but is not an obligate requirement. Quiescent infections occur on immature fruit of *P. alata* and *P. edulis* f. *flavicarpa*, whereby infections stop development after appressorium formation (Jeffries *et al.*, 1990). The mechanisms that regulate the dormancy of appressoria are not known.

MANAGEMENT Interspecific hybrids between *P. mollissima* and *P. tripartita*, and *P. mixta* and *P. cumbalensis* have exhibited stable resistance. Studies are needed on the agronomic characteristics of their fruit before they could be used in production (Sanudo-Sotelo and Zunga-Ravelo, 1991).

Control involves the use of pathogen-free seeds or seedlings, adequate spacing, planting in areas that are free of strong or constant winds (wind breaks are helpful), and avoiding injury to the plant and its fruit. Fruit should not be harvested during wet conditions, unduly exposed to the sun, or kept for long in the absence of refrigeration. In areas where there is no history of bacterial spot, pruning to eliminate affected areas and improve ventilation and light conditions helps control the disease. Pruning should be done when plants are dry, and should be followed with applications of a protective fungicide. Applications of mixed formulations of protective and curative fungicides are needed during disease conducive conditions.

Brown spot

Brown spot has been recorded in Australia, Canada, Indonesia, Kenya, Mauritius, New Zealand, New Guinea, South Africa, Tanzania, the USA and Zambia (Reid, 1938; Holliday, 1980; Lutchmeah, 1993; Persley, 1993; de Goes, 1998). *P. quadrangularis*, *P. edulis* f. *edulis* and *P. edulis* f. *flavicarpa* are susceptible, and the incidence on the latter taxon in areas of high rainfall can be as high as 98%.

SYMPTOMS The two most common brown spot agents produce distinct symptoms. *Alternaria passiflorae* causes reddish brown spots on leaves, ~5 mm in diameter (Plate 107) (Holliday, 1980; de Goes, 1998). They enlarge and become zonate under high humidity. Conidia and conidiophores of the causal fungus are found in the centre of the lesion. Since a single lesion can cause abscission, much of the plant may defoliate. Small, dark, elongated lesions form on branches next to the insertion point of the leaf petioles, and drying and death of branches may occur during conducive conditions. Slightly circular spots develop on fruit. They are reddish brown, exceed 1 cm in diameter, form on fruit at an advanced stage of development, and damage the fruit's commercial value (Plate 108).

In contrast, *A. alternata* causes smaller spots, 1–5 mm in diameter, with chlorotic haloes on leaves (Plate 109) (K.G. Pegg, personal communication). The stem lesions it causes rarely kill vines. Spots on fruit have dark green, greasy margins (Plate 110). *A. alternata* is more virulent on hybrids on which it induces rapid defoliation (K.G. Pegg, personal communication).

CAUSAL AGENTS Nine species of *Alternaria* have been described as pathogens of passion fruit. After *A. passiflorae* and *A. alternata* (Holliday, 1980; de Goes, 1998), the less common species are *A. macrospora* (Ram *et al.*, 1977); *A. aliena*, *A. aragakii*, *A. hawaiiensis*, *A. tenuissima* and *A. tropica* (Simmons, 1993); and *A. guangxiensis* and *A. bannaensis* (Chen and Zhang, 1977). *A. tomato* has also been associated with the disease.

A. alternata is described in Chapter 1. *A. passiflorae* produces solitary conidia on its host, but chains of up to five in culture (Ellis, 1971). They are pale to mid brown, smooth or occasionally minutely verrucose, straight to slightly curved, and obclavate or with the body of the conidium ellipsoidal tapering to the simple or branched beak which is usually the same length or longer than the body (Fig. 18.4). They are 100–250 μm long and 14–29 μm wide at the widest point, have 5–12 transverse and a few longitudinal or oblique septa, and are constricted at septa. Conidiophores are produced singly or in groups, usually simple, straight or flexuous, pale to mid brown, smooth, with several conidial scars, and up to 120 μm long and 6–10 μm wide.

A. macrospora produces solitary or sometimes chains of two conidia that are mid to mid-dark reddish brown, straight or curved, usually minutely verrucose, and obclavate or with the body of the conidium ellipsoidal tapering abruptly to a very narrow simple beak which is usually the same length or twice as long as the body (Fig. 18.5) (Ellis, 1971). They are 90–180 (134) μm long and 15–22 (17.7) μm wide at the widest point, have 4–9 transverse and several longitudinal or oblique septa, and are slightly constricted at septa. Conidiophores are produced singly or in groups, simple, straight or flexuous, pale to mid brown, smooth, with one to several conidial scars, and up to 80 μm long and 4–9 μm wide.

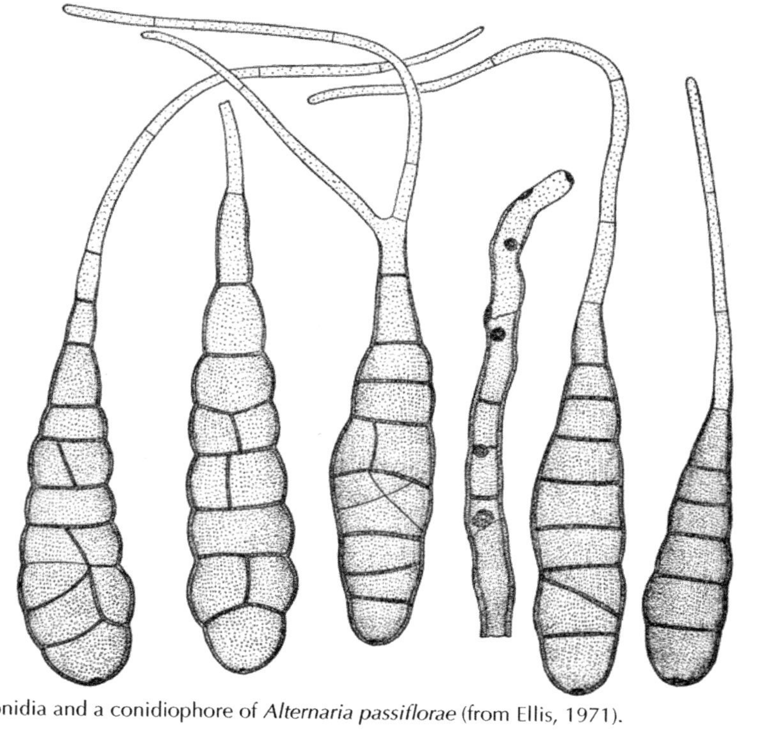

Fig. 18.4. Conidia and a conidiophore of *Alternaria passiflorae* (from Ellis, 1971).

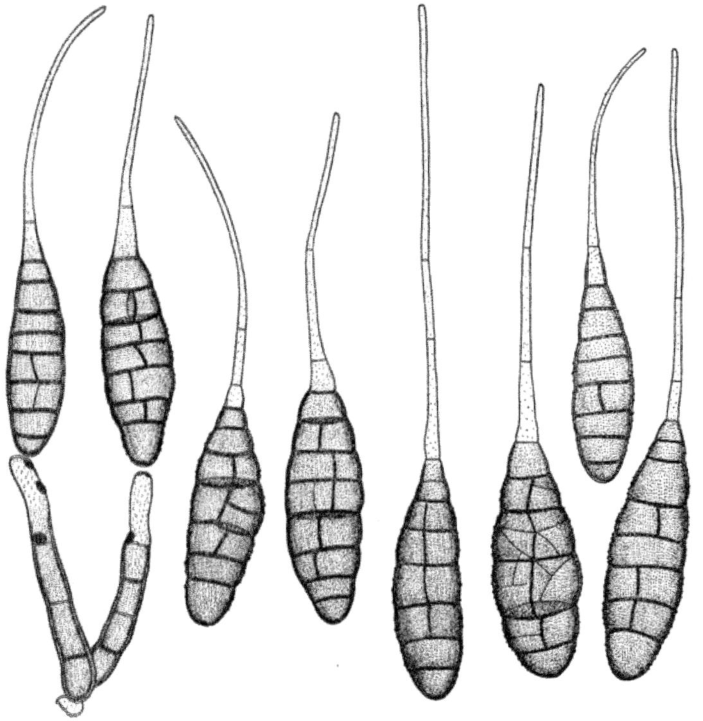

Fig. 18.5. Conidia and conidiophores of *Alternaria macrospora* (from Ellis, 1971).

A. tenuissima produces solitary or short chains of conidia that are pale to mid-clear golden brown, straight or curved, sometimes minutely verrucose, and obclavate or with the body of the conidium ellipsoidal tapering gradually to the beak which is up to half the length of the conidium (Fig. 18.6) (Ellis, 1971). They are 22–95 (54) μm long and 8–19 (13.8) μm wide at the widest point, have 4–7 transverse and several longitudinal or oblique septa, and are slightly or not constricted at septa. Conidiophores are produced singly or in groups, simple or branched, straight or flexuous, pale or mid pale brown, smooth, with one to several conidial scars, and up to 115 μm long and 4–6 μm wide.

EPIDEMIOLOGY The conidia are dispersed by wind, irrigation water and rain, and, occasionally, by infected seedlings. Little is known about the pre- and post-infection events in the host and physiological specialization in the pathogen.

MANAGEMENT Thinning vines to increase ventilation and penetration by fungicides can reduce disease pressure (de Goes, 1998). The most efficient fungicides include the carbamates, iprodione and copper compounds. Mancozeb should be applied every 2 weeks during the period most favourable for disease development, and over greater intervals when conditions are less favourable. In Australia, good control has been obtained with dicarboximide fungicides, although they become ineffective as resistance to them develops. Hybrids between *P. edulis* f. *edulis* and f. *flavicarpa* have been selected in Australia, most of which are resistant to *A. passiflorae* but susceptible to *A. alternata* (K. Pegg, personal communication).

Scab (Cladosporium rot)

Scab, which is also known as Cladosporium rot, has been reported in Australia and Brazil

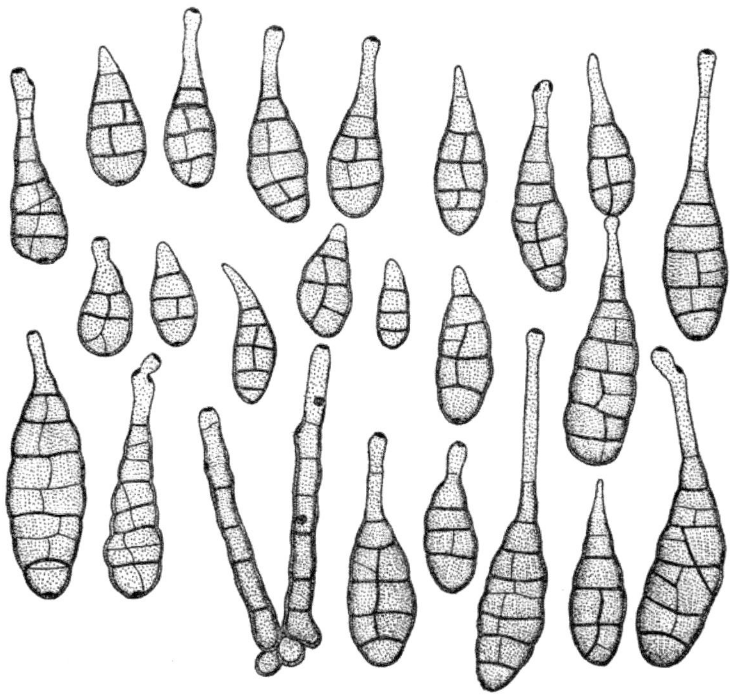

Fig. 18.6. Conidia and conidiophores of *Alternaria tenuissima* (from Ellis, 1971).

(Persley, 1993; de Goes, 1998). When scab is severe, especially on branches, it can delay flowering and production, as well as reduce the commercial quality of fruit. Passion fruit of the Amazon, *P. nitida* and *P. edulis* f. *flavicarpa* are susceptible.

SYMPTOMS Symptoms develop on young leaves, branches, tendrils, flower buds and fruit (Fig. 18.7, Plate 111) (Persley, 1993; de Goes, 1998). Small spots, 3–5 mm in diameter, form on leaves. They are light brown, translucent, round and become necrotic. Lesions can perforate leaves or, when they occur on veins, cause them to be deformed; they often cause abscission. Small brown spots form on branches and tendrils, and become elongated and sunken in the form of a canker. As scar tissue forms, branches become weakened and break in the wind.

Lesions on flower buds are 5 mm in length, brown and contain greenish grey powdery signs. High numbers of lesions on flower buds or on peduncles can greatly reduce the number of flower buds.

Circular spots, up to 5 mm in diameter, are produced on fruit. They are translucent, and are later covered with a brown, rough tissue that rises several millimetres above the fruit surface (Pio-Ribeiro and Mariano, 1997). Several lesions may form on the same fruit, causing it to be deformed and stunted. The fungus does not directly affect the internal parts of the fruit. On fruit of *P. nitida*, dark brown lesions are restricted to the skin. In large numbers, they coalesce and seriously affect the fruit's growth.

CAUSAL AGENTS *Cladosporium oxysporum* is responsible in Australia, and *C. cladosporioides* and *C. herbarum* in Brazil (Persley, 1993; Barreto *et al.*, 1996). *C. cladosporioides* and *C. herbarum* are described in Chapter 6. *C. oxysporum* produces macronematous, straight or slightly flexuous conidiophores that are nodulose, pale or mid pale brown, smooth, and up to 500 μm long and 3–5 μm wide, with terminal and intercalary swellings 6–8

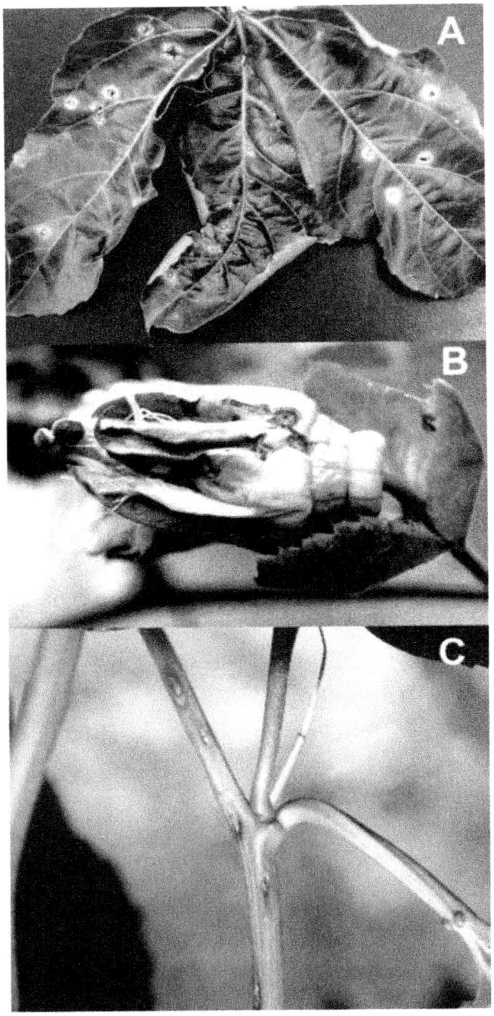

Fig. 18.7. Scab symptoms on a passion fruit (A) leaf, (B) flower and (C) stem (photos: A. de Goes).

μm in diameter (Fig. 18.8) (Ellis, 1971). Conidia are cylindrical and rounded at the ends, ellipsoidal, limoniform or subspherical, subhyaline or pale olivaceous brown, smooth, 5–30 × 3–6 μm, and arise simply or in branched chains from terminal swellings, which later become intercalary.

EPIDEMIOLOGY The optimum temperature for the agents varies: *C. oxysporum*, 19.5–24°C; *C. cladosporioides*, 20–28°C; and *C. herbarum*, 28–30°C (Domsch *et al.*, 1980; K.G. Pegg, personal communication). In

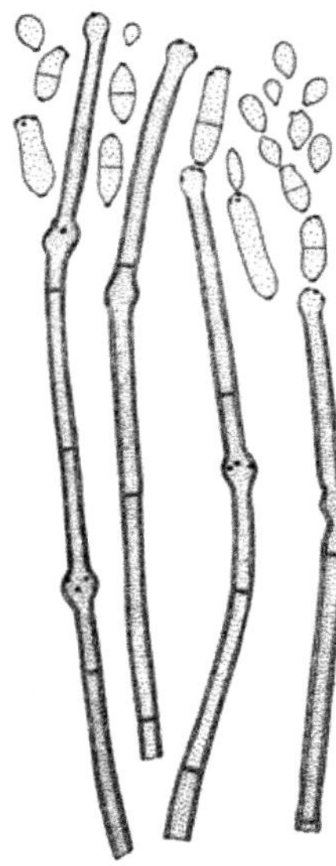

Fig. 18.8. Conidia and conidiophores of *Cladosporium oxysporum* (from Ellis, 1971).

Australia, scab is a winter disease (K.G. Pegg, personal communication). Elsewhere, a complex of scab symptoms develops under mild conditions on leaves, branches, tendrils, flower buds and fruit, whereas hot conditions result in more frequent development on the external parts of the floral organs, especially the bracts and the buds.

Dissemination occurs through infected seedlings, and by wind and sprinkler water. Although conidia are found frequently on seeds, there is no evidence for seed dissemination.

MANAGEMENT The measures that are recommended for anthracnose are also effective against scab. Regular fungicide applications are needed when disease-conducive conditions prevail, especially during periods of intense growth and flowering.

Septoria blotch (spot)

Septoria blotch, which is also known as Septoria spot, has been recorded in Australia, Florida, Central and South America, India, Oceania, Mauritius and South Africa (Inch, 1978; Holliday, 1980;

Lutchmeah, 1993; Alfieri *et al.*, 1994). It occurs on *P. edulis* f. *flavicarpa*, *P. quadrangularis*, and weed species such as *P. mollissima* (Trujillo *et al.*, 1994; Norman and Trujillo, 1995). The disease damages both seedlings and adult plants, but is less severe and prevalent than anthracnose.

SYMPTOMS Symptoms are most common on leaves, and occur less commonly on fruit and branches (Holliday, 1980). Leaf lesions are necrotic, light brown, 2–8 mm in diameter, slightly rounded, and normally accompanied by a yellow halo (Plate 112). Pycnidia of the pathogen occur in the centre of lesions. A single lesion per leaf is sufficient to cause abscission, and even leaves without visible symptoms may fall prematurely. Under disease-conducive conditions, leaves of all ages may fall, resulting in drying of the branches and, at times, plant death.

According to Inch (1978), infection of fruit may occur at any stage of development. Lesions are light brown with dark green borders (Plate 113). Lesions may coalesce and cover extensive areas of a fruit, affecting its development or maturation.

CAUSAL AGENTS Three different species have been reported, although the literature is confused on their identity. *Septoria fructigena* has been reported in Kenya, South Carolina and perhaps other locations (Nattrass, 1939; Farr *et al.*, 1989) but, based on its description is apparently a species of *Phomopsis* (M. Priest, personal communication). *S. passiflorae* has been described for isolates from diseased passion fruit in South Africa. Shortly after it was described Louw (1941) reported a new species in South Africa but, because he was not aware of the earlier report, also used the epithet *S. passiflorae*. In 1980, Punithalingam reported that these fungi were distinct, and renamed Louw's species *S. passifloricola*. Although *S. passifloricola* appears to be more widely spread than *S. passiflorae*, unfortunately it is called *S. passiflorae* in much of the recent literature.

S. passifloricola produces dark, spherical and subepidermic pycnidia that are 50–160 μm in diameter (Fig. 18.9) (Holliday, 1980). They may erupt and become ostiolate. The short conidiophores bear at their extremities conidia that are 2 × 23 μm, one- to four-celled, and filiform at either end or slightly obtuse and round. The conidia are released in hyaline cirri and are agglutinated by a mucilaginous substance. The presence of pycnidia on the lesion surface is an important identifying characteristic for the disease.

EPIDEMIOLOGY Conidia contained in cirri are moved by water and insects. The fungus survives in infected tissues, and mucilage in the cirrus is thought to aid survival. Prolonged rains and mild temperatures favour disease development.

MANAGEMENT The use of carbamate and benzimidazole fungicides, with measures to control anthracnose and brown spot, are effective against Septoria blotch. To reduce the buildup of resistance in the latter fungicides, they should be mixed or alternated with those with different modes of action (Peterson, 1977).

Postharvest Diseases

Postharvest diseases often result from infections of immature or maturing fruit in the field. Latent infection of fruit is increased by prolonged exposure to rain and high humidity.

Typically, producers harvest fruit only after they have abscised. This is done at variable intervals, depending mainly on the quantity produced.

Some of the most frequent postharvest diseases are anthracnose, Fusarium patch (caused by *Fusarium pallidoroseum*), Phomopsis rot and stem-end rot (*Phomopsis tersa*; Fig. 18.10). Occurring less frequently are Diplodia rot (*Diplodia theobromae*), Fusarium dry rot (*F. proliferatum*), Penicillium rot (*Penicillium expansum*), Pestalotiopsis brown spot (*Pestalotiopsis* sp.) and soft rot (*Rhizopus stolonifer*). The most common pathogens on *P. alata* are *C. gloeosporioides*, *Sclerotium rolfsii*, *D. theobromae*, *Gliocladium roseum*, *Alternari passiflorae* and *R. stolonifer*. *D. theobromae* causes a black rot of the skin and the pulp. *G. roseum* and *R.*

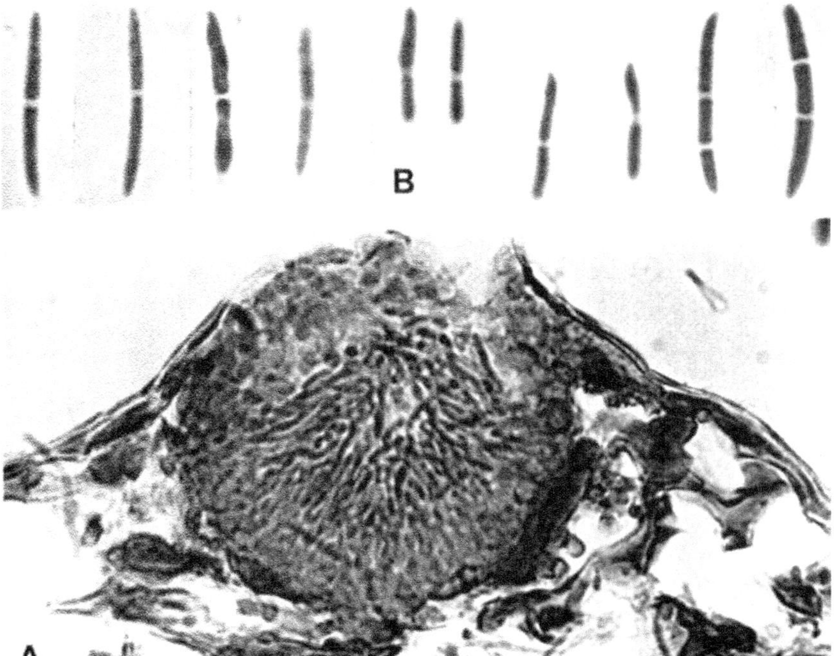

Fig. 18.9. (A) Pycnidium and (B) conidia of *Septoria passifloricola* (from CMI description no. 670).

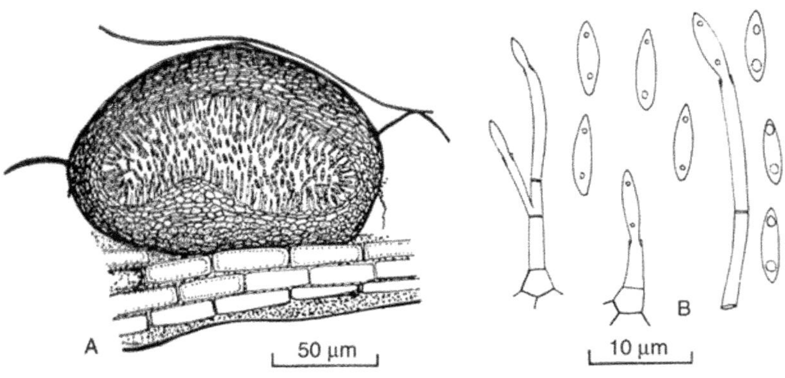

Fig. 18.10. (A) Vertical section of stromatic conidiomata and (B) conidiophores, conidiogenous cells and conidia of *Phomopsis tersa* (from CMI description no. 1169).

stolonifer initiate infection via the peduncular region, whereas *S. rolfsii* may initiate it on any part of the fruit. *A. passiflorae* causes spots and rotting (Anselmo and Junqueira, 1997). Simultaneous infections by other pathogens often occur.

Control of this complex of diseases involves adequate care of plants and suppression of disease-predisposing factors. Careful harvest, transport and storage are indispensable measures to improve the longevity of the fruit. Good anthracnose control has been obtained by treating fruit at 47.5°C for 5–10 min, or at 54°C for 1–5 min (Jesus *et al.*, 1994).

Miscellaneous Aboveground Diseases

Vine dieback differs from that which results from the interaction between *F. solani* and *C. gloeosporioides* (Power and Verhoeff, 1984), in that symptoms are not restricted to sprouts and it does not cause plant death. According to Cedeño *et al.* (1995), it is caused by *D. theobromae*. The fungus causes lesions on the internodes of main branches and where leaves and secondary branches emerge, later invading the internal tissues and inducing dieback. The disease is associated with structural damage caused by the mite, *Tetranychus urticae*, and with water stress.

A flower rot caused by *R. stolonifer* resulted in losses of up to 63% of the flowers of *P. alata* in Brazil (de Goes, 1998). More recently, a similar disease on flowers of *P. edulis* f. *flavicarpa* has been associated with a non-identified species of *Rhizopus*. Flower rot occurs mainly in the summer, during periods of prolonged rain. Little is known about its geographic distribution, incidence, impact and control. Symptoms include dark spots that are water soaked on the interior of the flower bud, especially on the sepals and petals (Fig. 18. 11). Later, spots occur over the entire flower bud, and dark grey mycelia and sporangia develop. The flower bud becomes putrid and abscises easily. Control has been difficult under conditions that favour for the disease. To date, only tebuconazole has been effective.

Soilborne Diseases

Fusarium wilt

Fusarium wilt was once a major disease of *P. edulis* f. *edulis* in Queensland and New South Wales, Australia. This vascular wilt was first reported by McKnight (1951), and was one of the most important constraints to production in Australia prior to the use of resistant rootstocks. The disease may also be present in Brazil (Carvalho-Dias *et al.*, 1998) and South Africa (Grech and Rijkenberg, 1991), but the incomplete nature of these reports makes it unclear whether Haematonectria canker or Fusarium wilt was reported. Although a clear distinction between Haematonectria canker and Fusarium wilt is made in this chapter, this is not the case in other publications (e.g. Cole *et al.*, 1992).

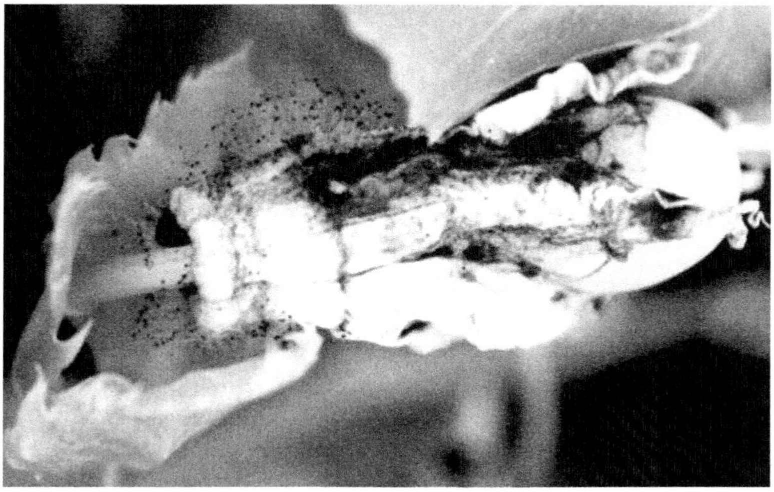

Fig. 18.11. Rot of passion fruit flower caused by *Rhizopus stolonifer*. Note the black sporangia of the pathogen covering the flower's base (photo: A. de Goes).

SYMPTOMS The first external symptoms of Fusarium wilt are a pale colour of new growth which proceeds to a wilt of one or more shoots as soon as 24–48 h after symptoms first appear (Holliday, 1980). Eventually, affected plants die. Symptom development may be unilateral or encompass the entire plant. Fruit shrivel and remain attached to the vine if they are immature. Unlike the somewhat localized vascular browning that occurs in plants affected by Haematonectria canker, internal symptoms of Fusarium wilt may extend to 2 m above the soil line.

CAUSAL AGENT *Fusarium oxysporum* f. sp. *passiflorae* causes Fusarium wilt of passion fruit. The species is described in Chapter 1.

Purss (1954) determined that *F. oxysporum* f. sp. *passiflorae* was a distinct forma speciales after showing that it was pathogenic on *P. edulis* f. *edulis* but not on other Fusarium wilt suscepts (i.e. tomato, watermelon and pea). Gardner (1989) demonstrated that *P. foetida*, *P. ligularis* and *P. mollissima* were also susceptible, but that *P. edulis* f. *flavicarpa*, *P. suberosa* and 24 miscellaneous monocots and dicots were resistant.

P. edulis is in the subgenus *Granadilla*, series *Incarnatae*. Since the susceptible species were not closely related (*foetida* is in subgenus *Dysomia*, *mollissima* in the subgenus *Tacsonia*, and *ligularis* in the subgenus *Granadilla*, series *Tiliaefoliae*), Gardner (1989) speculated that the host range of *F. oxysporum* f. sp. *passiflorae* probably included more species of *Passiflora* than were identified in his work.

MANAGEMENT Considerable effort was made in Australia to identify or breed resistant rootstocks during the 1950s and 1960s. Due to the widespread and damaging nature of Fusarium wilt, resistant rootstocks are now viewed as essential for the production of passion fruit in Australia (Winks *et al*.., 1988).

In New South Wales, Cox and Kiely (1961) determined that *P. edulis* f. *flavicarpa*, *P. coerulea* (sic = *caerulea*), *P. incarnata* and *P. herbertiana* were highly resistant. However, since *P. herbertiana* had a slender stem and *P. caerulea* and *P. incarnata* had a suckering growth habit, they were deemed unsuitable as rootstocks. Their results agreed with those of Groszmann and Purss (1958) in Queensland who indicated that *P. edulis* f. *flavicarpa* was generally tolerant of the disease. Groszmann and Purss (1958) identified a superior wilt-resistant selection of *P. edulis* f. *flavicarpa* that had the added attributes of resistance to nematodes and Phytophthora blight; it was still a standard in commercial production 30 years later (Winks *et al.*, 1988).

More recently, interform and interspecific hybrids have been evaluated for Fusarium wilt resistance and other traits (Winks *et al.*, 1988). A third generation cross between *P. edulis* and *P. incarnata*, 3–19/F3, grew to a graftable size more quickly and was more vigorous than *P. edulis* f. *flavicarpa*. However, it was intermediate in its susceptibility between *P. edulis* f. *flavicarpa* and *P. edulis* f. *edulis* and was not recommended for release to the industry. It was hoped that useful hybrids would be produced in crosses with new accessions of *P. incarnata* and *P. caerulea*.

Haematonectria canker, sudden wilt, collar rot, crown canker, base rot

Haematonectria canker has been identified in Australia, Florida (USA), Suriname, Taiwan, Uganda, Venezuela and Zimbabwe (Emchembe and Mukiibi, 1976; Power and Verhoeff, 1984; Lin and Chang, 1985; Cedeño *et al.*, 1990; Ploetz, 1991; Cole *et al.*, 1992; K.G. Pegg, personal communication). Based on symptoms and the fungi that have been recovered from affected plants, it may also occur in New Zealand, Réunion, Samoa and South Africa (Simmonds, 1938; Young, 1970; Gerlach, 1983; Terblanche *et al.*, 1986; Aubert, 1987).

Several different names are used to describe this serious disease including, unfortunately, Fusarium wilt (Cole *et al.*, 1992). Haematonectria canker and Fusarium wilt produce many of the same symptoms on passion fruit, and are both caused by species of *Fusarium*. Thus, it is not surprising that these diseases are frequently confused.

SYMPTOMS The first aboveground symptom is a dramatic decrease in the rate of shoot elongation, beginning within 3 months of planting (Power and Verhoeff, 1984). In Suriname, it was reported that affected terminals may wilt and/or die back, or may continue sporadic growth during subsequent rainy seasons. In most cases, however, a sudden and terminal collapse of vines is most common (Emchembe and Mukiibi, 1976). Terminal wilting of vines begins to occur shortly after they are trellised (Cole *et al.*, 1992) or after they begin to set fruit (Ploetz, 1991). All leaves in an affected vine lose turgor and become chlorotic, but are usually not shed. Fruit on bearing vines also remain attached and become shrivelled after vine death. Within 3–5 years, all vines in an affected planting die.

Wilting and dieback of the canopy result from the damage which the disease causes to stems and roots, and the corresponding reduction of the host's capacity to extract and conduct water (Fig. 18.12) (Cole *et al.*, 1992). Cankers that develop at the root collar or at physically wounded locations along the stem ultimately can girdle the entire stem (Plate 114) (Ploetz, 1991). Considerable root decay also occurs, and in some locations this is the only symptom reported from beneath the canopy (Power and Verhoeff, 1984). In Australia, root damage is associated with swelling of the stem at the soil line and adventitious root formation (Fig. 18.13) (K.G. Pegg, personal communication).

A reddish brown to brown discoloration of the host vascular system occurs in association with cankers and root decay, but usually does not progress more than 0.5 m from the affected area. Emchembe and Mukiibi (1976) examined affected tissues microscopically, and reported that the pathogen invaded and colonized the cortical and xylem parenchyma.

CAUSAL AGENTS Haematonectria canker is caused by homothallic strains of *Haematonectria haematococca* (anamorph: *Fusarium solani*). The fungus usually produces crimson coloured perithecia, singly or in groups, on the surface of cankers and affected roots. They are ~200 μm in diameter, and produce unitunicate asci, 80 μm long, in which eight bicellular ascospores, 14 μm long, are produced (Fig. 18.14) (Hanlin, 1990). Colonies of the pathogen grow 7.5 mm in

Fig. 18.12. (A) Stem canker and (B) root necrosis of passion fruit cuttings caused by artificial inoculation with *Haematonectria haematococca* (photos: R.C. Ploetz and A.W. Cooke, respectively).

Fig. 18.13. Swelling of a passion fruit stem at the soil line caused by *Haematonectria haematococca*. This symptom is common in Australia (photo: A.W. Cooke).

diameter day^{-1} on oatmeal agar, with abundant aerial mycelium and, eventually, numerous sporodochia; they are cream, aqua or blue, but never orange (Domsch *et al.*, 1980; Nelson *et al.*, 1983). The anamorph produces micro- and macroconidia on branched and non-branched monophialides. Microconidia are sparse to abundant, usually one-celled, oval to kidney shaped, and have thicker cell walls than those that are produced by *F. oxysporum*. Macroconidia are abundant, cylindrical, thick-walled and stout, with rounded, foot-shaped or notched basal and blunt or rounded apical cells. Chlamydospores are often abundant and form singly or in pairs. Both ascospores and conidia of the fungus are pathogenic (Emchembe and Mukiibi, 1976).

Other species of *Fusarium* and their respective teleomorphs may cause crown canker and sudden wilt of passion fruit. Young (as reported by Emchembe and Mukiibi (1976)) '...observed a crown rot of passion fruit after invasion by *F. sambucinum*...'. Sale (1987) indicated that *F. redolens*, *F. avenaceum*, *Gibberella baccata* (anamorph: *F. lateritium*) and *G. saubinetii* (anamorph: *F. graminearum*) were also 'known to be involved' in New Zealand. It is not clear whether Koch's postulates were completed with these fungi.

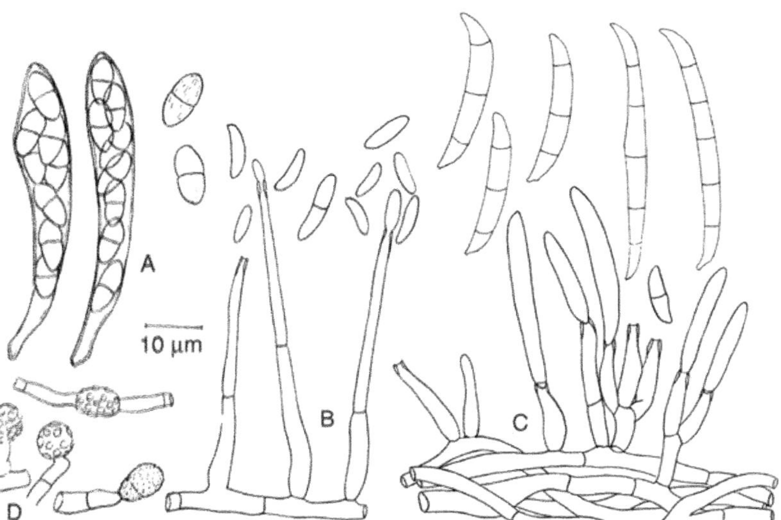

Fig. 18.14. (A) Asci and ascospores of *Haematonectria haematococca*, and (B) microconidia and conidiophores and (C) macroconidia and conidiophores of its anamorph, *Fusarium solani* (from CMI description no. 29).

EPIDEMIOLOGY Artificial inoculation studies indicate that *H. haematococca* is not a particularly aggressive pathogen on passion fruit, and that wounding has a profound effect on disease development. Several authors have shown that the onset of symptoms is hastened, and the incidence and severity of the disease is increased, if vines are wounded mechanically prior to inoculation (Emchembe and Mukiibi, 1976; Lin and Chang, 1985; Ploetz, 1991; Cole *et al.*, 1992). Emchembe and Mukiibi (1976) increased the percentage of affected plants by 100% by simply hoeing in between rows, a practice that presumably injured roots, thereby providing entry points for the pathogen. Ants of the genera *Solenopsis* and *Cromatogaster* may play a role in disease development in Venezuela (Cedeño *et al.*, 1990), and the disease is known to interact with Phytophthora blight, particularly where *P. edulis* is grown (Lin and Chang, 1985; Cole *et al.*, 1992).

Ploetz (1991) demonstrated that the pathogen colonizes the host vascular system in advance of visible external or internal symptoms, and that the pathogen could be recovered from asymptomatic plants in a commercial nursery. Thus, the pathogen could be moved in 'healthy' nursery stock or by cuttings from asymptomatic plants. Cuticle thickness or composition on the host may be important, since the disease can be reproduced in the absence of wounding when the cuticle is removed from stems with acetone prior to inoculation (R.C. Ploetz, unpublished).

Resistance to Haematonectria canker increases as plants age. Emchembe and Mukiibi (1976) reported that 76–100% of the 10-week-old plants that were root-dip or wound inoculated wilted, whereas only 28–48% of 12-month-old plants inoculated in the same manner succumbed.

MANAGEMENT When cuttings are used to propagate the crop, it is imperative that pathogen-free stock be used. No effective chemical measure has been described. Ssekeyewa *et al.* (1999) reported that biweekly drenches of copper oxychloride reduced the numbers of plants that developed collar rot in a small experiment in Uganda, but the experiment was not repeated and no statistical evidence for the significance of this treatment was presented. The same authors reported that three taxa, *P. edulis* f. *flavicarpa*, *P. maliformis* and *P.* ex Queensland ('Queensland Purple') were significantly more resistant than the susceptible *P. edulis* f. *edulis*. Resistance has also been reported among selections of *P. edulis* f. *flavicarpa* in Taiwan. During 1981–1982, an 800 ha plantation was established using the selections as rootstocks (Lin and Chang, 1985).

Phytophthora root and crown rot

This destructive disease, which is also called Phytophthora blight, has been reported from Australia, India, Sarawak, South Africa, Taiwan and Venezuela (Simmonds, 1959b; Van den Boom and Huller, 1970; Turner, 1974; Hseih, 1979; Erwin and Ribiero, 1996; Gonzalez *et al.*, 1999). Due to the wide distribution and host range of the causal agent (Erwin and Ribiero, 1996), the disease may occur in other countries.

Symptoms

The pathogen causes diverse symptoms on *P. edulis*. Affected plants first exhibit yellow or scorched leaves throughout the canopy (Plate 115). Vine apices die and are black. Subsequently, necrosis of leaf veins, wilting, and defoliation occur (Fig. 18.15). In its terminal phases, branches die back and extensive bark and cortical decay develop at the root collar, which thickens to three to five times its normal diameter. Grey green, water-soaked lesions develop on immature fruit, which then usually abscise (Plate 116).

Causal agent

Phytophthora nicotianae causes the above symptoms on *P. edulis* (Erwin and Ribiero, 1996). It is described in Chapter 1.

Epidemiology

P. nicotianae is a 'watermould' that requires moist conditions to complete many phases of its life cycle. The general epidemiology of the diseases its causes are discussed in Chapter 1.

Management

New plantings should be made in freely drained soil, preferably on a site with enough slope to encourage runoff of rainfall and irrigation. In South Africa, *P. edulis* f. *flavicarpa* is used as a rootstock for the susceptible *P. edulis* f. *edulis*, and bud unions of at least 0.6 m above the soil line are recommended to discourage disease development on the scion due to soil splash (Brodrick *et al.*, 1976). Although *P. edulis* f. *flavicarpa* is resistant to root and crown rot in Australia (K.G. Pegg, personal communication), this taxon is reported to be susceptible in Venezuela (Gonzalez *et al.*, 1999). Fungicides decrease the spread of the disease, and canopies should be thinned when they are used to ensure good penetration.

PASSION FRUIT DISEASES THAT ARE CAUSED BY NEMATODES

Nematodes are important pathogens in many different growing regions (dos Santos, 1998). The reniform nematode, *Rotylenchus reniformis*, and at least three species of root-knot nematode, *Meloidogyne arenaria*, *M. incognita* and *M. javanica*, are important. They can cause economic losses of fruit and reduce the life of a plantation.

R. reniformis is widely distributed and has a host range of 140 species, including *P. edulis*. In Fiji, the nematode was reported to cause foliar chlorosis and was found in 84% of the areas that were surveyed (Kirby, 1978). In Brunei, the nematode was reported to predispose *P. edulis* f. *flavicarpa* to root rot caused by *Phytophthora* spp. (Perregrine and Yunton, 1980).

Nematodes are disseminated in infested soil, water and equipment, as well as infected plant materials. They are best controlled with strict exclusion and sanitation measures. In general, the registrations of various nematicides are being lost rapidly due to their high toxicity and adverse environmental impacts.

PASSION FRUIT DISEASES THAT ARE CAUSED BY VIRUSES

Passiflora spp. are affected by viruses in several different families. Although many of the diseases they cause are localized or are laboratory curiosities, others impact production over

Fig. 18.15. Large-scale damage in a passion fruit planting caused by *Phytophthora nicotianae*. Note the dry, necrotic foliage in the foreground and the numerous skips throughout the planting (photo: A.W. Cooke).

large areas. Undoubtedly, a key reason for the importance of these pathogens is the extent to which this crop is propagated vegetatively.

Worldwide, the most important virus-induced disease of passion fruit is woodiness and other similar diseases, all of which are caused by potyviruses. Although vines are rarely killed, production is severely affected and the crop can become unprofitable in its first year.

Symptoms are not very useful to identify viruses that affect *Passiflora* spp. since many are caused by more than one virus and mixed infections are common. Furthermore, detailed symptom descriptions are often lacking in the literature. Commercially available enzyme-linked immunosorbent assay (ELISA) kits enable the identification of viruses in the *Potyvirus* genus.

Whilst it is somehow comforting to know the exact virus that affects a crop, there is usually little that can be done to rectify a situation once symptoms appear. For these diseases, prevention is the preferred means of control. This and other control measures are discussed at the end of this section. Due to its importance, extensive coverage of the woodiness disease is given below, whereas brief overviews of other, less important virus-induced diseases follow.

Woodiness

Woodiness was first described in southern Australia, where it was first recognized in the 1890s and seriously affected production by 1930 (Magee, 1948). The disease has also been reported in Brazil, Nigeria, South Africa and Taiwan and, based on symptoms alone, was also diagnosed in Hawaii, India, Italy, Japan, Kenya, Malaysia and the Philippines.

The term 'woodiness' describes the classical symptom of this disease, the distortion, hardening and thickening of the fruit pericarp. As work was conducted on this problem, it became clear that several different causal agents were responsible for this and other symptoms that were associated with the disease. Although Magee (1948) ascribed the disease to *Cucumber mosaic virus* (CMV), subsequent reports indicted different potyviruses, in particular, *Passionfruit woodiness virus* (PWV). The latter agents now appear to play more important roles in the development of this disease than CMV. Since the diseases that are caused by the different viruses are so intertwined, historically and in practice, they will be dealt with together below.

Symptoms

The pericarp is malformed, thickened and hardened with a much reduced pulp cavity and fewer seeds. Orange brown gum pockets are common in the rind (Plate 117). The lumpiness and distortion can be detected at an early stage, but may be confused with fruit fly stings. In a given vineyard, an outbreak of woody fruit can be preceded or followed by normal-appearing fruit that may be smaller or have slightly thickened pericarps. Ring spots, 3–8 mm in diameter, may occur on woody fruit, and can fuse to produce irregular blotching. Fruit may also be finely stippled.

Temperature influences the development of fruit and leaf symptoms. Symptoms are more common on fruit that set during autumn and winter and ripen from late winter to early summer, and on leaves are less pronounced in summer.

Leaf mosaic, mottle and ringspot symptoms are often associated with fruit woodiness. Mosaic symptoms appear as dark green, raised blisters on a centimetre scale on a yellow–green background (Fig. 18.16). Leaves are hardened and frequently distorted and puckered by the differential growth rates of the infected tissue. Where young terminal growth is affected, leaves have translucent yellow–white membranous areas where cells have collapsed to cause major distortions or missing lobes. Vein clearing may occur. Growth is slowed, internodes shorten and the terminals have a bunchy appearance. Mottle appears as numerous, small, yellow spots or flecks on young, mature leaves. These can be up to 3 mm in diameter, round and bright yellow, or the small spots can coalesce into yellow patches. The mottle may disappear as the leaf ages. Ringspots on leaves are rare.

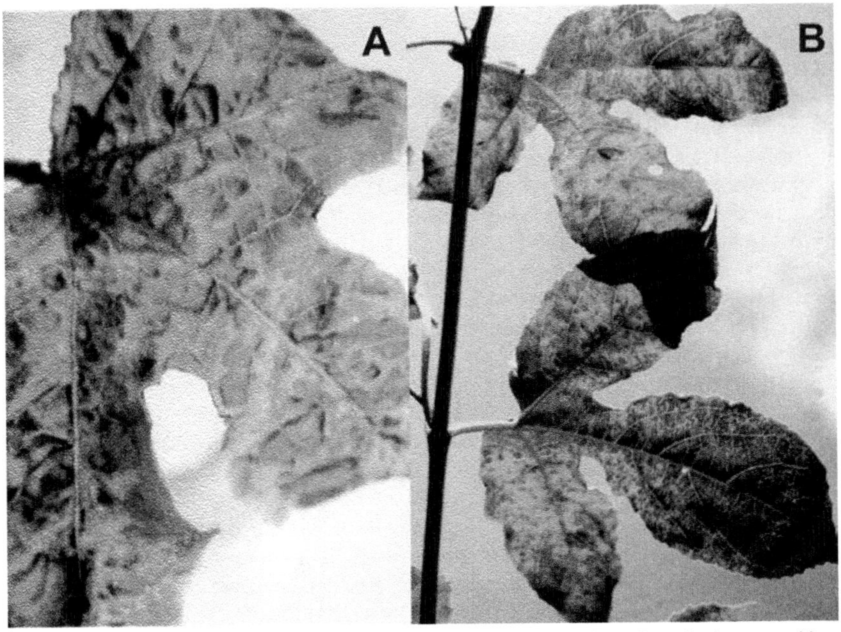

Fig. 18.16. (A) Dark green, raised blisters and (B) vein clearing, and small, yellow flecks caused by passion fruit woodiness (photos: A.W. Cooke and E.W. Kitajima, respectively).

Several authors indicated that dual infections resulted in increased symptom severity, even leading to vine dieback (Taylor and Kimble, 1964; Pares *et al.*, 1985; Kitajima *et al.*, 1986; Brand, 1992). Teakle *et al.* (1963) reported that CMV caused leaf mottling, distortion and stunted growth, whereas Kitajima *et al.* (1986) observed only local lesions on yellow passion fruit and no fruit woodiness with the same virus. Taylor and Kimble (1964) wrote that CMV caused vein clearing and epinasty of the terminal leaves of small seedlings of *P. edulis* which disappeared as plants matured, that PWV caused a systemic mosaic which persisted, and that dual infection led to tip necrosis, leaf fall and severe stunting and yellowing.

Causal agents

Magee (1948) first reported that CMV caused woodiness in Australia. CMV is the type species of the genus *Cucumovirus*, family *Bromoviridae*. It has a single-stranded, positive-sense RNA genome consisting of three unique RNAs, 1, 2 and 3, each of which is encapsidated separately in the same capsid protein. The particles with RNA3 also contain a subgenomic coat protein messenger, RNA4. The presence of all three particles is required for infection. The particles are ~30 nm in diameter, icosahedrally symmetric and consist of 180 capsid units.

In Queensland, McKnight (1953) was unable to associate CMV with woodiness symptoms. Taylor and Kimble (1964) later demonstrated that two viruses were associated with woodiness, CMV and a long flexuous virus, PWV. They suggested that the major cause of woodiness in Victoria and New South Wales was CMV, but in Queensland was PWV.

PWV is a member of the genus *Potyvirus*, family *Potyviridae*, the largest plant virus family. Virions are flexuous rods, 680–900 × 11–15 nm, that contain single-stranded, positive-sense RNA of ~9–10 kb, encapsidated with ~2000 helically arranged coat protein units.

The genome and protein products of the potyviruses have been characterized extensively (Dougherty and Carrington, 1988). They induce cytoplasmic 'pinwheel' inclusions in infected cells that are diagnostic for these viruses. Their morphology is independent of the host and has been used to divide the *Potyviridae* into subgroups. Classical serology does not assess the relatedness of different potyviruses reliably, but antibodies that are specific for the N-terminus of the coat protein and monoclonal antibodies can be used for this purpose. Potyvirus taxonomy currently is investigated with RNA and deduced amino acid sequences.

Mild, severe and tip blight strains of PWV exist. They are serologically inseparable and their coat protein sequences are at least 96% homologous. However, these strains are only 85% homologous with a strain from a hybrid between *P. edulis* f. *flavicarpa* and *P. edulis* f. *edulis* (Shukla *et al.*, 1988; Gough and Shukla, 1992).

It is possible that early workers who described CMV in passion fruit were, in fact, dealing with mixed infections. The only time CMV has been confirmed in passion fruit by itself was by Teakle *et al.* (1963) in California; with electron microscopy, they observed CMV virions in *P. caerulea* and *P. alata-caerulea*. Magee (1948) did not have the facilities needed to detect dual infections, and Taylor and Kimble (1964) did not test their CMV isolates for the presence of flexuous virus particles.

In South Africa, Brand (1992) occasionally found *Tobacco necrosis virus* (TNV; now split into two viruses *Tobacco necrosis virus A* and *Tobacco necrosis virus D*) and always found CMV in association with a potyvirus in woodiness-affected plants. In contrast, Da Graça (1976) found only a potyvirus in affected plants. It was isolated from *P. edulis* in Mpumalanga Province and had a modal length of 650 nm, whereas the virus characterized by Brand (1992) was isolated from *P. caerulea* in Kwa-Zulu Natal and had a modal length of 670 nm. The coat protein region of the latter was sequenced, and found to be distinct from PWV in Australia. It was termed PWV-SA but, based on sequence data, has since been identified as a strain of *Cowpea aphid-borne mosaic virus* (CABMV) and renamed CABMV-SAP (McKern *et al.*, 1994). Recently, Gerhard Pietersen serologically identified PWV in South Africa (J.E. Thomas, personal communication).

In Brazil, CMV is regarded as a minor pathogen, causing yellow rings or spots on leaves, but no systemic mosaic or fruit woodiness (Kitajima *et al.*, 1986). However, severe damage is observed in co-infections with PWV. By itself, PWV causes severe losses. It appears to be restricted to northwestern production areas of the country and is related, serologically, to other potyviruses. Symptoms in Brazil are identical to those in Australia, but sequence comparisons between isolates from both countries have not been conducted.

Woodiness was also reported from Nigeria (Martini, 1962). Although typical leaf mottling and distortions were described, fruit woodiness was not mentioned. CMV was not observed, and *Passion fruit mosaic virus* was the name proposed for the causal agent. The virus was not characterized sufficiently, and is now considered a synonym of PWV.

In Taiwan, woodiness is also a serious disease. The disease is widespread on *P. edulis* f. *flavicarpa*, reducing yields from 35 t ha^{-1} to 5 t ha^{-1} (C.A. Chang, 1992, personal communication). PWV often occurs in conjunction with *Passion fruit mottle virus* (Chang, 1992).

In summary, the classic woodiness symptoms on fruit are caused by a number of different potyviruses. The available evidence suggests that CMV is not a prime cause of the disease, but that co-infection with potyviruses is common and enhances the damage they cause on passion fruit. It is also possible that other viruses (see below) cause woodiness under cool conditions. Leaf mosaic and mottle seem to be consistent characteristics of potyvirus infection. Work is warranted to determine the relationships between potyviruses from different countries and the Australian PWV type strain of this pathogen.

Epidemiology

Members of the genus *Potyvirus* are vectored by aphids, and are also transmitted mechanically and by grafting (Nattrass, 1944). PWV reportedly is not seedborne (Chang, 1992).

CMV is transmitted by aphids in a non-persistent fashion. It can also be transmitted mechanically, and seed transmission is common.

OTHER POTYVIRUSES

Passionfruit ringspot virus

De Wijs (1974) described a disease of *P. edulis* f. *flavicarpa* in the Ivory Coast. The most common symptoms were yellow spots and mottling on otherwise normal leaves, and ringspots on younger leaves, especially where they were shaded. A few plants had severe mosaic and malformation on some of the leaves, and severely affected plants were stunted and bore few fruit. Woodiness was not seen, but fruit were small and shrivelled before ripening. Young, developing leaves showed yellow veins and epinasty after inoculation, followed by mottling.

The cause was identified as a potyvirus, *Passionfruit ringspot virus* (PRV), which was serologically related, but not identical, to PWV. It could be transmitted non-persistently by the aphids *Aphis gossypii* and *A. spiraecola*, but not by seed. De Wijs (1975) later demonstrated that an indigenous member of the *Passifloraceae*, *Adenia lobata*, was a native source of the virus. In the wild, infected *A. lobata* often was symptomless. Infection of cultivated plants depended on aphid incursions from these sources.

Passionfruit mottle virus

Chang (1992) described passionfruit mottle in Taiwan. Symptoms included mottling of leaves and the skin of the fruit, but not woodiness. On the basis of particle morphology, serology, production of cytoplasmic inclusions, aphid transmissibility (*Myzus persicae* Sulzer), and host range studies, the virus was classified as a potyvirus, *Passionfruit mottle virus* (PaMV). It had a different host range and caused different symptoms than PWV, and was serologically related to PWV, *Bean common mosaic virus* (BCMV), *Blackeye cowpea mosaic virus* (synonym: BCMV), *Watermelon mosaic virus 2* (WMV2) and *Soybean mosaic virus* (SMV). In Taiwan, PaMV was as widely spread as PWV and often occurred in mixed infections with this virus.

Soybean mosaic virus

Benscher *et al.* (1996) described a severe disease of *P. edulis*, *P. ligularis* and *P. quadrangularis* in Colombia. Symptoms included mosaic, chlorosis and epinasty that was followed by hardening of the leaves, defoliation and, ultimately, plant death. Prominent ringspots were present on fruit, and on *P. edulis* mild mottle symptoms and red coloration of the leaves in an oak-leaf pattern were also observed.

Viruses isolated from *P. edulis* and *P. ligularis* were transmitted mechanically, by the aphids *A. gossipii* and, less efficiently, *Toxoptera citricida*, but not by seed. Based on host range, symptomatology and serology, the isolates were similar to SMV. Further confirmation of the isolates as SMV was based on their 98% homology with the amino acid sequence of the coat protein and dot-blot DNA complementation hybridization with the 3′ terminus of the coat protein-coding region of a strain of SMV.

Passionfruit Sri Lankan mottle virus

Dassanayake and Hicks (1992) described a potyvirus that affected *P. edulis* f. *flavicarpa* in Sri Lanka. Symptoms began as vein yellowing and mild crinkling of new leaves, developing later into numerous yellow spots and a yellow–green mottle and mosaic. Fruit were shrunken and showed slightly depressed green spots on ripe fruit. The virus caused flower break symptoms on artificially infected *P. mollisima*.

The virus was non-persistently aphid (*M. persicae*, *A. gossypii*, *A. craccivora* and *A spiroecola*), graft and mechanically transmissible, and possessed the characteristic morphology of potyviruses. The virus produced typical pinwheel inclusions in plants. It was serologically related to, but distinct from, PRV, PWV and *Potato virus Y*, and reacted only weakly with WMV2 and BCMV.

Incompletely described potyviruses

Chang *et al.* (1996) reported another potyvirus in Taiwan, provisionally designated *Passionfruit crinkle potyvirus*. It produced crinkle symptoms on leaves but did not affect fruit. Serologically, it was indistinguishable from a virus recorded by Escudero *et al.* (1988)

from Puerto Rico that was later named *Puerto Rican passionfruit virus* (PRPV). PRPV appeared to occupy a taxonomic position between PWV and WMV2, and may be a strain of BCMV-US1 (Benscher *et al.*, 1993).

Based on coat protein sequences, another potyvirus from the Dominican Republic appeared to be closely related to CABMV, whereas a strain from Thailand was distinct from these and the SMV strains from Colombia (Benscher *et al.*, 1993, 1996). Finally, *Bean yellow mosaic virus* has been recorded on *P. caerulea* with CMV (Pleše and Wrisher, 1984).

Carlavirus

Passionfruit latent virus

Passionfruit latent virus (PLV) is a carlavirus. Virions are filamentous, flexuous rods, 600–700 × 12–13 nm, and contain single-stranded RNA. They are transmitted by aphids in a non-persistent manner, by grafting and by mechanical inoculation, but not by contact between plants. They occasionally are transmitted by seed and rarely by pollen.

PLV appears to be a problem primarily on vegetatively propagated ornamental *Passiflora* spp. It was first identified in glasshouse plants of *P. caerulea* and *P. suberosa* in Germany (Brandes and Wetter, 1963), and caused concern in The Netherlands in the late 1970s where *P. caerulea* was an important ornamental (Hakkart and Versluys, 1981). PLV was found later in Florida in the ornamental species *Passiflora* × Incense (St Hill *et al.*, 1992), and was common in European collections of ornamental *Passiflora* spp. (Hicks *et al.*, 1996). The virus has also been found in imported ornamental *Passiflora* spp. in South Africa.

Pares *et al.* (1997) reported that PLV was found widely in Australia in cultivated *P. edulis* that had been graft propagated. Seedlings only rarely were infected. *P. edulis* f. *flavicarpa* was also infected, but only where it was grown in association with infected *P. edulis*.

PLV causes an inconspicuous systemic foliar mosaic. In cooler weather, older leaves are mottled. PLV is transmitted by grafting and mechanical means, but not through seed or by insects.

Rhabdoviruses

The plant-associated rhabdoviruses are divided into two genera, *Cytorhabdovirus* and *Nucleorhabdovirus*, but those that affect passion fruit have not been assigned. The virions have a characteristic envelope, are 100–430 × 30–130 nm, and contain single-stranded RNA. Transmission is by aphids and Cicadellidae in a persistent manner, and the virus is transmitted to progeny of the vectors. These viruses can be transmitted by grafting but do not pass through seed. Only some members of these genera can be transmitted mechanically.

Passion fruit vein-clearing rhabdovirus

A widespread disease in Brazil was first described by Batista and co-workers in 1981 on *P. edulis* f. *flavicarpa* (Kitajima *et al.*, 1986). Affected plants exhibited shortened internodes, small crinkled leaves and vein clearing. Branches lignified and fruit on heavily infected plants was, similar to woodiness, malformed with reduced pulp and a hard pericarp. Production was severely affected but plants did not die. The disease was transmitted by grafting, but not seed. Spread in the field appeared to involve a vector, but tests excluded certain aphids and mites as candidates. A rhabdovirus was observed by electron microscopy mostly in the vascular parenchyma.

Pares *et al.* (1983) recorded a rhabdovirus in a mixed infection with PWV in *P. edulis* in garden plants in New South Wales, Australia, but not in commercial fields. It was not possible to separate the viruses for further study, and it is not known whether this virus, provisionally termed *Passion fruit nucleorhabdovirus*, is related to the Brazilian rhabdovirus.

Green spot

Kitajima *et al.* (1997) reported a severe disease in São Paulo state, Brazil, that killed

P. edulis. It caused necrotic lesions on stems, green spots on fruit and leaf senescence. Death occurred when stems were girdled by confluent lesions. The disease was associated with heavy infestations of the mite, *Brevipalpus phoenicis*, which could transmit the disease. Electron microscopy revealed typical bacilliform rhabdovirus particles.

Other Viruses

Passionfruit yellow mosaic virus

Passion fruit yellow mosaic was first recorded from the state of Rio de Janeiro in Brazil (Kitajima *et al.*, 1986). Symptoms started with vein clearing that became a brilliant yellow, net-like pattern. Chlorosis extended to the interveinal areas to form patches and, in severe cases, only the main veins retained a green border. There was some crinkling of the leaves. Fruit was unaffected and, although plants were unthrifty, overall effects on yield could not be determined since a limited number of plants were affected.

The cause was shown to be a tymovirus, *Passionfruit yellow mosaic virus*. It is isometric, 28 nm in diameter, and was transmitted experimentally by the beetle *Diabrotica speciosa* in a non-persistent manner. It was also easily transmitted mechanically and by grafting, but did not pass through seed. The host range is limited to the genus *Passiflora* (Crestani *et al.*, 1986). A tymovirus forming part of a complex has also been recorded from Colombia (Varón de Agudelo *et al.*, 1992).

Purple granadilla mosaic virus

Purple granadilla mosaic currently is limited to a small area in the state of São Paulo in Brazil where it was discovered during the course of a survey in the early 1980s (Kitajima *et al.*, 1986). Symptoms are a mild mosaic, yellowing of major veins, woodiness of fruits and reduced yield. Symptoms are more severe during cooler months. An isometric virus, 30 nm in diameter, was found but could not be related serologically to other isometric viruses and is as yet unclassified. It was readily transmitted mechanically by grafting and experimentally by the beetle *D. speciosa* in a non-persistent manner. Its host range appeared to be limited to the *Passifloraceae*.

Citrus tristeza virus

Passiflora is the only genus outside the *Rutaceae* that is affected by *Citrus tristeza virus* (CTV). Müller *et al.* (1974), following up a report by Dr A. Osores, managed to infect *P. gracilis* with CTV by aphid transmission. This was repeated, with difficulty, by Roistacher and Bar-Joseph (1987) with *P. caerulea*, who were then able to transmit CTV by grafting to other species.

Symptoms in *P. gracilis* were severe stunting, interveinal yellowing followed by total chlorosis and death. Faint vein clearing was only seen on very young leaves. *P. caerulea* was severely stunted, with small leaves, but death did not ensue. Other species, notably *P. edulis*, were not affected. Another closterovirus has been found in a complex of viruses affecting *P. edulis* in Colombia (Varón de Agudelo *et al.*, 1992).

Miscellaneous Viruses

Tomato ringspot virus has been recorded from *P. edulis* f. *flavicarpa* in Peru (Koenig and Fribourg, 1986) in association with a virus for which the name *Maracuja mosaic virus* (MrMV) was later proposed (Fribourg *et al.*, 1987). MrMV is a tobamovirus that causes systemic mosaic, is transmitted mechanically, but is only distantly related to other tobamoviruses. A serologically related strain of MrMV has also been found in *P. incarnata* in Florida (St Hill *et al.*, 1992).

Jatropha mosaic virus, a geminivirus, was transmitted by the whitefly *Bemesia argentifolii* (formerly the 'B' biotype of *B. tabaci*) to cultivated *P. edulis* in Puerto Rico (Brown and Bird, 1995).

Management

P. edulis is self-pollinated and genetically fairly homogeneous. Since potyviruses are rarely transmitted through seed, seed selection from good producers can eliminate these pathogens as factors in the production of this crop. Unfortunately, seed transmission of CMV can be as high as 50%. Thus, rigorous selection in the nursery is needed to eliminate this virus. Symptoms can be evanescent, which further complicates matters.

P. edulis f. *flavicarpa* is generally self-incompatible. Since seed progeny of this form segregate, it is desirable to propagate superior selections used for production clonally, a practice that increases the risk of propagating and spreading viruses. The use of clonal rootstocks incurs the same risks. A case in point is the use of *P. caerulea* as a rootstock for *P. edulis* in South Africa in the 1970s. Although it was tolerant of Fusarium wilt, the clonally propagated stocks were latently infected with PWV and CMV. The spread and use of grafted plants across the country brought the industry to its knees.

In Australia, a certification scheme using mild strain cross-protection was used for the control of woodiness (Simmonds, 1959a). Good producing clones that contain mild strains of PWV are used as budwood sources. *P. edulis* f. *flavicarpa* and *P. edulis* f. *flavicarpa* × *P. edulis* f. *edulis* hybrids are also more resistant than *P. edulis*. Various small breeding programmes for virus tolerance have been launched without great success.

Passiflora spp. are not favoured by aphids when other hosts are available. Thus, significant infestations of the vectors of the major virus pathogens are not common. Visiting alates, however, are capable of initiating these diseases in the field. Field control of aphids is rarely justified economically.

Where practised, pruning operations can transmit these viruses. Sterilization of pruning implements can be easily applied and is effective. Suitable sterilants or inhibitors of transmission are sodium hypochlorite (which aggressively rusts tools), formaldehyde, skim milk and soap. Tools should be dipped in these products as often as possible.

Roguing of infected plants is best applied where a limited percentage of plants are affected. The same applies when a plantation becomes uneconomic and needs to be destroyed before re-establishment. Elimination of weeds in and around the plantation will assist in reducing alternative hosts of the viruses. In the special cases of green spot and jatropha mosaic, directed control of the insect vectors is indicated.

Acknowledgements

The authors thank Ken Pegg and John Thomas for useful editorial comments on this chapter, and Cornelia Büchen-Osmond for current information on virus taxonomy and nomenclature.

References

Abeysinghe, A. (1973) Commercial passionfruit cultivation, processing and marketing. *Journal of the National Agricultural Society of Ceylon* 9, 87–111.

Alfieri, S.A. Jr, Langdon, K.R., Kimbrough, J.W., El-Gholl, N.E. and Wehlburg, C. (1994) *Diseases and Disorders of Plants in Florida*. Bulletin No. 14, Florida Department of Agriculture and Consumer Services, Contribution No. 680.

Anselmo, R.M. and Junqueira, N.T.V. (1997) Doenças de maracujá-doce (*Passiflora alata* Dryand) em pós-colheita. *Fitopatologia Brasileira* 22, 244.

Aubert, B. (1987) The cultivation of granadilla in Reunion. Prospects and constraints. *Fruits* 42, 717–723.

Barreto, R.W., Requia, A.C. and Casa, R.T. (1996) Queima de mudas de maracujazeiro *Passiflora edulis* causada por *Cladosporium cladosporioides*. *Fitopatologia Brasileira* 21, 348.

Benscher, D., Pappu, S.S., Niblett, C.L., Rybicki, E.P. and Bird, J. (1993) Biological and molecular characterisation of potyviruses from *Passiflora*. *Phytopathology* 83, 1422.

Benscher, D., Pappu, S.S., Niblett, C.L., Aguledo, F.V., Morales, F., Hodson, E., Alvarez, E., Acosta, O. and Lee, R.F. (1996) A strain of soybean mosaic virus infecting *Passiflora* spp. in Colombia. *Plant Disease* 80, 258–262.

Bradbury, J.F. (1986) *Guide to Plant Pathogenic Bacteria*. CAB International, Wallingford, UK.

Brand, R.J. (1992) Viruses implicated in the woodiness disease of South African passion fruit and the molecular characterization of a new potyvirus. PhD thesis, University of Cape Town.

Brandes, J. and Wetter, C. (1963) Studies on the characteristics and relationships of Passiflora latent virus. *Phytopatologiese Zeitschrift* 49, 61–70.

Brodrick, H.T., Milne, D.L., Wood, R. and Mulder, N.J. (1976) Control of Phytophthora stem-rot of granadillas (*Passiflora edulis*) in South Africa. *Citrus and Subtropical Fruit Journal* 508, 15–17.

Brown, J.K. and Bird, J. (1995) Introduction of an exotic whitefly (*Bemesia*) vector facilitates secondary spread of jatropha mosaic virus, a geminivirus previously vectored exclusively by the 'Jatropha' biotype. In: *Bemesia 1995: Taxonomy, Biology, Damage Control and Management*. Intercept Ltd, Aldershot, UK, pp. 351–353.

Carvalho-Dias, M.S., Peres, N.A.R., Souza, N.L. and Kuramae-Izioka, E.E. (1998) Pathogenic, morphological and molecular aspects of *Fusarium* spp. from passion fruit plants with symptoms of premature death disease. Abstract 6.112, Offered Paper Abstracts – Volume 3. 7th International Congress of Plant Pathology, Edinburgh, Scotland, August 9–16, 1998.

Cedeño, L.R., Prü, E.L.P., Marquez, N.J. and Tavira, M.E. (1990) *Nectria haematococca*, agente causal de la muerte repentina de la parchita en Venezuela. *Fitopatologîa Venezolana* 3, 15–18.

Cedeño, L., Mohali, S. and Palacios-Prü, E. (1993) Antracnose causada por dos cepas de *Glomerella cingulata* em frutos de parchita. *Fitopatologîa Venezolana* 6, 30–33.

Cedeño, L., Carrero, C., Mohali, S. and Palacios-Prü, E. (1995) Muerte regresiva en parchita por *Lasiodiplodia theobromae* en Venezuela. *Fitopatologîa Venezolana* 8, 7–10.

Chan, H.T. (1980) Passion fruit. In: Nagy, S. and Shaw, P.E. (eds) *Tropical and Subtropical Fruits*. AVI Publishing, Westport, Connecticut, pp. 300–315.

Chang, C.A. (1992) Characterization and comparison of passion fruit mottle virus, a newly recognized potyvirus, with passion fruit woodiness virus. *Phytopathology* 82, 1358–1363.

Chang, C.A, Chen, C.C., Deng, T.C. and Zettler, F.W. (1996) Characterization of passion fruit crinkle potyvirus – a newly found virus infecting passion fruit. *Plant Protection Bulletin of Taipei* 38, 339–354 (in Chinese).

Chen, W.Q. and Zhang, T.Y. (1977) Two new species of *Alternaria* from China. *Mycological Research* 101, 1257–1258.

Cole, D.L., Hedges, T.R. and Ndowora, T. (1992) A wilt of passion fruit (*Passiflora edulis* f. *edulis* Sims) caused by *Fusarium solani* and *Phytophthora nicotianae* var. *parasitica*. *Tropical Pest Management* 38, 362–366.

Cox, J.E. and Kiely, T.B. (1961) Fusarium resistant rootstocks for passion vines. *The Agricultural Gazette, New South Wales* 72, 314–318.

Crestani, O.A., Kitajima, E.W., Lin, M.T. and Marinho, V.L.A. (1986) Passion fruit yellow mosaic virus, a new Tymovirus found in Brazil. *Phytopathology* 76, 951–955.

Da Graça, J.V. (1976) Studies on woodiness disease of passion fruit, *Passiflora edulis*, in South Africa. *Phytophylactica* 8, 37–40.

Dassanayake, E.M. and Hicks, R.G.T. (1992) Sri Lankan passion fruit mottle virus, a potyvirus infecting golden passion fruit in Sri Lanka. *Annals of Applied Biology* 120, 459–469.

de Goes, A. (1998) Doenças fúngicas da parte aérea da cultura de Maracujá. In: *Simpósio Brasileiro Sobre a Cultura do Maracujazeiro*. Faculdade de Ciências Agrárias e Veterinárias, Universidade Estadual Paulista, Jaboticabal, São Paulo, Brazil, pp. 208–216.

De Wijs, J.J. (1974) A virus causing ringspot of *Passiflora edulis* in the Ivory Coast. *Annals of Applied Biology* 77, 33–40.

De Wijs, J.J. (1975) The distribution of passion fruit ringspot virus in its main host plants in Ivory Coast. *Netherlands Journal of Plant Pathology* 81, 144–148.

Dias, S.C. (1990) Morte precoce do maracujazeiro amarelo (*Passiflora edulis* f. *flavicarpa*) causada por patógenos que afetam a parte aérea da planta. MSc thesis, Universidade de Brasília, Brasília, Brazil.

Dodd, J.C., Estrada, A. and Jeger, M.J. (1992) Epidemiology of *Colletotrichum gloeosporioides* in the tropics. In: Bailey, J.A. and Jeger, M.J. (eds) *Colletotrichum: Biology, Pathology and Control*. CAB International, Wallingford, UK, pp. 308–325.

Domsch, K.H., Gams, W. and Anderson, T.-H. (1980) *Compendium of Soil Fungi*, Vol. 1. Academic Press, New York.

dos Santos, J.M. (1998) Nematóides. In: *Simpósio Brasileiro Sobre a Cultura do Maracujazeiro*. Faculdade de Ciências Agrárias e Veterinárias, Universidade Estadual Paulista, Jaboticabal, São Paulo, Brazil, pp. 204–207.

Dougherty, W.G. and Carrington, J.C. (1988) Expression and function of potyviral gene products. *Annual Review of Phytopathology* 26, 123–143.

Ellis, M.B. (1971) *Dematiaceous Hyphomycetes*. Commonwealth Mycological Institute, Kew, UK.

Emchembe, A.M. and Mukiibi, J. (1976) Nectria collar and root rot of passion fruit in Uganda. *Plant Disease Reporter* 60, 227–231.

Erwin, D.C. and Ribiero, O.K. (1996) *Phytophthora Diseases Worldwide*. APS Press. St Paul, Minnesota.

Escudero, J., Monllor, A.C., Bird, J. and Zettler, F.W. (1988) Mosaic of passion fruit (*Passiflora edulis*) in Puerto Rico. *Phytopathology* 78, 857.

Farr, D.F., Bills, G.F., Chamuris, G.P. and Rossman, A.Y. (1989) *Fungi on Plant and Plant Products in the United States*. APS Press, St Paul, Minnesota.

Francisco Neto, E., Nakamura, K. and Oliveira, J.C. (1994) Influência de algunas fatores na germinação de conídios, no crescimento micelial e na esporulação de alguns isolados de *Colletotrichum gloeosporioides*, obtidos de *Passiflora*. *Summa Phytopathologica* 20, 96–100.

Fribourg, C.E., Koenig, R. and Lesemann, D.E. (1987) A new tobamovirus from *Passiflora edulis* in Peru. *Phytopathology* 77, 486–491.

Gardner, D.E. (1989) Pathogenicity of *Fusarium oxysporum* f. sp. *passiflorae* to banana poka and other *Passiflora* spp. in Hawaii. *Plant Disease* 73, 476–478.

Gerlach, W.W.P. (1983) Observations on passionfruit collar rot in Western Samoa. *Alafua Agricultural Bulletin* 8, 76–78.

Gonzalez, M.S., Suarez, Z., Rosales, C. and Parra, D. (1999) Collar rot and wilt of yellow passion fruit in Venezuela. *Plant Disease* 70, 1038.

Gough, K.H. and Shukla, D.D. (1992) Major sequence variations in the N-terminal region of the capsid protein of a severe strain of passionfruit woodiness virus. *Archives of Virology* 124, 389–396.

Grech, N.M. and Rijkenberg, F.H.J. (1991) Laboratory and field evaluation of the performance of *Passiflora caerula* as a rootstock tolerant to certain fungal pathogens. *Journal of Horticultural Science* 66, 725–729.

Groszmann, H.M. and Purss, G.S. (1958) Beating passionfruit wilt. *Queensland Agricultural Journal* 84, 214–216.

Hakkaart, F.A. and Verslys, J.M.A. (1981) Virus in *Passiflora caerulea* door meristeemcultuur. *Vakblad voor de Bloemisterij* 1, 24–25.

Hanlin, R.T. (1990) *Illustrated Genera of Ascomycetes*. APS Press, St Paul, Minnesota.

Hicks, R.G.T., Mohamed, M.E. and Blakesley, D. (1996) Passiflora latent carlavirus in European collections of ornamental *Passiflora*. *Journal of Phytopathology* 144, 203–205.

Holliday, P. (1980) *Fungus Diseases of Tropical Crops*. Cambridge University Press, Cambridge.

Hseih, H.J. (1979) New diseases caused by *Phytophthora* species. *Phytophthora Newsletter* 7, 39–40.

Inch, A.J. (1978) Passion fruit diseases. *Queensland Agricultural Journal* 104, 479–484.

Jeffries, P., Dodd, J.C., Jeger, M.J. and Plumbley, R.A. (1990) The biology and control of *Colletotrichum* species on tropical fruits crops. *Plant Pathology* 39, 343–366.

Jesus, W.C. Jr, Benato, E.A. and Souza, N.L. (1994) Postharvest hot water treatment of passion fruit (*Passiflora edulis* f. *flavicarpa*) for control of anthracnose (*Colletotrichum gloeosporioides*). *Fitopatologia Brasileira* 19, 283.

Killip, E.P. (1938) *The American Species of Passifloraceae*. Botanical Series Vol. XIX. Field Museum of Natural History, Chicago.

Kirby, M.F. (1978) Reniform and root knot nematodes on passionfruit in Fiji. *Nematropica* 8, 21–25.

Kitajima, E.W., Chagas, C.M. and Crestani, O.A. (1986) Enfermidades de etiologia viral e associadas a organismos do tipo micoplasma em maracujazeiros no Brasil. *Fitopatologia Brasileira* 11, 409–432.

Kitajima, E.W., Rezende, J.A.M., Rodrigues, J.C.V., Chiavegato, L.G., Piza-Junior, C.T. and Morozini, W. (1997) Green spot of passion fruit, a possible viral disease associated with infestation by the mite *Brevipalpus phoenicis*. *Fitopatologia Brasileira* 22, 555–559.

Knight, R. Jr (1980) Origin and world importance of tropical and subtropical fruit crops. In: Nagy, S. and Shaw, P.E. (eds) *Tropical and Subtropical Fruits*. AVI Publishing, Westport, Connecticut, pp. 1–120.

Koenig, R. and Fribourg, C.E. (1986) Natural occurrence of tomato ringspot virus in *Passiflora edulis* from Peru. *Plant Disease* 70, 244–245.

Lin, Y.S. and Chang, H.J. (1985) Collar rot of passion fruit possibly caused by *Nectria haematococca* in Taiwan. In: Parker, C.A., Rovira, A.D., Moore, K.J., Wong, P.T.W. and Kollmorgen, J.F. (eds) *Ecology and Management of Soilborne Plant Pathogens*. APS Press, St Paul, Minnesota, pp. 41–44.

Louw, A.J. (1941) Studies on *Septoria passiflorae* n.sp. occurring on passion fruit with special reference to its parasitism and physiology. *Scientific Bulletin of the South African Department of Agriculture* 229.

Lutchmeah, R.S. (1993) Common field and postharvest diseases of passion fruit (*Passiflora edulis* f. *flavicarpa*) and the associated fungi in Mauritius. *Revue Agricole et Sucriere de I'le Maurice* 72, 55–59.

Magee, C.J. (1948) Woodiness or mosaic disease of passion fruit. *The Agricultural Gazette* April 1, 199–202, 208.

Malavolta, V.A. (1998) Bacterioses do maracujazeiro. In: *Simpósio Brasileiro Sobre a Cultura do Maracujazeiro*. Faculdade de Ciências Agrárias e Veterinárias, Universidade Estadual Paulista, Jaboticabal, São Paulo, Brazil, pp. 217–229.

Martin, F.W. and Nakasone, H.Y. (1970) The edible species of *Passiflora*. *Economic Botany* 24, 333–343.

Martini, C.K.H. (1962) Some properties of the virus causing 'woodiness' of passion fruit in Western Nigeria. *Annals of Applied Biology* 50, 163–168.

McKern, N.M., Strike, P.M., Barnett, O.W., Dijkstra, J., Shukla, D.D. and Ward, C.W. (1994) Cowpea aphid borne mosaic virus – Morocco and South African *Passiflora* virus are strains of the same potyvirus. *Archives of Virology* 136, 207–217.

McKnight, T. (1951) A wilt disease of the passion vine (*Passiflora edulis*) caused by a species of *Fusarium*. *Queensland Journal of Agricultural Science* 8, 1–4.

McKnight, T. (1953) The woodiness disease of the passion vine. *Queensland Journal of Agricultural Science* 10, 4–35.

Müller, G.W., Costa, A.S., Kitajima, E.W. and Camargo, I.J.B. (1974) Additional evidence that tristeza virus multiplies in *Passiflora* spp. In: Weathers, L.G. and Cohen, M. (eds) *Proceedings of the 6th Conference of the International Organisation of Citrus Virologists*. University of California, Riverside, pp. 75–78.

Nattrass, R.M. (1939) Annual Report of the Senior Plant Pathologist. *Report of the Department of Agriculture, Kenya, 1938*, pp. 42–47.

Nattrass, R.M. (1944) The transmission of the virus of the 'woodiness' disease of passion fruit (*Passiflora edulis*) by single leaf grafts. *Annals of Applied Biology* 31, 310–311.

Nelson, P.E., Toussoun, T.A. and Marasas, W.O. (1983) *Fusarium Species. An Illustrated Guide for Identification*. Pennsylvania State University Press.

Norman, D.J. and Trujillo, E.E. (1995) Development of *Colletotrichum gloeosporioides* f. sp. *clidemiae* and *Septoria passiflorae* into two mycoherbicides with extended viability. *Plant Disease* 79, 1029–1032.

Pares, R.D., Martin, A.B. and Morrison, W. (1983) Rhabdovirus like particles in passion fruit. *Australasian Plant Pathology* 12, 51–52.

Pares, R.D., Martin, A.B. and Fitzell, R.D. (1985) Virus-induced tip necrosis of passionfruit (*Passiflora edulis* Sims). *Australasian Plant Pathology* 14, 76–78.

Pares, R.D., Gunn, L.V., Keskula, E.N., Martin, A.B. and Teakle, D.S. (1997) Occurrence of Passiflora latent carlavirus in cultivated and wild *Passiflora* species in Australia. *Plant Disease* 81, 248–350.

Perregrine, W.T.H. and Yunton, B.A. (1980) A preliminary note on nematode pests in Brunei. *Tropical Pest Management* 26, 416–419.

Persley, D. (ed.) (1993) *Diseases of Fruit Crops*. Department of Primary Industries, Indooroopilly, Queensland.

Peterson, R.A. (1977) Benomyl resistance in *Septoria passiflora* Louw. *APPS Newsletter* 6, 3–4.

Pio-Ribeiro, G. and Mariano, R. de L.R. (1997) Doenças do maracujazeiro. In: Kimati, H., Amorim, L., Bergamin Filho, A., Camargo, L.E.A. and Rezende, J.A.M. (eds) *Manual de Fitopatologia. Doenças das Plantas Cultivadas*, 3rd edn. Editora Agronômica Ceres Ltda, São Paulo, Vol. II, pp. 525–534.

Pleše, N. and Wricher, M. (1984) A mixed infection of *Passiflora caerulea* L. with two viruses. *Acta Botanica Croatia* 43, 1–6.

Ploetz, R.C. (1991) Sudden wilt of passionfruit in southern Florida caused by *Nectria haematococca*. *Plant Disease* 75, 1071–1073.

Power, R.H. and Verhoff, K. (1984) Dieback of passion fruit in Surinam. *Phytopathologische Zeitschrift* 110, 336–345.

Purseglove, J.W. (1968) *Tropical Crops. Dicotyledons*. Longman, London.

Purss, G.S. (1954) Identification of the species of *Fusarium* causing wilt in passion vines in Queensland. *Queensland Journal of Agricultural Science* 11, 79–81.

Ram, B., Naidu, R. and Singh, H.P. (1977) *Alternaria macrospora* Zimm. a new record on passion fruit (*Passiflora edulis* Sims) from India. *Current Science* 46, 165.

Reid, W.D. (1938) Grease spot of passion fruit. *New Zealand Journal of Science and Technology* 20, 260–265.

Rezende, D.V. and Junqueira, N.T.V. (1997) Queda de flores de maracujá-doce (*Passiflora alata* Dryand) provocada por *Rhizopus stolonifer* Sac. *Fitopatologia Brasileira* 22, 301.

Rínderman, R.S. and Cruz, M.A.G. (1997) *El Maracuyá Fruta de la Pasión*. Universidad Autónoma Chapingo.

Roistacher, C.N. and Bar-Joseph, M. (1987) Transmission of citrus tristeza virus by *Aphis gossypii* and by graft inoculation to and from *Passiflora* spp. *Phytophylactica* 19, 179–182.

Ruggiero, C. (ed.) (1998) *Maracujá do Plantio a Colheita*. FUNEP.

Sale, P.R. (1987) *Passionfruit Culture*. Ministry of Agriculture and Fisheries, Wellington, New Zealand.

Sanudo-Sotelo, B. and Zuniga-Ravelo, B. (1991) Hibrido interspecificos de curuba resistentes a la anthracnosis *Colletotrichum gloeosporioides* (Penz.) Sacc. en el departamento de Narino. *ASCOLFI Informa* 17, 9–10.

Shukla, D.D., McKern, N.M. and Ward, C.W. (1988) Coat protein of potyviruses. 5. Symptomology, serology, and coat protein sequences of three strains of passion fruit woodiness virus. *Archives of Virology* 102, 221–232.

Simmonds, J.H. (1938) Passion vine diseases. *Queensland Agricultural Journal* 45, 322–330.

Simmonds, J.H. (1959a) Mild strain protection as a means of reducing losses from the Queensland woodiness virus in the passion vine. *Queensland Journal of Agricultural Science* 16, 371–380.

Simmonds, J.H. (1959b) *Report of the Plant Pathology Section for 1958–1959*. Department of Agriculture, Queensland, pp. 49–50.

Simmons, E.G. (1993) *Alternaria* themes and variations. *Mycotaxon* 46, 171–199.

Ssekeyewa, C., Fina Opio, A., Swinburne, T.R., Van Damme, P.L.J. and Abubakar, Z.M. (1999) Sustainable management of collar rot of passion fruits in Uganda. *International Journal of Pest Management* 45, 173–177.

St Hill, A.A., Zettler, F.W., Elliot, M.S., Petersen, M.A., Li, R.H. and Bird, J. (1992) Presence of passiflora latent virus and a serologically distinct strain of maracuja mosaic virus in *Passiflora* spp. in Florida. *Plant Disease* 76, 843–847.

Taylor, R.H. and Kimble, K.A. (1964) Two unrelated viruses which cause woodiness of passion fruit (*Passiflora edulis* Sims). *Australian Journal of Agricultural Research* 15, 560–570.

Teakle, D.S., Gill, C.C., Taylor, R.H. and Raabe, R.D. (1963) Cucumber mosaic virus in *Passiflora* in California. *Plant Disease Reporter* 47, 677–678.

Terblanche, J.H., Grech, N., Frean, R., Crabbe, F. and Joubert, A. (1986) Good news for passion fruit industry. *CSFRI Information Bulletin* 164.

Trujillo, E.E., Norman, D.J. and Killgore, E.M. (1994) Septoria leaf spot, a potential biological control for banana poka vine in forests of Hawaii. *Plant Disease* 78, 883–885.

Turner, G.J. (1974) Phytophthora wilt and crown blight of *Passiflora edulis*. *Transactions of the British Mycological Society* 62, 59–63.

Van den Boom, T. and Huller, I.M. (1970) Phytophthora stem rot of passion fruit, *Passiflora edulis*, in South Africa. *Phytophylactica* 2, 71–74.

Varón de Agudelo, F., Castaño, M., Arroyave, J.A., Velasco, A.C., Vuillaume, C. and Morales, F.J. (1992) Complejo viral que affecta plantaciones de maracuyá (*Passiflora edulis* Sims) en el Valle del Cauca. *Fruits* 47, 321–328.

Winks, C.W., Menzel, C.M. and Simpson, D.R. (1988) Passionfruit in Queensland 2. Botany and cultivars. *Queensland Agricultural Journal* July–August, 217–224.

Wolcan, S., and Larran, S. (2000) First report of anthracnose cause by *Glomerella cingulata* on passion fruit in Argentina. *Plant Disease* 84, 706.

Young, B.R. (1970) Root rot of passion vine (*Passiflora edulis*) in the Auckland area. *New Zealand Journal of Agriculture* 13, 119–125.

1

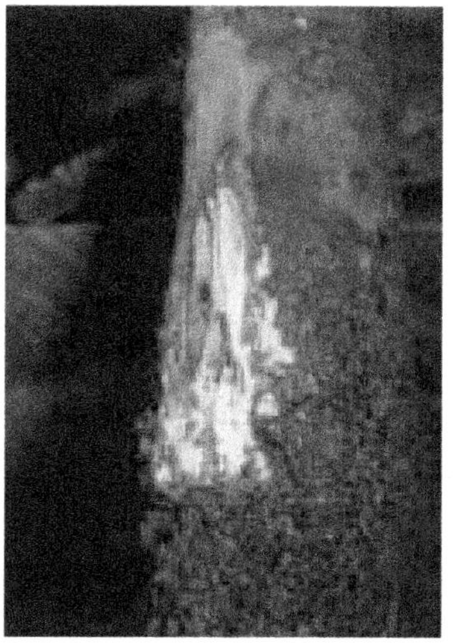

2

3

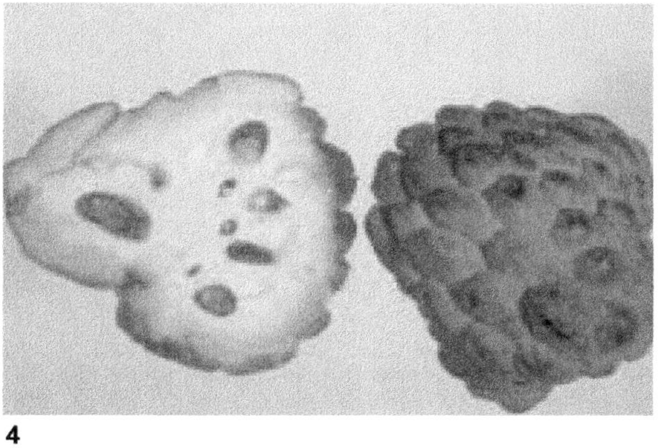

4

5

Plate 1. A cluster of basidiomes (mushrooms) of *Armillaria mellea* at the base of an avocado tree affected by Armillaria root rot. Note the shaggy, skirt-like ring surrounding the stem that is produced by *A. mellea*, but not *A. socialis* (photo: J. Menge, UCR).

Plate 2. Mycelial plaque of *Armillaria mellea* formed beneath the bark of an avocado tree affected by Armillaria root rot (photo: J. Menge, UCR).

Plate 3. Discolored vascular system of a custard apple tree affected by bacterial wilt, caused by *Ralstonia solanacearum* (photo: A.W. Cooke, QDPI).

Plate 4. External and internal symptoms of black canker on sugar apple, caused by *Phomopsis annonacearum*. Note the superficial nature of the lesion (photo: A.W. Cooke, QDPI).

Plate 5. A sugar apple affected by Botryodiplodia fruit rot, caused by *Diplodia theobromae* (photo: A.W. Cooke, QDPI).

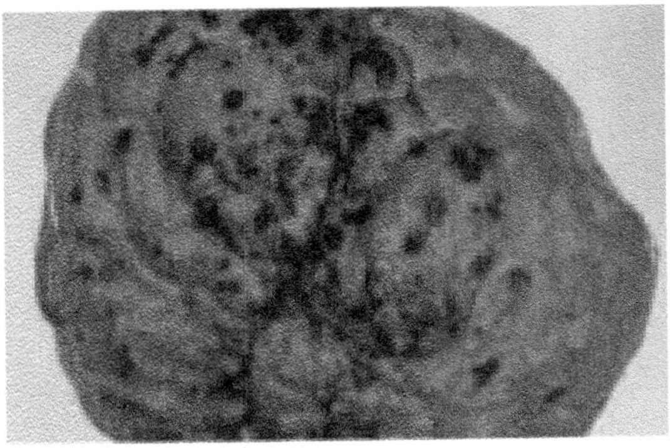

6

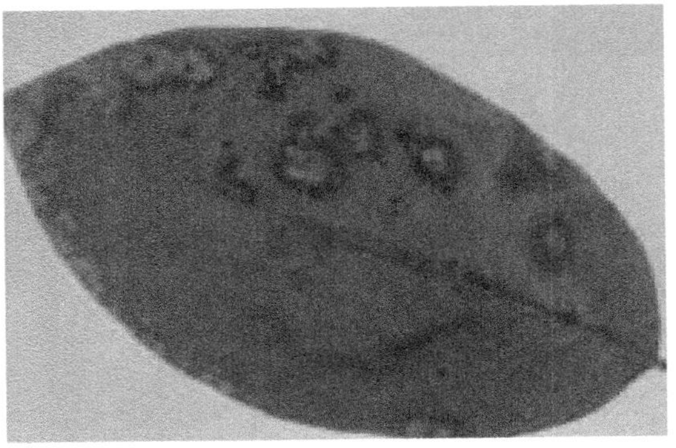

7

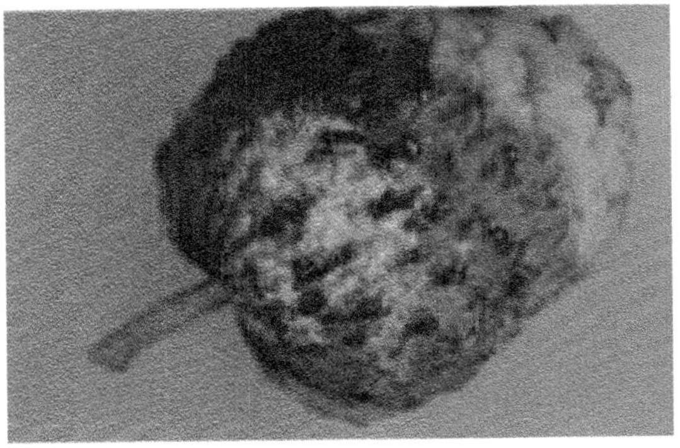

8

9

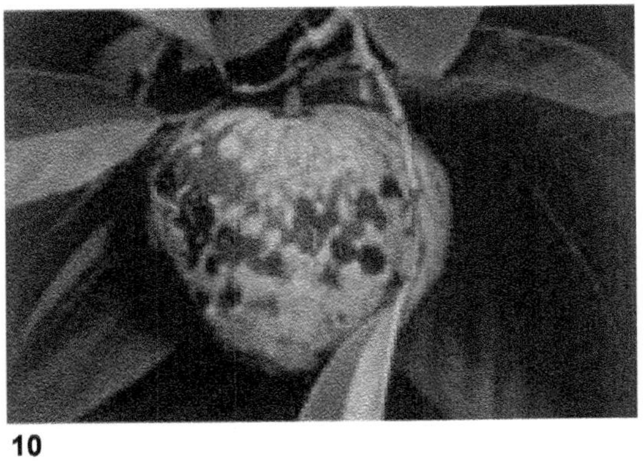

10

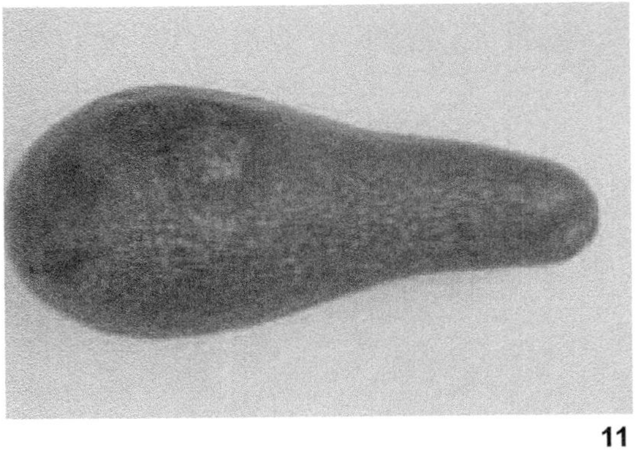

11

Plate 6. Cylindrocladium fruit spot on atemoya (photo: A.W. Cooke, QDPI).
Plate 7. Cylindrocladium leaf spot on sugar apple (photo: A.W. Cooke, QDPI).
Plate 8. Purple blotch, caused by *Phytophthora palmivora*, on an atemoya fruit (photo: A.W. Cooke, QDPI).
Plate 9. Pink disease on custard apple, caused by *Erythricium salmonicolor* (photo: A.W. Cooke, QDPI).
Plate 10. Pseudocercospora fruit spot on custard apple, caused by *Pseudocercospora* sp. (photo: A.W. Cooke, QDPI).
Plate 11. Dark anthracnose lesion on avocado fruit. Note sporulation of the causal fungus, *Colletotrichum gloeosporioides*, on the lesion surface (photo: D. Prusky, ARO).

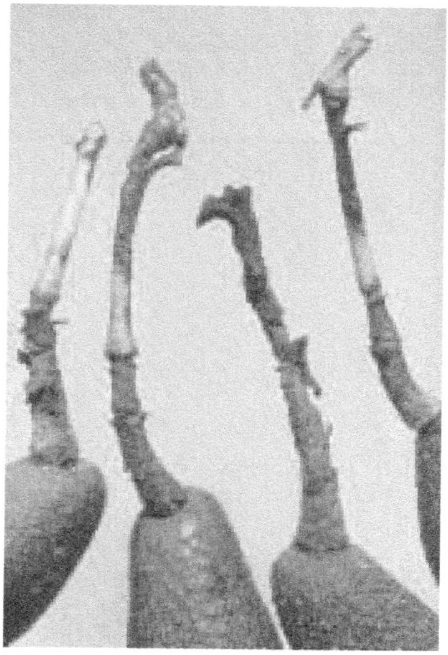

12

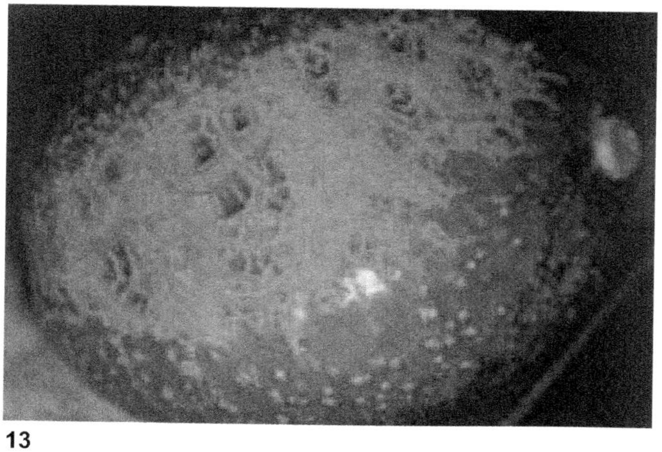

13

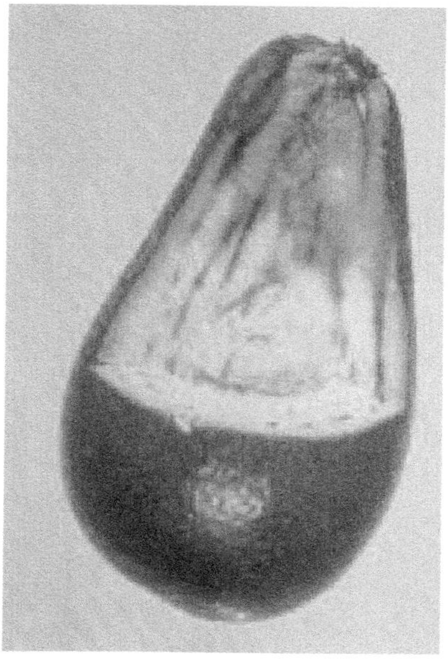

14

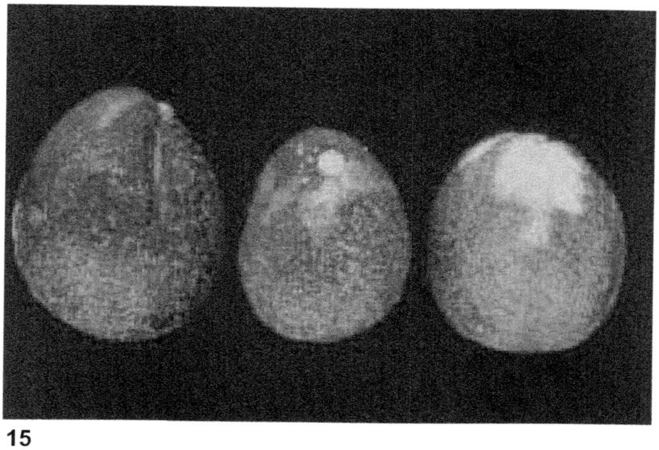

15

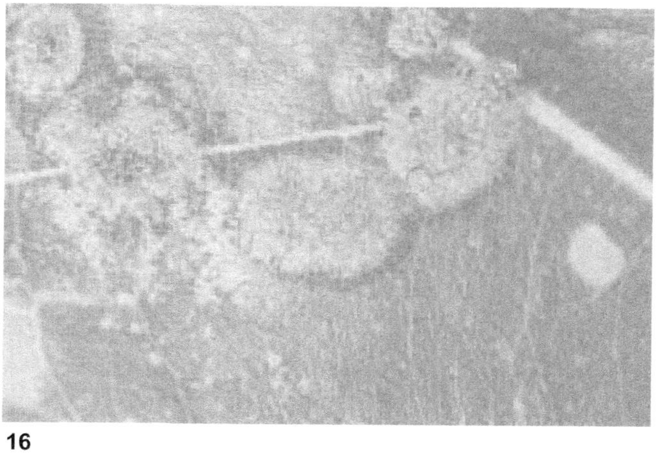

16

Plate 12. Symptoms of ringneck on avacado fruit from a 'Haas' seedling (photo: A. Whiley, QDPI).
Plate 13. Rough, corky scab lesions on avocado fruit caused by *Sphaceloma perseae* (photo: J. Menge, UCR).
Plate 14. Internal symptoms of Dothiorella stem-end rot (photo: A. Whiley, QDPI).
Plate 15. Sunblotch symptoms on fruit of 'Fuerte' avocado (photo: G.A. Zentmyer, UCR).
Plate 16. Symptoms of algal leaf spot on an avocado leaf (photo: R.C. Ploetz, UF).

17

18

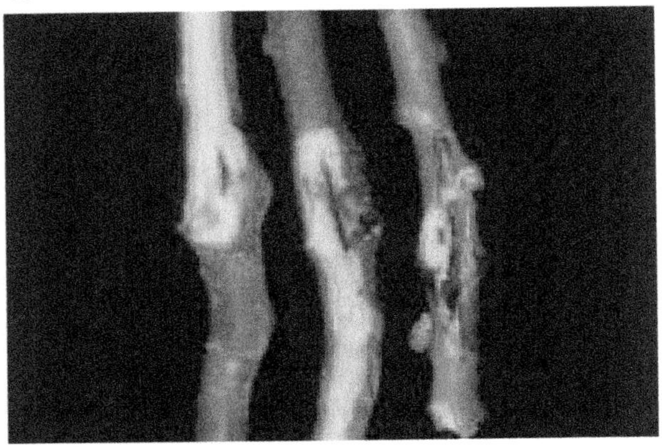

19

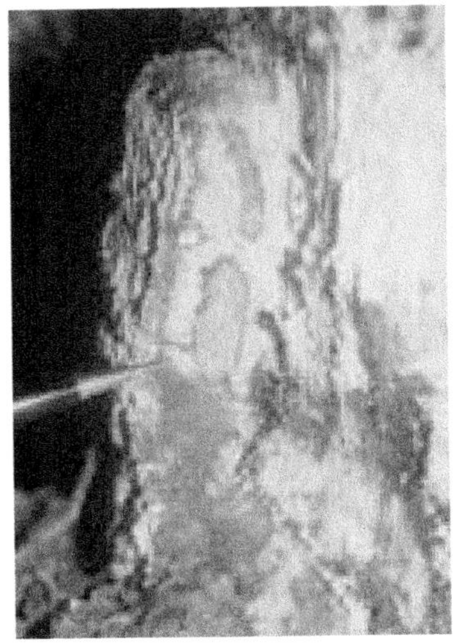

20

21

22

Plate 17. Powdery sugar exudate associated with black streak disease of avocado (photo: R.C. Ploetz, UF).
Plate 18. Bacterial canker lesions, caused by *Xanthomonas campestris,* on avocado tree in California (photo: H.D. Ohr, UCR).
Plate 19. Brown discoloration associated with Dothiorella stem canker in graft unions of avocado seedlings (photo: J. Menge, UCR).
Plate 20. Reddish-brown symptoms in avocado cambium caused by *Phytophthora citricola* (photo: J. Menge, UCR).
Plate 21. Healthy avocado seedling roots (right) compared with those affected by Phytophthora root rot (left), caused by *Phytophthora cinnamomi* (photo: G.A. Zentmyer, UCR).
Plate 22. Internal symptoms of Moko disease on 'Bluggoe' ABB (photo: J. Parrado, UF).

23

24

25

26

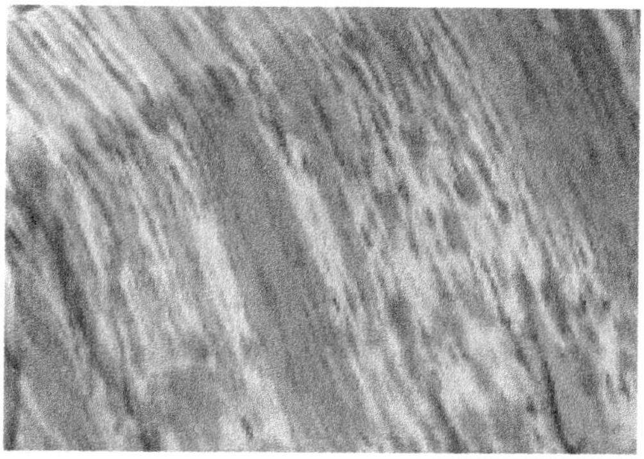

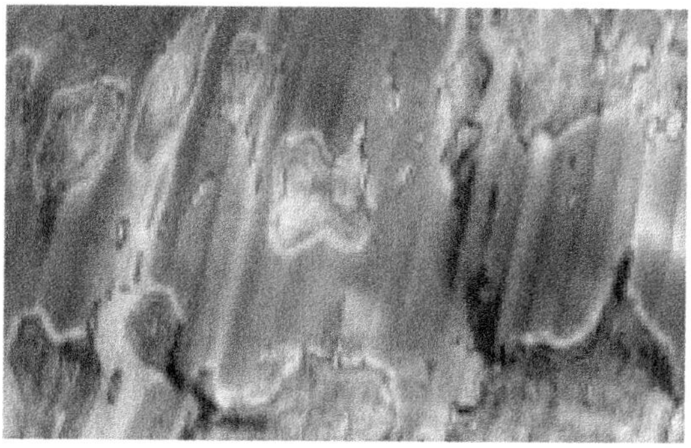

27

Plate 23. Symptoms of black cross leaf spot on the underside of a leaf of *Musa acuminata* ssp. *banksii* in Papua New Guinea (photo: D.R. Jones, QDPI).
Plate 24. Rust-coloured streak symptom of black Sigatoka. These streaks on the upperside of a leaf of 'Dwarf Cavendish' AAA have matured and coalesced on the right to produce enlarged, blackened lesions; note their moist appearance and chlorosis that has begun to form on their borders (photo: R.C. Ploetz, UF).
Plate 25. Severe black Sigatoka damage on 'Dwarf Cavendish' AAA in Malawi. In the absence of fungicide applications, complete defoliation can occur on plants that have fruited (photo: R.C. Ploetz, UF).
Plate 26. Symptoms of Cladosporium speckle on a highland AAA cultivar in Uganda (photo: R.C. Ploetz, UF).
Plate 27. Extensive lesions of Cordana leafspot that have formed around smaller yellow Sigatoka lesions. Note the greyish caste of, and the dark borders of the Sigatoka lesions in, the Cordana lesions (photo: R.C. Ploetz, UF).

28

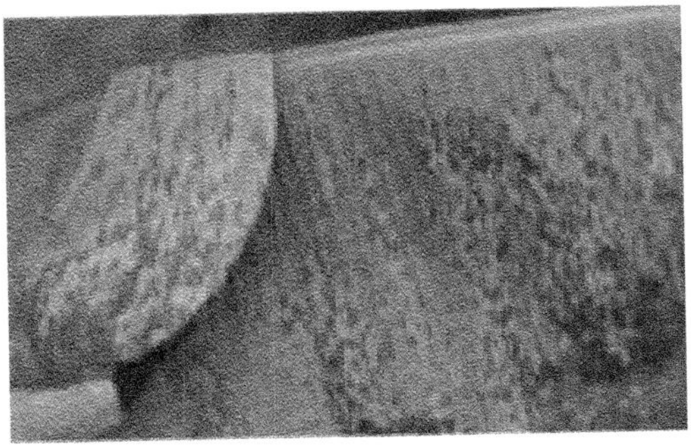

29

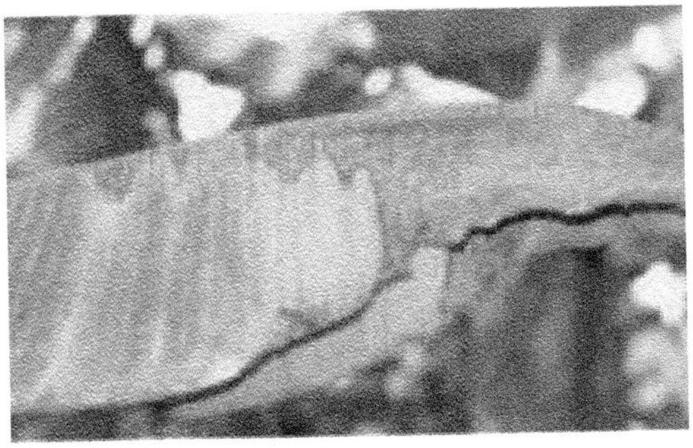

30

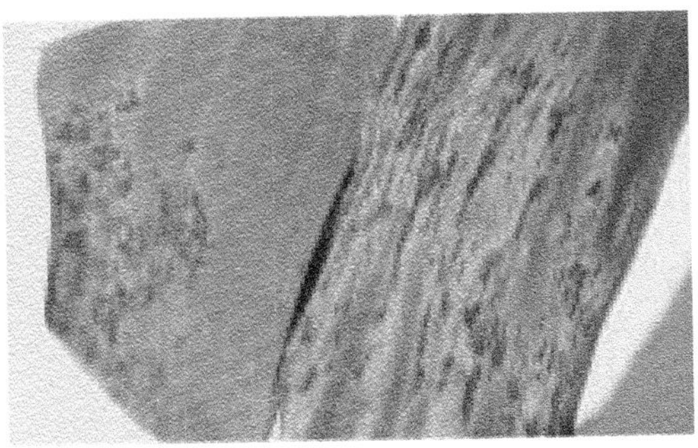

31

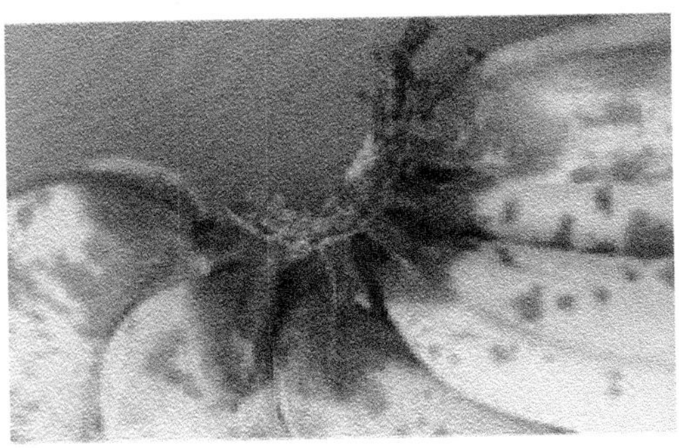

32

33

Plate 28. Symptoms of Eumusae leaf spot, caused by *Mycosphaerella eumusae* (anamorph: *Pseudocercospora eumusae*) (photo: R.C. Ploetz, UF).

Plate 29. Symptoms of freckle, caused by *Guignardia musae*, on 'Giant Cavendish' AAA in Taiwan. Note the blackish pycnidia of the anamorph, *Phyllosticta musarum*, along the midrib in the necrotic portion of the leaf; they give the leaf a rough texture (photo: R.C. Ploetz, UF).

Plate 30. Symptoms of Mycosphaerella speckle (left), caused by *Mycosphaerella musae*, and yellow Sigatoka (right), caused by *M. musicola*, on 'Williams' AAA in South Africa (photo: R.C. Ploetz, UF).

Plate 31. Anthracnose and crown rot on a hand of 'Grand Nain' AAA fruit (photo: R.C. Ploetz, UF).

Plate 32. Internal symptoms of Panama disease (photo: R.C. Ploetz, UF).

Plate 33. Symptoms of mosaic, caused by *Cucumber mosaic virus* (CMV), on 'Dwarf Cavendish' AAA in Egypt (photo: R.C. Ploetz, UF).

34

36

38

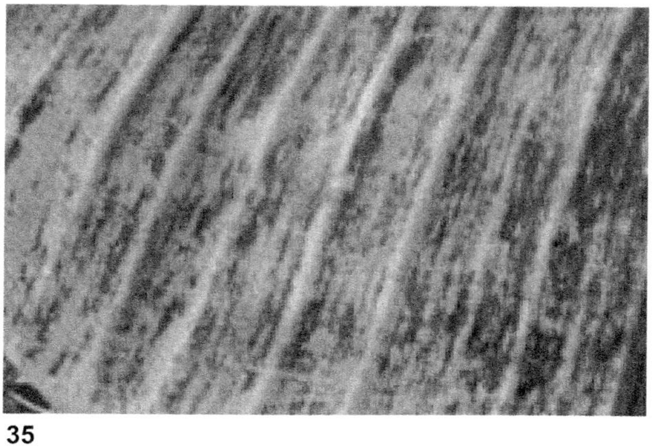

35

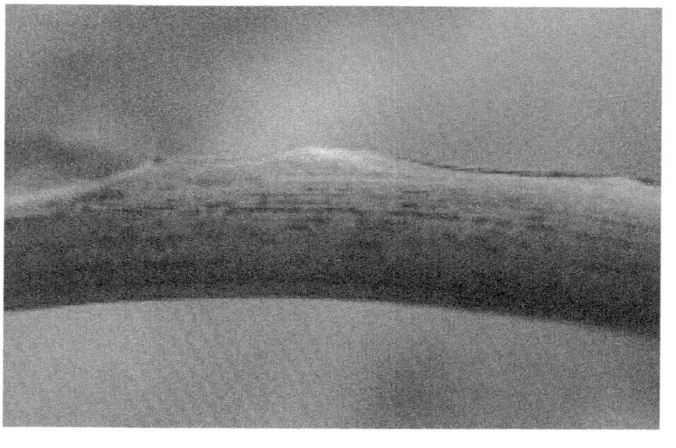

37

Plate 34. Initial fleck symptoms of banana streak, caused by *Banana streak virus* (BSV), on 'Mysore' AAB (photo: R.C. Ploetz, UF).
Plate 35. Dark necrotic streaks on 'Mysore' AAB, caused by *Banana streak virus* (BSV) (photo: R.C. Ploetz, UF).
Plate 36. Rosetting and stunting of leaves, caused by *Banana bunchy top virus* (BBTV) in Burundi (photo: R.C. Ploetz, UF).
Plate 37. Dot – dash Morse code symptom of banana bunchy top on a petiole of a 'Dwarf Cavendish' AAA leaf in Egypt (photo: R.C. Ploetz, UF).
Plate 38. Anthracnose on young breadfruit associated with insect injury (photo: A.W. Cooke, QDPI).

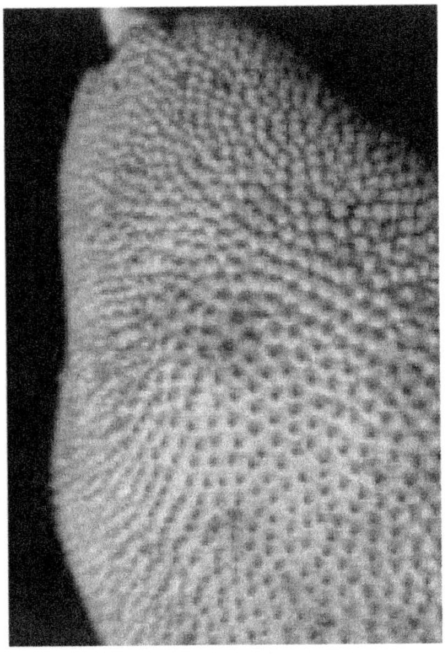

39

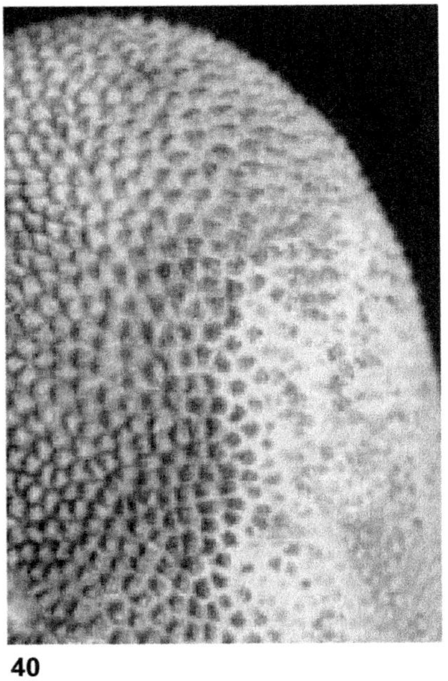

40

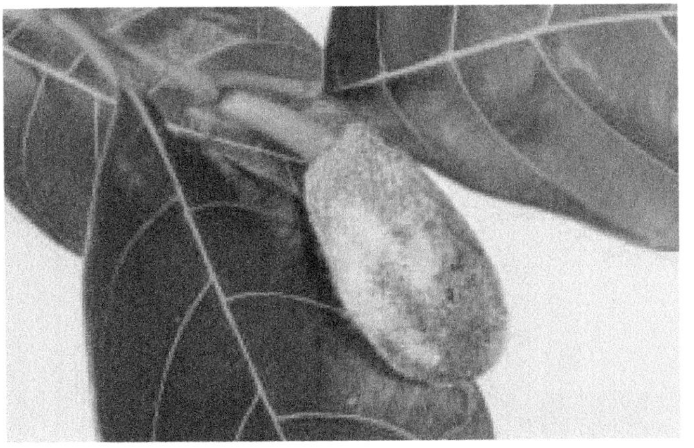

41

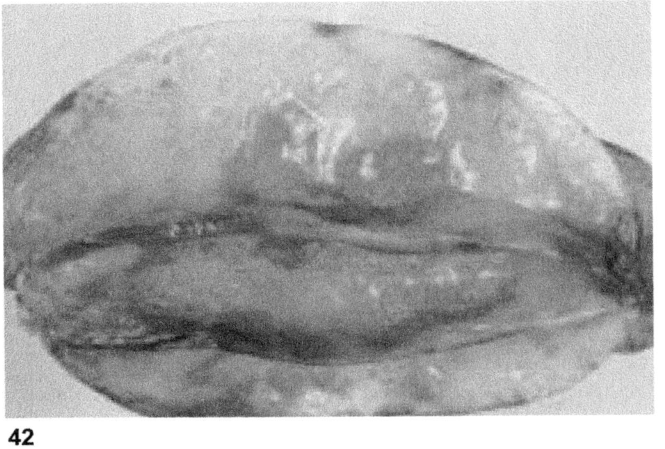

42

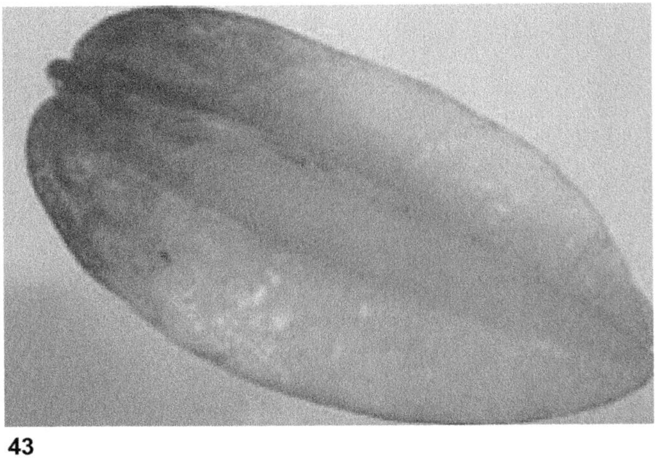

43

Plate 39. Anthracnose on mature jackfruit (photo: Somsiri Sangchote, KU).
Plate 40. Symptoms of Diplodia fruit rot on jackfruit (photo: Somsiri Sangchote, KU).
Plate 41. Symptoms of Rhizopus rot on young jackfruit (photo: A.W. Cooke, QDPI).
Plate 42. Advanced symptoms of anthracnose on a carambola fruit, caused by *Colletotrichum acutatum* (photo: A. W. Cooke, QDPI).
Plate 43. Dothiorella rot on a carambola fruit, caused by *Dothiorella* sp. (photo: A. W. Cooke, QDPI).

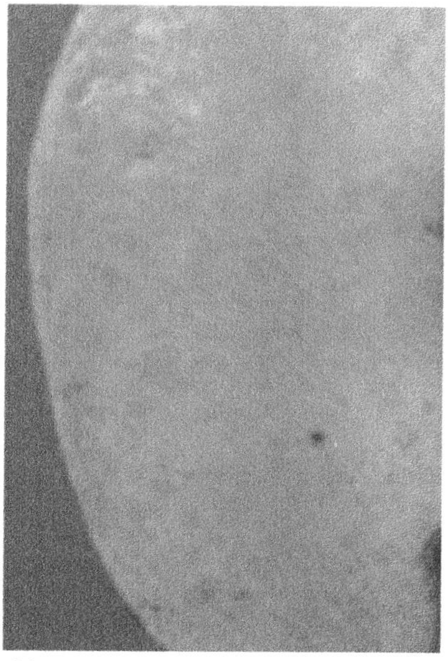
44

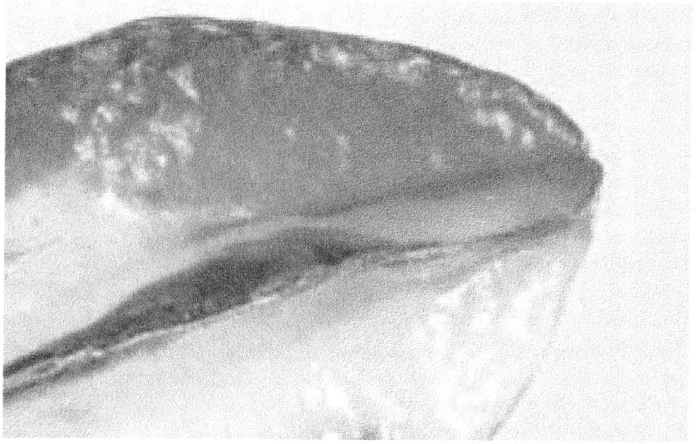
45

46

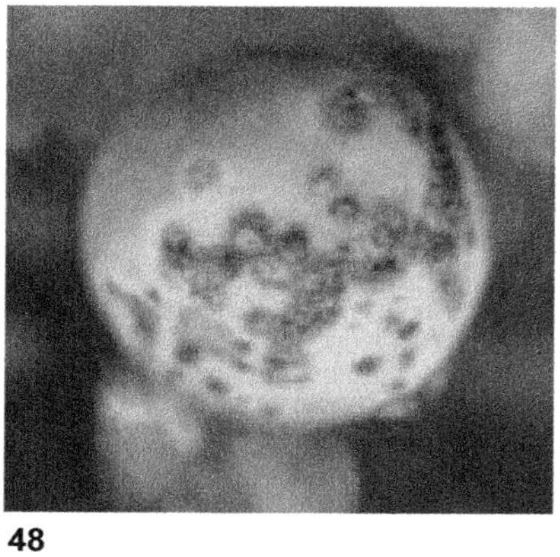
48

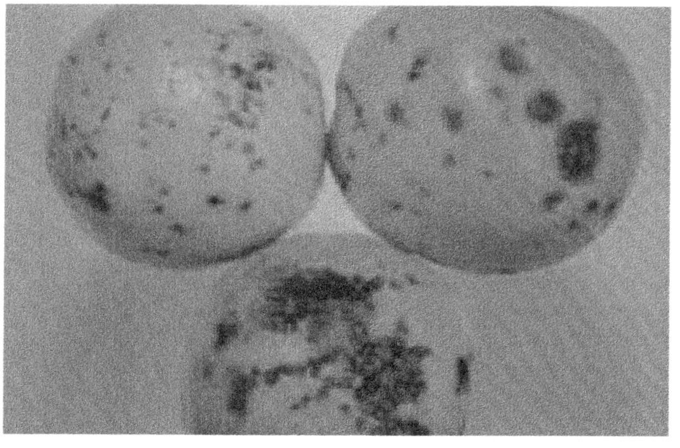
47

Plate 44. Sooty blotch on the surface of an 'Arkin' carambola fruit. Note the symptoms of flyspeck at the lower right-hand corner of the plate (photo: R.C. Ploetz, UF).
Plate 45. Phomopsis rot on a 'Fwang Tung' carambola fruit. This disease usually develops near the stem and stylar ends, but also occurs on wounded areas and particularly on the ribs (photo: A.W. Cooke, QDPI).
Plate 46. Lesions on flower petals affected by postbloom fruit drop, caused by *Colletotrichum acutatum* (photo: L.W. Timmer, UF).
Plate 47. Lesions of black spot, caused by *Guignardia citricarpa*, on 'Valencia' sweet orange in Australia. Symptoms range from small freckle spots to expanding virulent lesions (photo: Lowan Turton).
Plate 48. Lesions of citrus canker, caused by *Xanthomonas axonopodis* pv. *citri*, on a 'Pera' sweet orange fruit in Parana, Brazil (photo: J.H. Graham, UF).

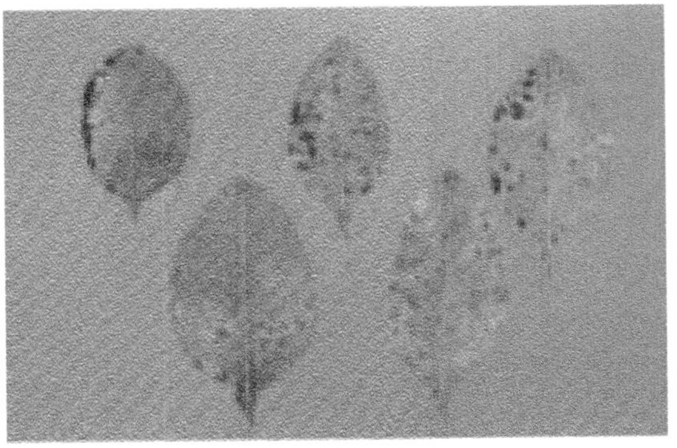

49

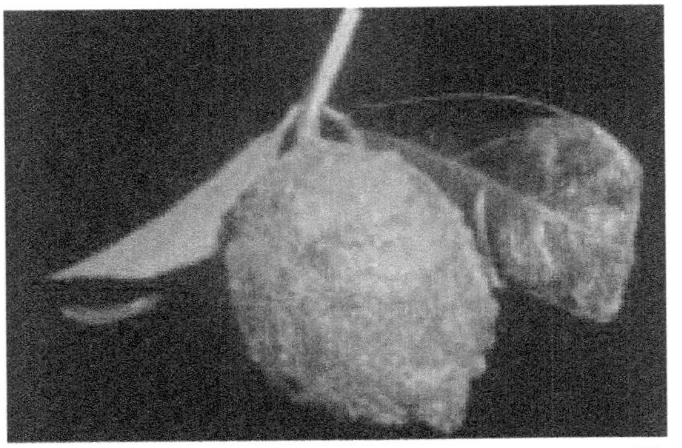

50

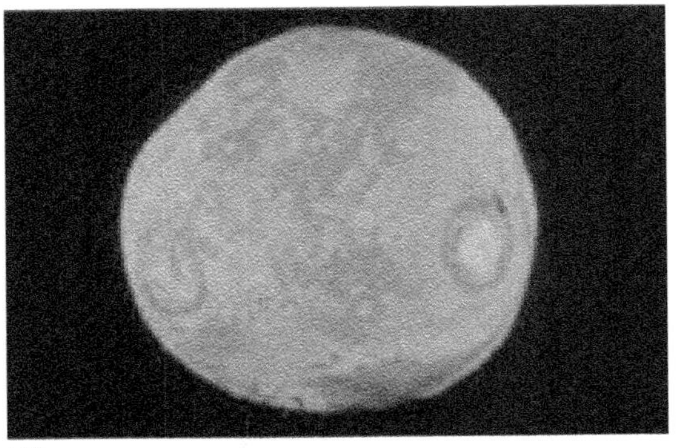

52

51

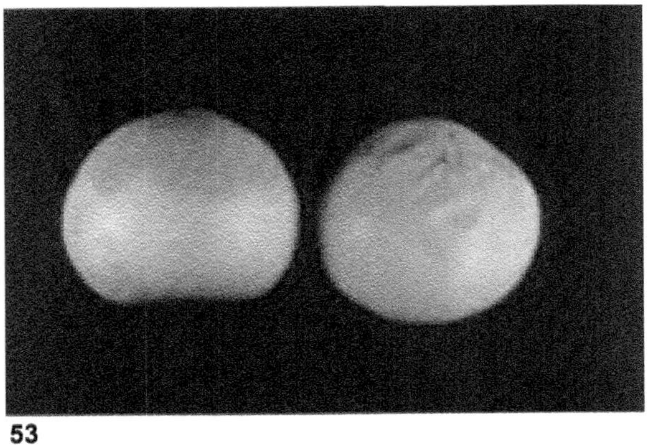

53

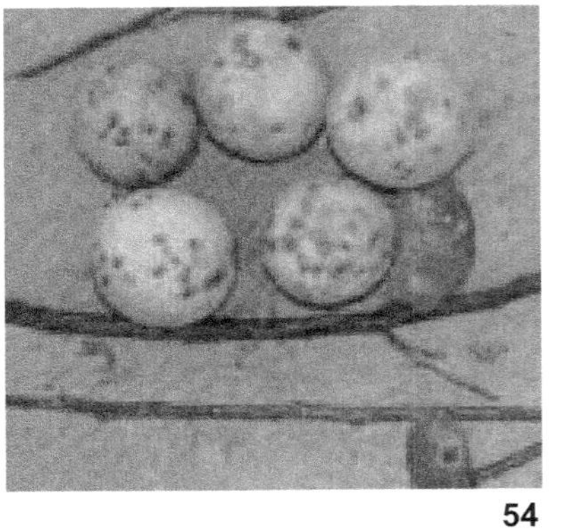

54

Plate 49. Lesions of greasy spot, caused by *Mycosphaerella citri*, on grapefruit leaves in Florida (photo: L.W. Timmer, UF).
Plate 50. Pustules of citrus scab, caused by *Elsinoë fawcettii*, on 'Eureka' lemon fruit in Australia (photo: Lowan Turton).
Plate 51. Gummosis (foot rot), caused by *Phytophthora* sp., on the trunk of a grapefruit tree on 'Cleopatra' mandarin rootstock (photo: S.M. Garnsey, UF).
Plate 52. A sweet orange fruit affected by green mould, caused by *Penicillium digitatum*, and blue mould, caused by *P. italicum* (photo: L. W. Timmer, UF).
Plate 53. Stem-end rot, caused by *Diplodia theobromae*, on 'Star Ruby' grapefruit (photo: L.W. Timmer, UF).
Plate 54. Symptoms of leprosis, caused by a rhabdovirus, on sweet orange leaves, twigs and fruit in Brazil (photo:

55

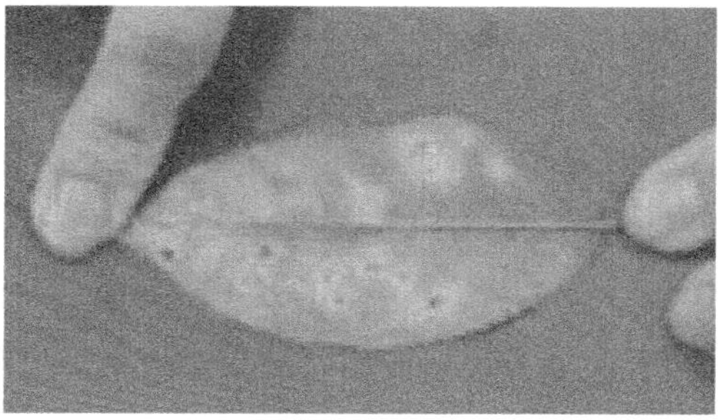

56

57

58

58

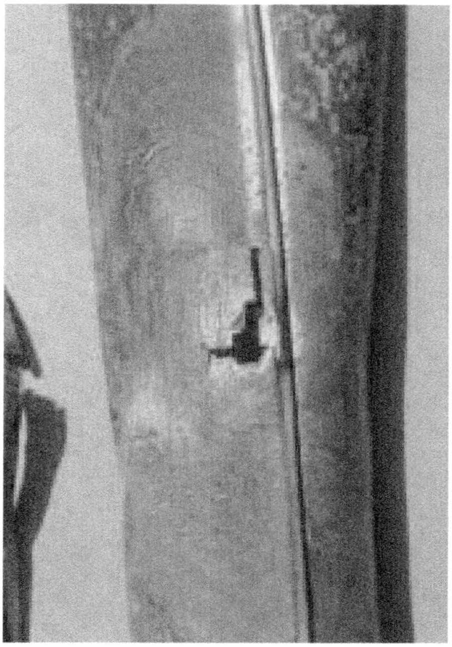

60

P. Broadbent).

Plate 55. Aerial view of a sweet orange grove on sour orange rootstock showing trees declining from citrus tristeza and new replacement trees in Florida (photo: S.M. Garnsey, UF).

Plate 56. Symptoms of citrus variegated chlorosis, caused by *Xylella fastidiosa*, on a sweet orange leaf in Uruguay. Note chlorosis, gum impregnation, and collapse of areas between the principal veins (photo: L.W. Timmer, UF).

Plate 57. Symptoms of huanglongbing (greening) disease, caused by Liberibacter asiaticus, on mandarins in Malaysia (photo: P. Broadbent).

Plate 58. Symptoms of citrus blight (declinamiento) in Argentina. Note wilt, twig dieback and small fruit (photo: L.W. Timmer, UF).

Plate 59. Early symptoms of bud rot on coconut in Hawaii, caused by *Phytophthora* sp. (photo: J.J. Ooka and J.Y. Uchida, UH).

61

62

63

64

65

Plate 60. Symptoms of grey leaf blight of coconut, caused by *Pestilotiopsis palmarum* leaf (photo: R.C. Ploetz, UF).
Plate 61. Symptoms of hartrot on coconut, caused by a uniflagellate protozoan in the genus *Phytomonas* (photo: Michel Dollet, IHRO).
Plate 62. Discoloration of a new inflorescence on a coconut palm affected by lethal yellowing (photo: N.A. Harrison, UF).
Plate 63. Various stages in the development of lethal yellowing on coconut palm (photo: N.A. Harrison, UF).
Plate 64. Destruction of a coconut plantation in Ghana by Cape St Paul wilt (photo: P. Jones, IACR).

66

67

69

68

70

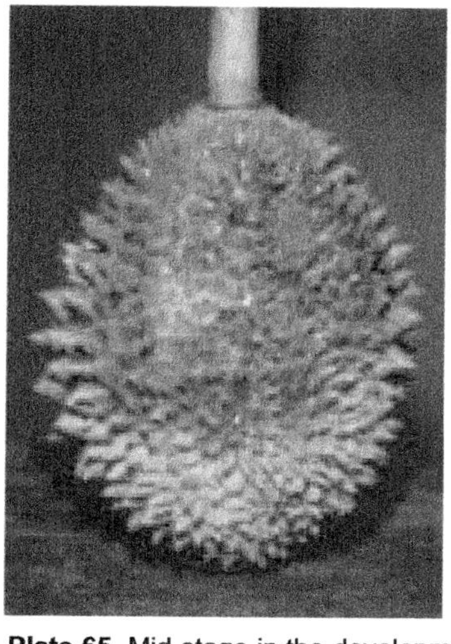

71

Plate 65. Mid-stage in the development of lethal decline of coconut in Tanzania (photo: P. Jones, IACR).
Plate 66. In longitudinal section, elongated nuts with poorly developed endosperm from coconut palms affected by Kalimantan wilt (left) compared with a healthy nut on the right (photo: P. Jones, ICAR).
Plate 67. Coconut palm affected by red ring. Note the relatively nondescript decline of the palm's canopy, absence of nuts and the diagnostic orange to red-brown ring in the trunk's interior (photo: K. Gerber).
Plate 68. Unilateral necrosis of a newly mature date palm leaf caused by bayoud. This is usually the first external symptom caused by this disease (photo: J. Carpenter, USDA).
Plate 69. Advanced symptoms of bayoud on several date palms. Eventually the disease kills all leaves in the canopy and the palm dies (photo: J. Carpenter, USDA).
Plate 70. Symptoms on durian fruit caused by *Phytophthora palmivora* (photo: T.K. Lim).

73

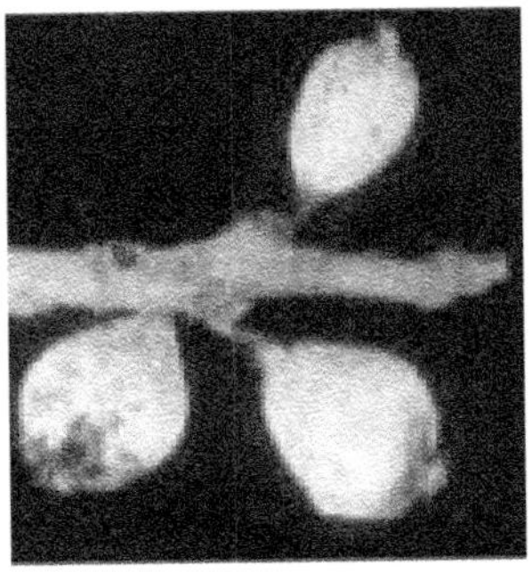

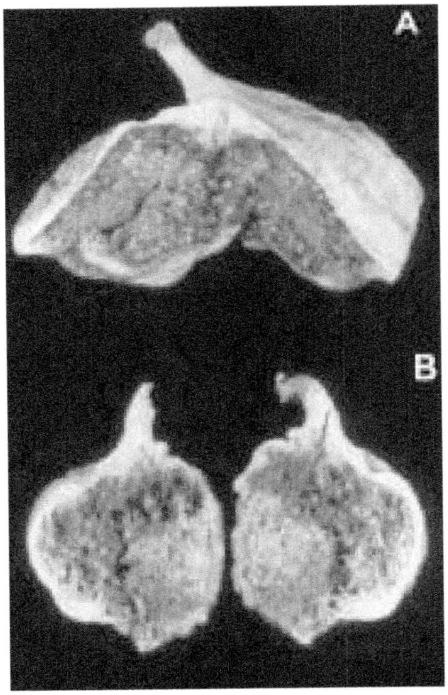

72

74

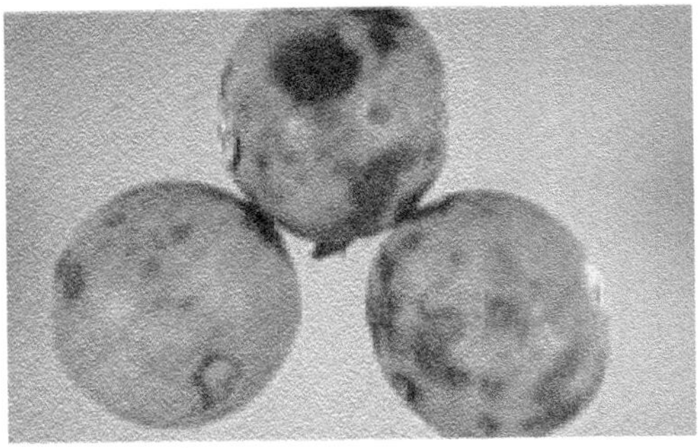

75

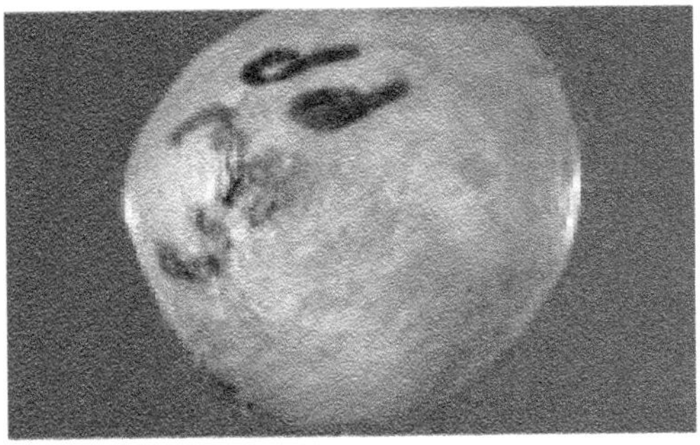

77

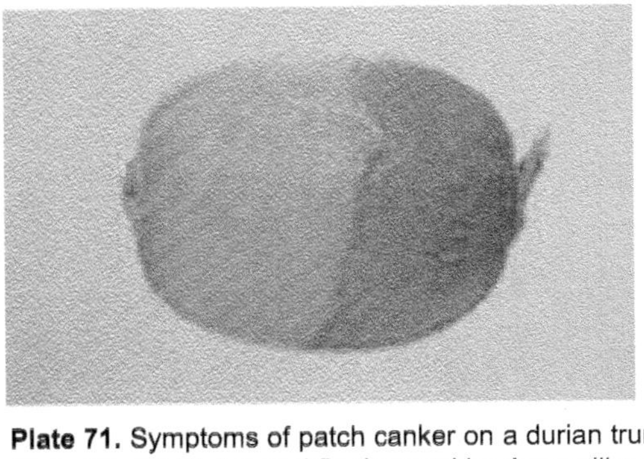

76

Plate 71. Symptoms of patch canker on a durian trunk, caused by *Phytophthora palmivora* (photo: T.K. Lim).
Plate 72. (A) 'Calimyrna' fig decayed by *Aspergillus flavus*, and (B) a fig decayed by *A. parasiticus* (photos: T.J. Michailides, UC).
Plate 73. Mummies of summer (mammoni) crop caprifigs infected by *Fusarium lactis* that have survived until April of the following year. The surface and the interior of these caprifigs are covered with the pathogen's mycelia and conidia (photos: T.J. Michailides, UC).
Plate 74. Symptoms of anthracnose on guava, caused by *Colletotrichum acutatum* (photo: A.W. Cooke, QDPI).
Plate 75. Symptoms of Rhizopus rot on guava, caused by *Rhizopus* sp. (photo: B. Brown).
Plate 76. Extensive canker that has developed at the base of the trunk of a kiwifruit vine due to temperatures below 0°C. The necrotic tissue was exposed after peeling off the bark. Note the healthy vine that has sprouted from beneath the damaged area (photo: B.A. Latorre).
Plate 77. Peeled kiwifruit affected by Botrytis stem-end rot. The affected flesh is conspicuously darker, and has a

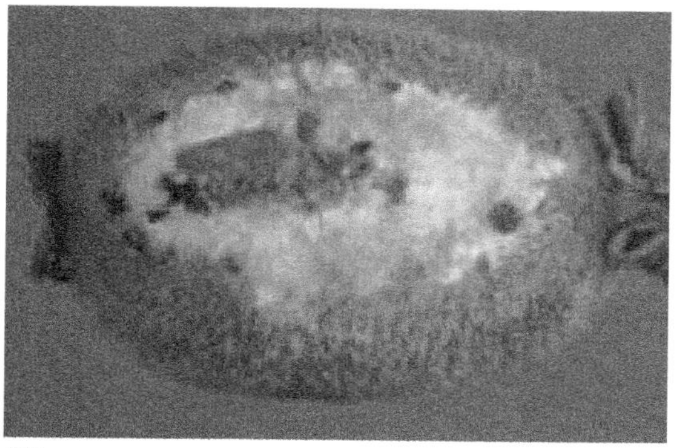

78

79

80

81

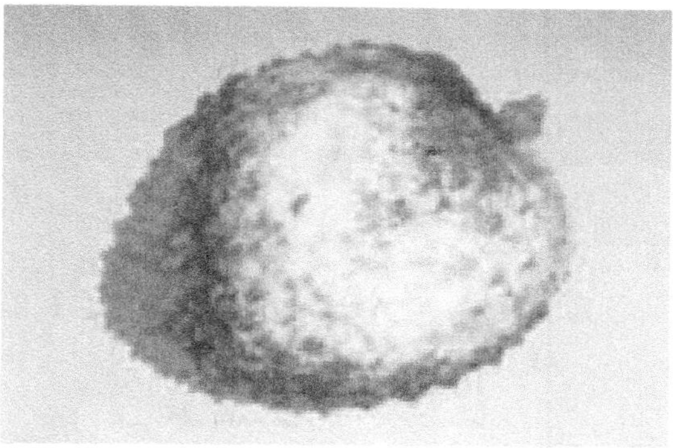

82

glassy, water-soaked appearance (photo: B.A. Latorre).

Plate 78. Sclerotinia rot of kiwifruit originating at the point of contact with an infected petal. Black sclerotia of the pathogen, *Sclerotinia sclerotiorum*, are clearly visible amongst the white mycelium (photo: B.A. Latorre).

Plate 79. Tree in the advanced stages of longan decline (photo: C. Sittigul, CMU).

Plate 80. Powdery mildew on young rambutan fruit (photo: S. Sanchote, KU).

Plate 81. Longan tree with witches' broom. Note the crowded panicles and broom-like appearance of the inflorescences (photo: C. Sittigul, CMU).

Plate 82. Anthracnose symptoms on a lychee fruit. Note the copious white mycelium and salmon-coloured masses

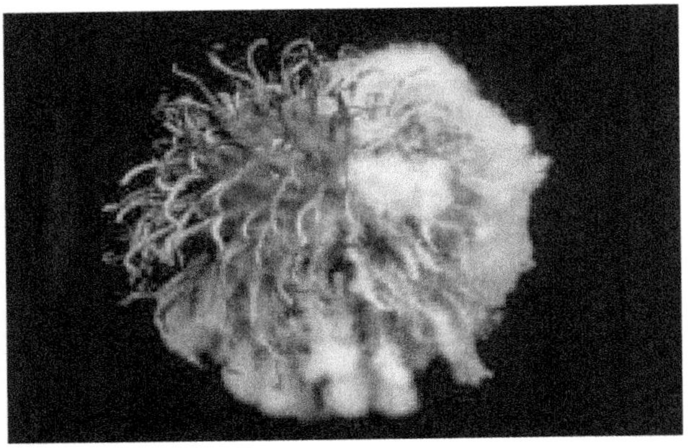

83

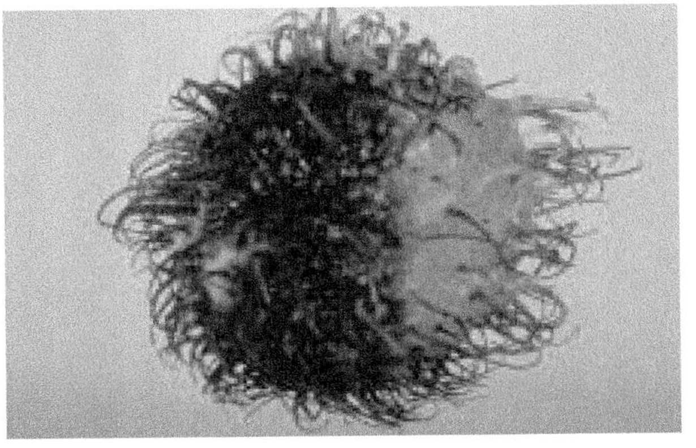

84

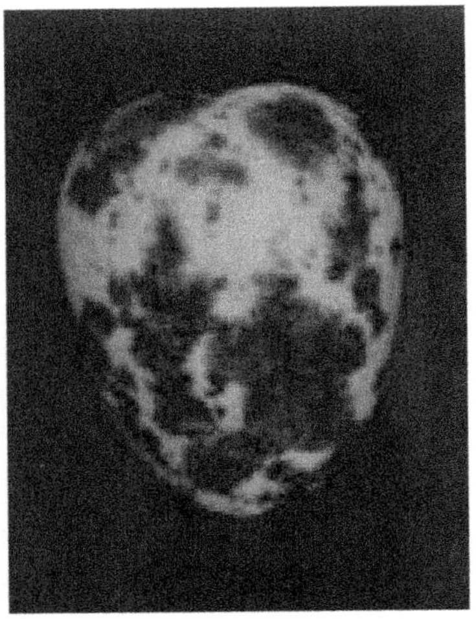

85

86

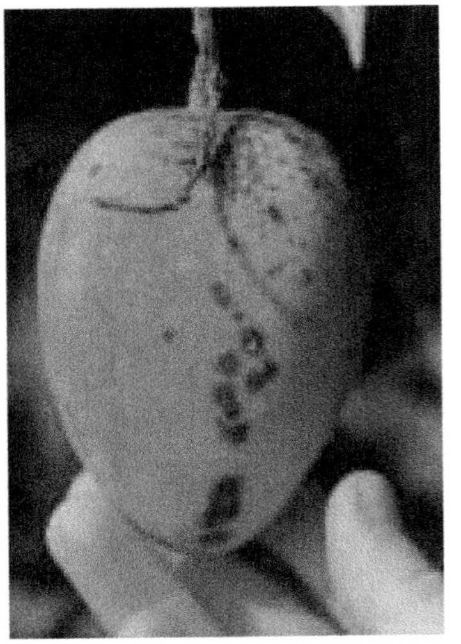

87

of the causal fungus, *Colletotrichum gloeosporioides* (photo: A.W. Cooke, QDPI).
Plate 83. Symptoms of Gliocephalotrichum rot on a rambutan fruit. Note the copious mycelium of the causal fungus covering the fruit surface (photo: S. Sanchote, KU).
Plate 84. Symptoms of Phomopsis stem-end rot on a rambutan fruit (photo: A.W. Cooke, QDPI).
Plate 85. Advanced symptoms of Alternaria rot on mango. This damage is darker and harder than that caused by anthracnose (photo: D. Prusky, ARO).
Plate 86. Anthracnose symptoms on newly formed leaves on an 'Edward' mango. Foliar damage this extensive occurs only when high levels of rainfall and inoculum occur during vegetative flushes (photo: R.C. Ploetz, UF).
Plate 87. Anthracnose symptoms on an 'Edward' mango fruit. Note the streak of lesions that originates at the fruit's stem end that result from conidia of the pathogen, *Colletrichum gloeosporiodes*, dripping from the necrotic

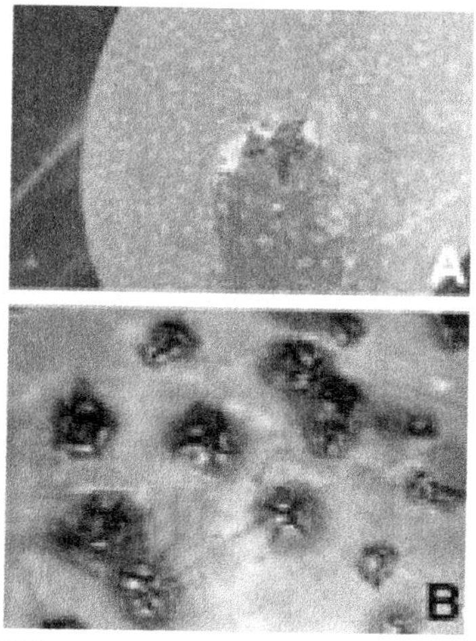

88

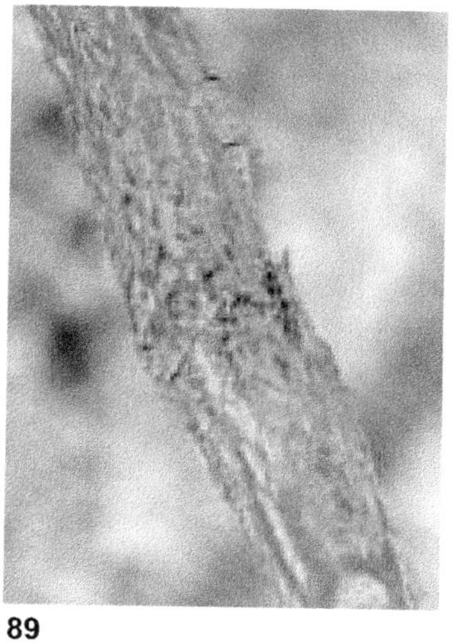

89

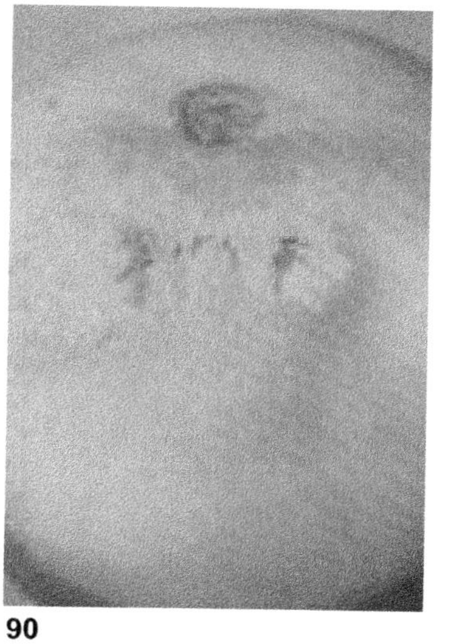

90

91

92

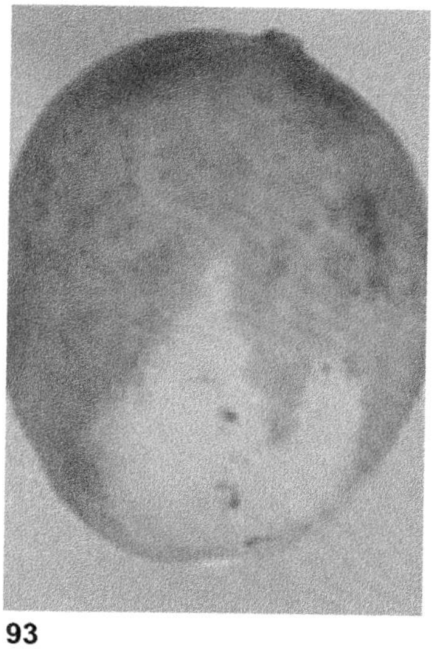

93

panicle (photo: R.C. Ploetz, UF).

Plate 88. Bacterial black spot (BBS) lesions on the surface of a mango fruit. (A) The liquid that has oozed from these lesions is laden with the causal bacterium, *Xanthomonas*. sp. pv. *mangiferaeindicae*. Note the lenticels beneath the lesion that have become infected and begun to develop symptoms. (B) Close up of BBS lesions showing the deep, cracked craters caused by this disease (photos: O. Pruvost, CIRAD).

Plate 89. A twig canker, caused by *Xanthomonas*. sp. pv. *mangiferaeindicae,* that is associated with a wound (photo: O. Pruvost, CIRAD).

Plate 90. Stem-end cavity on a 'Kensington' mango fruit in Australia (photo: P. Scholefield).

Plate 91. Symptoms of vegetative malformation of mango in Egypt (photo: R.C. Ploetz, UF).

Plate 92. Stem bleeding on trunk and scaffold limbs of a 'Tommy Atkins' mango tree in Florida. The tree was exposed to temperatures of -4°C during the previous winter. Freezing temperatures are one of many predisposing factors associated with the decline syndrome (photo: R.C. Ploetz, UF).

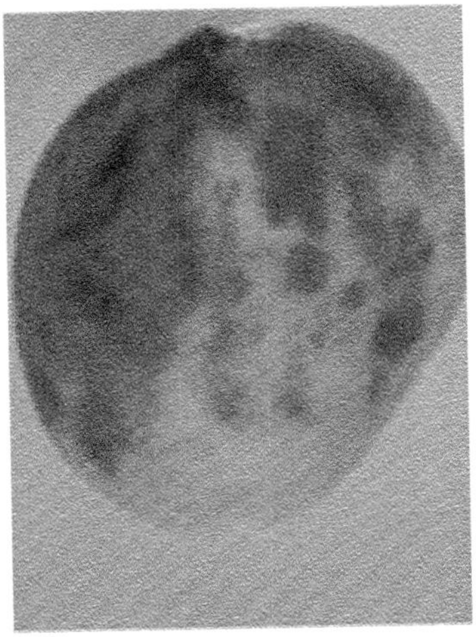

94

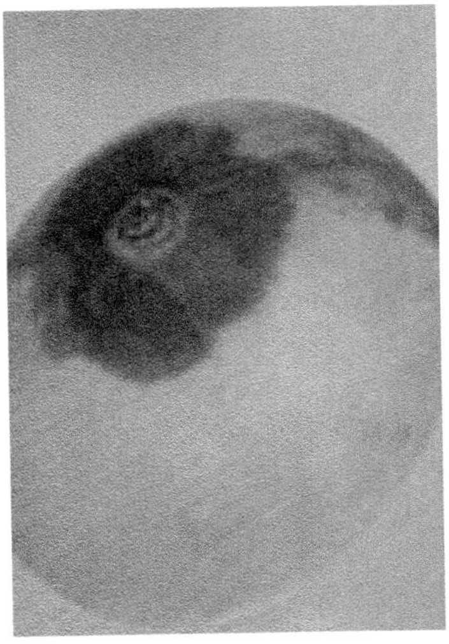

95

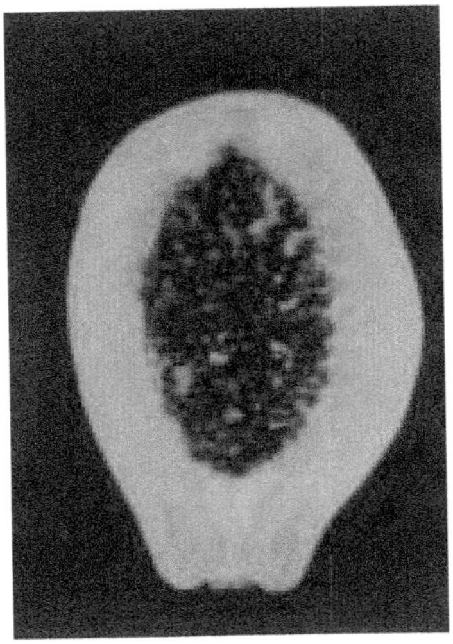

97

96

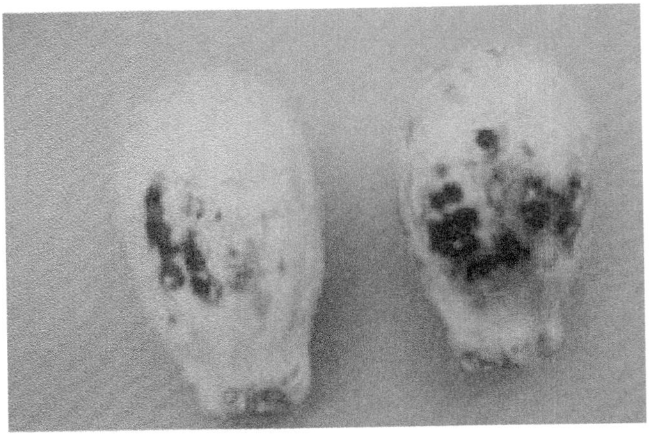

98

Plate 93. External symptoms of mango stem-end rot caused by *Fusicoccum aesculi* (photo: A.W. Cooke, QDPI).
Plate 94. External symptoms of mango stem-end rot, caused by *Fusicoccum mangiferum* (photo: A.W. Cooke, QDPI).
Plate 95. External symptoms of mango stem-end rot, caused by *Phomopsis* sp. (photo: A.W. Cooke, QDPI).
Plate 96. Internal symptoms of mangosteen fruit rot, caused by *Diplodia theobromae* (photo: A.W. Cooke, QDPI).
Plate 97. Internal yellowing of a papaya fruit, caused by *Enterobacter cloacae* (photo: K.A. Nishijima, USDA).

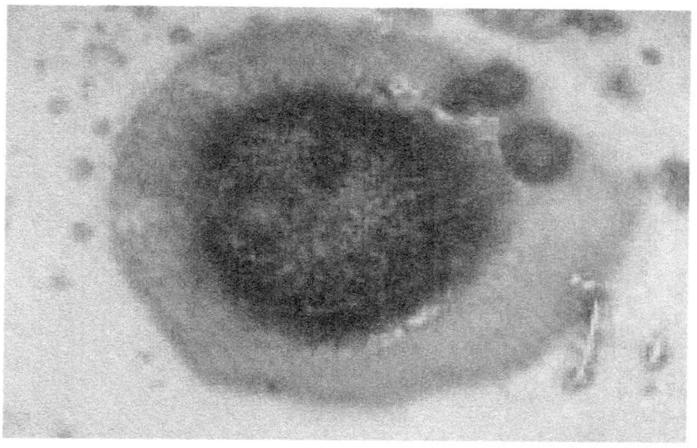

100

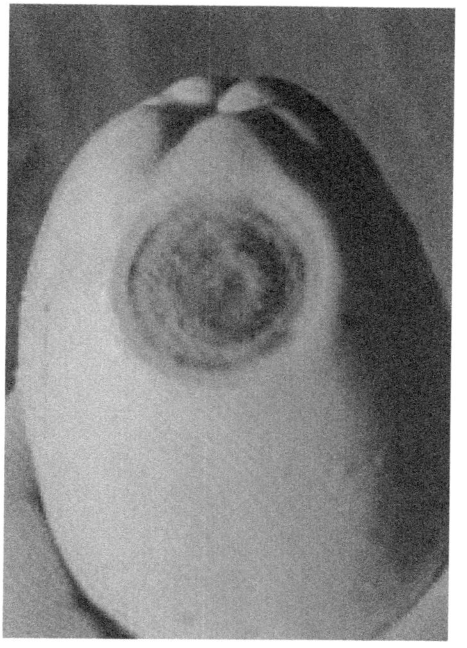

99

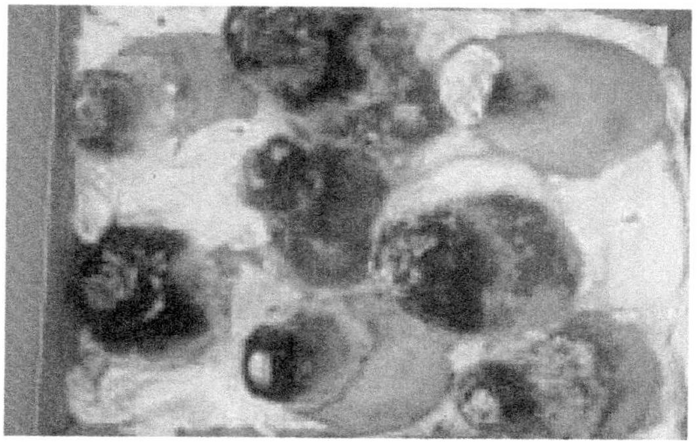

101

102

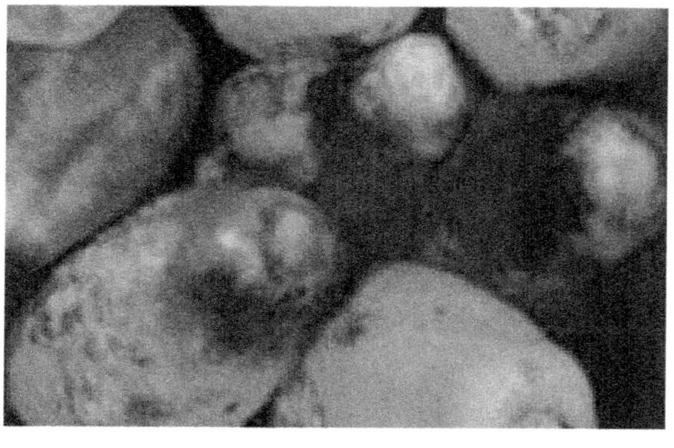

103

104

Plate 98. Symptoms of Alternaria fruit spot on papaya (photo: W.T. Nishijima, UH).
Plate 99. Typical anthracnose lesion on papaya (photo: W.T. Nishijima, UH).
Plate 100. Black rot lesion on papaya, caused by *Mycosphaerella caricae*. This fungus is also a common cause of stem-end rot (photo: A.W. Cooke, QDPI).
Plate 101. Fruit and stem-end rot of papaya, caused by *Diplodia theobromae* (photo: W.T. Nishijima, UH).
Plate 102. Symptoms of Phytophthora fruit rot on an immature papaya fruit (photo: A.W. Cooke, QDPI).
Plate 103. Rhizopus soft rot of papaya fruit (photo: W.T. Nishijima, UH).

105

106

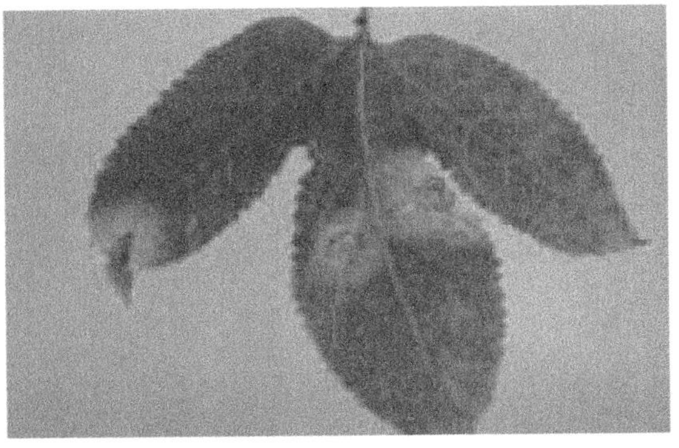

107

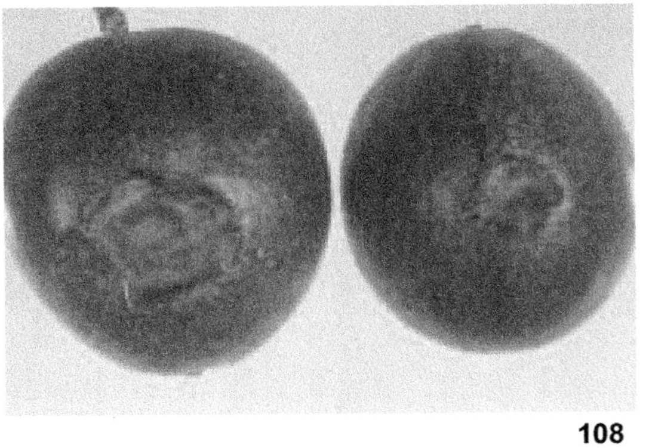

108

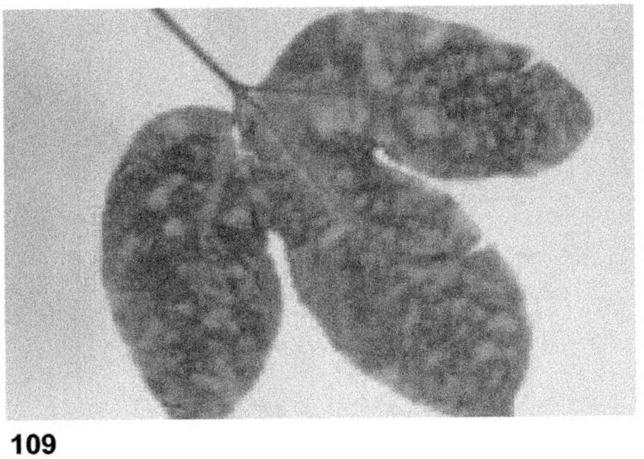

109

Plate 104. Symptoms of yellow crinkle on papaya (A.W. Cooke, QDPI).
Plate 105. Strapleaf symptoms on papaya leaves, caused by *Papaya ringspot virus* (photo: A.W. Cooke, QDPI).
Plate 106. Large anthracnose lesion on a passion fruit, caused by *Colletotrichum gloeosporioides* (photo:A.W. Cooke, QDPI).
Plate 107. Symptoms of brown spot on a passion fruit leaf, caused by *Alternaria passiflorae* (photo: A.W. Cooke, QDPI).
Plate 108. Sunken brown spot lesions on a passion fruit, caused by *Alternaria passiflorae* (photo: A.W. Cooke, QDPI).

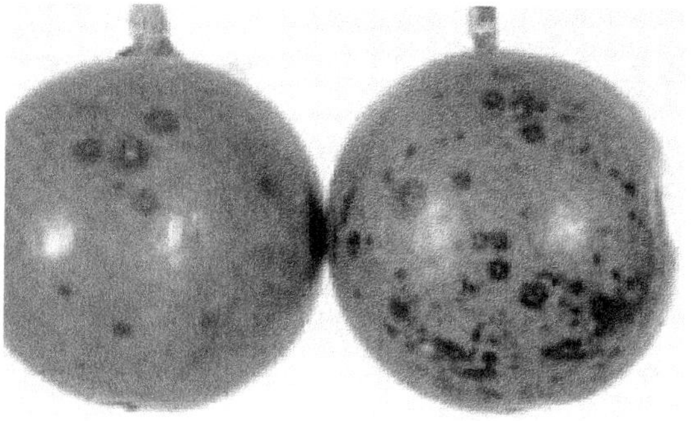

110

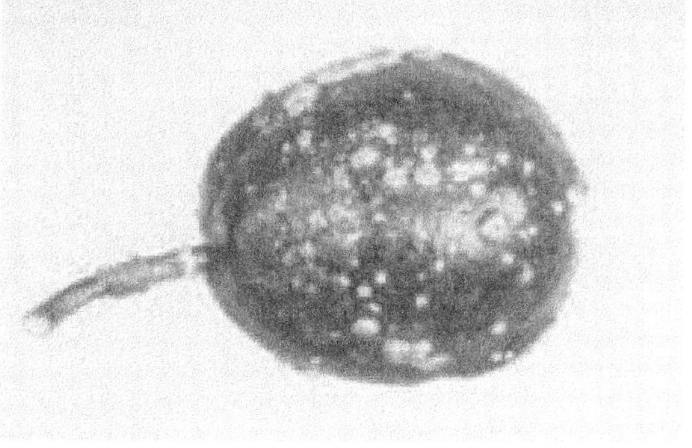

111

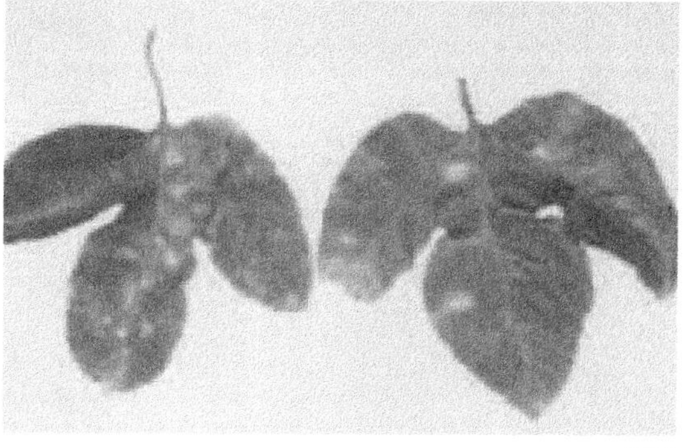

112

113

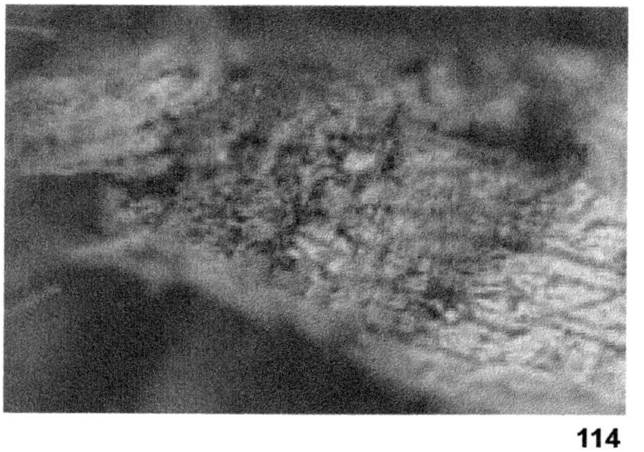

114

Plate 109. Foliar lesions on passion fruit, caused by *Alternaria alternata* (photo: A.W. Cooke, QDPI).
Plate 110. Lesions on passion fruit, caused by *Alternaria alternata*. Note the oily, greasy appearance of the lesion margins (photo: A.W. Cooke, QDPI).
Plate 111. Scab lesions, caused by *Cladosporium herbarum*, on a passion fruit (photo: A.W. Cooke, QDPI).
Plate 112. Septoria blotch symptoms on passion fruit leaves (photo: A.W. Cooke, QDPI).
Plate 113. Septoria blotch symptoms on passion fruits (photo: A.W. Cooke, QDPI).
Plate 114. Basal canker on a passion fruit vine that has rotted through the entire stem. Note the crimson perithecia

115

116

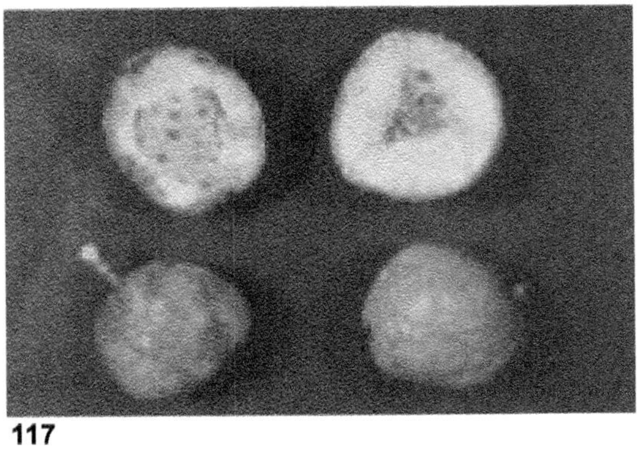

117

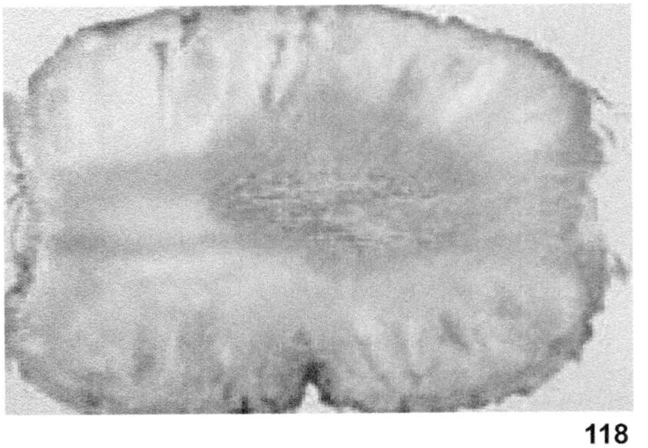

118

119

of the causal fungus, *Haematonectria haematococca*, at the stem's base (photo: R.C. Ploetz, UF).

Plate 115. Necrosis of passion fruit leaves, stems and fruit, caused by *Phytophthora nicotianae* (photo: A.W. Cooke, QDPI).

Plate 116. Extensive lesion on an immature passion fruit, caused by *Phytophthora nicotianae* (photo: R.C. Ploetz, UF).

Plate 117. Distortion and internal discoloration of passion fruits, cause by passion fruit woodiness (photo: A.W. Cooke, QDPI).

Plate 118. Symptoms of black rot or water blister in the interior of a pineapple fruit (photo: A.W. Cooke, QDPI).

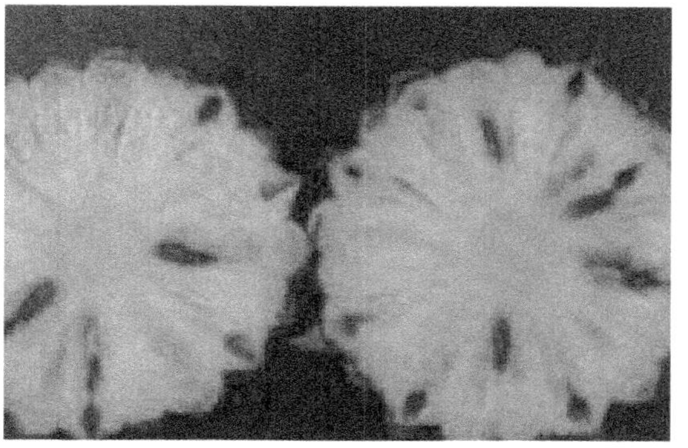

120

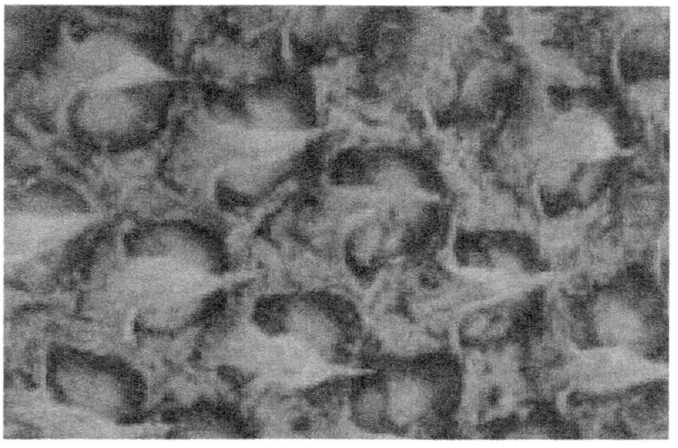

121

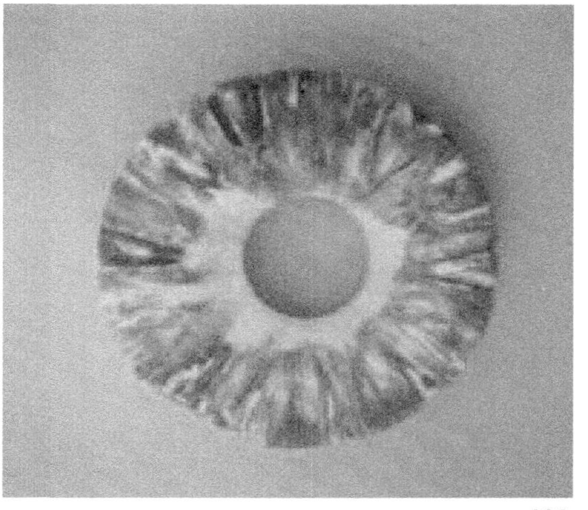

122

123

124

Plate 119. External symptoms of fruitlet core rot (photo: A.W. Cooke, QDPI).
Plate 120. Internal symptoms of fruitlet core rot (photo: A.W. Cooke, QDPI).
Plate 121. External symptoms of interfruitlet corking, caused by *Penicillium funiculosum* (photo: A.W. Cooke, QDPI).
Plate 122. Severe internal symptoms of marbling on pineapple (photo: A.W. Cooke, QDPI).
Plate 123. A pineapple fruit core affected by pink disease (left) compared with one not affected by the disease (right) (photo: A.W. Cooke, QDPI).

125

126

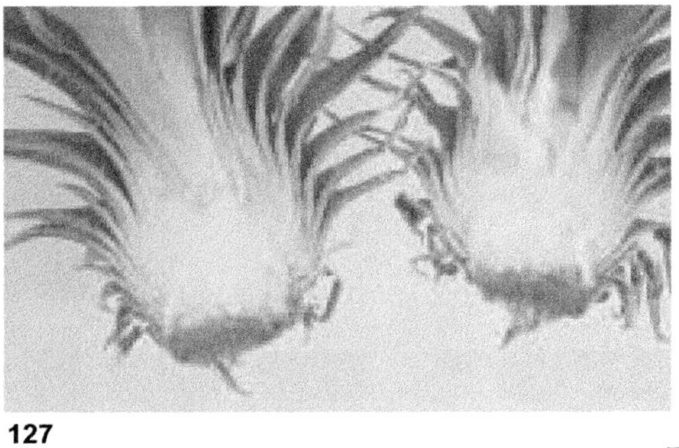

127

128

Plate 124. Yellow spot symptoms on young pineapple plants (photo: A.W. Cooke, QDPI)
Plate 125. Foliar symptoms of mealybug wilt on pineapple. Although they can be caused by other factors, they are diagnostic if they occur at the margins of fields and in association with mealybugs (photo: A.W. Cooke, QDPI).
Plate 126. Above ground symptoms of root rot on pineapple, caused by *Phytophthora cinnamomi* (photo: A.W. Cooke, QDPI).
Plate 127. Butt rot on pineapple suckers, caused by *Chalara paradoxa* (photo: A.W. Cooke, QDPI).

19 Diseases of Pineapple

K.G. Rohrbach[1*] and Donald Schmitt[2]
[1] *University of Hawaii, College of Tropical Agriculture;* [2] *University of Hawaii, Plant Pathology Department, Honolulu, Hawaii, USA*

Introduction

An estimated 12,562,440 t of pineapple, *Ananas comosus* (family: *Bromeliaceae*), were produced in 1998. The major producing countries are, in descending order, Thailand, the Philippines and Brazil (Fig. 19.1). Fruit for export, either fresh or processed, are grown on large plantations due to the need for year-round production and the large investment of capital that is involved. In contrast, locally consumed fruit are usually produced and marketed by small farmers.

Botanical and commercial terminology for pineapple is described in Table 19.1. Commercial production consists of a series of fruit cycles, the number depending on the effectiveness of pest and disease management. The first cycle is termed the 'plant crop' whereas subsequent cycles are 'ratoons'. In Hawaii, the plant crop fruits in ~18–20 months, depending on temperature (elevation), and subsequent ratoons take ~10–12 months (Rohrbach, 1986).

Several characteristics of the pineapple plant affect the types of diseases that impact commercial production. Pineapple plant growth occurs between 21 and 35°C, and the plant's phenotype varies depending on the production area's mean temperature. Roots of the pineapple plant are adventitious and will not regenerate if damaged. Thus, loss of functional roots has a major impact on plant and fruit growth. The pineapple inflorescence emerges out of a rosette of leaves that collect water. Several microorganisms and arthropods colonize this area and contaminate the emerging inflorescence. Anthesis of the inflorescence occurs over a 2–4 week period, and individual flowers are open for 1 or 2 days. Thus, on a given plant, susceptible flowers are available for a relatively long time (Rohrbach and Schmitt, 1994).

Several comprehensive reviews on pineapple diseases and pests have been published (Lim, 1985; Rohrbach and Apt, 1986; Broadley *et al.*, 1993; Rohrbach and Schmitt, 1994). Criteria for disease measurement are seasonal frequency (occurrence of disease), the proportion of the population that is affected (incidence) and the effect of disease on each plant part (severity). Severity ranges from a reduction in plant size and fruit weight to plant mortality. Major pineapple diseases that are covered in this chapter are listed based on the host organ that is affected (Rohrbach and Apt, 1986; Rohrbach and Schmitt, 1994).

*New address: 40900 South Applefield Circle, Elizabeth, CO 80107, USA.

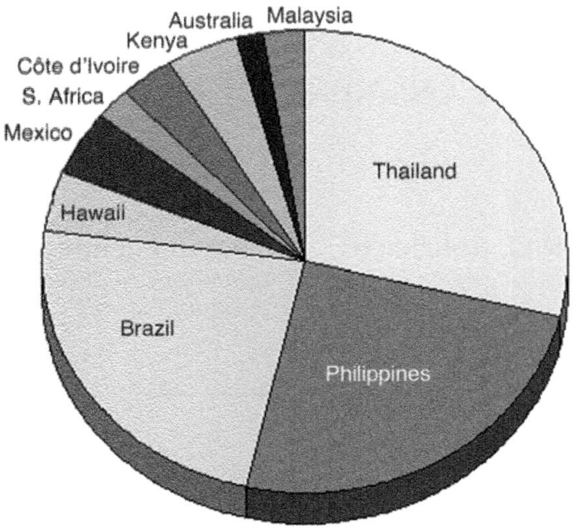

Fig. 19.1. Relative pineapple production in the world's leading producing countries (FAO, 1998).

Table 19.1. Pineapple production terminology.

Production term	Botanical term
Roots	Adventitious roots
Stem	Stem
Leaf	Leaf
Heart leaves	Rosette of leaves
Open heart. Two stages: 1.25 cm and 2.5 cm (Rohrbach and Taniguchi, 1984)	Emergence of inflorescence
Cone. Three stages: early, mid and late cone (Rohrbach and Taniguchi, 1984)	Emergence of inflorescence
Inflorescence. Three stages: early, mid and late flower (Rohrbach and Taniguchi, 1984)	Multiple cone-shaped sessile, complete trimerous, flowers and subtending bracts
Fruit	Multiple fruit
Crown	Apical vegetative shoot
Slip	Lateral vegetative shoot on the peduncle borne on the apical end of vestigial fruits
Sucker	Axillary vegetative shoot

Fruit Diseases

Fruit pathology is complex due to the wide range of microorganisms that are found on the exterior and interior of the developing inflorescence. Determining the aetiology of many of the fruit diseases has been difficult due to their sporadic occurrence. Economically significant fruit diseases are black rot, fruit collapse, fruitlet core rot, fusariosis, internal browning, marbling and pink disease.

A pineapple fruit disease index for measuring the proportion of fruitlets with external and internal symptoms has been devised for use in the cannery (Rohrbach and Taniguchi, 1984; Paull and Rohrbach, 1985; Rohrbach and Schmitt, 1994). Since a single affected fruitlet may require an entire slice to be downgraded, the scale is weighted to recognize the high economic impact of minor damage (1=1–2%, 2=3–5%, 3=6–10%, 4=11–25%, 5=26–50% and 6=51–100%).

Bacterial fruit collapse

Bacterial fruit collapse was first recorded in Malaysia in 1935 (Thompson, 1937). It also occurs in Brazil, Costa Rica and the Philippines (Rohrbach, 1983). The intensive use of the susceptible 'Singapore Spanish' cultivar may account for the high incidence of disease in Malaysia.

Symptoms

Fruit collapse usually appears on fruit 2–3 weeks before normal ripening (Johnston, 1957a,b). Infected fruit exude juice and release gas typical of fermentation. The fruit shell becomes olive green as cavities develop within the fruit.

Causal agent

Erwinia chrysanthemi causes fruit collapse (Johnston, 1957a,b). It is a facultative anaerobe that affects a wide range of tropical and subtropical plants. The bacterium is a motile, Gram-negative, 0.6–0.9 × 1.5–3.0 μm rod, with peritrichous flagella. Virulence of different strains is related to the ability to produce large quantities of endopolygalacturonic transeliminase and grow at higher temperatures than other soft rot bacteria (Lim, 1974).

Epidemiology

The epidemiology of bacterial fruit collapse was studied in Malaysia in the 1970s (Lim, 1978; Lim and Lowings, 1979a,b). The primary source of inoculum is exudates from infected fruit. The pathogen is thought to be vectored to open pineapple flowers by several insect species, mostly ants, and remains latent until sugar levels rise and polyphenoloxidase levels decline in the ripening fruit (Rohrbach, 1989).

Management

Control strategies focus on reducing inoculum and vector activity, and proper cycling of flowering. Infected fruit should be removed from the field. Control of insects during flowering should be beneficial (Lim and Lowings, 1977). Significant control has been obtained with ethylene products that are used to inhibit flower opening and reduce nectar flow (Lim and Lowings, 1979a). Forcing flowering to ensure that it does not coincide with fruiting in adjacent plantings can also reduce disease development. Where they are adapted, the resistant 'Smooth Cayenne' and 'Smooth Cayenne' × 'Red Spanish' selections could be used rather than the susceptible Spanish types (Lim and Lowings, 1979b).

Black rot

Black rot, which is also called water blister or soft rot, was one of the first pineapple diseases to be studied. Major losses occurred when fruit were held for more than a few days before shipment (Larsen, 1910). Losses are now minimal due to refrigeration (Rohrbach and Phillips, 1990).

Black rot is a worldwide problem in fresh fruit. Its severity depends on the degree of bruising or wounding during harvesting and packing, the level of inoculum on the fruit at harvest, and the storage temperature during transportation and marketing. Black rot is not significant in processed fruit because it is handled quickly, and does not occur in the field unless fruit are over-ripe or injured (Rohrbach and Apt, 1986).

Symptoms

Black rot causes a soft watery rot (Plate 118). Affected tissue turns dark in the later stages of development due to the dark coloured mycelium and chlamydospores of the pathogen (Larsen, 1910). Eventually, the entire fruit can be destroyed.

Causal agent

Pineapple black rot is caused by *Chalara paradoxa*. Characteristics of the fungus are listed in Chapter 1.

Epidemiology

Fruit are infected through wounds that result from natural growth cracks and bruises. Both green and ripe fruit are susceptible (Larsen, 1910). Conidia are produced on affected tissues under conditions of high humidity, and are disseminated by wind to fruit in the field. Inoculum levels on fruit at harvest vary according to the environmental conditions prior to harvest. The high correlation between rainfall prior to harvest and disease following harvest has resulted in the name 'water blister' (Plate 118).

Management

Fruit injury should be minimized during harvest and handling. Fruit are dipped in an appropriate fungicide 6–12 h after harvest, and storage at 9°C retards development of the disease (Rohrbach and Phillips, 1990).

The 'Red Spanish' cultivar generally is more resistant than 'Smooth Cayenne'. Some hybrids developed by the Pineapple Research Institute of Hawaii were much more resistant than 'Smooth Cayenne' (K.G. Rohrbach, unpublished data).

Fruitlet core rot, leather pocket and interfruitlet corking

Fruitlet core rot (FCR), black spot, fruitlet brown rot and eye rot are terms that have been used to describe brown to black diseased centres of individual pineapple fruitlets. Leathery pocket (LP) and interfruitlet corking (IFC) are additional symptoms that develop as FCR continues to develop. These widespread problems have complex aetiologies.

Causal agents and symptoms

The fungi *Penicillium funiculosum* and *Fusarium guttiforme* (formerly *F. moniliforme*), the round yeast *Candida guilliermondi*, the pineapple fruit mite, *Steneotarsonemus ananas*, and the pineapple red mite, *Dolichotetranychus floridanus*, are associated with FCR (Rohrbach and Schmitt, 1994; O'Donnell *et al.*, 1998). Although no cause and effect relationship has been established for *C. guilliermondi*, flower infections by *P. funiculosum* and *F. guttiforme* are involved. They presumably are enabled or exacerbated by feeding wounds that are made by, respectively, *S. ananas* and *D. floridanus* (Oxenham, 1962; Rohrbach and Pfeiffer, 1976a; Lim and Rohrbach, 1980; Mourichon, 1991; K.G. Rohrbach, unpublished). LP and IFC have only been associated with *P. funiculosum*.

The importance of *P. funiculosum* and *F. guttiforme* varies among different production areas. In Brazil, *F. guttiforme* is the predominant cause of FCR, whereas *P. funiculosum* is the most common cause of FCR and LP in South Africa (Rohrbach, 1980; Rohrbach and Taniguchi, 1984). In Hawaii, both *P. funiculosum* and *F. guttiforme* cause FCR symptoms, while only *P. funiculosum* causes IFC and LP.

In Hawaii, three strains of *P. funiculosum* are found on pineapple, but only one of these, P1, is pathogenic. In Malaysia, six strains of *P. funiculosum* have been associated with FCR, LP and cork spot (CS) symptoms. Descriptions of the strains are presented elsewhere (Lim and Rohrbach, 1980; Lim, 1985).

Cultures of *F. guttiforme* are floccose, greyish white becoming purple on the bottom side of the culture (Rohrbach, 1994; Nirenberg and O'Donnell, 1998). Microconidia are aseptate, hyaline, ovoid to obclavate, 8–12 × 2.5–3 μm and produced on mono- and polyphialides. Macroconidia are thin walled, falcate, 3–5 septate and 32–53 × 3.0–4.5 μm. Chlamydospores are not formed. Despite its obvious affinities for *G. fujikuroi*, no teleomorph has been reported for this pathogen. *F. guttiforme* also causes fusariosis in South America, a much more destructive disease than FCR.

S. ananas is light brown. The adult male is oval with an average length of 0.20 mm and width of 0.07 mm. The adult female is slightly larger than the male, with an average length of 0.24 mm and width of 0.10 mm. Males have a fourth pair of legs that are robust and claw-like, whereas female legs are long and thin (Petty, 1975, 1978).

D. floridanus is the largest phytophagous mite found on pineapple, and is conspicuous because of its bright orange to red colour. The adult male is 0.3–0.4 mm long and 0.1 mm wide (Petty, 1975, 1978).

Early symptoms that *P. funiculosum* causes in closed flowers are necrosis of the anthers and pistil, blue–green sporulation on ovules and locule walls, and cork formation on the locule surface (Plates 119 and 120). As FCR symptoms progress, septa between locules become dark to medium brown, and the discoloration may extend into adjacent noncarpellary tissue (Plate 121) (Larsen, 1910). Further corking of locules as fruit mature results in LP (Hepton and Anderson, 1968; Petty, 1977; Rohrbach and Schmitt, 1994). IFC develops on the fruit surface between affected fruitlets that do not enlarge as rapidly as healthy fruitlets; as a result, fruit shape becomes distorted (Hepton and Anderson, 1968). Symptoms identical to IFC can also be caused by boron deficiency (Rohrbach and Schmitt, 1994).

F. guttiforme causes a light to dark brown discoloration of septa that may extend down the entire fruitlet core. White to pinkish mycelium and sporulation of the pathogen occur in locules (Larsen, 1910).

FCR symptoms that are associated with yeast infections are usually light brown.

Epidemiology

The P1 strain of *P. funiculosum* is found in soil and on pineapple plant residue (Lim and Rohrbach, 1980). Conidia require exogenous nutrients (simple sugars) and a pH of 3.5 to germinate. The optimum temperature for infection is 16–21°C during the 6 weeks after plants are treated with ethylene to induce flowering. Rainfall during this period does not appear to be critical (Rohrbach and Taniguchi, 1984).

The pineapple fruit mite that is associated with *P. funiculosum*-induced FCR feeds on trichomes on the basal portion of leaves in the plant heart and on bracts and sepals of the developing flowers. Trichome injury results in brown patches on the leaves and bracts. Populations of the pineapple fruit mite are most abundant between the time of flower induction and the 2.5 cm open-heart stage of inflorescence development; they peak when the inflorescence emerges 6–7 weeks after flower induction (see Table 19.1) (K. Sakimura, unpublished data). Infection occurs through the developing flower 1–2 weeks prior to normal anthesis (Rohrbach and Namba, 1983).

F. guttiforme is soilborne. Conidia survive for 6–13 weeks in soil, depending on moisture and temperature, with survival being highest in dry soils. Survival in pineapple residue in soil is ~10 months. The pathogen develops in the heart of the plant and on the inflorescence, and infection apparently takes place only through the open flower (Bolkan *et al.*, 1979; Rohrbach and Taniguchi, 1984).

Very high populations of red mite have been associated with epidemics of *F. guttiforme*-induced FCR in Hawaii (K.G. Rohrbach, unpublished data). Populations increase only under dry conditions and most commonly on the basal leaves of the developing crown and on stored seed material. The mite feeds on the epidermis, causing necrotic sunken areas at the leaf bases, and high populations on developing crowns can result in necrosis of the basal leaves (K. Sakimura, unpublished data).

Little is known of the epidemiology of the associated yeasts and bacteria.

Management

P. funiculosum-induced FCR, LP and IFC can be reduced with endosulfan applied at flower induction and 3 weeks after forcing under low to moderate disease levels (Rohrbach *et al.*, 1981). Fungicides have not been effective except when applied directly into the opening of the terminal leaves that is created by the emerging inflorescence. No controls for *F. guttiforme*-induced FCR have been developed.

Cultivars vary in their response to *P. funiculosum* and *F. guttiforme*. In Hawaii, 'Smooth Cayenne' artificially inoculated with either pathogen develops low levels of FCR, and with *P. funiculosum* moderate levels of IFC and LP. In contrast, the PRI hybrid '53–116' was very susceptible to *P. funiculosum*-induced LP, IFC and FCR, but not very susceptible to *F. guttiforme*, and the hybrid '58–1184' (Hawaii Germplasm Repository Number 159) was very susceptible to FCR induced by both pathogens, but only slightly susceptible to IFC (Rohrbach and Pfeiffer, 1976b).

Fusariosis

Fusariosis was first observed in Argentina in 1954. It was first reported in southern Brazil in 1964 (Kimati and Tokeshi, 1964), and within 10 years was recognized throughout the country (Laville, 1980). Fusariosis is now also known in Paraguay and Uruguay, and recently was introduced to Bolivia via infected slips from Brazil (de Matos *et al.*, 1992; Ventura *et al.*, 1993). Fusariosis has not been reported outside South America.

Symptoms

Fusariosis affects virtually all parts of the pineapple plant, but is most conspicuous and damaging on fruit (Pissarra *et al.*, 1979; Rohrbach, 1994). Initial symptoms are an off-colour of fruitlets and exudation of gum, which can be confused with exudation caused by feeding of the pineapple fruit caterpillar, *Thecla basilides* (de Matos, 1987). Fruitlets become sunken, light to dark brown and covered with light pink to greyish mycelium and sporulation of the causal fungus. Individual fruitlets or large areas of the fruit surface may be affected, and damage can range from superficial to extending to the fruitlet core.

Affected plants exhibit a bent or dead stem apex, shortened stem, disrupted phylotaxy, and general stunting and chlorosis (Pissarra *et al.*, 1979). Secondary symptoms include: a bending or curvature of the plant; rosetting of the leaves, which includes an increase in the number of leaves per spiral; reduction of leaf size and length; death of the terminal growing point; and chlorosis (Pissarra *et al.*, 1979; de Matos and Mourichon, 1993).

Causal agent

Fusariosis is caused by *F. guttiforme* (Rohrbach, 1994; Ventura, 1994; Nirenberg and O'Donnell, 1998; O'Donnell *et al.*, 1998). It is described in the previous section on fruitlet core rot.

Epidemiology

Fusariosis is the most serious fruit disease in Brazil, where major losses occur on 'Pérola', 'Jupi' and 'Smooth Cayenne' (Laville, 1980). Fruit losses may reach 80% under severe disease conditions (Robbs *et al.*, 1965; de Matos 1987). Losses also result from the infection of propagative materials (primarily slips). In fields in which 75–80% of the fruit are infected, 30–40% of the propagative materials may be infected and, when these are planted in new fields, 20% of the plants may die (Laville, 1980; Aguilar, 1981).

Infections of the inflorescence and fruit occur primarily via injuries caused by insects, particularly the pineapple caterpillar, *T. basilides*. Infected propagative materials are the primary means of spread to newly planted fields, and infected fruit provide inoculum for stem infections particularly in the cultivar 'Pérola' with its 'collar-of-slips' habit.

Since the pathogen does not survive longer than 10 months in infested crop residues or soil, infested soils are not considered a primary source of inoculum (Maffia, 1980). Insects and wind are the primary means by which the fungus spreads from infected to healthy plants. Spore dispersal occurs throughout the crop cycle, with peak production in mid afternoon (de Matos and Costa, 1997). In addition, environmental conditions and associated insects may enhance disease development (de Matos, 1987).

Management

Fusariosis is controlled by planting disease-free propagation material, controlling insects, and by protecting the inflorescence and developing fruit with fungicides (Ventura *et al.*, 1993). Hot benomyl (54°C for 90 min) is effective for disinfestation of slips and suckers, but retards growth and kills up to 50% of the plants (Maffia and Chaves, 1978). More often, a rapid propagation method is used whereby healthy stumps are divided, dipped in benomyl and grown in a greenhouse. Plants that develop symptoms are quickly discarded and only healthy plants are used.

Applications of fungicides and insecticides at 20-day intervals, from forcing and through harvest, provided good fruit rot control in Brazil (de Matos, 1987). Fungicides work best if fruiting is initiated during periods of low rainfall. Fluazinan retarded growth of benomyl-resistant strains of the fungus *in vitro*, and captan was most effective against the disease in the field (Ventura *et al.*, 1995).

In Brazil, Ventura (1994) reported fusariosis incidences of 87% on 'Pérola', 80% on 'Smooth Cayenne' and 29% on 'Primavera'. In another experiment, 'Perolera', 'Piña Negra', 'Rondon', 'Tapiracanga', 'Amapá', 'Amarelo-de-Uaupés', 'Cabeçona', 'Turi verde' and 'Ver-o-peso' were considered resistant after 2 years of evaluation (Ventura *et al.*, 1993). Under greenhouse conditions in France, de Matos *et al.* (1991) evaluated the reactions of 45 accessions to artificial inoculation. Six pineapple cultivars, 'Blanca', 'Samba', 'Angelita 1', 'Iris 1', 'BR 189' and 'Tapiricanga', were resistant, as were the 'Ananas São Bento', 'Branco do Mato' and 'BR 123' accessions of *A. bracteactus* and the 'VE 64' accession of *A. parguazensis*. Resistance to fusariosis is also found in the genus *Pseudoananas* (Rohrbach, 1994). In the future, somatic hybridization via protoplast fusion may enable the incorporation of fusariosis resistance from some of the primitive materials into desirable cultivars (Pinho Guedes *et al.*, 1997).

Internal browning

Internal browning, also termed black heart or endogenous browning, was described in several early pineapple company reports.

Symptoms

The initial symptom of a small greyish translucent zone at the base of the fruitlet adjacent to the fruit core occurs ~4 days after fruit refrigerated at 7°C is removed from storage (Paull and Rohrbach, 1985). The zone later turns brown to black, and in severe cases the entire interior of the fruit darkens, giving rise to the name black heart.

Aetiology

Linford (1932a) established that chilling induces internal browning. The physiological disorder is of major significance in Australia, Kenya, South Africa and Taiwan, where fruit are grown and harvested under cool conditions (0–10°C). It is also important when fresh fruit are refrigerated to extend shelf life.

Epidemiology

Internal browning development is associated with high levels of polyphenol oxidase, and low ascorbic acid and temperatures below 21°C prior to harvest.

Management

Internal browning can be reduced by coating the fruit with paraffin-polyethylene waxes (wax:water ratios of 1:4–9). Waxing increases internal CO_2 concentrations, thereby lowering O_2 concentrations and reducing polyphenol oxidase activity. Cultivars that have higher than normal ascorbic acid contents were reported to resist internal browning (Paull and Rohrbach, 1985), but a recent report from Australia disputes this association (Sanewski and Giles, 1997). In the latter study, PRI hybrid 53–116 remained symptomless after storage at 10°C for 14 days followed by 20°C for 8 days.

Marbling

Marbling of pineapple fruit was first described by Serrano (1928) in the Philippines as bacterial brown rot. It has been reported in most pineapple production areas of the world (Rohrbach, 1983).

Symptoms

Infected tissues are yellowish to dark, dull brown and may be speckled; they generally become hardened, granular and brittle (Keetch, 1977a; Rohrbach and Schmitt, 1994). Symptom development, occurring during the last month of fruit maturation, may involve single fruitlets or the entire fruit including core tissues (Plate 122) (Rohrbach, 1989). Externally, severely infected fruit may be identified by a 'woody' sound when tapped (Rohrbach and Schmitt, 1994).

Causal agents

Acetobacter peroxydans, *Acetobacter* sp. and *Pantoea ananatis* (formerly *Pantoea ananas*) cause marbling (Rohrbach, 1989; Mergaert *et al.*, 1993; Trüper and de Clari, 1997). They are Gram-negative, motile, catalase-positive bacteria. They oxidize ethanol to acetate, grow on Hoyer's medium, do not produce dihydroxyacetone from glycerol, and contain refractile, sudanophilic inclusions (Rohrbach and Apt, 1986; Rohrbach and Schmitt, 1994).

Epidemiology

Marbling is most severe in the lowland tropics where temperatures remain above 21–27°C. In Thailand, the disease can occur on 5–20% of the slices in October and November and can stop canning operations (Rohrbach and Schmitt, 1994).

Infection usually occurs through the open flower, but may also occur through fruit surface growth cracks during the later stages of fruit development. In Hawaii, the bacteria appear to be ubiquitous since the application of surfactants prior to and during flowering significantly increases disease. Flower infections remain latent until ~1 month before fruit maturity. Low fruit acidity and brix are associated with high levels of the disease (Rohrbach, 1989).

Management

Marbled fruit are identified at the cannery by their woody sound or appearance, and are discarded (Rohrbach and Schmitt, 1994). Differences in cultivar susceptibility have been noted, with 'Smooth Cayenne' being moderately resistant (Rohrbach and Schmitt, 1994).

Pink disease

Pink disease is primarily a problem in canned products and is of little importance in fresh fruit. It was first reported in Hawaii by Lyon (1915), and is now also known in Australia, the Philippines, South Africa and Taiwan (Rohrbach, 1983). It should not be confused with the pink disease of other crops in this volume that is caused by the basidiomycete, *Erythricium salmonicolor*.

Symptoms

The term 'pink disease' refers to the pinkish discoloration of affected fruit flesh (Plate 123). Severely infected fruit in the field may be slightly pink and have desiccated bracts and fruitlet surfaces. Depending on the causal bacterium, uncooked infected fruit may be symptomless, light pink to brown, or have a cantaloupe-like odour (Rohrbach and Pfeiffer, 1976c). When the fruit is heated during the canning process, infected tissue turns brown to black. Because the can is sealed during processing, infected pineapple slices can only be detected using quality control procedures. When disease levels are low, affected slices may reach the consumer and influence future purchases (Rohrbach and Schmitt, 1994).

Causal agents

Bacteria have always been associated with pink disease (Buddenhagen and Dull, 1967; Rohrbach and Pfeiffer, 1975; Kontaxis and Hayward, 1978; Cho *et al.*, 1980; Gossele and Swings, 1986; Kageyama *et al.*, 1992).

Pantoea citrea (formerly *Erwinia herbicola*) is a non-motile, Gram-negative, glucose-fermenting rod. The bacterium is catalase positive, Kovacs' oxidase negative, nitrate reducing, and does not produce dihydroxyacetone from glycerol. The optimum *in vitro* pH for growth is 6.8 (Cho *et al.*, 1980). The pathogen does not normally cause symptoms in fresh fruit (Rohrbach and Pfeiffer, 1975). The process by which it induces pink discoloration in fruit has been described (Pujol and Kado, 1998).

Gluconobacter oxydans is a motile, Gram-negative rod with polar flagella that produces acid from ethanol on yeast–dextrose–calcium carbonate (YDC) media. Growth is optimum at a pH of 6, but can initiate at pHs as low as 4.0–4.5, and requires some vitamins and amino acids (Kontaxis and Hayward, 1978; Cho *et al.*, 1980).

Acetobacter aceti is a motile, Gram-negative rod with peritrichous flagella that produces acid from ethanol. Growth is optimal at pH 5, but will occur at 4.0–4.5. Strains that cause pink disease are brown, over-oxidize ethanol, produce catalase, grow on Hoyer's medium, and produce acid from glucose and dihydroxyacetone from glycerol (Kontaxis and Hayward, 1978; Cho *et al.*, 1980). Gosselle and Swings (1986) identified these brown-pigmented strains as *A. liquefaciens*.

Epidemiology

In Hawaii and Taiwan, epidemics occur from February to April, and in the Philippines from August to September. In general, pink disease does not occur in the lowland tropics (Rohrbach, 1989). The pathogens are vectored to pineapple flowers by nectar-feeding insects and mites. The nectar or yeast that is present in the nectar may be utilized by the bacteria (Hine, 1976). Bacteria inside the flower remain latent in the nectar gland or stellar canal and locule until the fruit matures and sugar concentrations increase (Rohrbach, 1989).

Management

Disulfoton application starting at the red bud stage followed by three additional applications at 5-day intervals through flowering has resulted in the highest level of control (Kontaxis, 1978). Seasonal peaks in disease potential can be avoided by selectively forcing flowering. Cultivars and hybrids vary from highly resistant to very susceptible, and 'Smooth Cayenne' is relatively resistant (Rohrbach and Pfeiffer, 1976c).

Leaf Diseases

White leaf spot

This disease was first reported by Larsen (1910) in Hawaii, but was first given the name white leaf spot by Oxenham (1957) in Australia. White leaf spot occurs worldwide (Rohrbach, 1983). Although it can cause serious symptoms, economic impacts have not been demonstrated (Keetch, 1977b).

White leaf spot starts as a rapidly advancing brownish wet rot, usually where the leaf has been injured. Spots become greyish brown with a dark brown margin prior to drying, and turn whitish and papery; thus, the term 'white leaf spot' (Fig. 19.2) (Larsen, 1910).

Pineapple white leaf spot is caused by *Chalara paradoxa* (Rohrbach and Schmitt, 1994). Characteristics of the fungus are listed in Chapter 1. *C. paradoxa* requires injuries, such as those caused by wind and grasshoppers, to infect leaves (Larsen, 1910; Keetch, 1977b). Epidemics occur following rainstorms with high winds. These conditions also favour production and distribution of conidia (Larsen, 1910). The loss of pineapple leaf area is minimal and does not result in economic losses (Larsen, 1910; Oxenham, 1953; Keetch, 1977b; Lim, 1985).

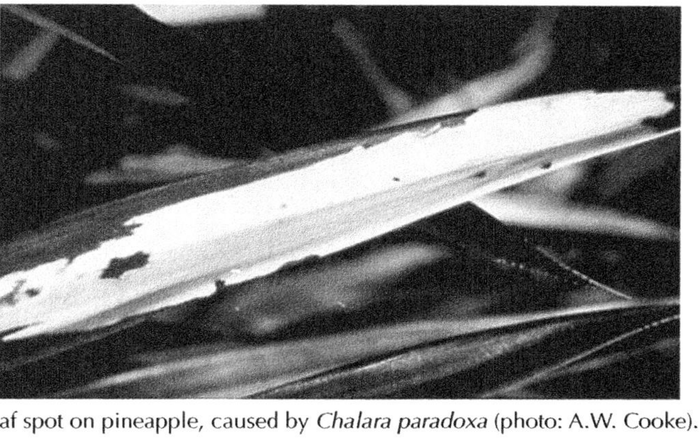

Fig. 19.2. White leaf spot on pineapple, caused by *Chalara paradoxa* (photo: A.W. Cooke).

'Red Spanish' and 'Queen' are generally more resistant to the disease than 'Smooth Cayenne' (Keetch, 1977b).

Yellow spot

Yellow spot was first described by Illingworth (1931) in Hawaii. It occurs in all production areas of the world.

Symptoms

The initial symptom is a slightly raised yellowish spot with a darkened centre on the upper surface of the leaf. Shortly thereafter, a characteristic chain of secondary spots develops that progress into a rot of the basal leaf and stem (Plate 124). Frequently, particularly on young plants, the rotting and cessation of growth on one side of the stem causes the plant to bend severely, eventually killing the entire plant. The disease can occur on the developing crown with the rot progressing into and distorting the fruit (Illingworth, 1931).

Causal agent

Yellow spot is caused by the *Tomato spotted wilt virus* (TSWV), known formerly as the *Yellow spot virus* (Sakimura, 1940).

Epidemiology

TSWV is vectored to pineapple by *Thrips tabaci* (Linford, 1932b). Infection occurs most frequently on plants during early growth, and crowns on developing fruit are infected occasionally (Linford, 1943). Pineapple is not a preferred host of *T. tabaci*. It moves into pineapple fields from adjacent weed hosts (Linford, 1932b), which in Hawaii is primarily *Emilia fosberyii*.

Infection is always fatal. Therefore, vegetative propagation does not transmit the virus to subsequent plantings (Rohrbach and Apt, 1986).

Management

Yellow spot has been controlled via weed control and adjusting the timing of operations in adjacent crops to minimize thrip movement (Rohrbach and Apt, 1986).

Root Diseases

Diseases caused by nematodes

Although nematodes probably affect pineapple in all production fields, only a few species are important. Their geographic distribution on pineapple is not well understood. Although variation in soils and temperatures do not adequately explain their occurrence, cultivars and management practices that are used in different regions are thought to be important factors.

Symptoms and causal agents

Meloidogyne incognita, *Meloidogyne javanica*, *Pratylenchus brachyurus* and *Rotylenchulus reniformis* are most important (Rohrbach and Schmitt, 1994; Singh, 1996). The foliar symptoms they cause are similar, and include stunting and chlorosis. High populations result in lower yields and unacceptable fruit size and quality.

Root-knot nematodes, *Meloidogyne* spp., are more widely distributed than any other major group of plant-parasitic nematodes (Sasser, 1977). *M. javanica* is a serious problem on pineapple throughout the world, whereas *M. incognita* is less important. *M. javanica* can be distinguished from *M. incognita* on the basis of perennial patterns (Chitwood, 1949), male head shapes (Eisenback, 1981) and esterase phenotypes (Esbenshade and Triantaphyllou, 1985). Several races of *M. incognita* are characterized cytologically and by host differential tests (Eisenback, 1981).

Root-knot nematodes cause galls on roots that vary in size and shape, depending on the species and population density of nematodes in the gall. Those induced by *M. javanica* and *M. incognita* are large and are not diagnostic. The morphological changes that these nematodes cause occur very quickly after infection (Kaplan, 1979), and major metabolic changes occur in the plant at the nematode feeding site (Gommers, 1977). Additional effects on the plant can result from interactions with soil microflora and microfauna.

The lesion nematode, *P. brachyurus*, also causes root stunting and necrosis. It was first described as a problem in Hawaii by Godfrey (1929), and is also important in Africa (Kehe, 1997), Brazil and probably other areas since the damage that it causes often is underestimated. *P. coffeae* also occurs on pineapple.

Lesion nematodes are migratory endoparasites. They are vermiform and range from 0.3 to 0.9 mm in length. The anterior end of the nematode is broadly flattened with an offset lip region. The terminus is bluntly rounded. They have a well-developed but rather short stylet, 15–20 μm long, with conspicuous basal knobs. In most species, the vulva of females is ~70–80% of the distance from the head to the terminus. Males are not common.

The reniform nematode, *R. reniformis*, is common on many plant species and is a significant pest of pineapple in Hawaii and the Philippines (Caswell, 1990). It is sexually dimorphic. Males and immature females are vermiform, and mature females are arcuate ventrally and become saccate in the shape of a kidney. The female stylet is somewhat delicate and ~16–18 μm long. The male has a poorly developed stylet and oesophagus with a reduced metacorpus and indistinct valve. In the female, the median oesophageal bulb is strongly developed and the basal bulb extends over the intestine, forming a large flattened lobe. Females have two opposed ovaries and the vulva is located 60–75% of the distance from the lip region to the terminus. Females produce a gelatinous matrix into which its eggs are deposited externally (Sivakumar, 1971). Although the root stunting and necrosis that is caused by *R. reniformis* is itself not diagnostic, the nematode can be identified by the soil-encrusted egg masses it produces on the root.

Epidemiology

The initial population density of *Meloidogyne* spp. is inversely proportional to plant growth and yield on pineapple. Standard regression models or the Seinhorst regression model fit data from soybean (Nardacci, 1979; Kinloch, 1982). Their life cycle is similar on most hosts. The second-stage juvenile penetrates the root near the tip, and giant cells are initiated as the juvenile begins to swell. After three moults, the nematode becomes an adult. The females extrude a gelatinous matrix into which eggs are deposited. In the two species that attack pineapple, males that emerge from roots after the fourth moult are not normally involved in reproduction. The optimum temperature range for reproduction of *M. incognita* is between 24 and 35°C (Dropkin, 1963; Nardacci, 1979).

P. brachyurus and *P. coffeae* reproduce best at 30°C (Acosta, 1979). Lesion nematodes have four juvenile stages and an adult stage. The second-, third- and fourth-stage juveniles and adults are infective and feed on root tissue. Populations increase with rainfall, attaining a peak 7–8 months after planting, and remain high thereafter (Dinardo-Miranda, 1997). In Côte d'Ivoire, numbers of *P. brachyurus* decreased as soil pH increased from 3.85 to 6.0 (Osseni, 1997).

The life history of *R. reniformis* on pineapple probably resembles that on other hosts, such as okra (Sivakumar, 1971). Immature females penetrate the root until about one-third of the anterior portion of their body is inside the root. The area of the body immediately surrounding the vulva bulges, and the female attains its characteristic reniform shape. It can then produce 12 eggs per day, but only ~45 are retained in the egg mass. The life cycle from egg to egg required from 20 to 25 days on different hosts (Birchfield, 1962; Sivakumar, 1971; Rebois, 1973). Population development and the effect of the nematode on root growth were greatest at 29°C (Rebois, 1973).

Management

Pre-plant control with fumigants is generally effective (Rabie, 1997; Sipes, 1997). Control of *M. javanica* and *R. reniformis* in Hawaii is achieved by soil fumigation with 1,3-dichloropropene or methyl bromide (Schneider, 1993). Fenamiphos and oxamyl commonly are applied after planting by drip irrigation to maintain the control achieved with the fumigants (Schneider, 1993). It is critical to apply the proper quantity of water to prevent excessive movement of these nematicides. A fallow period is often used between plantings to decrease nematode population densities before fumigation.

In South Africa, *P. brachyurus* was controlled by spraying the butt of suckers with 2.5 kg fenamiphos ha^{-1} or dipping in 600 p.p.m. of fenamiphos for 10 min followed by foliar applications of the same after planting and every 3 months thereafter (Keetch, 1977b). Pre-plant fumigation plus post-plant applications with non-fumigants also maintained low population densities of this nematode (Sarah, 1980).

Alternative measures include maintaining soil pH between 5.5 and 6.0 to control *P. brachyurus* (Osseni, 1997), and the use of sugarcane and pangola grass as cover crops between pineapple cropping cycles to control *R. reniformis* and *M. incognita* (Ayala, 1971). In Hawaii, *R. reniformis* population densities were reduced in all covercrop treatments during the intercycle period (Wang, 1998).

Mealybug wilt

Mealybug wilt is a universal problem. The strong association of mealybugs with several species of ants makes control of mealybug wilt very difficult. When vegetative seed material (crowns and slips) is taken from affected plants, it can be a source of inoculum and eventually develop symptoms (Rohrbach *et al.*, 1988).

Symptoms

The first symptoms are a reddening of the leaves, usually at the margins of fields (Plate 125). These symptoms are caused by a cessation of root growth and a collapse of the root system (Carter, 1933b), but the same symptoms may also result from drought (Carter, 1933a), nematode damage and root rot. Leaf symptoms are not diagnostic unless high levels of mealybugs are also present or they are associated with field edges. When a pineapple plant with early mealybug symptoms recovers, the term 'terminal mottle' is used to denote the recovered plant (Rohrbach *et al.*, 1988). If the disease progresses, however, leaves wither and brown, and the plant eventually dies. Because the root system is severely affected, plants can be pulled from the soil easily (Rohrbach *et al.*, 1988).

Associated agents and epidemiology

Ants, mealybugs and one or more viruses are associated with the disease. Ants protect the mealybugs from parasites and predators, as well as remove honeydew that they produce (Rohrbach *et al.*, 1988). An early hypothesis that wilt symptoms resulted from toxins in mealybug saliva has been discounted.

At least three viruses are present in affected plants: two closteroviruses, *Pineapple mealybug wilt-associated virus*-1 (PMWaV-1) and *Pineapple mealybug wilt-associated virus*-2 (PMWaV-2) and an unnamed badnavirus (Gunasinghe and German, 1989; Ullman *et al.*, 1989; Wakman *et al.*, 1995; Hu *et al.*, 1997; Sether *et al.*, 1998; Melzer *et al.*, 2001). A causal role for PMWaV-2 was reported recently (Sether and Hu, 2002). In this work, it was shown that mealybug feeding played a critical role. Plants that were infected with PMWaV-2 and infested with mealybugs developed symptoms of the disease, but no symptoms developed on plants that were only infected or infested. Although PMWaV-1 and the badnavirus were often detected in symptomatic plants in the field, they did not cause mealybug wilt symptoms alone or in combination with mealybug feeding.

The pink mealybug, *Dysmicoccus brevipes*, and the grey mealybug, *D. neobrevipes*, vector PMWaV-2 (Sether and Hu, 2002). *D. brevipes* is usually found on roots and the lower stem of pineapple, as well as perennial grasses including sugarcane, whereas *D. neobrevipes* is found on the upper stem, inflorescence and fruit of pineapple, and on sisal (Beardsley *et al.*, 1982; Beardsley, 1993). Both are found on banana and other hosts.

Mealybug populations generally do not develop in pineapple fields where mealybug parasites and predators occur, unless ants are present. In Hawaii, three species of ants are found in different production zones. The big-headed ant, *Pheidole megacephala*, occurs below 500 m, the Argentine ant, *Iridomyrmex humills*, above 600 m, and the fire ant,

Solenopsis geminata, in the dry lowlands (Rohrbach *et al.*, 1988). The bigheaded ant predominates (Sulaiman, 1997).

Management

When ants are controlled and mealybug predators and parasites are present, mealybug wilt is usually not a problem. Ants can be controlled with insecticides, and hydramethylon baits have been effective on certain species (Rohrbach *et al.*, 1988; Petty and Manicom, 1995). Some cultivars are resistant but are generally not used (Carter and Collins, 1947).

Root rot

Pineapple root rot was first studied in Hawaii by Sideris and Paxton (1930). Prior to their work, this disease was confused with mealybug wilt.

Worldwide losses due to root rot are highly variable. Since its symptoms are subtle, its importance frequently is underestimated. As with nematodes and mealybug wilt, losses result from a suppression of plant growth and yield. Production areas with the greatest root rot potential are Australia, Hawaii, the Philippines, South Africa and Thailand. In cooler environments with high rainfalls, root rot can eliminate the ratoon crop.

Symptoms

Since aboveground symptoms are similar to those that are caused by nematodes, mealybug wilt and low levels of O_2 in the soil, they are not diagnostic. Leaf growth slows or stops, leaves redden, and leaf tips and margins yellow and eventually become necrotic (Plate 126). Symptom development is relatively slow. If conditions become dry following infection, affected plants may appear reddish as if under severe drought stress. As with severe mealybug wilt and root-knot nematode damage, root damage can be extensive (Fig. 19.3), and affected plants can be pulled from the soil easily.

Causal agents

Considerable confusion appeared in the early literature on the identification and nomenclature of the root rot pathogens (Sideris and Paxton, 1931). Current binomials for the major causes of root rot are *Phytophthora cinnamomi*, *P. nicotianae* and *Pythium arrhenomanes* (Klemmer and Nakano, 1964; Erwin and Ribiero, 1996).

Fig. 19.3. Severe root rot on the pineapple plant on the left, caused by *Phytophthora cinnamomi*, compared with a healthy plant on the right (photo: A.W. Cooke).

Epidemiology

P. cinnamomi is a problem in cooler production areas of Australia, Hawaii and South Africa. Optimum soil temperatures for disease development range from 19 to 25°C. Nematodes, especially the lesion nematode, are associated with root rot.

P. cinnamomi and *P. nicotianae* are heterothallic and generally form oospores only when the A1 and A2 mating types are present (Erwin and Ribiero, 1996). Thus, chlamydospores are more significant survival structures for these pathogens. In turn, zoospores are the important infective propagules. Since their production and release, and infection and disease development are all favoured by high soil moisture, root rot is especially prevalent in poorly drained soils (Boher, 1976). Both species are described in Chapter 1.

Root rot of pineapple caused by *Pythium arrhenomanes* has not been well studied. It has been associated with root-wounding organisms, such as nematodes, Anomela beetle larvae and symphylids.

Management

Many factors and agents contribute to root decay. Raised beds and other measures to improve surface drainage generally reduce root rot. Fosetyl-Al gives excellent control of the root rots caused by *Phytophthora* spp. (Rohrbach and Schenck, 1985). The soil fumigant 1–3 dichloropropene reduced root rot caused by *Py. arrhenomanes*, due presumably to the control of root-damaging pests (Anderson, 1966).

Stem Diseases

Bacterial heart rot

Bacterial heart rot is significant in Malaysia, and also occurs in Brazil, Costa Rica and the Philippines (Rohrbach, 1983; Lim, 1985). Economic loss is attributed largely to the development of axillary buds, and asynchronous flowering and fruit development. The intensive use of the susceptible 'Singapore Spanish' cultivar may account for the high incidence of the disease in Malaysia (Lim, 1985).

Symptoms

Water-soaked lesions first appear on the white basal portion of the leaves of the central whorl. As they develop, the green midportion of the leaves assume an olive green and bloated appearance, symptoms that distinguish bacterial heart rot from Phytophthora heart rot (Lim, 1985).

Causal agent

Johnston (1957) initially identified *Erwinia carotovora* as the cause of bacterial heart rot, but it was later renamed *E. chrysanthemi* (Lim, 1974). It is described under bacterial fruit collapse.

Epidemiology

Primary inoculum is thought to come from infected fruit or plants. Infection occurs through the stomata of young or injured leaves. Inoculum is dispersed by wind, wind-blown rain and insects, most often the bigheaded and Argentine ants. The souring beetles *Haptoncus ocularis* and *Carpophilus foveicollis* are minor vectors. Urease from bacteria in dirty spray water may enhance disease by releasing phytotoxic ammonia especially on 4- to 8-month-old plants in the first crop. Under optimum conditions for disease development, the entire cycle may occur in 1–2 weeks (Lim, 1985).

Management

Infected fruit should be removed from the field, and crowns or slips from plants with fruit symptoms should not be used. Mechanical leaf damage resulting from crop lodging or weed control during periods of early growth must be avoided (K.G. Rohrbach, personal observations). Partial control of bacterial heart rot has been obtained with miticides and ant control. Where they are adapted, the resistant 'Smooth Cayenne' and a 'Smooth Cayenne' × 'Red Spanish' selection can be used rather than the susceptible Spanish types (Lim, 1985).

Butt rot

Butt rot is a common stem disease. The term 'base rot' was first used by Larsen (1910), but the term 'butt rot' has replaced it in Australia (Broadley *et al.*, 1993) and Hawaii (Rohrbach and Apt, 1993).

Symptoms

The basal portions of crowns, slips and suckers become soft and black (Plate 127) (Lyon, 1915). Under disease-conducive conditions, the entire propagule may rot.

Causal agent

The pathogen, *Chalara paradoxa*, is described in Chapter 1.

Epidemiology

C. paradoxa survives as chlamydospores in soil and decaying pineapple residues. When nutrients are available, they germinate to initiate infection (K.G. Rohrbach, unpublished data). Since *C. paradoxa* requires wounds to infect, crowns, slips or suckers rot where they are attached to the mother plant. Infection is most common when these propagules are stored on wet, infested soil or planted in infested soil soon after detachment from the mother plant (Cho *et al.*, 1977). Conidia are produced under conditions of high humidity and may be dispersed by wind (Rohrbach and Schmitt, 1994).

Management

Storing newly detached crowns, slips or suckers on mother plants during dry weather where there is good air circulation and minimal exposure to infested soil is effective (Rohrbach and Apt, 1986; Rohrbach and Schmitt, 1994). Freshly detached materials must be treated with an appropriate fungicide within 12 h if they are planted immediately after harvest (Cho *et al.*, 1977; Rohrbach and Phillips, 1990). 'Red Spanish' is generally more resistant than 'Smooth Cayenne'.

Phytophthora heart rot

Heart rot, which is also known as top rot in Australia, was first described in the 1920s (Ashby, 1920). Sideris and Paxton (1930) provided the first in-depth report on the causes of the disease. Worldwide losses are highly variable, and are caused by plant mortality rather than reduced fruit size or quality (Fig. 19.4). Heart rot has been reported in

Fig. 19.4. Extensive mortality in a pineapple planting caused by Phytophthora heart rot (photo: A.W. Cooke).

Australia, Côte d'Ivoire, Fiji, Hawaii, Jamaica, Martinique, the Philippines, South Africa, Taiwan and Thailand (Rohrbach, 1983).

Symptoms

Young leaves of affected plants fail to elongate and become chlorotic (Plate 128). The terminal whorl of leaves leans to one side of the plant and is easily removed (Fig. 19.5).

Causal agents

Heart rot is caused by *Phytophthora cinnamomi*, *P. nicotianae* and *P. palmivora*; *P. cinnamomi* and *P. nicotianae* are most common. All three species are described in Chapter 1. Little information is available on the role of *P. palmivora* as a pineapple pathogen (Klemmer and Nakano, 1964). The three species usually do not occur together. *P. nicotianae* and *P. palmivora* are found at lower elevations where temperatures range from 25 to 36°C, and development of *P. nicotianae* is slowed when temperatures drop below 25°C. In contrast, *P. cinnamomi* causes heart rot at cooler, higher elevations in Hawaii and in cooler production areas such as Australia. Optimum soil temperatures for disease development are 19–25°C.

Epidemiology

The disease is limited to fine textured, high pH soils under wet environmental conditions. Malaysia, where production is on low pH peat soils, is the only major pineapple production area where heart rot has not been observed.

Chlamydospores of the three species are primary inoculum, and can survive for years in soil or infested plant debris. They germinate in moist soil to produce hyphae that can infect roots or immature leaf and stem tissue directly, or terminate in the formation of sporangia. Chlamydospores and sporangia can be disseminated in moving or rainsplashed soil or by wind. Free water is required for sporangium formation and zoospore release.

Fig. 19.5. A pineapple plant, affected by Phytophthora heart rot, from which the central core of leaves has been removed easily (photo: A.W. Cooke).

Infection by zoospores of *P. nicotianae* is most common through leaf axils during the first 3–4 months following planting, and introduction of infested soil into leaf axils by deep planting or heavy rains may increase infection. Infection by *P. cinnamomi* is mostly through root tips. It grows up the root, through the older part of the stem and eventually to the stem apex, causing the heart rot symptom. Little evidence exists for secondary spread from infected crowns.

Infection by *P. nicotianae* is probably less dependent on high soil moisture than *P. cinnamomi*. High soil moisture and poor drainage increase infection by *P. cinnamomi* and suppress root growth. Soil water content below 15% reduces germination of *P. cinnamomi* chlamydospores. Damage by *P. nicotianae* is most severe at pHs of 7 and above, but will occur at pHs as low as 5.

P. cinnamomi affects well over 1000 plant species, and the host ranges of *P. nicotianae* and *P. palmivora* each include several hundred species (Erwin and Ribiero, 1996). However, host specialization occurs in *P. nicotianae*.

Management

A combination of strategies is used. Most important is the improvement of soil drainage with increased surface flow and planting on raised beds (Sideris and Paxton, 1930). Although mulching with plant residues generally increases disease, plastic mulch has reduced disease severity. Plants should not be planted too deeply. Excessive lime applications should be avoided since they increase disease (Sideris and Paxton, 1930). If *Thiobacillus* is present, the application of sulphur to soil decreases heart rot (Pegg, 1977). Pre-plant dips and post-plant foliar applications of fosetyl-Al are also effective. Metalaxyl is effective, but is registered only as a pre-plant dip (Rohrbach and Schenck, 1985). Tolerance to metalaxyl can develop after prolonged use.

'59-656', a cultivar that was developed at the Pineapple Research Institute of Hawaii, is resistant to both *P. cinnamomi* and *P. nicotianae*, and is available at the USDA-ARS Germplasm Repository in Hilo, Hawaii. It has fruit characteristics, quality and yield potential that are similar to those for 'Smooth Cayenne'.

References

Acosta, N. and Malek. R.B. (1979) Influence of temperature on population development of eight species of *Pratylenchus* on soybean. *Journal of Nematology* 11, 229–232.

Aguilar, J.A.E. (1981) *Fusariose do Abacaxizeiro*. Cruz das Almas, Bahia, Brazil, EMBRAPA-CNPMF.

Anderson, E.J. (1966) 1–3 Dichloropropene, 1–2 dichloropropane mixture found active against *Pythium arrhenomanes* in field soil. *Down to Earth* 22, 23.

Ashby, S.F. (1920) Notes on two diseases of the coco-nut palm in Jamaica caused by fungi of the genus *Phytophthora*. *West Indian Bulletin* 18, 71.

Ayala, A. (1971) Rotation of pangola grass and sugar cane beneficial for pineapple production in Puerto Rico. *Nematropica* 1, 6–7.

Beardsley, J.W. (ed.) (1993) The pineapple mealybug complex; taxonomy, distribution, and host relationships. First International Pineapple Symposium. *Acta Horticulturae* 334, 383–386.

Beardsley, J.W., Su, T.H., McEwen, F.L. and Gerling, D. (1982) Field interrelationships of the big-headed ant, the gray pineapple mealybug, and pineapple mealybug wilt disease in Hawaii. *Proceedings of the Hawaiian Entomological Society* 24, 51–67.

Birchfield, B. (1962) New hosts and non host of reniform nematode. *Plant Disease Reporter* 46, 683–685.

Boher, B. (1976) Pineapple heart rot. Study of infection by *Phytophthora nicotianae* var. *parasitica* and *Phytophthora palmivora*. Penetration of underground organs. *Fruits* 31, 365–371.

Bolkan, H.A., Dianese, J.C. and Cupertino, F.P. (1979) Pineapple flowers as principal infection sites for *Fursarium moniliforme* var. *subglutinans*. *Plant Disease Reporter* 63, 655–657.

Broadley, R.H., Wasserman, R.C. and Sinclair, E. (eds) (1993) *Pineapple Pests and Disorders*. Department of Primary Industries, Brisbane, Queensland, Australia.

Buddenhagen, I.W. and Dull, G.G. (1967) Pink disease of pineapple fruit caused by strains of acetic acid bacteria. *Phytopathology* 57, 806.

Cabral, J.R.S., Matos, A.P and Souto, G.F. (1985) Reacao de germoplasma de abacaxi a inoculacao com *Fusarium moniliforme* var. *subgutinans*. *Pesq. agropec. bras., Brasilia* 20, 787–791.

Carter, W. (1933a) A wilt of pineapple similar to mealybug wilt but caused by drought. *Pineapple Quarterly* 3, 81–184.

Carter, W. (1933b) The pineapple mealybug, *Pseudococcus brevipes*, and wilt of pineapples. *Phytopathology* 23, 207–242.

Carter, W. and Collins, J.L. (1947) Resistance to mealybug wilt of pineapple with special reference to a Cayenne–Queen hybrid. *Phytopathology* 37, 332–348.

Caswell, E.P., Sarah, J.L. and Apt, W.J. (1990) Nematode parasites of pineapple. In: Luc, M., Sikora, R.A. and Bridge, J. (eds) *Plant Parasitic Nematodes in Subtropical and Tropical Agriculture*, CAB International, Wallingford, UK, pp. 519–537.

Cha, J.S., Pujol, C., Ducusin, A.R., Macion, E.A., Hubbard, C.H. and Kado, C.I. (1997) Studies on *Pantoea citrea*, the causal agent of pink disease of pineapple. *Journal of Phytopathology* 145, 313–319.

Chitwood, B.G. (1949) Root-knot nematodes. Part I. A revision of genus *Meloidogyne* Goeldi, 1887. *Proceedings of the Helminthological Society of Washington* 16, 340–352.

Cho, J.J., Rohrbach, K.G. *et al.* (1977) Induction and chemical control of rot caused by *Ceratocystis paradoxa* on pineapples. *Phytopathology* 67, 700–703.

Cho, J.J., Hayward, A.C. *et al.* (1980) Nutritional requirements and biochemical activities of pineapple pink disease bacterial strains from Hawaii. *Antonie Van Leeuwenhoek* 46, 191–204.

de Matos, A.P. (1987) Pineapple fusariosis in Brazil: an overview. *Fruits* 42, 417–422.

de Matos, A.P. and Costa, J.L. (eds) (1997) Patterns of diurnal and seasonal airborne spore concentrations of *Fusarium subglutinans* on a pineapple orchard. *Acta Horticulturae* 425, 515–522.

de Matos, A.P. and Mourichon, X. (eds) (1993) Development of resistance to infection by *Fusarium moniliforme* var. *subglutinans* in wounds of pineapple plantlets. *Acta Horticulturae* 334, 423–428.

de Matos, A.P., Mourichon, X. and Lapeyre, F. (1991) Reaction of pineapple accessions to inoculations with *Fusarium moniliforme* var. *subglutinans*. *Fruits* 46, 647–652.

de Matos, A.P., Mourichon, X. and Lapeyre, F. (1992) Occurrence of *Fusarium moniliforme* var. *subglutinans* on pineapple in Bolivia. *Fruits* 47, 33.

Dinardo-Miranda, L.L., Spironello, A. and Martins, A.L.M. (1997) Dinamica populacional de nematoides fitoparasitos em cultura de abacaxi. *Nematologia Brasileira* 21, 49–57.

Dropkin, V.H. (1963) Effect of temperature on growth of root-knot nematodes in soybeans and tobacco. *Phytopathology* 53, 663–666.

Eisenback, J.D., Hirschmann, H., Sasser, J.N. and Triantaphyllou, A.C. (1981) A guide to the four common species of root-knot nematodes (*Meloidogyne* spp.) with a pictorial key. Raleigh, NC, North Carolina State University.

Erwin, D.C. and Ribiero, O.K. (1996) *Phytophthora Diseases Worldwide*. APS Press, St Paul, Minnesota.

Esbenshade, P.R.T. and Triantophyllou, A.C. (1985) Use of enzyme phenotypes for identification of *Meloidogyne* species. *Journal of Nematology* 17, 6–20.

Frossard, P. (1964) Influences of temperature and acidity on the growth in culture of *Thielaviopsis paradoxa*, a pineapple parasite. *Fruits* 19, 461–463.

Godfrey, G.H. (1929) A destructive root disease of pineapples and other plants due to *Tylenchus brachyurus*, n. Sp. *Phytopathology* 19(7), 611–629.

Gommers, F.J. and Dropkin, V.H. (1977) Quantitative histochemistry of nematode-induced transfer cells. *Phytopathology* 67, 869–873.

Gossele, F. and Swings, J. (1986) Identification of *Acetobacter liquefaciens* as causal agent of pink-disease of pineapple fruit. *Journal of Phytopathology* 116, 167–175.

Gunasinghe, U.B. and German, T.L. (1989) Purification and partial characterization of a virus from pineapple. *Phytopathology* 79, 1337–1341.

Hepton, A. and Anderson, E.J. (1968) Interfruitlet corking of pineapple fruit, a new disease in Hawaii. *Phytopathology* 58, 74–78.

Hine, R.B. (1976) Epidemiology of pink disease of pineapple fruit. *Phytopathology* 66, 323–327.

Hu, J.S., Sether, D.M., Ullman, D.E and Lockhart, B.E. (1997) Mealybug wilt of pineapple: pineapple viruses and two-step treatment of pineapple crowns. *Acta Horticulturae* 425, 485–492.

Illingworth, J.F. (1931) Yellow spot of pineapples in Hawaii. *Phytopathology* 21, 865–880.

Johnston, A. (1957a) Bacterial heart rot of the pineapple. *Malay Agricultural Journal* 40, 2–8.

Johnston, A. (1957b) Pineapple fruit collapse. *Malay Agricultural Journal* 40, 253–263.

Kageyama, B., Nakae, M., Yagi, S. and Sonoyama, T. (1992) *Pantoea punctata* sp. nov., *Pantoea citrea* sp. nov. and *Pantoea terrea* sp. nov. isolated from fruit and soil samples. *International Journal of Systematic Bacteriology* 42, 203–210.

Kaplan, D.T., Thomason, I.J. and Van Gundy, S.D. (1979) Histological study of the compatible and incompatible interaction of soybeans and *Meloidogyne incognita*. *Journal of Nematology* 11, 338–343.

Keetch, D.P. (1977a) H.7 Marbling in pineapples. *Pine Series H. Diseases and Pests*. Government Printer, Republic of South Africa.

Keetch, D.P. (1977b) H.3 White leaf spot, base, and soft rot in pineapples. *Pine Series H. Diseases and Pests*. Government Printer, Republic of South Africa.

Kehe, M., Gnonhouri, P. and Adiko, A. (1997) Time course of infestation by *Hanseniella ivorensis* (symphilid) and *Pratylenchus brachyurus* (nematode) on pineapple crop in Cote d'Ivoire. *Acta Horticulturae* 425, 465–474.

Kimati, H. and Tokeshi, H. (1964) Note on the occurrence of a species of *Fusarium* causing gummosis in pineapples. *Review of Agriculture, Piracicaba* 39, 131–133.

Kinloch, R.A. (1982) The relationship between soil populations of *Meloidogyne incognita* and yield reduction of soybean in the coastal plain. *Journal of Nematology* 14, 162–167.

Klemmer, H.W. and Nakano, R.Y. (1964) Distribution and pathogenicity of *Phytophthora* and *Pythium* in pineapple soils in Hawaii. *Plant Disease Reporter* 48, 848–852.

Kontaxis, D.G. (1978) Control of pink disease of pineapple fruit with disulfoton in the Philippines. *Plant Disease Reporter* 62, 172–173.

Kontaxis, D.G. and Hayward, A.C. (1978) The pathogen and symptomatology of pink disease, *Acetobacter aceti, Gluconobacter oxydans*, on pineapple fruit in the Philippines. *Plant Disease Reporter* 62, 446–450.

Larsen, L.D. (1910) *The Diseases of Pineapple*. Hawaii Sugar Planters Association Experiment Station Bulletin. Hawaii Sugar Planters Association, Honolulu 10, 1–72.

Laville, E. (1980) Fusarium disease of pineapple in Brazil. I. Review of current knowledge. *Fruits* 35, 101–113.

Lim, W.H. (1974) The etiology of fruit collapse and bacterial heart rot of pineapple. *MARDI Research Bulletin* 2, 11–16.

Lim, W.H. (1978) *Studies on the Etiology, Epidemiology, Ecology and Control of Pineapple Fruit Collapse*. Cambridge University Press, Cambridge.

Lim, W.H. (1985) Diseases and disorders of pineapples in peninsular Malaysia. MARDI Report No. 97, Malaysian Agricultural Research and Development Institute (MARDI) 97, 53.

Lim, W.H. and Lowings, P.H. (1977) Insects associated with collapsed pineapple fruits, heart rot plants and inflorescences. *MARDI Research Bulletin* 5, 153–157.

Lim, W.H. and Lowings, P.H. (1979a) Effects of ethephon on anthesis and 'fruit collapse' disease in pineapple. *Experimental Agriculture* 15, 331–334.

Lim, W.H. and Lowings, P.H. (1979b) Pineapple fruit collapse in Peninsular Malaysia: symptoms and varietal susceptibility. *Plant Disease Reporter* 63, 170–174.

Lim, T.K. and Rohrbach, K.G. (1980) Role of *Penicillium funiculosum* strains in the development of pineapple fruit diseases. *Phytopathology* 70, 663–665.

Linford, M.B. (1932a) Endogenous or non-parasitic brown spot. *Pineapple Quarterly* 11, 46–58.

Linford, M.B. (1932b) Transmission of the pineapple yellow-spot virus by *Thrips tabaci*. *Phytopathology* 22, 301–324.

Linford, M.B. (1943) Influence of plant populations upon incidence of pineapple yellow spot. *Phytopathology* 33, 408–410.

Lyon, H.L. (1915) A survey of pineapple problems. *Planters Record* 13, 125–139.

Maffia, L.A. (1980) Persistence of *Fusarium moniliforme* var. *subglutinans* Wr. and Rg. in the soil and on plant remains and elimination of this pathogen from pineapple suckers by hot water treatment. *Fruits* 35, 217–243.

Maffia, L.A. and Chaves, G.M. (1978) Eradication of *Fusarium moniliforme* Sheld. var. *subglutinans* Wr. and Rg. by thermotherapy and chemotherapy of pineapple slips. *Fitopatologia Brasileira* 3, 119–120.

Melzer, M.J., Karasev, A.V., Sether, D.M. and Hu, J.S. (2001) Nucleotide sequence, genome organization and phylogenetic analysis of pineapple mealybug wilt-associated virus-2. *Journal of General Virology* 82, 1–7.

Mergaert, J.L., Verdonck, L. and Kersters, K. (1993) Transfer of *Erwinia ananas* (synonym, *Erwinia uredovora*) and *Erwinia stewartii* to the genus *Pantoea* emend. as *Pantoea ananas* (Serrano 1928) comb. nov. and *Pantoea stewartii* (Smith 1898) comb. nov., respecctively, and description of *Pantoea stewartii* subsp. *indologenes* subsp. nov. *International Journal of Sysstematic Bacteriology* 43, 162–173.

Mourichon, X. (1991) A study of fruit diseases: black spot and leathery pocket. *Fruits* 46, 390–394.
Nardacci, J.F. and Barker, K.R. (1979) The influence of temperature on *Meloidogyne incognita* on soybean. *Journal of Nematology* 11, 62–70.
Nirenberg, H.I. and O'Donnell, K. (1998) New *Fusarium* species and combinations within the *Gibberella fujikuroi* species complex. *Mycologia* 90, 434–458.
O'Donnell, K., Cigelnik, E. and Nirenberg, H.I. (1998) Molecular systematics and phylogeography of the *Gibberella fujikuroi* species complex. *Mycologia* 90, 465–493.
Osseni, B., Sarah, J.L. and Hugon, R. (1997) Effect of soil pH on the population development of *Pratylenchus brachyrus* (Godfrey) in pineapple roots and on the growth and yield of the plant. *Acta Horticulturae* 425, 423–433.
Oxenham, B.L. (1957) Diseases of the pineapple. *Queensland Agriculture Journal* 83, 13–26.
Oxenham, B.L. (1962) Etiology of fruitlet core rot of pineapple in Queensland. *Queensland Journal of Agricultural Science* 19, 27–31.
Paull, R.E. and Rohrbach, K.G. (1985) Symptom development of chilling injury in pineapple fruit. *Journal of the American Society for Horticultural Science* 110, 100–105.
Pegg, K.G. (1977) Soil application of elemental sulphur as a control of *Phytophthora cinnamomi* root and heart rot of pineapple. *Australian Journal of Experimental Agriculture and Animal Husbandry* 17, 859–865.
Petty, G.J. (1975) Pineapple mites. *Citrus and Subtropical Fruit Journal* 498, 15–18.
Petty, G.J. (1977) *H2. Pineapple Pests: Leathery Pocket in Pineapples.* Government Printer, Republic of South Africa.
Petty, G.J. (1978) *H16. Pineapple Mites. Pineapple Series H. Diseases and Pests.* Government Printer, Republic of South Africa, pp. 1–4.
Petty, G.J. and Manicom, B.Q. (1995) Control of big-headed ant, *Pheidole megacephala*, in pineapple plantations with proprietary bait Amdro. *Fruits* 50, 343–346.
Pinho Guedes, N.M., Maria, J., Zambolin, L. and Ventura, J.A. (1997) Protoplast isolation of *Ananas comosus* (L.) Merr. cv. 'Perolera.' *Acta Horticulturae* 425, 259–265.
Pissarra, T.B., Chaves, G.M. and Ventura, J.A. (1979) Sintomatologia da fusariose (*Fusarium moniliforme* Sheld. var. *subglutinans* Wr. & Reink.) do abacaxizeiro. *Fitopatologia Brasileira* 4, 255–263.
Pujol, C.J. and Kado, C.I. (1998) Characterization of pUCD5000 involved in pink disease color formation by *Pantoea citrea*. *Plasmid* 40, 169–173.
Rabie, E.C. and Tustin, H.A. (1997) Pre-plant control of nematodoes in pineapple. *Acta Horticulturae* 425, 435–442.
Rebois, R.V. (1973) Effect of soil temperature on infectivity and development of *Rotylenchulus reniformis* on resistant and susceptible soybeans, *Glycine max*. *Journal of Nematology* 5, 10–13.
Rohrbach, K.G. (1980) Climate and fungal pineapple fruit diseases. *Second Southeast Asian Symposium on Plant Diseases in the Tropics*, October 20–26, 1980, Bangkok, Thailand.
Rohrbach, K.G. (1983) Pineapple diseases and pests and their potential for spread. *Exotic Plant Quarantine Pests and Procedures for Introduction of Plant Materials.* K.G. Singh, ASEAN Plant Quarantine Centre and Training Institute, Serdang, Selangor, Malaysia, pp. 145–171.
Rohrbach, K.G. (1986) *Pineapple: the Plant and its Culture.* HITAHR: Hawaii Institute of Tropical Agriculture and Human Resource, University of Hawaii at Manoa, Hawaii Agricultural Experiment Station, Honolulu, Hawaii.
Rohrbach, K.G. (1989) Unusual tropical fruit diseases with extended latent periods. *Plant Disease* 73, 607–609.
Rohrbach, K.G. (1994) Fusariosis. In: Ploetz, R.C., Zentmyer, G.A., Nishijima, W.T., Rohrbach, K.G. and Ohr, H.D. (eds) *Compendium of Tropical Fruit Diseases.* APS Press, St Paul, Minnesota, p. 49.
Rohrbach, K.G. and Apt, W.J. (1986) Nematode and disease problems of pineapple. *Plant Disease* 70, 81–87.
Rohrbach, K.G. and Apt, W.J. (1993) Common names for plant diseases – pineapple (*Ananas comosus* [L.] Merr.) diseases. *Plant Disease* 77, 320–321.
Rohrbach, K.G. and Namba, R. (1983) Population dynamics of the pineapple fruit mite and *Penicillium* induced fruit diseases. *10th International Congress of Plant Protection 1983.* Proceedings of a Conference held at Brighton, UK, November 20–25, 1983. British Crop Protection Council, Croydon, UK.
Rohrbach, K.G. and Pfeiffer, J.B. (1975) The field induction of bacterial pink disease in pineapple fruits. *Phytopathology* 65, 803–805.

Rohrbach, K.G. and Pfeiffer, J.B. (1976a) Field induction of pineapple interfruitlet corking, leather pocket, and fruitlet core rot with *Penicillium funiculosum*. *Phytopathology* 66, 392–395.
Rohrbach, K.G. and Pfeiffer, J.B. (1976b) Susceptibility of pineapple cultivars to fruit diseases incited by *Penicillium funiculosum* and *Fusarium moniliforme*. *Phytopathology* 66, 1386–1390.
Rohrbach, K.G. and Pfeiffer, J.B. (1976c) The interaction of four bacteria causing pink disease of pineapple with several pineapple cultivars. *Phytopathology* 66, 396–399.
Rohrbach, K.G. and Phillips, D.J. (1990) Postharvest diseases of pineapple. *Acta Horticulturae* 269, 503–508.
Rohrbach, K.G. and Schenck, S. (1985) Control of pineapple heart rot, caused by *Phytophthora parasitica* and *P. cinnamomi*, with metalaxyl, fosetyl Al, and phosphorous acid. *Plant Disease* 69, 320–323.
Rohrbach, K.G. and Schmitt, D.P. (1994) Pineapple. In: Ploetz, R.C., Zentmyer, G.A., Nishijima, W.T., Rohrbach, K.G. and Ohr, H.D. (eds) *Compendium of Tropical Fruit Diseases*. APS Press, St Paul, Minnesota, pp. 45–55.
Rohrbach, K.G. and Taniguchi, G. (1984) Effects of temperature, moisture, and stage of inflorescence development on infection of pineapple *Ananas comosus* by *Penicillium funiculosum* and *Fusarium moniliforme* var. *subglutinans*. *Phytopathology* 74, 995–1000.
Rohrbach, K.G., Namba, R. and Taniguchi, G. (1981) Endosulfan for control pineapple interfruitlet corking, leathery pocket and fruitlet core rot. *Phytopathology* 71, 1006.
Rohrbach, K.G., Beardsley, J.W., German, T.L., Reimer, N.J. and Sandford, W.G. (1988) Mealybug wilt, mealybugs, and ants of pineapple. *Plant Disease* 72, 558–565.
Sakimura, K. (1940) Evidence for the identity of the yellow-spot virus with the spotted-wilt virus: experiments with the vector, *Thrips tabaci*. *Phytopathology* 30, 281–399.
Sanewski, G.M. and Giles, J. (1997) Blackheart resistance in three clones of pineapple [*Ananas comosus* (L.) Merr.] in subtropical Queensland. *Australian Journal of Experimental Agriculture* 37, 459–461.
Sarah, J.L. (1980) Use of systemic nematicides for the control of *Pratylenchus brachyurus* in pineapples. I. Preventive and curative effects of foliar applications on root infestations. *Fruits* 35, 745–757.
Sasser, J.N. (1977) Worldwide dissemination and importance of the root-knot nematodes *Meloidogyne* spp. *Journal of Nematology* 9, 26–29.
Schneider, R.C., Green, R.E. and Apt, W.J. (1993) Management of drip-applied nematicides in pineapple. *Acta Horticulturae* 334, 351–360.
Serrano, F.B. (1928) Bacterial fruitlet brown rot of pineapple. *Philippine Journal of Science* p. 36.
Sether, D.M. and Hu, J.S. (2002) Closterovirus infection and mealybug exposure are necessary for the development of mealybug wilt of pineapple disease. *Phytopathology* 92, 928–935.
Sether, D.M., Ullman, D.E. and Hu, J.S. (1998) Transmission of pineapple mealybug wilt-associated virus by two species of mealybugs (*Dysmicoccus* spp.). *Phytopathology* 88, 1224–1230.
Sideris, C.P. and Paxton, G.E. (1930) Heart rot of pineapple plants. *Phytopathology* 20, 951–958.
Sideris, C.P. and Paxton, G.E. (1931) Pathological, histological, and symptomatological studies on pineapple root rots. *American Journal of Botany* 18, 465–498.
Singh, M. and Khan, E. (1996) Five new species under subfamily *Longidorinae* (*nematodea*) associated with fruit crops from north and north-eastern India, with comment on the geneus, *Neologidorus* Khan, 1986. *Indian Journal of Nematology* 26, 158–171.
Sipes, B.S. (1997) Pre-plant and post-plant pesticides for nematode control in pineapple. *Acta Horticulturae* 425, 457–464.
Sivakumar, C.V. and Seshadri, A.R. (1971) Life history of the the reniform nematode, *Rotylenchus reniformis* Linford and Oliera, 1940. *Indian Journal of Nematology* 1, 7–20.
Sulaiman, S.F.M. (1997) Impact of weed management on ant density and fruit yield in the control of pineapple wilt disease. *Acta Horticulturae* 425, 475–484.
Thompson, A. (1937) Pineapple fruit rots in Malaya. A preliminary report on fruit rots of the Singapore canning pineapple. *Malaysian Agriculture Journal* 25, 407–420.
Trüper, H.G. and de'Clari, L. (1997) Taxonomic note: necessary correction of specific epithets formed as substantives (nouns) 'in apposition'. *International of Systematic Bacteriology* 47, 908–909.
Ullman, D.E., German, T.L., Gunasinghe, U.B. and Ebesu, R.H. (1989) Serology of a closterovirus-like particle associated with mealybug wilt of pineapple. *Phytopathology* 79, 1341–1345.
Ventura, J.A. (1994) Pineapple fusariosis: characterization of the pathogen, epidemiology of the disease, resistance and micropropagation of the host *in vitro*. PhD thesis. University of Viçosa.
Ventura, J.A., Zambolin, L. and Chaves, G.M. (1993) Integrated management system for pineapple Fusarium disease control. *Acta Horticulturae* 334, 439–453.

Ventura, J.A., Fanton, C.J. and Costa, H. (1995) Effect of new fungicides on pineapple fusariosis control. In: *Book of Abstracts, International Seminar on Fusarium Mycotoxins, Taxonomy and Pathogenicity.* Martina Franca, Italy, May 9–13, 1995. p. 103.

Wakman, W., Teakle, D.S., Thomas, J.E. and Dietzgen, R.G. (1995) Presence of a clostero-like virus and a bacilliform virus in pineapple plants in Australia. *Australian Journal of Agricultural Research* 46, 947–958.

Wang, K.H. and Sipes, B.S. (1998) Suppression of reniform nematode, *Rotylenchulus reniformis*, on pineapple with tropical cover crops. *Acta Horticulturae* 529, 247–253.

20 Management of Tropical Fruit Diseases: Current Overview and Future Outlook

Randy C. Ploetz[1], L.W. Timmer[2] and Steve M. Garnsey[2]
[1] *University of Florida, Tropical Research and Education Center, Homestead, Florida, USA;* [2] *University of Florida, Citrus Research and Education Center, Lake Alfred, Florida, USA*

Introduction

The preceding chapters should make it abundantly clear that diseases are major impediments to the production of tropical fruit. In the future, the numbers and types of diseases that affect each of these crops undoubtedly will change. New diseases will enter production areas, and only with effective quarantine and pathogen detection measures will it be possible to minimize these threats. Producers and researchers must be aware of, and be ready to deal with, new diseases that lurk outside given production areas.

The ways in which tropical fruit diseases are managed are bound to change. New approaches to germplasm improvement have the potential to revolutionize the production of many of these crops and their protection from important diseases. New pesticides will be developed that are more environmentally benign, but they will need to be used rationally and effectively. Resistance to pesticides will continue to be a problem but, with disease forecast models, enhanced understandings of how and why resistance develops to different compounds, better application technologies and improved resistance in the host crops to these diseases, the effective life of these pesticides should be increased. In this chapter, we summarize the status of, and future trends and issues that surround, the management of these diseases.

Germplasm Improvement

New cultivars with improved quality, yield, appearance and disease resistance are always in high demand. They can reduce production costs, improve consumer acceptance, reduce environmental risks and facilitate the use of biological control agents.

Improved rootstocks for the grafted crops are also desirable since they can significantly impact yield and resistance to soilborne diseases. They also affect the quality of, and disease development in, fruit that are produced by the scion (Willingham *et al.*, 2001; A.W. Whiley, personal communication, 2002). Although the latter phenomenon has only been described in avocado and citrus, its existence in other tropical fruit crops should be investigated.

When evaluating rootstocks, the rootstock–scion combination must be considered in total. Performance of the rootstock alone does not predict how it will interact with a given scion, and resistance to a given root disease does not ensure that the scion will resist other, less important root diseases. In the latter cases, minor diseases can assume an increased importance (e.g. *Phytophthora citricola*-induced cankers on *P. cinnamomi*-resistant avocado rootstocks). Scions may also not produce high yields when grafted on disease-resistant rootstocks (e.g. 'Haas' avocado grafted on the 'Martin Grande' rootstock). Certain citrus combinations have unique disease susceptibilities that are not present in the separate components, and there may be an adverse effect that one component has on the other, such as changes in cold hardiness. Moreover, physiological incompatibility can occur in combinations that are graft compatible (A.W. Whiley, personal communication, 2002).

The deployment of new genotypes as a means of disease control is a more costly and long-term solution for perennial crops than for annual crops. With the exceptions of passion fruit and pineapple, the crops discussed in this volume are often grown for 10 years or more before replanting is considered. Replacing susceptible cultivars as individual plants die or decline is less expensive than converting an entire orchard or plantation but, in the interest of uniform crop management or disease containment, complete conversion may be needed.

Conventional approaches

How the above objectives of improvement are met depends on the crop and the resources that are available. Improvement for many has occurred primarily through the selection of naturally occurring genotypes. Little usually is known about the parentage of such germplasm and, even with open-pollinated seedlings, both the female and male parent may be unknown. Advances have been made recently in deciphering some of the ambiguous parental backgrounds (Carreel, 1994; Schnell *et al.*, 1995; Nicolosi *et al.*, 2000).

Conventional breeding programmes that have been established for tropical fruit crops are confronted with several problems. Parents that are used are often heterozygous and lack well-defined genetic traits (e.g. avocado, citrus and mango). Desirable parents may be sterile (e.g. banana), sexually incompatible (e.g. avocado and citrus) or lack resistance to important diseases (e.g. banana and citrus). Improving the crops for which fruit set is low and fruit produce a single seed is also difficult. For example, only one in 1000 avocado flowers and three in 1000 mango flowers produce mature fruit (Bergh, 1976; Singh, 1976). Controlled pollinations of these crops have been time consuming, expensive and produced few useful hybrids. Finally, long periods of juvenility and the large areas of land that are needed to evaluate meaningful populations of hybrids (essentially all of the woody crops in this volume) add to the expense of traditional breeding schemes. Clearly, there are ample reasons for using non-conventional approaches to improve these crops.

Non-conventional approaches

The molecular characterization of these crops assists in improvement efforts. Phylogenetic studies enable the reconstruction of complex, natural hybrids (Carreel, 1994; Nicolosi *et al.*, 2000). Linkage maps exist or are being developed for avocado, banana, citrus and mango (Fauré *et al.*, 1993; Gmitter *et al.*, 1996; Kaemmer *et al.*, 1997; Kijas *et al.*, 1997; Sharon *et al.*, 1997; Kashkush *et al.*, 2001). As they become saturated, they have the potential to direct conventional breeding programmes in two key ways. Crossing strategies could be guided to avoid undesirable linkages and favour specific genetic combinations (Gonzáles de Leon and Fauré, 1993). Furthermore, markers could be developed for marker-assisted selection (MAS) of resistance and other traits. As an example, work is in progress to locate the gene for resistance to *Citrus tristeza virus* (CTV) in trifoliate orange (Gmitter *et al.*, 1996). Finally, a recently initiated consortium to sequence the genome of banana ultimately could provide an enormous database for improving this crop (Anonymous, 2001).

Molecular characterization of pathogens also assists improvement programmes. Genetic transformation of these crops for resistance has only begun and will probably become more important in the future. To date, this is most common with the virus-induced diseases, especially those that affect banana, citrus and papaya. The small and relatively simple genomes of viruses are usually well understood, and this information has been used to devise disparate pathogen–gene-mediated approaches to disease resistance. Viral coat protein genes were the first type of genes to be used, but other genes or versions of genes for virus replication, movement within the host and vector transmission have also been tested.

Although genetic transformation is the most prevalent non-traditional approach, additional strategies have been used, including mutation breeding, selection of somaclonal variants and somatic hybridization.

Genetic transformation

Genetic transformation is a powerful tool for genetically manipulating crops. Although several of the crops discussed in this volume have been transformed with foreign DNA, substantial progress has been made with only three: banana, citrus and papaya.

For banana, the major disease focus has been banana bunchy top. Minimal or no natural resistance to this disease is found in the edible clones (see Chapter 4). Current transformation strategies use diverse genes from the *Banana bunchy top virus* (BBTV). Black Sigatoka and Panama disease, two diseases that are caused by fungi, are also targets, and genes for antifungal proteins from other plants have been used.

Transformation of citrus with several CTV genes is in progress. Other transformation approaches involve genes that will trigger a resistance response in the plant when infection is initiated by the pathogen, or genes that will allow the plant to express products that either repel, or are toxic to, insects that are pests or pathogen vectors.

To date, the greatest success has been with papaya. In 1992, a team from Hawaii and Cornell University reported the development of the first transgenic papaya plants that resisted papaya ringspot in the field (Fitch *et al.*, 1992). Two cultivars developed by the team, 'SunUp' (a transformant of 'Sunrise' that is homozygous for a coat protein transgene) and 'Rainbow' (a cross between 'Sunrise' and the non-transformed 'Kapoho' that is hemizygous for the transgene) have been credited with saving the papaya industry on the island of Hawaii (Ferreira *et al.*, 2002). Groups elsewhere recently have modified protocols used by the above team to produce ringspot-resistant transformants of their own (Gonsalves, 1998; Yeh *et al.*, 1998; Ying *et al.*, 1999; Lines *et al.*, 2002).

While there are many positive long-term aspects of genetic engineering for crop improvement, in many cases expectations have been overly optimistic. Efficient tissue culture and transformation systems are not available for all cultivars or crops. A case in point is the transformation of mango somatic embryos with reporter and selection genes (e.g. Mathews *et al.*, 1992; Cruz-Hernández and Litz, 1997), even though somatic embryos of this crop have not produced plants that can be taken to the field. There are also relatively few genes that are well characterized and useful for transformation, although this will surely change with time (Roose, 1996). Expression of foreign genes after successful transformation is uncertain, and large populations may need to be screened to identify transformants that stably express a given gene, are disease resistant and have an otherwise unchanged phenotype. Given the considerable effort and resources that are needed to genetically engineer a resistant crop, and the intractable nature of many of the crops in tissue culture, it is clear that genetic transformation will be a viable option for only some tropical fruit crops in the near future.

Numerous factors govern the use of genetically modified organisms (GMOs) in agriculture. Government regulations for the release of food crops are particularly involved, especially in Europe and North America. Food safety issues, both real and imagined, are overriding concerns for governments and consumers, and once a transgenic crop has been released there is also no assurance that the general public will accept it. Distrust and non-acceptance of GMOs is greatest in Europe, and it is clear that GM crops will not be marketable there until major changes occur in the public's perception of these products. For food crops that are close relatives of weeds, there is also concern that transgenes will move from the crop into weeds via pollen, resulting in disease-, insect- or herbicide-resistant weeds. Finally, licensing restrictions affect the commercialization of new cultivars since many of the most efficient transformation vectors and promoters are patented and not freely available.

Mutation breeding

Ionizing radiation (e.g. X-rays, γ-rays and neutrons) and chemical mutagens have been used to induce useful mutations in diverse crops. As above, the absence of reliable *in vitro* regeneration protocols often are key stumbling blocks in the successful use of this approach.

Recently, Ahloowalia and Maluszynski (2001) indicated that >1800 cultivars of seed-propagated crops had been developed either directly from, or after crosses with, mutant lines. An additional 465 mutants of vegetatively propagated crops were found in the database of the International Atomic Energy Agency, including banana, date palm and pineapple. Disease resistance is among the traits that are sought (e.g. Bhagwat and Duncan, 1998).

Somatic hybridization

Somatic hybridization has been used most successfully in the improvement of citrus (Grosser and Gmitter, 1990; Guo and Deng, 2001). There are very useful traits among citrus relatives including resistance to tristeza, exocortis, greening, the citrus nematode and *Phytophthora*-induced diseases. Unfortunately, polyembryony, pollen and ovule sterility, sexual and graft incompatibility, and long juvenility prevent the introduction of these traits into useful lines via conventional breeding.

These barriers have been overcome by fusing protoplasts of diverse taxa with polyethylene glycol or electroporation. With this approach, 60 different sexually compatible and incompatible intergeneric combinations of *Citrus* spp. and citrus relatives have been produced, most of which have come from a group led by Jude Grosser in Florida (Guo and Deng, 2001). In general, only hybrids from sexually compatible combinations have been useful. At this time, it appears that these hybrids have their greatest potential as rootstocks. However, because some have flowered and set fruit, it is possible that some could be useful as scions or without grafting.

Somatic hybridization of avocado and avocado relatives has been attempted recently in order to produce Phytophthora root rot-resistant rootstocks. Unfortunately, the combinations between *Persea americana* and red bay, *P. borbonia*, that have been attempted have not survived.

Somaclonal selection

Somaclonal variation has been recognized in numerous crop species, and has been used to improve many. Among the crops in this volume, it has been used most widely in banana. In the 1980s, a group in Taiwan began evaluating somaclonal variants of 'Giant Cavendish' AAA for resistance to race 4 of Panama disease (Hwang and Ko, 1987). Although the initial selections possessed poor bunch characteristics, recent products of this programme produce excellent bunches and have considerable resistance to this important disease (Hwang *et al.*, 1994). Recently, somaclonal resistance has been reported in 'Williams' AAA to yellow Sigatoka (Vidal and Garcia, 2000) and in plantains AAB to black Sigatoka (Nwauzoma *et al.*, 2002).

Pesticide Usage: Human and Environmental Health Issues

Pesticides often are necessary tools in the production of tropical fruit. Without them, economic yields of many crops would not be possible, nor would the production of a high quality and aesthetically pleasing product.

Unfortunately, there are well-known downsides to the use of some pesticides. Two important issues are associated with tropical fruit production. First, some of the compounds that are or have been used are quite toxic. Although more benign materials are used now than in the past, undesirable non-target effects are still recognized (Smith, 2000). Secondly, in many of the poor countries in which tropical fruit production occurs, environmental degradation that can result from pesticide use may be perceived as either not important or simply the cost of doing business (Colburn, 1997). In addition, agricultural labourers may be viewed as 'politically superfluous' and may not be protected from occupational hazards to the extent of other sectors in these societies (Sass, 2000). In these situations, the health of workers and the environment are of a lesser importance than profitable fruit production. Although this mindset is changing and not widespread, substantial progress is still needed in some areas (Smith, 2000).

The environmental and human health impacts of pesticide use vary widely for the crops discussed in this volume and, on some (e.g. smallholder banana and coconut production), use is negligible. However, in other situations, pesticide use can be high. That in the export banana plantations is an extreme example.

Among all crops worldwide, the amounts of pesticides that are used to produce export bananas are exceeded only by those in cotton production. In a given year, >55 kg of the active ingredient of pesticides and petroleum oils can be applied per hectare in a given plantation. Some of these are deemed highly toxic by the US Environmental Protection Agency (Fig. 20.1).

Historically, the nematicide DBCP (1,2-dibromo-3-chloropropane) is most notorious. Its use was phased out after it was determined to be a carcinogen and male reproductive toxicant. None the less, it caused male sterility and cancer in banana plantation workers (Sass, 2000; Smith, 2000).

The numerous pesticides that are still used on export bananas are of concern (Sass, 2000; Smith, 2000). Of these, the nematicides terbufos and phenamiphos and the herbicide paraquat are most dangerous. In banana plantations in Costa Rica, they are responsible for >50% of the pesticide poisonings that are reported each year (Smith, 2000). Plastic bunch covers that are impregnated with the insecticide chlorpyriphos are used

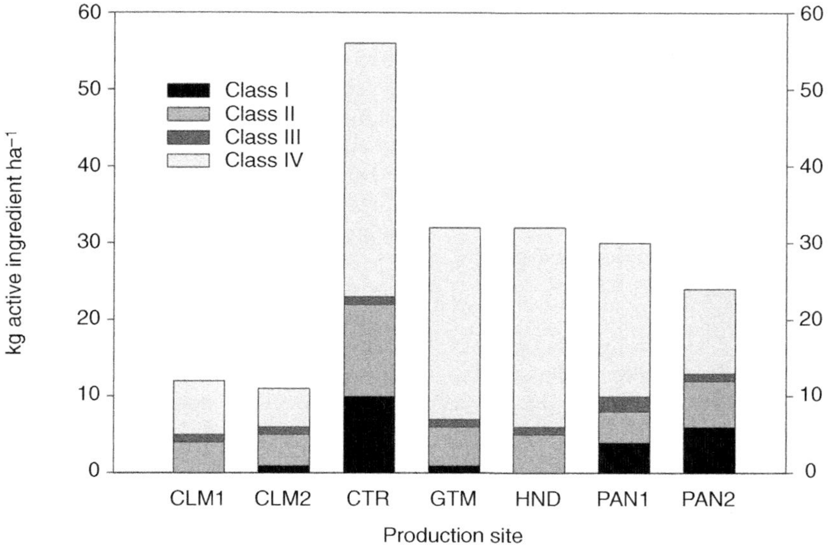

Fig. 20.1. Pesticide usage in Chiquita banana plantations in 2000. Pesticide classes are those of the US Environmental Protection Agency, with Class I being the most and Class IV the least toxic. Production sites are: CLM1 = Santa Marta, Colombia; CLM2 = Turbo, Colombia; CTR = Costa Rica; GTM = Guatemala; HND = Honduras; PAN1 = Armuelles, Panama; and PAN2 = Bocas, Panama. Figures are found at the following URL: http://www.chiquita.com/chiquitacr.1/3perform/cgr20.asp

to protect fruit in the field and are a conspicuous, non-biodegradable trash in plantations. Finally, diverse fungicides are used in pre- and postharvest operations. Although they are far less toxic than the nematicides (e.g. Class III or IV versus Class I), their use is pervasive. Fruit at packing stations is dipped in thiabenzadole or imazalil fungicides before shipment to foreign markets. Most importantly, several fungicides are used to combat black Sigatoka and other leaf diseases (see Chapter 4). They are applied by aircraft as many as 30–40 times per year, especially in high rainfall environments (e.g. the Atlantic coast of Costa Rica). In the recent past, applications were often made when workers were harvesting or conducting maintenance in plantations. In these cases, direct exposure to fungicides often caused dermal and respiratory problems. Worker exposure to these fungicides has decreased, but aquatic and other life forms are still affected. For example, sterol-inhibiting fungicides, such as propiconazole and tridemorph, have been implicated in significant mortality in Ecuador's shrimp industry (i.e. the Taura syndrome), and other fungicides, such as azoxystrobin, are toxic to fish (Colburn, 1997).

As awareness of such pesticide issues increase, there will be an increased demand for safer crop protection practices. Meeting these demands are now major goals of the large international producers of tropical fruit.

Development of Fungicide-resistant Strains

The appearance of fungicide-resistant strains is a continuing problem. Many of the cases that have occurred involve benomyl. *Colletotrichum gloeosporioides* (McMillan *et al.*, 1989), *Mycosphaerella citri* (Whiteside, 1980a), *M. fijiensis* and *M. musicola* (Stover, 1990) developed resistance to benomyl quickly. Scab fungi have also developed resistance to benomyl in many areas, although resistant strains have not spread as rapidly because the spores are not airborne (Whiteside, 1980b). With few exceptions (e.g. Peres *et al.*, 2002b), benomyl is no longer used to control tropical fruit diseases. *Phytophthora nicotianae* has also developed resistance to metalaxyl in Florida, and resistance in *P. cinnamomi* and its enhanced degradation in soil have been reported in California (Timmer *et al.*, 1998; Chapter 3). Recently, a 500-fold increase in resistance to a strobilurin fungicide, trifloxystrobin, was recorded in *M. fijiensis* in Costa Rica (Chin *et al.*, 2001). In contrast, tolerance to the sterol-inhibiting fungicides develops quite differently, in that highly tolerant populations of the pathogens do not predominate. For example, in banana plantations, wide ranges in sensitivity are observed among individual isolates of *M. fijiensis*. It is possible that isolates of this fungus that are highly tolerant to these fungicides are less fit than sensitive strains (Romero and Sutton, 1997).

A key to avoiding resistance problems is the development of products with diverse modes of action. These fungicides could then be alternated or mixed to help avoid problems with resistance. Fungicides should be applied preventively or when inoculum levels are low to reduce the potential for selecting resistant strains. Another important consideration is to avoid distribution of resistant strains. Nurseries are an important source of resistant strains due to their disease-conducive environments and the long distances they can spread once stock is moved to the field. Sanitation and control of environmental conditions, rather than fungicides, should be used to control diseases in nurseries whenever possible.

Development of New Pathotypes

New pathotypes are continuing threats to production. Unfortunately, their development is often poorly understood. Single mutations in viroids can result in a change in symptom expression or in the species or cultivars that are affected (Schnell *et al.*, 2001), and strains of CTV have probably also arisen via mutation (Pappu *et al.*, 1993). The existence of mixtures of CTV strains that differ in virulence and aphid transmission complicates the recognition of new strains of this virus.

The development of new pathotypes of other pathogens should be investigated. With certain diseases, host specificity is easily identified. Tangerine cultivars that are susceptible to Alternaria brown spot are readily identified, and those that are resistant are essentially immune (Kohmoto *et al.*, 1979; Whiteside, 1976). There are indications that the tangerine strain may have developed from saprophytic strains or possibly from the rough lemon pathotype. With citrus scab, distinct pathotypes exist and differential hosts have been determined. On banana, there is good circumstantial evidence for the existence of two pathotypes of *Guignardia musae*, and severe and mild strains of BBTV, *Banana streak virus* and *Cucumber mosaic virus* have been described (Jones, 2000). Recognizing and understanding such variation is needed in order to combat the problems that it presents.

Growers usually choose cultivars according to their market potential, rather than resistance to diseases. Although host resistance is taken into consideration during cultivar development, it has been difficult to incorporate genes for resistance. For rootstocks, efforts have been directed more specifically towards locating and evaluating naturally occurring hybrids for tolerance to soilborne and systemic diseases as well as abiotic factors.

Development and Use of Predictive Models for Disease Forecasting

Models have been developed to forecast the development of several diseases in this volume. In general, they enable fewer pesticide applications, thereby reducing control costs and minimizing environmental impact.

Several forecast models have been developed for black Sigatoka control, and they are now used routinely to determine application schedules in the export banana plantations (Stover, 1990). In general, these systems incorporate data on disease severity and environmental factors that are known to affect infection and disease development. A system that was created by Ganry and Laville (1983) for use in the French Caribbean and modified by Fouré (1990) for use in Cameroon, and by Marín and Romero (1992) for use in Central America is most prevalent. Other forecasting systems for export banana plantations were developed by Chuang and Jeger (1987) and Wielemaker (1990), and for plantain by Bureau (1990) and Lescot *et al.* (1998).

Timmer and Zitko (1996) developed a model for postbloom fruit drop of citrus that helps predict disease outbreaks and determine the need for and timing of fungicide application in Florida. Peres *et al.* (2000a) developed a more broadly based advisory system for postbloom fruit drop that is useful in all citrus-growing areas. Timmer *et al.* (2000) developed the Alter-Rater model for timing sprays for the control of Alternaria brown spot, and weather-based models could be developed for other citrus diseases such as Septoria spot and Phytophthora brown rot.

Empirical models have been developed for use on mango. They have been used to reduce the number of applications of a systemic fungicide to control blossom blight (Fitzell *et al.*, 1984) and anthracnose (Dodd *et al.*, 1991).

In the future, fungicide applications will be timed more precisely and control will improve with less and more specific fungicide usage. Moreover, as better, less expensive, and easily used weather monitoring equipment becomes available, predictive models are becoming more feasible. Global satellite information will also improve weather forecasts and thus aid in disease predictions. Essential ingredients for the use of such models are efficient sprayers that require low volumes of water and can cover large areas quickly, and fungicides with 'kick-back' or post-infection activity.

New Disease Threats

As world travel has increased, so too has the movement of exotic pathogens. New diseases can cause unpredictable losses and negatively impact the exchange of germplasm. Although the reasons for the appearance of some diseases often are understood, at times it is not possible to determine why and where new problems will originate.

Recent studies of some causal agents have revealed that similar symptoms on some hosts can be caused by previously unrecognized pathogens. On banana, the newly described species *Mycosphaerella eumusae* was discovered during a survey of the Sigatoka leaf spots in Asia (Crous and Mourichon, 2002). Although the fungus causes symptoms that resemble closely those of black Sigatoka, it is distinct from *M. fijiensis*. Whether its host range differs from those of the Sigatoka pathogens, and what kind of threat it poses outside Asia are not known. In a similar way, a new species, *Fusarium sterilihyphosum*, was described during studies on the fungus that causes mango malformation, *F. mangiferae* (Britz *et al.*, 2002). The threat that this fungus poses to mango production is not known.

Other newly recognized diseases may have arisen on another host or may have been present previously at low and previously undetected levels. Examples on citrus are the whitefly-transmitted chlorotic dwarf in Turkey (Korkmaz *et al.*, 1996), witches' broom in Oman (Garnier *et al.*, 1991) and citrus variegated chlorosis in Brazil (Chang *et al.*, 1993), and on date palm are slow decline and white tip dieback in Sudan (Cronjé *et al.*, 2000a, b). Strict quarantine measures must be observed to ensure that these and other diseases of restricted distribution do not spread (Broadbent and Timmer, 1997).

Diseases rarely are static. There will be continued dissemination of pathogens, pests and vectors to new areas, and modern air transport and global commerce will allow production areas to be exposed to pests and diseases found elsewhere. Expansion of the geographic range of diseases is dependent on favourable weather conditions and susceptible hosts. For example, fungal and bacterial pathogens that require humid tropical conditions are unlikely to become problems in arid areas. Conversely, diseases such as citrus stubborn, that have a pathogen:vector complex that is adapted to arid areas, may never become problems in the humid tropics. Other pathogens and diseases that pose a threat if they become more widely distributed are listed in Table 20.1.

New pathotypes will continue to arise via mutation and selection, and cultivars that currently are not affected by existing strains or pathotypes may be affected suddenly. The appearances of new strains of the Alternaria leaf spot and scab pathogens on citrus are examples.

Finally, when new scion and rootstock cultivars are deployed, they may be susceptible to new or previously unimportant pathogens. For example, a previously unknown disease, stem pitting of 'Duke 6', was recognized only after this root-rot-resistant avocado rootstock was exported to South Africa. In citrus, the adoption of new rootstocks in areas affected by tristeza decline has often revealed other problems, such as viroids and blight, which did not affect trees on sour orange rootstock. Genetic and virus-induced bud union incompatibilities may occur with a change in either scion or rootstock.

Many tropical fruit pathogens are closely related to those found on other plant species. In citrus, the exocortis and cachexia viroids, citrus variegation and citrus leaf rugose ilarviruses, citrus tatterleaf, citrus yellow mosaic badnavirus, stubborn and citrus variegated chlorosis are all examples of this phenomenon. Several of the viruses that affect passion fruit have primary hosts other than this crop. On banana, *Ralstonia solanacearum* biovar 1, race 2 (cause of Moko disease) evolved on *Heliconia* spp., *Cucumber mosaic virus* evolved on cucurbits, and the *Banana streak virus* probably arose on sugarcane.

The existence of alternative hosts for these pathogens has important implications for the management of these diseases. If the number of primary infections is low and there is no secondary spread by vectors, the basic risk is only from secondary spread via propagation. However, if secondary spread occurs within the crop, the potential for damage is greater. Alternative hosts create additional problems for establishing effective quarantines since they must be identified and regulated.

Improved Detection Methods

Extensive progress in detection methods has occurred as protein and nucleic acid compo-

Table 20.1. Pathogens of tropical fruit crops that have limited distributions and pose major threats to the safe movement of germplasm.

Hosts	Pathogens/diseases
Annonaceous crops	None
Avocado	*Avocado sunblotch viroid*, black streak (cause unknown), Duke 6 stem pitting (cause unknown)[a]
Banana	*Banana bract mosaic virus*, *Banana bunchy top virus*, *Banana streak virus*, blood disease (unclassified bacterium related to *Ralstonia solanacearum*), *Cucumber mosaic virus* (severe strains), *Fusarium oxysporum* f. sp. *cubense* (tropical race 4), *Guignardia musae*, *Mycosphaerella fijiensis*, *Mycosphaerella eumusae*, *Phyllachora musicola*, *Pratylenchus goodeyi*, *Ralstonia solanacearum* (biovar 1, race 2)
Breadfruit	Pingelap disease (cause unknown)
Carambola	None
Citrus	Canker, black spot, Phaeoramularia leaf and fruit spot[b], and diverse viruses, viroids and phytoplasmas
Coconut	*Bursaphelenchus cocophilus*, *cadang-cadang viroid*, *Coconut foliar decay virus*, hart rot (uniflagellate protozoan), phytoplasmas (diverse), *tinangaja viroid*
Date	*Fusarium oxysporum* f. sp. *albedinis*, phytoplasmas (lethal yellowing, slow decline and white tip dieback)
Durian	None
Fig	*Fig mosaic virus*
Guava	Guava wilt (*Penicillium vermoesini*?), *Puccinia psidii*
Jackfruit	None
Kiwifruit	*Pseudomonas savastanoi*
Lychee	*Peronophythora litchii*
Longan	Witches' broom (phytoplasma?)
Mango	*Denticularia mangiferae*, *Fusarium mangiferae*, *Xanthomonas* sp. pv. *mangiferaeindicae*
Mangosteen	*Gliocephalotrichum bulbilium*, *Marasmiellus equicrinus*, *Marasmiellus scandens*
Papaya	Bunchy top (rickettsial bacterium), *Papaya ringspot virus*, phytoplasmas (diverse)
Passion fruit	*Fusarium oxysporum* f. sp. *Passiflorae*, *Passionfruit nucleorhabdovirus*, *Passionfruit ringspot virus*, *Passionfruit yellow mosaic virus*, *Passionfruit woodiness virus*, *Xanthomonas campestris* pv. *passiflorae*
Pineapple	*Fusarium guttiforme*
Rambutan	*Gliocephalotrichum bulbilium*, *Gliocephalotrichum microchlamydosporum*, *Greeneria* sp.

[a]This disease has not been detected after it was originally described and may have been eradicated.
[b]See Frison and Taher (1991).

nents of pathogens have been characterized (Matthews, 1993). Application of molecular technologies to the development of sensitive and reliable assays for plant pathogen detection have helped elucidate the aetiology of previously enigmatic diseases, such as cadang-cadang and New Hebrides disease of coconut. Enzyme-linked immunosorbent assay (ELISA) and polymerase chain reaction (PCR) have enabled major advances (Jagoueix *et al.*, 1996). However, many problems remain, and the absence of totally effective detection methods still hampers control efforts for numerous diseases. A number of virus-like pathogens are still poorly characterized, and molecular and serological measures for detecting these pathogens will not be developed until they are better understood. Several important prokaryotic pathogens remain uncultured, and the sketchy knowledge of their properties has been deduced only with difficulty. Irregular distribution and low or variable titres that occur with seasonal conditions, growth stage and cultivar present additional problems. The aetiology of other diseases is not known, making reliable detection either impossible (e.g. Pingelap disease of breadfruit) or dependent upon indirect measures (e.g. citrus decline).

Pathogen detection often entails two divergent needs. In some cases, it is important to detect all strains of a given pathogen regardless of pathogenicity and all pathogens regardless of identity. This is often paramount in certification programmes where the absence of pathogens, not their identity, is of primary interest. Thus, the specificity of serological tests and most PCR-based assays may not allow a broad range of isolates to be detected. In this case, it will be necessary to broaden the scope of existing probes by using mixtures of antibodies that recognize all strains of a pathogen or degenerate primers that detect highly conserved areas of the pathogen genome. PCR-based detection is especially relevant to viroids, even though the small size of the viroid genome hampers efforts to develop generic probes. The use of biological indicators and generic molecular assays such as detection of double-stranded RNAs and sequential polyacrylamide gel electrophoresis (sPAGE) for viroids will continue to have applications for broad scale assays.

In other cases, there is a need to discriminate between strains of a pathogen that are closely related, but differ in pathogenicity. Several 'selective' probes have been developed (Permar *et al.*, 1990; Gillings *et al.*, 1996), but most of these are based on apparent correlations between molecular and biological properties that have been derived empirically, and not from exact knowledge of the nucleotide sequence that actually affects symptom expression. Development of such information will require extensive research on pathogen genes that are responsible for symptom expression.

The cost and expertise required for sophisticated detection methods must be reconciled with the practical requirements for tests that are quick, inexpensive and accurate. While nucleotide-based tests such as PCR offer the best hope for specificity, serological tests such as ELISA are simpler and less expensive. Immunocapture PCR, whereby the pathogen is trapped selectively for amplification without nucleic acid extraction, is one approach. The effort required to transfer detection technology from a research to a practical level may be justified only for large-scale applications, such as budwood schemes or surveys for regulatory purposes.

International Trade, Plant Quarantine, Exclusion, Eradication and New Threats

International trade

Trade liberalization is needed to ensure a balanced and equitable development of exports and earnings from fresh and processed tropical fruit. The Uruguay Round of Negotiations for the General Agreement on Tariffs and Trade (GATT) presented opportunities for all countries to benefit from greater access to world markets by facilitating competitive and fair trade. The Agreement on Agriculture aims to improve market access, reduce domestic support and export subsidies, remove trade barriers, and facilitate a fair and market-oriented agricultural trading system. GATT included an agreement on Sanitary and Phytosanitary matters (the SPS Agreement) to reduce the use of plant quarantine as a trade barrier (GATT, 1994).

Plant quarantine

In 1995, the World Trade Organization (WTO) replaced GATT and, at the same time, the SPS Agreement took effect. GATT nominated the International Plant Protection Convention (IPPC) as the international standard on plant quarantine. In accepting the principle of managed risk, i.e. that some risk always exists, countries can no longer prohibit importations through a nil-risk policy. Pest risk analysis (PRA) is now a cornerstone of decision making in plant quarantine. Commodity import restrictions imposed on countries of equivalent phytosanitary risk cannot be more stringent than actions that are taken at the national level (Anonymous, 1995, 1996).

The current IPPC definition of a quarantine pest is 'a pest of potential economic importance to the area endangered thereby and not yet present, or present and not widely distributed and being officially controlled'. Difficulties with this definition are the absence of reliable information on the potential economic effects of a pest and of pest records in some countries.

Knowledge of variability in a pathogen is required for the development of rational regulations on the movement of fruit and plant material. There are numerous strains of CTV, but those that pit grapefruit, orange and mandarin are not uniformly distributed worldwide. Furthermore, severe strains can be 'hidden' among more benign strains and only revealed by aphid transmissions (Moreno *et al.*, 1993; Broadbent *et al.*, 1996). Clearly, disastrous economic consequences could result if they were introduced into new areas. Whether other variants present an important risk is not known. For example, there are three scab pathogens of citrus that differ in host range but not in morphology (Timmer *et al.*, 1996). The problem lies in whether the detected variation represents new and significant quarantine threats (Phillips and Chandrashekar, 1992). Without such information, it may not be possible to prevent potentially hazardous introductions from occurring. Recent, failed efforts by the Australian banana industry to prohibit imports from the Philippines due to the unverified presence in that country of severe strains of *Guignardia musae* are a case in point (D.R. Jones, personal communication, 2002).

IPPC is developing international standards for phytosanitary measures. For example, the standard for citrus canker describes methodologies for general surveillance, and detection, monitoring and delimiting surveys. This will enable countries with canker to delineate and designate disease-free areas, such as a citrus-growing area in Argentina. 'The Principles of Plant Quarantine as Related to International Trade' endorse the concept of 'pest-free areas', i.e. an area in which a specific pest does not occur as demonstrated by scientific evidence and in which, where appropriate, this condition is being officially maintained.

Exclusion

Efforts to exclude a pathogen or pathogen strain from a given production area are usually more cost effective than those that are required for eradication once an introduction has occurred. A hallmark of exclusion efforts is the establishment of pathogen-free germplasm collections. The extent to which such collections are available and used varies considerably for the crops discussed in this volume.

International exchange of germplasm

There are at least two international *in vitro* collections of banana germplasm. Accessions in both have been indexed for the important virus pathogens of this crop and, because they are derived from meristem cultures, they are also free of bacterial, fungal and nematode pathogens. The International Transit Centre at the Katholieke Universiteit in Lueven, Belgium (Laboratory of Tropical Crop Improvement, Kardinal Mercierlaan 92, 3001 Heverlee) is sponsored by the International Network for the Improvement of Banana and Plantain (INIBAP) and contains diverse improved and primitive genotypes, as well as related taxa, whereas the International Institute of Tropical Agriculture (IITA, c/o L.W. Lambourn & Company, Carolyn House, 26 Dingwall Road, Croydon CR9 3EE, UK) contains mainly plantain accessions. Given the serious threat that is posed by viruses and other pathogens of banana that are disseminated in traditional seed pieces (see Chapter 4 and Table 20.1), it is imperative that virus-indexed *in vitro* materials be used when new *Musa* germplasm is introduced into a given area (Diekman and Putter, 1996).

Unfortunately, international repositories of pathogen-free germplasm of crops other than banana generally are not maintained. Thus, caution is required when importing germplasm of other crops discussed in this volume. Guidelines have been published for the dissemination of citrus (Frison and Taher, 1991) and coconut (Frison *et al.*, 1993), and these enable the safe international exchange of germplasm of these crops. For other crops, producers should know whether the germplasm comes from an area that is affected by either an important pathogen or an insect pest. Carambola, durian, jackfruit and the annonaceous crops have no pathogens that pose major threats to the international exchange of germplasm (Table 20.1), but at least one of these, durian, is affected by several threatening insect pests (Subhadrabandhu and Ketsa, 2001). Obviously, when threatening pathogens are an issue, planting materials should not be imported from countries in which they are found unless they come from certified pathogen-free areas or nurseries. In all cases, budwood and other traditional propagation materials should be cleaned before shipment and be free of extraneous soil and leaf litter.

Clean nursery stock

The importance of using pathogen-free propagation materials when establishing new plantings cannot be overstated. The productivity and longevity of a planting are increased greatly when it is started with clean stock. Pathogen-free materials have been generated with a variety of techniques including meristem culture, *in vitro* shoot-tip grafting, thermotherapy and improved indexing techniques.

The establishment of local accreditation schemes for healthy budwood and nursery plants has been successful for citrus (Broadbent and Timmer, 1997). Government agencies usually assume responsibility for horticultural evaluations of new cultivars, virus elimination, pathogen indexing and even the maintenance of virus-free clones. Industry usually takes responsibility for commercial increase of budwood and seed. The best schemes include components where research is conducted on improvement of detection techniques.

Budwood schemes often need to change their mode of operation to meet changing economic or environmental conditions and disease status. Maintenance of foundation trees under insect-proof screening has become more frequent as threats from vector-based pathogens have increased. In Australia, the Australian Citrus Improvement Program has implemented the importation of new cultivars and their early release to nurserymen. It has multiplied rapidly new cultivars that are in short supply.

Some citrus production programmes have included mild strain cross-protection against CTV as a part of the improvement strategy. Pre-inoculation of virus-free scions with mild protective isolates of CTV has given reasonable protection to grapefruit in Australia (Broadbent *et al.*, 1991) and South Africa (Van Vuuren *et al.*, 1993), and to Pera oranges in Brazil (Costa and Muller, 1980). Pre-inoculated foundation trees are maintained in insect-proof screenhouses.

Eradication

Eradication campaigns to eliminate introduced pathogens from production regions are always expensive and have met with mixed success. Diseases of banana and citrus provide the best examples.

The rapid growth of Australia's banana industry and its use of suckers that were affected by banana bunchy top almost ruined the trade by 1925; production in some areas was reduced by 90–95% (Magee, 1927). Only with a thorough quarantine and eradication programme and its strict enforcement was the industry able to rebound and become as viable as it is today (see Fig. 4.30). In contrast, where eradication efforts have not been enforced as strictly, bunchy top has had a devastating and ongoing impact on production (Ferreira *et al.*, 1989; Khalid and Soomro, 1993; Ploetz, 1994). How successful Australia's banana quarantine will be in eliminating two other diseases remains to be seen. Black Sigatoka has been found periodically in the Torres Straight and continues to reappear after eradication due probably to movement of the pathogen from neighbouring Papua New Guinea. In the late 1990s, tropical race 4 of Panama disease was found in the Northern Territory. Areas within the affected plantations have been taken out of production, but it is too early to determine whether the race has been eliminated or will reappear.

Citrus canker has received the most widespread and intense eradication efforts of any disease discussed in this volume. It has been eradicated in Australia, New Zealand and South Africa, and active eradication or suppression programmes are underway in Brazil, Florida and Uruguay. Some programmes, like one in Argentina, failed, and canker eventually became endemic. However, even when these campaigns were not successful, they provided a period of time during which control measures were not needed. Thus, markets for fruit often remained open and control costs for growers in disease-free areas were reduced.

When the threat posed by CTV was recognized in Israel in the 1950s, affected trees were identified and destroyed. Nevertheless, the disease was found to be widespread by 1986 and eradication was discontinued. While eradication was not totally effective, the disease was suppressed, and citrus production on sour orange rootstock was extended by 5–10 years (Bar-Joseph *et al.*, 1989).

In areas where overall CTV incidence is low, management by suppression of the virus with the long-term aim of eradication may be a viable option if infected trees are removed faster than new infections occur. Management of CTV by eradication has been practised by the Citrus Tristeza Eradication Agency (CCTEA) in three counties in the Central Valley of California.

The success of eradication campaigns is dependent on the rapid establishment and delimitation of the infested area. That, in turn, is dependent on the availability of speedy, accurate detection techniques to confirm visual observations. Failure of campaigns is usually due to the lack of information on a disease's distribution.

Eradication campaigns will successfully combat new disease outbreaks only if they are detected early. Countries that are interested in excluding pests and diseases will need to maintain intense survey and detection programmes for the most important problems.

Conclusion

Significant challenges lie ahead for tropical fruit producers. In industrialized nations where banana and citrus were once the only tropical fruits that were available, a wide array of tropical fruits has now entered the marketplace. Competition in national and global markets and rising expectations from increasingly sophisticated consumers will become ever more important as producers attempt to meet demands for high quality and economical fruit. Clearly, producers must embrace new strategies to managing diseases whenever they become available. Just as these diseases are dynamic and ever changing, so too must a producer's approach be towards their management.

References

Ahloowalia, B.S. and Maluszynski, M. (2001) Induced mutations – a new paradigm in plant breeding. *Euphytica* 118, 167–173.

Anonymous (1995) Principles of plant quarantine as related to international trade. *International Standards for Phytosanitary Measures Publication Number* 1. Food and Agricultural Organization, Rome.

Anonymous (1996) Requirements for the establishment of pest free areas. *International Standards for Phytosanitary Measures Publication Number 4*. Food and Agricultural Organization, Rome.

Anonymous (2001) Genome busting – launching the global *Musa* genomics consortium. *Info*Musa 10, I.

Bar-Joseph, M., Marcus, R. and Lee, R.F. (1989) The continuous challenge of citrus tristeza virus control. *Annual Review of Phytopathology* 27, 291–316.

Bergh, B.O. (1976) Avocado. In: Simmonds, N.W. (ed.) *Evolution of Crop Plants*. Longman, London, pp. 148–151.

Bhagwat, B. and Duncan, E.J. (1998) Mutation breeding of Highgate (*Musa acuminata*, AAA) for tolerance to *Fusarium oxysporum* f. sp. *cubense* using gamma irradiation. *Euphytica* 101, 143–150.

Britz, H., Steenkamp, E.T., Coutinho, T.A., Wingfield, B.D., Marasas, W.F.O. and Wingfield, M.J. (2002) Two new species of *Fusarium* section *Liseola* associated with mango malformation. *Mycologia* 94, 722–730.

Broadbent, P. and Timmer, L.W. (1997) Challenges of modern citriculture: citrus disease. *Proceedings of the International Society of Citriculture* 1, 196–198.

Broadbent, P., Bevington, K.B. and Coote, B.G. (1991) Control of stem-pitting of grapefruit in Australia by mild strain cross protection. In: Brlansky, R.H., Lee, R.F. and Timmer, L.W. (eds) *Proceedings of the 11th Conference of the International Organization of Citrus Virologists*, Riverside, California, pp. 64–70.

Broadbent, P., Brlansky, R.H. and Indsto, J. (1996) Biological characterization of Australian isolates of citrus tristeza virus and separation of sub-isolates by single aphid transmissions. *Plant Disease* 80, 329–333.

Bureau, E. (1990) Adoption of a forecasting system to control black Sigatoka (*Mycosphaerella fijiensis* Morelet) in plantain plantations in Panama. *Fruits* 45, 329–338.

Carreel, F. (1994) Etude de la diversté génétique des bananiers (genre *Musa*) à l'aide des marqueurs RFLP. PhD thesis, Institut National Agronomique Paris-Grignon, France.

Chang, C.J., Garnier, M., Zreik, L., Rossetti, V. and Bové, J.M. (1993) Culture and serological detection of the xylem-limited bacterium causing citrus variegated chlorosis and its identification as a strain of *Xylella fastidiosa*. *Current Microbiology* 27(3), 137–142.

Chin, K.M., Wirz, M. and Laird, D. (2001) Sensitivity of *Mycosphaerella fijiensis* from banana to trifloxystrobin. *Plant Disease* 85, 1264–1270.

Chuang, T.-Y. and Jeger, M.J. (1987) Predicting the rate of development of black Sigatoka (*Mycosphaerella fijiensis* var. *difformis*) disease in southern Taiwan. *Phytopathology* 77, 1542–1547.

Colburn, F.D. (1997) Shrimp or bananas. *Journal of Business Research* 38, 97–103.

Costa, A.S. and Muller, G.W. (1980) Tristeza controlled by cross protection. A United States–Brazil cooperative success. *Plant Disease Reporter* 64, 538–541.

Cronjé, P., Dabek, A.J., Jones, P. and Tymon, A.M. (2000a) First report of a phytoplasma associated with a disease of date palms in North Africa. *New Disease Reports*. http://www.bspp.org.uk/ndr/2000/2000-4.htm

Cronjé, P., Dabek, A.J., Jones, P. and Tymon, A.M. (2000b) Slow decline: a new disease of mature date palms in North Africa associated with a phytoplasma. *New Disease Reports*. http://www.bspp.org.uk/ndr/2000/2000-7.htm

Crous, P.W. and Mourichon, X. (2002) *Mycosphaerella eumusae* and its anamorph *Pseudocercospora eumusae* spp. nov.: causal agent of eumusae leaf spot disease of banana. *Sydowia* 54, 35–43.

Cruz-Hernández, A. and Litz, R.E. (1997) Transformation of mango somatic embryos. In: Lavi, U., Degami, C., Gazit, S., Lahar, E., Pess, E., Prusky, D., Tower, E. and Wysoki, M. (eds) *Proceedings of the 5th International Mango Symposium*, Vol. 1. ISHS, Leuven, Belgium, pp. 292–298.

Dodd, J.C., Estrada, A.B., Matcham, J., Jeffries, P. and Jeger, M.J. (1991) The effect of climatic factors on *Colletotrichum gloeosporioides*, causal agent of mango anthracnose, in the Philippines. *Plant Pathology* 40, 568–575.

Diekman, M. and Putter, C.A.J. (1996) *FAO/IPGRI Technical Guidelines for the Safe Movement of Germplasm. No. 15.* Musa, 2nd edn. Food and Agricultural Organization of the United Nations, International Plant Genetics Resources Institute, Rome.

Fauré, S., Noyer, J.-L., Horry, J.-P., Bakry, F., Lanaud, C. and Gonzalez de Leon, D. (1993) A molecular marker-based linkage map of diploid bananas (*Musa acuminata*). *Theoretical and Applied Genetics* 87, 517–526.

Ferreira, S.A., Trujillo, E.E. and Ogata, D.Y. (1989) *Bunchy top disease of bananas*. Information leaflet prepared by the Hawaii Cooperative Extension Service, Hawaii Institute of Tropical Agriculture and Human Resources, University of Hawaii at Manoa. Commodity Fact Sheet BAN-4(A), FRUIT.

Ferreira, S.A., Pitz, K.Y., Manshardt, R., Zee, F., Fitch, M. and Gonsalves, D. (2002) Virus coat protein transgenic papaya provides practical control of *Papaya ringspot virus* in Hawaii. *Plant Disease* 86, 101–105.

Fitch, M.M.M., Manshardt, R.M., Gonsalves, D., Slightom, J.L. and Sanford, J.C. (1992) Virus resistant papaya derived from tissues bombarded with coat protein of *Papaya ringspot virus*. *Bio-Technology* 10, 1466–1472.

Fitzell, R.D., Peak, C.M. and Darnell, R.E. (1984) A model for estimating infection levels of anthracnose disease of mango. *Annals of Applied Biology* 104, 451–458.

Fouré, E. (1990) La lutte integrée contre le cercosporiose noire des bananiers au Cameroun. L'avertissement biologique et son évolution de 1985 à 1988. In: Fullerton, R.A. and Stover, R.H. (eds) *Sigatoka Leaf Spot Diseases of Bananas*, Proceedings of an International Workshop held in San José, Costa Rica, March 28–April 1, 1989. INIBAP, Montpellier, France, pp. 124–134.

Frison, E.A. and Taher, M.M. (1991) *FAO/IPGRI Technical Guidelines for the Safe Movement of Citrus Germplasm*. Food and Agricultural Organization of the United Nations, International Board for Plant Genetic Resources, Rome.

Frison, E.A., Putter, C.A.J. and Diekman, M. (1993) *FAO/IPGRI Technical Guidelines for the Safe Movement of Coconut Germplasm*. Food and Agricultural Organization of the United Nations, International Board for Plant Genetic Resources, Rome.

Ganry, J. and Laville, E. (1983) Les cercosporiosis du bananier et leurs traitments. Evolution des méõhodes de traitment. 1. Traitments fungicides. 2. Avertissement. *Fruits* 3, 3–20.

Garnier, M., Zreik, L. and Bové, J.M. (1991) Witches' broom, a lethal mycoplasmal disease of lime trees in the Sultanate of Oman and the United Arab Emirates. *Plant Disease* 75, 546–551.

General Agreement on Tariff and Trade (GATT) (1994) *Results of the Uruguay Round of Multilateral Trade Negotiations*, the legal texts. Agreement on the application of sanitary and phytosanitary measures. General Agreement on Tariff and Trade Secretariat, Geneva, pp. 69–84.

Gillings, M., Broadbent, P. and Indsto, J. (1996) Restriction analysis of amplified CTV coat protein cDNA is sensitive and rapid method for monitoring and controlling CTV infections. In: da Graça, J.V., Moreno, P. and Yokomi, R.K. (eds) *Proceedings of the 13th Conference of the International Organization of Citrus Virologists*, Riverside, California, pp. 25–37.

Gmitter, F.G. Jr, Xiao, S.Y., Huang, S., Hu, X.L., Garnsey, S.M. and Deng, Z. (1996) A localized linkage map of the citrus tristeza virus resistance gene region. *Theoretical and Applied Genetics* 92, 688–695.

Gonsalves, D. (1998) Control of *Papaya ringspot virus* in papaya: a case study. *Annual Review of Phytopathology* 36, 415–437.

Gonzáles de Leon, D. and Fauré, S. (1993) Genetic mapping of the banana diploid genome. In: *Biotechnological Applications for Banana and Plantain Improvement*. INIBAP, Montpellier, France, pp. 29–46.

Grosser, J.W. and Gmitter, F.G. Jr (1990) Protoplast fusion and *Citrus* improvement. *Plant Breeding Reviews* 8, 339–374.

Grosser, J.W. and Gmitter, F.G. Jr (1996) New cultivars in the citrus improvement pipeline. *Proceedings of the International Society of Citriculture*, pp. 31–34.

Guo, W.W. and Deng, X.X. (2001) Wide somatic hybrids of *Citrus* with its related genera and their potential in genetic improvement. *Euphytica* 118, 175–183.

Hwang, S.C. and Ko, W.H. (1987) Somaclonal variation of bananas and screening for resistance to Fusarium wilt. In: Persley, G.J. and DeLanghe, E.A. (eds) *Proceedings of an International Workshop in Cairns, Australia, October 13–17, 1986*. pp. 151–156.

Hwang, S.C., Ko, W.H. and Chao, C.P. (1994) GCTCV-215-1: a promising Cavendish clone resistant to race 4 of *Fusarium oxysporum* f. sp. *cubense*. *Plant Protection Bulletin (Taiwan)* 36, 281–291.

Jagoueix, S., Bové, J.M. and Garnier, M. (1996) PCR detection of the two liberobacter species associated with greening disease of citrus. *Molecular Cellular Probes* 10, 43–50.

Jones, D.R. (ed.) (2000) *Diseases of Banana, Abacá and Enset*. CAB International, Wallingford, UK.

Kaemmer, D., Fisher, D., Jarrett, R.L., Baurens, F.-C., Grapin, A., Dambier, D., Noyer, J.-L., Lanaud, C., Kahl, G. and Lagoda, P.J.L. (1997) Molecular breeding in the genus *Musa*: a strong case for STMS marker technology. *Euphytica* 96, 49–63.

Kashkush, K., Jinggui, F., Tomer, E., Hillel, J. and Lavi, U. (2001) Cultivar identification and genetic map of mango (*Mangifera indica*). *Euphytica* 122, 129–136.

Khalid, S. and Soomro, M.H. (1993) Banana bunchy top disease in Pakistan. *Plant Pathology* 42, 923–926.

Kijas, J.M.H., Thomas, M.R., Fowler, J.C.S. and Roose, M.L. (1997) Integration of trinucleotide microsatellite into a linkage map of *Citrus*. *Theoretical and Applied Genetics* 94, 701–706.

Kohmoto, K., Scheffer, R.P. and Whiteside, J.O. (1979) Host selective toxins from *Alternaria citri*. *Phytopathology* 69, 667–671.

Korkmaz, S., Kersting, U., Ertugrul, B. and Cinar, A. (1996) Transmission and epidemiology of citrus chlorotic dwarf (CCD) disease in the eastern Mediterranean region of Turkey. *Journal of Turkish Phytopathology* 25, 71–76.

Lescot, T., Simonet, H., Fages, O. and Escalant, J.V. (1998) Developing a biometeorological forecasting system to control leaf spot disease of plantains in Costa Rica. *Fruits* 53, 3–16.

Lines, R.E., Persley, D.M., Dale, J.L., Drew, R.A. and Bateson, M.F. (2002) Genetically engineered immunity to papaya ringspot virus in Australian papaya cultivars. *Molecular Breeding* 10, 119–129.

Magee, C.J.P. (1927) *Bulletin No. 30. Investigation on the Bunchy Top Disease of the Banana*. Council for Scientific and Industrial Research, Melbourne.

Marín, D.H. and Romero, R.A. (1992) *El Combate de la Sigatoka Negra*. Boletín No. 4. Departamento de Investagaciones, CORBANA, San José, Costa Rica.

Mathews, H., Litz, R.E., Wilde, H.D., Merkle, S.A. and Wetztein, H.W. (1992) Stable integration and expression of β-glucuronidase and NPT-II genes in mango somatic embryos. In vitro *Cellular and Developmental Biology* 28, 172–178.

Matthews, R.E.F. (1993) *Diagnosis of Plant Virus Diseases*. CRC Press, Boca Raton, Florida.

McMillan, R.T., Moss, M.A., Bowling, C.R. and Stempel, L. (1989) Variation in tolerance to benomyl among *Colletotrichum gloeosporioides* isolates from mango. *Proceedings of the 3rd International Mango Symposium*, CSIRO Division of Horticulture, Darwin, Australia, p. 53 (abstract).

Moreno, P., Guerri, J., Ballaster-Olinos, J.F., Fuertes-Polo, C., Albiach, R. and Martinez, M. (1993) Variations in pathogenicity and double-stranded RNA (dsRNA) patterns of citrus tristeza virus isolate induced by host passage. In: Moreno, P., da Graça, J.V. and Timmer, L.W. (eds) *Proceedings of the 12th Conference of the International Organization of Citrus Virologists*. Riverside, California, pp. 8–15.

Nicolosi, E., Deng, Z.N., Gentile, A., La Malfa, S., Continella, G. and Tribulato, E. (2000) Citrus phylogeny and genetic origin of important species as investigated by molecular markers. *Theoretical and Applied Genetics* 100, 1155–1166.

Nwauzoma, A.B., Tenkouano, A., Crouch, J.H., Pillay, M., Vuylsteke, D. and Daniel Kalio, L.A. (2002) Yield and disease of plantain (*Musa* spp., AAB group) somaclones in Nigeria. *Euphytica* 123, 323–331.

Pappu, H.R., Pappu, S.S., Manjunah, K.L., Lee, R.F. and Niblett, C.L. (1993) Molecular characterization of a structural epitope that is largely conserved among severe isolates of a plant virus. *Proceedings of the National Academy of Sciences USA* 90, 3641–3644.

Peres, N.A.R., Kim, S., Beck, H.W., Souza, N.L. and Timmer, L.W. (2002a) A fungicide application decision (FAD) support system for postbloom fruit drop of citrus (PDF). *Plant Health Progress* doi:10.1094/PHP-2002-0731-01-RV.

Peres, N.A.R., Souza, N.L., Zitko, S.E. and Timmer, L.W. (2002b) Activity of benomyl for control of postbloom fruit drop of citrus caused by *Colletotrichum acutatum*. *Plant Disease* 86, 620–624.

Permar, T.A., Garnsey, S.M., Gumpf, D.J. and Lee, R.F. (1990) A monoclonal antibody that discriminates strains of citrus tristeza virus. *Phytopathology* 80, 224–228.

Phillips, D. and Chandrashekar, M. (1992) Scientific issues arising from the use of the International Plant Convention definition of a quarantine pest. *Bulletin European and Mediterranean Plant Protection Organization* 22, 597–606.

Ploetz, R.C. (1994) Banana production in Egypt. *Info*Musa 3, 16–18.

Romero, R.A. and Sutton, T.B. (1997) Sensitivity of *Mycosphaerella fijiensis*, causal agent of black Sigatoka of banana, to propiconazole. *Phytopathology* 87, 96–100.

Roose, M.L. (1996) The impact of biotechnology on citriculture. *Proceedings of the International Society of Citriculture*, pp. 41–45.

Saas, R. (2000) Agricultural 'killing fields': the poisoning of Costa Rican banana workers. *International Journal of Health Services: Planning, Administration, Evaluation* 30, 491–514.

Schnell, R.J., Ronning, C.M. and Knight, R.L. Jr (1995) Identification of cultivars and validation of genetic relationships in *Mangifera indica* L. using RAPD markers. *Theoretical and Applied Genetics* 90, 269–274.

Schnell, R.J., Kuhn, D.N., Olano, C.T. and Quintanilla, W.E. (2001) Sequence diversity among avocado sunblotch viroids isolated from single avocado trees. *Phytoparasitica* 29, 451–460.

Sharon, D., Cregan, P.B., Mhameed, S., Kusharsja, M., Hillele, E., Lahav, E. and Lavi, U. (1997) An integrated genetic linkage map of avocado. *Theoretical and Applied Genetics* 95, 911–921.

Singh, L.B. (1976) Mango. In: Simmonds, N.W. (ed.) *Evolution of Crop Plants*. Longman, pp. 7–9.

Smith, A. (2000) The modern banana plantation – still a 'green prison'. *Pesticide News* 48, 9.

Stover, R.H. (1990) Sigatoka leaf spots: thirty years of changing control strategies: 1959–1989. In: Fullerton, R.A. and Stover, R.H. (eds) *Sigatoka Leaf Spot Diseases of Banana*. Proceedings of an International Workshop held in San José, Costa Rica, March 28–April 1, 1989. INIBAP, Montpellier, France, pp. 66–74.

Subhadrabandhu, S. and Ketsa, S. (2001) *Durian: King of Tropical Fruit*. CAB International, Wallingford, UK.

Timmer, L.W. and Zitko, S.E. (1996) Evaluation of a model for prediction of postbloom fruit drop of citrus. *Plant Disease* 80, 380–383.

Timmer, L.W., Priest, M., Broadbent, P. and Tan, M.K. (1996) Morphological and pathological characterization of *Elsinoe* spp. causing citrus scab diseases. *Phytopathology* 86, 1032–1038.

Timmer, L.W., Graham, J.H. and Zitko, S.E. (1998) Metalaxyl-resistant isolates of *Phytophthora nicotianae*: occurrence, sensitivity, and competitive parasitic ability on citrus. *Plant Disease* 82, 254–261.

Timmer, L.W., Darhower, H.M., Zitko, S.E., Peevel, T.L., Ibáñez, A.M. and Bushong, P.M. (2000) Environmental factors affecting severity of Alternaria brown spot of citrus and their potential use in timing fungicide applications. *Plant Disease* 84, 638–643.

Van Vuuren, S.P., Collins, R.P. and da Graça, J.V. (1993) Evaluation of citrus tristeza virus isolates for cross protection of grapefruit in South Africa. *Plant Disease* 77, 24–28.

Vidal, M.C. and Garcia, E. (2000) Analysis of a *Musa* sp. somaclonal variant resistant to yellow Sigatoka. *Plant Molecular Biology Reporter* 18, 23–31.

Whiteside, J.O. (1976) A newly recorded Alternaria-induced brown spot disease on Dancy tangerines in Florida. *Plant Disease Reporter* 60, 326–329.

Whiteside, J.O. (1980a) Tolerance of *Mycosphaerella citri* to benomyl in Florida citrus groves. *Plant Disease Reporter* 64, 300–302.

Whiteside, J.O. (1980b) Detection of benomyl-tolerant strains of *Elsinoe fawcettii* in Florida citrus groves and nurseries. *Plant Disease Reporter* 64, 871–872.

Wielemaker, F. (1990) Practical notes on black Sigatoka control. In: Fullerton, R.A. and Stover, R.H. (eds) *Sigatoka Leaf Spot Diseases of Bananas*, Proceedings of an International Workshop held in San José, Costa Rica, 28 March–1 April, 1989. INIBAP, Montpellier, France, pp. 107–114.

Willingham, S.L., Pegg, K.G., Cooke, A.W., Coates, L.M., Langdon, P.W.B. and Dean, J.R. (2001) Rootstock influences postharvest anthracnose development in 'Hass' avocado. *Australian Journal of Agricultural Research* 52, 1017–1022.

Yeh, S.D., Gonsalves, D., Wang, H.L., Namba, R. and Chiu, R.J. (1988) Control of *Papaya ringspot virus* by cross protection. *Plant Disease* 72, 375–380.

Ying, Z., Xia, Y. and Davis, M.J. (1999) New method for obtaining transgenic papaya plants by *Agrobacterium*-mediated transformation of somatic embryos. *Proceedings of the Florida Horticultural Society* 112, 201–205.

Appendices

Appendix I Microbe Taxa, Authorities and Synonyms

Binomial	Synonyms
Eukaryotes	
Algae	
Cephaleuros virescens Kunze	*Cephaleuros parasiticus* Karst *Cephaleuros mycoidea* Karst
Trentepohlia aurea (L.) Martius	
Trentepohlia arborueum (C. Ag.) Hariot	
Trentepohlia monile De Wildeman	
Chromistans	
Peronophythora litchii Chen ex Ko, Chang, Su, Chen and Leu	
Phytophthora arecae (Coleman) Pethybridge	*Phytophthora omnivora* var. *arecae* Coleman
Phytophthora boehmeriae Sawada	
Phytophthora botryosa Chee	
Phytophthora cactorum (Lebert and Cohn) Schröter	
Phytophthora capsici Leonian	
Phytophthora cinnamomi Rands	
Phytophthora citricola Sawada	
Phytophthora citrophthora (R.E. Smith and E.H. Smith) Leonian	
Phytophthora cryptogea Pethybridge and Lafferty	
Phytophthora drechsleri Tucker	
Phytophthora gonapodyides (Petersen) Buisman	
Phytophthora heveae Thompson	
Phytophthora katsurae Ko and Chang	*Phytophthora castaneae* Katsura & Uchida
Phytophthora lateralis Tucker and Milbrath	
Phytophthora megasperma Drechsler	
Phytophthora nicotianae Breda de Hahn	*Phytophthora nicotianae* var. *parasitica* (Dastur) G.M. Waterhouse *Phytophthora parasitica* Dastur

Binomial	Synonyms
	Phytophthora parasitica Dastur var. *macrospora* Ashby
Phytophthora palmivora (E.J. Butler) E.J. Butler	*Phytophthora carica* Hara *Phytophthora faberi* Maublanc *Phytophthora fici* Hori *Phytophthora omnivora* de Bary
Phytophthora tropicalis Aragaki and J.Y. Uchida	*Phytophthora capsici* Leonian
Pythium aphanidermatum (Edson) Fitzp.	
Pythium arrhenomanes Drechsler	
Pythium splendens Braun	
Pythium ultimum Trow var. *ultimum*	*Pythium debaryanum* sensu de Bary *Pythium haplomitrii* Lilienfeld
Pythium vexans de Bary	*Pythium complectans* Braun
Fungi	
Acremonium diospyri (Crandell) W. Gams	
Acremonium stromaticum W. Gams and R.H. Stover	
Acrodontium simplex (G. Mangenor) de Hoog	
Albonectria rigidiuscula (Berk. and Broome) Rossman and Samuels anamorph: *Fusarium decemcellulare* C. Brick	*Nectria rigidiuscula* Berk. and Broome
Alternaria aliena E.G. Simmons	
Alternaria alternata (Fr.) Keissler	*Alternaria fasciculata* (Cooke and Ellis) L. Jones and Grout *Alternaria tenuis* Nees *Macrosporium fasciculatum* Cooke and Ellis
Alternaria aragakii E.G. Simmons	
Alternaria bannaensis sp. nov. Chen and Zhang	
Alternaria citri Ell. and Pierce	
Alternaria guangxiensis sp. nov. Chen and Zhang	
Alternaria hawaiiensis E.G. Simmons	
Alternaria macrospora Zimm.	
Alternaria passiflorae J.H. Simmonds	
Alternaria tenuissima (Kunze ex Pers.) Wiltshire	
Alternaria tomato (Cooke) Jones	
Alternaria tropica E.G. Simmons	
Armillaria luteobubalina Watling and Kile	
Armillaria mellea (Vahl:Fr.) P. Kumm.	*Armillariella mellea* (Vahl ex fr.) P. Karst
Armillaria novae-zelandiae (Stev.) Herink	
Armillaria socialis (DC:Fr.) Herink.	*Armillaria tabescens* (Scop.) Dennis, Orton and Hora *Clitocybe tabescens* (Scop.) Bres.
Aspergillus alliaceus Thom and Church	
Aspergillus carbonarius (Bainier) Thom	
Aspergillus flavus Link:Fr.	
Aspergillus fumigatus Fresen.	
Aspergillus japonicus Saito var. *aculeatus*	
Aspergillus japonicus Saito var. *japonicus*	
Aspergillus melleus Yukawa	
Aspergillus niger van Tieghem	
Aspergillus niger van Tieghem var. *awamori* (Nakazawa) Al-Musallam	

Aspergillus niger van Tieghem var. *phoenicis* (Corda) Al-Musallam	*Aspergillus phoenicis* (Corda) Thom
Aspergillus ochraceus K. Wilh.	
Aspergillus parasiticus Speare	
Aspergillus sclerotiorum Huber	
Aspergillus tamarii Kita	
Asperisporium caricae (Speg.) Maubl.	
Athelia rolfsii (Curzi) Tu and Kimbrough	*Corticium rolfsii* Curzi
anamorph: *Sclerotium rolfsii* Saccardo	
Bipolaris incurvata (C. Bernard) Alcom	
Botryosphaeria cocogena Subileau	
anamorph: *Diplodia theobromae* (Pat.) W. Nowell[a]	
Botryosphaeria disrupta (Berk. and M.A. Curtis) Arx and E. Muller	
Botryosphaeria dothidea (Mougeot ex Fries) Cesati and de Notaris	*Physalospora perseae* Doidge
anamorph: *Fusicoccum aesculi* Corda	*Dothiorella dominicana* Petr. and Cif. *Dothiorella gregaria* Saccardo
Botryosphaeria obtusa (Schwein.) Shoemaker	
anamorph: *Sphaeropsis malorum* Peck	
Botryosphaeria quercuum (Schwein.) Saccardo	
Botryosphaeria rhodina (Cooke) Arx	*Physalospora rhodina* Cooke
anamorph: *Diplodia theobromae* (Pat.) W. Nowell[a]	*Diplodia natalensis* Pole-Evans *Lasiodiplodia theobromae* (Pat.) Griffon and Maubl.
Botryosphaeria ribis Grossenb. and Duggar	
anamorph: *Fusicoccum parvum* Pennycook and Samuels	
Botryotinia fuckeliana (de Bary) Whetzel	*Sclerotinia fuckeliana* (de Bary) Fuckel
anamorph: *Botrytis cinerea* Pers. ex. Fr.	*Peziza fuckeliana* de Bary
Brooksia tropicalis Hansf.	
Calonectria colhounii Peerally	
anamorph: *Cylindrocladium colhounii* Peerally	
Calonectria ilicicola Boedijn and Reitsma	*Calonectria crotalariae* (C.A. Loos) D.K. Bell and Sobers
anamorph: *Cylindrocladium parasiticum* Crous, M.J. Wingfield	*Cylindrocladium crotalariae* (Loos) Bell and Sobers and Alfenas
Calonectria pteridis Crous, M.J. Wingfield and Alfenas	
anamorph: *Cylindrocladium pteridis* F.A. Wolf	
Calonectria leguminum (Rehm) Crous	*Nectria leguminum* Rehm *Calonectria quinqueseptata* Figueiredo and Namek.
anamorph: *Cylindrocladium leguminum*	*Cylindrocladium quinqueseptatum* Boedijn and Reitsma
Calonectria spathiphylli El-Gholl *et al.*	
anamorph: *Cylindrocladium spathiphylli* Shoulties, El-Gholl and Alfieri	*Cylindrocladium musae* Semer, D.J. Mitchell, M.E. Mitchell and Alfenas
Candida guilliermondi (Dastellani) Langeron and Guerra	
teleomorph (uncommon): *Pichia guilliermondi* Wickerham	
Capitorostrum cocoes K.D. Hyde and Philemon	
Capnodium moniliforme Fraser	
Ceratocystis adiposa (E.J. Butler) C. Moreau	
anamorph: *Chalara* sp.	
Ceratocystis fimbriata Ellis & Halst	
anamorph: *Chalara* sp.	
Ceratocystis paradoxa (Dade) C. Moreau	
anamorph: *Chalara paradoxa* (De Seynes) Saccardo	*Thielaviopsis paradoxa* (De Seynes) Höhn.

Binomial	Synonyms
Cercospora averrhoae Petch.	
Cercospora hayi Calp.	
Cercospora mamaonis Viégas & Chupp	
Cercospora papayae Hansf.	
Cercospora wellesiana (Welles) Chupp	
Cerotelium fici (Butler) Arthur	*Kuehneola fici* Butler *Physopella fici* (Cast.) Arthur
Chaetomium globosum Kunze:Fr.	
Chaetothyrina musarum (Speg.) Theiss	
Choanophora cucurbitarum (Berk. & Ravenel) Thaxt.	
Chondrostereum purpureum (Pers.:Fr.) Pouzar	
Citromyces ramusus Bain. and Sart.	
Cladosporium cladosporioides (Fresen.) de Vries	
Cladosporium herbarum (Pers) Link ex Gray	
Cladosporium musae E.W. Mason apud Martyn	
Cladosporium oxysporum Berk and M.A. Curtis	
Cochliobolus eragrostidis (Tsuda and Ueyama) Sivan. anamorph: *Curvularia eragrostidis* (Henn.) J.A. Meyer	
Cochliobolus setariae anamorph: *Drechslera setariae* (Sawada) Subramanian and Jain	
Cochliobolus spicifer R.R. Nelson anamorph: *Bipolartis spicifera* (Bainier) Subramanian	
Cochliobolus tuberculatus Sivan. anamorph: *Curvularia tuberculata* P.C. Jain	
Colletotrichum capsici (Syd.) E.J. Butler and Bisby	
Colletotrichum crassipes (Speg.) V. Arx	
Colletotrichum gloeosporioides var. *minor* Simmonds	
Coprinus micaceus Fr.	
Cordana johnstonii M.B. Ellis	
Cordana musae (A.W. Zimmerm.) von Höhnel	
Corticium penicillatum Petch.	
Corynespora cassiicola (Berk. & M.A. Curtis) C.T. Wei	*Cercospora melonis* Cooke *Cercospora vignicola* E. Kawamura *Helminthosporium cassiicola* Berk. & M.A. Curtis *Helminthosporium vignae* Olive in Olive, Bain, & Lefebvre *Helminthosporium vignicola* (E. Kawamura) Olive
Curvularia carica-papayae Srivastava and Bilgrami	
Curvularia lunata (Wakk.) Boedijn	
Cylindrocarpon musae C. Booth and R.H. Stover	
Cylindrocarpon tonkinense Bugn.	
Cylindrocladiella parva (P.J. Anderson) Boesewinkel	*Cylindrocladium parvum* P.J. Anderson
Cylindrocladium gracile (Bugnicourt) Boesewinkel	*Cylindrocarpon gracile* Bugnicourt
Cylindrocladium scoparium Morg.	
Cytosphaera mangiferae Died.	
Cytospora palmarum Cooke	
Deightoniella torulosa (Syd.) M.B. Ellis	
Denticularia mangiferae (Bitancourt and Jenkins) comb. nov. teleomorph?: *Elsinoë mangiferae* Bitancourt and Jenkins	*Sphaceloma mangiferae* (Bitancourt and Jenkins)
Diaporthe actinidiae Sommer and Beraha anamorph: *Phomopsis* sp.	

Diaporthe cinerascens Saccardo	
anamorph: *Phomopsis cinerascens* (Saccardo) Bubak.	
Diaporthe citri Wolf	
anamorph: *Phomopsis citri* Fawcett	
Diaporthe perniciosa Marchal	
Diplodia phoenicum (Sacc.) H. Fawc. and L.J. Klotz	
Diplodia recifensis Batista	
Drechslera gigantea (Heald and F.A. Wolf) S. Ito	*Helminthosporium giganteum* Heald and F.A. Wolf
Drechslera halodes (Drechsler) Subram. and Jain	
Drechslera musae-sapientum (Hansford) M.B. Ellis comb. nov.	
Drechslera rostrata (Drechsler) Richardson and Fraser	
Elsinoë annonae Bitancourt and Jenkins	
Elsinoë australis Bitancourt and Jenkins	
anamorph: *Sphaceloma australis* Bitancourt and Jenkins	
Elsinoë fawcettii Bitancourt and Jenkins	
anamorph: *Sphaceloma fawcettii* Jenkins	
Epicoccum nigrum Link	*Epicoccum purpurascens* Ehrenb.
Erysiphe cichoracearum De Candolle	
Erythricium salmonicolor (Berk. and Broome) Burdsall	*Corticium salmonicolor* Berk. and Broome
anamorph: *Necator decretus* Massee	*Phanerochaete salmonicolor* (Berk. and Broome) Jülich
Eurotium amstelodami Mangin	
anamorph: *Aspergillus amstelodami* (Mangin) Thom and Church	
anamorph: *Setosphaeria rostrata* K.J. Leonard	
Exserohilum rostratum (Drechs.) K.J. Leonard and E.G. Suggs	
Fusarium acuminatum Ellis and Everhart	
Fusarium avenaceum (Fr.:Fr.) Saccardo	
Fusarium concentricum Nirenberg and O'Donnell	
Fusarium dimerum Penzig	*Fusarium episphaeria* (Tode) W.C. Snyder and H.N. Hansen Penz. in Saccardo
Fusarium equiseti (Corda) Saccardo	
Fusarium guttiforme Nirenberg and O'Donnell	*Fusarium subglutinans* f. sp. *ananas* *Fusarium subglutinans* *Fusarium moniliforme* Sheldon var. *subglutinans*
Fusarium lactis Pirotta and Riboni	*Fusarium moniliforme* Sheldon var. *fici* Caldis
Fusarium mangiferae Britz, Wingfield and Marasas sp. nov.	
Fusarium moniliforme Sheldon	
Fusarium oxysporum Schlechtend.:Fr. f. sp. *albedinis* (Kill. and Maire) Melencon	
Fusarium oxysporum Schlechtend.:Fr. f. sp. *canariensis*	
Fusarium oxysporum Schlechtend.:Fr. f. sp. *cubense* (E.F. Smith) W.C. Snyder and H.N. Hansen	
Fusarium oxysporum Schlechtend.:Fr. f. sp. *passiflorae* Gordon apud Purss	
Fusarium oxysporum Schlechtend.:Fr. f. sp. *psidii*	
Fusarium pallidoroseum (Cooke) Saccardo	*Fusarium semitectum* Berk. and Rav.
Fusarium proliferatum (T. Matsushima) Nirenberg	*Fusarium moniliforme* var. *intermedium* Neish and Legget
Fusarium redolens Wollenweber	
Fusarium sterilihyphosum Britz, Marasas and Wingfield sp. nov.	

Binomial	Synonyms
Fusicoccum luteum Pennycook and Samuels	*Dothiorella aromatica* (Saccardo) Petr. and Syd.
Fusicoccum mangiferum	*Dothiorella mangiferae* H. and P. Syd. and But. *Hendersonia creberrima* Syd. *Hendersonula toruloidea* Nattras *Natrassia mangiferae* (Nattras) Sutton and Dyko
Fusicoccum sp.	*Dothiorella* 'long'
Galactomyces citri-aurantii Butler and Peterson	
anamorph: *Geotrichum citri-aurantii* (Ferraris) Butler	
Ganoderma applanatum (Pers.) Pat.	
Ganoderma boninense Pat.	*Ganoderma miniatocinctum* Steyaert
Ganoderma brownii (Murrill) R.L. Gilbertson	
Ganoderma lucidum (W. Curt.:Fr.) P. Karst	
Ganoderma philippi (Bres. and P. Henn.) Bres.	*Ganoderma pseudoferreum* (Wakerf.) Van Over. and Steinm. *Fomes pseudoferreum*
Ganoderma tornatum (Pers.) Bresad.	
Ganoderma zonatum Murrill	*Ganoderma sulcatum* Murrill *Polyporus lucidus* var. *zonatus* (Murrill) Overholts
Geastrumia polystigmatis Batista and M.L. Farr	
Geotrichum candidum Link	
Geotrichum ludwigii (Hanson) Sin-Fang *et al.*	
Gibberella baccata (Wallr.) Saccardo	
anamorph: *Fusarium lateritium* Nees	
Gibberella fujikuroi (Sawada) S. Ito in S. Ito and K. Kimura	
Gibberella pulicaris (Fr.:Fr.) Saccardo	
anamorph: *Fusarium sambucinum* Fuckel	
Gibberella saubinetii	*Gibberella zeae* (Schwein.) Petch
anamorph: *Fusarium graminearum* Schwabe	
Gilbertella persicaria (E.D. Eddy) Hasseltine	*Choanephora persicaria* Eddy
Gliocephalotrichum bulbilium J.J. Ellis and Hesseltine	
Gliocephalotrichum microchlamydosporum (J. Meyer) Wiley and Simmons	
Gloeodes pomigena (Schwein.) Colby	
Glomerella acutata J.C. Guerber and J.C. Correll	
anamorph: *Colletotrichum acutatum* J.H. Simmonds	
Glomerella cingulata (Stoneman) Spaulding and von Schrenk	
anamorph: *Colletotrichum gloeosporioides* (Penz.) Penz. and Sacc.	
Glomerella fructigena (Clinton) Saccardo	*Glomerella cingulata* (Stoneman) Spaulding and von Schrenk
anamorph: *Colletotrichum caricae* Stevens and Hall	*Colletotrichum gloeosporioides* Penz.
Glomerella sp.	*Glomerella musarum* Petch
Colletotrichum musae (Berk. and M.A. Curtis) Arx	*Gloeosporium musarum* Cooke and Massee
Grallomyces portoricensis Stevens	
Graphiola phoenicis (Moug.) Poit.	
Guignardia citricarpa Kiely	
anamorph: *Phyllosticta citricarpa* McAlp	
Guignardia musae Racib.	
anamorph: *Phyllosticta musarum* (Cooke) van der Aa	*Phyllostictinia musarum* (Cooke) Petrak

Haematonectria haematococca (Berk. and Broome) Samuels and Nirenberg	*Nectria haematococca* Berk. and Broome
anamorph: *Fusarium solani* (Mart.) Appel and Wollenweber emend Snyder and Hansen	
Haplobasidion musae M.B. Ellis	
Hypocrea ceramica Ellis and Everh.	
anamorph: *Trichoderma koningii* Oudem.	
Junghuhnia vincta (Berk.) Hood and M. Dick	
Khuskia oryzae H.J. Hudson	
anamorph: *Nigrospora oryzae* (Berk. and Broome) Petch	
Leptodontium elatius (G. Mangenot) De Hoog	
Leptosphaeria musarum Saccardo and Berl.	
Leptothyrium pomi (Mont. and Fr.) Saccardo	
Leptoxyphium sp. Speganizzi	
Leveillula taurica (Lév.) G. Arnaud	
anamorph: *Oidiopsis taurica* Salmon	
Macrophomina phaseolina (Tassi) Goidanich	
Macrosporium cocos Pass.	
Magnaporthe grisea (Hebert) Barr	
anamorph: *Pyricularia grisea* (Cooke) Saccardo	
Marasmiellus cocophilus Pegler	
Marasmiellus equicrinus Mull.	*Marasmius equicrinus* Mull.
Marasmiellus inoderma (Berk.) Singer	*Marasmius semiustus* Berk. and M. A. Curt.
Marasmiellus scandens (Mass.) Dennis and Reid	*Marasmius scandens* *Marasmius byssicola* Petch
Mauginiella scaettae Cav.	
Meliola durionis Hansford	
Meliola garcinae Yates	
Meliola mangiferae Earle	
Meliola nephelii var. *singalensis* Hansf.	
Mucor hiemalis Wehmer	
Mucor piriformis Fischer	
Mycosphaerella caricae H. and P. Sydow	
anamorph: *Phoma caricae-papayae* (Tarr.) Punithalingam	*Ascochyta carica* Pat. *Ascochyta caricae-papayae* (Tarr)
Mycosphaerella citri Whiteside	
anamorph: *Stenella citri-grisea* (Fisher) Sivanesan	
Mycosphaerella eumusae Crous and X. Mourichon	
anamorph: *Pseudocercospora eumusae* Crous and X. Mourichon	
Mycosphaerella fijiensis M. Morelet	
anamorph: *Pseudocercospora fijiensis* (M. Morelet) Deighton	*Paracercospora fijiensis* (M. Morelet) Deighton
Mycosphaerella musae (Speg.) Syd. and P. Syd.	
Mycosphaerella musicola J.L. Mulder in J.L. Mulder and R.H. Stover	
anamorph: *Pseudocercospora musae* (Zimm.) Deighton	
Mycosphaerella palmicola Chaudhury and P.N. Rao, emend.	
Mycosphaerella perseae Stevens	
anamorph: *Septoria* sp.	
Mycosphaerella tassiana (De Not.) Johans.	
anamorph: *Cladosporium herbarum* (Per.:Fr.) Link	
Myxosporium psidii Sawada and Kurosawa	
Nectria foliicola Berk. and M.A. Curtis	
Nectria ochroleuca (Schwein.) Berk.	
anamorph: *Gliocladium roseum* Bainier	
Nectria pseudotrichia Berk. and M.A. Curtis	*Thyronectria pseudotrichia* (Berk. and M.A. Curtis) Seeler

Binomial	Synonyms
anamorph: *Tubercularia laterita* (Berk.) Seifert	
Nigrospora sphaerica (Saccardo) E.W. Mason	
Oidium citri (Yen) U. Braun	*Oidium erysiphoides* f. *citri* Yen
Oidium mangiferae Berthet	
Oidium nephelii Hadiwidjaja ex U. Braun.	
Oidium tingitaninum Carter	*Acrosporium tingitaninum* (Carter) Subramanian
Omphalia pigmentata Bliss	
Omphalia tralucida Bliss	
Ovulariopsis papayae van der Bilz	
Oxyporus latemarginatus (Durieu, and Mont. Ex Mont.) Donk	*Poria latemarginatus* (Durieu and Mont.) Cooke
Pellicularia filamentosa (Pat.) Rogers	
Pellicularia koleroga Cooke	
Peltaster fructicola Johnson	
Penicillium chrysogenum Thom	
Penicillium digitatum Saccardo	
Penicillium expansum Link ex Gray	
Penicillium funiculosum Thom	
Penicillium italicum Wehmer	
Penicillium lilacinum Thom	
Penicillium ulaiense Hsieh, Su and Tzean	
Penicillium vermoesini (Biourge) Thom.	
Periconiella cocoes M.B. Ellis	
Pestalotiopsis disseminata (Thüm.) Steyaert	
Pestalotia elasticola Hennings	*Pestalotiopsis elasticola*
Pestalotiopsis flagisettula	*Pestalotia flagisettula*
Pestalotiopsis guipini (Desm.) Steyaert	
Pestalotiopsis leprogena (Speg.) Steyaert	*Pestalotia leprogena* Speg.
Pestalotiopsis mangiferae (Henn.) Steyaert	
Pestalotiopsis palmarum (Cooke) Steyaert	*Pestalotia palmarum* Cooke
teleomorph: *Rhynchosphaeria* sp.	
Pestalotiopsis psidii (Pat.) Mordue	
Pestalotiopsis versicolor (Spreg.) Steyaert	
Phaeoramularia angolensis (DeCarvalho and Mendez) P.M. Kirk	*Cercospora angolensis* Car. and Men.
Phaeoseptoria musae Punithalingham	
Phakopsora cherimoliae (Lagerh.) Cummins	
Phellinus lamaoensis	
Phellinus noxius (Corner) Cunn.	*Fomes noxius* Corner
Phoma epicoccina Punithalingham, Tulloch and C.M. Leach	
Phoma exigua Desmazières	
Phoma macrostoma Montagne	
Phoma nigricans Johnston and Boerema	
Phoma tracheiphila (Petri) Kantsch. and Gik.	
Phomopsis annonacearum Bondartseva-Monteverde	
Phomopsis artocarpi H. Sydow	
Phomopsis artocarpina	
Phomopsis caricae-papayae Petr. and Cif.	
Phomopsis durionis Syd.	
Phomopsis mangiferae Ahmad	
Phomopsis sp.	*Septoria fructigena* Berk. and Curt.
Phomopsis persae Zerova	
Phomopsis psidii Nag Raj and Ponnappa	

Phomopsis tersa (Saccardo) comb. nov.	
Phragmocapnias betle (Sydow and Butler) Theissen and Sydow emend. Reynolds	
Phyllachora anonicola Chardon	
Phyllachora gratissima Rehm.	
Phyllachora musicola C. Booth and D.E. Shaw	
Phyllachora torrendiella (Batista) Subi Leau	
Phyllosticta caricae-papayae Allesch.	
Phyllosticta palmetto Ellis and Everh.	
Polychaeton sp. (Pers.) Lev.	
Pseudocercospora artocarpi (H. Sydow and P. Sydow) Deighton	*Cercospora artocarpi* (H. Sydow and P. Sydow)
Pseudocercospora nephelii Sutton and Peregrine	
Pseudocercospora purpurea (Cooke) Deighton	*Cercospora purpurea* Cooke
Pseudoepicoccum cocos (F. Stevens) M.B. Ellis	
Puccinia psidii Winter.	
Pucciniastrum actinidiae Hiratsuka	
Pyricularia oryzae Cavara	
Ramichloridium musae (Stahel ex M.B. Ellis) de Hoog	*Chloridium musae* Stahel *Periconiella musae* Stahel ex M.B. Ellis *Veronaea musae* Stahel ex M.B. Ellis
Ramularia necator Massee	
Rhizopus arrhizus A. Fischer	*Rhizopus oryzae* Went and Prinsen Geerligs
Rhizopus artocarpi Racib.	
Rhizopus microsporus van Tieghem	
Rhizopus stolonifer (Ehrenb.:Fr.) Vuill.	*Rhizopus nigricans* Ehrenb.
Rigidoporus lignosus (Klotzsch) Imazeki	*Fomes lignosus* (Klotzsch) Lloyd *Leptoporus lignosus*
Rigidoporus ulmarius (Sow.: Fr.) Imazeki	
Rigidoporus vinctus (Berk.) Ryv.	
Rosellinia bunodes (Berk. and Broome) Saccardo	
Rosellinia necatrix (Hartig) Berl. ex Prill. anamorph: *Dematophora necatrix* R. Hartig	
Rosellinia pepo Pat.	
Schizothyrium pomi (Mont. and Fr.) Arx anamorph: *Zygophiala jamaicensis* Mason	
Sclerotinia sclerotiorum (Libert) de Bary	
Scorias spongiosa (von Schweinitz) Fries emend. Reynolds	
Septobasidium bogoriense Pat.	
Septoria passiflorae Sydow	
Septoria passifloricola Punith.	*Septoria passiflorae* A.J. Louw non Syd.
Sphaceloma perseae Jenk.	
Sphaerodothis acrocomiola (Montagne) Von Arx and Muller	*Coccostroma palmicola* (Speg.) Von Arx and Muller
Sphaerotheca caricae-papayae S. Tanda and U. Braun anamorph: *Oidium caricae* F. Noack	 *Oidium caricicola* J. Yen and Wang
Sphaerotheca fuliginea (Schlecht.) Poll.	
Sphaerotheca humuli (D.C.) Burr.	
Stemphylium floridanum Hannon and G.F. Weber	
Stemphylium lycopersici (Enjoji) W. Yamamoto	*Thyrospora lycopersici* Enjoji
Stigmina mangiferae (Koorders) M.B. Ellis	*Cercospora mangiferae* Koorders
Stigmina palmivora (Sacc.) S.J. Hughes	

Binomial	Synonyms
Syncephalastrum racemosum Cohn ex J. Schröt.	
Thanatephorus cucumeris (Frank) Donk.	
anamorph: *Rhizoctonia solani* Kühn	
Trachysphaera fructigena Tabor and Bunting	
Trichomerium grandisporum (Ellis and Martin) Bat. and Cif.	
Trichopeltheca asiatica Bat., Costa and Cif.	
Trichothecium roseum Link	
Trichoderma hamatum (Bonord.) Bainier	
teleomorph: *Hypocrea* sp.	
Trichoderma harzianum Rifai	
teleomorph: *Hypocrea* sp.	
Tripospermum sp. Speganizzi	
Uncinula actinidiae Miyabe ex Jacz.	
Uredo musae Cummins	
Uromyces musae Henn.	
Verticillium albo-atrum Reinke and Berth.	
Verticillium dahliae Kleb.	
Verticillium theobromae (Turconi) E.W. Mason and S.J. Hughes	
Nematodes	
Belonolaimus longicaudatus Rau	
Bursaphelenchus cocophilus (Cobb) Baujard	*Rhadinaphelenchus cocophilus*
Helicotylenchus dihystera Cobb	
Helicotylenchus multicinctus Cobb	
Hemicriconemoides mangiferae Siddiqi	
Meloidogyne arenaria (Neal) Chitwood	
Meloidogyne hapla Chitwood	
Meloidogyne incognita (Kofoid and White) Chitwood	
Meloidogyne javanica (Treub) Chitwood	
Pratylenchus brachyurus (Godfrey) Filipjev and Schuurmans Stekhoven	
Pratylenchus coffeae (Zimm.) Filip. and Schuur.	
Pratylenchus goodeyi Sher. and Allen	
Pratylenchus penetrans (Cobb) Filipjev and Schuurmans-Stekhoven	
Pratylenchus vulnus (Allen and Jensen)	
Radopholus similis (Cobb) Thorne	
Rotylenchulus parvus (Williams) Sher	
Rotylenchulus reniformis Linford and Oliveira	
Tylenchulus semipenetrans Cobb.	
Xiphinema brevicolle Lordello and da Costa	
Prokaryotes	
Bacteria	
Acetobacter aceti (Pasteur) Beijennek	
Acetobacter liquefaciens (Asai) Gosselé *et al.*	
Acetobacter peroxydans Visser 't Hooft	
Agrobacterium radiobacter (Beijerinck and van Delden) Conn	
Agrobacterium tumefaciens (E.F. Smith and Townsend) Conn	
Agrobacterium tumefaciens (E.F. Smith and Townsend) Conn, biovar 3	*Agrobacterium vitis* Ophel and Kerr
Bacillus subtilis (Ehrenberg) Cohn	

Candidatus Liberobacter africanum Jagoueix *et al.*	*Bacillus ananas* Serrano
Candidatus Liberobacter asiaticum Jagoueix *et al.*	*Erwinia herbicola* var. *ananas* (Serrano) Dye
Enterobacter cloacae (Jordan) Hormaeche & Edwards	
Erwinia carotovora (Jones) Holland	
Erwinia carotovora ssp. *carotovora*	
Erwinia chrysanthemi Buckholder, McFadden and Dimock	
Erwinia cypripedii (Hori) Bergey *et al.*	*Pectobacterium cypripedii* (Hori) Brenner *et al.*
Erwinia herbicola Löehnis Dye	
Erwinia herbicola var. *ananas* (Serrano) Dye	*Erwinia ananas*
Erwinia mangiferae (Doidge) Begey *et al.*	*Erwinia herbicola*
Gluconobacter oxydans (Henneberg) Deley	
Pantoea annanatis (Serrano) Mergaert *et al.*	
Pantoea citrea Kageyama *et al.*	*Erwinia herbicola*
Pseudomonas carica-papayae Robbs.	
Pseudomonas fici (Cav.) Krasil'nikov.	
Pseudomonas savastanoi	
Pseudomonas syringae van Hall	
Pseudomonas syringae pv. *actinidia* Takikawa *et al.*	
Pseudomonas syringae pv. *passiflorae*	
Pseudomonas syringae pv. *syringae* van Hall	
Pseudomonas viridiflava (Burkholder) Dowson	
Ralstonia solanacearum (Smith) Smith	*Pseudomonas solanacearum* *Burkholderia solanacearum*
Xanthomonas axonopodis pv. *citri* (Hasse) Vaut.	*Xanthomonas citri* *Xanthomonas campestris* pv. *citri (Hasse) Dye*
Xanthomonas axonopodis pv. *citrumelo* (Hasse) Vauterin *et al.*	
Xanthomonas campestris (Pammel) Dowson	
Xanthomonas campestris pv. *passiflorae* (Pereira) Dye	
Xanthomonas nepheliae Barr.	
Xanthomonas sp. pv. *mangiferaeindicae*	*Xanthomonas campestris* pv. *mangiferaeindicae* (Patel, Moniz and Kulkarni) Robbs, Ribiero and Kimura *Pseudomonas mangiferae-indicae* Patel, Moniz and Kulkarni
Xylella fastidiosa Wells *et al.*	
Mollicutes	
Candidatus Phytoplasma australasia White *et al.*	
Candidatus Phytoplasma australiense Davis *et al.*	
Spiroplasma citri Saglio *et al.*	

[a]The synonym *Botryodiplodia theobromae* Pat. recently was declared a *nomen dubium* (Crous and Palm, 1999).

Reference

Crous, P.W. and Palm, M.E. (1999) Reassessment of the anamorph genera of *Botryoplodia, Dothiorella* and *Fusicoccum*. *Sydowia* 52, 167–175.

Appendix II Plant Taxa, Authorities and Common Names

Taxa [Synonym]	Common name (Synonym)
Abelmoschus esculentus L.	Okra (Bhindi)
Actinidia chinensis Planch.	Mihoutao
Actinidia deliciosa (A. Chevalier) C.F. Liang and A.R. Ferguson [*A. chinensis* Planch.]	Kiwifruit
Agave sisalana Perrine	Sisal
Anacardium occidentale L.	Cashew
Ananas comosus (L.) Merrill	Pineapple
Annona glabra L.	Pond apple
Annona cherimola Mill.	Cherimoya
Annona diversifolia Saff.	Ilama
Annona muricata L.	Soursop (Guanábana)
Annona squamosa L.	Sugar apple (Sweetsop)
Artocarpus camansi Blanco	Kamangsi
Artocarpus heterophyllus Lam.	Jackfruit
Artocarpus lakoocha Roxb.	Monkey jack
Artocarpus lingnanensis Merr.	Kwai muk
Artocarpus odoratissimus Blanco [*A. tarap* Becc.]	Morang (Marang, Tarap)
Artocarpus rigida Blume	Monkey jack
Artocarpus rotundus Merr.	Monkey jack
Annona reticulata L.	Custard apple (Bullock's heart)
Artocarpus altilis (Parkinson) Fosberg [*A. communis* Forster]	Breadfruit
Artocarpus hirsuta Lam.	Aini
Artocarpus integer (Thunb.) Merr. [*A. champeden* (Loffr.) Stokes]	Chempedek (tibadak)
Artocarpus elastica Reinw. ex Blume	
Artocarpus gomezianus Wall. Ex Trec.	Tapang (Tampang)
Artocarpus mariannensis Trec.	Dugdug (Cheibiei)
Averrhoa bilimbi L.	Bilimbi
Averrhoa carambola L.	Carambola
Blighia sapida Koenig	Akee
Bougainvillea spp.	Bougainvillea
Capsicum spp.	Pepper
Carica papaya L.	Papaya
Cassytha filiformis L.	Dodder laurel
Casuarina equisetifolia J.R. Forst. and G. Forst.	Australian pine
Cinnamomum zeylandicum Blume (*C. verum* J.S. Presl)	Cinnamon
Citrus aurantifolia (Christm.) Swing.	Mexican lime (Key lime)
Citrus aurantium L.	Sour orange
Citrus grandis (L.) Osb. [*C. maximus* (Burm.) Merrill]	Pummelo
Citrus hystrix DC	Combava
Citrus jambhiri Lush	Rough lemon
Citrus latifolia Tan.	Tahiti (Persian) lime
Citrus limon (L.) Burm. f.	Lemon
Citrus limonia Osb.	Rangpur lime
Citrus medica L.	Citron
Citrus mitis Blanco	Calamondin
Citrus paradisi Macf.	Grapefruit
Citrus reticulata Blanco	Mandarin
Citrus sinensis (L.) Osb.	Sweet orange
Cocos nucifera L.	Coconut
Coffea spp.	Coffee

Cuscuta campestris Yuncker	Dodder
Cynodon dactylon (L.) Pers.	Bermuda grass
Dimocarpus longan Lour. [*Euphoria longan* (Lour.) Steud.]	Longan
Durio zibethinus Murray	Durian
Echinochloa colona (L.) Link	Marsh grass
Eleusine indica (L.) Gaertener	Goosegrass
Emilia fosberyii D.H. Nicholson	Tassel flower
Eriobotrya japonica (Thunb.) Lindl.	Loquat
Ficus benjamina L.	Weeping fig
Ficus carica L.	Fig
Ficus palmata Forsok	Fagwara
Garcinia mangostana L.	Mangosteen
Hevea brasiliensis (Willd. ex A. Juss.) Muell. Arg.	Para rubber tree (Rubber)
Hibiscus tiliaceous L.	
Lawsonia inermis L.	Henna
Litchi chinensis Sonn.	Lychee
Lycopersicum esculentum Mill.	Tomato
Mangifera indica L.	Mango
Manilkara zapota (L.) van Royen [*Achras zapota* L.]	Sapodilla
Metrosideros collina A. Gray	Ohia
Medicago sativa L.	Lucerne
Morus alba L.	White mulberry
Morus nigra L.	Black mulberry
Murraya paniculata (L.)	Orange jasmine
Musa acuminata Colla ssp. *banksii* (F. Muell) Simmonds	Banana and plantain relative
Musa acuminata Colla ssp. *burmannica* Simmonds	Banana and plantain relative
Musa acuminata Colla ssp. *malaccensis* (Ridley) Simmonds	Banana and plantain relative
Musa acuminata Colla ssp. *microcarpa* (Beccari) Simmonds	Banana and plantain relative
Musa acuminata Colla ssp. *siamea* Simmonds	Banana and plantain relative
Musa acuminata Colla ssp. *zebrina* (van Houtte)	Banana and plantain relative
Musa balbisiana Colla	Banana and plantain relative
Musa lolodensis E.E. Cheesman	Fe'i banana relative
Musa maclayi F.J.H. von Mueller	Fe'i banana relative
Musa peekelii C.(K.)A.G. Lauterbach	Fe'i banana relative
Musa schizocarpa Simmonds	Banana and plantain relative
Musa textilis Née	Manila hemp
Nephelium lappaceum L.	Rambutan
Nephelium mutabile Blume	Pulasan
Passiflora edulis Sims f. *edulis* Sims	Purple passion fruit
Passiflora edulis Sims f. *flavicarpa* Degner	Yellow passion fruit
Passiflora foetida L.	Stinking passion flower
Passiflora ligularis Juss.	Sweet granadilla
Passiflora mollissima (HBK) Bailey	Curubá (Banana passion fruit)
Passiflora quadrangularis L.	Giant granadilla
Persea americana Miller	Avocado
Persea borbonia (L.) K. Spreng.	Red bay
Phoenix canariensis Hort. ex Chabaud	Canary Island date palm
Phoenix dactylifera L.	Date palm
Piper nigrum	Black pepper
Poncirus trifoliata (L.) Raf.	Trifoliate orange
Pongamia pinnata (L.) Pierre	Pongam
Psidium guajava L.	Guava
Rollinia pulchrinervis DC [*R. deliciosa* Saff.]	Biriba
Saccharum officinarum L.	Sugarcane
Schinus terebinthifolius Raddi	(Brazil) pepper tree
Solanum tuberosum L.	Potato
Spondias mangifera Willd.	Imra

Taxa [Synonym]	Common name (Synonym)
Spondias mombin L.	Yellow mombin
Syzygium aqueum (Burm. f.) Alston	Water apple
Syzygium samarangense (Blume) Merrill and L.M. Perry	Java apple
Theobroma cacao L.	Cacao
Viburnum odoratissimum Ker.	China laurestine

Appendix III Insect and Acarid Taxa, Authorities and Common Names

Binomial [synonym]	Common name
Aceria dimocarpi (Kuang)	Four-legged mite
Aceria ficus Cotte	Eriophyid mite
Aceria mangiferae Sayed [*Eriophyes mangiferae* (Sayed)]	Mango bud mite
Adoxophyes privatana Walk.	
Aphis gossypii Glover	Cotton or melon aphid
Aphis spiraecola Patch	Spirea aphid
Archips tabescens Walk.	
Bactrocera dorsalis [*Dacus dorsalis* Hend. complex]	Fruit fly
Blastophaga psenes L.	Fig wasp
Carpophilus foveicollis Murr.	Souring beetle
Ceroplastes pseudoceriferus Green	Longan wax scale
Coccus hesperidum L.	Soft brown scale
Cosmopolites sordidus Germar	Weevil borer
Cornegenapsylla sinica Yang and Li	Longan psyllid
Dacus dorsalis Hendel	Oriental fruit fly
Diacrotricha fasciola Zell	Fruit borer
Diaphorina citri Kuwayama	Asian psyllid
Diaprepes abbreviatus L.	Diaprepes weevil
Diastrombus mkurangai	Planthopper
Dolichotetranychus floridanus Banks	Pineapple red mite
Drepanococcus chiton (Green)	Soft scale
Drosophila melanogaster Meig	Vinegar fly
Dynamis borassi	Sugarcane weevil
Dysmicoccus brevipes Cockerell	Pink mealybug
Dysmicoccus neobrevipes Beardsley	Grey mealybug
Empoasca papayae Oman	Leafhopper
Empoasca stevensi Young	Leafhopper
Haptoncus ocularis Fairm.	Souring beetle
Helopeltis theobromae Miller	Cocoa mirid bug
Hypocryphalus mangiferae Stebb.	Scolytid beetle
Indarbela disciplaga Swinch	Stem borer
Iridomyrmex humills Mayer (Fluker and Beardsley)	Argentine ant
Metamasius hemipterus sericeus (Oliver)	Sugarcane weevil
Myndus adiopodoumeensis	Planthopper
Myndus crudus Van Duzee	Planthopper
Myndus taffini Bonfils	Planthopper
Myzus persicae Sulzer	Green peach aphid
Pentalonia nigronervosa (Coquerel)	Banana aphid
Pheidole megacephala Fabricius	Bigheaded ant
Pheidole megacephala Risso	Coastal brown ant
Porthesia (Euproctis) scientillans Walk.	
Rhynchophorus palmarum L.	American palm weevil
Solenopsis geminata Fabricius	Fire ant
Steneotarsonemus ananas Tryon	Pineapple fruit mite
Stephanitis typica (Distant)	Lace-bug
Tessaratoma papillosa Drury	Lychee stinkbug
Thecla basilides (Geyer)	Pineapple fruit caterpillar
Thrips tabaci Lindeman	
Toxoptera citricida (Kirkaldy)	Brown citrus aphid
Trioza erytreae (Del Guercio)	Citrus psyllid
Xyleborus ferrugineus (F.)	Beetle

Index

Bold page numbers indicate where relevant figures are located, and aka = also known as.